26th Extended European Cosmic Ray Symposium 2018

Journal of Physics: Conference Series Volume 1181

Barnaul-Belokurikha, Russia
6 - 10 July 2018

Editors:

Anatoly Lagutin
Igor Moskalenko
Mikhail Panasyuk

ISBN: 978-1-5108-8341-3
ISSN: 1742-6588

Printed from e-media with permission by:

Curran Associates, Inc.
57 Morehouse Lane
Red Hook, NY 12571

Some format issues inherent in the e-media version may also appear in this print version.

Institute of Physics
Dirac House, Temple Back
Bristol BS1 6BE UK

Phone: 44 1 17 929 7481
Fax: 44 1 17 920 0979

techtracking@iop.org

Additional copies of this publication are available from:

Curran Associates, Inc.
57 Morehouse Lane
Red Hook, NY 12571 USA
Phone: 845-758-0400
Fax: 845-758-2633
Email: curran@proceedings.com
Web: www.proceedings.com

TABLE OF CONTENTS

PLENARY SESSION

SOLAR AND HELIOSPHERIC COSMIC RAYS

COSMIC RAYS BELOW THE KNEE. NEW PHENOMENA AND THEIR INTERPRETATION

COSMIC RAYS ABOVE THE KNEE (E<1017 EV)

UHE COSMIC RAYS (E>1017 EV)

ACCELERATION OF COSMIC RAYS

COSMIC RAYS ANISOTROPY

CR INTERACTIONS WITH THE INTERSTELLAR MEDIUM AND ENIGMATIC PHENOMENA OBSERVED IN X-RAY AND GAMMA-RAY RANGES

DIRECT COSMIC-RAY OBSERVATION

HE/UHE GAMMA RAYS

HE/UHE MUONS AND NEUTRINOS

COSMIC RAYS AT EARTH AND PLANETS: APPLIED ASPECTS OF COSMIC RAYS

IOP Conf. Series: Journal of Physics: Conf. Series **1181** (2019) 011001 doi:10.1088/1742-6596/1181/1/011001

26th E+CRS

Preface

Particle astrophysics that is also called astroparticle physics combines efforts and interests of two communities, particle physicists and astrophysicists, and presents the broadest field of research that covers experiments involving man-made machines as well as experiments that are running for billions of years by Nature itself. These involve all types of known and yet unknown particles and emissions in the broadest energy range possible, from nano-eV to 10^{21} eV, far exceeding the capabilities of the terrestrial and extraterrestrial laboratories. Last decade was rich on discoveries and breakthrough results in many areas of traditional astroparticle physics, and gave birth to two brand new branches of astronomy: gravitational and neutrino. The hopes are that this rapid and exciting progress will continue in years to come and the key factors to ensure its sustainability are the face to face meetings and exchange of enriching ideas between the communities with different backgrounds.

The 26th Extended European Cosmic Ray Symposium (E+CRS) took place at Altai State University (Barnaul-Belokurikha, Russia, 610 July 2018), and was organized by Altai State University in partnership with the Skobeltsyn Institute of Nuclear Physics of the Lomonosov Moscow State University and Novosibirsk State University.

Since the first symposia, which started 50 years ago with two parts held in 1968 in Lodz, Poland (high energy, extensive air showers and astrophysical aspects) and in Bern (solar and heliospheric phenomena), the ECRS became one of the major conferences in the field of astroparticle physics with the emphasis on meeting the demands of young scientists and European collaborations.

Based on the decision of the International Advisory Committee (IAC), adopted during the 25th ECRS (Turin, Italy, 2016), the 26th European Cosmic Ray Symposium was scheduled to be held in 2018 in conjunction with the 35th Russian Cosmic Ray Conference at Altai State University (Barnaul, Russia). Conducting a joint conference in the geographical center of Eurasia was aimed at the expansion of the geography of participants, including young scientists from different countries. In addition, the IAC decided to modify the format of the symposium (following the COSPAR International Assembly model) in order to expand the representation of key global research programs in the areas of the symposium. A change in the name and the abbreviation of the symposium into the 26th Extended European Cosmic Ray Symposium (E+CRS) reflects these goals.

The 26th E+CRS covers a wide range of scientific issues divided into the following sections supervised by leading world experts:

- Solar and heliospheric cosmic rays (SH)
- Cosmic rays at Earth and Planets. Applied aspects of cosmic rays (GEO)
- Direct cosmic-ray observation (CRD)
- Cosmic rays below the knee. New phenomena and their interpretation (CR I)
- Cosmic rays above the knee ($E < 10^{17}$ eV) (CR II)
- CR interactions with the interstellar medium and enigmatic phenomena observed in X-ray and gamma-ray ranges (ISM)
- UHE cosmic rays ($E > 10^{17}$ eV) (CR III)
- HE/UHE muons and neutrinos (MN)
- HE/UHE gamma rays (GR)
- Cosmic rays anisotropy (ANI)
- Acceleration of cosmic rays (ACR)
- Future measurements of cosmic rays: new instrumentations and methods (FUT)
- Workshops section I (WSS I) LPM effect and its applications for extremely high energy cosmic ray research
- Workshops section II (WSS II) Hadron Interactions at superhigh energies (colliders and cosmic rays)

The conference was attended by 155 people representing 20 countries: Belgium, China, the Czech Republic, Denmark, Finland, France, Germany, India, Italy, Japan, Kazakhstan, the Netherlands, Norway, Poland, Portugal, Russia, South Africa, Spain, Switzerland, and the United States. All the world's leading collaborations in the areas covered by the symposium have presented their latest results.

The conference organizers express deep sincere gratitude to all participants, especially the section supervisors, who did a great job, to Sir Arnold W. Wolfendale, co-founder and long-time Chairman of the International Advisory Committee of European Cosmic Ray Symposium, Professor J. Nishimura and Academician G. Krymsky for their inspiring greetings to the Symposium participants. The Local Organizing Committee thanks the Russian Foundation for Basic Research for the support of the symposium.

Editors
Anatoly Lagutin (Altai State University)
Igor Moskalenko (Stanford University)
Mikhail Panasyuk (Moscow State University)

IOP Conf. Series: Journal of Physics: Conf. Series **1181** (2019) 011002 doi:10.1088/1742-6596/1181/1/011002

Peer review statement

All papers published in this volume of *Journal of Physics: Conference Series* have been peer reviewed through processes administered by the proceedings Editors. Reviews were conducted by expert referees to the professional and scientific standards expected of a proceedings journal published by IOP Publishing.

IOP Conf. Series: Journal of Physics: Conf. Series **1181** (2019) 012001 doi:10.1088/1742-6596/1181/1/012001

Fermi bubbles, their origin and possible connection to cosmic rays near the Earth

D. Chernyshov[1], K.S. Cheng[2], V. Dogiel[1], C.M. Ko[3]

[1] I.E.Tamm Theoretical Physics Division of P.N.Lebedev Institute of Physics, Leninskii pr. 53, 119991 Moscow, Russia

[2] Department of Physics, University of Hong Kong, Pokfulam Road, Hong Kong, China

[3] Institute of Astronomy, Department of Physics and Center for Complex Systems, National Central University, Jhongli District, Taoyuan City, Taiwan 320, R.O.C.

E-mail: chernyshov@lpi.ru

Abstract. Discovery of two giant structures seen in gamma-rays and radio, located above and below the Galactic center, known as Fermi bubbles, can be considered as one of the most interesting phenomena observed by Fermi-LAT. Their position and total energy content suggest a very strong energy outburst happened in the Galactic center in the past. We point out some limitations of theoretical models of Fermi bubbles. We also discuss interconnection between Fermi bubbles and Galactic cosmic rays. In particular, we point out that gamma-ray emitting electrons in the bubbles are most likely originated from cosmic ray electrons. In addition we show that Fermi bubbles may be responsible for the formation of the spectrum of cosmic rays above the "knee".

1. Introduction

Bipolar gamma-ray structures in our Galaxy correlated with the Galactic center also known as Fermi bubbles belong to the most remarkable discoveries made by Fermi-LAT. Initially discovered by two independent groups [23] and [39] and later confirmed by Fermi collaboration [3], Fermi bubbles remain as one of the most interesting astrophysical phenomena. They exhibit a very large size of order of 1 sr, making them the largest gamma-ray objects observed on the sky.

Fermi bubbles also are observed in other wavelengths. In particular observations in microwave band show a very good correlation of so-called "WMAP haze" with gamma-ray emission [4, 25]. There are some indications on the hot plasma inside the Fermi bubbles observed by a ROSAT as a narrow envelope with very sharp edges [10]. This structure is explained as a fast wind with a velocity $u_w \sim 10^8$ cm/s driving a shock into the halo gas. However subsequent observations of Fermi bubbles edges by Suzaku did not find any evidences of a strong shock there [30].

Since Fermi bubbles are very faint structures it is very difficult to observe exactly the same phenomena in other galaxies. Recently there was a claim of presence of similar objects in the galaxy M31 [36], but it was not confirmed by the Fermi collaboration. However some much more powerful objects with similar properties and probably similar nature are observed in some galaxies with active nuclei. For example even more giant structures are clearly seen in the direction of Cen-A in GHz radio [24, 29], GeV [43] and TeV [7] gamma-ray ranges. Giant X-ray and radio lobes (bubbles) were found also in the galaxies NGC 3801 [22], Mrk 6 [33] and Circinus Galaxy [34].

The nature of Fermi bubbles is being intensively discussed and a lot of phenomenological models were suggested (see review of [42] and references therein). With rather arbitrary parameters of the models it is possible to reproduce characteristics of emission from the bubbles. However, plausibility of these model is in many cases an open question.

In general, observations of powerful nonthermal emission the bubble from the bubbles indicate that they are source of CRs and may contribute into the total flux of CRs in the Galaxy. In this paper we estimate below this contribution into the total flux of CR electrons and protons.

2. Important properties of Fermi bubbles

Fermi bubbles have several peculiar properties that should be taken into account in theoretical models [3, 39].

- The surface brightness of the bubbles is uniform. Therefore their volume emissivity is non-uniform and edges of the bubbles should emit more gamma rays in comparison to the central part. Total gamma-ray flux is $F_\gamma \approx 4 \times 10^{37}$ erg/s.
- Edges of the bubbles are sharp. Transition region between the bubble interior and exterior occupies less than 2°-3° and is very small in comparison to the size of the bubble of 60°.
- Spectrum of the gamma-ray emission is the same everywhere in the bubble. The spectral variations suggested by [27] and [45] was not confirmed by analysis of [3]. The spectral index of gamma-ray emission equals -2 in energy range between 1 GeV and 100 GeV. Below 1 GeV there is a cut-off or spectral hardening and above 100 GeV there is an indication on spectral softening or a possible high-energy cut-off.

Theoretical models of Fermi bubbles can be separated into two major groups. There are hadronic models, where gamma-ray emission generated by proton-proton collisions and leptonic models, where gamma rays are produced by inverse-Compton scattering of relativistic electrons on soft (background) Galactic photons. At present neither of them can be completely ruled out. Below we present a preliminary analysis of the bubble emission, which may give ideas how to differentiate between these models if more detailed analysis or better observational data are available.

The first indication may come from radio observations of the Fermi bubbles. Indeed observations of WMAP and subsequent observations by Planck showed that a micro wave flux came from the region of Fermi Bubbles with a power-law spectrum:

$$\Phi_\nu^{FB} \propto \nu^{-\alpha}, \tag{1}$$

where α is the spectral index of emission which is between 0.5 and 0.63 at GHz frequency range. If the emission is synchrotron emission, then spectrum of the radio emitting electrons is close to a power-law,

$$N_e \propto E_e^{-(2\alpha+1)}. \tag{2}$$

Therefore the spectral index of radio emitting electrons should be close to -2.

These observations make the hadronic model problematic. If the gamma-ray emission is produced entirely by proton-proton collisions, then the spectral index of relativistic protons equals -2. The spectrum of secondary electrons produced by these protons has the spectral index -3 because of inverse-Compton and synchrotron losses, that is too steep to reproduce the observed radio emission. To make radio emission consistent with observational data one need to assume that there is an additional component of primary electrons responsible for the radio emission.

IOP Conf. Series: Journal of Physics: Conf. Series **1181** (2019) 012001 doi:10.1088/1742-6596/1181/1/012001

This model was analyzed in [18] and it was shown that under ideal background parameters hadronic emission may be responsible only less than 80% of the total gamma-ray flux. And at least 20% of the flux should be produced by primary electrons.

The observed spectral features at $E_\gamma < 1$ GeV and at $E_\gamma > 100$ GeV of the gamma-ray spectrum provide also restrictions for bubbles models. The feature at 1 GeV in the case of hadronic origin may be related to a threshold of the neutral pion production in the proton-proton reaction. In leptonic models explanation of this hardening is quite different. It is usually assumed that spectrum of electrons has a spectral index of -2 and a cut-off at several hundred GeV. Therefore spectrum of leptonic gamma-ray emission produced by inverse-Compton scattering of monoenergetic soft photons should be rather hard with spectral index of -1.5. On the other hand, above the Galactic plane we have at least 3 types of soft photons: relic, infra-red and optical. Their combination mimics a power-law spectrum with the index of -2. At low enough energies however only scattering on relic photons dominates that provides the spectrum hardening.

Taking this into account one can predict that spectrum of leptonic emission will most likely continue with index of -1.5 to low energies. For hadronic emission the situation is slightly different. Due to emission of secondary electrons with soft spectrum the overall spectrum gamma-ray emission should experience an upturn at low energies (see, e.g. [41]). The absence of the emission from secondary electrons at low energies can be a strong argument in favor of leptonic model, however the opposite is not necessary true. To test this scenario observation below 100 MeV are necessary.

In leptonic models high energy cut-off should be at fixed position since fine adjustment is required to mimic the spectral index of -2. Hadronic models on the other handare more relaxed. Current observations by Fermi-LAT [3] and HAWC [1] do not provide sufficient limitations for leptonic models, but in the future can potentially help to differentiate the origin of Fermi bubbles [37].

In the next two sections we are going to briefly discuss some complications of both hadronic and leptonic models.

3. Restrictions on the hadronic wind model

Hadronic wind model suggested by [21] assumes that relativistic protons generated in star formation regions within the central 200 pc radius are carried into the Galactic halo by plasma outflow and somehow trapped there for the time of proton-proton collisions. The total energy release in the central region was estimated as 10^{40} erg/s which is significantly larger than total gamma-ray brightness of the Fermi bubbles.

The problem of the model is the necessity to confine protons stored inside the Fermi bubbles for a very long time of proton-proton collisions, $\tau_{pp} = (n_0 \sigma_{pp} c)^{-1} \approx 4 \times 10^{10}$ yrs where $n_0 \approx 10^{-3}$ cm^{-3} is the gas density in the Fermi bubbles [30] and cross-section of proton-proton collisions is $\sigma_{pp} \approx 3 \times 10^{-26}$ cm^2. Then we should assume mirroring magnetic walls at the bubble edges [28]. The energy density of magnetic field there should exceed the energy density of CR protons in the bubbles. The latter can be estimated based on the surface brightness of the Fermi bubbles $I_\gamma \approx 5 \times 10^{-7}$ GeV cm^{-2} s^{-1} sr^{-1} as

$$\epsilon_p = \frac{4\pi I_\gamma}{2 \times 0.075 \sigma_{pp} n_0 c \times 2L} \ln\left(\frac{E_{max}}{E_{min}}\right) \approx 18 \log_{10}\left(\frac{E_{max}}{E_{min}}\right) \text{eV cm}^{-3}, \quad (3)$$

where $L \approx 1$ kpc is the characteristic size of the gamma-ray emitting region in the bubbles and 0.075 is an average fraction of the energy transferred from the proton to gamma-ray photon. E_{max} and E_{min} are maximum and minimum energy of protons.

To confine protons with energy density of 18 eV cm^{-3} it is necessary for the magnetic field to be at least of order of 20 μG. This value is very high even in comparison to the magnetic field strength in the Galactic disk and can potentially create some problems for explanation of the radio-emission [18]. The other problem is that observations of gamma-ray emission of molecular clouds give an estimate of CR energy density below 10 eV cm^{-3} [44].

Another problem may come from the fact that regular magnetic field cannot confine particles forever. Even in absence of turbulence there will be Hall diffusion or drift across the field lines with diffusion coefficient of order of $D_b \approx r_L c/3$, where r_L is particle gyroradius. The flux of escaping protons can be estimated as $j_{esc} = D_b \frac{\partial N_p}{\partial r} \approx D_b \frac{N_p}{\delta L}$, where N_p is the density of the relativistic protons and $\delta L \leq 0.3$ kpc is the characteristic size of the transitional region between the bubble and halo [3].

Therefore one can compare life time of the protons due to escape through the bubble walls with life time of proton-proton collisions:

$$\frac{\tau_{pp}}{\tau_{esc}} \approx \frac{r_L}{3\sigma_{pp} n_0 L \times \delta L} = 0.7 \times \left(\frac{B}{20\ \mu\text{G}}\right)^{-1} \times \left(\frac{E}{1\ \text{TeV}}\right). \tag{4}$$

It interesting to note that escape of relativistic protons starts at about 1 TeV, which corresponds to a high-energy turn-over in the gamma-ray spectrum.

The difficulties of confinement of protons can be resolved in hadronic jet model which does not assume stationary state and uses a single event to power the bubbles. However the total energy in this event should reach enormous value of $W \approx F_\gamma t_{pp} \approx 10^{56}$ erg.

4. Acceleration of electrons inside the bubbles

The energy of relativistic electrons needed to produce 100 GeV gamma-ray emission is about or more than several hundred GeV. Lifetime of these electrons is quite short in comparison with that of protons ($\tau_e \sim 3 \times 10^6$ yrs). Therefore, these electrons should be produce directly inside or nearby the emission region of the bubbles, i.e. nearby the bubble edges. Two different mechanism of electron acceleration were suggested. Thus, [16] assumed that these electrons were accelerate by a shock/shocks penetrating into the halo, which are produced by energy release in the GC due to a tidal disruption of a star on the central black hole. The spectrum of electrons accelerated by a shock is $\propto E^{-2}$ as needed for the observed gamma-ray from the bubbles. This model with a set of free parameters (like the number of accelerated particles or the coefficient of spacial diffusion near the shock etc.) is able to reproduce the observed gamma-ray and microwave emission from the bubble. The maximum energy of accelerate electrons is derived from the equality condition between the time of shock acceleration and the time of energy losses [12]. Although a shock is not observed directly, there are some indication for supersonic fluxes of plasma there (see [26]).

Alternatively, electrons can be accelerated by a magnetic turbulence excited behind the shock as suggested by [31]. This type acceleration is similar to classical stochastic acceleration suggested by Fermi in 1954, which produce hard spectra ($\propto E^{-1}$), However , the arbitrary assumed spatial dependence of a cut-off in the spectrum makes the gamma-ray flux close to E^{-2} because of integration along the line of sight.

The number of accelerated particles depends on the process of efficiency of particle injection into the acceleration regime. The latter is usually omitted since correct description of the injection requires careful analysis of the spectrum of turbulence at very short wavelengths. However this information is crucial for determination of the magnitude of the accelerated particles spectrum. Also, the knowledge about particle injection is also important for understanding if stochastic acceleration is effective for generation of nonthermal spectra or the interaction between charged particles with turbulence results only in heating of the background plasma, i.e. no acceleration [20, 35].

Since Fermi bubbles are located in the Galactic halo it is necessary to take into account that injection of electrons for any type of acceleration can happen from two main sources of particles:

either from thermal plasma or from the cosmic ray electrons. The thermal plasma has much more seed electrons therefore potential magnitude of the spectrum can be very high. On the other hand, cosmic ray electrons have much higher energy therefore it is easier to accelerated them.

Therefore for low acceleration rate is most likely that injection from cosmic ray electrons will be higher than injection from thermal plasma. As acceleration rate increases, the thermal plasma contribution will increase and at some point it may overcome the contribution from cosmic rays. However for the parameters of Fermi bubbles acceleration rate is not enough for thermal plasma to dominate and the main sources of the seed electrons are CRs [19], despite this fact is usually ignored for the studies of acceleration of particles in Fermi bubbles.

The major advantage of using cosmic rays as seed particles for acceleration is the fact that injection happens in the energy range where acceleration totally dominates over losses. E.g. in the case of stochastic acceleration if we assume that turbulence have a power-law spectrum, a power-law tail of particles should be formed. The reacceleration process can be roughly described as a smooth attachment of a power-law tail to the background distribution function i.e. the reaccelerated spectrum looks like

$$f_r(E) = f_0(E)\Theta(E_c - E) + f_0(E_c)\left(\frac{E}{E_c}\right)^{-\delta_1}\Theta(E - E_c)\,, \tag{5}$$

where $f_0(E)$ is the initial spectrum of particles before the reacceleration and E_c is determined from the continuity of the particle flow in the energy space. In most cases E_c can be estimated from the continuity of the derivative $\partial f/\partial E$. In the case of Galactic CR electrons the spectrum $f_0(E)$ have a sharp spectral break near 2.2 GeV so it is reasonable to assume that $E_c \approx 2$ GeV.

The intensity of the inverse Compton emission can be estimated in the following way

$$I_{IC} = \frac{\Delta x}{4\pi}\int\limits_{E_{IC}^{(min)}}^{E_{IC}^{(max)}} f_r(E)\left(\frac{dE}{dt}\right)_{IC} dE\,, \tag{6}$$

where $E_{IC}^{(min)}$ and $E_{IC}^{(max)}$ is energy of the electrons corresponding to maximum and minimum energy of gamma-ray photons accordingly and $(dE/dt)_{IC}$ are inverse Compton energy losses. Using for simplicity non-relativistic approximation $(dE/dt)_{IC} = \beta_{IC}E^2$ and assuming that $\delta < 3$ one can obtain that the thickness of the gamma-ray emitting region is

$$\Delta x = I_{IC}\frac{4\pi(3-\delta)}{\beta_{IC} f_0(E_c)}\left(\frac{E_{IC}^{(max)}}{E_c}\right)^{\delta-3} E_c^{-3}\,. \tag{7}$$

According to the one of the publicly available GALPROP simulations $S_S Z_8 R_{20} T_{\rm inf} C_2$ [2] the density of electrons at 2 GeV is $f_0(E_c) = 8.7\times 10^{-16}$ cm^{-3}MeV^{-1}. The cut-off in the gamma-ray spectrum requires $E_{IC}^{(max)} \approx 0.3$ TeV [39] and $\delta = 2.1$ as it is required by radio observations [4]. Therefore the thickness of FERMI bubble walls is only $\Delta x = 30$ pc. This value is much smaller than the value of 1-2 kpc required to reproduce the spatial shape of the bubble and also require an extremely small spatial diffusion coefficient to confine relativistic electrons in a thin shell. A finer $\Delta x(\delta)$ relations based on the correct expressions for the gamma-ray and radio emission together with some restriction from observation data are presented in Fig. 1.

One can conclude that reacceleration process overproduces relativistic electrons. This effect does not depend on the nature of the reacceleration mechanism as long as it produces power-law spectrum of electrons. To avoid this problem it is necessary to introduce another kind of losses

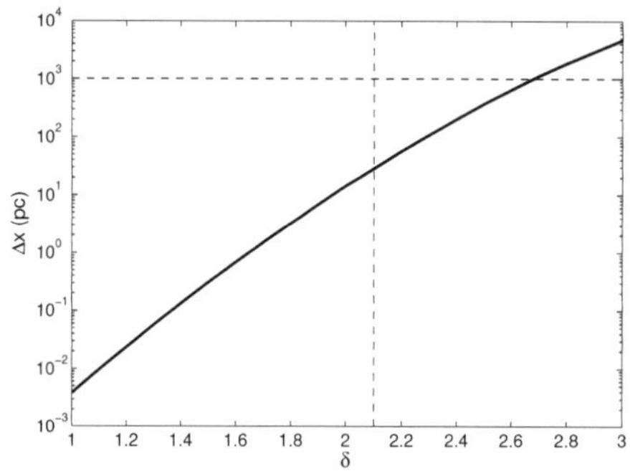

Figure 1. Thickness of the bubble walls Δx necessary to produce observed gamma-ray flux (thick line) depending on the spectral index of relativistic electrons inside the bubble δ. Vertical dashed line marks the expected spectrum of electrons based on radio data [4]. Horizontal dashed line marks the expected thickness based on the spatial structure [39].

which should be effective in GeV energy range. One of the possible solutions is introduction of adiabatic losses. Indeed if the plasma flow in the Galactic halo in the vicinity of Fermi bubbles is non-uniform particles start to lose energy with a rate

$$\frac{dE}{dt} = \frac{E}{3}\nabla \cdot \mathbf{u}, \tag{8}$$

where $\mathbf{u}$ is the flow velocity. According to [11] the velocity of the Galactic wind can be described as linear function of the altitude $u(z) = 3v_0 z$ and the gradient value v_0 can be as high $v_0 \simeq 10^{-15}\mathrm{s}^{-1}$. With this rate adiabatic losses become essential in GeV energy range and are able to significantly reduce the value of $f(E_0)$ in (7) and soften the restrictions on Δx and δ. As we showed in [19] this assumption indeed helps to reproduce observed spectrum of the gamma-ray and radio emission using reacceleartion of the CR electrons.

Another way to avoid this problem is to take into account anisotropic diffusion of electrons and structure of the global magnetic field in the vicinity of Fermi bubbles [32]. Since diffusion of the particles across the magnetic field lines is much weaker than the diffusion along the field lines, injection of the CR electrons into the acceleration zone may be limited. Careful analysis of the acceleration may solve the problem of particles overproduction.

5. Influence of Fermi bubbles on the spectrum of cosmic rays near the Earth

Since size of the Fermi bubbles is huge, one can expect that Fermi bubbles should also affect the population of CRs near the Earth. The effect of reacceleration should appear as a "bump" in the spectrum of electrons at energies of about 10-100 GeV. Indeed, Fermi bubbles act as a source of electrons with spectral index close to -2, while Galactic sources should produce electrons with spectral index close to -2.4 [2]. Therefore combination of these two sources should produce characteristic features in the spectrum of CR electrons. This component may be related to the spectral hardening seen by AMS-02 [6] and CALET [5] in the spectrum of electrons. However since electrons are strongly affected by energy losses, a careful detailed analysis required to estimate the possible influence of Fermi bubbles on the spectrum of CR electrons near the Earth.

Even stronger effects are expected if we take into account reacceleration of Galactic protons and nuclei, which are almost unaffected by energy losses during their propagation in the Galaxy. The important point of the CR spectrum is sudden steepening around 3×10^{15} eV which indicates on the change of the acceleration or propagation mechanism. Smooth attachment of the spectra above and below this energy and sharpness of the transition indicates that we are dealing with sole spectrum rather than with sum of two distinct components. The review of different models suggested to explain this phenomenon can be found in the original paper [17].

In summary, it is generally agreed that supernova remnants (SNR) shocks can only accelerate particles to energies $E < 10^{15}$ eV. Acceleration beyond this energy by SNRs is problematic due to restriction on the maximum energy for a typical SNR of age T expanding with velocity of u_{sh}

$$E_{max} \sim Ze\beta_{sh}BTu_{sh}\,, \tag{9}$$

where $\beta_{sh} = u_{sh}/c$, B is the magnetic field strength at the shock and the term $\mathcal{E} = \beta_{sh}B$ in this case can be interpreted as an effective electric field.

However Fermi bubbles with age and size exceeding that of typical SNR by 3 orders of magnitude can easily accelerate particles to much higher energies. Using Eq. (9) one can estimate that the bubbles have a potential to form the spectrum of CR in between 3×10^{15} eV and 10^{18} eV. We do not consider the possibility that Fermi bubbles can form a whole spectrum of CR both below and above 3×10^{15} eV. The reason for this is that the total power needed for the luminosity of CRs in our Galaxy $L_{CR} \sim 10^{41}$ erg/s [8] and the average energy release toward the Fermi bubbles is also 10^{41} erg/s [16]. Thus our model can be described in the following way: SNR in the disk accelerate particles with power-law distribution up to energies of 3×10^{15} eV and Fermi bubbles further re-accelerate this particles up to 10^{18} eV.

In our model for simplicity we assume that CRs only consist of protons. Despite actual spectrum of CRs is more complicated, we only want to use a toy model as a proof of concept. We note that since acceleration properties of nuclei depend only on their rigidity, application of the model to the multi-component spectrum of CRs is straightforward. However detailed information about the spectra of each CR specimen produced by SNRs is required.

In the model of Fermi bubbles origin we suggested that the activity responsible for the formation of the bubbles was due to stellar capture and tidal disruption by a central black hole [16, 17]. The average time between two successive captures in the Galaxy is between 10^4 yrs and 10^5 yrs [40]. Thus the activity is periodic and characteristic period between two event is shorter than the characteristic lifetime of the Fermi bubbles. Periodic energy releases in the Galactic center should form series of shocks propagating through the halo.

The problem of particle acceleration in conditions of supersonic turbulence (multi-shock structure) was extensively analyzed before. In series of papers by [14, 13, 15] as applied to acceleration processes in OB associations,which is quite similar to the structure of the Bubble, they introduce a nondimensional parameter characterizing acceleration regime as

$$\psi = \frac{l_{sh}}{l_D} = \frac{ul_{sh}}{D} \sim \frac{ul_{sh}}{cr_L}\,. \tag{10}$$

The corresponding energy E_1 that separates different regimes can be estimated from the condition $\psi \sim 1$ or $l_D(E_1) \sim l_{sh}$ which for the conditions of the Fermi bubble is

$$E_1 = eBl_{sh}u/c = 10^{15}\left(\frac{B}{5\ \mu G}\right)\left(\frac{l_{sh}}{30\ \mathrm{pc}}\right)\left(\frac{u}{10^8\ \mathrm{cm/s}}\right)\ \mathrm{eV}. \tag{11}$$

In the case of $\psi \gg 1$ or $l_{sh} \gg l_D$ analyzed in [14, 13], there is a combined effect of a fast particle acceleration by a single shock, which generates the spectrum E^{-2} and relatively slow transformation of this spectrum due to interaction with other shocks (stochastic Fermi acceleration) into a hard E^{-1} spectrum in the intershock medium at relatively low energies. However it is unclear if such slow transformation can be completed within the life time of the shocks in the Bubble. Furthermore the hard E^{-1} spectrum requires the power which significantly exceeds 10^{41} erg/s which the bubbles can not supply. It is reasonable to assume that Fermi bubbles do not affect the spectrum of CR below E_1.

For $\psi \ll 1$ or $l_{sh} \ll l_D$ the acceleration regime changes to a pure stochastic acceleration by a supersonic turbulence. In the stationary case the equation for accelerated CRs can be presented in the form [14]

$$\frac{\partial}{\partial z}D(\rho)\frac{\partial f}{\partial z} + \frac{1}{\rho}\frac{\partial}{\partial \rho}D(\rho)\rho\frac{\partial f}{\partial \rho} + \frac{1}{p^2}\frac{\partial}{\partial p}\kappa(\rho)p^2\frac{\partial f}{\partial p} = 0 \tag{12}$$

where ρ and z are the cylindrical spatial coordinates, p is the particle momentum, $D(\rho)$ is the spatial diffusion coefficient and κ is the momentum diffusion coefficient.

We present the bubble region as a cylinder extending above and below the Galactic plane from $z = 0$ to $z = \pm H$ with a radius $\rho = \rho_B$. As the boundary conditions we put the density of particles f equaled zero at the Galactic halo surface.

The diffusion coefficients inside and outside the bubble are supposed to be different

$$\begin{aligned} D(\rho) &= D_B\theta(\rho_B - \rho) + D_G\theta(\rho - \rho_B) \\ \kappa &= \kappa_B\theta(\rho_B - \rho)\Theta(E - E_1) \end{aligned} \tag{13}$$

where $D_B = Lc/3$ is the coefficient inside the bubble due to interactions with a supersonic turbulence and D_G is the average diffusion coefficient in the Galaxy defined e.g. in [9]. The momentum diffusion coefficient is $\kappa_B = p^2u^2/9D_D$ and we assume that there is little or no stochastic acceleration outside the bubble.

The momentum dependence of f can be presented by a power-law function, $f(p) \propto p^{-\gamma}$, where γ is determined by balance between acceleration inside the bubble and escape. From Eq.(12) it can be approximated as

$$\gamma \approx \frac{3}{2} + \sqrt{\frac{9}{4} + \pi^2\frac{D}{\rho_B^2\kappa}} = \frac{3}{2} + \sqrt{\frac{9}{4} + \frac{\pi^2 D_B^2}{u^2\rho_B^2}}\,. \tag{14}$$

Similarly to the reacceleration of the CR electrons one can expect that inside the bubbles spectra produced by SNRs and reaccelerations should be smoothly connected and described by Eq. (5) with corresponding choice of spectral indexes. However spectrum near the Earth should be distorted by propagation and escape from the Galaxy. Since sources of CR above the knee and below the knee are located in the different parts of the Galaxy, it is not obvious that smooth connection of the spectra will remain.

To test this idea, we work out a concrete numerical model. Essentially, we solve the stationary state CR transport equation (12) in our Galaxy with two Fermi Bubbles (one on each side of the Galactic plane). We modeled our Galactic halo as a cylinder of radius $\rho_G = 20$ kpc, and the top and bottom at ± 10 kpc from the mid-plane. Each Fermi Bubble is also a cylinder of the same height ± 10 kpc, but with a radius $\rho_B = 3$ kpc. The spatial diffusion coefficient are different inside and outside the bubble as described by Eq. (13).

Since we expect the average separation between shock in the Fermi bubbles to be of order of 100 pc [17] we consider a constant spatial diffusion coefficient and adopt $D_B = 2.08 \times 10^{30}$ cm^2 s^{-1}. Outside the bubble, we take into account the energy (or momentum) dependence of the spatial diffusion coefficient and adopt $D_G = D_0(pc/4\,\mathrm{GeV})^{0.6}$, $D_0 = 6.2 \times 10^{28}$ cm^2 s^{-1} [2].

For Galactic SNRs adopt the distribution suggested by [38] and modified it with a Gaussian thickness profile. We adopt the idea that SNRs inject energetic particles in the form of a power law with a high-energy cutoff at $p_{\max}c \approx 3 \times 10^{15}$ eV. Therefore the source function is

$$\begin{aligned} Q(\rho, z, p) &= Q_0\left(\tfrac{p}{p_{\max}}\right)^{-\mu}\exp\left(-\tfrac{p}{p_{\max}}\right)\left(\tfrac{\rho}{R_\odot}\right)^{1.2} \\ &\times\exp\left(-\tfrac{3.22\rho}{R_\odot}\right)\exp\left(-\tfrac{z^2}{h^2}\right), \end{aligned} \tag{15}$$

IOP Conf. Series: Journal of Physics: Conf. Series **1181** (2019) 012001 doi:10.1088/1742-6596/1181/1/012001

where we take $h = 100$ pc, $R_\odot = 8$ kpc.

The spectrum evaluated at Earth's position is the solid line shown in Figure 2. The model fits the data reasonably well and it is not coincident that the spectra join smoothly at the knee.

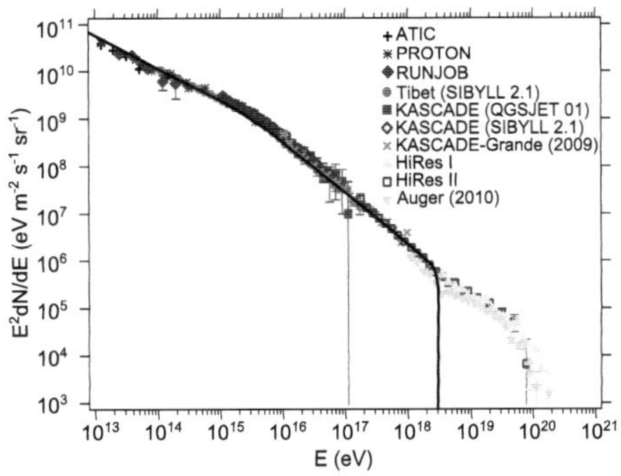

Figure 2. CR spectrum at the Earth as a combination of the SNR contribution (in the Galactic disk) and the stochastic acceleration in the Fermi Bubbles. The black solid line is the spectrum from our numerical model. For the references to the experimental data see the original paper [17].

6. Conclusion

We analyzed two groups of theoretical models of Fermi bubbles: hadronic and leptonic ones. We showed that hadronic models may experience some difficulties since long confinement of protons inside the bubbles is necessary.

Although at present we do not have reliable arguments observational evidences in favor of leptonic or hadronic origin of the gamma-ray emission from Fermi bubbles, it may be available in future from observations in the ranges below 100 MeV and above 1 TeV.

We also discussed connections between CRs in the bubbles and that seen at the Earth.

- If gamma-ray and radio emission from Fermi bubbles is produced by electrons accelerated in-situ, we cannot exclude that these electrons are emitted by SNRs which are re-accelerated in the halo. The point is that SN electrons propagate into halo. Processes of re-acceleration there provide particles with harder spectrum that explains the origin of gamma-ray and microwave emission from the bubbles. The advantage of this mechanism is that the energy of electrons is increased in two orders of magnitude only (cf. for acceleration of background electrons eleven orders of magnitude is needed). This source of accelerated particles is often overlooked despite being very efficient. Moreover, spatially uniform model involving reacceleration of the CR electrons overproduces gamma-ray emission from the Fermi bubbles. This effect does not depend on the particular choice of the reacceleration mechanism and only depends on the propagation model for the CR electrons. This fact may indicate on strong outflow in the Galactic center region or on the importance of the anisotropic diffusion in the region of Fermi bubbles or in the whole Galactic halo.
- As potential giant accelerators with age and size exceeding those of typical SNR by 3 orders of magnitude, Fermi bubbles can significantly distort spectrum of CR protons and nuclei observed near the Earth. in particular they may be responsible for formation of the spectrum of CRs beyond the "knee".

Acknowledgments
VAD and DOC are supported in parts by the grant RFBR 18-02-00075. DOC is supported in parts by foundation for the advancement of theoretical physics and mathematics "BASIS". CMK is supported in part by the Ministry of Science and Technology of Taiwan under grants MOST 104-2923-M-008-001-MY3 and MOST 105-2112-M-008-011-MY3. KSC is supported by the GRF Grant under HKU 17310916.

References
[1] Abeysekara, A. U.; Albert, A.; Alfaro, R. 2017, ApJ, 842, 85
[2] Ackermann, M., et al., 2012, ApJ, 750, 3
[3] Ackermann, M., Albert, A., Atwood W. B. et al. 2014, ApJ, 793, 64
[4] Ade, P. A. R., Aghanim, N., Arnaud, M. et al.(Planck Collaboration) 2013, A&A, 554, 139
[5] Adriani, O., Akaike, Y., Asano, K. et al. 2018, Phys. Rev. Lett., 120, 261102
[6] Aguilar, M., Aisa, D., Alpat, B. et al. 2014, Phys. Rev. Lett., 113, 221102
[7] Aharonian, F., Akhperjanian, A. G., Anton G. et al. 2009, ApJL, 695, 40
[8] Berezhko, E. G., Yelshin, V. K. & Ksenofontov, L. T. 1994, Astropart. Phys., 2, 215
[9] Berezinskii, V. S., Bulanov, S. V., Dogiel, V. A., Ginzburg, V. L., & Ptuskin, V. S. 1990, *Astrophysics of Cosmic Rays*, ed. V.L.Ginzburg, (Norht-Holland, Amsterdam)
[10] Bland-Hawthorn, J., & Cohen, M. 2003, ApJ, 582, 246
[11] Bloemen, J. B. G. M., Dogiel, V. A., Dorman, V. L., & Ptuskin, V. S. 1993, A&A, 267, 372
[12] Bulanov, S.V. & Dogiel V.A., 1979, Sov. Astr. Lett. 5, 278
[13] Bykov, A. M. & Fleishman, G. D. 1992, MNRAS, 255,269
[14] Bykov, A. M. & Toptygin, I. N. 1993, Physics Uspekhi, 36, 1020
[15] Bykov, A. M. & Toptygin, I. N. 2001, Astronomy Letters, 27, 625
[16] Cheng, K.-S., Chernyshov, D. O., Dogiel, V. A., Ko, C.-M., & Ip, W.-H. 2011, ApJ, 731, L17
[17] Cheng, K.-S., Chernyshov, D. O., Dogiel, V. A., Ko, C.-M., Ip, W.-H., Wang, Y. 2012, ApJ, 746, 116
[18] Cheng, K.-S., Chernyshov, D. O., Dogiel, V. A., Ko, C.-M. 2015, ApJ, 799, 112
[19] Cheng, K.-S., Chernyshov, D. O., Dogiel, V. A., Ko, C.-M. 2015, ApJ, 804, 135
[20] Chernyshov, D. O., Dogiel, V. A., Ko, C. M. 2012, ApJ, 759, 113
[21] Crocker, R. M., & Aharonian, F. 2011, PRL, 106, 101102
[22] Croston, J. H., Kraft, R. P. & Hardcastle, M. J. 2007, ApJ, 660, 191
[23] Dobler, G., Finkbeiner, D. P., Cholis, I., et al. 2010, ApJ, 717, 825
[24] Feain, I. J., Cornwell, T. J., Ekers, R. D. et al. 2011 ApJ, 740, 17
[25] Finkbeiner, D. P. 2004, ApJ, 614, 186
[26] Fox, A. J.; Bordoloi, R.; Savage, B. D.et al, 2015, ApJ, 799, L7
[27] Hooper, D., Slatyer, T. R. 2013, PDU, 2, 188
[28] Jones, D. I., Crocker, R. M., Reich, W., Ott, J., & Aharonian, F. A. 2012, ApJ, 747, L12
[29] Junkes, N., Haynes, R.F., Harnett, J. L., & Jauncey, D.L. 1993, A&A, 269, 29
[30] Kataoka, J., Tahara, M., Totani, T. et al. 2013, ApJ, 779, 57
[31] Mertsch, P. & Sarkar, S. 2014, PhRvD, 90, 061301
[32] Mertsch, P., Petrosian, V. 2018, arXiv:1808.00330
[33] Mingo, B., Hardcastle, M. J., Croston, J. H. et al. 2011, ApJ, 731, 21
[34] Mingo, B., Hardcastle, M. J., Croston, J. H. et al. 2012, ApJ, 758, 95
[35] Petrosian V. 2012, Space Sci. Rev., 173, 535
[36] Pshirkov, M. and Vasiliev, V. & Postnov, K. 2016, EPJWC, 125, 03010
[37] Razzaque, S. & Yang, L. 2018, Galaxies, 6, 47
[38] Stecker F. W., & Jones, F. C. 1977, ApJ, 217, 843
[39] Su, M., Slatyer, T. R., & Finkbeiner, D. P. 2010, ApJ, 724, 1044
[40] Syer, D. & Ulmer, A. 1999, MNRAS, 306, 35
[41] Yang, R., Kafexhiu, E., Aharonian, F. 2018, A&A, 615, 108
[42] Yang, H.-Y. and Ruszkowski, M. and Zweibel, E. 2018, Galaxies, 6, 29
[43] Yang, R.-Z., Sahakyan, N., de Ona Wilhelmi, E., et al. 2012, A&A, 542, 19
[44] Yang, R.-z., de Oña Wilhelmi, E., & Aharonian, F. 2014a, A&A, 566, A142
[45] Yang, R.-Z., Aharonian, F., Crocker, R. 2014, A&A, 567, 19

IOP Conf. Series: Journal of Physics: Conf. Series **1181** (2019) 012002 doi:10.1088/1742-6596/1181/1/012002

Astrospheres and Cosmic Rays

A. Struminsky[1,2] **and A. Sadovski**[1,3]

[1] Space Research Institute, 84/32 Profsouznaya str., Moscow, 117997, Russia
[2] Moscow Institute of Physics and Technology, 9 Institutskiy per., Dolgoprudny, Moscow Region, 141701, Russia
[3] National Research University Higher School of Economics, 20 Myasnitskaya ulitsa, Moscow, 101000, Russia

E-mail: asadovsk@iki.rssi.ru

Abstract. Ionization plays a key role in the formation of stars, planets and planetary atmospheres. Cosmic rays (CR) are the main source of ionization, therefore it is important to estimate fluxes of galactic and stellar cosmic rays (GCR and SCR) for different stages of evolution of stars and planetary systems. Radiation conditions on exoplanets might be important for the origin and development of life. We present a review of the current state of knowledge of astrospheres and their interactions with GCR and SCR. We pay special attention to estimates of radiation conditions near exoplanets recently discovered in a habitable zone of their hosting stars.

1. Introduction—astrophysics of low energy CR

Traditionally the astrophysics of cosmic rays (OG section of CR conferences) considers CR above 10 GeV since CR of lower energies have their flux modulated in the heliosphere by the solar magnetic field and solar wind, some part of low CR rays are accelerated inside the heliosphere in solar flares or by shock waves driven by Coronal Mass Ejections (CME). Problems of low energy CR physics were discussed at SH (Sun and Heliosphere) section of CR conferences [1]. Recent years have seen the birth of low energy CR astrophysics, which is tightly connected to astrochemistry, radioastronomy and solar physics. At present, the astrophysics of low energy CR consider CR in three locations—molecular clouds, protoplanetary discs and astrospheres.

The cosmic-rays dominate in the ionization and heating of the interstellar medium (ISM). Recent measurements, made beyond the heliopause by the Voyager I spacecraft, have provided data of lower energy cosmic-ray protons and electrons down to energies about 3 MeV [2]. Nevertheless, it remains unclear whether the cosmic-ray fluxes reach their unmodulated values even at the current location of Voyager I. However, the ionization cross-sections for atomic and molecular hydrogen have a maximum at 0.01 MeV, so an extrapolation of unobserved energies is still required to determine the implied cosmic ray ionisation rate, which remains quite uncertain even in the solar neighbourhood [3].

Remarkably, the spectra of both proton and electron CRs in the local interstellar medium (ISM)— at least down to particle energy of a few MeVs— are now known with some confidence thanks to the recent data collected by the Voyager probe at large distances from the Sun [2]. Whether or not such spectra are representative of the average Galactic spectra, especially for MeV CRs, is still not clear. However, the analysis of gamma rays from Molecular clouds (e.g., [4]) seems to indicate that at least the spectrum of proton CRs of energy above a few GeV is quite homogeneous in our Galaxy. The intensity of CRs in the local ISM as revealed by Voyager measurements is too weak to explain the level of ionization rate observed in clouds. Possible solutions to this problem include the presence of another source of ionization or a non-uniform intensity of low energy CRs throughout the Galaxy [5].

One of the possible sources of CR was suggested by [6]. They hypothesized the existence of an acceleration mechanism for both cosmic-ray protons and electrons through the diffusive shock acceleration near protostars. This mechanism may explain the high ionisation rate and the synchrotron emission observed. Also, such internal sources of energetic particles should have a strong influence on the ionisation of the protostellar disc, on star and planet formation processes, and on the formation of pre-biotic molecules [6].

As in the solar system where solar wind can inhibit the CRs propagation, also stellar winds can modulate CR spectra within the circumstellar environment and subsequently into the disk. In [7] a two-dimensional protoplanetary disk model of a T-Tauri star system was constructed and ionization from stellar and interstellar far ultraviolet, stellar X-ray photons, and CRs was found. According to their results stellar winds can conform a heliosphere-like analog, i. e., using the term of the authors T-Tauriosphere decreases CR ionization rates by several orders of magnitude at low to moderate CR energies (~ 1 GeV). The implications of a diminished CR ionization rate on the physics of the disk was performed by investigation of the magnetorotational instability: if winds are so efficient at CR modulation, than the major source of ionization would be short-lived radionuclides [7]. However, the authors didn't consider the stellar CR.

The Chandra X-ray satellite observed the large flares on T Tauri stars which can give a significant amount of energetic particles and recently the Herschel Space Observatory provided evidence of their possible contribution to the disk ionization of young stars. Authors of [8] modeled ionization of stellar energetic particles in protoplanetary disks around T Tauri stars using a particle energy distribution derived from solar flare observations and an enhanced stellar particle flux proposed for T Tauri stars. They have shown that stellar energetic particles can be an important ionization agent for disk chemistry.

Test-particle numerical simulations of energetic protons propagating into a realistic T Tauri stellar wind have shown that in the energy range 0.1–10 GeV they are consistent with expectations from Chandra and the Herschel Space Observatory observations [9]. Also in [9] was found that the disk ionization is dominated by X-rays over much of its area, except for some narrow regions where particles are channeled by the turbulent magnetic field, in contrast with a previous theoretical study.

Moreover, a proper description of particle transport is essential to compute the ionization rate since the electron and positron differential fluxes depend sensitively on the fluxes of both protons and photons [10]. Their results show that the CR ionization rate in high-density environments, such as the inner parts of collapsing molecular clouds or the mid-plane of circumstellar disks, is higher than previously assumed.

Exoplanet discovery gives a great impetus to investigate both habitability conditions and possible habitable planets. However, from the investigation of the Earth environment it seems to be clear that the radiation environment should have a great influence on any planet and its atmosphere. Despite their importance, CRs and their influence on radiation conditions near planets have dropped out of the discussion, and, perhaps, only one group actively investigates the influence of CRs on exoplanets (see for example [11]). The influence of Galactic cosmic rays (GCRs) on planets depend on planetary magnetosphere and atmosphere and has been considered in several papers (see, e.g., [12, 13]). In paper [14] the influence of CRs on the atmosphere of an Earth-like exoplanet was investigated and it was suggested that the GCR flux might be considered as an isotropic and nearly constant one. However the authors did not consider GCR modulation.

Such problems are similar to the investigations of archean Earth in the young Sun environment. For example, the quantitative model developed in [15] demonstrated that a change of the interstellar medium surrounding the heliosphere triggers significant changes of planetary environments caused by enhanced fluxes of neutral atoms as well as by increased cosmic ray fluxes for 10^4–10^6 years. In paper [16] the 3D MHD models for solar wind and a 2D model for CGR transferring near archean Earth were used. It was showed that the GCR flux has greatly reduced with comparison to present conditions. The reason mainly being the shorter solar rotation period and tighter winding of the Parker spiral, and to the different surface distribution of the more active solar magnetic field.

In contrast to Galactic cosmic rays, the detection (or separation from the total flux) of SCR is impossible far from the parent star, and because of this they are the "stepsons" of CR physics usually mentioned as a possible CR component approximately once every ten years [17, 18, 19, 20, 21, 22, 23]. Nevertheless, stellar CRs are considered an important factor of space weather in the habitable zone of a star in many papers [24, 25, 23]. Unfortunately, the spectrum of stellar CRs cannot be determined and the well known spectra of solar CRs [25, 24] are used to model stellar CRs. Another approach is based on general physical laws and also uses the solar–stellar analogy, but without using near-Earth solar CR observations [23].

Here we review methods proposed in our previous papers [23, 26, 27] showing the Sun-stellar similarities and scaling of stellar parameters which allows us to obtain the correct estimates for different stars by the order-of-magnitude. The previous paper was devoted to the Proxima Centauri system [26] and for TRAPPIST-1 system [27]. Below we present some our results for TRAPPIST-1 and new unpublished estimates for σ Ori E. The next section is devoted to the methods using for calculations and their discussion. The section 3 presents results for TRAPPIST and σ Ori E.

2. Methods

2.1. Stellar wind and astrosphere

Solar wind velocity, particle density, magnetic field and dimension of the heliosphere are essential parameters for propagation and modulation of CR observed near the Earth. Analogous parameters of stellar wind and astrosphere would be important for CR near exoplanets of some chosen star.

For the stellar wind velocity, we use the Parker's model [28], which depends only on the coronal temperature and star radius and is universal for any star system. For a given coronal temperature T_e we may estimate the sound speed $u_{cr} = (2k_B T_e/m_p)^{-1/2}$ and critical distance, where the wind speed became equal to the sound speed, $r_{cr} = GM_*/u_{cr}$ (here M_* is a star mass, G is the gravitational constant). The stellar wind speed for the distance much larger than critical may be estimated as $V_{\rm SW} = u_{cr}\log(r/r_{cr})$.

We may get a coronal temperature from X-ray observations of a particular star or may use estimate of a maximal possible value of the coronal temperature

$$T_{\rm cor}^{\rm max} = G\frac{M_* m_p}{4kR_*}(\sqrt{3}-1)$$

determined by the condition that a radius of critical point is equal to the stellar radius plus the coronal height

$$H_{\rm cor} = \frac{2kT_{\rm cor}m_p R_*^2}{GM_*}.$$

Here R_* is the star's radius.

The number density of stellar wind may be estimated from thermal coronal loss. Let us suppose that the heat flux from corona is $Q = -(8\pi/7)R_* k(T_e)T_e$, where $k(T_e) = 6\times10^{-6}T_e^{5/2}$ erg $\cdot$ cm$^{-1}\times$ s$^{-1}\cdot$ K^{-1} is the thermal conductivity coefficient for fully ionized gas. Assuming that all the heat flux Q goes for the coronal spherical symmetrical expansion, we find that

$$Q \approx 4\pi r^2 m_p \frac{N(r)V_{\rm SW}}{2}(V_{\rm SW}^2 + V_{\rm esc}^2),$$

where $V_{\rm esc} = (2GM_*/R_*)^{1/2}$ is the escape velocity) and stellar wind number density N at some distance r is

$$N(r) = \frac{48\pi}{7} R_* \frac{10^{-6}T_e^{7/2}}{2\pi r^2 m_p V_{\rm SW}\left(V_{\rm SW}^2 + V_{\rm esc}^2\right)}$$

Also now it is possible to find the mass loss rate $M_r = 4\pi r^2 N m_p V_{\rm SW}$ and the astrosphere radius

$$R_{\rm as} = R\sqrt{\frac{m_p N V_{\rm SW}^2}{P_{\rm ISM}}},$$

where $P_{\rm ISM} = 1$ eV/cm^3 is the pressure of interstellar medium.

For estimates of the magnetic field strength in some point of the astrosphere we may use the Parker spiral. From the magnetic flux conservation follows $B(r) = B_* \left(R_*/r\right)^2$, where B_* is the stellar magnetic field at photospheric level. The radius of the first turn of Parker spiral equals to $r_1 = V_{\rm SW}T(\varphi/2\pi)$, where T is the stellar rotation period.

2.2. CR modulation

According to [29] a modulated flux of galactic cosmic rays is

$$j_0(\eta) = j_\infty(\eta)\exp\left\{-\frac{12V_{\rm SW}(r_1 - r_2)\, l Z^2 e^2 B^2(\eta+1)}{\pi^2 m^2 c^5 \left[\eta(\eta+2)\right]^{3/2}}\right\},$$

Where η is a ratio of kinetic energy to the rest particle energy, Ze is a particle charge, m is the particle mass, l is a scale of magnetic field fluctuations ($l = 2 \times 10^6$ km for the solar wind), $(r_1 - r_2)$ is a dimension of modulation shell. In our estimates we assume that a dimension of modulation shell is a difference between radius of point of interest and a radius of first Parker turn. In this case a radius of astrosphere should be greater than the radius of the first Parker turn and the planet orbit should be within the first turn. Therefore to estimate CR modulation near exoplanets we need to know the magnetic field strength and the rotation period from observations of the hosting star.

2.3. Stellar cosmic rays

To estimate the SCR influence let us assume that in stellar flares, the mechanism of proton acceleration is the same as in solar flares and that the main building blocks in star and solar coronas are magnetic loops [23].

Our main suggestion is that in the stable loop magnetic field energy should be in equipartition with thermal plasma energy. From pressure balance $B_0^2/8\pi = nkT = Gm_pM_*H/R_*^2$ assuming the mean free path $H = (n\sigma_T)^{-1}$ (where σ_T is the Thompson scattering cross-section) we find an estimate of the photospheric magnetic field in the form

$$B_0 = \frac{1}{R_*}\sqrt{\frac{8\pi G m_p M_*}{\sigma_T}}.$$

Balona [30] showed that the characteristic scale of an active region is about ten percent of star radius, $L = \alpha R$, magnetic field over an active region is about one tenth of photospheric

$B = \beta B_0$. Here coefficients $\alpha\beta < 1$. Using these parameters, we can estimate the flare energy as

$$E_{fl} = \frac{B^2 L^3}{8\pi} = 2.3 \times 10^{37} \alpha^3 \beta^2 \left(\frac{R}{R_\odot}\right) \left(\frac{M}{M_\odot}\right).$$

In big solar flares approximately 10% of flare energy is the energy of proton acceleration. Assuming similar processes for the stars we estimate amount of accelerated protons with an average energy E_p as $N - 0.1 E_{fl}/E_p$.

The electric field over an active region is $E = uB/c$, where u is velocity of reconnection, and the maximum energy of protons accelerated in flares would be $E_{\max} = (\alpha\beta/c)uB_0R_*$.

The largest fluxes of solar protons $j_{\max}$ observed near the Earth correspond to the arrival of strong magnetic field disturbances and $j_{\max} = F_{\max}/(2\pi\tau)$, where $\tau = r/V_{SW}$ is a propagation time of solar wind disturbances to 1 AU. Let us assume a similar value for stellar cosmic rays, i. e., assume that convection plays a main role in the propagation of stellar CR for strong magnetic field. Note that the solar magnetic field is rather weak in comparison with the magnetic field of several tens to several thousand Gauss observed at magnetic active stars. Considering that stellar CR propagate within spatial angle of 60×60 degrees we will get fluencies $F = 9N/\pi r^2$ and maximal intensities $j_{\max} = F/(2\pi\tau)$, which might be observed at distance r in the astrosphere.

2.4. Discussion of methods

According to the dynamo theory O–B, A stars have not developed convective zone, as opposed to F–M stars and, therefore, they should not have a magnetic field. If there are no magnetic fields, then we would see an absence of stellar activity, flares, hot corona and hot stellar wind. However magnetic fields [31, 32], X-ray emission (see [33] and references therein) and flares [34] are observed at O–B stars, flares [35] and stellar spots [36] are observed at A-stars. The nature of magnetic fields is still unknown [37], X-ray emission is described by shock waves propagating in cold stellar wind [38] and stellar flares are attributed to an invisible cold companion [34, 39, 40] or wind-shock X-ray emission [34]. We don't discuss here the magnetic field nature but we assume that presence of rather strong magnetic fields supposes that magnetic structures like arcs and filaments could be formed at O–B, A stars, then we may discuss expanding hot coronas (stellar wind) and their X-ray emission, stellar flares, as well as make estimates using solar-stellar analogous [26].

3. Results

3.1. Estimates for TRAPPIST 1

In the year 2016 the planetary system of TRAPPIST-1 with 7 exoplanets was discovered, four of them are in a possible habitable zone [41]. TRAPPIST-1 is an M8 star at a distance of 10 light years, its parameters are $M = (0.089 \pm 0.006)M_\odot$, radius $R = (0.121 \pm 0.003)R_\odot$, effective temperature $T = 2516$ K. The rotation period is $P_{\rm rot} = 3.295 \pm 0.003$ days [42]. The TRAPPIST-1 star has a hot corona with a ratio of X-ray to bolometric luminosity $L_x/L_{\rm bol} = (2–4) \times 10^{-4}$ [43], i.e., should have a hot stellar wind similar to the solar wind. Its flare activity was observed by the MOST satellite [42] a frequency of flares with energy $\sim 10^{33}$ erg is $f = 1.2 \times 10^{-7}$ s^{-1}. The measured average magnetic field was estimated as 600 G [44], but it is too weak of a source to get ZDI magnetic field maps by using modern instruments. In this paper we will not discuss the possibility of generating such a large field, which is doubtful.

In our recent paper [27] we presented our view on radiation conditions in the TRAPPIST 1 system. Here we show some of our estimates for TRAPPIST 1d only, one of four exoplanets in the habitable zone.

Modelling characteristic values for stellar wind of TRAPPIST 1 are shown in Tab. 1 for different values of the coronal temperature.

Table 1. Stellar wind parameters for the TRAPPIST-1 d ($r_d = 0.0214$ AU)

T, MK	Q, erg/s	$n_{\rm SW}$, cm^{-3}	u_{cr}, km/s	r_{cr}/R$_*$	$V_{\rm SW}$, km/s	$r_{\rm Park}$, AU	$R_{\rm AS}$, AU
1	1.813×10^{26}	1614	131	4.0	294/366	0.6	61.5
2	2.051×10^{27}	6267	186	2.0	545/647	1	224
3	8.480×10^{27}	12510	228	1.4	760/884	1.5	442
4	2.321×10^{28}	19710	263	1.0	953/1097	1.8	696

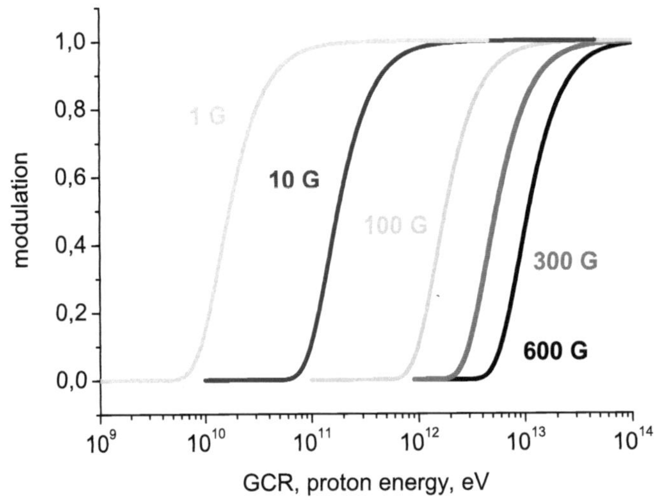

Figure 1. The dependence of CR modulation for TRAPPIST 1d for different values of stellar magnetic field and stellar wind velocity 545 km/s ($T_{\rm cor} = 2$ MK)

Figure 1 illustrates a possible effect of GCR modulation for hypothetical values of TRAPPIST 1d magnetic field of 1–600 G and coronal temperatures of 2 MK. This value of coronal temperature corresponds to the solar coronal temperature and is about twice as small as the maximum possible temperature for TRAPPIST 1 corona. From Fig. 1 it is clear, that CR with energies less 1 TeV should be swept out from the astrosphere in a case of magnetic field > 300 G. Moreover as $R_{AS} \gg r_1$ the real modulation GCR will be greater. However it is possible, that the real magnetic fields will be much smaller than measured. The modulation effects even for 1 G are more than an order of magnitude higher than for the Earth (Fig. 1).

Assuming the following parameters of an active region: $L = \alpha R = \alpha 8.5 \times 10^9$ cm, $B = \beta B_0 = 3000\beta$ G, $V = 100$ km/s, we find a flare energy of $E_{fl}\alpha^3\beta^2 = 2.2 \times 10^{35}\alpha^3\beta^2$ erg $= 10^{33}$ erg which is a reasonable value for TRAPPIST 1 flares. A frequency of such flares is $f = 10^{-2}$ year^{-1} $= 1.2^{-7}$ s^{-1}. The number of protons accelerated in one flare and the average rate of their generation as well as their density and flux within the first Parker spiral turn are presented in Tab. 2.

Maximum proton fluences and intensities for possible extreme events of stellar CR, Trappist-1d are presented in Tab. 3.

Table 2. The number of protons N accelerated in one flare and average rate of their generation, density and flux of protons within the first Parker spiral turn (1 AU)

E_p, MeV	N, protons	fN, proton/s	$n = 3N/4\pi r^3_{\text{park}}$, proton/cm^3	$nV_{\text{SW}}/(2\pi)$, proton/(cm$^2 \cdot$ s $\cdot$ sr)
30	2.1×10^{36}	2.5×10^{29}	1.5×10^{-4}	1300
200	3.1×10^{35}	3.7×10^{28}	2.2×10^{-5}	191

Table 3. Maximum fluences and proton intensity for $V_{\text{SW}} = 545$ km/s in extreme events near TRAPPIST 1 d ($r_d = 0.0214$ AU)

E_p , MeV	N, protons	$F = 9N/\pi r^2$, cm^{-2}	$j_{\max} = F/(2\pi\tau)$ (cm $\cdot$ s $\cdot$ sr)$^{-1}$
30	2.1×10^{36}	5.8×10^{13}	1.6×10^{9}
200	3.1×10^{35}	8.6×10^{12}	2.4×10^{8}

3.2. Estimates for σ Ori E

The magnetic helium-strong star σ Ori E is famous for its outstanding characteristics: the surface magnetic field strengths are at least 10 kG [45], X-ray emission $L_X/L_b = 10^{-7}$ [46] and $L_X/L_b = 3.9 \times 10^{-7}$, flare activity [34, 39], a weak cold stellar wind of some $10^{-10} M_\odot$ year^{-1} [48]. Fundamental parameters of σ Ori E are $T_{\text{eff}} = 22500$ K, $R = 3.77 R_\odot$, $M = 8.3 M_\odot$, the rotation period $P = 1.190847$ days [47]. The estimated overall energy of X-ray from a flare observed at σ Ori E was between 5.3×10^{35} and 2.9×10^{36} [34].

Applying the methods presented above we may get values of X-ray luminosity L_x and flare energy $\alpha^3\beta^2 E_{fl}$, which are very close to those observed. The mass rate of hot stellar wind appeared to be about 3–4 orders less than the observed mass rate of cold stellar wind $10^{-10} M_\odot$ year^{-1} [48], that may explain why the hot stellar wind of σ Ori E is not observed. An electron density in the corona n_e was a free parameter in our calculations.

Table 4. Calculated parameters of the hot corona of σ Ori E

T, MK	n_e, cm^{-3}	H_{cor}/R_*	L_X/L_b	M_r ($M_\odot$, year^{-1})	B$_0$, G	$\alpha^3\beta^2 E_{fl}$, erg
4.6	2.9×10^{8}	0.18	1.4×10^{-7}	4.0×10^{-14}	980	4.9×10^{38}
8.9	2.6×10^{8}	0.37	1.3×10^{-7}	1.7×10^{-13}	980	6.9×10^{38}

Table 5. Stellar wind parameters for σ Ori E in the habitable zone ($\sim$ 14 AU)

T, MK	Q, erg/s	n_{SW}, cm^{-3}	u_{cr}, km/s	r_{cr}/R_*	V_{SW}, km/s	r_{park}, AU	R_{AS} AU
4.6	1.2×10^{30}	0.5	218	2.6	1609	1.1	4051
8.9	1.2×10^{31}	1.6	392	1.4	2497	1.7	10770

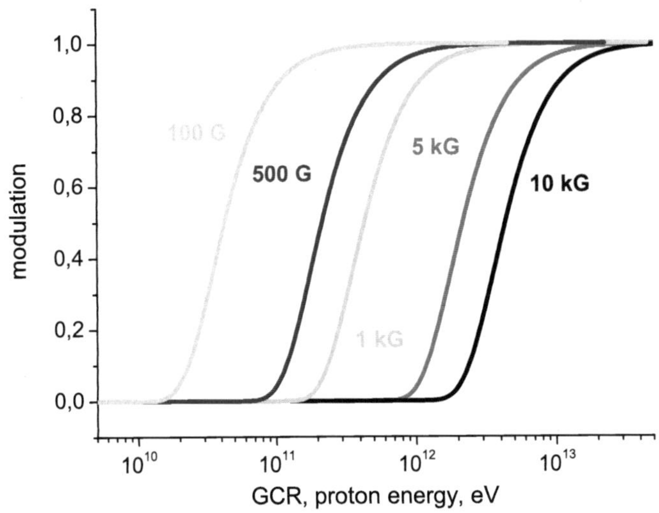

Figure 2. The dependence of CR modulation in the habitable zone of σ Ori E for hypothetical values of magnetic field of 100 G–10 kG and coronal temperatures of 4.6 MK

Figure 2 illustrates a possible effect of GCR modulation for hypothetical values of σ Ori E magnetic field of 100 G–10 kG and coronal temperatures of 4.6 MK. This value of coronal temperature is about twice as smaller as the maximum possible temperature for σ Ori E corona. Since the radius of the habitable zone is $R_H \approx 14$ AU $\gg r_{\rm park} \sim 1$ AU, we assumed a dimension of modulation region is about 30 AU. From Fig. 2 it is clear, that CR with energies less 1 TeV should be swept out from the astrosphere in a case of magnetic field > 5 kG. Moreover as $R_{\rm AS} \gg 30$ AU the real modulation GCR will be greater. However it is possible, that real values of the stellar magnetic field will be much less than measured. Flare observations may help to estimate their real values.

Assuming the following parameters of our active region: $L = H_{\rm cor} = \alpha R = 0.18R(0.37R)$, $B = \beta B_0 = 980\beta$ G, $V = 100$ km/s, we find a flare energy of $E_{fl}\alpha^3\beta^2 = 2.8 \times 10^{36}\beta^2$ $(3.4 \times 10^{37}\beta^2)$ erg, which is very close to the observed flare energy 5.3×10^{35}–2.9×10^{36} erg [34], i.e., $\beta = 0.44$–0.29. The number of protons accelerated in such a flare as well as their density and flux within the radius of habitable zone are presented in Tab. 6. These values are in agreement with our previous estimates and conclusions [23].

Table 6. Number of protons, their maximum fluencies and intensities in habitable zone of σOri E ($R_H = 14$ AU, flare energy 2×10^{36} erg)

E_p , MeV	N, protons	$F = 9N/\pi r^2$, cm^{-2}	$j_{\rm max} = F/(2\pi\tau)$, cm $\cdot$ s $\cdot$ sr^{-1}
30	4.2×10^{40}	2.7×10^{12}	3.2×10^5
200	6.2×10^{39}	4.1×10^{11}	4.9×10^4

It is possible that the observed magnetic field at σ Ori E $B_* = 10$ kG might be not the average stellar magnetic field but the local one (magnetic field of active region). In this case we may right $\beta_* B_* = \beta B_0$ and $\beta_* = \beta B_0/B_* = 0.04$–$0.03$. An average magnetic field should be much less than 10 kG, about several tens of Gauss, that is below the observational threshold.

4. Conclusions

It may be said recent measurements by the Voyager I spacecraft gave birth to qualitative low energy CR astrophysics and have provided data of lower-energy cosmic-ray particles out of the heliosphere [49]. This give us a reference point to evaluate low-energy CR intensities obtained by other methods in molecular clouds, protoplanetary discs and astrospheres. Our knowledge of the low CR interaction environment is at the same level as our knowledge of the heliosphere in the 1950s. So for the initial estimates we can use the Sun-stellar similarities and the methods developed for CR interaction with heliospheric medium. The most unknown parameter in these estimates is the stellar magnetic field which can vary by several orders of magnitude in comparison with solar magnetic field. The main problem is the measured stellar field may be the magnetic field of active regions, i.e., the local one. We believe that the developed approach can be used for any magnetic star with hot corona. Results for any planetary system should coincide with those obtained by more advanced models by the order of magnitude.

The estimates for stellar wind parameters, GCR and SCR parameters were found for the TRAPPIST 1 system. We obtain that for a coronal temperature of 2 MK the scale of astrosphere is $\sim$ 224 AU and radius of the first Parker spiral is $\sim$ 1 AU. Particles should be swept out by the stellar wind and the main factor of GCR modulation is the stellar wind magnetic field. As a result of modulation GCR less than 1 TeV should be absent near the planet. However, flare frequencies and energies allow us to state that the radiation environment determined by SCR.

For σ Ori E applying the methods presented above we obtain the values of X-ray luminosity L_X and flare energy $\alpha^3\beta^2 E_{fl}$, which are very close to that observed. The mass rate of hot stellar wind appeared to be about 3–4 orders less than the observed mass rate of cold stellar wind $10^{-10} M_\odot$ year^{-1} [48], which may explain why the hot stellar wind of σ Ori E is not observed. From the flare energy estimations it is possible that the observed magnetic field at σ Ori E might not be the average stellar magnetic field but the local one.

e-ASTROGAM is a concept of a breakthrough observatory carrying a γ-ray telescope for the study of the non-thermal Universe in the photon energy range from 0.15 MeV to 3 GeV. The gamma-rays of such energy give us a new source for understanding of low-energy CR.

The gamma-rays of energies from 0.15 MeV to 3 GeV may give us information for better understanding low-energy CR in their different locations in the Galaxy. e-ASTROGAM is a concept of a breakthrough observatory carrying a γ-ray telescope for the study of the non-thermal Universe in this new photon energy range [50, 51].

Acknowledgments

The work is supported by Russian Foundation for Basic Research grant 16-02-00328.

References

[1] Ginzburg V L 1988 Sov. Phys. Usp. **31** 491
[2] Cummings A C, Stone E C, Heikkila B C, Lal N, Webber W R, Jóhannesson G, Moskalenko I. V, Orlando E, and Porter T A 2016 *Astrophys. J.* **831** 18
[3] Neufeld D A, Wolfire M G 2017 *Astrophys. J.* **845** 163
[4] Yang R-z, de Oa Wilhelmi E, and Aharonian F 2014 *A&A* **566** A142
[5] Phan V H M, Morlino G, Gabici S 2018 *Monthly Notices Roy. Astronom. Soc.* **480** 5167
[6] Padovani M, Marcowith A, Hennebelle P, and Ferrière K 2016 *A&A* **590** A8
[7] Cleeves L I, Adams F C, and Bergin E A 2013 *Astrophys. J.* **772** 5
[8] Rab Ch, G udel M, Padovani M, Kamp I, Thi W.-F, Woitke P, and Aresu G 2011 *A&A* **603** A96
[9] Fraschetti F, Drake J J, Cohen O, Garraffo C 2018 *Astrophys. J.* **853** 112
[10] Padovani M, Ivlev A V, Galli D, and Caselli P 2018 *A&A* **614** A111
[11] Grießmeier J-M, Stadelmann A, Motschmann U, Belisheva N K, Lammer H, Biernat H K 2005 *Astrobiology* **5** 587
[12] Atri D, Hariharan B, Grießmeier J-M 2013 Astrobiology **13** 910
[13] Grießmeier J-M, Tabataba-Vakili F, Stadelmann A, Grenfell J L, Atri D 2015 *Astrophys. J.* **581** A44

[14] Grießmeier J-M, Tabataba-Vakili F, Stadelmann A, Grenfell J L, Atri D 2016 *Astrophys. J.* **587** A159
[15] Scherer K, Fichtner H, Stawicki O 2002 *J. Atmospheric Solar-Terrestrial Phys.* **64** 795
[16] Cohen O, Drake J J, Kóta J 2012 *Astrophys. J.* **760** 85
[17] Unsöld A 1957 *Proc. 4th IAU Symp.* p 238
[18] Edwards P J and McQueen M 1971 *Proc. 12th ICRC* **1** 323
[19] Lovell A C B 1974 *Phill. Trans. Roy. Soc. A* **277** 489
[20] Mullan D J 1979 *Astrophys. J.* **234** 588
[21] Kopysov Yu S and Stozhkov Yu I 2005 *Proc. 29th ICRC* **3**, 141
[22] Stozhkov Yu I 2011 *Bulletin of the Russian Academy of Sciences. Physics* **75** 323
[23] Struminsky A. and Sadovski A 2017 *Stellar Cosmic Rays in a Habitable Zone, Stars: From Collapse to Collapse, Proc. conf. Special Astrophysical Observatory, Nizhny Arkhyz, Russia 3–7 October 2016* Ed Yu Yu Balega, D O Kudryavtsev, I I Roma-nyuk, and I A Yakunin (San Francisco: Astronomical Society of the Pacific) p 105
[24] Tabataba-Vakili F, Grenfell J L, Grießmeier J-M, Rauer H 2016 *A&A* **585** A96
[25] Atri D 2017 *MNRAS* **465** L34
[26] Sadovski A M, Struminsky A B, and Belov A 2018 *Astron Lett* **44** 324
[27] Struminsky A B, Sadovski A M, and Zharikova M S 2018 *Geomagnetism Aeronomie* in print
[28] Parker E N *Astrophys. J.* **128** 664
[29] Parker E N 1958 *Phys. Rev.* **110** 1445
[30] Balona L A 2015 *MNRAS* **447** 2714
[31] Berdyugina S V 2009 *Proceedings IAU Symp.* **259** 323
[32] Linsky J L and Schöller M 2015 *Space Science Reviews* **191** 26
[33] Güdel M 2004 *A&ARv* **12** 71
[34] Groote D and Schmitt J H M M 2004 *A&A* **418** 235
[35] Balona L A 2012 *MNRAS* **423** 3420
[36] Balona L A 2017 *MNRAS* **467** 1830
[37] Hubrig S, Schöller M, Fossati L, Morel T, Castro N, Oskinova L M, Przybilla N, Eikenberry S S, Nieva M-F, Langer N 2015 *A&A* **578** L3
[38] Babel J and Montmerle T 1997 *A&A* **323** 121
[39] Mullan D J 2009 *Astrophys. J.* **702** 759
[40] Pedersen M G, Antoci V, Korhonen H, White T R, Jessen-Hansen J, Lehtinen J, Nikbakhsh S, Viuho J 2017 *MNRAS* **466** 3060
[41] Gillon, M, Demory B.O, Van Grootel V, Motalebi F, Lovis C, Cameron A C, Charbonneau D, Latham D, Molinary E, Pepe F A, Segransan D, Sasselov D, Udry S, Mayor M, Micela G, Piotto G, and Sozzetti A 2017 *Nature* **542** 456
[42] Vida K, Kővári Zs, Pál A, Oláh K, Kriskovics L 2017 *Astrophys. J.* **841** 124
[43] Wheatley P J, Louden T, Bourrie V, Ehrenreich D, and Gillo M 2017 *MNRAS* **465L** 74W
[44] Reiners A and Basri G A 2010 *Astrophys. J.* **710** 924
[45] Landstreet J D and Borra E F 1978 *Astrophys. J.* **224** L5
[46] Pallavicini R, Golub L, Rosner R, Vaiana G S, Ayres T, Linsky J L 1981 *Astrophys. J.* **248** 279
[47] Townsend R H D, Rivinius Th, Rowe J F, Moffat A F J, Matthews J M, Bohlender D, Neiner C, Telting J H, Guenther D B, Kallinger T, Kuschnig R, Rucinski S M, Sasselov D, and Weiss W W 2013 *Astrophys. J.* **769** 33
[48] Groote D and Hunger K 1982 *A&A* **116** 64
[49] Stone E C, Cummings A C, McDonald F B, Heikkila B C, Lal N, Webber W R 2013 *Science* **341** 150
[50] *Science with e-ASTROGAM* arXiv:1711.01265v4 [astro-ph.HE]
[51] Orlando E 2018 *MNRAS* **475** 2724

IOP Conf. Series: Journal of Physics: Conf. Series **1181** (2019) 012003 doi:10.1088/1742-6596/1181/1/012003

The CALorimetric Electron Telescope (CALET) on the International Space Station: Results from the First Two Years On Orbit

Y Asaoka[1,2], O Adriani[3,4], Y Akaike[5,6], K Asano[7], M G Bagliesi[8,9], E Berti[3,4], G Bigongiari[8,9], W R Binns[10], S Bonechi[8,9], M Bongi[3,4], A Bruno[11], P Brogi[8,9], J H Buckley[10], N Cannady[12], G Castellini[13], C Checchia[14,15], M L Cherry[12], G Collazuol[14,15], V Di Felice[16,17], K Ebisawa[18], H Fuke[18], T G Guzik[12], T Hams[5,19], N Hasebe[1], K Hibino[20], M Ichimura[21], K Ioka[22], W Ishizaki[7], M H Israel[10], K Kasahara[1], J Kataoka[1], R Kataoka[23], Y Katayose[24], C Kato[25], N Kawanaka[26,27], Y Kawakubo[28], K Kohri[29], H S Krawczynski[10], J F Krizmanic[19,5], T Lomtadze[9], P Maestro[8,9], P S Marrocchesi[8,9], A M Messineo[30,9], J W Mitchell[6], S Miyake[31], A A Moiseev[32,19], K Mori[1,18], M Mori[33], N Mori[4], H M Motz[34], K Munakata[25], H Murakami[1], S Nakahira[35], J Nishimura[18], G A De Nolfo[11], S Okuno[20], J F Ormes[36], S Ozawa[1], L Pacini[3,13,4], F Palma[16,17], V Pal'shin[28], P Papini[4], A V Penacchioni[8,37], B F Rauch[10], S B Ricciarini[13,4], K Sakai[19,5], T Sakamoto[28], M Sasaki[19,32], Y Shimizu[20], A Shiomi[38], R Sparvoli[16,17], P Spillantini[3], F Stolzi[8,9], S Sugita[28], J E Suh[8,9], A Sulaj[8,9], I Takahashi[39], M Takayanagi[18], M Takita[7], T Tamura[20], N Tateyama[20], T Terasawa[35], H Tomida[18], S Torii[1,40], Y Tsunesada[41], Y Uchihori[42], S Ueno[18], E Vannuccini[4], J P Wefel[12], K Yamaoka[43], S Yanagita[44], A Yoshida[28], and K Yoshida[45]

[1] Waseda Research Institute for Science and Engineering, Waseda University, Japan
[2] JEM Utilization Center, Human Spaceflight Technology Directorate, Japan Aerospace Exploration Agency, Japan
[3] Department of Physics, University of Florence, Italy
[4] INFN Sezione di Florence, Italy
[5] Department of Physics, University of Maryland, USA
[6] Astroparticle Physics Laboratory, NASA/GSFC, USA
[7] Institute for Cosmic Ray Research, The University of Tokyo, Japan
[8] Department of Physical Sciences, Earth and Environment, University of Siena, Italy
[9] INFN Sezione di Pisa, Italy
[10] Department of Physics, Washington University, USA
[11] Heliospheric Physics Laboratory, NASA/GSFC, USA
[12] Department of Physics and Astronomy, Louisiana State University, USA
[13] Institute of Applied Physics (IFAC), National Research Council (CNR), Italy
[14] Department of Physics and Astronomy, University of Padova, Italy
[15] INFN Sezione di Padova, Italy
[16] University of Rome "Tor Vergata", Italy
[17] INFN Sezione di Rome "Tor Vergata", Italy

IOP Conf. Series: Journal of Physics: Conf. Series **1181** (2019) 012003 doi:10.1088/1742-6596/1181/1/012003

[18] Institute of Space and Astronautical Science, Japan Aerospace Exploration Agency, Japan
[19] CRESST and Astroparticle Physics Laboratory NASA/GSFC, USA
[20] Kanagawa University, Japan
[21] Faculty of Science and Technology, Graduate School of Science and Technology, Hirosaki University, Japan
[22] Yukawa Institute for Theoretical Physics, Kyoto University, Japan
[23] National Institute of Polar Research, Japan
[24] Faculty of Engineering, Division of Intelligent Systems Engineering, Yokohama National University, Japan
[25] Faculty of Science, Shinshu University, Japan
[26] Hakubi Center, Kyoto University, Japan
[27] Department of Astronomy, Graduate School of Science, Kyoto University, Japan
[28] College of Science and Engineering, Department of Physics and Mathematics, Aoyama Gakuin University, Japan
[29] High Energy Accelerator Research Organization, Japan
[30] University of Pisa, Italy
[31] Department of Electrical and Electronic Systems Engineering, National Institute of Technology, Ibaraki College, Japan
[32] Department of Astronomy, University of Maryland, College Park, Maryland 20742, USA
[33] Department of Physical Sciences, College of Science and Engineering, Ritsumeikan University, Japan
[34] Global Center for Science and Engineering, Waseda University, Japan
[35] RIKEN, Japan
[36] Department of Physics and Astronomy, University of Denver, USA
[37] ASI Science Data Center (ASDC), Italy
[38] College of Industrial Technology, Nihon University, Japan
[39] Kavli Institute for the Physics and Mathematics of the Universe, The University of Tokyo, Japan
[40] School of Advanced Science and Engineering, Waseda University, Japan
[41] Division of Mathematics and Physics, Graduate School of Science, Osaka City University, Japan
[42] National Institutes for Quantum and Radiation Science and Technology, JAPAN
[43] Nagoya University, Japan
[44] College of Science, Ibaraki University, Japan
[45] Department of Electronic Information Systems, Shibaura Institute of Technology, Japan

E-mail: `yoichi.asaoka@aoni.waseda.jp`

Abstract. The CALorimetric Electron Telescope (CALET) is a high-energy astroparticle physics space experiment installed on the International Space Station (ISS), developed and operated by Japan in collaboration with Italy and the United States. The CALET mission goals include the investigation of possible nearby sources of high-energy electrons, of the details of galactic particle acceleration and propagation, and of potential signatures of dark matter. CALET measures the cosmic-ray electron+positron flux up to 20 TeV, gamma-rays up to 10 TeV, and nuclei with Z=1 to 40 up to 1,000 TeV for the more abundant elements during a long-term observation aboard the ISS. Starting science operation in mid-October 2015, CALET performed continuous observation without major interruption with close to 20 million triggered events over 10 GeV per month. Based on the data taken during the first two-years, we present an overview of CALET observations: 1) Electron+positron energy spectrum, 2) Nuclei analysis, 3) Gamma-ray observation including a characterization of on-orbit performance. Results of the electromagnetic counterpart search for LIGO/Virgo gravitational wave events are discussed as well.

1. Introduction

The CALorimetric Electron Telescope (CALET) [1, 2] is a high-energy astroparticle physics mission on the ISS, with development and operation conducted by Japan in collaboration with Italy and the United States. The detector was launched to orbit aboard the unmanned H2

Transfer Vehicle (HTV) atop the Japanese H2-B carrier rocket on August 19, 2015. After arrival of the HTV at the ISS, it was installed on the Japanese Experiment Module–Exposed Facility (JEM-EF). The initial mission duration was two years, extendable to five years or more. The image in Fig. 1 shows the JEM with CALET attached at port #9 of the JEM-EF, which features a mostly unobstructed field-of-view of 45° from zenith. The schematic drawing in Fig. 1 gives an overview of the CALET payload.

Figure 1. JEM-EF and the CALET payload attached at the port #9. The inset is the CALET instrument package showing the main calorimeter and CALET Gamma-ray Burst Monitor (CGBM) subsystems [3].

CALET's main instrument is a very thick calorimeter incorporating imaging and total absorption calorimeters. The overall thickness of CALET for normal incidence angle is 30 radiation length, corresponding to ~1.3 proton interaction length. Installed on the ISS, CALET carries out long term observation with a large area detector, thus providing a high-statistics measurement. CALET is designed to discover signatures of nearby cosmic-ray accelerators and potentially dark matter in the all-electron (electron+positron) spectrum, which is measured with high precision over a wide energy range from 1 GeV to 20 TeV, as well as in the gamma-ray spectrum measured up to 10 TeV. Protons, helium, and heavier nuclei through iron, are the main components of cosmic rays, and can be measured in the range to a PeV. The detailed mechanism and parameters governing propagation and acceleration of the galactic cosmic-rays will be investigated based on nuclei spectra measurements. CALET is expected to extend the limits of direct measurements.

2. Instrument

The CALET calorimeter (see the right-hand panel of Fig. 2) consists of a Charge Detector (CHD), which identifies the charge of the incident particle [4, 5], an IMaging Calorimeter (IMC), which reconstructs the track of the incident particle and records the initial shower development with fine resolution, and a Total AbSorption Calorimeter (TASC), which absorbs the entire energy of the incoming particle and identifies the particle species using hodoscopic lead-tungstate crystal arrays. The components and read-out sensors are summarized in the left panels of Fig. 2.

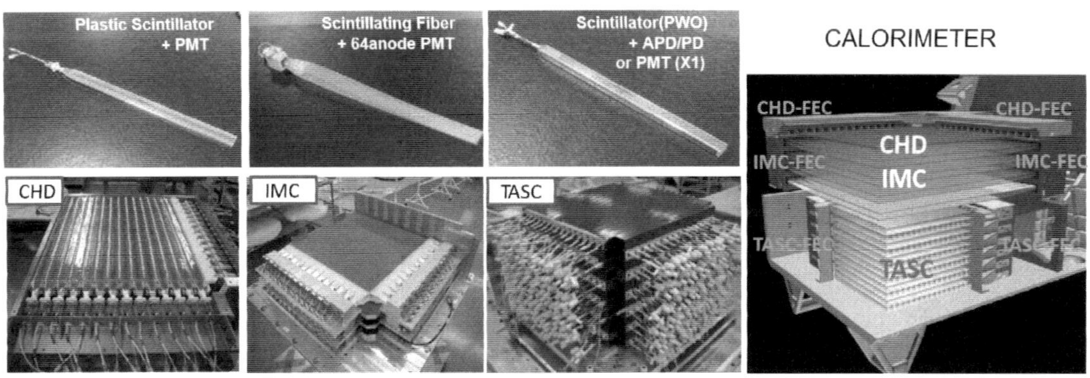

Figure 2. CALET calorimeter consisting of three sub detectors, i.e., a CHarge Detector (CHD), an IMaging Calorimeter (IMC), and Total AbSorption Calorimeter (TASC).

Combining these sub-detectors as well as the trigger system and data acquisition system, the CALET instrument features (1) a proton rejection factor larger than 10^5, (2) a 2% energy resolution above 20 GeV for electrons, (3) very wide dynamic range from 1 GeV to 1 PeV, (4) charge resolution of 0.1–0.3 electron charge unit from protons to above iron (up to $Z = 40$), (5) an angular resolution of 0.1 to 0.5°, and (6) a geometrical factor of the order of 0.1 m^2sr.

Figure 3 summarizes the CALET capability of particle identification. Top left, top right, bottom left, and bottom right panels show 3 TeV electron candidate, proton candidate with equivalent shower energy, iron candidate with shower energy of 9.3 TeV, gamma-ray candidate with 44 GeV reconstructed energy, respectively. The calorimeter with 30 radiation-length on-axis thickness absorbs the full electron shower energy even in the TeV range. Charge measurement using CHD and IMC separates each of the elements from $Z = 1$ to 26 and above. Gamma-rays are identified as charge zero because they do not produce any signal before the pair creation. While both of electrons and protons have $Z = 1$, they can be separated using the differences in their shower shapes. Because of the continuing showering activity in the lower part of the TASC due to subsequent interactions of secondary pions, electrons, and protons are easily separated by a simple cut even in the TeV region. In addition to this, various parameters characterizing the shower shape can be utilized to improve the separation power [6, 7].

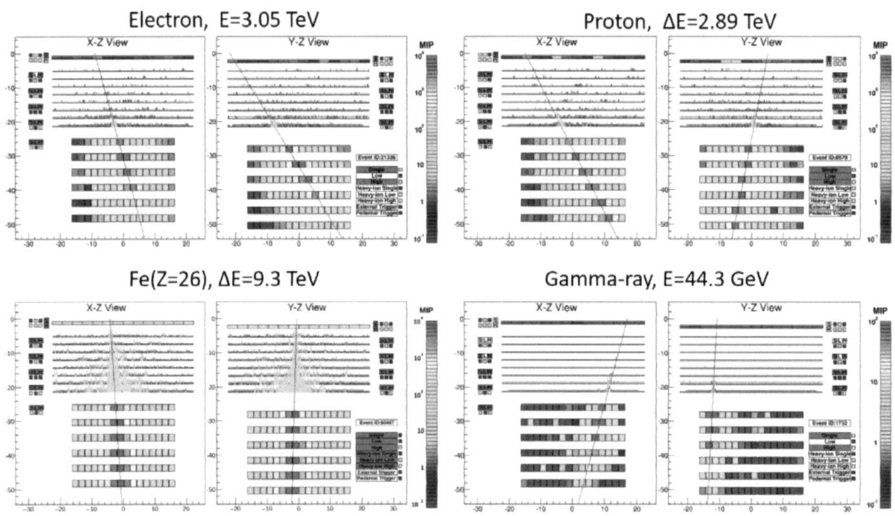

Figure 3. Event examples of high-energy showers collected by CALET onboard the ISS. (*Top Left*) a 3 TeV electron candidate, (*Top Right*) a proton candidate with equivalent shower energy, (*Bottom Left*) an iron candidate with shower energy of 9.3 TeV, and (*Bottom Right*) a 44 GeV gamma-ray candidate.

3. On-Orbit Operations

For on-orbit operation of CALET, JAXA Ground Support Equipment (JAXA-GSE) and the Waseda CALET Operations Center (WCOC) were set up at Tsukuba Space Center and Waseda University, respectively. The data taken by the CALET instrument are transferred from the ISS to JAXA using NASA's data relay system. The scientific operations of CALET are planned at WCOC [8] based for example on the variation of the geomagnetic rigidity cutoff depending on ISS position. Accordingly is the observation mode of CALET controlled by scheduled command sequences. These sequences define the time profile of calibration and data acquisition tasks as for example the recording pedestal and penetrating particle events, as well as switching on/off observation modes such as a low-energy electron trigger operating at high geomagnetic latitude, a low-energy gamma-ray (LE-γ) trigger operating at low geomagnetic latitude, and an almost continuously active ultra heavy trigger mode, during each ISS orbit. Maximum exposure to high-energy electrons and other high-energy shower events is ensured by an always active high-energy (HE) trigger mode.

As of May 31, 2018, the total observation time is 962 days with a live time fraction of ~84% relative to the total time. Close to 630 million events were taken with the HE (E >10 GeV) trigger mode. Figure 4 shows the accumulated live time for the HE trigger. Cumulative observation time has increased without significant interruption since scientific operation start in October 2015. Data transmission from JAXA-GSE to WCOC and data processing for scientific analysis at WCOC also proceeded smoothly.

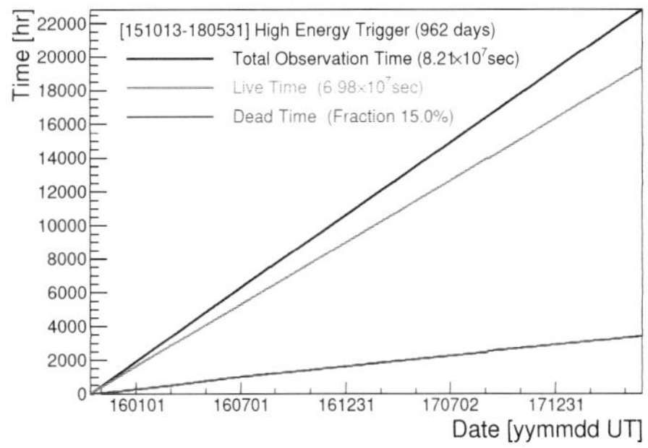

Figure 4. Cumulative of observation time (black line) for HE trigger. The red and blue lines indicate the live and dead times, respectively.

4. Calibration

Energy calibration is a key procedure for CALET as a calorimeter instrument to achieve high precision and accurate measurements. While excellent energy resolution inside the TeV region is one of the most important features of a thick calorimeter instrument like CALET, calibration errors must be carefully assessed and taken into account in the estimation of the actual energy resolution.

Our energy calibration [9] includes the evaluation of the conversion factors between ADC units and energy deposits, ensuring linearity over each gain range (TASC has four gain ranges for each channel), and provides a seamless transition between neighboring gain ranges. Temporal gain variations occurring during long time observations are also corrected for by the calibration procedure [6].

The errors at each calibration step, such as the correction of position and temperature dependence, consistency between energy deposit peaks of non-interacting protons and helium, linear fit error of each gain range, and gain ratio measurements, as well as slope extrapolation, are included in the estimation of the energy resolution.

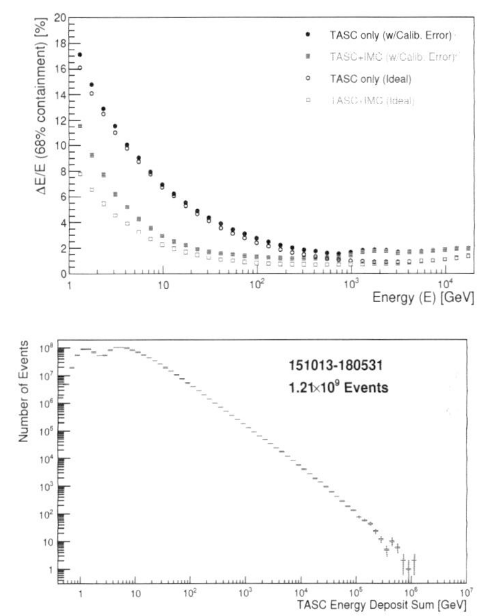

Figure 5. (*Top*) Energy resolution for electron measurements [9], (*Bottom*) TASC energy deposit spectrum using all triggered events up to May, 2018.

As a result, a very high resolution of 2% or better is achieved above 20 GeV [9] as shown in the top panel of Fig. 5. It should be noted that even with such a detailed calibration, the determining factor for the energy resolution is the calibration uncertainty, as the intrinsic resolution of CALET is ~1%. Intrinsic resolution refers to the detector's capability by design, taking advantage of the thick, fully-active total absorption calorimeter. Also important is the fact that the calibration error in the lower gain ranges is crucial for the spectrum measurements in the TeV range.

The bottom panel of Fig. 5 shows the TASC energy deposit spectrum using all triggered events through the end of May, 2018. The first bump is due to low-energy triggered events, while the second bump is caused by high-energy triggered events and the tail at high energy reflects the power-law nature of the cosmic-ray spectrum. The spectrum spans more than six orders of magnitude in energy with highest energy past a PeV, and the lowest energy below 1 GeV. This clearly demonstrates the CALET capability to observe cosmic rays over a very wide dynamic range.

5. Results

5.1. All-Electron Spectrum

A precise measurement of the all-electron (electron + positron) spectrum in the TeV region might reveal interesting spectral features to provide the first experimental evidence of the presence of a nearby cosmic-ray source [10, 11]. In addition, the unexpected increase of the positron fraction above 10 GeV established by PAMELA [12] and AMS-02 [13] may require a primary source component for positrons in addition to the generally accepted secondary origin. Candidates for such primary sources range from astrophysical (pulsar) to exotic (dark matter). Since these primary sources emit electron-positron pairs, it is expected that the all-electron spectrum would exhibit a spectral feature, near the highest energy range of the primary component.

The CALET collaboration published its first result on electrons in the energy range from 10 GeV to 3 TeV [6]. Subsequently, the DArk Matter Particle Explorer (DAMPE) collaboration published their all-electron spectrum in the energy range from 25 GeV to 4.6 TeV [14]. The latter publication was followed by many papers speculating about the origin of a peak-like structure near 1.4 TeV in the DAMPE data.

Recently, an updated version of the CALET all-electron spectrum using 780 days of flight data and the full geometrical acceptance was published in the energy range from 11 GeV to 4.8 TeV [7]. Figure 6 shows the updated all-electron spectrum obtained with CALET using the same energy binning as in our previous publication [6], except for adding one extra bin at the high energy end. The error bars along horizontal and vertical axes indicate bin width and statistical errors, respectively. The gray band is representative of the quadratic sum of statistic and systematic errors. The exhaustive study on the systematic uncertainties was performed and is described in Refs. [6, 7] including their Supplemental Materials.

There are four important implications from the current status on the all-electron spectrum measurements. First, CALET's spectrum is consistent with AMS-02 below 1 TeV. As both detectors have enough capability to identify electrons up to 1 TeV, and the detection principles are largely different (calorimeter versus magnet spectrometer), the agreement is an important indication. Secondly, there are two group of measurements: AMS-02 + CALET vs Fermi−LAT + DAMPE, indicating the presence of unknown systematic errors. Thirdly, CALET observes flux suppression consistent with DAMPE within errors above 1 TeV. No peak-like structure was found at 1.4 TeV in CALET data, irrespective of energy binning as shown in Fig. 7.

In the left-hand plot of Fig. 7, we have adopted exactly the same energy binning as DAMPE to show our spectrum. The flux is inconsistent between the two experiments with a 4 σ significance; the CALET data does not show any significant excess in the 1.4 TeV bin. The significance includes the systematic errors quoted from both experiments. Furthermore, possible binning related effects in the CALET all-electron spectrum are studied with a shift of one fourth of the bin width as shown in the right-hand plot of Fig. 7. The solid curves in the figure show the energy dependent systematic uncertainty band. The deviation due to binning is well below our energy dependent systematic uncertainty or statistical fluctuations. Therefore, bin-to-bin migration and related effects are negligible compared to our estimated systematic uncertainties, in accordance with the estimated CALET energy resolution of 2% above 20 GeV.

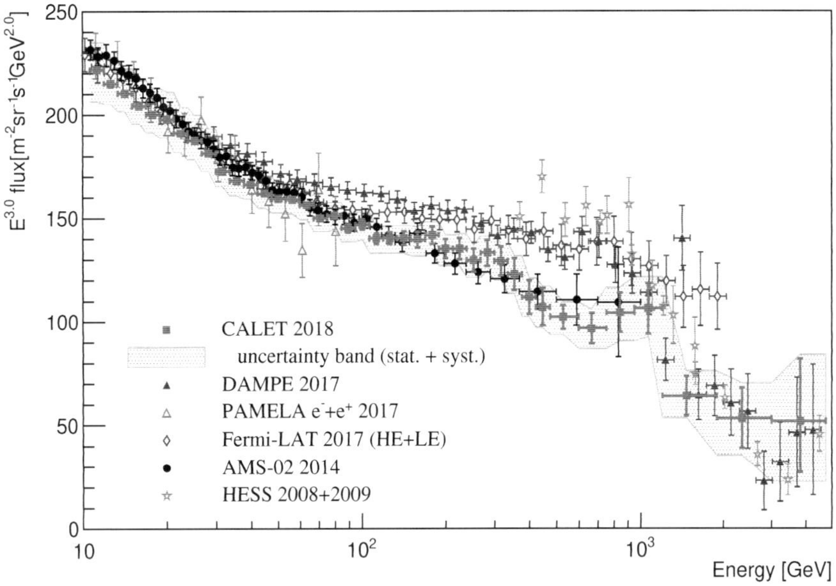

Figure 6. Cosmic-ray all-electron spectrum measured by CALET from 10.6 GeV to 4.75 TeV [7], where the gray band indicates the quadratic sum of statistical and systematic errors (not including the uncertainty on the energy scale). Also plotted are direct measurements in space [14, 15, 16, 17] and from ground-based experiments [18, 19].

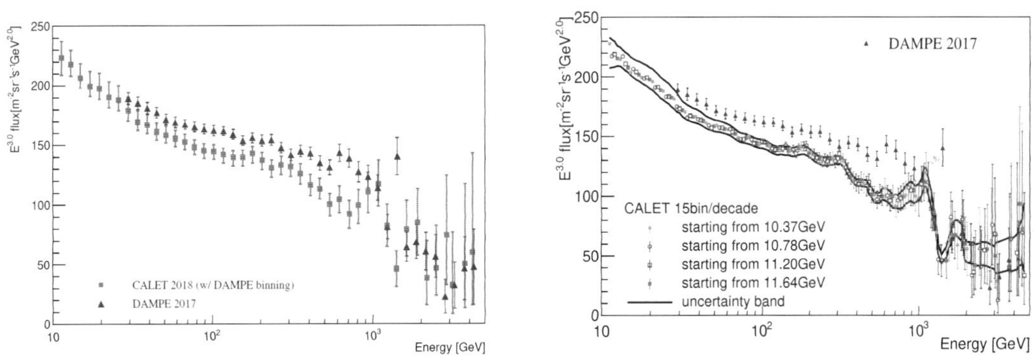

Figure 7. (*Left*) Cosmic-ray all-electron spectrum measured by CALET from 10.6 GeV to 4.57 TeV [7], using the same energy binning as DAMPE's result [14] and compared with it. The error bars indicate the quadratic sum of statistical and systematic errors. (*Right*) Study of possible binning related effects in the CALET all-electron spectrum [7].

5.2. Hadrons

Direct measurements of the high-energy spectra of each species of cosmic-ray nuclei up to the PeV energy scale provide information complementing all-electron observations and provides more detailed insight into the general conditions of cosmic-ray acceleration and propagation. A possibly charge-dependent cutoff in the nuclei spectra is hypothesized to explain the "knee" in the all-particle spectrum. This hypothesis could only be investigated by a space experiment with sufficient exposure. The acceleration limit of supernova remnants calculated with nominal parameters is typically found to be far smaller than the energy of the "knee" [20] observed indirectly by ground detectors. Therefore, precise direct observation of the proton and helium spectra up to PeV energy is highly important. Also the spectral hardening observed in spectra of various nuclei calls for an careful investigation, with CALET's wide dynamic range from GeV

IOP Conf. Series: Journal of Physics: Conf. Series **1181** (2019) 012003 doi:10.1088/1742-6596/1181/1/012003

to PeV energies allowing to study the feature unaffected by systematics from combination of spectra measured by different experiments. Another detailed study will be done on the spectral behavior of heavier elements, including secondary-to-primary ratios up to 1 TeV/n energy region, which should yield important information about propagation parameters such as the diffusion coefficient.

Figure 8 illustrates the charge identification capability of CALET based on the CHD data only, showing clear separation of protons, helium and nuclei up to iron and nickel. The IMC provides an independent and about equally powerful charge separation capability for light elements. Until now preliminary results on protons [21] and heavier nuclei [22, 23] have been presented, taking advantage of the accurate charge determination and wide energy range. The first results on the nuclei spectra will be published soon, including a detailed assessment of systematic uncertainties.

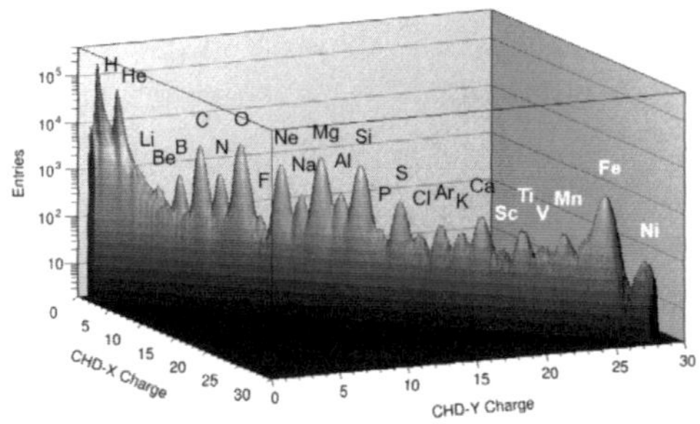

Figure 8. CHD charge separation.

5.3. Gamma-Rays

With a fully active 30 radiation-length thick calorimeter, CALET is capable of measuring gamma-rays up to the TeV region. In addition to HE trigger, CALET uses LE-γ trigger to be sensitive to gamma rays with primary energies down to 1 GeV. To avoid large dead-time fraction, however, LE-γ trigger is activated only at low geomagnetic latitudes or following gamma-ray burst triggered onboard by CALET gamma-ray bust monitor (CGBM).

The first 24 months of on-orbit scientific data provide valuable characterization of the performance of the calorimeter based on the analyses of the gamma-ray data set [24]. It includes optimization of event selection criteria, calculation of effective area, determination of point spread function, confirmation of absolute pointing accuracy, observation of bright point sources and study of diffuse components. Based on the developed analysis method, CALET gamma-ray sky seen by LE-γ trigger is shown in the left-hand panel of Fig. 9, where galactic emission and bright gamma-ray sources are clearly identified. In the right-hand panel of Fig 9, projection of the observed and expected number of photons onto galactic latitude for the galactic plane region $|l| < 80^o$ is shown. Expected number of photons are calculated using Fermi/LAT flux map and CALET's exposure in the same region of the sky. The very good consistency confirmed our sensitivity. It should be noted, that it is important to take the effects of ISS structures in the field-of-view into account.

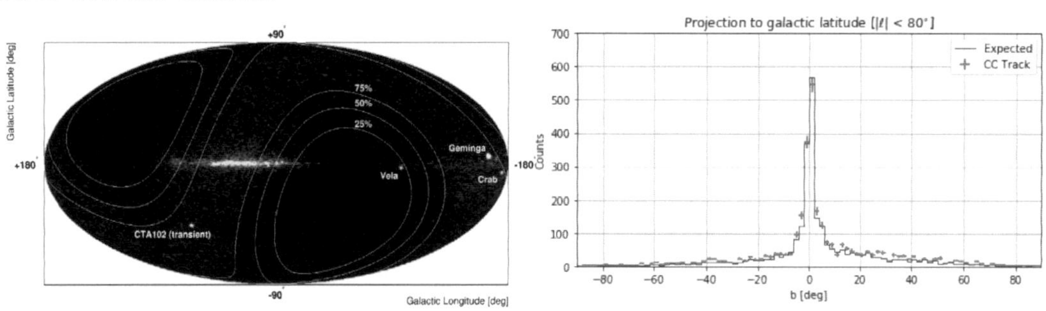

Figure 9. (*Left*) Gamma-ray sky map shown in a Mollweide projection of galactic coordinates. White contours show the relative level of exposure compared to the maximum on the sky. (*Right*) Projection of the observed and expected number of photons onto galactic latitude for the galactic plane region $|l| < 80^o$ for the energy range from 1 to 100 GeV.

Gamma-ray transients are also an important observational target for CALET. CGBM detected nearly 60 GRBs (~20% short GRB among them) per year in the energy range of 7 keV–20 MeV, as expected [25]. To search for GeV-energy counterpart emission from such sources also detected by other instruments, we check the CAL data at the reported trigger times for gamma-ray candidates. For events checked based on CGBM, Swift, and Fermi/GBM triggers, no significant counterparts have been detected at this stage for timescales ranging from 1 s to 1 hr [24]. Regarding the counterpart search for gravitational wave events, combined analyses of CGBM and calorimeter were performed for GW151226, resulting in our upper limits set on X-ray and gamma-ray counterparts [26]. Furthermore, complete search results of CALET calorimeter on the LIGO/Virgo's Observation Run 2 has been published recently [27].

6. Summary and Prospects

CALET was successfully launched on Aug. 19, 2015, and detector performance for scientific observation has been continuously very stable since Oct. 13, 2015. Careful calibrations using "MIP" signals of the non-interacting protons and helium events have been successfully carried out, and the linearity of the energy measurements up to 10^6 MIPs was established based on observed events [8, 9].

The all-electron spectrum [6] has been published with extended energy range from 11 GeV to 4.8 TeV [7] and improved statistics. Figure 10 presents the current situation of all-electron spectrum direct measurements in space. Five years or more of observations with CALET will triple the statistics, which together with a reduction of systematic errors based on a better understanding of the detector with an increased amount of the flight data will lead to significantly improved precision. This will allow a refined study of the possible fine structures around a few hundred GeV and ~1 TeV, which are currently not significant, which might shed light on the origin of the positron excess. An extension of the high-energy reach using improved statistics and electron separation methods might for the first time reveal a charged cosmic-ray signature of a local accelerator.

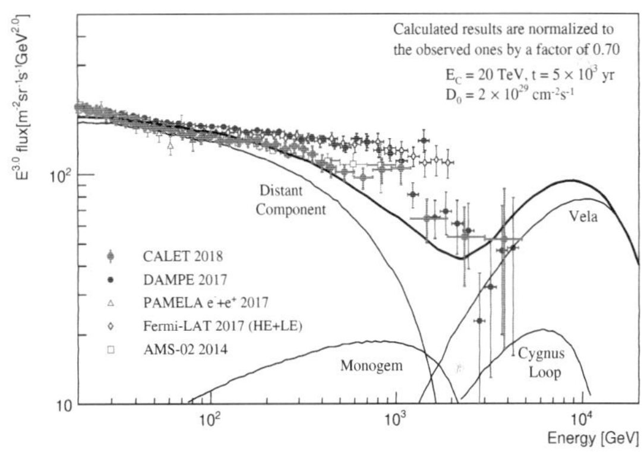

Figure 10. Current situation of all-electron spectrum directly measured in space, together with the model calculation [11], where main model parameters are shown in the plot.

Preliminary results on protons [21], primary and secondary nuclei up to $Z = 26$ and their ratios (for example, boron to carbon) [22, 23] were presented so far, demonstrating CALET's wide dynamic range of energy measurement from 1 GeV to 1 PeV and the accuracy of its charge determination capability. Their publication is foreseen in the near future, and will address important questions in cosmic-ray physics, such as the charge dependence of the acceleration limit in supernova remnants, the universality of the widely observed spectral hardening, and the energy dependence of the diffusion coefficient. The relative abundance of the ultra heavy nuclei up to $Z = 40$ is also analyzed [28].

There is also significant progress in CALET's gamma-ray analysis. Based on the science data taken in two years on orbit, the performance of the gamma-ray measurements has been characterized [24]. These results confirm the capability of CALET to observe gamma rays in the energy range from ~1 GeV to over 100 GeV. CALET's current results on electromagnetic counterpart search for gravitational wave events [26, 27] show the great potential of follow-up observations during the upcoming LIGO/Virgo's third observation run (Observation Run 3). The continuous GeV gamma-ray sky observation with CALET complements the coverage by other missions and may help to identify unexplored high-energy emissions from future transient events. Watching for various transient phenomena including those in gamma rays, but not limited to them is an important task for CALET as an on-orbit observation strategy. Through the detection of many events of MeV electrons originating from the radiation belt [29], a phenomenon called relativistic electron precipitation, space weather was added as an additional observational target for CALET after the start of on-orbit operations.

The so far excellent performance of CALET and the outstanding quality of the data suggest that a 5-year (or more) observation period will most likely provide a wealth of interesting new results.

Acknowledgment

We gratefully acknowledge JAXA's contributions to the development of CALET and to the operations onboard the International Space Station. We also wish to express our sincere gratitude to ASI and NASA for their support of the CALET project. This work was supported in part by JSPS Grant-in-Aid for Scientific Research (S) (No. 26220708), JSPS Grant-in-Aid for Scientific Research (B) Number 17H02901, JSPS Grant-in-Aid for Scientific Research (C) Number 16K05382, and by the MEXT-Supported Program for the Strategic Research Foundation at Private Universities (2011-2015) (No. S1101021) at Waseda University. The CALET effort in the United States is supported by NASA through Grants No. NNX16AB99G, No. NNX16AC02G, and No. NNH14ZDA001N-APRA-0075.

References

[1] Torii S *et al.* (CALET Collaboration) 2015 *Proceeding of Science (ICRC2015) 581*
[2] Torii S *et al.* (CALET Collaboration) 2017 *Proceeding of Science (ICRC2017) 1092*
[3] Yamaoka K *et al.* 2013 *Proc. 7th Huntsville Gamma-Ray Burst Symposium, GRB 2013* p 41
[4] Marrocchesi P S *et al.* 2011 *Nucl. Instrum. Methods Phys Res., Sect. A* **659** 477–483
[5] Marrocchesi P S *et al.* (CALET Collaboration) 2013 *Proc. of 33rd ICRC* p 362
[6] Adriani O *et al.* (CALET Collaboration) 2017 *Phys. Rev. Lett.* **119** 181101
[7] Adriani O *et al.* (CALET Collaboration) 2018 *Phys. Rev. Lett.* **120** 261102
[8] Asaoka Y, Ozawa Y, Torii S *et al.* (CALET Collaboration) 2018 *Astropart. Phys.* **100** 29
[9] Asaoka Y, Akaike Y, Komiya Y, Miyata R, Torii S *et al.* (CALET Collaboration) 2017 *Astropart. Phys.* **91** 1
[10] Nishimura J *et al.* 1980 *ApJ* **238** 394–409
[11] Kobayashi T, Komori Y, Yoshida K and Nishimura J 2004 *Astrophys. J.* **601** 340–351
[12] Adriani O *et al.* 2009 *Nature* **458** 607
[13] Accardo L *et al.* (AMS Collaboration) 2014 *Phys. Rev. Lett.* **113** 121101
[14] Ambrosi G *et al.* (DAMPE Collaboration) 2017 *Nature* **552** 63
[15] Adriani O *et al.* 2017 *La Rivista del Nuovo Cimento* **40** 473
[16] Abdollahi S *et al.* (The Fermi-LAT Collaboration) 2017 *Phys. Rev. D* **95** 082007
[17] Aguilar M *et al.* (AMS Collaboration) 2014 *Phys. Rev. Lett.* **113** 221102
[18] Aharonian F *et al.* (H.E.S.S. Collaboration) 2008 *Phys. Rev. Lett.* **101** 261104
[19] Aharonian F *et al.* (H.E.S.S. Collaboration) 2009 *Astron Astrophys.* **508** 561
[20] Bell A R, Schure K M, Reville B and Giacinti G 2013 *MNRAS* **431** 415
[21] Marrocchesi P S *et al.* (CALET Collaboration) 2017 *Proceeding of Science (ICRC2017) 156*
[22] Akaike Y *et al.* (CALET Collaboration) 2017 *Proceedings of Science (ICRC2017) 181*
[23] Akaike Y *et al.* (CALET Collaboration) 2018 *at this conference*
[24] Cannady N, Asaoka Y *et al.* (CALET Collaboration) 2018 *ApJS* **238** 5
[25] Yamaoka K *et al.* (CALET Collaboration) 2017 *Proceeding of Science (ICRC2017) 613*
[26] Adriani O *et al.* (CALET Collaboration) 2016 *ApJL* **829** L20
[27] Adriani O *et al.* (CALET Collaboration) 2018 *ApJ* **863** 160
[28] Brian R, Akaike Y *et al.* (CALET Collaboration) 2017 *Proceeding of Science (ICRC2017) 180*
[29] Kataoka R *et al.* 2016 *Geophys. Res. Lett.* **43** doi:10.1002/2016GL068930

IOP Conf. Series: Journal of Physics: Conf. Series **1181** (2019) 012004 doi:10.1088/1742-6596/1181/1/012004

The Dipole Anisotropy of Galactic Cosmic Rays

Markus Ahlers

Niels Bohr International Academy & Discovery Centre, Niels Bohr Institute, University of Copenhagen, Blegdamsvej 17, DK-2100 Copenhagen, Denmark

E-mail: markus.ahlers@nbi.ku.dk

Abstract. The arrival directions of Galactic cosmic rays exhibit anisotropies up to the level of one per-mille over various angular scales. Recent observations of TeV–PeV cosmic rays show that the dipole anisotropy has a strong energy dependence with a phase-flip around 100 TeV. We argue that this behavior can be well understood by the combination of various effects: the anisotropic diffusion of cosmic rays, the presence of nearby sources, the Compton-Getting effect from our relative motion and the reconstruction bias of ground-based observatories.

1. Introduction

The arrival directions of Galactic cosmic rays are highly isotropic. This is expected from the presence of turbulent magnetic fields in our Galaxy with a root-mean-square field strength at the level of micro-Gauss. For cosmic ray nuclei in the TeV–PeV energy range and charge Z the maximal gyroradius can be estimated as $r_g \simeq 1.1\, Z\, \mathcal{R}_{\rm PV}\, B_{\mu{\rm G}}^{-1}$ pc, where $\mathcal{R}_{\rm PV}$ is the rigidity in units of peta-Volts and $B_{\mu{\rm G}}$ the magnetic field strength in units of micro-Gauss. This length scale is much smaller than the typical distance of cosmic ray sources, presumably supernova remnants (SNRs) [1], and a cosmic ray nucleus is therefore expected to encounter many magnetic scattering centers before arrival at Earth. This diffusive process can also account for the observed steepness of the cosmic ray power-law spectrum, in contrast to the hard spectrum expected from diffusive shock acceleration [2, 3].

However, cosmic ray diffusion predicts a weak anisotropy that is related to the distribution of sources. In recent years, many experiments achieved the necessary level of statistics to be able to find anisotropies at the level of one per-mille and below, see *e.g.* Refs. [4, 5]. These data revealed that, besides a dominant dipole anisotropy expected from diffusion theory, there is also evidence for medium- and small-scale structures in the maps of cosmic ray arrival directions, that extend down to angular scales of 10 degrees. Figure 1 shows a recent (preliminary) anisotropy map from a combined analysis of HAWC and IceCube data that illustrates the complexity of the pattern [6].

The appearance of medium- and small-scale features is most likely related to non-diffusive cosmic ray streaming on length scales much smaller than the effective diffusion length. Various authors have considered the effect of the heliosphere [7, 8, 9, 10], the local realization of turbulence [11, 12, 13, 14, 15, 16] or modifications of cosmic ray transport via non-uniform pitch-angle diffusion [17, 18]. We refer to the recent review [5] for further details.

In this brief review we will focus on the interpretation of the Galactic dipole anisotropy. Diffusion theory predicts that this large-scale anisotropy is proportional to the cosmic ray gradient and the effective diffusion length. Naïvely, one would expect that its orientation is towards the Galactic center and its amplitude is rising as a simple power-law according to the

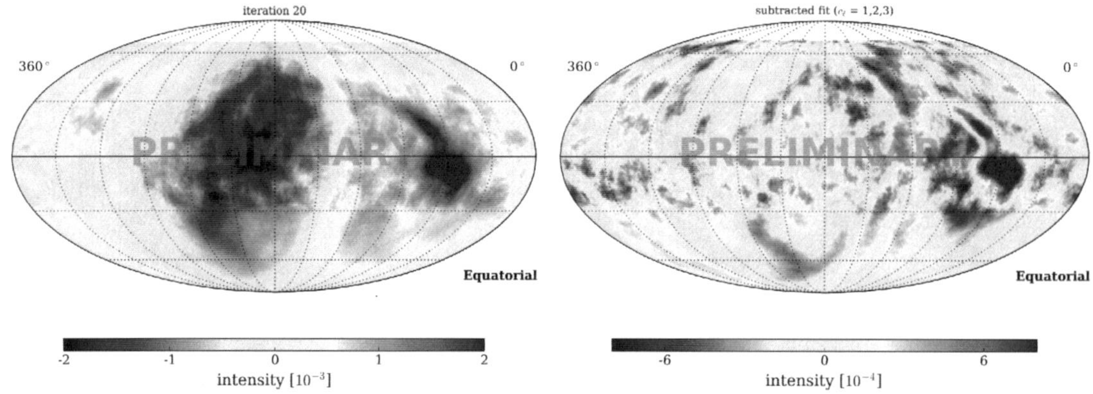

Figure 1. Preliminary results of the 10 TeV cosmic ray anisotropy from a combined analysis of HAWC ($-30° \leq \delta \leq 64°$) and IceCube ($-90° \leq \delta \leq -20°$) data (from Ref. [6]). The analysis is based on an iterative maximum-likelihood method introduced in Ref. [19]. The left map show the full anisotropy in the equatorial coordinate system. The right map shows the same anisotropy after removal of the dipole ($\ell = 1$), quadrupole ($\ell = 2$) and octopole ($\ell = 3$).

rigidity dependence of the diffusion length. However, this is not supported by recent TeV–PeV cosmic ray data. Nevertheless, we will show that the data can be understood in the context of diffusion theory if one accounts for additional effects, namely, the anisotropic diffusion of cosmic rays, the presence of nearby SNRs as the sources of cosmic rays, the Compton-Getting effect from our relative motion in the rest frame of diffusion and a reconstruction bias of ground-based observatories.

2. Anisotropy Reconstruction

We assume in the following that the TeV–PeV cosmic ray flux is constant over the lifetime of observatories. The angular distribution can then be expressed as a function of celestial longitude α (right ascension) and latitude δ (declination) as,

$$\phi(\alpha, \delta) = \phi^{\rm iso} I(\alpha, \delta)\,, \tag{1}$$

where $\phi^{\rm iso}$ corresponds to the isotropic flux level (in units of $\rm cm^{-2}\,s^{-1}\,sr^{-1}$) and $I(\alpha, \delta)$ is the relative intensity of the flux as a function of position in the sky. The anisotropy is defined as the deviation $\delta I = I - 1 \ll 1$. The reconstruction of Galactic cosmic ray anisotropies at the level of one per-mille or less requires the collection of hundreds of millions of events with detectors that are precisely calibrated. It is challenging to satisfy both of these conditions simultaneously. Large event numbers can be achieved by ground-based observatories, that detect cosmic ray air showers via surface arrays or underground detectors [4]. Since the atmosphere is part of the detector, a precise determination of the detector response including small variations in the local acceptance is difficult to achieve from simulations alone. However, one can overcome some of these limitations by using the same cosmic ray data also for calibration.

In the local coordinate system of the ground-based observatory the arrival direction of a cosmic ray is parametrized by its azimuth angle φ (from the north, increasing to the east) and zenith angle θ,

$$\mathbf{n}' = (\cos\varphi\sin\theta, -\sin\varphi\sin\theta, \cos\theta)\,. \tag{2}$$

The arrival direction in the local horizontal coordinate system is related to the arrival direction in the right-handed equatorial coordinate system via a time-dependent rotation, $\mathbf{n} = \mathbf{R}(t)\mathbf{n}'$, where the unit vector $\mathbf{n}$ is given as

$$\mathbf{n} = (\cos\alpha\cos\delta, \sin\alpha\cos\delta, \sin\delta)\,. \tag{3}$$

The rotation matrix $\mathbf{R}$ depend on the local sidereal time t of the observation, which is equivalent to the right ascension angle of the zenith, $\alpha_{\rm zen}(t) = \omega t$, where the sidereal angular frequency $\omega = \omega_{\rm solar} + \omega_{\rm orbit}$ is a combination of the solar angular frequency $\omega_{\rm sol} = 2\pi/24{\rm h}$ and Earth's orbital angular frequency $\omega_{\rm orbit} = 2\pi/1{\rm yr}$. For an experiment located at a geographic latitude Φ, the transformation is then given by

$$\mathbf{R}(t) = \begin{pmatrix} -\cos\omega t\sin\Phi & -\sin\omega t\sin\Phi & \cos\Phi \\ \sin\omega t & -\cos\omega t & 0 \\ \cos\omega t\cos\Phi & \sin\omega t\cos\Phi & \sin\Phi \end{pmatrix}\,. \tag{4}$$

At any time, the ground-based observatory covers an instantaneous field of view which is typically characterized by a maximal zenith angle, $\theta \leq \theta_{\rm max}$. After many sidereal days of operation, the detector will then observe an integrated field of view characterized by a declination band, $\delta_{\rm min} < \delta < \delta_{\rm max}$, bounded by $\delta_{\rm min} = \max(-90^\circ, \Phi - \theta_{\rm max})$ and $\delta_{\rm max} = \min(90^\circ, \Phi + \theta_{\rm max})$.

As indicated earlier, the local detector exposure $\mathcal{E}$ is expected to depend on atmospheric and other time-dependent effects and is difficult to determine from first principles at a level of one per-mille or less. However, after observation over many sidereal days, the events collected during a fixed sidereal time bin $[t, t+\Delta t]$ are related to a time-integrated detector exposure, where local variations in the detector acceptance can be approximated by the average relative acceptance. To be more concrete, it is assumed that the time-integrated local detector exposure $\mathcal{E}$ (accumulated over many sidereal days) can be expressed as a product of its angular-integrated exposure E per sidereal time (units of ${\rm cm}^2\,{\rm sr}$) and relative acceptance $\mathcal{A}$ (units of ${\rm sr}^{-1}$ and normalized as $\int {\rm d}\Omega\mathcal{A}(\Omega) = 1$):

$$\mathcal{E}(t, \varphi, \theta) \simeq E(t)\mathcal{A}(\varphi, \theta)\,. \tag{5}$$

This ansatz assumes that the relative acceptance of the detector does not strongly depend on sidereal time and time-variations can be compensated by an overall rescaling of the angular-integrated exposure. Note that this is a typical assumption of anisotropy reconstruction methods, like the methods of direct integration [20] or time-scrambling [21].

The expected number of cosmic rays at a sidereal time t from an azimuth angle φ and zenith angle θ can now be expressed as

$$\mu(t, \varphi, \theta) = \Delta t\Delta\Omega\, I(\alpha(t, \varphi, \theta), \delta(t, \varphi, \theta))\, \mathcal{N}(t)\, \mathcal{A}(\varphi, \theta)\,, \tag{6}$$

where $\mathcal{N}(t) \equiv \phi^{\rm iso}E(t)$ gives the expected rate of isotropic background events at sidereal time t. The previous relation allows us to simultaneously reconstruct the relative cosmic ray intensity I, isotropic background rate $\mathcal{N}$ and relative acceptance $\mathcal{A}$ via a fit to the data $n(t, \varphi, \theta)$. For mid-latitude detectors, where the instantaneous field of view is rapidly changing, one can derive the anisotropy via an iterative maximum-likelihood analysis, as introduced and validated in Ref. [19].

Finally, the reconstructed anisotropy δI can be analyzed in terms of its spherical harmonics expansion defined via

$$\delta I(\alpha, \delta) = \sum_{\ell\geq 1}\sum_{m=-\ell}^{\ell} a_{\ell m}Y^{\ell m}(\pi/2 - \delta, \alpha)\,, \tag{7}$$

where the complex expansion coefficients are related by $a_{\ell\text{-}m} = (-1)^m a^*_{\ell m}$. Unfortunately, due to the observatory's limited field of view ($\delta_{\rm min} \leq \delta \leq \delta_{\rm max}$), this expansion (7) can not be inverted unless we introduce an *a priori* truncation scale in the multipole expansion. However, there is evidence of small-scale anisotropy in cosmic ray data and this truncation seems not to be well motivated. We therefore expect that the multipole reconstruction reported by various observatories in the North and South can show strong variation introduced by the cross-talk of multipoles in the analysis.

There is an additional obstacle related to the analysis method of ground-based observatories where the cosmic ray data is used for the calibration of the detector. Since this is an important limitation that is often overlooked, we provide in the following section a rigorous mathematical derivation of this effect (see also Appendix of Ref. [22]).

3. Observational Bias

The events recorded at a fixed position (φ, θ) in the local coordinate system can only probe the cosmic ray flux along a constant declination $\delta(\varphi, \theta)$, *i.e.*, only variations of the flux with respect to right ascension $\alpha(t, \varphi, \theta)$ as the sidereal time increases. Hence, the expectation values (6) are invariant under the simultaneous rescaling

$$\begin{aligned} I(\alpha,\delta) &\rightarrow I'(\alpha,\delta) \equiv I(\alpha,\delta)/a(\delta)/b\,, & (8)\\ \mathcal{N}(\varphi,\theta) &\rightarrow \mathcal{N}'(\varphi,\theta) \equiv \mathcal{N}(\varphi,\theta)bc\,, & (9)\\ \mathcal{A}(\varphi,\theta) &\rightarrow \mathcal{A}'(\varphi,\theta) \equiv \mathcal{A}(\varphi,\theta)a(\delta(\varphi,\theta))/c\,, & (10) \end{aligned}$$

where $a(\delta)$ is an arbitrary function of declination. The normalization factors b and c are defined by the normalization conditions $\int \mathrm{d}\Omega\mathcal{A}'(\Omega) = 1$ and $\int \mathrm{d}\Omega\delta I'(\Omega) = 0$. The symmetry transformations in Eqs. (8), (9) and (10) imply that the relative intensity I can only be determined up to a declination-dependent function. As we will show in the following, this makes it impossible to infer azimuthally-symmetric ($m = 0$) anisotropies, if the cosmic ray data are also used for detector calibration.

We can solve Eq. (8) for the rescaled anisotropy as

$$\delta I'(\alpha,\delta) = \frac{1+\delta I(\alpha,\delta)}{a(\delta)b} - 1\,, \tag{11}$$

where the normalization factor b has the explicit form

$$b = \frac{1}{4\pi}\int \mathrm{d}\Omega\frac{1+\delta I(\alpha,\delta)}{a(\delta)}\,. \tag{12}$$

The family of solutions implied by the transformation (11) also include the true anisotropy, denoted by $\delta\widehat{I}(\alpha,\delta)$ in the following. In order to extract a particular member of the family we have to introduce an additional condition on δI that is consistent with the normalization condition, $\int \mathrm{d}\Omega\delta I(\alpha,\delta) = 0$, but breaks the symmetry (8). The natural choice of this "fixing" condition is

$$\int \mathrm{d}\alpha\delta I^{\mathrm{fix}}(\alpha,\delta) = 0\,. \tag{13}$$

Note that this condition singles out a unique solution, as one can infer from the azimuthally-averaged Eq. (11): two solutions, $\delta I'^{\mathrm{fix}}$ and δI^{fix}, imply $a(\delta)b = 1$ and are therefore identical. On the other hand, if we had found a solution δI that does not obey condition (13), we can find the fixed solution δI^{fix} by the transformation (11) using the rescaling function

$$a(\delta)b = \frac{1}{2\pi}\int \mathrm{d}\alpha(1+\delta I(\alpha,\delta))\,. \tag{14}$$

Now, the true anisotropy, $\delta\widehat{I}$, can be expanded into spherical harmonics as in Eq. (7) with coefficients $\widehat{a}_{\ell m}$. Applying the fixing condition (13) to $\delta\widehat{I}$ implies the transformation (11) with scaling function

$$a(\delta)b = 1 + \sum_{\ell\geq 1}\widehat{a}_{\ell 0}\sqrt{\frac{2\ell+1}{4\pi}}P_\ell(\sin\delta)\,. \tag{15}$$

But this expression is simply the azimuthally-averaged relative intensity, $\langle\widehat{I}\,\rangle(\delta)$. From Eq. (11) we therefore arrive at the (unique) solution

$$\delta I^{\rm fix}(\alpha,\delta) = \frac{1}{\langle\widehat{I}\,\rangle(\delta)}\sum_\ell\sum_{m\neq 0}\widehat{a}_{\ell m}Y^{\ell m}(\pi/2-\delta,\alpha)\,. \tag{16}$$

Since the azimuthally-averaged relative intensity is dominated by the monopole, $\langle\widehat{I}\,\rangle(\delta)\simeq 1$, this expression approximates the spherical harmonic expansion of the true anisotropy, $\delta\widehat{I}$, except for azimuthally symmetric components with $m=0$.

In summary, the anisotropy reconstruction by ground-based observatories that use the same cosmic ray data to determine the local detector acceptance is insensitive to the $m=0$ components of the spherical harmonics expansion of the anisotropy in the equatorial coordinate system. This observational bias was first pointed out in Ref. [23] and has important consequences for the interpretation of cosmic ray anisotropies.

4. Dipole Anisotropy

The dipole anisotropy is the leading-order effect predicted from cosmic ray diffusion. The dipole term is usually parametrized as

$$\delta I_{\rm dipole}(\alpha,\delta) = \boldsymbol{\delta}\cdot\mathbf{n}(\alpha,\delta)\,, \tag{17}$$

where $\mathbf{n}$ is the unit vector introduced in Eq. (3). The dipole vector $\boldsymbol{\delta}$ can be expressed in terms of the expansion coefficients of the spherical harmonics (7) as

$$\boldsymbol{\delta} \equiv \big(\delta_{\rm 0h},\delta_{\rm 6h},\delta_{\rm N}\big) = \sqrt{\frac{3}{2\pi}}\big(-\Re(a_{11}),\Im(a_{11}),a_{10}\big)\,. \tag{18}$$

Here, we introduced the notation $\delta_{\rm 0h}$ and $\delta_{\rm 6h}$ corresponding to the dipole components parallel to the equatorial plane and pointing to the direction of the local hour angle 0h ($\alpha=0^\circ$) and 6h ($\alpha=90^\circ$) of the vernal equinox, respectively. We also introduce $\delta_{\rm N}$ as the dipole component pointing to north.

As discussed in the previous section, ground-based observatories only report the dipole components aligning with the equatorial plane (cf. Eq. (18)) as

$$\big(\delta_{\rm 0h},\delta_{\rm 6h}\big) = (A_1\cos\alpha_1, A_1\sin\alpha_1)\,. \tag{19}$$

The observed dipole amplitude corresponds to the projection of the true dipole onto the equatorial plane, $A_1=\cos\delta_1 A=\sqrt{3/2\pi}|a_{11}|$. Only the phase α_1 corresponds to the true right ascension angle. Figure 2 shows a summary of the projected dipole amplitude and dipole phase from various experiments. One can notice a phase-flip and a drop in the amplitude around at an energy of about 100 TeV. We will argue in the following that this behavior can be understood in terms of standard cosmic ray diffusion, after accounting for strong anisotropic diffusion in our local environment and the presence of local cosmic ray sources.

Fick's law of diffusion theory predicts that the dipole anisotropy is proportional to the spatial gradient of the CR density $\nabla n^\star$ and the diffusion tensor $\mathbf{K}$,

$$\boldsymbol{\delta}^\star = \frac{3\mathbf{K}\cdot\nabla n^\star}{n^\star}\,. \tag{20}$$

In general, the diffusion tensor is expected to be invariant under rotations along the orientation of the local ordered magnetic field and can be written in the form

$$K_{ij} = \kappa_\parallel\widehat{B}_i\widehat{B}_j + \kappa_\perp(\delta_{ij}-\widehat{B}_i\widehat{B}_j) + \kappa_A\epsilon_{ijk}\widehat{B}_k\,. \tag{21}$$

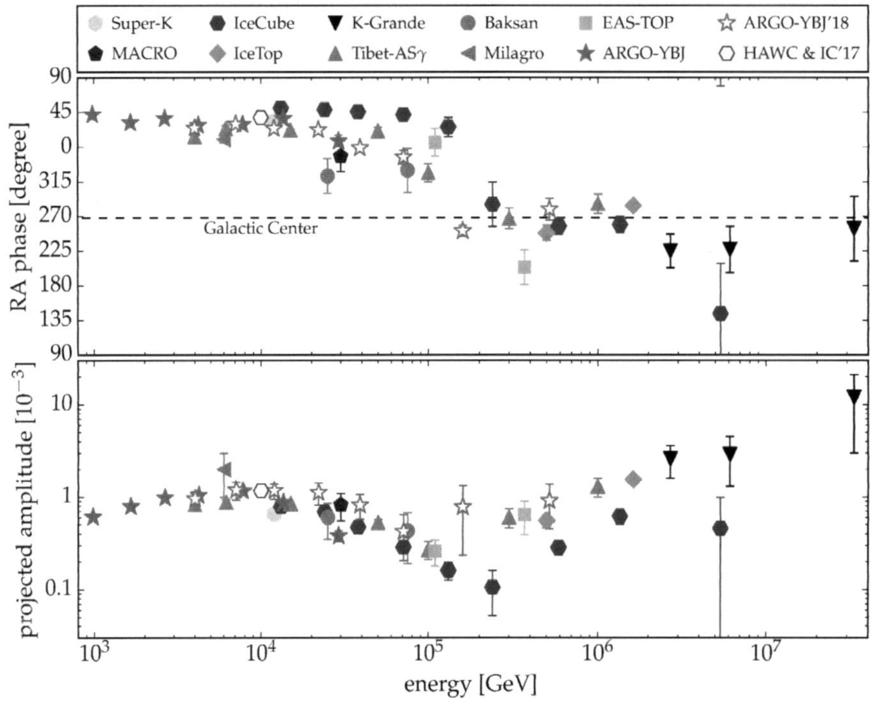

Figure 2. The phase (top panel) and amplitude (bottom panel) of the projected dipole anisotropy from recent cosmic ray measurements [5, 24, 6]. The dashed line in the top panel shows the right ascension angle of the Galactic center.

Here, $\hat{\mathbf{B}}$ is a unit vector pointing in the direction of the regular magnetic field, $\kappa_\parallel$ and $\kappa_\perp$ denote the diffusion strength along and perpendicular to the magnetic field, respectively, and κ_A is the axial diffusion strength (see, *e.g.*, Ref. [25]).

Strictly speaking, the prediction (20) only holds in the rest frame of diffusion (denoted by starred quantities). The relative motion of the solar system with respect to the rest frame of diffusion introduces an apparent dipole anisotropy, which is generally referred to as the Compton-Getting effect [28]. The dipole in the observer's frame can then be written as the sum

$$\boldsymbol{\delta} = \boldsymbol{\delta}^\star + (2+\Gamma)\mathbf{v}/c\,, \tag{22}$$

where $\mathbf{v}$ is the relative velocity of the solar system through the rest frame of the magnetic scattering centers and $\Gamma \simeq 2.7$ is the cosmic ray spectral index [29]. The Compton-Getting effect of the sidereal dipole anisotropy has large uncertainties, related to the ambiguity of identifying the rest frame of cosmic ray diffusion. One possible choice is the local standard of rest (LSR) corresponding to our motion towards the solar apex. The velocity vector in Galactic coordinates has been inferred to point towards $l \simeq 47.9^\circ \pm 2.9^\circ$ and $b \simeq 23.8^\circ \pm 2.0^\circ$ with absolute velocity $v_{\rm LSR} \simeq 18.0 \pm 0.9$ km/s [30]. We will use this estimate in the following as a nominal value. Another choice corresponds to the relative velocity through the local interstellar medium (ISM) towards $l \simeq 5.25^\circ \pm 0.24^\circ$ and $b \simeq 12.0^\circ \pm 0.5^\circ$ with absolute velocity $v_{\rm ISM} \simeq 23.2 \pm 0.3$ km/s [31]. Note that, while the precise value of the sidereal Compton-Getting effect is uncertain, the predicted amplitude is at the level of 3×10^{-4} and almost an order of magnitude smaller than the observed dipole anisotropy at 10 TeV (see Fig. 2).

The field strength of the local ordered magnetic field is expected to be comparable to the turbulent contribution. This implies that the expansion coefficients of the diffusion tensor, Eq. (21), are hierarchical with $\kappa_\perp \ll \kappa_\parallel$ and $\kappa_A \ll \kappa_\parallel$. Therefore, the local diffusion tensor

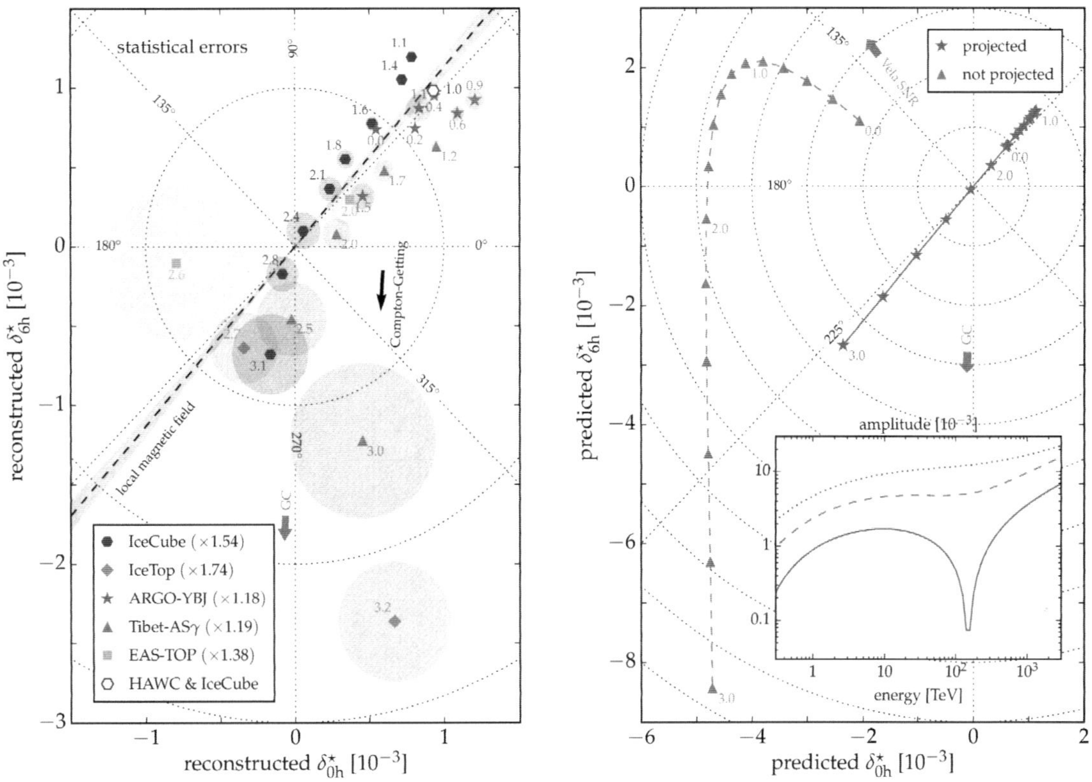

Figure 3. Left Panel: Summary plot of the reconstructed TeV–PeV dipole components $\delta^\star_{0h}$ and $\delta^\star_{6h}$ in the equatorial plane (from Ref. [26]). The black arrow indicates the Compton-Getting effect from the solar motion with respect to the local standard of rest that we subtracted from the data following Eq. (22). In most cases, the reported dipole amplitude was rescaled to compensate for a bias from the experiment's limited field of view and/or reconstruction techniques. The numbers attached to the data indicate the median energy of the bins as $\log_{10}(E_{\rm med}/{\rm TeV})$. The colored disks show the statistical 1σ error range. The dashed line and gray-shaded area indicate the magnetic field direction and its uncertainty (projected onto the equatorial plane) inferred from IBEX observations [27]. We also indicate the direction towards the Galactic center (GC). **Right Panel:** Same as the left panel but now showing the predicted dipole anisotropy in a model that accounts for the contribution of local SNRs (from Ref. [26]). The red line shows the prediction assuming isotropic diffusion, where the dipole follows the local cosmic ray gradient. The green line shows the same configuration assuming anisotropic diffusion. The inset plot highlights the drop in the dipole amplitude related to the phase-flip around 100 TeV.

(21) reduces to a projector onto the magnetic field direction [32, 33, 34, 26]. The local ordered magnetic field on distance scales less than 0.1 pc can be inferred from the emission of energetic neutral atoms (ENA) from the outer heliosphere observed by the *Interstellar Boundary Explorer* (IBEX) [35]. The emission of ENA is enhanced along a circular ribbon that defines a magnetic field axis along $l \simeq 210.5^\circ$ and $b \simeq -57.1^\circ$ with an uncertainty of $\sim 1.5^\circ$ [27]. This agrees with the magnetic field direction inferred from polarization measurements of local stars [36].

The left panel of Figure 3 shows the reconstructed dipole anisotropy in the equatorial plane (cf. Eq. (19)) inferred from TeV–PeV dipole data. The data are corrected by the predicted Compton-Getting dipole from our relative motion in the LSR. The data in the 1-100 TeV energy region show a strong alignment with the orientation of the local magnetic field projected

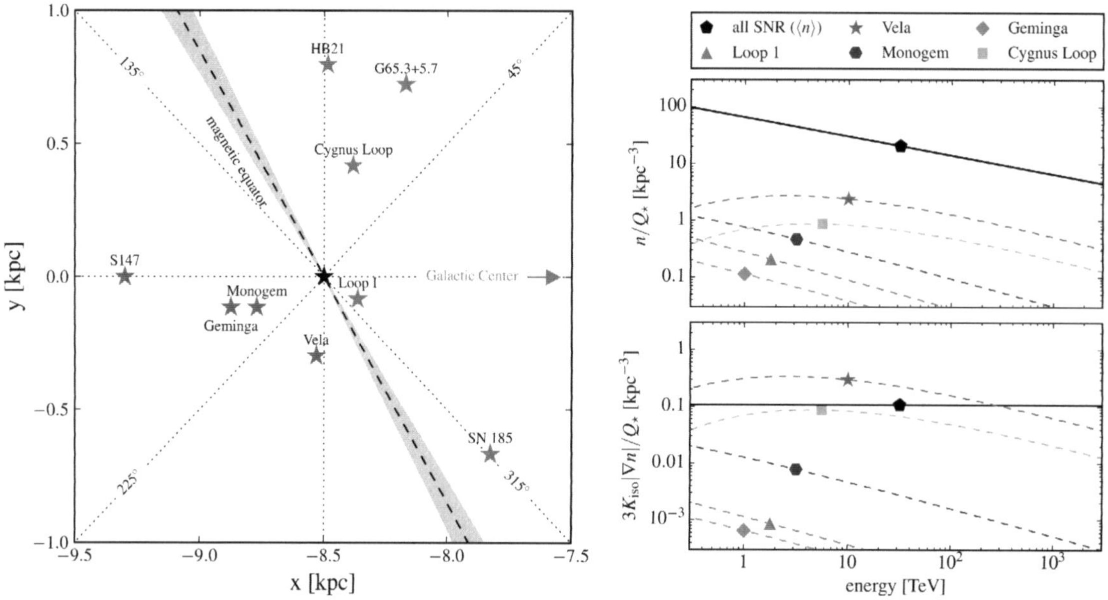

Figure 4. Left Panel: The position of local SNRs in the vicinity of the solar system (black star). The dashed line shows the intersection of the magnetic equatorial plane with the Galactic plane. A cosmic ray gradient aligned with the green (red) sources would be visible at a right ascension phase $\alpha \simeq 49°$ ($\alpha \simeq 229°$) in the model of strong anisotropic diffusion. **Right Panel:** The contribution of the five closest SNRs to the flux (upper panel) and gradient (lower panel) compared to the average contribution of the sum of all Galactic SNRs.

onto the equatorial plane. This is expected from anisotropic diffusion. Note that the data are only shown with statistical errors. As indicated in the previous section, the limited field of view of observatories is expected to introduce cross-talk between multipole moments that can be responsible for the large scatter in the data. The plot also shows the result of the recent combined analysis of IceCube and HAWC [6] (cf. Fig. 1). This measurement is not expected to suffer from strong cross-talk because the combined field of view covers about 95% of the sky. After correcting for the predicted Compton-Getting shift from our motion in the LSR, the dipole orientation agrees very well with the orientation of the local magnetic field inferred by IBEX.

The projection of the cosmic ray gradient onto the local magnetic field axis leaves only two possible dipole orientations, depending on the relative location of the gradient with respect to the magnetic equator. In the left panel of Fig. 4 we show the intersection of the magnetic equatorial plane with the Galactic plane together with the location of known nearby SNRs. Assuming that the cosmic ray gradient is aligned with the position of one of these potential cosmic ray sources, the dipole would be oriented towards $\alpha_1 \simeq 49°$ for SNRs indicated in green and $\alpha_1 \simeq 229°$ for SNRs indicated in red. The observed dipole phase below 100 TeV is oriented towards $\alpha_1 \simeq 49°$, which favors a strong contribution from the local SNRs Vela, Monogem or Geminga. The phase-flip at 100 TeV would indicate the transition of the gradient into the opposite magnetic hemisphere, that also includes the location of the Galactic center.

One can estimate the relative contribution of local SNRs to the cosmic ray flux (monopole) and gradient (dipole), taking into account the position and age of these candidate cosmic ray sources [37, 38, 39, 40, 41, 42]. The calculation of Ref. [26] assumes impulsive cosmic ray emission at the beginning of the Sedov phase (after about 100 yrs) and cosmic ray diffusion with an (effectively) isotropic diffusion coefficient $K_{\rm iso} \simeq 4 \times 10^{28}(E/3{\rm GeV})^{1/3}{\rm cm}^2/{\rm s}$. The average contribution of all Galactic SNRs assumes a spatial distribution following Ref. [43] with a source rate of $1/30\,{\rm yr}^{-1}$ and a vertical diffusion height of 3 kpc. The results of the cosmic ray flux and gradient are shown in the right panel of Fig. 4. In this model, the flux of local SNRs is subdominant compared to the total contribution of Galactic SNRs, whereas the gradient is dominated below 100 TeV by the Vela SNR.

The right panel of Fig. 3 shows the corresponding dipole projection of this dipole model onto the equatorial plane. The red dashed line shows the predicted dipole under the assumption that the local diffusion tensor is the same as the effective isotropic diffusion tensor assumed for the calculation of the flux and gradient. This is clearly inconsistent with the data shown in the left panel of Fig. 3. However, anisotropic diffusion along the local magnetic field orientation predicts a dipole following the green solid line, in agreement with the behavior seen in the dipole data. The inset plot in the right panel of Fig. 3 compares the energy dependence of the amplitude of the full isotropic dipole (blue dotted line), the projected isotropic dipole (red dashed line) and the projected anisotropic dipole (green solid line).

In summary, the observed energy dependence of the dipole amplitude and phase can be well reproduced with diffusion theory after the effect of anisotropic diffusion, local sources, Compton-Getting effect and reconstruction bias are taken into account. A natural candidate for the local source responsible for the phase-flip in the data is the Vela SNR [26].

5. Conclusions

The arrival directions of TeV–PeV Galactic cosmic rays show a rich structure of anisotropies up to a level of one per-mille on various angular scales. In this brief review we have shown that the observed dipole data is consistent with the expectation from diffusion theory, if one accounts for the combined effect of one or more local sources, the presence of a strong ordered magnetic field in our local environment and the reconstruction bias of ground-based observatories.

The sidereal Compton-Getting effect from our relative motion with respect to the rest frame of diffusion is typically much smaller than the experimental sensitivity if one accounts for systematic uncertainties related to the reconstruction of spherical harmonic coefficients in the limited field of view. However, a recent combined analysis [6] of IceCube and HAWC data with an integrated field of view that covers about 95% of the sky allows to keep multipole cross-talk under control. The inferred dipole at a median cosmic ray energy of 10 TeV aligns well with the local orientation of the magnetic field inferred from IBEX after accounting for our relative motion in the local standard of rest.

The presence of anisotropies at smaller angular scales can be related to the cosmic ray transport in the local (turbulent and ordered) magnetic field within one diffusion length. Here, the impact of the heliosphere and the random realization of magnetic turbulence have been invoked to account for hotspots in the anisotropy map (cf. Fig. 1) and the excess power down to angular scales of at least 10° in the angular power spectrum [5].

Acknowledgements

I would like to thank the organizers of the *26th Extended European Cosmic Ray Symposium 2018* at Altai State University for their invitation to present this work. I am grateful for many fruitful discussions with and the support of my collaborators Segev Y. BenZvi, Paolo Desiati, Juan Carlos Díaz-Vélez, Daniel W. Fiorino, Philipp Mertsch and Stefan Westerhoff. This work was supported by Danmarks Grundforskningsfond (project no. 1041811001) and VILLUM FONDEN (projects no. 18994).

References

[1] Baade W and Zwicky F 1934 *Proceedings of the National Academy of Science* **20** 259–263
[2] Bell A R 1978 MNRAS **182** 147–156
[3] Blandford R D and Ostriker J P 1978 ApJ **221** L29–L32
[4] Di Sciascio G and Iuppa R 2013 *Homage to the Discovery of Cosmic Rays, the Meson-Muon and Solar Cosmic Rays* ed Pérez-Peraza J A (New York: Nova Science Publishers) chap 9, pp 221–257 (*Preprint* 1407.2144)

[5] Ahlers M and Mertsch P 2017 *Prog. Part. Nucl. Phys.* **94** 184–216 (*Preprint* 1612.01873)
[6] Velez J C D *et al.* (IceCube, HAWC) 2018 *PoS* **ICRC2017** 539 (*Preprint* 1708.03005)
[7] Lazarian A and Desiati P 2010 ApJ **722** 188–196 (*Preprint* 1008.1981)
[8] Desiati P and Lazarian A 2013 *Astrophys.J.* **762** 44 (*Preprint* 1111.3075)
[9] Drury L O 2013 *Proceedings of the ICRC 2013* (*Preprint* 1305.6752)
[10] Zhang M, Zuo P and Pogorelov N 2014 *Astrophys. J.* **790** 5
[11] Battaner E, Castellano J and Masip M 2011 A&A **527** 5 (*Preprint* 1006.2346)
[12] Giacinti G and Sigl G 2012 *Phys.Rev.Lett.* **109** 071101 (*Preprint* 1111.2536)
[13] Ahlers M 2014 *Phys.Rev.Lett.* **112** 021101 (*Preprint* 1310.5712)
[14] Pohl M and Rettig R 2016 *PoS* **ICRC2015** 451
[15] López-Barquero V, Farber R, Xu S, Desiati P and Lazarian A 2016 *Astrophys. J.* **830** 19 (*Preprint* 1509.00892)
[16] Ahlers M and Mertsch P 2015 *Astrophys. J.* **815** L2 (*Preprint* 1506.05488)
[17] Malkov M, Diamond P, Drury L and Sagdeev R 2010 ApJ **721** 750–761 (*Preprint* 1005.1312)
[18] Giacinti G and Kirk J G 2017 *Astrophys. J.* **835** 258 (*Preprint* 1610.06134)
[19] Ahlers M, BenZvi S Y, Desiati P, Díaz-Vélez J C, Fiorino D W and Westerhoff S 2016 *Astrophys. J.* **823** 10 (*Preprint* 1601.07877)
[20] Atkins R W *et al.* (Milagro) 2003 ApJ **595** 803–811 (*Preprint* astro-ph/0305308)
[21] Alexandreas D E *et al.* 1993 *Nucl. Instrum. Meth. A* **328** 570–577
[22] Ahlers M 2018 *Astrophys. J.* **863** 146 (*Preprint* 1805.08220)
[23] Iuppa R and Di Sciascio G 2013 ApJ **766** 9286 (*Preprint* 1301.1833)
[24] Bartoli B *et al.* (ARGO-YBJ) 2018 *Astrophys. J.* **861** 93 (*Preprint* 1805.08980)
[25] Bhatnagar P, Gross E and Krook M 1954 *Phys.Rev.* **94** 511–525
[26] Ahlers M 2016 *Phys. Rev. Lett.* **117** 151103 (*Preprint* 1605.06446)
[27] Funsten H O *et al.* 2013 ApJ **776** 30
[28] Compton A H and Getting I A 1935 *Physical Review* **47** 817–821
[29] Forman M A 1970 Planet. Space Sci. **18** 25–31
[30] Schönrich R, Binney J and Dehnen W 2010 MNRAS **403** 1829–1833 (*Preprint* 0912.3693)
[31] McComas D J *et al.* 2012 *Science* **336** 1291
[32] Jones F C 1990 ApJ **361** 162–172
[33] Mertsch P and Funk S 2015 *Phys.Rev.Lett.* **114** 021101 (*Preprint* 1408.3630)
[34] Schwadron N A *et al.* 2014 *Science* **343** 988–990
[35] McComas D J *et al.* 2009 *Science* **326** 959
[36] Frisch P C *et al.* 2015 ApJ **814** 112 (*Preprint* 1510.04679)
[37] Erlykin A D and Wolfendale A 2006 *Astropart.Phys.* **25** 183–194 (*Preprint* astro-ph/0601290)
[38] Blasi P and Amato E 2012 *JCAP* **1201** 010 (*Preprint* 1105.4521)
[39] Blasi P and Amato E 2012 *JCAP* **1201** 011 (*Preprint* 1105.4529)
[40] Pohl M and Eichler D 2013 ApJ **766** 4 (*Preprint* 1208.5338)
[41] Sveshnikova L G, Strelnikova O N and Ptuskin V S 2013 *Astropart. Phys.* **50-52** 33–46 (*Preprint* 1301.2028)
[42] Kumar R and Eichler D 2014 ApJ **785** 129
[43] Case G L and Bhattacharya D 1998 ApJ **504** 761 (*Preprint* astro-ph/9807162)

Features of the cosmic ray 27-day variation within 2014 November-December

V E Sdobnov, M V Kravtsova, S V Olemskoy

Institute of Solar-Terrestrial Physics of Siberian Branch of Russian Academy of Sciences, Irkutsk, RF

E-mail: *sdobnov@iszf.irk.ru*

Abstract. From the ground-based measurements at the global world network of neutron monitors and from the GOES-15 satellites, investigated was the cosmic ray (CR) intensity 27-day variation within 2013, 2014, and 2015 November–December.

Significant changes in the energy losses as particles move in the heliosphere regular electromagnetic fields were shown to be the determining factor for the considerable difference in the amplitudes of the CR 27-day variation within 2014 November-December (as compared with that of 2013 and 2015 November–December). Within 2014 November–December, there was a long-living corotating trap under the action of a vast coronal hole in the Sun south in the interplanetary space. This trap configuration facilitated the energy loss of CRs with the ~3–20 GV rigidity, due to which ground-based neutron monitors measured an abnormally large amplitude of the 27-day variation.

1. Introduction

An important feature of the interplanetary magnetic field (IMF) is its structure. This means that, in the interplanetary space, there may be an even number of the sectors with different IMF radial component direction in the ecliptic plane.

The IMF sector structure is related to the existence of the neutral current sheet dividing the hemispheres, in which the solar magnetic field radial component has the opposite direction (because the magnetic flux through any closed surface is equal to zero, the magnetic field in different regions of space should be directed in opposite directions). Every 22 years, the Sun magnetic field changes its sign (polarity reversal). The current sheet is approximately in the solar equator plane and has a corrugated structure: it alternately declines north- and southward of the equator. As a result of the Sun rotation, the current sheet folds twist in spirals.

An observer on the Earth, who moves in the plane, is either below, or above the current sheet. Due to this, he passes through the sectors with different direction of the IMF radial component. Because of longitudinal and latitudinal solar wind (SW) velocity gradients that occur near the Sun, there emerge radial velocity gradients (when moving away from the Sun) that lead to the production of collisionless shocks near the sector boundaries. Such waves originate at >1 AU, and are traced at several AU.

As long as an active region exists within several rotations of the Sun, then it will corotate with the Sun, because the magnetic inhomogeneity is, as though, bound to a certain field line "leaving" the given meridian. The cosmic ray (CR) propagation in the heliosphere in the presence of such inhomogeneities causes a 27-day variation. There are numerous papers that study the effect of corotating SW structures on the CR behavior, for example [1–3].

Normally, the amplitude of the CR neutron component 27-day variation measured at the sea-level does not surpass ~0.7–1.0 % [4].

Starting with the 2014 second half until 2015 March, the world network CR stations recorded the 27-day variation of an abnormally large amplitude [5]. Thus, at the Irkutsk CR station (threshold rigidity R=3.66 GV) located at 433 m above the sea-level, the amplitude of the CR 27-day variation was ~8% in 2014 November–December. The amplitudes of the CR 27-day variation in 2013 November-December, and in 2015 did not show any abnormal values.

Unlike the situation in the heliosphere within 2013 November–December and in 2015, the existence of a huge coronal hole near the South Pole of the Sun [6] in the 2014 second half led to an asymmetric magnetic configuration at the mid- and high-heliolatitudes. The 2014 November–December current sheet inclination was ~52° [7]. Besides, during this period, there was a polarity reversal on the Sun [7].

In this study, we interpret the abnormally large amplitude of the 2014 November–December CR 27-day variation.

2. Data and method

For the analysis, we used the data of ground-based measurements with world-network neutron monitors. The data were averaged over hour intervals. For 2013 November–December, there were data from 45 stations, for 2014 November–December – from 39 stations, for 2015 November–December – from 41 stations. Also, the GOES-15 satellite data were involved [8].

Through the spectrographic global survey [9, 10], we obtained the information on variations in the primary CR angular and energy distributions outside the Earth magnetosphere, as well as on the variations in the geomagnetic cutoff rigidity planetary system for each observational hour. The modulation amplitudes were counted off from the 2013 November 19 background level (for 2013 November–December), from the 2014 November 28 background level (for 2014 November–December), and from the 2015 November 24 – for 2015 November–December. From the world-network CR station data, and from the GOES-15 data on protons in the Earth orbit [10], we calculated the CR proton spectra at individual instants of the investigated events. To calculate the CR spectra, we used the expression obtained within the model for the CR modulation by heliospheric regular electromagnetic fields [11].

3. Results and conclusions

The existence of a huge coronal hole in the 2014 second half and early 2015 [6] led to the presence of quasi-stationary high-velocity flows of solar plasma in the interplanetary space. When propagating in the interplanetary space, the high-velocity flow overtakes the slow one, and, as a result of coupling between the two flows, there forms a magnetic trap that corotates with the Sun at the 27-day periodicity, thus causing the 27-day variation in the CR intensity.

According to [12], the CR 27-day variations imply the presence of the ordered IMF. Therefore, the CR energy variation (when moving in the heliospheric regular electromagnetic fields) may be? to a great extent, an important factor in the CR modulation with a 27-day periodicity. The energy variation, in its turn, is determined by the value and by the nature of the interplanetary medium electric fields, and by the time of particle interaction with those fields [11]. That time is notably determined by the IMF structures of "magnetic trap" type. At such an approach, the energy variation is described by the expression [11]

$$\Delta\varepsilon = \frac{ze\Omega B_0 r_0^2}{c}(1-\cos\lambda_E) = zeU,$$

where $U = \Omega B_0 r_0^2(1-\cos\lambda_E)$ is the electric field potential at the λ_E heliolatitude, Ω is the Sun rotation angular velocity, B_0 is the strength of the mean magnetic field at r_0. The CR rigidity spectrum in the Earth orbit is described by the expression [11]

IOP Conf. Series: Journal of Physics: Conf. Series **1181** (2019) 012005 doi:10.1088/1742-6596/1181/1/012005

$$J(R) = A\frac{\left(\varepsilon^2 - \varepsilon_0^2\right)^{3/2}}{\varepsilon\left[(\varepsilon+\Delta\varepsilon)^2 - \varepsilon_0^2\right]}\left(\frac{\varepsilon+\Delta\varepsilon}{T_0+\varepsilon_0}\right)^{-\gamma},$$

where ε is the total energy of particles with the rigidity R; $\Delta\varepsilon$ is its variations within the heliospheric electromagnetic fields; ε_0 is the rest energy; T_0 is the kinetic energy of particles, at which the CR intensity corresponding to the rigidity in the Galaxy is A; γ are the spectral index of the galactic spectrum.

Apparently, the particle energy losses at the potential electric field do not depend on the particle rigidity, and, at the IMF characteristic strength (~5 nT), are ~200 MeV.

In that case, when the IMF has the structure different from spiral (for example, in the presence of loop-like structures or "magnetic clouds", i.e. "magnetic traps"), the $\Delta\varepsilon$ values will depend on the strength of the magnetic fields of the provided structures and on the SW velocity, and may surpass 200 MeV by almost an order of magnitude [11].

Figure 1 presents the IMF strength modulus (|B|), the IMF orientation longitudinal angle (Long), and the solar wind velocity (V) for 2013, 2014, and 2015 November–December. Within the presented intervals, one can clearly see the periods, when the IMF was directed away from the Sun, and when the magnetic field direction turned to the opposite one, i.e., one can see a sectoral IMF structure. This figure also provides: the CR variation amplitudes ($\Delta I/I$, blue line) at the Irkutsk Station (Rc=3.66 GV); the CR variation amplitude with the 10 GV rigidity in the Earth orbit ($\Delta J/J$, red line); the CR geomagnetic cutoff rigidity variations (ΔR, blue line) in Irkutsk; the Dst-index (red line); $\Delta\varepsilon$ being the particle energy variations within the heliospheric electromagnetic fields during the indicated periods.

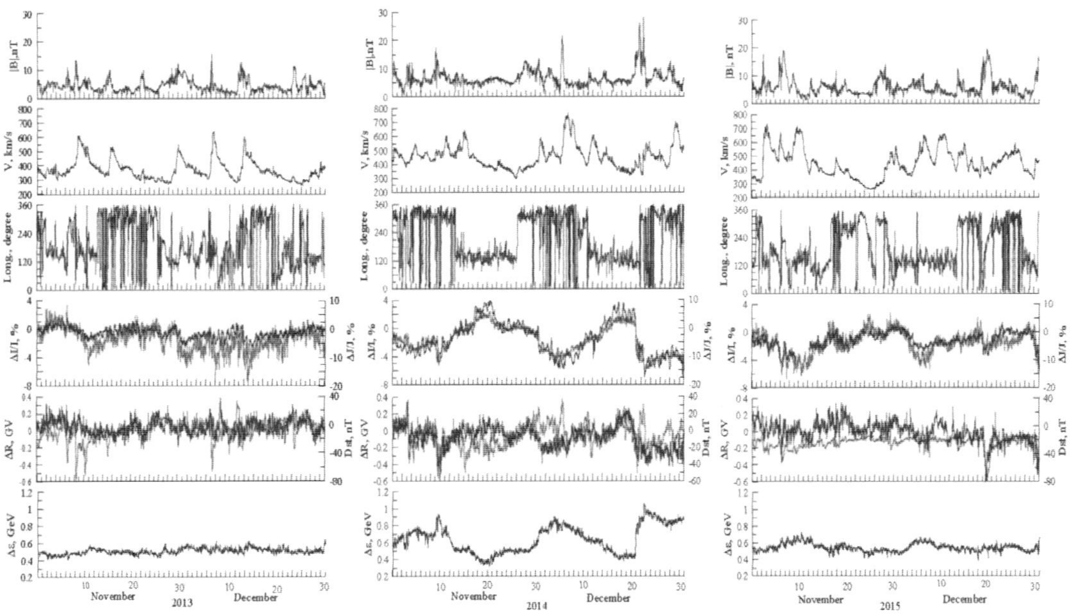

Figure 1. |B| is the IMF strength modulus, Long is the IMF orientation longitudinal angle, V is the solar wind velocity, $\Delta I/I$ is the amplitude of the CR variations at the Irkutsk Station (Rc=3.66 GV, blue line), $\Delta J/J$ is the amplitude of the variations in the 10 GV rigidity CRs in the Earth orbit (red line), ΔR is the CR geomagnetic cutoff rigidity variation in Irkutsk (blue line), Dst-index (red line); $\Delta\varepsilon$ is the particle energy variation within the heliospheric electromagnetic fields.

From the satellite measurements of SW parameters [13], the IMF modulus mean value in the sector with the magnetic field sunward direction in 2014 was more (~7.2 nT), than this value in the sector with the magnetic field direction away from the Sun (~6.2 nT). In 2013 and 2015, the IMF modulus mean value in the sector with the magnetic field sunward direction (~4.6 and ~6.6 nT, respectively) was less, than this value in the sector with the magnetic field direction away from the Sun (~6.8 and ~6.7 nT, respectively).

From Figure 1, one can see that the amplitude of the CR neutron component 27-day variation in Irkutsk over 2014 November–December is much more, than that over the same period in 2013 and 2015. Such a great difference in the variation amplitudes is provided by the different values of the CR particle energy losses in the heliospheric electromagnetic fields during the indicated periods. Thus, in 2013 November–December, the $\Delta\varepsilon$ value varied between ~0.4 and ~0.6, in 2014 November–December ~ 0.3 through ~1.1, in 2015 November–December ~0.4 through ~0.7. From the above, it follows that the determining factor of the significant difference in the galactic CR flow variations in the Earth orbit over 2014 November–December (as compared with 2013 and 2015 November–December) is essential variations in the particle energy losses, as they move in the regular heliospheric electromagnetic fields. We also note that the particle energy losses over 2014 November–December in the sector, where the IMF was directed away from the Sun, are much less, than those in the sector, where the IMF was sunward.

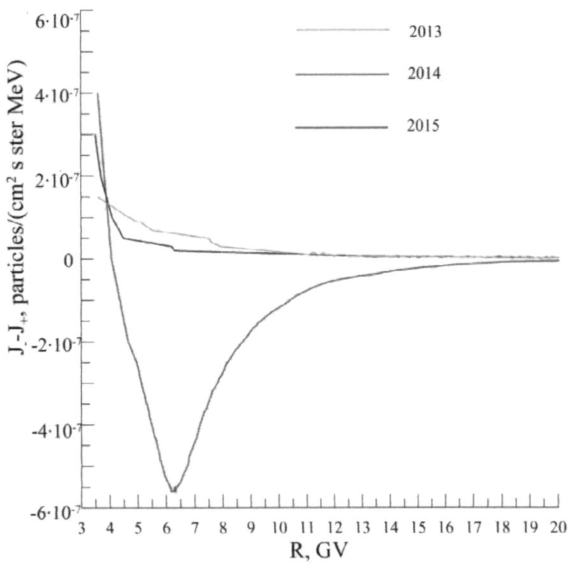

Figure 2. (J_--J_+) is the difference in the mean CR rigidity spectra in the sectors with the IMF direction toward (J_-) and away from (J_+) the Sun.

To reveal, in which rigidity range the maximal modulation in different IMF sectors occurs, we calculated the CR mean rigidity spectra in the sectors with the IMF direction toward (J_-) and away from (J_+) the Sun for 2013, 2014, 1nd 2015 November–December. Figure 2 presents the difference in such spectra (J_-J_+) for the above periods. Apparently, within the ~3–20 GV rigidity range, where neutron monitors are most sensitive, the 2014 November–December modulation in the sectors with the IMF direction toward the Sun prevailed over the modulation in the sectors with the IMF direction away from the Sun. Presumably, over this period, there emerged a long-lived corotating trap of such a configuration, in which the ~3–20 GV CRs were losing their energy most efficiently under the effect of the huge coronal hole in the south of the Sun. Due to that, the ground-based CR measurements with neutron monitors displayed an abnormally large amplitude of the 27-day variation.

The study was done within the 2018 State Assignment 007-00163-18-00 of 2018 January 12 by using the research equipment of the "Angara" Multi-Access Center (Angara MAC) and of the "Russian National Ground-Based Network of Cosmic Ray Stations" (CRS Network) Unique Research Facility (URF).

References

[1] Lee C O, Luhmann J G, I. de Pater, Mason G M, Haggerty D, Richardson I G, Cane H V,· Jian L K, Russell C T and Desai M I 2010 *Solar Phys.* 263 239-61 DOI 10.1007/s11207-010-9556-x
[2] Modzelewska R and Alania M V 2013 *Solar Phys.* 286 593-607 DOI 10.1007/s11207-013-0261-4A
[3] Gil A and Alania M V 2016 *Solar Phys.* 291 1877-86 DOI 10.1007/s11207-016-0924-z
[4] Dorman L I 1963 *Cosmic ray variations and space exploration* (Moscow: USSR AS Publishers) p 1028
[5] Gil A and Mursula K 2015 *Proc. 34th Int. Cosmic Ray Conf.* (Netherlands: Hague) p 149
[6] http://www.solarmonitor.org
[7] http://wso.stanford.edu/
[8] http://satdat.ngdc.noaa.gov./sem/goes/data/new_avg/
[9] Dvornikov V M et al 1983 *Proc.18th Int. Cosmic Ray Conf.* 3 (India: Bangalore) p 249
[10] Dvornikov V M and Sdobnov V E 2002 *Int. J. Geomagn. Aeron.* 3 3 217-28
[11] Dvornikov V M, Kravtsova M V, Sdobnov V E 2013 *Geomagnetism and Aeronomy* 53 4 430-40
[12] Krymsky G *F* 1969 *Modulation of cosmic rays in the interplanetary space* (Moscow, Nauka Science Publishers) p 152
[13] https://omniweb.gsfc.nasa.gov/ow.html

Analysis of sub-GLE and GLE events using NM data: space weather applications

A. Mishev[1,2] **and I. Usoskin**[1,2]
[1]Space Climate Research Unit, University of Oulu, Finland
[2]Sodankylä Geophysical Observatory, University of Oulu, Finland

E-mail: alexander.mishev@oulu.fi

Abstract. Detailed study of solar energetic particle events is important to understand their acceleration and propagation in the interplanetary space and allows one to model various processes related to space weather and space climate e.g. radiation environment at flight altitudes and atmospheric ionization. A special class of events, called ground level enhancements, registered by ground based detectors, can dramatically change the radiation environment in the Earth's atmosphere. New recently installed high-altitude polar neutron monitors NMs, namely DOMC/B and historically installed South Pole, have made the worldwide neutron monitor network more sensitive to strong solar energetic particle events, registered at ground level. The two high-altitude polar stations are able to detect lower energy events, which most likely would not be registered by the other (near sea level) neutron monitor stations. Here, using the worldwide neutron monitor database (NMDB) records and an optimization procedure combined with simulations, we assess the spectral and angular characteristics of a new class of events sub-GLEs. With the estimated spectral characteristics as an input, we evaluate the effective dose rate in polar and sub-polar regions at typical commercial flight altitude during such events as well as during GLEs.

1. Introduction

Detailed study of solar energetic particle events (SEP) events is important in order to clarify their acceleration and transport mechanisms in the interplanetary space [1, 2]. In addition, possessing information about their spectra and angular distributions, makes it possible to assess processes related to space weather, such as exposure to radiation, henceforth exposure, at flight altitudes [3, 4]. The exposure in the vicinity of Earth and in the Earth's atmosphere is variable and highly dynamic. It is governed by the galactic cosmic rays (GCRs). The Earth is constantly hit by high-energy particles, which collide with nucleus of the atmosphere and produce new, energetic particles, which also collide with nuclei etc...Each collision adds a number of particles to the developing a cascade – an extensive air shower. Another important but sporadic source, which makes the radiation environment in the vicinity of Earth and within its atmosphere highly dynamical is related to eruptive solar processes, namely solar flares and coronal mass ejections, leading to production of SEPs [5, 6]. A special class of SEP events, called ground level enhancements (GLEs) [6, 7], registered by ground based detectors e.g. neutron monitors (NMs), can dramatically change the radiation environment in the Earth's atmosphere [8]. GLEs can be studied using the worldwide NM network [9, 10].

New recently installed high-altitude polar NMs, namely DOMC/B and historically installed both South Pole NMs, have made the worldwide NM network more sensitive to strong SEPs and GLEs. The two high-altitude polar NM stations are able to detect lower energy SEP events, which most likely would not be registered by the other (near sea level) NMs. Here, on the basis of NM records and an optimization procedure, we assess the spectral and angular characteristics of GLE and sub-GLE particles and subsequently evaluate the exposure at typical commercial flight altitude during such events.

2. Analysis of GLE and sub-GLEs using NM data

According to recently adjusted definition, a GLE event is registered when there is a simultaneous statistically significant increase of the count rate of at least two differently located NM stations accompanied with a statistically significant increase of SEP flux observed by a space-borne instrument(s). A new sub-class of events sub-GLEs, accordingly is registered when there are simultaneous statistically significant increase of the count rates of at least two differently located high-altitude NMs with corresponding enhancement in SEP flux measured by a space-borne in-instrument(s), but no statistically significant enhancement in the count rates of NMs near to the sea level, for details see [11]. For the analysis of such events we employ a method similar to that applied to earlier GLEs [12, 13], which detailed description is given in [14, 15, 16].

The method consists of: computation of asymptotic viewing cones and cut-off rigidities of the NMs used for the analysis of the corresponding event; initial guess of the optimization procedure by assuming the apparent source position along the interplanetary magnetic field (IMF) line derived from ACE satellite measurements, the latter being estimated from the 20-minute averaged measurements explicitly considering the time shift of the field direction at the nose of the Earth's bow shock [16]; optimization over modelled and measured NM responses assuming a corresponding set of unknown parameters.

The relative count rate increase of a given NM can be expressed as:

$$\frac{\Delta N(P_{cut})}{N} = \frac{\int_{P_{cut}}^{P_{max}} J_{||SEP}(P,t)Y(P)G(\alpha(P,t))A(P)dP}{\int_{P_{cut}}^{\infty} J_{GCR}(P,t)Y(P)dP} \tag{1}$$

where $J_{||SEP}(P,t)$ is the rigidity spectrum of the primary SEPs with a given rigidity P in the direction of the maximal flux along the IMF, $J_{GCR}(P,t)$ is the rigidity spectrum of GCR at a given time t with the corresponding modulation, $Y(P)$ is the NM yield function, $G(\alpha(P,t))$ is the pitch angle distribution (PAD) of SEPs, $A(P)$ is a discrete function with $A(P)$=1 for allowed trajectories (proton with rigidity P can reach the station), accordingly $A(P)$=0 for forbidden trajectories (proton with rigidity P cannot reach the station), N is the count rate due to GCR, $\Delta N(P_{cut})$ is the count rate increase due to SEPs, P_{cut} is the lower cut-off rigidity of the station and P_{max} is the maximal rigidity of SEPs considered in the model to be 20 GV, respectively. In this study we used a newly computed NM yield function, which provides good agreement with experiments and models [17, 18, 19]. In our model the rigidity spectrum of SEPs is described by a modified power law:

$$J_{||}(P) = J_0 P^{-(\gamma+\delta\gamma(P-1))} \tag{2}$$

where $J_{||}(P)$ is the particle flux with given rigidity P in [GV] arriving from the Sun along the axis of symmetry whose direction is defined by geographic coordinate angles Ψ and Λ (latitude and longitude), γ is the power-law spectral exponent at rigidity P = 1 GV, $\delta\gamma$ is the rate of the spectrum steepening. In Eq. (1) we can consider also an exponential spectrum similarly:

$$J_{||}(P) = J_0 \exp(-P/P_0). \tag{3}$$

where $J_{||}$ is defined as in Eq. (2) and P_0 is a characteristic proton rigidity. The PAD is assumed to be a superposition of two Gaussians, which allows to model a complicated particle flow:

$$G(\alpha(P)) \sim \exp(-\alpha^2/\sigma_1^2) + B.\exp(-(\alpha - \pi)^2/\sigma_2^2) \tag{4}$$

where α is the pitch angle, σ_1 and σ_2 are parameters corresponding to the width of the pitch angle distribution, B is a parameter corresponding to the contribution of the particle flux arriving from the anti-sun direction. The optimization is performed using the Levenberg-Marquardt method [20, 21] with variable regularization [22, 23, 24]. The magnetospheric computations are performed with the MAGNETOCOSMICS code [25] using a combination of Tsyganenko 1989 (external) [26] and IGRF (internal) [27] geomagnetic models, which allows one to compute straightforwardly the cut-off rigidity and asymptotic directions with reasonable precision [28, 29, 30].

In the case of sub-GLEs we possess information only from two NMs with statistically significant count rate increase. Therefore, for the analysis of sub-GLEs we simplify the model by assuming a simple power law or exponential rigidity spectrum of SEPs, with one directional Gaussian PAD. We also assume the apparent source position to be along the IMF derived from ACE satellite measurements, but not as a free parameter. Subsequently we derive a set of solutions with similar quality and with the mean we performed a simulation of the global NM network response. The best fit of the global NM network response from the forward modelling is assumed as the final apparent source position. Therefore, by optimization over a limited set of parameters and simulation of the global NM network response (fixing one and/or several of the parameters, but varying the others) after several iterations we assessed the spectral and angular characteristics of sub-GLE particles and accordingly their spectral constraints, the details are given in [16].

3. Examples of sub-GLE analysis

As an example, we demonstrate the analysis of a sub-GLE event on 29 October 2015 (Fig.1). Details for the analysis of this event as well several other sub-GLEs is given elsewhere [16].

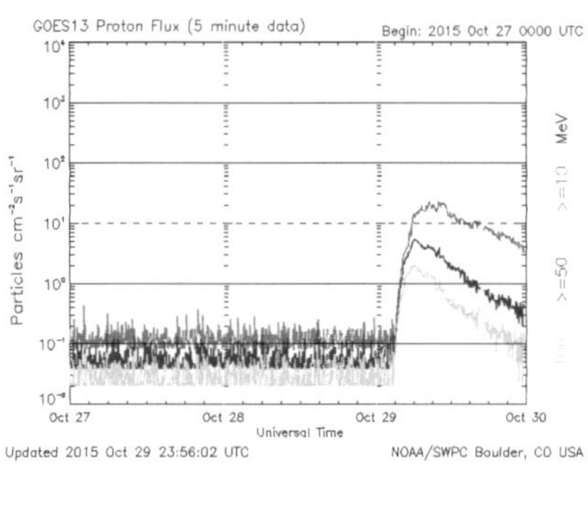

Figure 1. Profile of the time variation of GOES 13 proton flux during sub-GLE event on 29 October 2015.

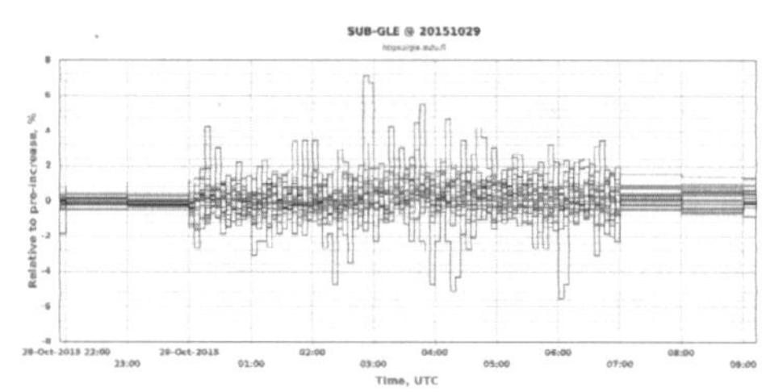

Figure 2. Profile of the time variation of NMs count rate during sub-GLE event on 29 October 2015
.

The derived spectra and PAD of the studied sub-GLEs are compared with those of a weak event – GLE 71 and with a moderately strong event – GLE 70, the results are presented in Fig.3a,b.

The derived rigidity spectra of sub-GLEs are softer than those of GLEs (Fig.3a). In addition, the proton flux at 1 GV during sub-GLE events is nearly an order lower compared to a weak event as GLE 71.

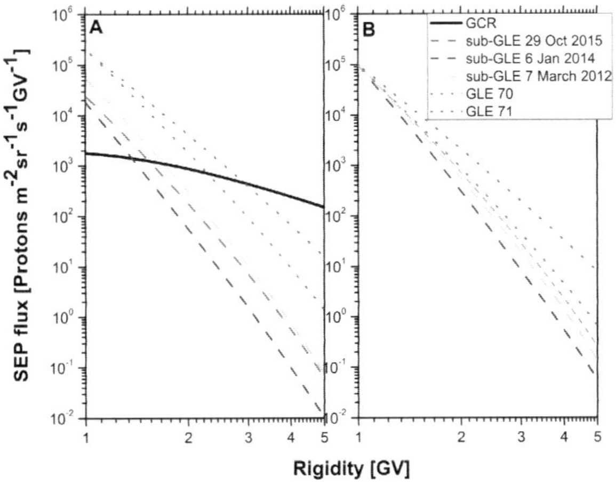

Figure 3. Rigidity spectra during several sub-GLE events compared with GLE 70 on 13 December 2006 and GLE 71 on 17 May 2012 as denoted in the legend. The black solid line in the left panel denotes the average GCR flux.

4. Computation of exposure during GLEs and sub-GLEs

The computation of the dose rate at flight altitudes due to CRs is rather challenging, because the large variability and extended energy range of the produced secondaries. During the last decade several models for assessment of the exposure due to CRs have been proposed e.g. [31, 32, 33]. Here, the exposure at flight altitudes is assessed using the derived spectral and angular parameters of SEPs and employing a recent numerical model based on pre-computed yield functions [34]. The effective dose rate at a given atmospheric depth h induced by a primary CR particle is computed by convolution of the yield function with a corresponding primary CR particle spectrum:

$$E(h) = \sum_i \int_{T_{cut}(P_c)}^{\infty} J_i(T) Y_i(T,h) dT \tag{5}$$

where $J_i(T)$ is the differential energy spectrum of the primary CR arriving at the top of the atmosphere for $i-$th component (proton or $\alpha-$particle) and Y_i is the effective dose yield function for this type of particles. The integration is over the kinetic energy T above $T_{cut}(P_c)$, which is defined by the local cut-off rigidity P_c for a nuclei of type i, $T_{cut,i} = \sqrt{\left(\frac{Z_i}{A_i}\right)^2 P_c^2 + E_0^2} - E_0$, where $E_0 = 0.938$ GeV/c^2 is the proton's rest mass.

Accordingly, the effective dose yield function Y_i is defined as:

$$Y_i(T,h) = \sum_j \int_{T^*} F_{i,j}(h,T,T^*,\theta,\varphi) C_j(T^*) dT^* \tag{6}$$

where $C_j(T^*)$ is the coefficient converting the fluence of secondary particles of type j (neutron, proton, γ, e^-, e^+, μ^-, μ^+, π^-, π^+) with energy T^* to the effective dose, $F_{i,j}(h,T,T^*,\theta,\varphi)$ is the fluence of secondary particles of type j, produced by a primary particle of type i (proton or $\alpha-$particle) with a given primary energy T arriving at the top of the atmosphere from zenith angle θ and azimuth angle φ.

According to our analysis the contribution of SEPs during sub-GLE event is comparable to the contribution of GCR, the details re given in [16]. In addition, we performed analysis of the GLE 72 on 10 September 2017, which appeared as a week event. Details about the GLE 72 analysis are given in [35] and these proceedings. The distribution of the effective dose rate as a function of the geographic coordinates at altitude of 35 kft due to high energy GLE and GCR particles during the peak phase of GLE 72 is given in Fig. 4.

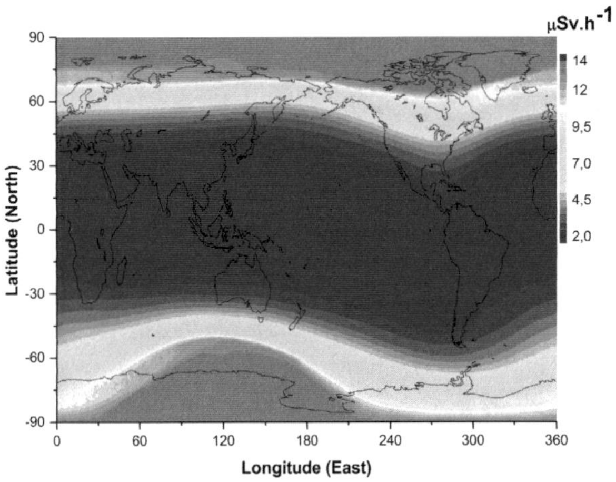

Figure 4. Distribution of the effective dose rate at altitude of 35 kft during GLE 72

5. Conclusions

In the work presented here we have studied several candidates for a new subclass of SEP events, sub-GLEs, which are of special interest for space weather applications. On the basis of data retrieved from different NMs and using a combination of a full simulation of the global NM network response and optimization procedure over a set of unknown parameters describing the SEP features, we have assessed the spectral and angular characteristics of sub-GLE particles. With the estimated spectral characteristics and using a previously developed model, we have assessed the effective dose rate in a polar and sub-polar region at commercial flight altitudes of 35 kft. During those computations a conservative scenario concerning the contribution of sub-GLE to the exposure was assumed [36]. It was shown that the contribution of sub-GLE particles to the exposure is at least comparable to the GCRs contribution. In addition, a similar analysis of GLEs was performed, their spectra derived and distribution of the exposure over the globe obtained. The derived spectra and assessed exposure during various stages of sub-GLEs and GLEs allow one to compute to quantify the radiation radiation environment in the troposphere and stratosphere [37, 38]. In this study, we demonstrated that the global NM network is a useful tool to estimate an important space weather effect, namely the exposure of aircrew due to CR of galactic and solar origin. The derived spectra and angular distributions will be integrated into the GLE database [39].

Acknowledgments

This work was supported by the Academy of Finland (project No. 272157, Center of Excellence ReSoLVE and project No. 267186). The operation of the DOMC/DOMB NM is supported by the French-Italian Concordia Station (IPEV program n903 and PNRA Project LTCPAA PNRA14 00091), projects CRIPA and CRIPA-X No. 304435 and Finnish Antarctic Research Program (FINNARP). We acknowledge NMDB and all of the colleagues and PIs from the neutron monitor stations who kindly provided the data used in this analysis, namely: AATB, APTY, ATHN, BKSN, BRBG, CALG, CAPS, DOMC, DRBS, FSMT, INVK, IRKT, JBGO, JUNG, KERG, KIEL, KGST, LMKS, MCMD, MGDN, MWSN, MXCO, MOSC, NAIN, NWRK, OULU, PWNK, PTFM, ROME, SOPO, TERA, THUL, TXBY.

References

[1] Tylka A and Dietrich W 2009 *Proc. of 31th ICRC Lodz, Poland, 7 -15 July 2009* 0273
[2] Klein K L and Dalla S 2017 *Space Science Reviews* **212** 1107–1136
[3] Pelliccioni M 2000 *Radiation Protection Dosimetry* **88** 279–297
[4] Mishev A 2014 *Advances in Space Research* **54** 528–535
[5] Reames D 2013 *Space Science Reviews* **175** 53–92
[6] Desai M and Giacalone J 2016 *Living Reviews in Solar Physics* **13** 3
[7] Aschwanden M 2012 *Space Science Reviews* **171** 3–21
[8] Vainio R, Desorgher L, Heynderickx D, Storini M, Flückiger E, Horne R, Kovaltsov G, Kudela K, Laurenza M, McKenna-Lawlor S, Rothkaehl H and Usoskin I 2009 *Space Science Reviews* **147** 187–231
[9] Hatton C 1971 *Progress in Elementary Particle and Cosmic-ray Physics X* (Amsterdam: North Holland Publishing Co.) chap 1
[10] Moraal H and McCracken K 2012 *Space Science Reviews* **171** 85–95
[11] Poluianov S, Usoskin I, Mishev A, Shea M and Smart D 2017 *Solar Physics* **292** 176
[12] Shea M and Smart D 1982 *Space Science Reviews* **32** 251–271
[13] Cramp J, Duldig M, Flückiger E, Humble J, Shea M and Smart D 1997 *Journal of Geophysical Research* **102** 24 237–24 248
[14] Mishev A, Kocharov L and Usoskin I 2014 *Journal of Geophysical Research* **119** 670–679
[15] Mishev A and Usoskin I 2016 *Solar Physics* **291** 1225–1239
[16] Mishev A, Poluianov S and Usoskin S 2017 *Journal of Space Weather and Space Climate* **7** A28
[17] Mishev A, Usoskin I and Kovaltsov G 2013 *Journal of Geophysical Research* **118** 2783–2788
[18] Gil A, Usoskin I, Kovaltsov G, Mishev A, Corti C and Bindi V 2015 *J. Geophys. Res.* **120** 7172–7178
[19] Mangeard P S, Ruffolo D, Sáiz A, Nuntiyakul W, Bieber J, Clem J, Evenson P, Pyle R, Duldig M and Humble J 2016 *Journal of Geophysical Research: Space Physics* **121** 11,620–11,636
[20] Levenberg K 1944 *Quarterly of Applied Mathematics* **2** 164–168
[21] Marquardt D 1963 *SIAM Journal on Applied Mathematics* **11** 431–441
[22] Tikhonov A, Goncharsky A, Stepanov V and Yagola A 1995 *Numerical Methods for Solving ill-Posed Problems* (Dordrecht: Kluwer Academic Publishers) ISBN 978-90-481-4583-6.
[23] Mavrodiev S, Mishev A and Stamenov J 2004 *Nuclear Instruments and Methods in Physics Research, Section A: Accelerators, Spectrometers, Detectors and Associated Equipment* **530** 359–366
[24] Mishev A, Mavrodiev S and Stamenov J 2005 *International Journal of Modern Physics A* **20** 7016–7019
[25] Desorgher L, Flückiger E, Gurtner M, Moser M and Bütikofer R 2005 *Internationl Journal of Modern Physics A* **20** 6802–6804
[26] Tsyganenko N 1989 *Planetary and Space Science* **37** 5–20
[27] Langel R 1987 *Geomagnetism* (London: J.A. Jacobs Academic Press) chap 1, pp 249–512
[28] Kudela K and Usoskin I 2004 *Czechoslovak Journal of Physics* **54** 239–254
[29] Mishev A and Usoskin I 2013 *Journal of Physics: Conference Series* **409** 012152
[30] Nevalainen J, Usoskin I and Mishev A 2013 *Advances in Space Research* **52** 22–29
[31] Sato T, Yasuda H, Niita K, Endo A and Sihver L 2008 *Radiation Research* **170** 244–259
[32] Mishev A, Adibpour F, Usoskin I and Felsberger E 2014 *Advances in Space Research* **55** 354–362
[33] Copeland K 2017 *Radiation Protection Dosimetry* **175** 419–431
[34] Mishev A and Usoskin I 2015 *Journal of Space Weather and Space Climate* **5** A10
[35] Mishev A, Usoskin I, Raukunen O, Paassilta M, Valtonen E, Kocharov L and Vainio R 2018 *Solar Physics* **in press** DOI:10.1007/s11207-018-1354-x
[36] Copeland K, Sauer H, Duke F and Friedberg W 2008 *Advances in Space Research* **42** 1008–1029
[37] Mishev A and Velinov P 2015 *Journal of Atmospheric and Solar-Terrestrial Physics* **129** 78–86
[38] Mishev A and Velinov P 2018 *Advances in Space Research* **61** 316–325
[39] Tuohino S, Ibragimov A, Usoskin I and Mishev A 2018 *Advances in Space Research* **62** 398–407

IOP Conf. Series: Journal of Physics: Conf. Series **1181** (2019) 012007 doi:10.1088/1742-6596/1181/1/012007

The rigidity spectrum of the long-term cosmic ray variations during solar activity cycles 19–24

V G Yanke, A V Belov, R T Gushchina, V N Zirakashvili

Pushkov Institute of Terrestrial Magnetism, Ionosphere and Radio Wave Propagation, RAS, IZMIRAN, Moscow, Russian, 108840.

yanke@izmiran.ru

Abstract. The monthly rigidity spectrum of cosmic ray (CR) variations in the 19-24 cycles of solar activity was obtained by the global survey method using the data of continuous ground and near-Earth monitoring of CRs, exempted from atmospheric and local effects.The changes of the spectrum, first obtained for such a long period, made it possible to reveal the features of large-scale effects in CR modulation, the presence of 22-year and 11-year CR variations in the spectrum, and confirm an abnormal spectrum change in the 70s. The paper assumes a rigidity spectrum of CRs, given in a three-parameter form. Analysis of the obtained long-term CR variations for particles with rigidity 10 GV shows that the amplitude of the 22-year wave in the CR intensity increases from cycle to cycle and reaches its maximum value at the minimum of 23/24th solar activity cycle. Softening of the spectrum at the cycle minima has been revealed for the negative polarity of the solar magnetic field (qA<0). The reasons for the abnormally high CR density at the minimum of the 24th cycle and the spectrum features in the 70s are discussed. The spectrum of long-term CR variations in the 19-24 solar activity cycles, determined from the experimental data, makes it possible to verify some conclusions of the theory of heliospheric CR modulation concerning the role of the magnetic drift of particles in cycles with the different polarity of the solar magnetic field. In particular, we propose the explanation for the observed R^{-2} spectrum of the variations in the minima of the negative solar activity cycles, related with the scattering of particles in the vicinity of the neutral current sheet.

1. Introduction

60 years of permanent observations of cosmic ray (CR) variations via the worldwide neutron monitor network permits to obtain the information about the influence of the long-term behavior of the solar activity on the CR intensity observed at the Earth and in the interplanetary medium. Different methods are used for the investigation of the cosmic ray modulation like an analysis of 11-year and other kinds of the solar cyclicity, empirical and theoretical modeling of CR modulation etc.

The long duration of the continuous registration of CR intensity by stable working detectors is the necessary and main condition for the investigation of rigidity spectra of long-term CR variations and for the basis of the modulation theory describing CR variations during the solar cycle.

The spectra of CR variations of the ground based and near Earth CR monitoring are generally found using the exempted from atmospheric and local effects CR variations obtained via the global survey method [1] and its modifications [2-5]

It is known that the rigidity dependence of long-term CR variations has a complex shape that is different from the power law one. In the present work we shall use a three-parametric form of the rigidity spectrum. Our task is to obtain the spectral parameters for every month for the long period of 6 solar cycles and to discuss the cyclic changes of the rigidity dependence of CR variations and its features at anomalous periods.

The rigidity spectrum of long-term CR variations in the 19-24 solar activity cycles, determined from the experimental data, makes it possible to verify some conclusions of the theory of heliospheric

CR modulation concerning the role of the magnetic drift of particles in cycles with the different polarity of the solar magnetic field [6-10]. In particular, we propose the explanation for the observed R^{-2} spectrum of the variations in the minima of the negative solar activity cycles, related with the scattering of particles in the vicinity of the neutral current sheet.

2. Data and method

The spectrum of long-term CR variations was determined for every month in years 1957-2016. For this purpose the data of worldwide neutron monitor network (41 neutron monitor (NM) stations), 3 stations of stratospheric observations [11](and the data of multi-directional meson telescope (Nagoya station, 17 arrival directions) [12] were used.

We determined the spectral characteristics of long-term CR variations based on 1) results of continuous CR observations in 1957-2016 years; 2) the global survey method; 3) three-parameter rigidity dependence of the spectrum $\delta N/N(R) = a/(b+R)^{\gamma}$; 4) the rigidity cut-off determined for every station in year 2016 [13].

The rigidity dependence is normalized to rigidity 10 GV that is close to the effective rigidity of neutron monitors. The model of variation for primary CRs is the following:

$$\delta = \Delta N / N = a_{10}(10+R)^{\gamma} / (b+R)^{\gamma} \quad (1)$$

We use the global survey method. The detectors mentioned above are considered as one unified multi-directional detector with high accuracy standard devices.

The spectral parameters a10, b and γ of long-term CR variations were determined for cycles 19-24 for rigidities R=1-25 GV. The calculation of the spectrum for every month during 60 years (1957-2016) was performed for the one basic period 01.2009-12.2009. This helps to avoid errors appearing from the different of base periods. An additional advantage is the high CR intensity in 2009 that is the highest during 60 years. This result in better accuracy of the parameters determined.

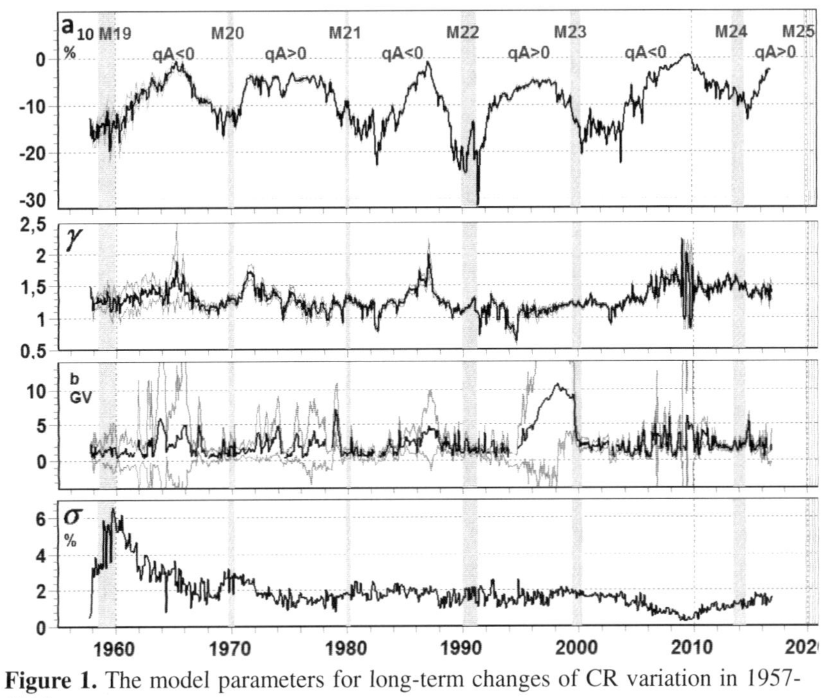

Figure 1. The model parameters for long-term changes of CR variation in 1957-2016:
the top panel - the amplitude a_{10} and its error (red),
the second panel - the spectral index γ and its error (red),
the third panel - the parameter b and its error (red),
the bottom panel - the standard deviation of the data and the model σ.

3. Parameters of rigidity spectrum of long-term CR variations

The parameters for the spectrum of galactic CR variations a10, γ, b in expression (1) and the standard statistical error σ are shown on figure 1.

The value of the parameter b was found between b=0 in 1994 and b=10.8 in 1998. This means that the spectrum of CR variations can be power law at some periods of time while the spectral shape is more complex in other periods of the solar activity. The results found for the parameter b are not robust because of the low accuracy. The good quality of the model applied is reflected by the small mean square deviation σ of experimental data from the model (see figure 1). Its value is below 2% after 1965 up to the present. The number of stations is about 40 to provide such a value of σ. The error is higher at the first years (1957-1965) of the worldwide network operation. The effective area of the ground based detectors was lower at the first years. The same is true for the statistical accuracy. However the high value of σ is probably not because of this. It is mainly explained by maximum time distance from the basic period, by influence of aperture drifts and large CR variations in 19-th cycle.

Significant deviations for the rigidity dependence of variations manifest themselves in the behavior of the parameter γ (see figure 1). The parameter γ is shown together with the mean statistical error. The value of the parameter γ was found in the interval 0.6(09.1994) – 2.2(02.2009). The statistical error for the parameter γ is small. Therefore we can make justified conclusions about γ behavior in the solar cycles with opposite directions of the polar magnetic field (qA>0 и qA<0 that is the standard notation used for the modulation of charged particles for negative and positive directions of the heliospheric magnetic field).

4. Cyclic changes of CR variation spectra in 1957-2016 calculation

Results of long continuous observations of CRs. solar and heliospheric magnetic fields were not available during early investigations of CR variations. Therefore it was difficult to reveal the long-term effects in CR modulation. The situation is different now. In particular it is possible to investigate how solar magnetic cycles influence on the rigidity dependence of the long-term CR variations.

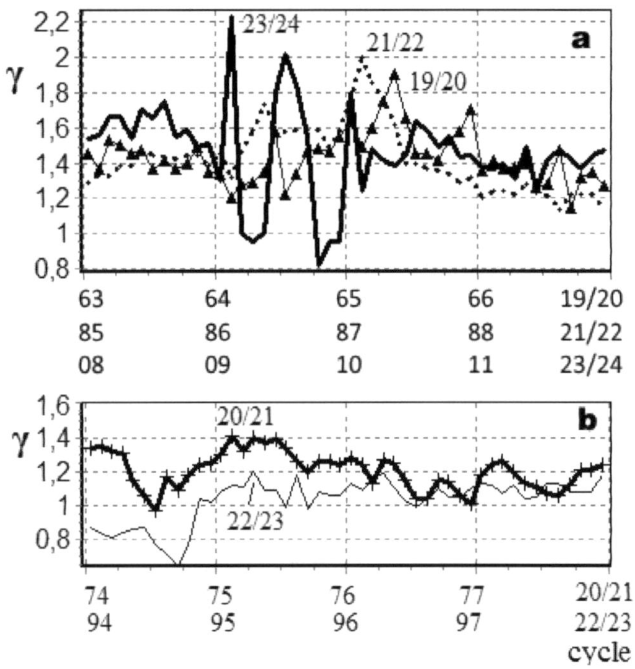

Figure 2.
a - Temporal changes γ during the solar minima 19/20 (1963-1966-triangles), 21/22 (1985-1988 -dots), 23/24 (2008-2011-thick curve);
b - 20/21 (1974-1977-crosses) and 22/23 (1994-1997-thin curve).

The dependence of the rigidity spectrum on the solar magnetic polarity and on the solar activity cycle was found in papers [14-15]. It was shown that the change of the magnetic polarity is accompanied by the change of the rigidity spectrum of long-term CR variations.

In this work we use the spectrum of CR variations obtained during the longer period of time. This allows to investigate the influence of solar magnetic cycles on CRs in more detail and also to consider the changes of the rigidity spectrum during anomalous periods like 70-s [4].

The periodicity found for the parameter γ for the used spectral model (1) in 6 cycles of solar activity (19-24) is related with the dependence of CR variations on the cycle phase and on the sign of the global solar magnetic field qA>0 and qA<0. It is shown in this work that there exist 22-year wave of CR intensity at the negative minima of the solar activity qA<0. The amplitude of this wave increases with time and is maximal in the last minimum of the solar activity between 23th and 24th cycle.

The periodicity found for the parameter γ for the used spectral model (1) in 6 cycles of solar activity (19-24) is related with the dependence of CR variations on the cycle phase and on the sign of the global solar magnetic field qA>0 and qA<0. It is shown in this work that there exist 22-year wave of CR intensity at the negative minima of the solar activity qA<0. The amplitude of this wave increases with time and is maximal in the last minimum of the solar activity between 23th and 24th cycle.

The change of the parameter γ points to the softening of the CR variation spectrum in three minima of 22-year cycles of the solar activity when the solar magnetic field was negative qA<0: 19/20 (γ≈1.9), 21/22 ((γ≈2.0) and 23/24 (γ≈2.2) (see figure 2a). This is not the case for the positive minima of cycles of the solar activity qA>0: 20/21 (γ≈1.4) и 22/23 (γ≈1.2) (see figure 2b).

We cannot obtain accurate values of γ in the minimum of 23/24 cycle because 2009 year is the basic period and CR variations are close to 0 in 2009. Nevertheless the parameter γ is high also before the CR maximum in 2009 and after 2009. At this time the spectrum of CR variations was soft with the parameter γ close to 1.5-1.7.

CR variations at high and low energies have a different behavior at this period. Low energy particles show a faster increase of intensity in comparison with high energy ones. We investigated the rigidity dependence in 24th cycle and compared it with CR intensity in minima of other cycles [16]. Time dependence of CR variations reflects the complex behavior of modulation for particles with different energies in 2009. In particular we found that low energy particles reach the maximum intensity faster contrary to the standard point of view. CRs of 10 GV rigidity reach the maximum two months later (09.2009) than 5 GV particles (07.2009).

The different behavior for the modulation of low energy and high energy particles during subsequent minima of solar activity was already mentioned in the paper [17]. It was explained by the difference of the drift motion of charged particles in the heliospheric magnetic field during the change of solar magnetic polarity. It seems that our results show a clear difference of the rigidity dependence of CR variations at opposite magnetic polarities. However this is not so clear at 70th years between 21 and 22 cycles. This special period is discussed below.

5. Anomalous behavior in 70th

It is clear from figure 2 that the spectral index of CR variations γ increases in 70th years and its behavior is strongly different in comparison with the next similar period of the magnetic polarity with qA>0 in 90th years. This is not the only one special feature of this period.

The influence of the magnetic drift on the CR intensity during 1970-1971 years was discussed in paper [18]. The observed peculiarity of the long-term CR modulation was explained by the same sign of the magnetic field at both solar poles. Such a magnetic configuration produced an additional penetration of galactic CR particles in near-equatorial regions at ±40° latitudes [18]. This extra penetration of particles resulted in the fast increase of CR intensity at the end of 1971.

The special behavior of CR variations was explained in (Belov et al., 1993) by consequences for observed CR variations of a so called “mini-cycle” in (1972-1974) [19 and references therein]. It was characterized by the rigidity dependence of CR modulation close to the one found in negative (qA<0) minima of the solar activity like in 1965 and 1987. This feature was mentioned in the paper [20] where it was related with the magnetic drift of particles in the framework of the heliospheric model of CR modulation in different solar cycles.

So the anomalous behavior of the index γ is related with the anomalous conditions in the heliosphere in 70th years. This does not contradict to our conclusion about 22-year periodicity for the spectrum of CR variations. The spectrum is softer for qA<0 than for qA>0. It is not excluded that this difference increases with time.

6. Drift effects and energy dependence of CR modulation

The spectrum of CR variations derived from the experimental data obtained during the long time period (19-24 cycles of solar activity) can be used to check some predictions of the theory of CR heliospheric modulation concerning the role of magnetic drifts in solar cycles with the opposite magnetic polarity.

In the case of the weak modulation the decrease of CR intensity in the heliosphere δN is determined by the energy losses ΔE [6]:

$$\delta N / N = (2+\gamma)\Delta E / E, \tag{2}$$

here γ is the slope of CR differential energetic spectrum. The energy losses in the idealized heliosphere with the flat neutral current sheet are given by the following expression [6]

$$\Delta E = |q\Phi| + p\int_r^R \frac{u}{\lambda} dr \tag{3}$$

Here u is the solar wind speed, λ, q and p are the free path length, the charge and the momentum of the particle respectively, Φ is some electric potential difference. During the period of positive magnetic polarity A>0 the galactic CR protons penetrate into heliosphere in polar regions, after that they drift in the nonuniform magnetic field from the poles to the equator and leave the heliosphere along the neutral current sheet. The magnetic drift from the pole to the equator is accompanied by energy losses corresponding to the potential difference Φ that is equal to the heliospheric potential Φ_0:

$$\Phi_0 = B_0 r_0^2 \Omega / c = 180MV \frac{u}{400km/c} \frac{B_e}{3\cdot 10^{-5} Gauss}, \tag{4}$$

that is determined by the solar magnetic field B_0, by the solar radius r_0 and by the angular rotation velocity of the Sun Ω. It can be also rewritten in terms of solar wind speed and azimuthal component of the interplanetary magnetic field B_e at the Earth orbit.

For negative magnetic polarity A<0 the protons enter the heliosphere along neutral current sheet, drift on latitude from the equator to poles and leave the heliosphere in polar regions. Therefore the potential difference Φ in expression (3) for the equator region is

$$\Phi = \Phi_0,\ qA>0;\ \Phi=0,\ qA<0 \tag{5}$$

This expression is written for protons. The dependence on the magnetic polarity is opposite for electrons.

For A<0 the proton modulation is determined by energy losses during the propagation in the equatorial region. They are described by the second term in Eq. (3). The dependence on the free path length λ does not mean that the particles reach the Earth along magnetic field lines. In the outer heliosphere the particles move across magnetic lines with the speed comparable with the speed of light moving along the neutral current sheet. Random scattering of particles destroy this motion along the current sheet and diminish the mean speed of particles in the radial direction.

The higher the scattering frequency the lower the mean speed and the longer time for the particle to reach the Earth. The same is true for the energy losses described by the second term in Eq. (3). For the opposite sign of magnetic polarity A>0 there are an additional energy loses related with the magnetic drift from the pole to equator. It is described by the first term in Eq. (3). This heliospherical potential about 180 MV corresponds to 8 percent of modulation for 10 GeV protons.

It is important that two terms in Eq. (3) have a different energy dependence. The particles with energies above several GeV are scattered by magnetic inhomogeneities with size that is smaller than the gyroradius of the particles. The theory predicts the quadratic dependence of the free parth length on momentum $\lambda \sim p^2$ [21] for this case.

The free path length have a weak energy dependence at lower energies.

Therefore during the minima of negative solar cycles when the tilt of the current sheet is small we expect to observe p^{-2} dependence for CR modulation and for long-term CR intensity variations of 10 GeV protons. In real situation the tilt makes the path of CR particles to the Earth longer and results in stronger modulation [21]. However Eq. (3) is not valid for large tilts. It is expected that in this case the mean speed is determined by the magnetic drifts which corresponds to p^{-1} dependence of modulation and CR variations.

During the minima of positive cycles A>0 the modulation is mainly determined by the first term in Eq. (3) and we expect to observe p^{-1} dependence of the modulation and CR variations. At this period the influence of the tilt on the modulation is minimal. During the maxima when the tilt is large the drift of particles in the vicinity of the neutral current sheet is directed to the outer heliosphere and can partially compensate fast inward drift motion of particles in polar regions. This will result in stronger modulation with the same p^{-1} energy dependence.

7. Conclusion

Rigidity spectrum of CR variations is obtained for every month during the last 60 years using the data of worldwide network detectors and applying the improved global survey method that provides the accuracy of the spectral parameters derived.

The shape of the rigidity spectrum of CR variations depends on the level of the solar activity and on the magnetic polarity. It is impossible to describe this shape with one set of parameters.

We observe the softening of long term CR variation spectrum in minima of the solar activity and change from R^{-1} to R^{-2} dependence after even cycles.

The spectrum of long-term CR variations in the 19-24 solar activity cycles, determined from the experimental data, makes it possible to verify some conclusions of the theory of heliospheric CR modulation concerning the role of the magnetic drift of particles in cycles with the different polarity of the solar magnetic field. In particular, we propose the explanation for the observed R^{-2} spectrum of the variations in the minima of the negative solar activity cycles, related with the scattering of particles in the vicinity of the neutral current sheet.

This work was partially supported by RFBR grant 17-02-00508, by the program of the Presidium of RAS №3, support the project UNU "Russian national network of ground stations of cosmic rays". We are grateful to all the staff of the World Network of cosmic ray stations http://cr0.izmiran.ru/ThankYou.

References

[1] Krymsky G F 1969 *Cosmic Rays Modulation in the Interplanetary Space* (Moscow: Nauka) P152
[2] Dorman L I 1975 *Variations of Galaxy Cosmic Rays* (Moscow: MGU) P214
[3] Dorman L I 1976 *Geomagnetism and Aeronomy* **16** 6 980
[4] Belov A V, Gushchina R T, Sirotina I V 1993 *Proc. 23rd ICRC (Calgary)* **3** 605
[5] Belov A V, Eroshenko E A, Yanke V G et al. 2018 *Solar Phys* **293** 4 23
[6] Kota J 1979 Proc. *16th ICRC (Kyoto)* **3** 13
[7] Jokipii J R, Thomas BT 1981 *Astrophys. J.* **243** 1115
[8] Kota J, Jokipii J R 1983 *Astrophys. J.* **265** 573 doi: 10.1086/160701
[9] Potgieter M S, Burger R A Ferreira S.E.S. 2001 *Space Science Reviews* **97** Issue 1–4 295
[10] Siluszyk M., Iskra K., Alania M V 2014 *Solar Phys* **289** 4297 doi:10.1007/s11207-014-0573-z.
[11] Stozhkov Yu I, Svirzhevsky N S, Bazilevskaya G A et al. 2007 *Cosmic ray fluxes in the maximum of absorption curve maximum in the atmosphere and on the atmospheric boundary FIAN N. 14* Moscow
[12] http://www.stelab.nagoya-u.ac.jp/ste-www1/div3/muon/muon3.html
[13] Gvozdevskii B B, Abunin A A, Kobelev P G et al. 2016 *Geomagnetism and Aeronomy* **56** 4 381
[14] Belov A V, Gushchina R T, and Sirotina I V 1995. *Proc. 24-th ICRC (Roma)* **4** 542
[15] Belov A V, Gushchina R T, Yanke V G 1998 *Geomagnetism and Aeronomy* **38** N4 131
[16] Gushchina R.T., Belov A.V., and Yanke V.G. 2013 *Bull. Russ. Acad. Sci.: Phys.* **77** N. 5 513
[17] Reinecke J. P. L., Potgieter M S 1994 *JGRA* **991** (A8) doi: 10.1029/94JA00792
[18] Alekseev V A, Ustinova G K 2013 *Doklady Akademii Nauk* **450** N 2 158 ISSN 0869-5652
[19].Alanko-Huotari K, Mursula K, Usoskin I G et al. 2006 Solar Phys. **238** 391

IOP Conf. Series: Journal of Physics: Conf. Series **1181** (2019) 012007 doi:10.1088/1742-6596/1181/1/012007

[20] Krymsky G F, Krivoshapkin P A, Mamrukova V P et al. 2007 J. Exp. Theor. Phys. **104** N 2 189
[21] Dolginov A Z 1966 JETP **51** 177

Planetary long term changes of the cosmic ray geomagnetic cut off rigidities

B B Gvozdevsky[1], A V Belov[2], R T Gushchina[2], E A Eroshenko[2], V G Yanke[2]

[1]Polar Geophysical Institute, RAS, PGI, Apatity, Russia, 184209
[2]Pushkov Institute of Terrestrial Magnetism, Ionosphere and Radio Wave Propagation, RAS, ИЗМИРАН, Moscow, Russian, 108840.

yanke@izmiran.ru

Abstract. The geomagnetic cutoff rigidities for vertical and inclined detectors of the World Network neutron monitors and muon telescopes were obtained with an annual resolution for the period 1950 2015 and for the forecast time up to 2050 by the method of trajectory calculations on the basis of the IGRF model. The results of the calculations indicate the manifestation of two World anomalies of the time dependence of the rigidity of geomagnetic cutoff and an irregular course for the north-western and south-eastern directions for detectors in the northern and southern hemispheres.

1. Introduction

The modern development of the experiment, both in the duration of observations, and in the accuracy of the experimental data obtained, requires more thorough studies of magnetospheric effects. This is due, for example, to the fact that during the sixty-year period of observations of cosmic rays the geomagnetic field has decreased by ~ 4%, and in different regions of the Earth the decrease is at a different rate. It was noted in [1] that while the dipole moment decreased significantly by 6.5% since 1900, the contribution of high field harmonics during the same period increased by ~ 30%. The center of the dipole has shifted by 200 km in the direction of the Pacific Ocean.

To assess the consequences of such a large adjustment of the magnetic field from the point of view of the magnetospheric effects in cosmic rays, it is necessary to obtain a planetary distribution of the rigidity of the geomagnetic cutoff for the entire observation period. In some regions the change in the main energy characteristic of incoming particles, the rigidity of geomagnetic cutoff, reaches ~ 2 GV, which is very important for the study of long-period cosmic ray variations.

An overview of the magnetospheric effects can be found in [2-7]. The most complete and systematic studies of the magnetospheric effects of cosmic rays, including their long-term changes, were carried out by M. Shea and D. Smart. Based on the geomagnetic field models for the respective epochs and trajectory calculations, global distributions of vertical geomagnetic cutoff rigidities were calculated in increments of 5°×15° in latitude and longitude for the epochs 1955, 1965, 1975, and 1980, [8-11] and 1990, 1995, 2000 [12-14]. Vertical rigidity of geomagnetic cutoff was obtained for all neutron monitors of the world network for the nine five-year epochs of 1955-1995 [15].

Shea M. and Smart D. [10], Storini M. [16], drew attention to the unevenness of changes in the planetary distribution of the geomagnetic cutoff rigidity in 20 years from 1955 to 1975, especially in the northern and southern waters of the Atlantic Ocean. In the southern part of the Atlantic Ocean, a decrease in rigidity was observed, while in the northern part a comparable increase was observed.

For the last four epochs, such calculations for the operating network of detectors have not been carried out. In addition, over the last 20 years, about a third of new neutron monitors have been included into operation and a network of multidirectional muon telescopes has been created for which the rigidities of the geomagnetic cutoff.

2. Method of calculation

It is known, there is no analytical solution of the equation of charge particle moving even in the field of a magnetic dipole (except of the trajectories in the equatorial plane).

The solution of the problem of the charged particle motion in the earth's magnetosphere was carried out by the method described in [17]. Integration is completed in three cases:

0 - the particle has moved beyond the magnetosphere - the trajectory is allowed;

1 - the particle penetrated to a depth less (R_{Earth} + 20) km (the particle returned to the atmosphere, the trajectory is forbidden): there are two options here - the rigidity of the particle is less than Stormer's rigidity or rigidity higher, but particle collided with the surface of the Earth);

2 - the integration is completed after a predetermined time (the particle is considered to be captured by the radiation belt)

As a result of the calculation, a discrete function of the penumbra g (R) is formed with the values 0, 1 or 2 for all rigidity values with the step 0.001 GV selected by us.

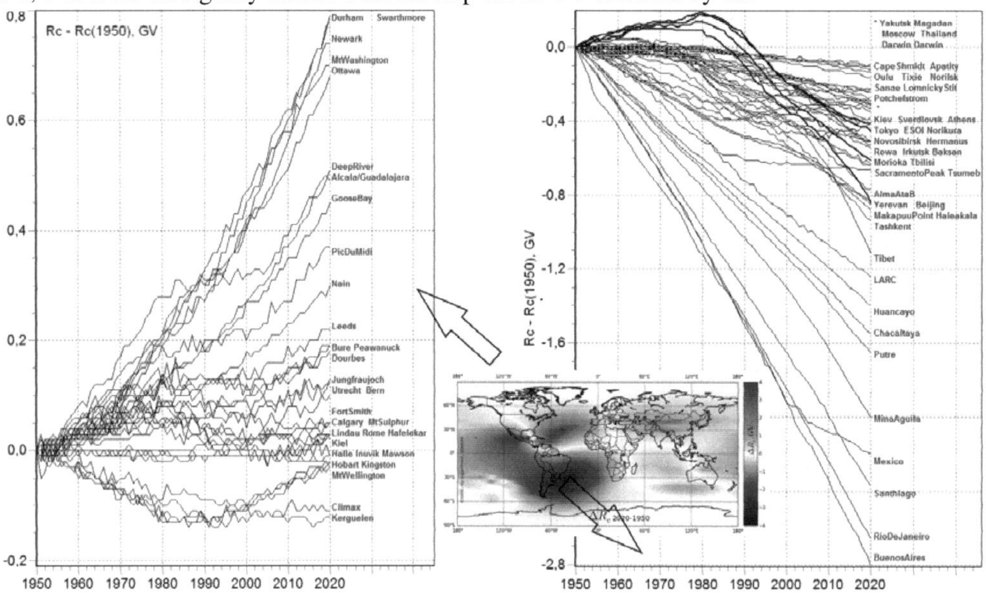

Figure 1. Changes with respect to the 1950 period of vertical rigidities of geomagnetic cutoff of the World Neutron Monitor Network. The insertion shows the changes in the planetary distribution of the vertical rigidities of the geomagnetic cutoff [20].

3. Discussion of the results. Neutron monitors

In work [18] for vertical directions of arrival of particles for every five epochs since 1900 by method of trajectory calculations the calculations of planetary distributions of geomagnetic cutoff rigidity (R_c) for grid 5° in latitude and 15° in longitude were performed. Geomagnetic cutoff rigidities were obtained using the International Geomagnetic Reference Field (IGRF) model [19]. Digital results for all epochs can be found on the server [20]. For the same period with an annual resolution, the geomagnetic cutoff rigidities were obtained for the World wide neutron monitor network (NMN). The results on the changes of the planetary distributions of the geomagnetic cutoff rigidity indicate the manifestation of two World anomalies (figure 1, insertion): in the zone of one, the rigidity decreases with time, while in the zone of the other, it increases with respect to the 1950 period, but globally rigidity of the geomagnetic cutoff decreases by approximately 3.4% for 50 years.

The time variations in the R_c for each station are determined by the influence of the anomaly in the zone of action of which the station is located. Figure 1 (left panel) shows the changes (increase) in the R_c of a group of stations that are in the zone of influence of the North Atlantic anomaly, and in figure 1 (right panel) the changes (decrease) of the R_c of a group of stations that are in the zone of influence South Atlantic anomaly. At peripheral stations, insignificant changes (~ 0.1 Gv) in the rigidity of the geomagnetic cutoff are observed. The accuracy of the obtained rigidity values is better than 0.05 GV.

4. Discussion of the results. Muon telescopes

To date, the world network of muon detectors has about 20 devices: super-telescopes, medium telescopes (~ 10m^2) and small telescopes (several m^2), as well as several hodoscopes. These are Nagoya, Hobart, Sao Martinho, hodoscope Kuwait, the short description of which can be found in [21], a giant hodoscope GRAPES-3 [22], hodoscope Moscow Nevod [23], spectrographYakutsk [24], Novosibirsk [25], Moscow-OPTO [26], Mustang [27-28], hodoscope Sierra Negra [29]. All detectors, with the exception of small, except vertical, detect particles from a large number of inclined directions. So, for the Nagoya telescope this is a vertical and 16 inclined directions.

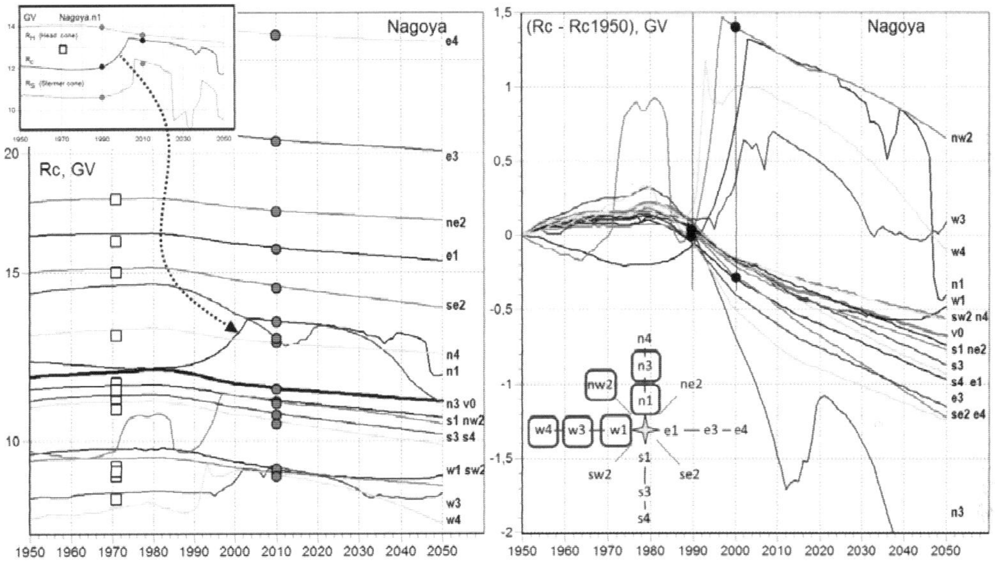

Figure 2. Absolute and relative (base 1950) temporary changes in the rigidity of the geomagnetic cutoff of the Nagoya telescope.

Calculations of temporal changes in the rigidity of geomagnetic cutoff were carried out for all muon detectors of the World Network. The result for the Nagoya telescope is shown in figure 2. For 1971 (square), the value of R_c obtained by the authors of a multichannel super-telescope Nagoya [30] using the model [31].

Attention is drawn to the irregular behavior of time dependences for some directions of arrival of particles, which is especially clearly seen when considering the relative changes in R_c (figure 2, right panel). First of all, in order to eliminate the code error, independent test calculations were performed [32], which are shown in figure 2 (circles). In addition, figure 2 (left panel, insert) shows the temporal stiffness for Nagoya.n1 for the lower R_S, upper R_H and effective R_{eff} of the geomagnetic cutoff stiffness. This detailed comparison of our and independent calculations shows that code errors can be eliminated.

Let us consider the behaviour of the rigidity of geomagnetic cutoff for the inclined directions of arrival of particles. Analyzing the time course of the cutoff rigidities for the inclined directions of the World Network muon detectors, two groups of detectors can be distinguished. Detectors of the first group, have an irregular course in time dependence only for inclined north-west directions for detectors of the northern hemisphere (in particular, Nagoya, in the sum of 17 detectors, figure 3, left). Detectors of the second group have an irregular course only for inclined directions south east (in particular, Hobart, in the sum of 6 detectors, figure 3, right) for detectors of the southern hemisphere.

The reasons for this behavior of the geomagnetic cutoff rigidity for inclined directions can be understood by considering the penumbra, its variations, and the particle trajectories themselves.

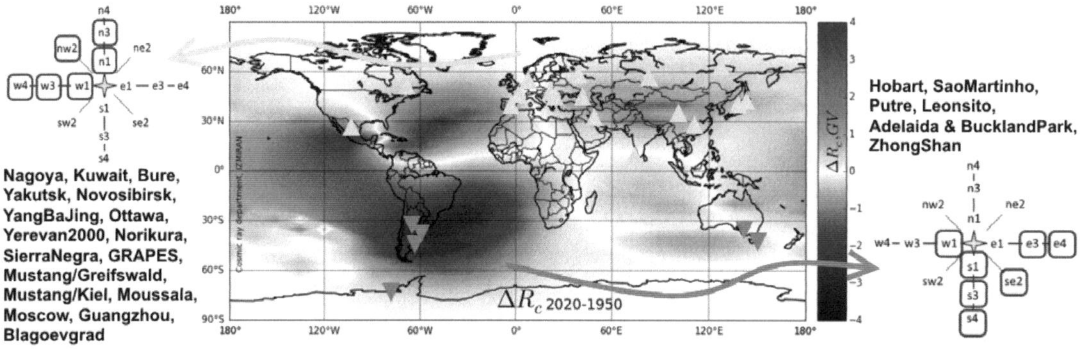

Figure 3. Relative changes in vertical rigidities of geomagnetic cutoff for 2 groups of cosmic ray stations relative to the 1950 period. The first group is the telescopes of the northern hemisphere (left side), the second is the telescopes of the southern hemisphere.(right side).

The region of R_c, according to the modern Stormer theory, taking into account the results of Lemaitre and Wallart [33], can be represented as follows:
Störmer region of rigidities, within which all trajectories are forbidden;
The main area of rigidity, within which all trajectories are allowed;
The area of semi-shadow "penumbra" between the Stormer and the main areas, but within this zone there are intermittent permitted and forbidden areas; it is important for us that the penumbra zone is formed by two regions or two types of trajectories:
The shadow rigidity region immediately following the Störmer boundary rigidity and actually adjacent to it is filled with particles crossing the Earth's surface; the latter region is characteristic of obliquely incident particles facing the north or south pole, and is of greatest importance at large zenith angles;
The region of rigidities of trapped particles to the left of and to the right of the Störmer boundary rigidity is the region formed by particles of small rigidities captured by radiation belts; The latter area is typical only for medium latitudes and in the case of low latitudes is not visible;

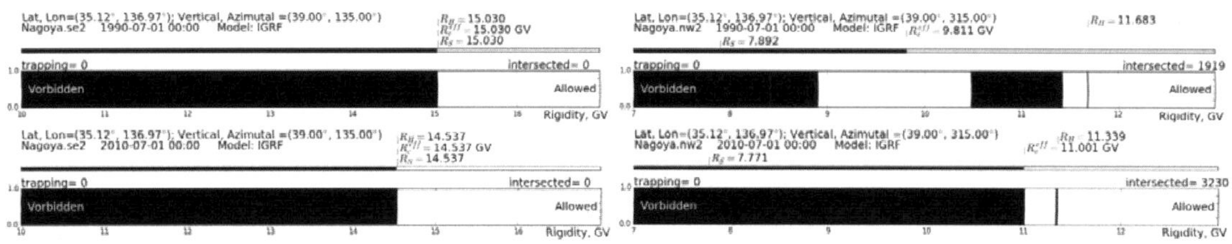

Figure 4. Penumbra for two types of trajectories (Nagoya.se2- left panel and Nagoya.nw2-right panel) and two epochs – 1990 и 2010.

Let us consider for 1990 and 2010 years the structure of the penumbra for the directions ***Nagoya.se2*** (figure 4, the left panel), the time course of the rigidity of the geomagnetic cutoff which behaves monotonically. For these directions there are neither captured (*trapping=0*) nor intersecting the Earth's surface (*intersected=0*) trajectories. Consequently, the region of the penumbra is absent.

A completely different structure has a penumbra for the directions ***Nagoya.nw2*** (figure 4, right panel), for which the time course of the rigidity of the geomagnetic cutoff behaves "unexpectedly strange". For these directions there are also no trapped (*trapping=0*) trajectories, but a large number of trajectories crossing the Earth's surface are observed, i.e. penumbra is formed only by trajectories that cross the Earth's surface. The width of the penumbra region for the 1990 period is 1.990 GV, and for the 2010 epoch 3.320 GV, i.e. for 20 years from 1990 to 2010, the rigidity of the geomagnetic cutoff for the direction ***Nagoya.nw2*** increases by 1.3 GV, which follows and from figure 2 (right panel).

Consequently, for inclined directions the penumbra region is very wide, and its strong variability with sufficiently small changes in the geomagnetic field leads to changes in the rigidity of the geomagnetic cutoff, sometimes even sharp, for inclined incident particles. Lemaitre and Vallarta [33] concluded that the shadow region contains particles with trajectories near the horizon from the north or south directions and have the largest values at large zenith angles. But for real detectors located not only on the equator, the increased shadow region of rigidities is characteristic for particles coming from north-west or south-east directions.

5. Results of digital calculations

For the practical use of the results of numerical calculations, an archive of geomagnetic cutoff rigidities data was created for all detectors of the World Network of neutron monitors and muon telescopes, which can be found on the Internet resource [20]. The archive also contains some graphical results and animations:

1) Digital and graphical results for all neutron monitors of the World network during the period 1950 - 2015 and prognosis to 2050 in the directory "R_c for all neutron monitors (Table and Graph)".
2) Digital and graphical results for all telescopes of the World network during the period 1950 - 2020 and prognosis to 2015 in the directory "R_c for all Telescopes (Table and Graph)" (see **guide.pdf**).
3) Animations of the trajectories, in particular, for the directions Nagoya.se2 and Nagoya.nw2 for 2010 are found in the section "Traces for some Trajectories (Graph)".
4) Penumbra and animations, in particular, for the directions Nagoya.se2 and Nagoya.nw2 for 1990 and 2010 – in the section "Penumbra for some Example (Graph)".
5) Tables of the planetary distribution of vertical R_c with a resolution of 5°×15° in latitude and longitude for the epochs from 1900 to 2050 with a step of 5 years in the section "Tables of planetary distributions R_c".
6) Calculator of the geomagnetic cutoff rigidity http://crsv.izmiran.ru/cutoff. It calculates the rigidity of the geomagnetic cutoff for a given date and a given geographic point. The program can calculate trajectories for one of the given model of the magnetosphere: dipole, IGRF, Tsyganenko IGRF+T89, IGRF+T96, IGRF+T02. The result of the program is the lower R_S, upper R_H and the effective Reff values of the geomagnetic cutoff rigidity and the penumbra.

6. Conclusions

1) The reason for the significant change in the rigidity of geomagnetic cutoff is the general decrease in the Earth's magnetic field, against which background its peculiar "contrast" is growing with the appearance of two anomalous zones: the North and South Atlantic anomaly and its plume.
2) The temporal changes in the R_c for vertically arriving particles at stations of the World Network of neutron monitors are completely consistent with the behavior dictated by these two anomalous zones.
3) Over the observational period of cosmic rays, the mean planetary R_c for the vertical arrival of particles decreased by 0.2 GV or 3.4%, which leads to an increase in the measured particle flux by ~1% over 50 years. In the epicenter of anomalies, during the same period, the flux changes to 12%.
4) The temporal changes in the R_c for oblique incoming particles for the muon telescopes of the World Network are also completely consistent with the behavior dictated by the two anomalous zones. But the second factor also acts, since the change in the magnetic field leads to a significant variability of the penumbra and, ultimately, to the instability trajectories of the particles from inclined directions.
5) The instability of particle trajectories from inclined directions leads to an irregular course of the time dependence of the R_c for certain directions, which are determined by the location of the detector relative to the geomagnetic equator. Thus, for detectors of northern hemisphere (**Nagoya, Kuwait,**

Yakutsk, GRAPES, YangBaJing, Yerevan2000, Norikura, SierraNegra, Mustang/Greifswald, Mustang/Kiel, Moscow, Moussala, Blagoevgrad, Guangzhou, Novosibirsk, Ottawa, Bure) irregularity of the temporal course of the rigidity of geomagnetic cutoff is observed for oblique north-west directions. For detectors of the southern hemisphere (**Hobart, SaoMartinho, Putre, Leonsito, Adelaida & BucklandPark, ZhongShan**) irregularity of the temporal course of the rigidity of geomagnetic cutoff is observed for oblique south-east directions.

This work was partially supported by the program of the Presidium of RAS №3, RFBR grant № 17-02-00508a, support the project USU "Russian national network of ground stations of cosmic rays". We are grateful to all the staff of the World Network of cosmic ray stations. http://cr0.izmiran.ru/ThankYou.

References

[1] Xu Wen-Yao, Wei Zigang, Ma Shizhuang 2000 *Chin. Sci. Bull* **45** 21 2013
[2] Hofmann D J, Sauer H H 1968 *Space Science Reviews* **8** 750
[3] Dorman L I, Smirnov V S, Tyasto M I 1971 *Cosmic rays in the Earth's magnetic field* (Moscow: Nauka) P400
[4] Smart D F, Shea M A, Flückiger E O 2000 *Space Science Reviews* **93** 1 305
[5] Kudela K, Bobik P 2004 *Solar Physics* **224** № 1-2 423
[6] Bobik P, Boella G, Boschini M J et al. 2006 *JGR* **111** № A5 CiteID A05205
[7] Dorman L 2009 *Cosmic Rays in Magnetospheres of the Earth* **358** (Springer) 770
[8] Shea M A, Smart D F 1967 *JGR* V **72** № 7 P 2021-2028
[9] Shea M A, Smart D F 1975 *AFCRL-TR-75-0381* № 524
[10] Shea M A, D F Smart 1975 *Proc. 14th ICRC* (Munich) **4** 1298
[11] Shea M A. Smart D F. 1983 *Proc. 18th ICRC* (Bangalore) **3** 514
[12] Smart D F, Shea M A 1997 *Proc. 25th ICRC* (Durban) **2** 401
[13] Smart D F, Shea M A 2007 *Proc. 30th ICRC* (Mexico) **1** 733
[14] Smart D F, Shea M A 2007 *Proc. 30th ICRC* (Mexico) **1** 737
[15] Shea M A, Smart D F 2001 *Proc. 27th ICRC* (Hamburg) 4063
[16] Storini M, Shea M A, Smart D F, Cordaro E G 1999 *Proc. 26th ICRC* (Salk Lake City) **7** 402
[17] Shea M A, Smart D F, McCracken K G 1965 *J. Geophys. Res.* **70** 4117
[18] Gvozdevskii B B, Abunin A A, Kobelev P G et al. 2016 *Geomagnetism and Aeronomy* **56** 4 381
[19] Thébault E., Finlay C.C., Toh H., Special issue "International Geomagnetic Reference Field—the twelfth generation" Earth, Planets and Space 2015 **67**:158/ doi 10.1186/s40623-015-0313-0; Model IGRF ePub 2015 http://www.ngdc.noaa.gov/IAGA/vmod/igrf.html
[20] Mag_Effect ePub 2016 ftp://crsb.izmiran.ru/MagEffect See *guide.pdf*
[21] Munakata K, Kozai M, Evenson P et al. 2018 *ApJ* **862** 170 9
[22] Gupta S K, Antia H M, Dugad S R 2009 *Nuclear Physics B* **196** 153
[23] Ampilogov N V, Astapov I I, Barbashina N S et al. 2016 Journal of Physics **675** 3 id 032042
[24] Starodubtsev S A, Grigoriev V G, Gololobov P Yu 2017 *Izvestiya RAS ser. fiz.* **81** 4 577
[25] Yanchukovsky V L, Grigoriev V G, Krymsky G F et al. 2016 *Solnechno-Zemnaya Fizika* **2** 1 76
[26] Kartyshov V G, Abunin A A, Klepach E G et al. 2018 *Proc. 26th ECRS* (Barnaul, Russia) GEO36
[27] Hippler R, Mengel A, Jansen F et al. 2008 Proc. 30th ICRC (Mexico) **1** (idSH980) 347
[28] Banjac S, Galsdorf D, Heber B et al. 2015 *Proc. 34th ICRC* (Hague) id1026
[29] Ortiz E, Valdes-Galicia J F, Matsubara Y et al. 2015 *Revista Mexicana de Fisica* **61** 6 466 doi:
[30] Nagoya Telescope 1970 http://www.stelab.nagoya-u.ac.jp/ste-www1/div3/muon 464
[31] Finch H P, Leaton B R 1957 O. Monthly Roy. Astron. Soc. Geophys. Suppl. No 7 314
[32] Gvozdevsky B B, Belov A V, Gushchina R T et al. 2017 *Nuclear Physics and Engineering* **8** 4 19
[33] Lemaitre G E, Vallarta M S 1936 *Phys. Rev.* **50** 493

IOP Conf. Series: Journal of Physics: Conf. Series **1181** (2019) 012009 doi:10.1088/1742-6596/1181/1/012009

On recurrent Forbush Decreases

A Melkumyan[1], A Belov[2], M Abunina[2], A Abunin[2], E Eroshenko[2], V Oleneva[2] and V Yanke[2]

[1] Gubkin Russian State University of Oil and Gas (National Research University), Moscow, Russia

[2] Pushkov Institute of Terrestrial Magnetism, Ionosphere and Radio Wave Propagation RAS (IZMIRAN), Troitsk, Moscow, Russia

E-mail: amelkumyan6@gmail.com

Abstract. The methods of nonparametric and multivariate statistics were used to study features of recurrent Forbush decreases (FDs) and to compare them with sporadic ones. The FEID database created and maintained in IZMIRAN allowed us to distinguish isolated FDs caused by high-speed streams from coronal holes (350 events) and by interplanetary disturbances from coronal mass ejections (207 events). Distributions of all parameters turned out to be more compact and to have smaller medians for recurrent events than for sporadic ones. Comparison of distributions of FD magnitude for different groups of events suggested that most FDs were related to coronal holes during the minimum between solar cycles 23 and 24. Multiple correlation coefficients between FD parameters and a set of solar wind parameters revealed that recurrent FDs depends less strongly on the characteristics of interplanetary disturbances than sporadic ones.

1. Introduction

Forbush decreases (FDs) are the results of influence of solar wind (SW) disturbances on background cosmic rays (CR) [1, 2, 3]. A relatively fast CR density decrease (with the biggest and extremely variable CR anisotropy) followed by a slower recovery on a time scale of several days are observed during FD. There are two different kinds of FDs associated with two types of solar sources: recurrent FDs caused by high-speed streams (HSS) from coronal holes (CHs) and sporadic FDs caused by interplanetary disturbances (ICMEs) from coronal mass ejections (CMEs) [4].

In the case of CHs, the region of interaction between HSS and ambient SW is characterized by greater density and stronger and more variable interplanetary magnetic field (IMF). The distinctive features of the main part of HSS are its higher speed and temperature, low density and weaker IMF [5, 6]. The influence of HSS on CR modulation has been studied in many works [e.g. 7, 8]. Some studies compare different types of FDs using statistical methods based on large number of events [9, 10, 11]. However, one can note some shortcomings of this area of research such as selecting groups of events by indirect features and/or using data observed at small number of stations and/or describing FD only by one parameter (usually, variations of CR density). The objective of our work is to study properties of recurrent FDs and to compare them with sporadic FDs using methods of nonparametric and multivariate statistics provided by large amount of data. We used FD characteristics (variations of CR density and CR anisotropy) obtained by global survey method (GSM) on the data of about 40 neutron monitors [12]. Recurrent and sporadic FDs were carefully selected with considering of various solar activity and interplanetary medium parameters [13].

2. Data and Methods

In this study we use the Forbush Effects and Interplanetary Disturbances (FEID) database {HYPERLINK "http://spaceweather.izmiran.ru/eng/dbs.html"} created and maintained in IZMIRAN. The database contains FDs parameters obtained from the data of worldwide neutron monitor network using GSM for particles with rigidity 10 GV. The FEID database also includes geomagnetic indices, SW parameters and solar geophysical data. The database contains more than 7000 FDs since 1957 (SW data since 1964). Our study of FDs associated with HSS from CHs assumes that each event is determined by a single solar source. We distinguished FDs created by only one source selecting the events with an onset no less 36 h prior to the next event and at least 60 h after the previous event (if the previous FD was greater than 1.4%).The next step – identification of recurrent and sporadic (for comparative analysis) FDs – required a joint analysis of interplanetary and solar data [13]. Analysis of SW parameters – velocity, density, temperature, IMF intensity – made it possible to refer FD to one of the two groups preliminary. Then the analysis of solar data was carried out which, as a rule but not always, confirmed the preliminary conclusions. At the last step, we compared the results to catalogs of CHs [14] and CMEs [15] and our conclusions mostly were confirmed. Since a confident identification of the event with a solar source requires complete and reliable data, then all selected recurrent FDs (CH group, 350 events) and sporadic FDs (CME group, 207 events) refer to the recent period (since 1997). We applied statistical analysis to study CR variations, geomagnetic activity and interplanetary disturbances in the CH group and to compare them with the CME group. To characterize each event, we used maximum (during the event) values: CR density variation (FD magnitude – A_F); absolute value of CR density hourly decrease (FD decrease rate – D_{max}); equatorial component of CR anisotropy (Axy_{max}); range of north–south component of CR anisotropy (Az_{range}); geomagnetic Ap index (Ap_{max}); absolute value of geomagnetic Dst index ($|Dst_{min}|$); SW velocity (V_{max}) and IMF intensity (B_{max}). Statistical analysis of these parameters included: analysis of distributions; calculation of descriptive statistics and Pearson correlation coefficients; multiple linear regression analysis [16, 17].

3. Results and Discussion

Table 1 presents the results of calculation of medians, mean values and interquartile ranges (a distance between 25% and 75% quartiles of the distribution) for eight parameters in the CH and CME groups. For all parameters (except SW velocity) medians and interquartile ranges are significantly less for recurrent FDs than for sporadic ones. Despite the great differences in HSS from CHs, the CH group turned out to be more compact and uniform as to events associated with CMEs. Median of SW velocity is a little more in the CH group although maximum velocity has been observed in the CME group; the interquartile ranges are approximately equal in both groups of events.

Figure 1 presents histograms of SW velocity in the CH and CME groups. Curve lines represent lognormal (left) and normal (right) distributions. Calculation of Kolmogorov-Smirnov statistics [16] reveals that the distribution of SW velocity does not fit to any of these laws in the CH group and corresponds to lognormal law in the CME group. Calculated value of skewness is statistically insignificant for recurrent FDs and significantly positive for sporadic ones. The difference in the distributions of SW velocity for recurrent and sporadic FDs could be explained by the fact that many FDs are caused by ICMEs with low speeds, but the FDs from low speed "high speed streams" are observed rarely (because they need even lower speed of the ambient SW). Figure 2 presents histograms of FD magnitude in the CH and CME groups. Calculation of Kolmogorov-Smirnov statistics and skewness reveals that both distributions do not fit to normal/lognormal laws and are right-skewed. The distribution is peaked, skewed, compact for recurrent FDs and highly skewed with a long "tail" in the region of large values for sporadic FDs. Figure 3 presents complementary cumulative distribution functions (*CCDF*) of FD magnitude for the CH and CME groups and for the events observed in the minimum between solar cycles 23 and 24. We can see that the distribution of FD magnitude in the minimum is very close to the distribution of recurrent FDs. This can be explained by the predominance of FDs associated with CHs in the period of this minimum.

Table 1. Descriptive statistics for FD and SW parameters and geomagnetic indices (Med – median, IQR – interquartile range)

	CH					CME				
	Med	Mean	IQR	Min	Max	Med	Mean	IQR	Min	Max
A_F, %	1.1	1.22 ± 0.03	0.7	0.4	3.6	2.1	3.03 ± 0.18	2.9	0.3	13.8
D_{max}, %/h	0.25	0.27 ± 0.01	0.09	0.11	1.02	0.41	0.58 ± 0.04	0.36	0.12	4.06
A_{xymax}, %	0.99	1.04 ± 0.02	0.35	0.47	2.43	1.61	1.79 ± 0.06	0.88	0.56	5.26
A_{zrange}, %	0.97	1.01 ± 0.02	0.35	0.53	2.52	1.76	2.04 ± 0.08	1.27	0.53	5.95
A_{pmax}	27	33.2 ± 1.2	31	1	154	48	64.7 ± 4.43	72	4	300
$\|D_{stmin}\|$	29	32.7 ± 1.04	21	1	119	47	64.4 ± 4.21	61	1	422
B_{max}, nT	11.2	11.7 ± 0.2	4.8	4.3	24.9	13.8	16.3 ± 0.65	10.5	4.7	65.6
V_{max}, km/s	564	567 ±6	162	331	907	492	522 ± 9	154	327	959

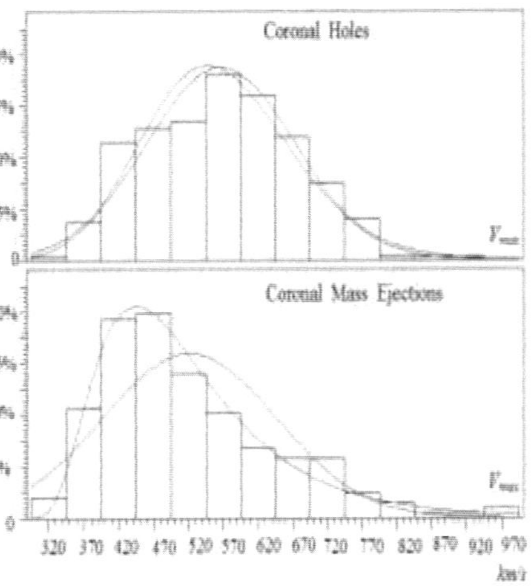

Figure 1. Histogram of SW velocity.

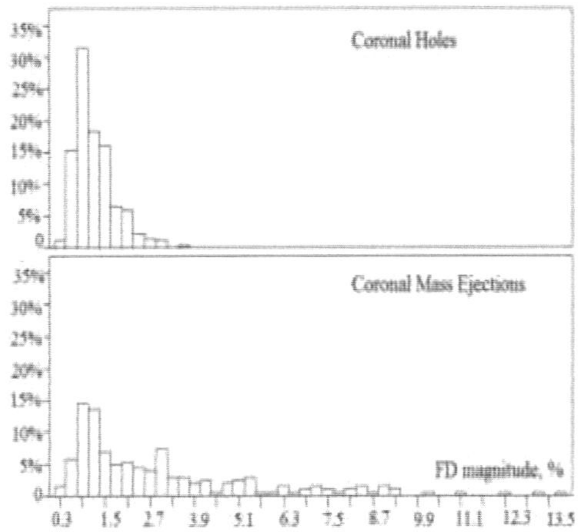

Figure 2. Histogram of FD magnitude.

Pearson paired correlation coefficients between the parameters were calculated for the CH and CME groups; z–statistic was applied to estimate the significance of the difference between correlation coefficients. The correlation coefficients between SW velocity and IMF intensity in the CH group (r = 0.52) and in the CME group (r = 0.55) differ statistically insignificantly. For both groups of events, SW disturbances with larger velocity result stronger compression of interplanetary matter and IMF; for large ICMEs, the correlation between SW velocity and IMF intensity can exist already in the solar sources [4]. In the CH group, the correlations between FD magnitude and SW parameters are moderate ($0.3 \leq |r| < 0.5$) while the correlations between three other FD parameters and SW parameters – weak ($|r| < 0.3$) or statistically insignificant. In the CME group, the correlations between FD magnitude and SW parameters are high ($0.7 \leq |r| < 0.9$) while the correlations between the rest of FD parameters and SW parameters are noticeable ($0.5 \leq |r| < 0.7$). Thereby, the correlations between FD parameters and SW parameters are significantly weaker in the CH group than in the CME group. Significant positive

IOP Conf. Series: Journal of Physics: Conf. Series **1181** (2019) 012009 doi:10.1088/1742-6596/1181/1/012009

correlation between FD parameters and SW parameters is consistent with the findings of e.g., [4] that when SW disturbance moves with faster speed and the magnetic field is stronger in it, the greater decrease of CR density occurs during the main phase of FD. The results of the multiple regression analysis (presented below) show that the relationship between FD parameters and geomagnetic indices is completely determined by their common dependence on SW parameters (partial correlation coefficients between FD parameters and geomagnetic indices are statistically insignificant).

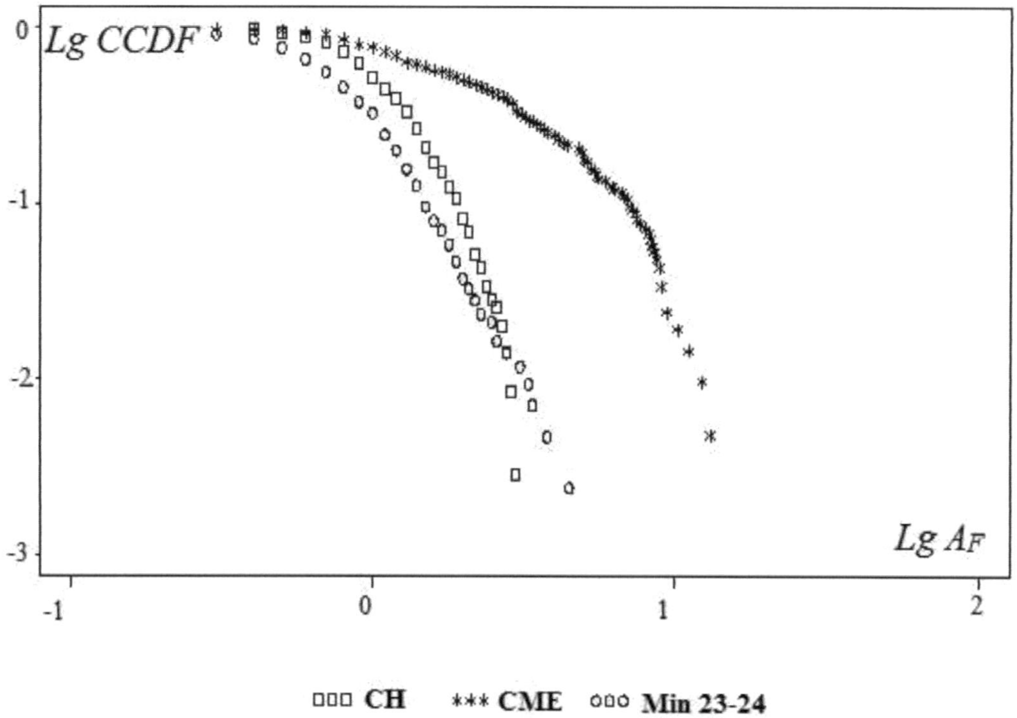

Figure 3. Complementary cumulative distribution function of FD magnitude on a log–log scale.

We estimated the dependence of FD parameters (predicted or dependent values) on the geomagnetic indices and SW parameters (a set of predictors) using multiple linear regression model [17]. The results are presented in table 2 for the dependent variables with confirmed (by F–test) overall statistical significance of linear regression model. In the CH group, the results are statistically significant only for FD magnitude and the multiple correlation coefficient indicates moderate correlation (R=0.48). We can see that the linear relationship of FD parameters on the predictors set much better corresponds to the observed data in the CME group: the multiple correlations are high for FD magnitude (R=0.81) and FD decrease rate (R=0.78) and noticeable for equatorial CR anisotropy (R=0.61). Standardized regression coefficients (or partial correlation coefficients) estimate a relative contribution of each predictor to the regression; only statistically significant coefficients are presented in table 2. The values of standardized regression coefficients in table 2 reveal that IMF intensity gives a larger part of contribution to the prediction of FD magnitude in the CH group. In the CME group, FD magnitude and FD decrease rate are predicted roughly equally by both SW parameters but equatorial CR anisotropy is predicted mostly by SW velocity. The results in table 2 also reveal that the multiple linear regression model gives statistically insignificant standardized regression coefficients for geomagnetic indices in both groups of events. We can conclude that both geomagnetic indices are related with CR density variations only through their common dependence on SW parameters.

Furthermore, geomagnetic activity is determined by the solar wind flowing around the Earth's magnetosphere, while GCR modulation is a result of the influence of the whole large–scale interplanetary disturbance [4].

Table 2. Standardized regression coefficients and multiple correlation coefficients for multiple linear regression model: dependent variables – A_F, Δ_{max}, Axy_{max}, Az_{range}; predictors – B_{max}, V_{max}, Ap_{max}, $|Dst_{min}|$. Only statistically significant regression coefficients are presented in the table.

Group of events		CH	CME		
Dependent variable		A_F	A_F	D_{max}	Axy_{max}
Standardized regression	B_{max}	0.38	0.46	0.38	–
coefficients	V_{max}	–	0.37	0.41	0.38
Multiple correlation coefficient		0.48	0.81	0.78	0.64

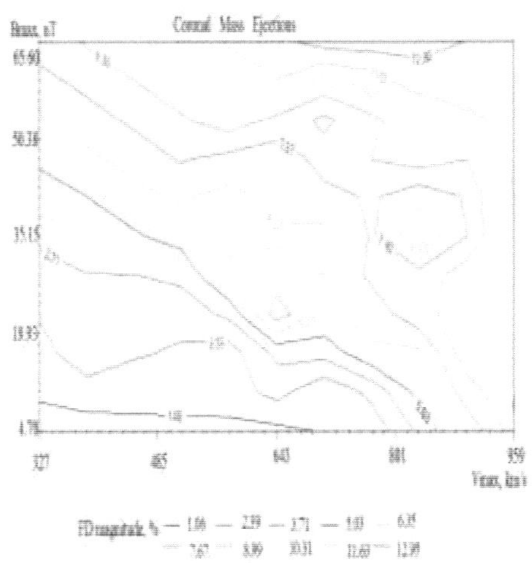

Figure 4. Dependence of FD magnitude (%) on SW parameters in the CME group.

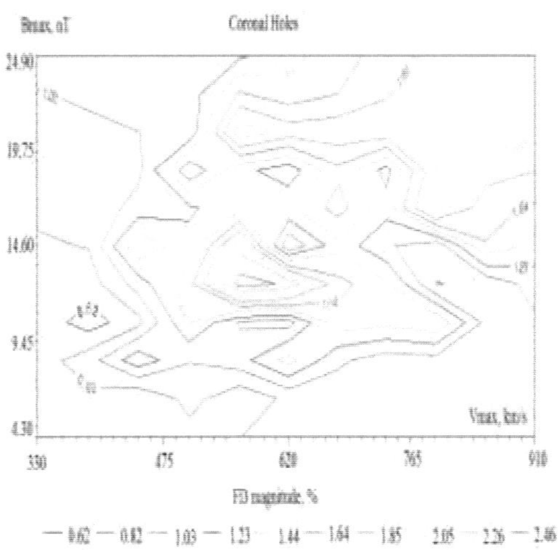

Figure 5. Dependence of FD magnitude (%) on SW parameters in the CH group.

It is clearly visible at the contour plots of A_F on the (V_{max}, B_{max}) plane (figure 4, 5) that the dependence of FD magnitude on SW parameters is much closer to the linear model in the CME group than in the CH group. On the contour plot for the CME group (figure 4), FD magnitude has minimum values for small V_{max} and B_{max} (the bottom left quadrant) and maximum values for the largest V_{max} and B_{max} (the upper right quadrant); the direction of FD magnitude gradient is almost unchangeable. The dependence of FD magnitude on SW parameters in the CH group is more complicated (figure 5). In this group, FD magnitude also has low values for small V_{max} and B_{max} and higher values for the largest V_{max} and B_{max}. However, a number of local maxima are singled out, indicating the existence of anomalously large Forbush decreases related with not strongly perturbed SW. For example, the anomalously large for the CH group FD magnitudes (up to 3.6%) is associated with values of SW parameters in the mean–third of the ranges (500 km/s < V_{max} < 650 km/s, 11.5 nT < B_{max} <16.5 nT).

4. Conclusions.

The FEID database allowed to select isolated (non–overlapping) FDs and to associate part of them (since 1997) to SW disturbances from CHs (350 recurrent FDs) and CMEs (207 sporadic FDs). Large amount of data enabled to use statistical methods (distribution analysis, correlation analysis, multiple linear regression analysis and descriptive statistics) to describe the properties of recurrent FDs and to compare them with sporadic FDs.

The distributions of FD parameters and of maximum (during the event) values of IMF intensity and geomagnetic indices (*Ap* and $|Dst|$) are unimodal, peaked and right-skewed for recurrent FDs. These distributions are more compact and have smaller medians than corresponding distributions for sporadic FDs. SW velocity distribution is almost symmetrical for recurrent FDs and right-skewed for sporadic ones. This can be explained by the fact that many of observed FDs are caused by ICMEs with low speeds, but the FDs from low speed "high speed streams" are registered rarely. Comparison of FD magnitude distributions for recurrent FDs, sporadic FDs and FDs registered during the minimum between solar cycles 23 and 24 revealed that FDs were mostly caused by HSS from CHs during this minimum. It is assumed that this is true for most of the periods of low solar activity.

For recurrent FDs, the correlation between SW parameters and FD parameters is moderate ($0.3 \le |r| < 0.5$) for FD magnitude and insignificant or weak ($|r| < 0.3$) for FD decrease rate and CR anisotropy. Corresponding correlations for sporadic FDs are significantly higher. Partial correlations between geomagnetic indices and FD parameters are statistically insignificant for both groups of FDs: the geomagnetic indices are related with CR density variations only through their common dependence on SW parameters.

Acknowledgements. The authors gratefully acknowledge all teams providing continued ground level cosmic ray monitoring: http://cr0.izmiran.ru/ThankYou and the NMDB database (www.nmdb.eu), founded under the European Union's FP7 program (contract no. 213007) for providing data.

References

[1] Forbush S E 1937 *Phys. Rev.* **51** 1108–09
[2] Lockwood J A 1971 *Space Sci. Rev.* **12** 685–715
[3] Iucci N, Parisi M, Storini M and Villoresi G *Nuovo Cimento C* **2** 1–52
[4] Belov A 2009 *Proc. IAU Symp.* **257** 119–130
[5] Wimmer-Schweingruber R F, von Steiger R and Paerli R 1997 *J. Geophys. Res.* 102 17407–417
[6] Cranmer S R 2009 *Living Rev. Sol. Phys.* 2009 **6** 3
[7] Dorman L I, Kaminer N S, Kuz'micheva A E and Mymrina N V 1985 *Proc. 19th ICRC* **6** 293
[8] Richardson I G 2004 *Space Sci. Rev.* **111** 267–376
[9] Abunin A A, Abunina M A, Belov A V, Eroshenko E A, Oleneva V A and Yanke V G 2012 *Geomagn Aeron.* **52** 292–299
[10] Dumbović M, Vršnak B, Čalogović J and Županı R 2012 *Astron. Astrophys.* **538** A28
[11] Badruddin K A 2016 *Sol. Phys.* **291** 559–580
[12] Belov A, Eroshenko E, Yanke V, Oleneva V, Abunin A, Abunina M, Papaioannou A and Mavromichalaki H 2018 *Sol. Phys.* **293** 68
[13] Melkumyan A A, Belov A V, Abunina M A, Abunin A A, Eroshenko E A, Oleneva V A and Yanke V G 2018 *Geomagn. Aeron.* **58** 154–168
[14] Tlatov A, Vasil'eva V and Tavastsherna K 2014 *Sol. Phys.* **289** 1349–58
[15] Richardson I and Cane H 2010 *Sol. Phys.* **264** 189–237
[16] Corder G W and Foreman D I 2009 *Nonparametric Statistics for Non-Statisticians* (New Jersey: John Willey & Sons) p 264
[17] Tabachnik B G and Fidell L S 2007 *Using Multivariate Statistics* (Boston: Pearson Education, Inc.) p 980

IOP Conf. Series: Journal of Physics: Conf. Series **1181** (2019) 012010 doi:10.1088/1742-6596/1181/1/012010

On Contribution of Poloidal Branch of Solar Activity to Heliosphere and GCR Modulation

Krainev M.B., Bazilevskaya G.A., Kalinin M.S., Svirzhevsky N.S.
Lebedev Physical Institute, RAS, Leninsky prosp., 53, 119991, Moscow, Russia
E-mail: mkrainev46@mail.ru

Abstract. The point of view is substantiated that during periods of low sunspot activity many heliospheric characteristics (the solar wind velocity, the form of the heliospheric current sheet, the strength of the heliospheric magnetic field) can be considered as the manifestations in the heliosphere of the poloidal branch of solar activity. As these characteristics strongly influence the process of the GCR transport through the heliosphere the contribution of the poloidal branch of solar activity to the GCR solar modulation should also be substantial. This conclusion is at odds with the wide-spread belief that the modulated GCR intensity is mainly determined by the sunspot cycle.

1. Introduction

When describing the cycle of solar activity on photosphere and in the underlying layers of the Sun two topologically different systems (or branches) of magnetic fields are isolated: toroidal, B^t_φ, and poloidal, $B^p_{r,\vartheta}$ [1]. These two systems are connected by motion of solar plasma, namely its differential rotation, convection etc. In the characteristics of these two systems of solar magnetic fields (SMF) observed on photosphere the toroidal branch corresponds to the active regions with the sunspots etc., while the poloidal branch corresponds to the coronal holes, high-latitude SMF, zonal unipolar magnetic regions etc. [1]. The properties of these two SMF branches are rather different both in their characteristics (scale and strength) and time behavior. Note that usually, speaking of the solar activity cycle, its maxima and minima, one has in mind the characteristics of toroidal branch of solar activity, because of its much more powerful manifestations (flares, coronal mass ejections) on the Sun and near the Earth.

However, the modulation of the GCR intensity takes place in the outer layer of solar atmosphere – heliosphere. For more than 60 years of the heliospheric and cosmic ray studies a huge amount of experimental data and theoretical ideas has been accumulated, but the picture of the 11-year cycle in sunspot activity as the root cause of long-term GCR variations is repeated from paper to paper (see review [2]). Small observed 22-year effects in GCR are usualy associated with variation of the polarity of heliospheric magnetic field (HMF). The causes of such situation are discussed in [3].

In [4] the point of view was suggested that during periods of low sunspot activity many principal heliospheric phenomena connected with the global heliospheric current sheet (GHCS) – the inhomogeneous solar wind (SW) velocity with respect to GHCS, the division of the heliosphere into two unipolar hemispheres with opposite polarity of the magnetic field – could be considered as the manifestations in the heliosphere of the poloidal branch of solar activity.

In this paper we first briefly consider the cause of different manifestation of two activity branches in the heliosphere and underlying spatial layers and then add the HMF strength to the list of heliospheric characteristics with the substantial contribution of the poloidal solar magnetic fields. As these characteristics strongly influence the process of the GCR transport through the heliosphere the contribution of the poloidal branch of solar activity to the GCR solar modulation should also be significant. More extended version of this paper will be published elsewhere.

2. Heliosphere as an outer layer of the Sun

It is wellknown that there is a counter-phase development of the toroidal and poloidal branches of solar activity [1]. Both of these SMF systems demonstrate the invertions of their magnetic fields. The azimuthal SMF of the toroidal branch changes its sign in the sunspot minimum, when the polarities of the preceeding and following spots of the bipole group reverse. The radial SMF of the poloidal branch in each polar regions changes sign in the sunspot maximum. However, in spite of the much more powerful toroidal SMF, the HMF polarity (which is important for GCRs) reflects that of the poloidal, but not toroidal branch of solar activity [5]. To understand this phenomenon we should consider the radial dependence of different types of the energy density.

As can be seen from the data in Fig. 1, there are roughly equal densities of thermal and magnetic energy at photosphere ($r = r_\odot$), but then, up to distances of the order of $r_{HS} \approx 10r_\odot$, the major dynamic factor is the magnetic field. At the same time in this layer, which we call the basement of the heliosphere, the SW velocity increases. Starting with the inner heliospheric boundary $r = r_{HS}$ and up to the heliopause $r = r_{HP}$ the heliosphere extends, where the major dynamic factor is supersonic and superalfvenic (except in the heliosheath) solar wind. Beyond the heliopause is a very local interstellar medium (VLISM). Such spherically symmetric geometry of the heliosphere is broken due to movement of the Sun relative to VLISM, as well as the latitude dependence of the SW characteristics. It leads to the complex structure of the heliosphere.

Thus, to clarify the issue of the relationship between the manifestations of two branches of activity on the Sun and in the heliosphere, the processes in the basement of the heliosphere are of crucial importance. In the model of potential SMF with a surface source worked out in Wilcox Solar Observatory (WSO) [7] and widely used to study GCR variations, the absence of currents is assumed in the basement the heliosphere, $r_\odot \leq r \leq r_{SS}$, where r_{SS} is the radius of so-called source surface. Then everywhere in the basement all three components of SMF, $B_{r,\vartheta,\varphi}$, can be calculated in potential approximation using the sets of the coefficients published in [7].

Note that partly due to the complete absence of currents and too small r_{SS} assumed in the model the absolute value and coordinate dependence of $B_r(\vartheta, \varphi)$ on the source surface, based on the WSO model, do not correspond to the observations [8] and are not commonly used. However, according to observations at spacecraft Ulysses, Pioneers, Voyagers the shape of the calculated neutral line $B_r(\vartheta, \varphi) = 0$ on this surface in the first approximation corresponds to the form of the heliospheric current sheet (HCS) [5, 9], and it is widely used in GCR studies. We make extensive use of two types of results of the WSO model: neutral isolines on the source surface and magnetic field flux through the photosphere and source surface. When calculating SMF fluxes we are also interested in the contribution of dipole SMF. Neutral isoline on the source surface is regarded as the base of the HCS. In [10], using the WSO model, we suggested the classification of the HMF polarity distributions and the meaning of the HMF inversion.

If using the WSO model we compare the distributions of SMF radial component on photosphere and source surface during periods of intermediate and low sunspot activity, we can see a complex distribution with numerous small islands of SMF of opposite polarities, neutral lines etc. on photosphere, while on the source surface there are only two hemispheres of opposite SMF polarity divided by single global neutral line. Hence the main function of the heliospheric basement is to filtrate the magnetic fields by their scales. So in the transition from photosphere

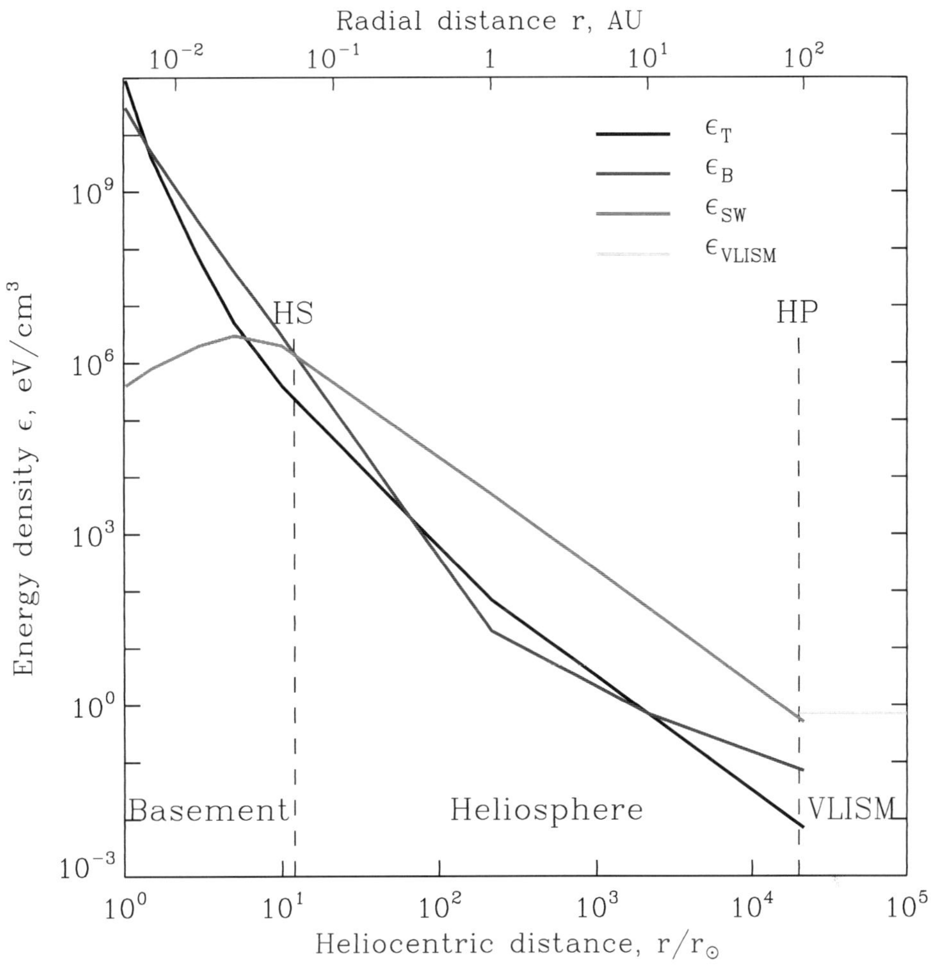

Figure 1. The heliocentric radial dependence of density of the different types of energy (thermal, magnetic, kinetic, interstellar), mainly based on [6]. The vertical dashed lines show the approximate positions of the heliospheric boundaries.

to heliosphere the poloidal SMFs due to their larger scale get the advantage over the toroidal ones. The second main function of the basement is that in this layer the form of the SMF tubes open to the heliosphere greatly affects the SW accelerating along these tubes [11].

3. Branches of solar activity and heliospheric characteristics important for GCR

It is wellknown that the heliospheric characteristics most important for the GCR propagation in the heliosphere are the SW velocity, the HMF strength and the form of the GHCS [2]. The detailed consideration of behaviour of these characteristics [4] leads to the conclusion that all main HMF and SW features important for GCR:

- the HMF separation into two unipolar hemispheres by the wavy GHCS
- the HMF inversion during sunspot maxima

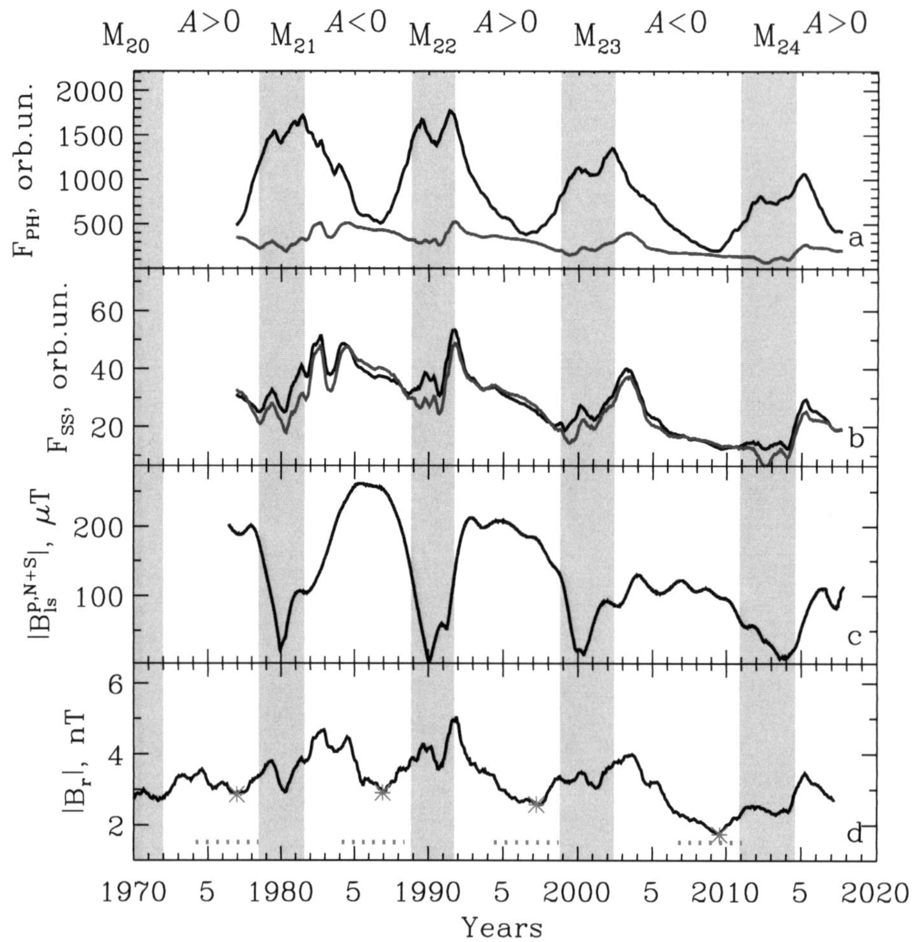

Figure 2. Magnetic field flux through different solar spheres and the poloidal field of the Sun in 1970-2018. All characteristics are 1-year smoothed. Panel a: the SMF flux through the photosphere (dark line) and its part from the dipole SMF (lighter/blue line); b: the same fluxes but through the source surface; c: sum of absolute values of the photospheric SMF line-of-sight components in both polar regions [7]; d: absolute value of the HMF radial component near the Earth [12]. Minimum values B_r^{min} are marked with asterisks, and doted horizontal lines in Figs. 2 and 3 show the periods when we associate HMF with the poloidal branch of solar activity.

- the strong dependence of SW velocity and hence the HMF azimuthal component on the angular distance from GHCS
- the independence of the HMF radial component on this distance

are due mainly to influence of the poloidal branch of solar activity on SW and HMF distributions in the heliospheric basement. And this influence is very significant: the SW velocity – the main dynamic factor in the heliosphere – changes by a factor of two in the vast volume. The regular HMF changes its direction in phase with the SMF of poloidal branch, and this variation in HMF is no less than that due to the cycle in the toroidal branch.

Now we consider which of two branches of solar activity accounts for the absolute value of $B_r(\vartheta, \varphi)$, constant in each of the unipolar "hemispheres" [5, 9]. Fig. 2 compares the time profiles of the SMF fluxes (both total flux and that due to dipole SMF) through the photosphere (panel a) and the source surface (b) with magnitudes of the line-of-sight-component of the high latitude photospheric SMF (c) and of the absolute value of HMF radial component (d). One can see that the magnetic flux through the photosphere behaves similarly to a toroidal branch (maxima in flux occur during sunspot maxima), and the contribution of the dipole fields is small. On the contrary, the magnetic field flux through a surface source (or an open SMF flux) is significantly shifted in time, its maxima being observed after sunspot maxima, and the contribution of dipole fields is essential.

It can be seen that an open SMF flux correlates well with the HMF radial component, but we believe that the main source of the HMF is different for different years of the periods between the HMF inversions, when the heliosphere is divided by the GHCS into two unipolar hemispheres. We call these periods the HMF dipole phases. In the early years of the HMF dipole phase the HMF characteristics are determined mainly by the decreasing but still large photospheric flux (i.e., the sunspot activity of the toroidal branch), and not the growing but still weak poloidal activity.

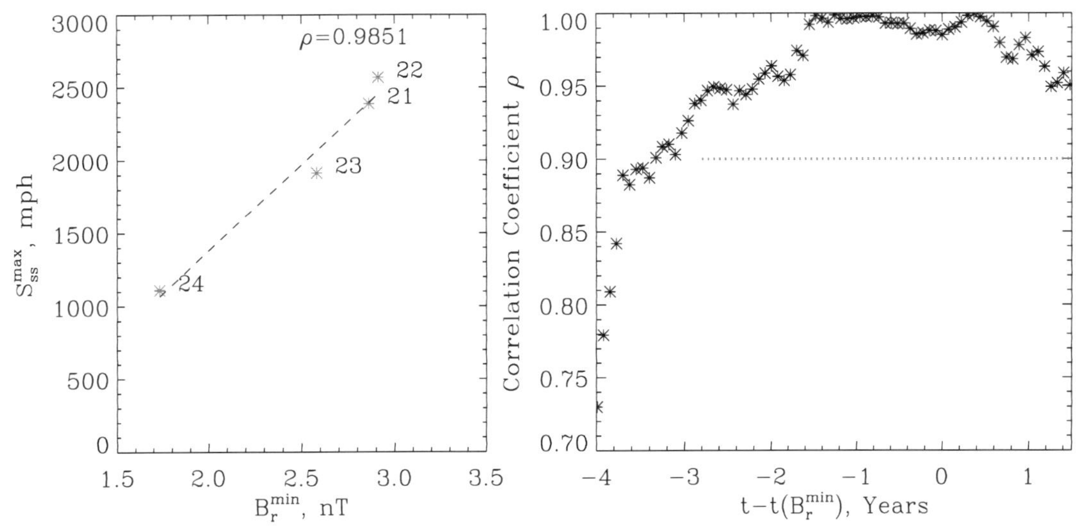

Figure 3. The correlation between the absolute value of the HMF radial components near the Earth in the periods of low sunspot activity and the maximum value of this activity S_{ss}^{max} in the next cycle for solar cycles 21-24. The explanations are in the text. The horizontal doted lines in this and previous figures show the periods when correlation coefficient ρ is greater than 0.95.

In the second half of the HMF dipole phase the situation is different. Here we use the idea that the characteristics of the poloidal branch of solar activity in minimum of sunspot cycle specify the next maximum of this cycle (see references in [13]). First, we consider the correlation between minimum value of the absolute value of the HMF radial components near the Earth, B_r^{min}, and maximum value of the sunspot area, S_{ss}^{max}, in the next sunspot cycle. Note that we calculated S_{ss}^{max} as the average value of the sunspot areas in two Gnevyshev peaks (see [14]). The left panel of Fig. 3 shows the regression between these characteristics for the last four solar cycles. You can see that correlation is very high. Then we calculate the correlation coefficient between

$|B_r|(t)$ at time t and S_{ss}^{max} and consider how this factor depends on time shift relative to time of minimum $|B_r|$. One can see that in period $t - t(B_r^{min}) \in [-2.5, 1.5]$ (years) the correlation coefficient is very high ($\rho > 0.95$), and before this period it decreases quite sharply. We believe that during this period of high correlation between $|B_r|$ and S_{ss}^{max} the HMF observed near the Earth depends mainly on the characteristics of the poloidal branch of the solar activity.

4. Conclusions

(i) In the formation and development of the heliospheric characteristics, important for GCR, essential roles are played by both the toroidal branch of solar activity (active regions, sunspots and related phenomena) and poloidal branch (coronal holes, high-latitude magnetic fields, zonal drift of unipolar magnetic field regions etc.), the role of each branch changing with the phase of the solar cycle. Outside the periods of heliospheric magnetic field inversion the poloidal branch of the solar activity forms the global heliospheric current sheet and connected with it distributions of HMF radial component and solar wind velocity. During periods of low sunspot activity the HMF radial component and strength are also the products of the poloidal branch of solar activity.

(ii) The relatively greater role of the poloidal branch of activity in the heliosphere than on and inside the Sun is due to processes in a layer between them, where the major dynamic factor is the solar magnetic field. As a result, in the transition from photosphere to heliosphere the poloidal solar magnetic fields get the advantage due to its larger scale over the toroidal ones.

(iii) As a result of the above heliospheric processes, the main observed long-term variation of the GCR intensity, 11-year cycle, as well as the 22-year cycle, are the manifestations of both branches of solar activity, the role of each branch varying with the phase of the solar cycle.

Acknowledgments
We thank for help the Russian Foundation for Basic Research (grants 17-02-00584a, 18-02-00585a) and the Program of the Presidium of the Russian Academy of Sciences "Fundamental Properties of Matter and Astrophysics".

References
[1] Charbonneau P 2010 Dynamo Models of the Solar Cycle *Living Rev. Solar Phys.* **7** 3
[2] Potgieter M S 2013 Solar modulation of cosmic rays *Living Rev. Solar Phys.* **10** 3
[3] Krainev M B Kota J Potgieter M S 2015 On the causes and mechanisms of the long-term variations in the GCR characteristics *PoS(ICRC2015)176*
[4] Krainev M B 2017 Causes of Long-Term Variations in Galactic Cosmic Ray Intensity in the Inner Heliosphere *Bulletin of the Russian Academy of Sciences: Physics* **81** 166-169
[5] Balogh A Jokipii J R 2009 The Heliospheric Magnetic Field and Its Extension to the Inner Heliosheath *Space Sci. Rev* **143** 85-110
[6] Rossi B Olbert S 1970 Introduction to the physics of space NY McGrow-Hill 454
[7] `http://wso.stanford.edu`
[8] Zhao X Hoeksema J T 1993 Unique determination of model coronal magnetic fields using photospheric observations *Solar Physics* **143** 41
[9] Smith E 2011 Solar cycle evolution of the heliospheric magnetic field: The Ulysses legacy *J. Atmosph.Solar-Terr. Phys.* **73** 277-289
[10] Krainev M B Bazilevskaya G A Kalinin M S et al. 2015 GCR intensity during the sunspot maximum phase and the inversion of the heliospheric magnetic field *PoS(ICRC2015)176*
[11] Owens M J Forsyth R J 2013 The Heliospheric Magnetic Field *Living Rev. Solar Phys.* **10** 5
[12] `ftp://spdf.gsfc.nasa.gov/pub/data/omni/low_res_omni/`
[13] Hathaway D H 2015 The Solar Cycle *Living Rev. Solar Phys.* **12** 4
[14] Bazilevskaya G A Krainev M B Makhmutov V S et al. Structure of the Maximum Phase of Solar Cycle 21 and 22 *Solar Physics* **197** 157

IOP Conf. Series: Journal of Physics: Conf. Series **1181** (2019) 012011 doi:10.1088/1742-6596/1181/1/012011

Behavior of zonal components of cosmic ray distribution and Dst-index of geomagnetic field during periods of geoeffective disturbances of solar wind

S A Starodubtsev, V G Grigoryev and P Yu Gololobov

Yu. G. Shafer Institute of Cosmophysical Research and Aeronomy of the Siberian Branch of the RAS (ShICRA), 31 Lenin ave., Yakutsk, Russia, 677980

E-mail: grig@ikfia.ysn.ru

Abstract. On a basis of processing of data of measurements of the world-wide neutron monitor network database NMDB by a global survey method in real-time, the behavior of the first two angular moments of cosmic ray distribution function are investigated during the periods of geomagnetic storms that were observed in 2017. It is shown that abrupt increases of positive amplitudes of zonal components, which can be considered as predictors of geomagnetic field disturbances with a probability around 75%, precede geomagnetic storms with Dst <-50 nT. The predictor's lead time is from a few hours to 1.5 days. A monitoring of geomagnetic disturbances is carried out in real-time mode and results of the forecasting is presented in the Internet through the link *http://www.ysn.ru/~starodub/SpaceWeather/global_survey_real_time.html.*

1. Introduction

As was shown in the works [1, 2, 3], investigation of the distribution of cosmic rays (CR) in real time on the basis of the data of world-wide network of neutron monitors and muon detectors could become one of the most effective instrument for short-term forecasting (from a few hours to days) of a beginning of major geomagnetic storms. Depending on the level of geomagnetic perturbation, the probability of forecasting can reach 75% and higher. The basic parameters, that react on arrival of geoeffective disturbances of the interplanetary medium to the Earth, are behaviors of amplitudes of zonal (north-south) components of isotropic intensity C_{00} and first two angular moments of CR distribution function C_{10} and C_{20} [4, 5, 6, 7]. For this components, we have determined the critical levels of their positive values C_{00}^+, C_{10}^+ and C_{20}^+ which were observed before the beginnings of magnetic storms. Moreover, their total factor $C^+ = C_{00}^+ + C_{10}^+ + C_{20}^+$ are also taken into account. The results of the CR monitoring are available in the Internet through the link: *http://www.ysn.ru/~starodub/SpaceWeather/global_survey_real_time.html.*

Below, on the basis of real-time realization of the global survey method using the NMDB database, we have carried an analysis of the predictor appearances and their relationship to geomagnetic disturbances that were observed in 2017 (WDC for Geomagnetism, Kyoto, *http://wdc.kugi.kyoto-u.ac.jp/index.html*).

2. Experimental data and method

For processing the measurements of the data of neutron monitors with the global survey method in real time [8], we use 1-hour data that are presented in database NMDB ($http://www01.nmdb.eu$). Characteristics of CR intensity observed by each device are defined by receiving characteristics of the registering device, which are unified under the concept of receiving vectors [9] and reflects its geometry, geographical position and the type of registered particles. Calculations of the receiving vectors of real devices are held with knowing of: coupling coefficients between the secondary $W(E)$ and primary particles, directional diagrams of the detector $N(\theta, \varphi)d\omega$, which reflects geometrical features of the device and also zenith-azimuthal sensitivity of particles, energy spectrum $f_n(E)$ of investigated CR variations and asymptotic angles of arrival of particles $\Psi(E, \theta, \varphi)$, $\Phi(E, \theta, \varphi)$.

The devices that has different receiving vectors $\vec{R}_n^m$ register CR intensity $\vec{I}$ that are defined by the following equation:

$$I = \sum_{n=0}^{\infty} \sum_{m=0}^{n} (a_n^m x_n^m + b_n^m y_n^m),$$

where x_n^m, y_n^m are components of receiving vector, a_n^m, b_n^m are the components of a multidimensional vector of CR distribuion $\vec{A}$. Then, for the selected devices we create the system of linear equations [9] which, in matrix form, has the next representation:

$$\vec{I} = M \cdot \vec{A},$$

where $\vec{I}$ is a vector column of observational data and M is a matrix of receiving vectors (coupling coefficients). The obtained system is solved by the last square method under the assumption that the decomposition of the CR distribution into the series of spherical functions is converges fast. So, usually, except an isotropic part, only first two harmonics of the distribution, which are well observed in the experiment, are taken into account. According to this, the first 9 components of the vector $\vec{R}$, including mentioned C_{00}, C_{10} and C_{20}, are need to be defined.

3. Results

As the analysis of the monitoring during 2017 has shown, the values of positive amplitudes of the components C_{00}^+, C_{10}^+ $C_{20}^+ > 0.8\%$ or the sum of their positive values $C^+ > 0.9\%$ can be considered as a predictor of the beginning of geomagnetic disturbance. If there is an excess of this critical values which indicates on oncoming of a magnetic disturbance, an appropriate warning is generated.

At the beginning of the geomagnetic disturbance, it starts the abrupt decrease of negative values of zonal harmonics C_{00}^-, C_{10}^- C_{20}^-. If their summary value $<$ -0.9%, then the message about the storm passing is generated. During the chosen period the predictors with different durations were observed 20 times. 2 of them were false and 3 had appeared after the beginning of disturbance. The other 15 predictors had appeared before the storm with the lead-times 5-16 hours and were associated with a passage of a large-scale disturbances of solar wind which cause the storms with Dst $<$ -30 nT.

As examples, in figures 1 and 2 the time dependencies of Dst-index and zonal components C_{00}, C_{10}, C_{20}, C^+ and C^- during two major geomagnetic storms in May and September 2017, when Dst-index has reached their minimal values -125 and -124 nT, respectively, are shown. The critical values for each component are shown by dotted lines. The up and down arrows shows the time moments when the components excess the positive and negative critical values, respectively.

As seen from the figures 1 and 2, the geomagnetic storms in 28th May and 8th September 2017 had the predictors in the values C_{20} and C^+ and were accompanied by decrease of value $C^- <$ -0.9% 6 hour after appearance of the predictor of disturbance.

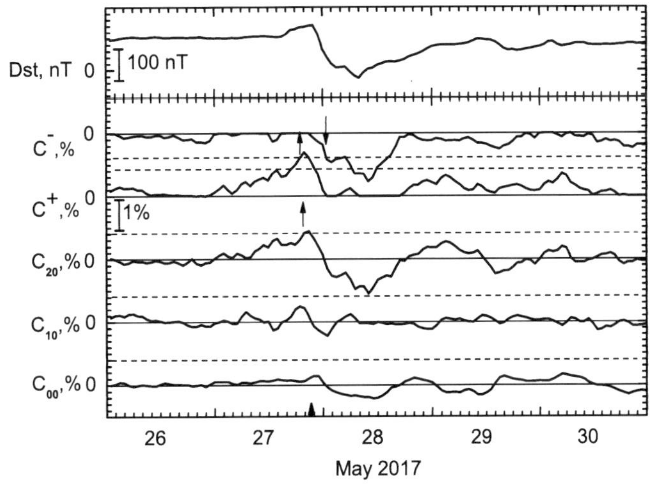

Figure 1. Behavior of the zonal harmonics C_{00}, C_{10}, C_{20} , the sum of their positive C^{+} and negative C^{-} values and Dst-index during the periods of geomagnetic disturbances in May 2017. Solid triangles at the bottom indicate on a sudden commencement of storm SSC

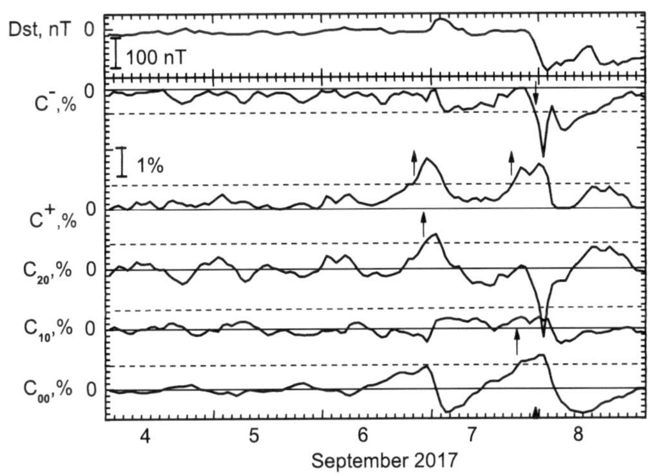

Figure 2. Behavior of the zonal harmonics C_{00}, C_{10}, C_{20} , the sum of their positive C^{+} an negative C^{-} values and Dst-index during the periods of geomagnetic disturbances in September 2017. Solid triangles at the bottom indicate on a sudden commencement of storm SSC.

4. Conclusion
The analysis of the results of monitoring of geomagnetic disturbances during 2017 carried out in ShICRA SB RAS have shown that the reliability of short-term forecasting of entering of the Earth into the geoeffective disturbances of solar wind is around 75%.

5. Acknowledgements
The work was carried out using the data of Unique scientific installation of the Russian national ground level network of cosmic ray stations and with the support of grant RFBR No. 18 – 42 – 140002 – r_a.

References

[1] Kuwabara T, Bieber J W, Clem J and et al 2006 *Space Weather* **4** S08001
[2] Rockenbach M, Lago A D, Gonzalez W D and et al 2011 *Geophys. Res. Lett.* **38** L16108
[3] Belov A V, Eroshenko E A, Yanke V G and et al 2018 *Geomag. Aeron.* **58** 356–372
[4] Grigoryev V G and Starodubtsev S A 2015 *Bull. of the RAS: Physics* **79** 649–653
[5] Grigoryev V G, Starodubtsev S A and Gololobov P Y 2015 *Proceedings of Science* **PoS(ICRC2015)** 076
[6] Grigoryev V G, Starodubtsev S A and Gololobov P Y 2017 *Bull. of the RAS: Physics* **81** 200–202
[7] Gololobov P Y, Grigoryev V G and Starodubtsev S A 2017 *Proceedings of Science* **PoS(ICRC2017)** 022
[8] Altukhov A M, Krymsky G F and Kuzmin A I 1970 *Proc. 11-th ICRC* **4** 457–460
[9] Krymsky G F, Kuzmin A I, Krivoshapkin P A and et al 1981 *Cosmic ray and solar wind* (Novosibirsk: Nauka) p 224

IOP Conf. Series: Journal of Physics: Conf. Series **1181** (2019) 012012 doi:10.1088/1742-6596/1181/1/012012

Investigation of mechanisms of formation of the second spherical harmonics of the galactic cosmic ray angular distribution

P. Yu. Gololobov, P. A. Krivoshapkin and G. F. Krymsky

Yu. G. Shafer Institute of Cosmophysical Research and Aeronomy of the Siberian Branch of the RAS (ShICRA), 31 Lenin ave., Yakutsk, Russia, 677980

E-mail: gpeter@ikfia.ysn.ru

Abstract. Mechanisms of formation of a second spherical harmonic of an angular distribution of galactic cosmic rays in the heliosphere are investigated. On the basis of data of Nagoya multidirectional muon telescope using a method of tensor anisotropy of cosmic rays time courses of symmetric diurnal, antisymmetric diurnal and semidiurnal variations of cosmic rays were obtained for the period 1971-2017. An abrupt offset of a phase of the semidiurnal variation to the earlier time during a positive polarity of the general magnetic field of the Sun, which closely correlates with the temporal change of a phase of the symmetric diurnal variations. It was shown the existence of annual variations of the tensor anisotropy components, which can be satisfactorely described by a single modulation factor - a screening of cosmic rays by the sector-structured interplanetary magnetic field. Moreover, the theory indicates on the existense of a stable north-south asymmetry of the interplanetary magnetic field - shift of the neutral current sheet to the south of solar equator.

1. Introduction

The heliospheric modulation of cosmic rays (CR) leads to the occurrence of the specific spatial angular distribution of CR which has an anisotropic character. This anisotropic flow existing in the interplanetary space reveals itself in data of ground-based CR detectors as diurnal and semidiurnal oscillations due to the rotation of the Earth around its own axis. At the same time, registration of the mentioned oscillations by a single CR detector reflects only small part of the full CR angular distribution. In order to obtain the full distribution it is necessary to use a huge number of differently directed CR detectors which could cover a sufficient area of the celestial sphere. Wherein, the number of detectors depends on how much detailed CR distribution is needed to be determined.

Our earlier calculations of the CR distribution using data of muon telescopes of Nagoya and Yakutsk stations and a world-wide network of neutron monitors allow us to obtain average parameters of the second spherical harmonic of the CR distribution, the so-called CR tensor anisotropy, which have been observed for the last 4 solar activity cycles [1, 2, 3]. As a theoretical interpretation of the obtained results it was suggested a model that consists of two CR modulation mechanisms which were able to create the necessary CR distribution: mechanisms of CR screening by the sector-structured interplanetary magnetic field (IMF) [4] and shear flow of solar wind [5]. Furthermore, juxtaposition of the model calculations and the experimental data allow us to reveal an important phenomenon - the north-south asymmetry of the IMF, which is today considered as a basic property of the heliosphere [6, 7]. Nevertheless, an approach to calculation of the tensor anisotropy and the theory itself had a number of shortcomings. Firstly, the influence of a temperature effect on muons was ignored. Secondly, the theoretical model based on two different mechanisms and this fact complicates the interpretation process. In the current work the further investigation of the temporal dynamic of the CR tensor anisotropy for the long time period 1971-2017 is carried out taking into account the mentioned backgrounds.

2. Investigation method

The angular distribution of CR is determined by processing of data of measurements of 17 independent directions of registration of the Nagoya multidirectiuonal muon telescope $http://www.stelab.nagoya-u.ac.jp/ste-www1/div3/muon/muon3.html$ using a method of receiving vectors [8, 9].The method allows us to take into account influences of the atmospheric and magnetospheric factors, except the temperature effect, on the registered CR intensity. The detailed description of the method is presented in [1, 10].

The angular distribution of CR in space $I(\theta,\varphi)$ is presented as series of spherical harmonics

$$I(\theta,\varphi)=\sum_{n=0}^{\infty}(a_n^m cos(m\varphi)+b_n^m sin(m\varphi))P_n^m(sin\theta)$$

, where θ and φ are latitude and longitude in a corresponding coordinate system, $P_n^m(sin\theta)$ is the associated Legendre's polinomials. If to represent the distribution $I(\theta,\varphi)$ as a multidimensional vector $\vec{A}=(a_n^m,b_n^m)$ then for each device (channel) one can define such a multidimensional vector $\vec{Z}=(x_n^m,y_n^m)$ that the CR intensity J registered by a device would be equal to a scalar product $J=A\cdot Z$. In the current work we use the receiving vectors which are presented in [11] with an assumption that spectra of vector f_1 and tensor f_2 anisotropies of CR have the next forms:

$$f_1=\begin{cases}E^1, & \text{if } E\leq E_1\\ 0, & \text{if } E>E_1\end{cases},\ f_2=\begin{cases}E^2, & \text{if } E\leq E_2\\ E^{-1}, & \text{if } E>E_2\end{cases},$$

where $E_1=100$ GeV and $E_2=70$ GeV, according to [11, 10].

If the number of devices (channels) are sufficient then the vector $\vec{A}$ can be defined by a solution of a system of linear equations which number is equals to the number of devices (channels). For example, the CR vector anisotropy requires at least 4 channels, while the tensor anisotropy - 9 channels.

3. Model of CR tensor anisotropy

Earlier in the works [3, 1] we developed the theory which can describe the experimental data on tensor anisotropy of CR. The theory was based on mechanism of CR screening by the sector-structure IMF which successfully explains an excess of CR in direction perpendicular to the IMF direction. Hence, the observed temporal dynamic of the tensor anisotropy appeared to be slightly complicated and the explanation required an additional mechanism. It was suggested that the second mechanism is the shear flow of solar wind which manifests itself during large scale disturbances of solar wind. On the other hand, we used average monthly data, so the model need to be stationary. Therefore, below is presented the new model which is based only on the CR screening mechanism.

Accoring to [10, 3] in the solar equatorial coordinate system, which polar axis is directed to the Sun's north pole and the zero longitude matches to the position of the Sun, the CR screening mechanism will have the tensor anisotropy with the next components:

$$\vec{A}_{Scr}=(1/2,0,0,\sqrt{3}/2\cos 2\varphi,\sqrt{3}/2\sin 2\varphi)$$

, where φ is an angle between the archimedian spiral and the Earth-Sun line, which in the model is equals to 45°.

Let us suggest that the amplitude of tensor anisotropy, that is due to the CR screening mechanism, depends on the heliolatitudinal position of the Earth (coefficient k) relatively to the plane of symmetry of the neutral of current sheet, which is deflected to the south or north from the solar equatorial plane on the angle α. Moreover, it is necessary to take into account the course of the heliolatitudinal position of the Earth over a year from +7.25° to −7.25°.

So, the vector $\vec{A}_{Scr}$ will have the next form:

$$\vec{A} = \vec{A}_{Scr}(\cdot 1 - \kappa \cos^4(30(\sin t - \alpha/7.25)))$$

, where $-14.5° \leq \alpha \leq 14.5°$ and t is a time of year which is counted from the day of the solstice - 21st June.

Transition from the solar heliographic coordinate system to solar geographical one, taking into account the possible north-south displacement of the plane of symmetry of the neutral current sheet, is carried out by the next transformations: $M^{long}(-t) \cdot M^{lat}(7.25°) \cdot M^{long}(t) \cdot M^{lat}(\alpha) \cdot \vec{A}$, where the rotation matrixes are presented in [9] and have the next forms:

$$M^{long}(\chi) = \begin{pmatrix} 1 & 0 & 0 & 0 & 0 \\ 0 & \cos\chi & \sin\chi & 0 & 0 \\ 0 & -\sin\chi & \cos\chi & 0 & 0 \\ 0 & 0 & 0 & \cos 2\chi & \sin 2\chi \\ 0 & 0 & 0 & -\sin 2\chi & \cos 2\chi \end{pmatrix},$$

$$M^{lat}(\xi) = \begin{pmatrix} 1 - \frac{3}{2}\sin^2\xi & 0 & -\frac{3}{2}\sin 2\xi & 0 & \\ 0 & \cos\xi & 0 & 0 & -\sin\xi \\ \frac{\sqrt{3}}{2}\sin 2\xi & 0 & \cos 2\xi & \sin 2\xi & 0 \\ -\frac{\sqrt{3}}{2}\sin^2\xi & 0 & -\frac{1}{2}\sin 2\xi & (1+\cos^2\xi) & 0 \\ 0 & \sin\xi & 0 & 0 & \cos\xi \end{pmatrix}.$$

The results of modeling of annual oscillations of the antisymmetric diurnal $\vec{R^1_{2,exp}}$ and semidiurnal $\vec{R^2_{2,exp}}$ components of the CR tensor anisotropy in the solar geographical coordinate system for k = 0.00, 0.15 0.30 and $\alpha = -10°$, 0°,10° are presened in figure 1. It is seen that during a year the component $\vec{R^1_{2,exp}}$ reveals oscillations strictly toward 9-21 h of LT and the vector skews to 9 h when the neutral sheet is tilted to the south from the solar equator, and vise versa, to 21 h when the sheet is tilted to the north. The effect rises with the value κ. The expected annual change of $\vec{R^2_{2,exp}}$ also depends on the value α. When the α increases, the semiannual oscillations of $\vec{R^2_{2,exp}}$ becomes complemented by annual oscillations, depending on κ. In the case when the neutral current sheet is shifted to the south (north) from the solar equator, the minimal values of the vector $\vec{R^2_{2,exp}}$ are expected in August-September (February-March) and the maximum values - in April-May (October-November).

4. Results and discussions

By the data of 17 directions of the multidirectional muon telescope Nagoya the decomposition of CR distribution into the vector $\vec{R^1_1} = (a^1_1, b^1_1)$ and the tensor $\vec{R^1_2} = (a^1_2, b^1_2)$, $\vec{R^2_2} = (a^2_2, b^2_2)$ anisotropy components is carried out for the period 1971-2017. The obtained values of the mentioned components are presented in figure 2. A detailed analysis of the components shows an occurrence of annual and semiannual periodic oscillation of the components a^1_2, b^1_2, a^2_2 and b^2_2. In calculations we have used the data are uncorrected for the temperature effect. On the other hand, using the method described in [12, 13] we have taken into account the temperature

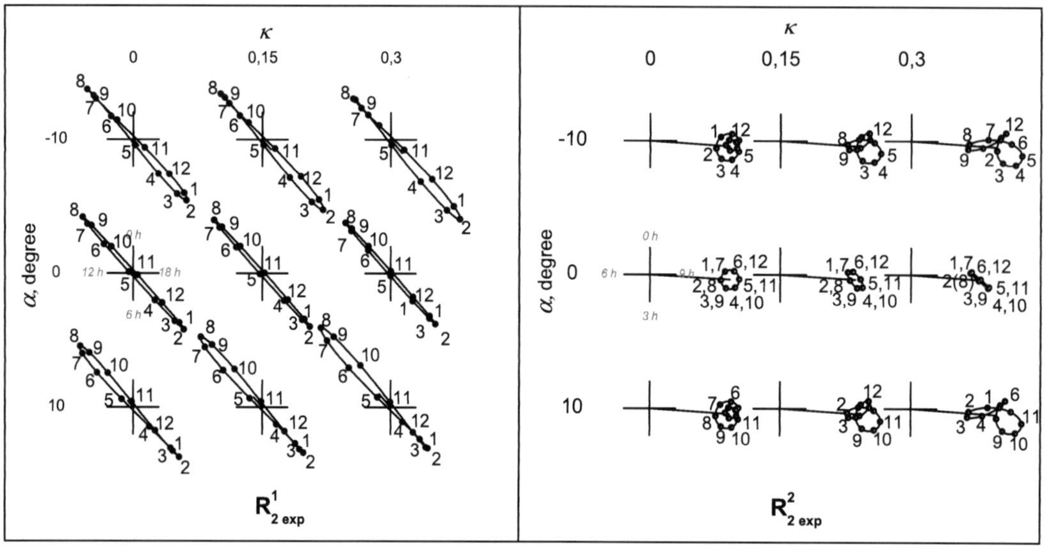

Figure 1. The vector diagram of annual variations of the antisymmetric diurnal $\vec{R_2^1}$ and semidiurnal $\vec{R_2^2}$ components of the CR tensor anisotropy that are expected by the model for different values of k and α. The numbers around the points denote months.

for the period 2006-2015. It is shown that for the selected period the average value of the vector anisotropy $|R_1^1|$ corrected for the temperature effect is equals to 0.29 % and time of maxima $T_{1,max}^1$ - 16.6 h LT while the values without corrections are 0.28 % and 15.3 h LT, respectively. It means that the temperature effect for the mentioned time period created a diurnal variation with an amplitude 0.1 % and phase 9.5 h LT. The temperature effect almost did not affect the anisymmetric diurnal and semidiurnal components of tensor anisotropy.

The obtained average yearly amplitudes and phases of vector and tensor anisotropies of CR for 1971-2017 are presented in figure 3. It is shown that the amplitudes $|R_1^1|$ and $|R_2^2|$ during the investigated period have changed synchronously respectively to the sunspot number - during the periods of solar activity minima the amplitudes decreases. The phases $T_{1,max}^1$ and $T_{2,max}^2$ have changed depending on a polarity of the general magnetic field of the Sun (GMFS) - during solar activity minima of positive polarity the phases shift to earlier time.

It is well known that during a 11-year solar activity cycle the geometry of solar wind in the Sun's meridional plane changes, the solar wind at solar activity minima becomes "flattened" and the level of regularity of the IMF increases. During the periods of positive GMFS the CR are mainly carried out away from the Sun through its equatorial plane. All this physical processes are well reflected in the observed amplitude-phase oscillations of the vector anisotropy [14, 15].

The phase $T_{2,max}^2$ in average is equals to 3 h of LT as it is expected by the CR screening mechanism, but during solar activity minima of positive polarity of the GMFS it shifts to the earlier time up to 1 h. Moreover, $T_{1,max}^1$ and $T_{2,max}^2$ are suspiciously well correlates to each other. The reason of this phenomenon is unknown.

It is necessary to note that the amplitude $|R_2^1|$ during the whole investigated period has low value and does not change with solar activity.

In figure 4 the average annual change of the components of the CR tensor anisotropy $\vec{R_{2,obs}^1}$ and $\vec{R_{2,obs}^2}$, which were observed during the investigated period, are shown. Also, in the figure there are presented $\vec{R_{2,exp}^2}$ and $\vec{R_{2,exp}^2}$ which are expected from the mechanism of CR screening by

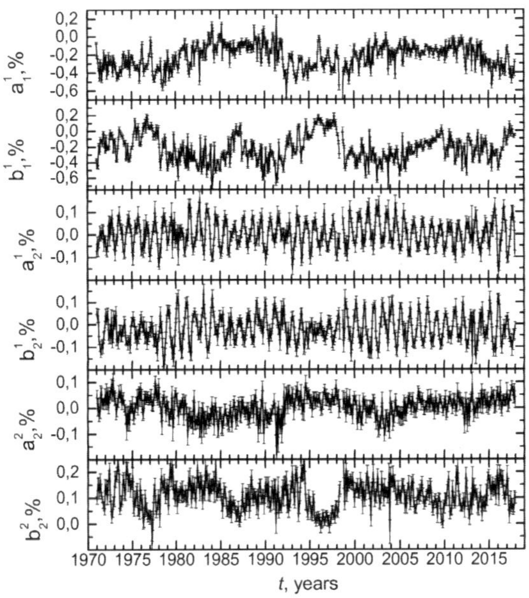

Figure 2. The observed average monthly values of the components of the vector (a_1^1, b_1^1) and the tensor $(a_2^1, b_2^1, a_2^2, b_2^2)$ anisotropies of CR which are obtained by the above mentioned method using the data of muon telescopes Nagoya for the period 1971-2017.

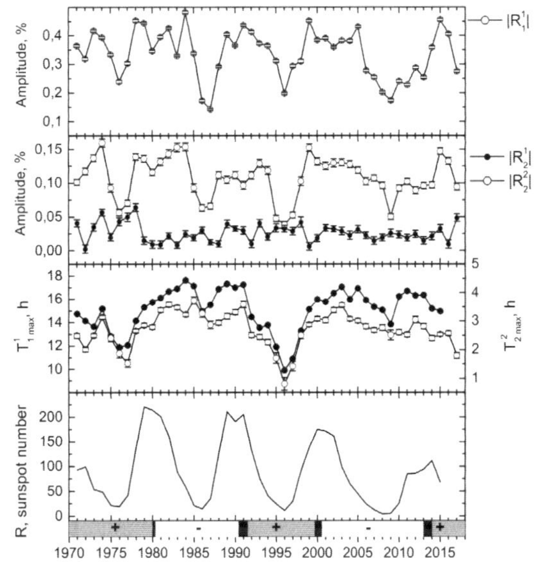

Figure 3. Average annual amplitudes of the vector $|\vec{R_1^1}|$ and the tensor $(|\vec{R_2^1}|, |\vec{R_2^2}|)$ anisotropies of CR and the maxima times $T^1_{1,max}$ and $T^2_{2,max}$, respectively, and also the sunsot number. The regions with positive and negative signs in the bottom of the figure reflect the GMFS polarity changes.

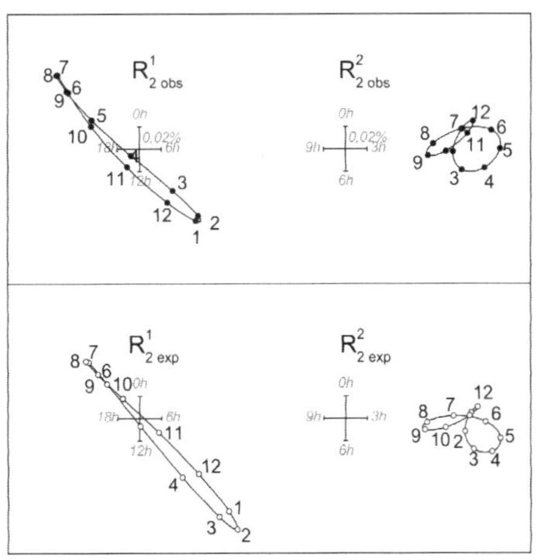

Figure 4. The vector diagrams of the average annual change of the antisymmetric diurnal $\vec{R^1_{2,obs}}$ and semidiurnal $\vec{R^2_{2,obs}}$ components of the CR tensor anisotropy that were observed by the Nagoya muon telescope during 1971-2017. And the same components expected by the model $\vec{R^2_{1,exp}}$, $\vec{R^2_{2,exp}}$ for $k = 0.3$ and $\alpha = -10°$.

the IMF for $\alpha = -10°$ and $\kappa = 0.3$. Comparison between the experimental data and the model calculations have shown that the suggested mechanism sufficiently describes the observed annual change of the tensor anisotropy. Moreover, the model indicates on the north-south asymmetric structure of the heliosphere - shift of the neutral sheet to the south from the solar equatorial plane.

It should be noted that another theoretical mechanisms (for example [16, 17, 18, 19]) which are able to create the same excess of particles in direction perpendicular to the IMF, can produce the same oscillations of CR tensor anisotropy.

5. Conclusions

Using the data of the multidirectional muon telescope Nagoya the vector and tensor anisotropies of CR for 1971-2017 were obtained. The components of the CR tensor anisotropy $(a_2^1, b_2^1, a_2^2, b_2^2)$ experiences annual and semiannual oscillations which are not due to the temperature effect. On the basis of the CR screening mechanism it was created the model which explains the observed oscillations of the CR tensor anisotropy. It is shown that the neutral current sheet is shifted to the south of the solar equator on $\sim 10°$.

5.1. Acknowledgments

The work has been carried out under the support of RFBR grants Nos. 18-32-00064-mol-a and 18-02-00451-a. We acknowledge the Cosmic Ray Section, Solar-Terrestrial Environment Laboratory, Nagoya University for providing the data.

References

[1] Gololobov P, Krymsky G and Krivoshapkin P 2017 *Proceedings of Science* **PoS(ICRC2017)** 021
[2] Gololobov P Y, Krymsky G F, Krivoshapkin P A, Gerasimova S K and Grigoryev V G 2015 *J. Phys.: Conf. Ser.* **632** 012056
[3] Krymsky G F, Krivoshapkin P A, Gerasimova S K and Gololobov P 2014 *Astron. Lett.* **40** 230–233
[4] Krivoshapkin P A, Krymsky G F, Kuzmin A I, Skripin G V and Metlyaeva E A 1970 *Acta Physics Academiae Scientiarum Hungaricae* **29**(2) 147–151
[5] Berezhko E G 1981 *JETP Letters* **33** 399–401
[6] Simpson J A, Zhang M and Bame S 1996 *The Astrophysical Journal* **465** L69–L72
[7] Krymsky G F, Krivoshapkin P A, Mamrukova V P and Gerasimova S K 2009 *Astronomy Letters* **35** 333–337
[8] Chirkov N P, Altukhov A M, Krymsky G F, Kuzmin A I and Skripin G V 1967 *Geomagnetism and aeronomiya* **7** 620–631
[9] Krymsky G F, Krivoshapkin P A and Kuzmin A I 1968 *Canadian Journal of Physics* **46** S959–S961
[10] Krymsky G F, Kuzmin A I, Krivoshapkin P A and et al 1981 *Cosmic rays and solar wind* (Novosibirsk, Russia: Nauka)
[11] Fujimoto K, Inoue A, Murakami K and et al 1984 *Coupling coefficients of cosmic ray daily variations for meson telescopes* (Nagoya, Japan: Rep. 9, Cosmic Ray Res, Lab., Nagoya University)
[12] Kostyuk M G, Petkov V B, Novoseltseva R V and et al 2013 *Bull. of. RAS: Phys.* **77** 569–571
[13] Osipenko A, Abunin A A, Berkova M D, Titova M A, Belov A V, Eroshenko E A, Yanke V G, Barabashina N S, Grigoryev V G, Starodubtsev S A, Kuzmenko V S and Yanchukovsky V L 2015 *Bull. of the RAS: Phys.* **75** 662–666
[14] Krymsky G F 1964 *Geomag. Aeron.* **4** 763–769
[15] Munakata K, Kozai M, Kato C and Kota J 2014 *The Astrophysical Journal* **791:22** 16
[16] Quenby J J and Lietti B 1968 *Planet. Space Sci.* **16** 1209–1219
[17] Sarabhai V, Pai G L and Wada M 1965 *Nature* **206** 703–704
[18] Nagashima K, Ueno H, Fujimoto K, Fujii Z and Kondo I 1972 *Rep. Ionos. Space Res. Japan* **26** 1–30
[19] Nagashima K, Fujimoto K, Fujii Z, Ueno H and Kondo I 1972 *Rep. Ionos. Space Res. Japan* **26** 31–68

IOP Conf. Series: Journal of Physics: Conf. Series **1181** (2019) 012013 doi:10.1088/1742-6596/1181/1/012013

Observation of a time lag in solar modulation of cosmic rays in the heliosphere

Miguel Orcinha[1], **Nicola Tomassetti**[2], **Fernando Barão**[1], **Bruna Bertucci**[2]

[1] Laboratório de Instrumentação e Física Experimental de Partículas, P-1000 Lisboa, Portugal
[2] Università degli Studi di Perugia & INFN-Perugia, I-06100 Perugia, Italy;

E-mail: miguel.orcinha@cern.ch

Abstract. The solar modulation effect of Galactic cosmic rays is a time-dependent phenomenon that is caused by the transport of these particles through the magnetized plasma of the heliosphere. Using a data-driven model of cosmic-ray transport in the heliosphere, in combination with a large collection of data, we report the evidence for a eight-month time lag between observations of solar activity and measurements of cosmic-ray fluxes in space. As we will discuss, this result enables us to forecast the cosmic ray flux at Earth well in advance by monitoring solar activity. We also compare our predictions with the new multi-channel measurements of cosmic rays operated by the AMS experiment in space.

1. Introduction

New-generation experiments of cosmic ray (CR) detection have reached an unmatched level of precision that is bringing transformative advances in astroparticle physics [9, 23]. Along with calculations of CR propagation in the galaxy, the interpretation of the data requires detailed modelling of the so-called *solar modulation effect*. Solar modulation is experienced by all CR particles that enter the heliosphere to reach our detectors near Earth. Inside the heliosphere, CRs travel through a turbulent magnetized plasma, the solar wind, which significantly reshapes their energy spectra. This effect is known to change with time, in connection with the quasi-periodical 11-year evolution of the solar activity, and to provoke different effects on CR particles and antiparticles [39].

Observationally, an inverse relationship between solar activity (often monitored by the number of solar sunspots) and the intensity of CRs at Earth is known for a long time. The effect of solar modulation in the low-energy CR spectra ($E \lesssim$ GeV) is measured by several experiments [13, 47]. Solar modulation is caused by a combination of basic particle transport processes such as diffusion, convection, adiabatic cooling, or drift motion, yet the underlying physical mechanisms and their associated parameters remain under active investigation.

Solar modulation models are dependent on two crucial factors: (i) precise knowledge of the interstellar spectra (LIS) of CRs outside the heliosphere; (ii) availability of time-series of CR data on different species. Recent accomplishments from strategic space missions have enabled us to make significant progress in this field. The entrance of Voyager-1 in the interstellar space provided us with the very first LIS data on CR protons and electrons [42, 18]. Long-duration space experiments PAMELA (on orbit since 2006) and AMS (since 2011) have been releasing a continuous stream of monthly-resolved data on CR particles and antiparticles [4, 3, 6, 7].

These measurements add to a large wealth of low-energy CR data collected in the last decades by space missions CRIS/ACE [48] IMP-7/8 [21], *Ulysses* [24], and more recently EPHIN/SOHO [27], as well as from ground data provided continuously by the neutron monitor (NM) worldwide network [33, 41].

In this proceeding, we present the work done on [35] which presents new calculations of CR fluxes near-Earth that account for the dynamics of CR modulation in the expanding heliosphere and expand on the understanding of the correlation between solar parameters and the CR fluxes. Using a large collection of modulated and interstellar CR data collected in space, we have constructed a predictive and measurement-validated model of solar modulation which depends only on direct solar-activity observables: the sunspot number (SSN) and the tilt angle of the heliospheric current sheet (HCS).

2. Methodology

The transport of CRs in heliosphere is described by the Parker equation for the omni-directional phase space density $\psi(t, p, \boldsymbol{r})$ expressed as function of time t, momentum p, and position $\boldsymbol{r}$ [38]:

$$\frac{\partial \psi}{\partial t} = -(\boldsymbol{V} + \boldsymbol{v_d}) \cdot \nabla\psi + \nabla \cdot (\boldsymbol{K} \cdot \nabla\psi) + \frac{1}{3}(\nabla \cdot \boldsymbol{V})\frac{\partial \psi}{\partial \ln p} \quad (1)$$

The various terms represent convection with a solar wind of speed $|\boldsymbol{V}| \cong 400\,\mathrm{km/s}$, drift motion with average speed $\boldsymbol{v_d}$, spatial diffusion with tensor $\boldsymbol{K}$, and adiabatic momentum losses.

We simulated this physical process in a minimal 2D description using $\boldsymbol{r} = (r, \theta)$, radius and heliolatitude [14]. In terms of diffusion coefficient, the parallel component was defined as $K_\parallel = \kappa^0 \frac{10^{22}\beta p/\mathrm{GeV}}{3B/B_0}$, in units of cm^2/s, where we have factorized an adimensional scaling factor, κ^0, of the order of unity. The perpendicular component is $K_\perp \cong 0.02\,K_\parallel$ [22].

The regular solar magnetic field (HMF) is modelled using the usual Parker structure, $B = \frac{B_0 A}{r^2}\sqrt{1+\Gamma}$ with $\Gamma = (\Omega r/V)\sin\theta$, where $\Omega = 2.866 \cdot 10^{-6}\,\mathrm{rad\,s^{-1}}$ is the angular rotation of the Sun, $B_0 \cong 3.4\,\mathrm{nT\,AU^2}$ is the field intensity at $r_0 = 1\,\mathrm{AU}$, and $A = \pm 1$ represents the magnetic polarity cycle of the Sun. The polarity is positive (negative) when the HMF points outward (inward) in the north hemisphere.

In this model we account for gradient and curvature drift effects. In particular, drift is important across the wavy layer of the HCS, *i.e.* the surface where magnetic polarity changes from north to south, the angular extension (and waviness) of which is described by the tilt angle α (see Figure 1). The drift velocity components v_r and v_θ are both proportional to $qA\frac{2\beta r p}{3B_0}$, so that the sign of $\boldsymbol{v_d}$ depends on the product qA [39].

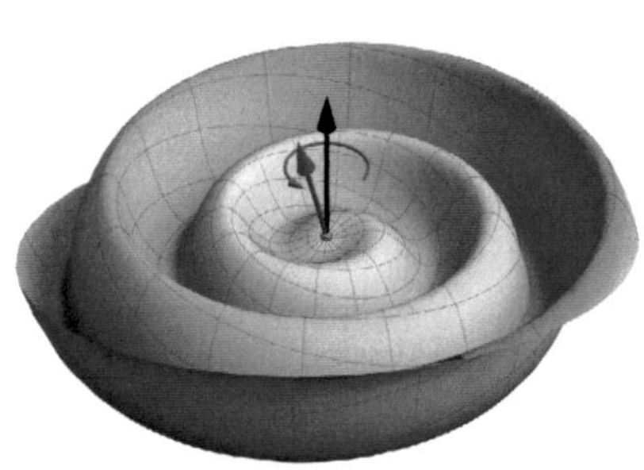

Figure 1. Graphical representation of the heliospheric current sheet in a Parker solar wind model. The black arrow represents the axis of rotation and the red one the magnetic dipole direction. Tilt angle is then defined as the angle between these two vectors.

With this setting, we compute the CR propagation from the termination shock (TS) to Earth's orbit using the stochastic differential equation approach of [26]. This method consists of a backwards-in-time propagation of a large number of pseudo-particles from Earth to the boundaries [40, 43, 8]. For a given particle type, steady-state solutions of Eq. 1 ($\partial\psi/\partial t = 0$) are then obtained by sampling. In our model we disregard re-acceleration effects occurring at the TS or modulation in the heliosheath [28, 29].

The LIS fluxes J^{IS} were calculated using an improved model of CR acceleration and propagation [45, 46, 19], being well constrained by Voyager-1 and AMS data [18, 5]. Positron LIS relies on secondary production calculations so it has larger uncertainties [19].

The role of positive and negative particles interchanges with polarity as can be seen in Figure 4. This phenomenon is due to sign-dependence of the the drift motion of particles. Depending on solar polarity CR particles and antiparticles sample different parts of the heliosphere. Along with qA, the interplay of the various physics processes depends on the levels of HCS waviness (see Figure 1) and HMF irregularities, that are here parametrized by α and κ^0. Tilt-angle measurements $\hat{\alpha}$ are provided on 10-day basis from the *Wilcox Solar Observatory*[1] using two reconstruction procedures: the "classic" model-L, and the improved model-R (both shown in Fig. 2). The latter was used for this work [25, 20]. A basic diagnostic for the HCS turbulence level is the manifestation of solar sunspots [37], so that SSNs and κ^0 are related each other [15, 14]. We used the the monthly-series of SSNs provided by the *Royal Observatory of Belgium*[2]. Both α and SSN were smoothly interpolated in order to create the functions $\hat{\alpha}$ and $\hat{S}$ For diffusion we adopted a simple two-coefficient relation $\kappa^0 \equiv a + b\log(\hat{S})$.

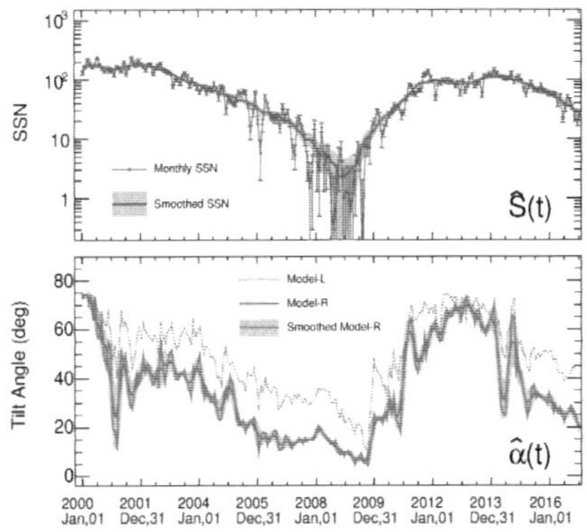

Figure 2. Reconstruction of the sunspot number ($\hat{S}(t)$) and tilt-angle of the heliospheric current sheet ($\hat{\alpha}(t)$) as function of time. Figure taken from [35].

The temporal behaviour of these quantities, shown in Fig. 2 from 2000 to 2017, is at the basis of the time-dependent nature of solar modulation. In practice, the problem is modelled under a *quasi-steady* fashion, *i.e.*, by providing a time-series of steady-state solutions corresponding to a time-series of input parameters. Due to the nature of the phenomenon, a finite amount of time is needed, in fact, for the properties observed in the solar corona to be transported in the outer heliosphere by the plasma. This motivated us to introduce a parameter ΔT in our calculation, describing a time-lag between the solar-activity indices of Fig. 2 and the medium properties of the modulation region, *i.e.*, the spatial region effectively sampled by CRs. Evidence for a lag of $\sim$6-12 months have been previously reported in NM-based studies [32, 12, 30, 36, 34, 16].

Our model is then specified by three free parameters only, a, b, and ΔT, that we constrain using a large amount of data. We use monthly-resolved proton data from the PAMELA experiment [4] collected between July 2006 and January 2010, and data from the EPHIN/SOHO space detector [27], yearly-resolved between 2000 and 2016. We also include data from the BESS Polar-I (Polar-II) mission from 13 to 21 December 2004 (from 23 December 2007 to 16 January 2008) [2, 1]. These measurements are given in terms of time-series of energy spectra $\hat{J}_{j,k} = \hat{J}(t_j, E_k)$, where each spectrum is a *snapshot* of the CR flux near-Earth at epoch t_j. Calculations $J(t_j, E_k)$ are performed using retarded functions of the physics inputs $\alpha_j = \hat{\alpha}(t_j - \Delta T)$ and $\kappa^0_j = a + b\log(\hat{S}(t_j - \Delta T))$. We then build a global χ^2-estimator:

$$\chi^2(a, b, \Delta T) = \sum_{j,k} \left[\frac{J(t_j, E_k; a, b, \Delta T) - \hat{J}_{j,k}}{\sigma_{j,k}} \right]^2 \tag{2}$$

The quantity $\sigma_{j,k}$ includes experimental errors in the data and model uncertainties due to finite statistics of the pseudo-particles simulation. The following sources of systematic

[1] http://wso.stanford.edu

[2] http://www.sidc.be

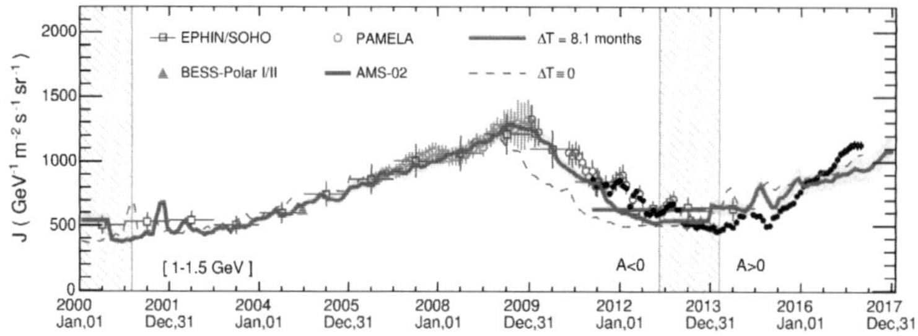

Figure 3. Time profile of the proton flux at $E = 1 - 1.5\,\text{GeV}$. Best-fit calculations are shown as thick solid line, along with the uncertainty band, in comparison with the data [4, 27, 2, 1, 5, 6, 31]. The most recent AMS data[6] is shown in black and PAMELA data[31] in dark blue. The shaded bars indicate the magnetic reversals of the Sun's polarity [44]. Figure taken from [35].

uncertainties are also accounted: (i) uncertainties in the LIS, from the constraints provided by Voyager-1 and AMS data; (ii) uncertainties on $\hat{\alpha}(t)$, from the discrepancy between L and R models and from the smoothing procedure; (iii) uncertainties on $\hat{S}(t)$ from the smoothed SSN variance, see Fig. 2. The free parameters are estimated by means of standard minimization techniques.

3. Results

The global fit has been performed to 3993 proton data points collected between 2000 and 2012 (in $A < 0$ conditions) at kinetic energy between 0.08 and 50 GeV. The best-fit parameters are $\hat{a} = 3.88 \pm 0.87$, $\hat{b} = \text{-}1.30 \pm 0.29$, and $\widehat{\Delta T} = 8.1 \pm 0.9$ months, giving $\chi^2/df = 2651/3990$. The fit was repeated after fixing $\Delta T \equiv 0$, *i.e.*, under a more conventional "unretarded" scenario. returning $\hat{a} = 3.36 \pm 0.76$, $\hat{b} = \text{-}1.08 \pm 0.24$, and $\chi^2/df = 4979/3991$. Results are shown in Fig. 3, illustrating the time and energy dependence of calculations in comparison with the data. The fits give satisfactory results at all energies and epochs. From Fig. 3, it can be seen that the model reproduces very well the time evolution of the proton flux, at $E = 1.5\,\text{GeV}$ and after the 2013 polarity reversal, the proton flux is predicted to increase with time.

It is also clear from the figure that the retarded scenario (with $\Delta T \equiv 8.1$ months, thick red line) allows for a much better description of the time evolution of the proton flux. Additionally, for comparison, recent AMS and PAMELA data were overlaid as well even though they are not used for the fit. The prediction shows great agreement with this recent data but some fine tuning is required. Due to the change of particle drift motion during and after a magnetic reversal, we expect that a new ΔT has to be determined for each epoch.

Additionally, we have also performed a time-series of fits to single energy spectra $\hat{J}_j$ by directly using the diffusion scaling as free parameter. This provided time-series of 62 $\hat{\kappa}^0_j$-values corresponding to various epochs t_j, which were done after fixing $\Delta T \equiv \widehat{\Delta T} = 8.1$ months. Inspecting the relation between $\hat{\kappa}^0_j$ and the "delayed" SSN revealed a correlation coefficient of $\rho_{\Delta T} = -0.89$ against the $\rho_0 = -0.66$ for the scenario of $\Delta T = 0$. Figures for both $\Delta T = 8.1$ months and $\Delta T = 0$ scenarios can be found in [35].

These findings explain why other authors, when proposing simple relations between κ^0 and SSN, had to adopt different coefficients for descending and ascending phases [14, 15]. Remarkably, this problem is naturally resolved in our model: once the time-lag is properly accounted, the κ^0-SSN relations can be described by a unique function.

Finally, our model was used to predict the time evolution of antimatter-to-matter ratios such as the e^+/e^-. Our calculations are shown in Fig. 4 for ΔT=8.1 months. The calculations were also done for ΔT=0 but were not shown in this figure. Measurements of the relative variation of the ratios are shown as reported by AMS [7] and PAMELA [3]. For the e^+/e^- ratio, we note that

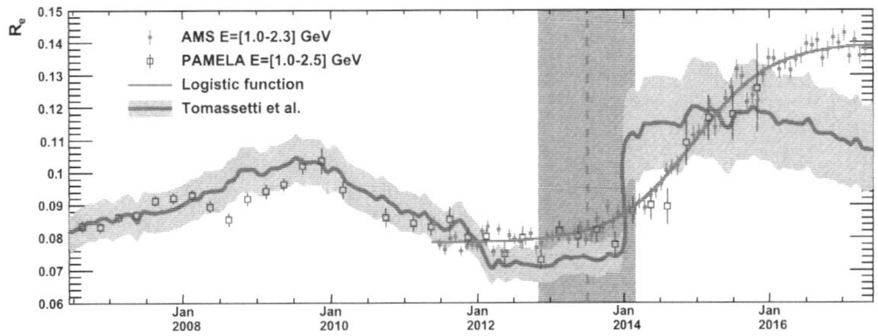

Figure 4. Time profile of the ratios R_e e^+/e^- at $E = 1 - 2.3$ GeV for AMS data[7] and at $E = 1 - 2.5$ GeV for PAMELA data. Model predictions and their corresponding uncertainties, are shown in comparisons with the data [3, 17]. The shaded bars indicate the magnetic reversals of the Sun's polarity [44]. Figure taken from the supplementary material of [7].

calculations within $\Delta T = 8.1$ month are favoured, although the data does not permit a resolute discrimination. Across the magnetic reversal, shown in the figure as shaded bars, a remarkable increase of the e^+/e^- ratio is predicted. It should be noted, however, that the dynamics of the transition could not be modelled during reversal, because the HMF polarity is not well defined. Nonetheless, a rise in the e^+/e^- ratio profile has been detected in both AMS [7] and PAMELA data [3], and this rise is found to occur a few months after completion of the Sun's polarity reversal.

4. Conclusions

Using new high-statistics measurements of CRs in space have determined the evolution of CR fluxes near Earth with unmatched time resolution. These data allow us to perform detailed studies of the solar modulation effect and its dynamical connection with the evolving solar activity. In this proceeding, we have reported new calculations of CR modulation based on a simple but physically consistent numerical model that accounts for particle diffusion, drift, convection and adiabatic cooling. We have adopted a simple formulation where the time-dependent physics inputs of the model consist only in SSN and HCS tilt angle. We have shown that this model reproduces well the time evolution of the Galactic proton spectra measured by AMS, PAMELA, EPHIN/SOHO, and BESS experiments. Our model is highly predictive once the correspondence between modulation parameters and solar-activity indices is established. Our study revealed an interesting aspect of the dynamics of CR modulation in the expanding wind, that is, the presence of time-lag ΔT between solar data and the condition of the heliosphere.

Using a large ensemble of CR proton data we found $\Delta T = 8.1 \pm 0.9$ month, which is in agreement with basic expectations and with recent NM based analysis [11, 34, 16]. An interesting consequence of this result is that the galactic CR flux at Earth can be predicted, at any epoch t, using solar-activity indices observed at the time $t - \Delta T$. This result is of great interest for real-time space weather forecast, which is an important concern for human space-flight.

In our results, the parameter ΔT has been determined using CR protons during negative polarity. This parameter has to be viewed as an effective quantity representing the average of several CR trajectories in the heliosphere during $A < 0$ conditions. Further elaborations may include the use of NM data, for larger observation periods, or the accounting for a latitudinal dependence in the wind profile or in the diffusion coefficient. Since CR particles and antiparticles sample different regions of the heliosphere, we expect slightly different time-lags, $\Delta T_{\pm}$ depending on the sign of qA (and in particular, $\Delta T_{-} \lesssim \Delta T_{+}$). With the precision of the existing data, we were unable to test this hypothesis. A detailed re-analysis of our model, in this direction, will be possible after the release of monthly-resolved data from AMS on CR particle and antiparticle fluxes.

Acknowledgments
We thank our colleagues of the AMS Collaboration for valuable discussions. MO acknowledges support from FCT - Portugal, *Fundação para a Ciência e a Tecnologia*, under grant SFRH/BD/104462/2014. NT and BB acknowledge the European Commission for support under the H2020-MSCA-IF-2015 action, grant No.707543 MAtISSE – *Multichannel investigation of solar modulation effects in Galactic cosmic rays.*

References
[1] Abe, K., *et al* , *Phys. Rev. Lett.* 108, 051102 (2012)
[2] Abe, K., *et al* , PLB 670, 103 (2008)
[3] Adriani, O., *et al* , *Phys. Rev. Lett.* 116, 241105 (2016)
[4] Adriani, O., *et al* , ApJ 765, 91 (2013)
[5] Aguilar, M., *et al* (a), *Phys. Rev. Lett.* 114, 171103 (2015)
[6] Aguilar, M., *et al* (c), *Phys. Rev. Lett.* 121, 051101 (2018)
[7] Aguilar, M., *et al* (d), *Phys. Rev. Lett.* 121, 051102 (2018)
[8] Alanko-Huotari, K., Usoskin, I. G., Mursula, K., Kovaltso, G. A., J. Geophys. Res. 112, A08101 (2007)
[9] Amato, E., & Blasi, P., ASR, in press (2017) [arXiv:1704.05696]
[10] Aslam, O. P. M., & Badruddin, Sol. Phys. 279, 269-288 (2012)
[11] Aslam, O.P.M. & Badruddin, Sol. Phys. 290, 2333-2353 (2015)
[12] Badruddin, Singh, M., Singh, Y. P., A&A 466, 697-704 (2007)
[13] Bindi, V., *et al* , ASR 60, 865-878 (2017)
[14] Bobik, P., *et al* , ApJ 745, 132 (2012)
[15] Boschini, M. J., *et al* , ASR (2017)
[16] Chowdhury, P., Kudela, K., Moon, Y.-J., Sol. Phys. 291, 581-602 (2016)
[17] Clem, J., & Evenson, P., J. Geophys. Res. 114, A10 (2009)
[18] Cummings, A. C., *et al* , ApJ 831, 18 (2016)
[19] Feng, J., Tomassetti, N., Oliva, A., PRD 94, 123007 (2016)
[20] Ferreira, S. E. S., & Potgieter, M. S., ApJ 603, 744 (2004)
[21] Garcia-Munoz, M., *et al* , ApJ SS 64, 269-304 (1987)
[22] Giacalone, J., & Jokipii, J. R., ApJ 520, 204 (1999)
[23] Grenier, I. A., Black, J. A., Strong, A. W., Annu. Rev. Astron. Astrophys., 53, 199–246 (2015)
[24] Heber, B., *et al* , ApJ 699, 1956-1963 (2009)
[25] Hoeksema, J. T., Space Sci. Rev. 72, 137-148 (1995)
[26] Kappl, R., Comp. Phys. Comm. 207, 386-399 (2016)
[27] Kühl, P., Gómez-Herrero, R., Heber, B., Sol. Phys. 291, 965-974 (2016)
[28] Langner, U. W., Potgieter, M. S., Webber, W. R., J. Geophys. Res. 108, A10, 8039 (2003)
[29] Langner, U. W. & Potgieter, M. S., J. Geophys. Res. 109, A01103 (2004)
[30] Lantos, P., Sol. Phys. 229, 373 (2005)
[31] Martucci, M., *et al* , ApJL854, L2 (2018)
[32] Mavromichalaki, H. & Petropoulos, B., Astrophys. Space Sci. 106, 61-71 (1984)
[33] Mavromichalaki, H., *et al* , ASR 47 2210-2222 (2011)
[34] Mishra, V. K. & Mishra, A. P., Indian J. Phys. 90, 1333-1339 (2016)
[35] N. Tomassetti, M. Orcinha, F. Barao, B. Bertucci, ApJL 849, L32 (2017)
[36] Nymmik, R. A., ASR 26, 1875-1878 (2000)
[37] Plunian, F., Sarson, G. R., Stepanov, R., MNRAS 400, L47-L51 (2009)
[38] Potgieter, M. S., Living Rev. Sol. Phys., 10, 3 (2013)
[39] Potgieter, M. S., ASR 53, 1415–1425 (2014)
[40] Raath, J. L., Potgieter, M. S., Strauss, R. S., Kopp, A., ASR 57, 1965–1977 (2016)
[41] Steigies, C., Proc. 34th ICRC - The Hague, PoS 225 (2015)
[42] Stone, E. S., *et al* , Science 341, 6142, 150 (2013)
[43] Strauss, R. D., Potgieter, M. S., Büshing, I., Kopp, A., Astrophys. Space Sci. 339, 223 (2012)
[44] Sun, X., Hoeksema, J. T., Liu, Y., Zhao, J., ApJ 798, 114 (2015)
[45] Tomassetti, N., PRD 92, 081301 (2015)
[46] Tomassetti, N., & Donato, F., ApJ 803, L15 (2015)
[47] Valdés-Galicia, J. F., & González, L. X., ASR 57, 1294-1306 (2016)
[48] Wiedenbeck, M., *et al* , Proc. 31st ICRC - Lodz, 0545 (2009)

IOP Conf. Series: Journal of Physics: Conf. Series **1181** (2019) 012014 doi:10.1088/1742-6596/1181/1/012014

Analysis of trajectories of primary particles and muons detected at the Earth's surface with different polarity of the Sun

A A Kovylyaeva, I I Astapov, N S Barbashina, A N Dmitrieva, V V Shutenko and I I Yashin

National Research Nuclear University MEPhI
(Moscow Engineering Physics Institute), Moscow, 115409, Russia

E-mail: AAKovylyaeva@mephi.ru

Abstract. The flux of cosmic rays passing through the heliosphere changes its characteristics (energy and angular distribution) due to the influence of the interplanetary magnetic field. Because of this, the muon flux produced by primary cosmic rays in the Earth atmosphere also changes its characteristics. Muons registered at the surface of the Earth are generated mainly by primary protons and helium nuclei with energies from 10 GeV to TeV. The trajectories of these particles in the heliosphere are significantly influenced by the interplanetary magnetic field. The construction of backward trajectories of particles from the Earth surface (detector) into the heliosphere region allows us to estimate the relationship between the trajectories of muons and primary particles. The trajectories of muons and parent protons and helium nuclei of various energies for positive and negative polarity of the Sun in a quiet period and under the influence of disturbances in the interplanetary magnetic field are presented. It is shown that the change in the polarity of the Sun and the influence of disturbances in the interplanetary magnetic field lead to qualitative changes in the regions of the heliosphere through which primary particles pass.

1. Introduction

The trajectories of charged primary cosmic rays (PCR) in the heliosphere are determined by the state of the interplanetary magnetic field (IMF) and the polarity of the Sun. Perturbations in the solar wind emanating from the Sun modulate the IMF and affect the PCR flux. With the help of ground-based cosmic ray detectors (neutron and muon), the variations in the PCR flux are investigated. In order to study these variations, it is necessary to know through which areas of the heliosphere pass the PCR (mainly protons and helium nuclei) which generate atmospheric neutrons and muons. Unlike neutron detectors, muon detectors can fix the direction of the registered muons. Knowing the direction of the muon in the detector, it is possible to build backward trajectories to the generation zone and then the trajectory of the proton or helium nucleus through the Earth's magnetosphere into the heliosphere. In this paper, the behavior of the primary protons trajectories in a quiet condition in the interplanetary magnetic field and under the influence of disturbances in the interplanetary magnetic field at negative and positive polarity of the Sun is discussed.

2. Atmosphere

Propagation of charged cosmic ray particles through the atmosphere of the Earth is accompanied by their interaction with nuclei of atoms of the atmosphere and energy losses. At interactions of primary particles, rapidly decaying pions and kaons are formed, they generate the atmospheric muon flux. The probability of interaction depends on the density of the atmosphere and the length of the way passed by particles. Energy losses - ionization and radiation - also depend on density of the atmosphere and the length of the way. The density of the atmosphere in its any point in geodetic coordinates WGS84 for the number of the day relative to the beginning of the year and the time of the day is given with empirical model of the atmosphere NRLMSISE-00 [1].

The altitude at which the interaction occurs (15 - 20 km) depends on the trajectory of the primary particle. At construction of the backward trajectory, this altitude can be estimated by means of numerical integration of the density along a direction of backward motion of a muon up to the atmosphere boundary, that is to calculate atmospheric depth (see figure 1). If the depth is less than a preset value (for example, 100 $g \cdot cm^{-2}$) muon is replaced with a parent particle (proton or helium nucleus).

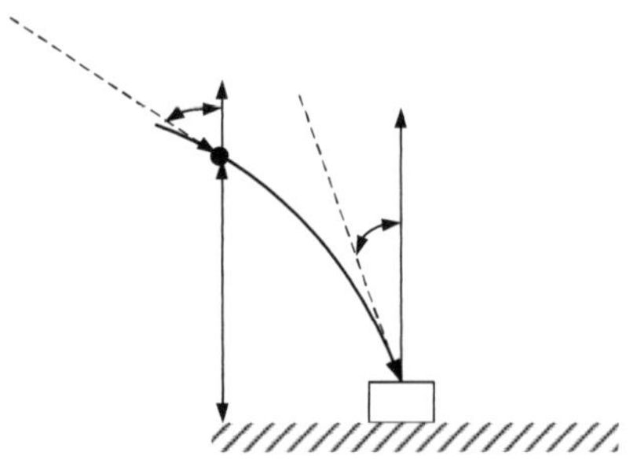

Figure 1. The altitude of the zone of muon generation is determined by atmospheric depth along the backward trajectory.

Ionization loss for muon, proton and helion, and also radiation loss of muon are taken into account according to known formulas [2]. The step along the trajectory Δr for muon must be chosen small enough, for example, $\Delta r = 10$ m or $\Delta r = (20\text{ m})/\cos\theta_0$, but not more than 150 m.

3. Earth magnetosphere

The magnetic field of the Earth has a complicated spatial distribution. For its description, various models are used. In our case, the necessary model should allow to get an estimation of a vector of a magnetic field in all magnetosphere regions for the preset time. Such opportunity provide the models of the magnetic field of the Earth by N.A.Tsyganenko TS05 (distant field) and GEOPACK-2008 (near-field) [3].

For the beginning, it is necessary to prepare input parameters describing a solar wind and geomagnetic conditions for the preset time. For this purpose, it is possible to take advantage of the data [4] or to calculate them with the use of database OMNI [5] by the algorithm described there. After preparation of input parameters, initialization with the use of subroutine RECALC_08 is carried out.

As the trajectory of particles can pass through regions with a big gradient of the magnetic field (cusp, magnetic anomalies) it is necessary to use a step along trajectory $\Delta r = 200$ м. At construction of the backward trajectory, it is necessary to check overrunning by the particle of the magnetosphere boundary. It is done with the help of subroutines GEOGSW_08 and SHUETAL_MGNP_08. After overrunning of the magnetosphere boundary, the particle appears in the interplanetary magnetic field.

4. Interplanetary magnetic field

Unlike the characteristics of the atmosphere and magnetosphere which can be approached in part to real, distribution of the magnetic field in the interplanetary space is not known. Only in the locations of space vehicles which are equipped with the corresponding equipment it is possible to get a real vector of the magnetic field. In the given work, the model of the interplanetary magnetic field considered in [6] was used. The following equations and parameters were used to calculate the magnetic field beyond the Earth's magnetosphere:

$$\begin{aligned}
&B_r(r,\theta,\varphi) = A\cdot S(\vec r)\cdot B_0\left(\frac{r_0}{r}\right)^2,\\
&B_\theta(r,\theta,\varphi) = 0;\\
&B_\varphi(r,\theta,\varphi) = -A\cdot S(\vec r)\cdot B_0\frac{\Omega}{V_{\mathrm{SW}}}(r-b)\left(\frac{r_0}{r}\right)^2\sin\theta,\\
&\cos\theta_{\mathrm{HCS}} = -\sin\alpha_{\mathrm{HCS}}\cdot\sin\left(\varphi+\varphi_{\mathrm{HCS}}+\frac{\Omega}{V_{\mathrm{SW}}}r+\varphi_{0\mathrm{HCS}}\right),\\
&\varphi_{0\mathrm{HCS}} = -\left(\varphi_{\mathrm{Earth}}+\frac{\Omega}{V_{\mathrm{SW}}}r_{\mathrm{Earth}}\right),\\
&S(\vec r)=\begin{cases}1, \text{ if } \cos\theta>\cos\theta_{\mathrm{HCS}},\\ -1, \text{ if } \cos\theta\le\cos\theta_{\mathrm{HCS}},\end{cases}
\end{aligned} \tag{1}$$

where $\vec r$, r, θ and φ are radius-vector and spherical coordinates in the HEEQ (Heliospheric Earth Equatorial) system [7]; B_r, B_θ and B_φ are magnetic field projections; $r_0 = 1$ a.u.; b = 0.005 a.u.; B_0 =5 nT is the magnetic field at distance r_0; $\Omega = 2.865\cdot10^{-6}$ radian·s^{-1} is the Sun's equatorial angular velocity; $V_{\mathrm{SW}} = 400$ km·s^{-1} is the solar wind speed; $A = \pm 1$ is the Sun polarity and $S(\vec r) = \pm 1$. If the point with coordinates (r, θ, φ) lies above the current sheet, $S(\vec r) = +1$; otherwise $S(\vec r) = -1$. Further, φ_{HCS} is the phase shift at the current sheet; $\varphi_{0\mathrm{HCS}}$ is the initial phase of the current sheet; r_{Earth} and φ_{Earth} give the Earth's position; θ_{HCS} is an operational variable; α_{HCS} is the slope of the current sheet.

5. Motion in magnetic field

Motion of the relativistic charged particles in the magnetic field is described by the formula:

$$m\frac{\mathrm{d}}{\mathrm{d}t}\gamma\vec v = q\left(\vec v\times\vec B\right),\quad \gamma=\frac{E}{E_0}+1,\quad \beta=\sqrt{1-\frac{1}{\gamma^2}}. \tag{2}$$

Here: m, q are mass and charge of the particle; γ is Lorentz factor; $\vec v$ is unit vector of velocity; E, E_0 are kinetic and rest energies; $\vec B$ is the vector of a magnetic field; t is the time.

Let us notice that movement in the magnetic field of particles of different types will be similar at equal values of $q/(m\gamma)$. Therefore, the motion of proton (rest energy E_{0p}) with kinetic energy E_p will be similar to motion of ${}^4_2$He (α-particle with rest energy $E_{0\alpha}$) with kinetic energy E_α:

$$E_\alpha = 2E_p + 2E_{0p} - E_{0\alpha} \approx 2E_p - 1.85\ [\mathrm{GeV}]. \tag{3}$$

In [8], detailed description of the construction of the backward trajectory and examples of trajectories of protons with kinetic energy 20 GeV for positive muon with kinetic energy 3 GeV registered in a vertical direction at three points on the surface of the Earth (Apatity, Khabarovsk, Moscow) and examples [9] of the trajectories (in HEEQ) with a sloped current sheet ($\alpha_{\mathrm{HCS}} = 10°$) are given.

IOP Conf. Series: Journal of Physics: Conf. Series **1181** (2019) 012014 doi:10.1088/1742-6596/1181/1/012014

6. Behavior of primary protons at different polarities of the Sun and under the influence of the disturbance in the interplanetary magnetic field

Below, the behavior of the trajectories of primary protons with energy 20 GeV in a quiet condition in the interplanetary magnetic field and when the Earth is in a closed region of anomalous vertical component of the interplanetary magnetic field *Bz* (the influence of the disturbance in the interplanetary magnetic field) on June 22 at 4 hours and 16 hours (when the asymptotic directions look towards and from the Sun, respectively) at negative (A < 0) and positive (A > 0) polarity of Sun are discussed. The region of anomalous *Bz* is at distance $R \approx 0.85$ - 0.9 and $R \approx 1.05$ - 1.1 a.u. from the Sun. At the chosen date, inclination and displacement of the current sheet, the Earth is located above the heliospheric current sheet. Figure 2 shows trajectories of primary protons in quiet condition of the interplanetary magnetic field on June 22 at 4 hours at negative (A < 0) (figure 2, a) and positive (A > 0) (figure 2, b) polarity of the Sun.

Figure 3 shows that on June 22 at 4 hours the region of anomalous *Bz* (the circle in the figure) is at distance $R \approx 0.85$ - 0.9 a.u. from the Sun and reflects the trajectory of primary protons for events: a) A < 0, $Bz = -30$ nT; b) A > 0, $Bz = -30$ nT; c) A < 0, $Bz = +30$ nT; d) A > 0, $Bz = +30$ nT.

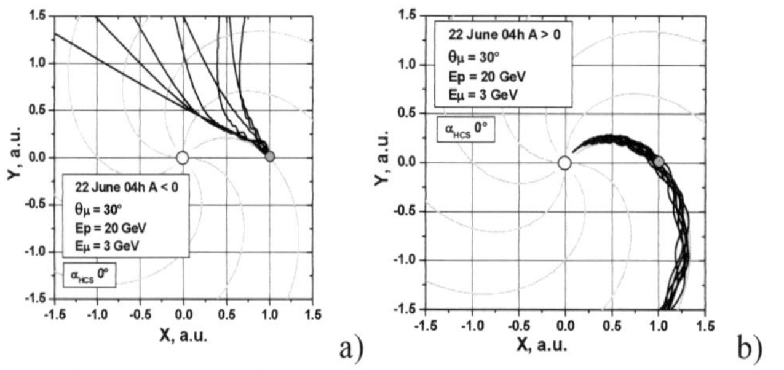

Figure 2. Trajectories of primary protons in quiet condition of the interplanetary magnetic field at negative (a) and positive (b) polarity of the Sun.

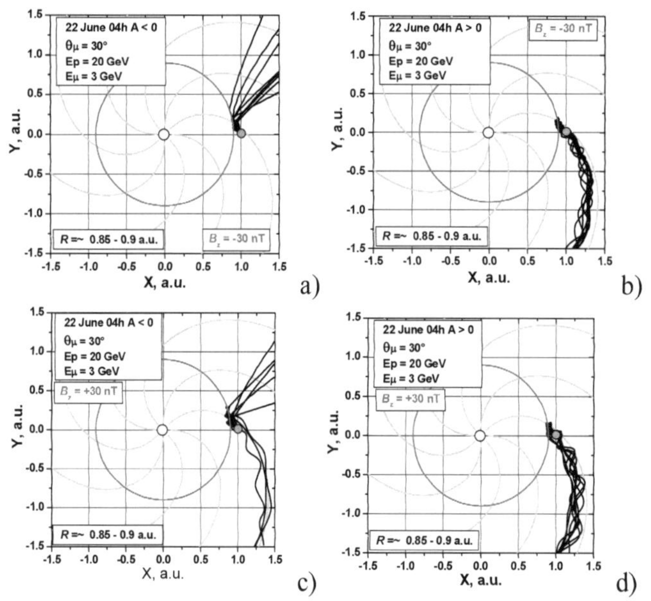

Figure 3. Trajectories of primary protons in the region of anomalous of *Bz* on June 22 at 4 hours at negative (A < 0) and positive (A > 0) polarity of the Sun at distance $R \approx 0.85$ - 0.9 a.u. from the Sun.

IOP Conf. Series: Journal of Physics: Conf. Series **1181** (2019) 012014 doi:10.1088/1742-6596/1181/1/012014

If the Earth is inside the closed region of anomalous *Bz* ($R \approx 1.05$ - 1.1 a.u.), this region limits access of primary protons from outside. Partially trajectories of primary protons come outside of the region of the anomalous *Bz* (at negative polarity) or completely do not come from outside (at positive polarity). The influence of the anomalous *Bz* region is stronger for events A > 0, *Bz* = -30 nT and A < 0, *Bz* = +30 nT, and weaker for other examples in figure 4 on June 22 at 4 hours and 16 hours: a, e) A > 0, *Bz* = -30 nT; b, f) A > 0, *Bz* = -30 nT; c, g) A < 0, *Bz* = +30 nT; d, h) A > 0, *Bz* = +30 nT.

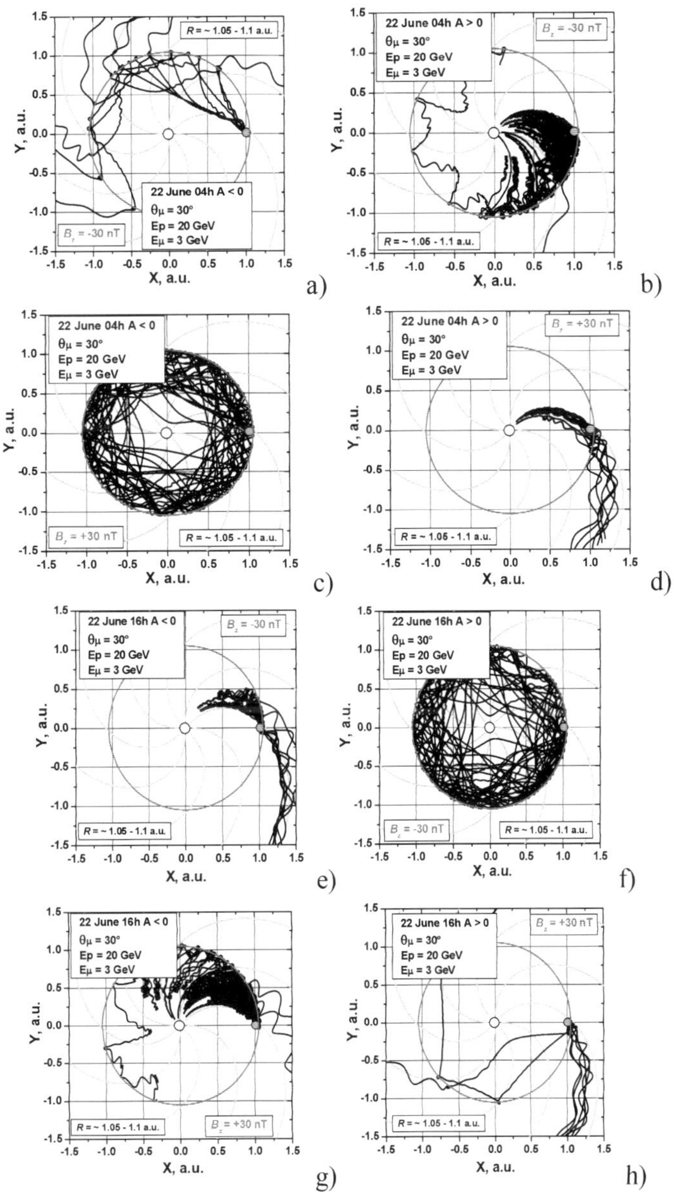

Figure 4. Trajectories of primary protons in the region of anomalous *Bz* on June 22 at 4 hours and 16 hours at negative (A < 0) and positive (A > 0) polarity of Sun at distance $R \approx 1.05$ - 1.1 a.u. from the Sun.

7. Conclusion

The described method of construction of backward-trajectories at different polarities of the Sun allows to study the behavior of trajectories of primary protons which give contribution to the muon flux registered at the Earth surface by the muon hodoscope URAGAN under the influence of disturbances in the interplanetary magnetic field. If the Earth is inside a closed region with an anomalous Bz, a decrease of the muon flux can be observed. The decrease of the muon flux depends on both the polarity of the Sun and the sign of Bz.

Acknowledgments

This work was performed at the Unique Scientific Facility "Experimental complex NEVOD" with the financial support of the MEPhI Academic Excellence Project (agreement with the Ministry of Science and Higher Education of the Russian Federation of August 27, 2013, project No. 02.a03.21.0005) and Russian Scientific Foundation (RSF) project No. 17-17-01215.

References

[1] http://www.brodo.de/space/nrlmsise/index.html
[2] Patrignani C *et al.* (Particle Data Group) 2016 *Chin. Phys.* C **40** 100001
[3] http://geo.phys.spbu.ru/~tsyganenko/modeling.html
[4] http://geo.phys.spbu.ru/~tsyganenko/TS05_data_and_stuff/
[5] ftp://spdf.gsfc.nasa.gov/pub/data/omni/
[6] Lipari P arXiv:1408.0431v1 [astro-ph.HE] 2 Aug 2014
[7] Hapgood M A 1992 *Planet. Space Sci.* **40** 711
[8] Shutenko V V 2017 *J. Phys.: Conf. Ser.* **798** 012046
[9] Shutenko V V, Astapov I I, Barbashina N S, Dmitrieva A N and Kovylyaeva A A 2017 *Bull. RAS: Phys.* **81** 194-6

27-day variation of galactic cosmic ray intensity by PAMELA experiment. Relationship with heliospheric parameters

R Modzelewska[1,2,3], A Mayorov[2], R Munini[3] on behalf of PAMELA collaboration

[1]Institute of Math. And Physics, Siedlce University, Siedlce, Poland
[2]National Research Nuclear University MEPhI, Moscow, Russia
[3]INFN, Sezione di Trieste, Trieste, Italy

email adress: renatam@uph.edu.pl

Abstract. We analyzed the dynamics of the 27-day variation of galactic cosmic rays (GCR) using PAMELA data in the solar cycle 24 covering the period of 2006-2015. We studied the time line of the proton and helium spectra for different rigidities measured by PAMELA and compared with neutron monitor (NM) observations. We studied the relationship between the 27-day variations of GCR and heliospheric parameters: solar wind velocity and heliospheric magnetic field (HMF). We found that the 27-day variation observed by PAMELA was changing over the solar cycle 24. Near the solar minimum in the 2007-2008 it was observed the stationary 27-day variation with stable amplitude and time of maximum for almost ~8 solar rotations. Solar wind velocity and Bx component of the HMF also showed the same clear 27-day periodicity. The recurrent changes of solar wind velocity were in strong antiphase with changes of GCR. In contrary in the rising phase of the solar cycle 24 the period of quasi-recurrent GCR variations was much more variable in time and in duration, in average 27-days, but also with the strong contamination of the higher harmonics. The antiphase relation between GCR and solar wind is not so clearly pronounced. Moreover, in the rising phase of the solar cycle 24 we established in all analyzed parameters some wave packages, periodically enhanced, taking place averagely each 1.5-2 years, suggesting the occurrence of the quasi biannual oscillations (QBOs).

1. Introduction

Flux of galactic cosmic ray (GCR) intensity contains various quasi-periodic changes with different time scales (from hours to several years), see e.g. [1,2] and references therein. On the background of the long term solar modulation: 22 years and 11 years, connected with the global heliospheric magnetic field (HMF) and the sun spot activity cycle, respectively, the short term modulation effects - sporadic (Forbush decreases) and recurrent (periodic) -also take place.

The 27-day GCR variation is connected with the heliolongitudinal asymmetry of heliospheric conditions during the solar rotation. The 27-day GCR variation is generally more typical and longer in duration during the minimum and near minimum epochs of solar activity. The amplitudes of the 27-day variations, and other GCR variations (e.g. Forbush decreases and solar diurnal anisotropies) in the range between few hours and few months vary, in general, in phase with the solar activity cycle, i.e. they reach the maximum values during solar activity maximum [3]. The 27-day variations of GCR are observed not only by ground based neutron monitors (NMs) [4], near Earth at L1 point by space missions (*SOHO, Stereo, ACE, WIND,* etc.) [5], in the inner heliosphere out of the ecliptic plane on *Ulysses* [6] and even on the *Voyager* spacecraft [7] in the outer heliosphere confirming them to be ubiquitous.

In spite of the progress has been made from the experimental (e.g. [8]) and theoretical (e.g. [9]) points of view in understanding the modulation processes governing the 27-day variation of the GCR intensity, it still remains much necessity to find new properties of the 27-day variation of GCR. Especially, up to now the relationship between the 27-day variation of GCR intensity and the heliospheric phenomena seems to be changeable over the solar cycle and is not completely understood requiring further investigation.

2. Data and methods

In this paper we study the dynamics of the 27-day variation of the GCR intensity in the period of 2006-2015 covering the main part of the solar cycle 24 and the minimum between the solar cycles 23 and 24 on the base of the PAMELA experiment [10]. PAMELA fills the largely unexplored energy gap between the particles detected in space (below few hundreds of MeV) and particles detected on the Earth (above few GeV). PAMELA data allow for the first time to study the time line of the 27-day variation of GCRs observed directly in space in a wide rigidity R range. Because of high statistics of PAMELA measurements and good enough accuracy it is possible to investigate the time- and rigidity-profile for different GCR particle species: proton and helium. Moreover it gives the possibility to discuss the relation of periodic GCR variations with heliospheric parameters through the solar cycle.
Here we use the daily proton and helium fluxes measured by PAMELA experiment:

- 17 rigidity bins for protons: 0.43 GV; 0.51 GV; 0.59 GV; 0.69 GV; 0.81 GV; 0.95 GV; 1.1 GV; 1.4 GV; 1.8 GV; 2.3 GV; 2.9 GV; 3.8 GV; 4.8 GV; 6.2 GV; 8.7 GV; 12.9 GV; 19.1 GV;
- 6 rigidity bins for helium: 0.89 GV, 1.2 GV, 1.7 GV, 2.4 GV, 3.4 GV, 4.7 GV

Since the main source of the 27-day GCR variation is the corotating interaction region (CIR), we also analyze the significant solar wind parameters: daily solar wind velocity V and the radial Bx component of the HMF. Figure 1 from top to bottom presents the daily proton (R=1.8 GV, Fig. 1a) and helium fluxes (R=1.7 GV, Fig. 1b) measured by PAMELA, Oulu NM count rate (geomagnetic rigidity cut-off Rc=0.8 GV, Fig. 1c), daily solar wind velocity V (Fig. 1d) and the radial Bx component of the HMF (Fig. 1e) for 2006-2015. As far the GCR intensity displays the long period trend connected with the solar activity, we normalize to the average of 29 days and exclude a long period trend by 29-days smoothing. Results are presented in Figure 2 for the PAMELA proton, helium fluxes and the Oulu NM, respectively. It allows to study the time line of the short period (27-days) GCR variation for 2006-2015.

For studying the periodic character of the data series we use the Cross Wavelet Transform (XWT) [11]. This method allows to study causal relationships in time-frequency space between two time series X and Y with corresponding continuous wavelet transform for each series: WXn (f) and WYn (f). Regions of high common power and consistent phase connection suggest causal relation between X and Y. The statistical significance of the Cross Wavelet Spectrum was estimated following [12] and [11].

IOP Conf. Series: Journal of Physics: Conf. Series **1181** (2019) 012015 doi:10.1088/1742-6596/1181/1/012015

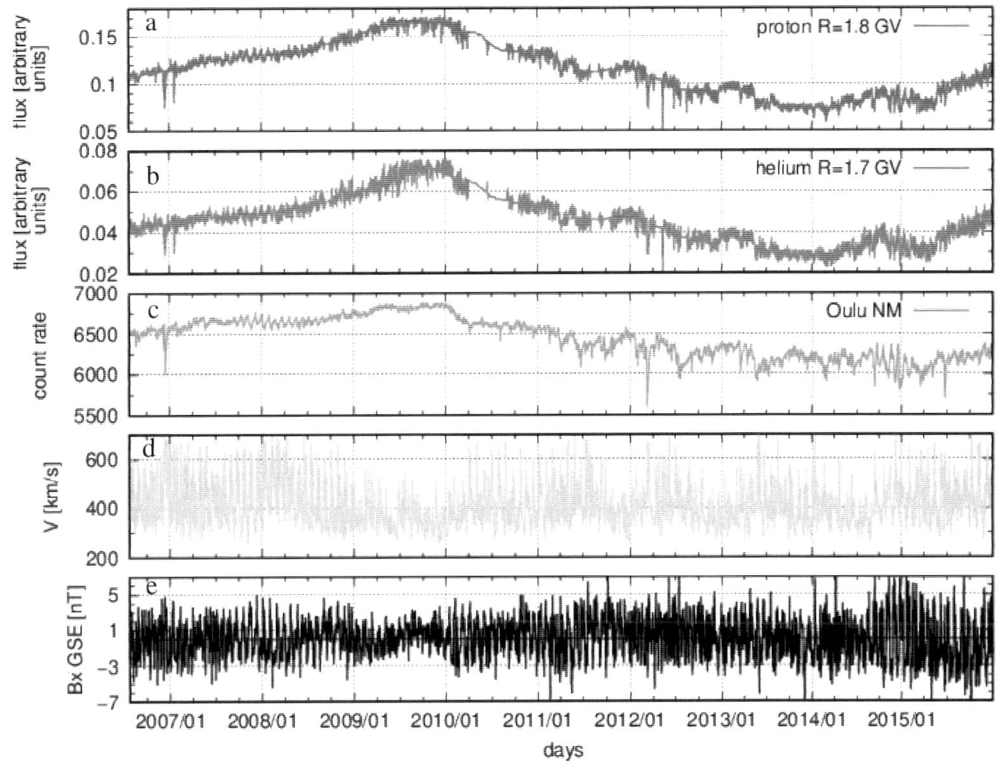

Figure 1. Daily PAMELA proton (R=1.8 GV), helium (R=1.7 GV) fluxes, Oulu NM (Rc=0.8 GV) GCR intensity count rates, solar wind velocity V and Bx component of the HMF for 2006-2015.

3. Analysis

First we crosscheck the normalized GCR intensity from Figure 2 measured by PAMELA with NMs data for 2006-2015. We use the NM observations with different geomagnetic cut-off rigidities R_c : Oulu (R_c=0.8 GV), Apatity (R_c=0.65 GV), Nain (R_c=0.3 GV), Moscow (R_c=2.43 GV), Rome (R_c=6.32 GV) and Potchefsroom (R_c=7 GV) NMs. In Figure 3ab we present the correlation coefficients vs rigidity R between NMs data and PAMELA each rigidity bin for 2006-2015 for proton and helium fluxes, respectively. Figure 3ab show that the correlation coefficients between PAMELA data and NMs observations are relatively high (~0.60±0.01) for protons and a bit lower but still high (~0.55±0.02) for helium for the rigidity interval from ~1GV up to ~5 GV. It seems that PAMELA data in this rigidity interval are in good correspondence to the GCR modulation effect registered by NMs. At the same time for rigidities below 1 GV the correlation coefficients for both, proton and helium, are relatively lower. It could suggest different contamination of solar cosmic rays in PAMELA data and in NMs. Time lines of the 27-day variations (Figure 2) observed by PAMELA proton (R=1.8 GV), helium (R=1.7 GV) fluxes and Oulu NM (Rc=0.8 GV) are similar in shape. However Figure 3ab show that the 27-day GCR intensity measured by PAMELA only for few rigidity bins between ~1GV and ~5 GV is comparable with the ground-based NMs data with effective rigidity ~15-20 GV responsible for solar modulation. Nevertheless it should be noted that we compare the differential GCR intensity as measured by PAMELA with an integrated GCR flux above the magnetic rigidity threshold typical for each NM.

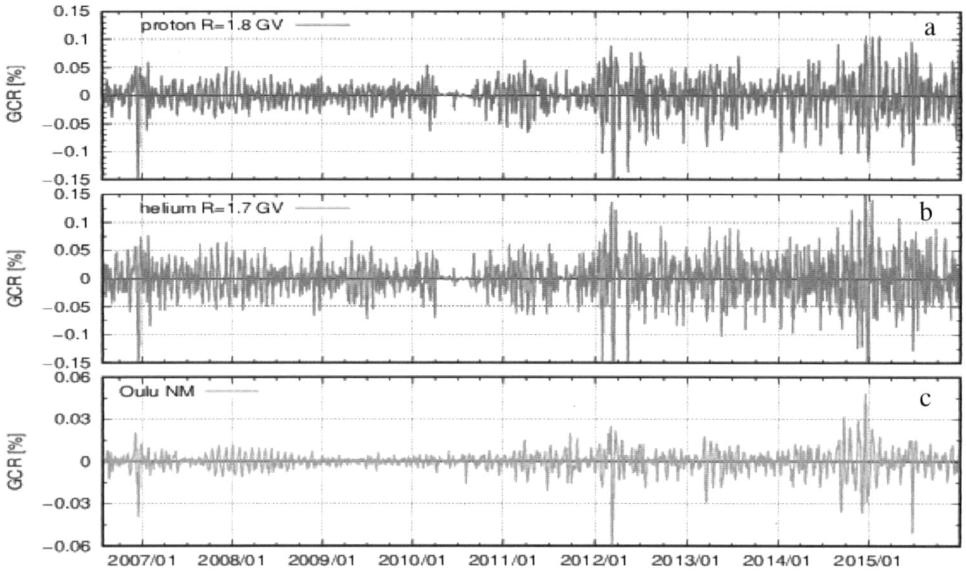

Figure 2. Daily PAMELA proton (R=1.8 GV), helium (R=1.7 GV) fluxes and Oulu NM (Rc=0.8 GV) GCR intensity normalized and detrended over 29 days smoothing for 2006-2015.

Next we analyze the relationship of quasi-periodic variations of GCR intensity and heliospheric parameters using the cross-wavelet time-frequency method (Grinsted et al., 2004). In Figures 4 we present the cross wavelet transformation XWT between (a) the PAMELA proton fluxes for R=1.8 GV and the Oulu NM data, (b) the PAMELA proton fluxes for R=1.8 GV and the PAMELA helium fluxes for R=1.7 GV, (c) the PAMELA proton fluxes for R=1.8 GV and solar wind velocity V and (d) the PAMELA proton fluxes for R=1.8 GV and Bx component of the HMF for 2006-2015. The line corresponds to the 95% confidence level of the wavelet spectra. The color-bar indicates the power of a period range (from blue-low power to red-high power). The arrows indicate the phase relationship between the data series in time-frequency space: (1) arrows pointing rightward mean the in-phase behavior of the two data series; (2) arrows pointing leftward indicate antiphase behavior; (3) arrows pointing downward indicate that the first dataset phase is shifted relative to the second one by 90 degrees. Figures 4 present the time-frequency variations of GCR intensity and solar wind parameters in the main part of the solar cycle 24 and prolonged solar minimum between solar cycles 23 and 24. Figures 4abc confirm the high positive correlation between PAMELA proton and helium fluxes and both with Oulu NM for the whole interval. In the time interval of 2006-2015 one can see two different modes of behaviors of the quasi-periodic variations, plotted in Figures 4. Mainly the first period 2007-2008 corresponds to the part of the prolonged solar minimum between solar cycles 23 and 24. During this period the stationary 27-day variation of GCR was observed with stable amplitude and phase for almost ~8 solar rotations. Heliospheric parameters: solar wind speed V and Bx component of the HMF also showed the same clear 27-day periodicity. The recurrent changes of solar wind velocity were in strong antiphase with changes of GCR. It is well known that this recurrent variation of solar wind speed are formed at CIR when a high speed solar wind stream overtakes and interacts with a slow speed stream, which was a case in 2007-2008. CIRs are especially prominent during the declining phase of the solar cycle and occur usually only at low latitudes, where the HMF has a well established sector structure and coronal holes spread to low latitudes (for a review see [13]).

The second period 2011-2015 represents the conditions of the rising phase of the solar cycle 24. Contrary to 2007-2008, in this period the quasi-recurrent GCR variations were much more variable in time and in duration, in average 27-days, but also with the strong contamination with the higher harmonics. The quasi-periodic GCR variations seem to be nonstationary [14]. The antiphase relation between GCR and solar wind V is not so clearly pronounced. Recently Gil and Mursula [15] compared the two 2007-2008 and 2014-2015 episodes of the 27-day variation using NMs data. They suggested that in both cases the source of periodic variations was a coronal hole, but the coronal holes in the two intervals were located quite differently: a strongly north-south asymmetric polar coronal hole existed in 2014-2015 and a transequatorial coronal hole governed the periodic variations during 2007-2008. Moreover, Figures 4 show that in the rising phase of the solar cycle 24 (2011-2015) one can observe in all parameters some wave packages with the periodic enhancements of the amplitude taking place averagely each 1.5-2 years. It suggests the existing of the knowing in literature the quasi biannual oscillations (QBOs) (for review see [2]). However this topic is out of the scope of this paper.

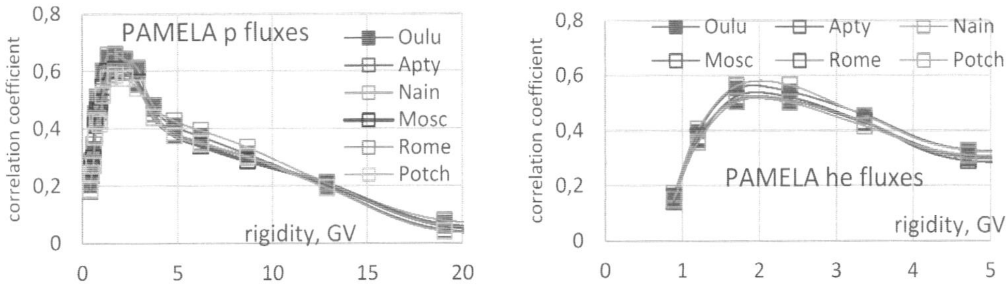

Figure 3ab. Correlation coefficient between the Oulu, Apatity, Nain, Moscow, Rome and Potchefsroom NMs data and the PAMELA each rigidity bin for proton and helium fluxes normalized and detrended over 29 days smoothing for 2006-2015.

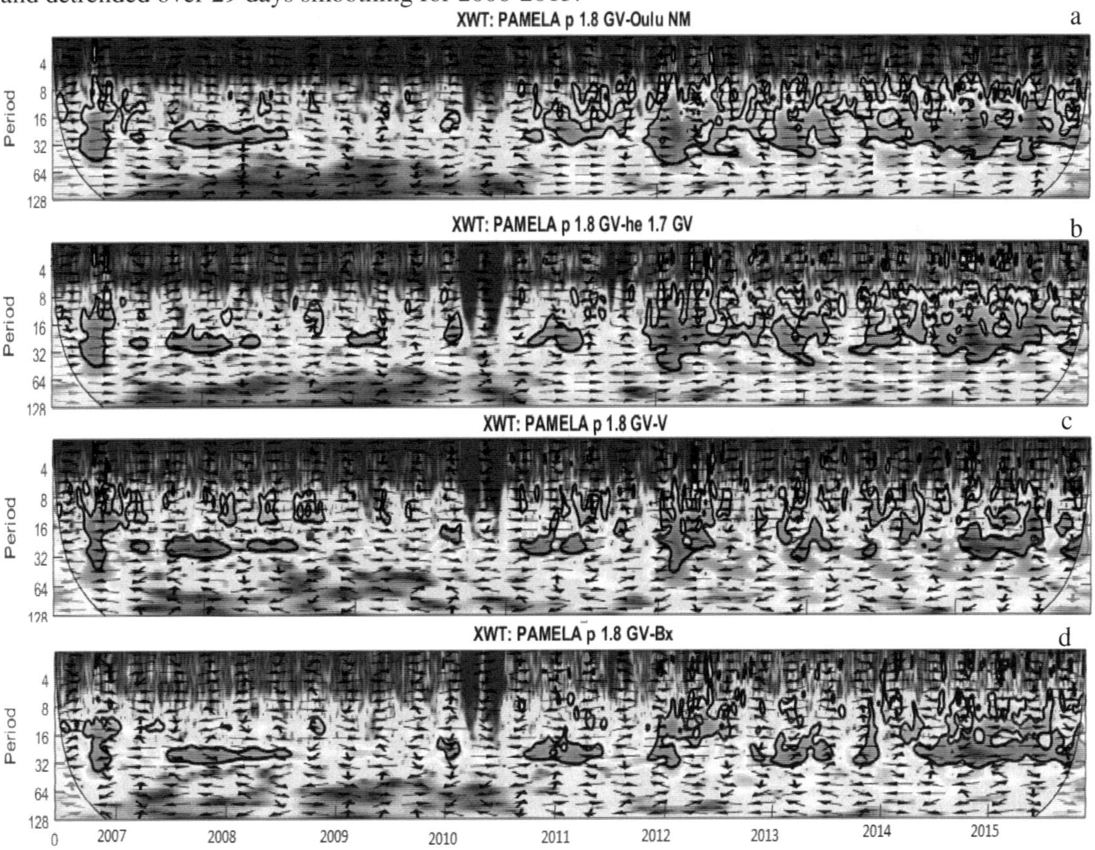

Figure 4. Cross wavelet transformation between (a) PAMELA proton spectra (R=1.8 GV) and Oulu NM data, (b) PAMELA proton spectra (R=1.8 GV) and helium spectra (R=1.7 GV), (c) PAMELA proton spectra (R=1.8 GV) and V and (d) PAMELA proton spectra (R=1.8 GV) and Bx for 2006-2015.

4. Discussion and conclusion

We studied the dynamics of the 27-day GCR variation observed by PAMELA over the main part of the solar cycle 24 and the solar minimum between solar cycles 23 and 24. We analyzed the relationship between the PAMELA proton and helium fluxes and the NMs data (normalized and detrended over 29 days smoothing) and found the high correlation for the rigidity interval ~1 GV-~5 GV. We showed that the time lines of the 27-day variations observed by PAMELA protons, helium and NMs are similar in shape. We studied the relationship between PAMELA data and solar wind velocity and heliospheric magnetic field over the main part of the solar cycle 24. We established that the 27-day variation observed by PAMELA was changing over the solar cycle 24. Near the solar minimum in the 2007-2008 the stationary 27-day variation was observed with stable amplitude and time of maximum for almost ~8 solar rotations. Solar wind velocity and Bx component of the HMF also showed the same clear 27-day periodicity. The recurrent changes of solar wind velocity were in strong antiphase with changes of GCR. In contrary, in the rising phase of the solar cycle 24 the period of quasi-cycling GCR variations was much more variable in time and in duration, in average 27-days, but also with the strong contamination with the higher harmonics. The antiphase relation between GCR and solar wind is not so clearly pronounced. Moreover, in the rising phase of the solar cycle 24, we found in all analyzed parameters some wave packages, with the periodic enhancements, taking place averagely each 1.5-2 years, suggesting the presence of the quasi biannual oscillations (QBOs).

Acknowledgments
We thank the PIs of Oulu, Apatity, Nain, Moscow, Rome and Potchefsroom NMs. Data of heliospheric parameters are from http://omniweb.gsfc.nasa.gov. Cross-Wavelet analysis tool is available at http://www.pol.ac.uk/home/research/waveletcoherence/ . R.M. acknowledges The Polish National Science Centre, decision number 2017/01/X/ST9/01023.

References

[1] Chowdhury P et al. 2016 A Study of Heliospheric Modulation and Periodicities of Galactic Cosmic Rays During Cycle 24 *Solar Physics* 291 pp 581-602
[2] Bazilevskaya G A 2014 A Combined Analysis of the Observational Aspects of the Quasi-biennial Oscillation in Solar Magnetic Activity *Sp. Sci. Rev.* 186 pp 359-386
[3] Bazilevskaya G A 2000 Observations of Variability in Cosmic Rays *Sp. Sci. Rev.* 94 pp 25–38
[4] Modzelewska R and Alania M V 2013 The 27-Day Cosmic Ray Intensity Variations During Solar Minimum 23/24 *Solar Physics* 286, pp 593–607
[5] Leske R et al. 2011 Observations of 27-Day Variations in Cosmic Ray Intensities During the Cycle 23/24 Solar Minimum *Procs of the 32nd ICRC2011* 11, pp 194
[6] McKibben R B et al. 1995 ULYSSES Out-of-Ecliptic Observations of ``27-day" Variations in High Energy Cosmic Ray Intensity *Sp. Sci. Rev.* 72 pp 403-408
[7] Decker J Y et al. 1999 Small-Scale Variations in ACR Intensities at Voyagers 1 and 2 in 1992-1998 *Procs of the 26th ICRC 1999*, 7 pp 512
[8] Gil A and Mursula K 2017 Hale cycle and long-term trend in variation of galactic cosmic rays related to solar rotation *Astronomy & Astrophysics* 599 pp A112
[9] Guo X and Florinsky V 2016 Galactic cosmic-ray intensity modulation by corotating interaction region stream interfaces at 1 AU. *Astrophys J* 826 13 pp
[10] Adriani O Barbarino G C Bazilevskaya G A, et al. 2014 *Phys. Rep.* 544 pp. 323-370
[11] Grinsted A Moore J C and Jevrejeva S 2004 *Nonlinear Processes in Geophysics* 11 pp 561-566
[12] Torrence C and Compo G P 1998 A Practical Guide to Wavelet Analysis *Bull Amer Meteor Soc* 79 pp 61-78
[13] Richardson I G 2018 Solar wind stream interaction regions throughout the heliosphere *Living Rev. Sol. Phys.* 15 95 pp
[14] Krainev M et al. 2017 *NFE* 8 pp 1-7
[15] Gil A and Mursula K 2018 Comparing two intervals of exceptionally strong solar rotation recurrence of galactic cosmic rays *J. Geophys Res.* (in press)

IOP Conf. Series: Journal of Physics: Conf. Series **1181** (2019) 012016 doi:10.1088/1742-6596/1181/1/012016

Passive Carrington Longitudes of Solar Cosmic Ray Sources in Solar Cycles 19–24

M V Podzolko

D. V. Skobeltsyn Institute of Nuclear Physics of M. V. Lomonosov Moscow State University (SINP MSU), Leninskie Gory, Moscow, 119991, Russian Federation

E-mail: spacerad@mail.ru

Abstract. Distribution of the fluences of solar proton events in solar cycles 19–24 over the Carrington longitude of their sources on the Sun has been considered. An interval of "passive" Carrington longitudes, extensive over the longitude (≈100–170°) and the lifetime (the whole period of observations) has been discovered, which very rarely produced the powerful solar proton events. It was earlier found in cycles 19–23 and now confirmed for the complete cycle 24 as well.

1. Introduction

In this study a distribution of the fluences of energetic protons of solar cosmic rays over the Carrington longitude of their sources on the Sun is considered.

A concept of the fluence (cumulative flux) of protons during the solar proton event (SPE) is standardly used in statistical models of solar cosmic rays. However, for the study of the longitudinal distribution of the solar activity typically the number of solar flares or number of sunspots or active regions are taken, which for the long time intervals demonstrate weak longitudinal dependency (see, for example, [1]).

We have carried out the analysis of the distribution of the fluences of energetic protons in SPEs over the Carrington longitude of their sources. A considerable inhomogeneity of this distribution has been found. In particular, an interval of "passive" Carrington longitudes, extensive over the longitude (≈100–170°) and the lifetime (the whole period of observations) was discovered, which very rarely produced the powerful solar proton events. It was earlier found in cycles 19–23 in the studies by I. V. Getselev et al. [2, 3], and now confirmed for cycle 24 as well.

2. Data

For this study a database on ≈450 solar proton events was used, which had been compiled from different sources and contained the information about the fluences of protons with energies >30 MeV during the SPEs and the heliolongitudes of the solar flares, during which these particles had been injected.

The data for solar cycles 19 and 20 were taken from different published catalogues [4–7]. For the 19th cycle the measurements were obtained using various methods (ground-based equipment, stratospheric balloons, geophysical rockets, first spacecraft); proton fluences were partially determined by indirect methods using the results of riometric measurements and registration of geomagnetic perturbations and peaks in solar radio emission. For the 20th cycle the data on proton fluences were obtained using the measurements onboard spacecraft.

For the period from 1970s till the current time we used the continuous sets of measurements of the interplanetary protons fluxes onboard IMP series spacecraft in the high-altitude near-Earth orbits and ACE spacecraft in the L1 libration point of the Sun-Earth system at a distance of ≈1.5 million km from the Earth, which are publically available [8]. Using these data the fluences of protons with energies >30 MeV in the SPEs were computed. During these computations the background fluxes of galactic cosmic ray protons were subtracted using the developed formalized method [9]. The links to the sources of particle injection on the Sun starting from 1976 were taken from the NOAA data base [10].

Using the data on the observed heliolongitudes of the solar proton event sources, the Carrington longitudes were computed for all SPEs in the data base using the constant Carrington rotation period.

3. Results of analysis

The results of the analysis of the distribution of the fluences of protons in SPEs over the Carrington longitude of their sources on the Sun are presented in Fig. 1.

In Fig. 1a the fluences of protons with energies >30 MeV for all considered SPEs during 19–24 solar cycles with the fluence higher than 10^7 cm^{-2} are shown related to the Carrington longitudes of their sources. In our study [3] it was shown, that during the period from 1956 till 2012 from a total of 60 most powerful SPEs with fluences of protons $\geq 10^8$ cm^{-2} only one SPE of November 4, 2001, generated by active region 9684, had the Carrington longitude of its source inside the interval of 100–170°. Taking into account the complete data on 24th solar cycle it appears that the last solar proton event of the cycle of September 10, 2017, generated by active region 12673, with the fluence value $8 \cdot 10^8$ cm^{-2} also had the Carrington longitude of its source inside the interval of 100–170°. Thus during the whole period of observation from 65 most powerful SPEs with fluences of protons with energies >30 MeV $\geq 10^8$ cm^{-2} only two SPEs had their sources inside the "passive" longitude interval. From another 85 "medium" SPEs with fluences 10^7–10^8 cm^{-2} in cycles 19–24 sources of only ≈12 SPEs were located in the passive longitude interval.

In Fig. 1b for each 10-degree interval of Carrington longitude a fraction of the fluences of protons with energies >30 MeV injected from this interval is computed, relative to the total fluence over all longitudes for all considered SPEs. From the plot it follows that the summarized fluence for the passive longitude interval amounts to only 6% from the total fluence for all SPEs during the whole investigated period of observations.

It should be mentioned, that the computations of the same distributions for each solar cycle separately showed, that the active Carrington longitudes, from which the major fraction of the proton fluences was injected, changed from one cycle to another. Whereas the discovered passive longitude interval can be determined for the whole 6 solar cycles.

Returning to Fig. 1a, one could see that the last powerful SPE on September 10, 2017 had its source near the border of the passive longitude interval. This led us to check the influence of the accuracy of the definition of Carrington period to the longitudinal distribution of the SPE fluences. The value of the sidereal period of Sun's rotation was determined by R. C. Carrington from the observations of low-latitude sunspots with the accuracy of 2 decimals equal to 25.38 days (corresponding synodic period 27.2753 days). We took the value of the sidereal period changed in the third digit to 25.379 days (corresponding synodic period 27.2742 days) and computed the distribution of the SPE fluences over the modified Carrington longitude values. To make the comparison of the initial and modified distributions more accessible in both cases for the time of the initial Sun's rotation the time of 1369s Carrington rotation 1956-01-07 13:00 UT was taken, which corresponds to the beginning of 19th solar cycle. The modified distribution of the SPE fluences over the Carrington longitude is shown in Fig. 2. As one could see from this figure, with the value of sidereal Carrington period $T = 25.379$ days only one "anomalous" powerful SPE of November 4, 2001 remained in the passive longitude interval; and the summarized value of the fluence of SPEs in the passive longitude interval decreased to only 4.5% from the total fluence for all the considered SPEs.

The changing of the Carrington period by notably larger value leads to disappearing of the inhomogeneity of SPE fluence distribution over the Carrington longitude.

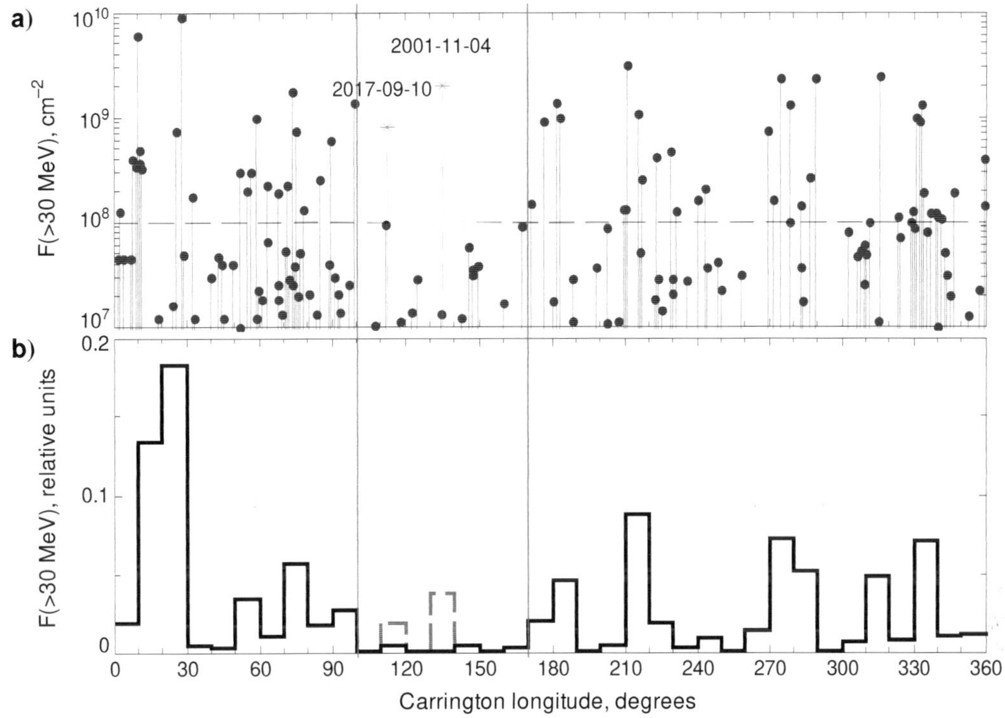

Figure 1. Distribution of the fluences of protons with energies >30 MeV in solar proton events in the solar cycles 19–24 over the Carrington longitude of their sources a) for the individual SPEs, b) summed over 10-degree intervals. Interval of "passive" longitudes 100–170° is emphasized. The two only "anomalous" strong SPEs during 62 years, which sources lay in the interval of passive longitudes, are marked by dash lines.

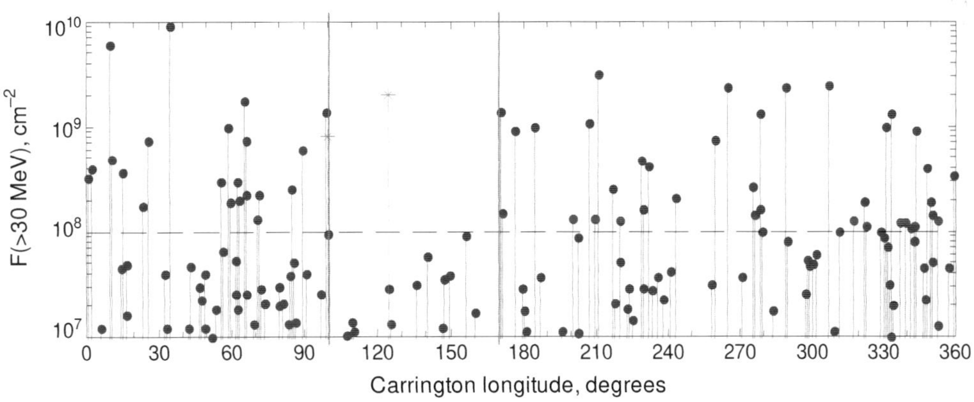

Figure 2. Distribution of the fluences of protons with energies >30 MeV in solar proton events in the solar cycles 19–24 over the Carrington longitude of their sources for the value of sidereal Carrington period $T = 25.379$ days instead of 25.38. Only one "anomalous" SPE of November 4, 2001 remained in the passive longitude interval.

4. Conclusions and discussion

On the basis of the study of the solar proton event fluence distribution over the longitude of their sources on the Sun the interval of "passive" Carrington longitudes, extensive over the longitude (≈100–170°) and the lifetime (6 solar cycles) has been discovered. From 65 most powerful SPEs during the whole period of observations with fluences of protons with energies >30 MeV $\geq 10^8$ cm^{-2} only 1–2 SPEs had their sources inside the passive longitude interval. The summarized value of the fluence of SPEs in the passive longitude interval amounts to only 4.5–6% from the total fluence for all the considered SPEs.

The existence of the interval of "passive" Carrington longitudes, extensive over the longitude and the lifetime, is the remarkable phenomenon in solar physics.

Author does not have the physical explanation of this phenomenon. Very speculatively, it may be the asymmetry of the solar magnetic dynamo.

It should be also noted, that such inhomogeneous longitudinal distribution of the sources of energetic proton fluences was received using the constant Carrington period of Sun's rotation, despite the fact that solar photosphere experiences the differential rotation. This fact may serve as a certain confirmation of the physical relevance of the Carrington period; and lead one to conjecture that the nature of the magnetic fields in the active regions, accelerating the particles during the most powerful solar flares is connected with the deeper layers of the Sun.

For the further research on the problem we intend to and investigate the longitudinal and temporal dependencies of the fluences of higher energy protons and the other indices of solar activity and the magnetic field structures on the Sun.

Proof of the existence of the passive interval of Carrington longitudes may have applications for the mid-term forecasting of the space radiation risks, for instance for determining the time window for sending the manned spacecraft to the Moon.

Acknowledgements
The research was supported by the grant No. 17-29-01022 of the Russian Science Foundation.

References

[1] Vernova E S, Mursula K, Tyasto M I and Baranov D G 2004 Long-term longitudinal asymmetries in sunspot activity: difference between the ascending and descending phase of the solar cycle *Solar Physics* **221** 151–65
[2] Getselev I V, Okhlopkov V P and Chuchkov E A 2003 Active longitudes in injection regions of solar cosmic rays *Bulletin of Russian Academy of Sciences: Physics* **67**(4) 532–4
[3] Getselev I V, Podzolko M V and Okhlopkov V P 2013 Passive longitudes of solar cosmic ray sources *J. Phys. Conf. Ser.* **409** 012203.
[4] Dolginova Yu N 1972 Catalogue of chromospheric flares and corresponding phenomena over the period from 1957 till 1965 *Sun-Earth Physics* vol. 2 (Moscow: IZMIRAN) (in Russian)
[5] Akinyan S T, Basilevskaya G A, Ishkov V N et al. 1982 *Catalogue of solar proton events 1970–1979* ed. Logachev Yu I (Moscow: IZMIRAN) (in Russian)
[6] King J H 1974 Solar proton fluences for 1977–1983 space missions *J. Spacecraft Rockets* **11** 401–8
[7] Feynman J, Armstrong T P, Dao-Gibner L and Silverman S 1990 A new interplanetary proton fluence model *J. Spacecraft Rockets* **27**(4) 403–10
[8] http://spdf.gsfc.nasa.gov/
[9] Getselev I V, Podzolko M V and Veselovsky I S 2009 Optimization of the interplanetary energetic proton flux database and its application in modeling radiation conditions *Sol. Syst. Res.* **43**(2) 136–42
[10] http://umbra.nascom.nasa.gov/SEP/

IOP Conf. Series: Journal of Physics: Conf. Series **1181** (2019) 012017 doi:10.1088/1742-6596/1181/1/012017

Underground neutron events at Tien Shan

A Shepetov[1], A Chubenko[1], O Kryakunova[1,2], O Kalikulov[3], S Mamina[1], K Mukashev[3], R Nam[1], V Piscal[1], V Ryabov[1], N Saduyev[1,3], T Sadykov[1,4], N Salikhov[1,2], E Tautaev[4], L Vil'danova[1], Zh Zhantayev[2], and V Zhukov[1]

[1] P N Lebedev Physical Institute of the Russian Academy of Sciences (LPI), 119991, Leninsky pr., 53, Moscow, Russia

[2] Institute of Ionosphere, 050020, Kamenskoye plato, Almaty, Kazakhstan

[3] Institute of Experimental and Theoretical Physics of al-Farabi Kazakh National University, 050040, al-Farabi pr., 71, Almaty, Kazakhstan

[4] Institute for Physics and Technology, Almaty, 050032, Ibragimova, 11, Kazakhstan

E-mail: ashep@tien-shan.org

Abstract. The events of multiple neutron production under a 2000 g/cm^2 thick rock absorber were studied at the Tien Shan mountain cosmic ray station, at an altitude of 3340 m above the sea level. From comparison of the experimental and Geant4 simulated neutron multiplicity spectra it follows that the great bulk of these events can be explained by interaction of cosmic ray muons with internal material of neutron detector. In synchronous operation of the underground neutron monitor with the Tien Shan shower detector system it was found that the characteristics of the muonic component of extensive air showers which is seemingly responsible for generation of neutron events underground do change noticeably within the energy range of the knee of primary cosmic ray spectrum. Some peculiar shower events were detected when the neutron signal reveals itself only ~100–1000 μs after the passage of the front of shower particles which probably means an existence of corresponding delay of the muon flux in such events.

1. Introduction

Investigation of the events of multiple neutron production in underground detectors of the Tien Shan mountain cosmic ray station was started about a decade ago. Phenomenological results which have been obtained then on the properties of these events were reported in [1, 2, 3]; in these publications it was stated that the original nature of neutron events observed underground still remains unclear. After completion of modification period of the complex detector system for cosmic ray studies at the Tien Shan station [4] and systematic introduction of the Geant4 package based simulation methods for determination of the properties of its particle detectors it became possible to return to the problem of underground neutron events at a new stage of experimental technique. In particular, the detection of neutron events in strict synchronization with shower installation now permits to study precisely the neutron bearing properties and temporal characteristics of the penetrative component of extensive air showers (EAS). An overview of the results newly obtained in simultaneous operation of the Tien Shan shower detector system with underground neutron detector is the subject of the present message.

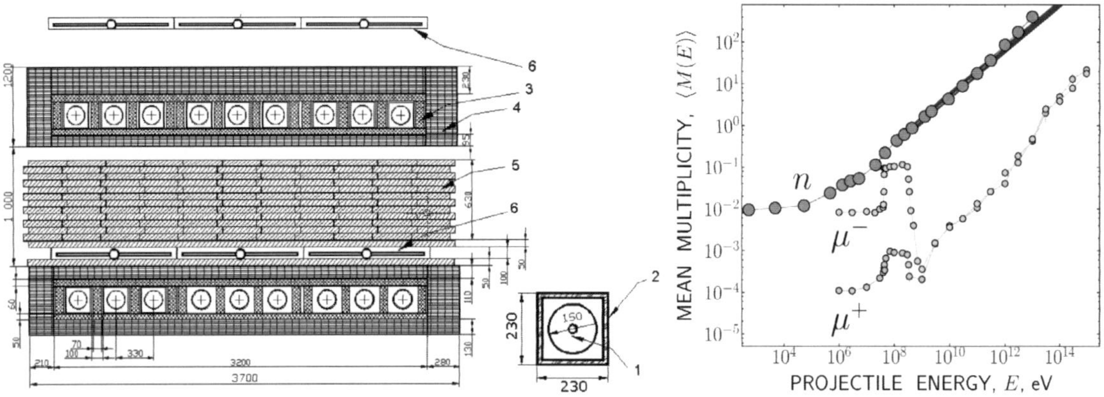

Figure 1. Left: internal arrangement of the underground neutron monitor installation in the present time: *1* – neutron counter, *2* – moderator, *3* – lead target, *4* – rubber (outer moderator and shielding), *5* – iron absorber, *6* – plastic scintillators. Dimensions are shown in millimeters. Right: Geant4 simulation results for multiplicity of the signals from neutron counters of a single monitor unit in dependence on the energy of incident neutrons and muons; continuous straight line beside the neutron data indicates the approximation of corresponding experimental results from the work [6].

2. The underground neutron monitor

Presently, the underground neutron detector at the Tien Shan station consists of a pair of separated units which have been made resembling the standard NM64 type neutron supermonitor [5], and were placed one above the other as it is shown in the left picture of figure 1. Both units include the layers of a heavy target absorber where the penetrative particles of cosmic radiation can experience nuclear interaction with lead nuclei. Evaporation neutrons which originate as a result of this interaction can be detected by the big $\varnothing 150 \times 2000$ mm^2 neutron counters with enriched $^{10}BF_3$ gas filling, so the detection of low-energy neutrons is possible there due to reaction $n(^{10}\mathrm{B},^{7}\mathrm{Li})\alpha$. Before the detection, the neutrons loose their initial MeV-order kinetic energy down to thermal level in multiple interactions with light nuclei within the sheets of internal moderator material which consists of wooden boxes surrounding the counters. Another sheets of the hydrogen enriched rubber (C_2H_2) which cover all the unit from outside play the role of external shielding to prevent the influence of environmental low energy neutron background on the measurement of the cosmic ray connected neutron signal.

Hence, the response of a neutron monitor unit to interaction of a nuclear-active cosmic ray particle is connected with a number of electric pulse signals obtained from its counters during some fixed time *gate* period after this interaction. Hereafter, sum number of such signals will be designated as the neutron *multiplicity* M. In [6] it was shown experimentally that for a monitor unit of considered construction the multiplicity M is nearly proportional to a square root of the energy deposit in primary interaction (in GeV order energy range), and typical duration of the gate time can be of the order of a few milliseconds. More precisely, the energy dependence $M(E)$ can be defined through detailed simulation both of the primary interaction and of the subsequent moderation and diffusion processes of originating neutrons within the monitor material which can be made on the basis of modern Geant4 toolkit [7]. The results of such calculations for hadronic (neutron) and muonic type primaries are presented in the right plot of the figure 1.

Same Geant4 simulation model can be of use to answer the question which was put in [1] on the nature of the underground neutron events. Taking into consideration the spectrum of energy deposits from penetrative cosmic ray particles which has been measured earlier in the

IOP Conf. Series: Journal of Physics: Conf. Series **1181** (2019) 012017 doi:10.1088/1742-6596/1181/1/012017

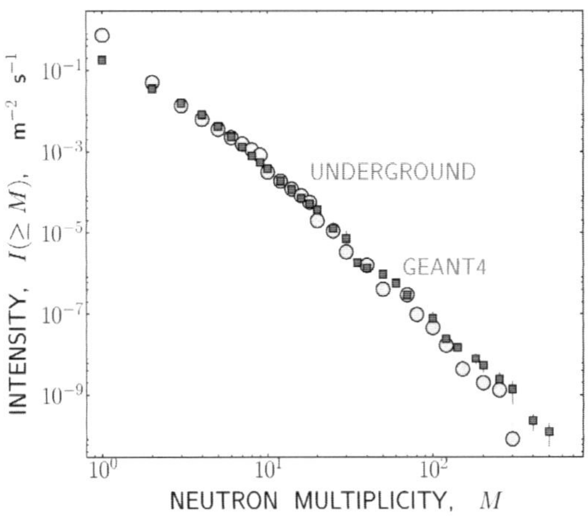

Figure 2. Experimental (circles) and Geant4 simulated (squares) multiplicity spectra of neutron events in the underground monitor.

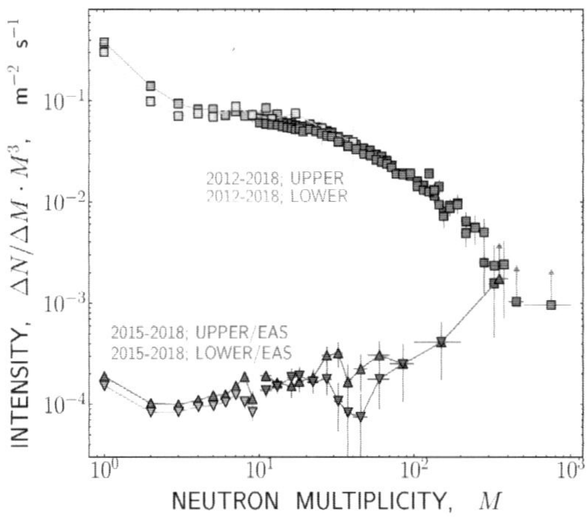

Figure 3. Multiplicity spectra of all neutron events seen underground (above), and of the events with EAS accompaniment only (below).

same underground room of Tien Shan station with ionization calorimeter [8], and using it as an input for Geant4 simulation series with $\mu^{\pm}$ type primaries one can obtain the expected neutron multiplicity spectrum of underground events. A comparison of such simulated and experimentally measured multiplicity spectra is made in the plot of figure 2 where it is seen that both spectra do agree rather well with each other. From this fact a conclusion can be drawn that in the case of underground monitor we deal mostly with the products of nuclear interaction caused by cosmic ray muons, so this monitor as a whole can be used as a specific detector of the cosmic ray muonic component, and particularly of the muons which accompany the passage of extensive air showers.

3. Underground neutron events and extensive air showers

During operation periods of the Tien Shan shower installation [4] in the years 2015-2018 the data taking process at the underground neutron monitor was fulfilled under the control of external trigger signal which was generated in the moments when an extensive air shower has been detected above the surface of Tien Shan station. The multiplicity spectra of the neutron events of muonic origin which have been registered simultaneously with an EAS are presented in figure 3, in comparison with analogous spectra calculated over the whole set of underground events. It is seen there that generally the intensity of EAS accompanied events occurs being 2–3 order of magnitude below the total flux of $M \lesssim 100$ events, but tends to match with the latter in the range of extremely high multiplicities. Seemingly, this difference can be explained by particulars of the shower trigger elaboration threshold and their influence on registration probability of the EAS connected neutron events underground.

To clarify more precisely what are the favorable conditions for detection of the muonic EAS component by the neutron signal from its nuclear interaction a correlation plot can be built between the multiplicity of underground neutron events M, and the core distance R and size N_e (total number of the charged particles) of accompanying showers. An example of such correlation is presented in the left plot of figure 4. The range of distance values here is limited from above with geometrical sizes of the central detector "carpet" of Tien Shan shower installation ($R^{max} \simeq 36$ m, see [4]) which was used for precise location of EAS axis in the measurement series when the considered experimental data have been obtained. From correlation plots of figure 4 it

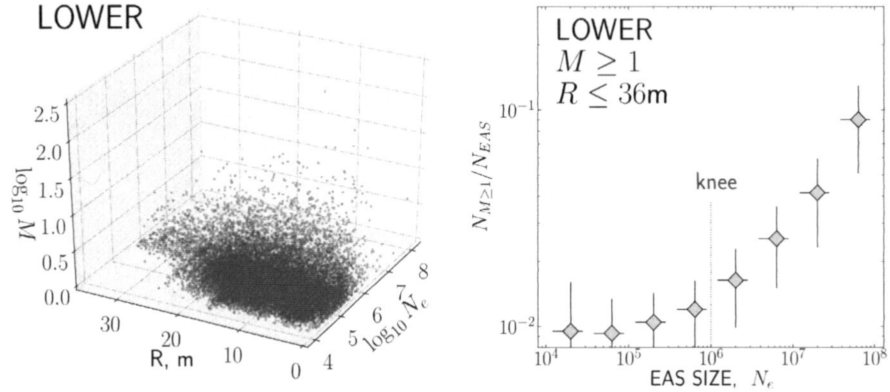

Figure 4. Left: correlation plots between the multiplicity of underground neutron events M, the shower size N_e, and the core distance R of accompanying EAS. Right: relative share of EAS with detected neutron signal from muonic interactions underground. (Here, there are plotted only the data obtained for the lower unit of the underground monitor; those for its upper unit are quite analogous to presented ones).

is seen that the events with non-zero multiplicity values were observed mostly in the cases when a shower axis has passed in relative vicinity ($R \lesssim 20 - 25$ m) to projection of the position of neutron detector to the surface of the ground, but the distance between EAS core and location of the monitor in such events can be somewhat bigger for the showers with $N_e \gtrsim 10^6$. As well, the events with comparatively high values of detected neutron multiplicity ($M \gtrsim 10 - 30$) were met only amongst EAS with $N_e \gtrsim 10^6$.

The dependence between the mean relative amount of EAS events with non-zero multiplicity of the underground neutron signal (which is normalized to the total statistics of registered extensive air showers N_{EAS}) and the average size of accompanying shower N_e is illustrated by the right plot of figure 4. It follows from this picture that for all EAS with $N_e \lesssim 10^6$ the share of $M \geqslant 1$ events remains at an approximately one and the same low level which statistically nearly fits zero, and generally can be explained by random coincidences of shower trigger with background neutron events in the monitor. Contrary to this, in the range of shower sizes $N_e \approx 10^6$ the relation $N_{M\geqslant1}/N_{EAS}$ starts to grow, and evidently can not be put into any agreement with its previous behaviour.

It should be noted that at the altitude of the Tien Shan mountain station the range of EAS sizes $N_e \approx 10^6$ corresponds to the primary energy of a cosmic ray particle $E_0 \approx 3 \cdot 10^{15}$ eV [9], i. e. to position of the well-known knee in the energy spectrum of cosmic rays. Since the revealed sharp increase in the relative share of the showers which were accompanied by the underground neutron events (and consequently, a corresponding increase in the average multiplicity or in the energy of EAS connected muons which are the original source for generation of these events) occurs nearly at the same point on energy scale, probably it could be added to a wide list of various peculiar effects which have been met up to date within the energy range around the knee of primary spectrum by many research groups, and particularly in the experiments which have been held previously at the Tien Shan cosmic ray station [4].

Another peculiarity found in behaviour of the underground neutron events concerns their temporal characteristics. Since the shower trigger provides a rather strict binding to the EAS passage it is possible to trace precisely the time distribution of subsequent neutron signals with regard to initial moment of nuclear interactions caused by the EAS muons in underground

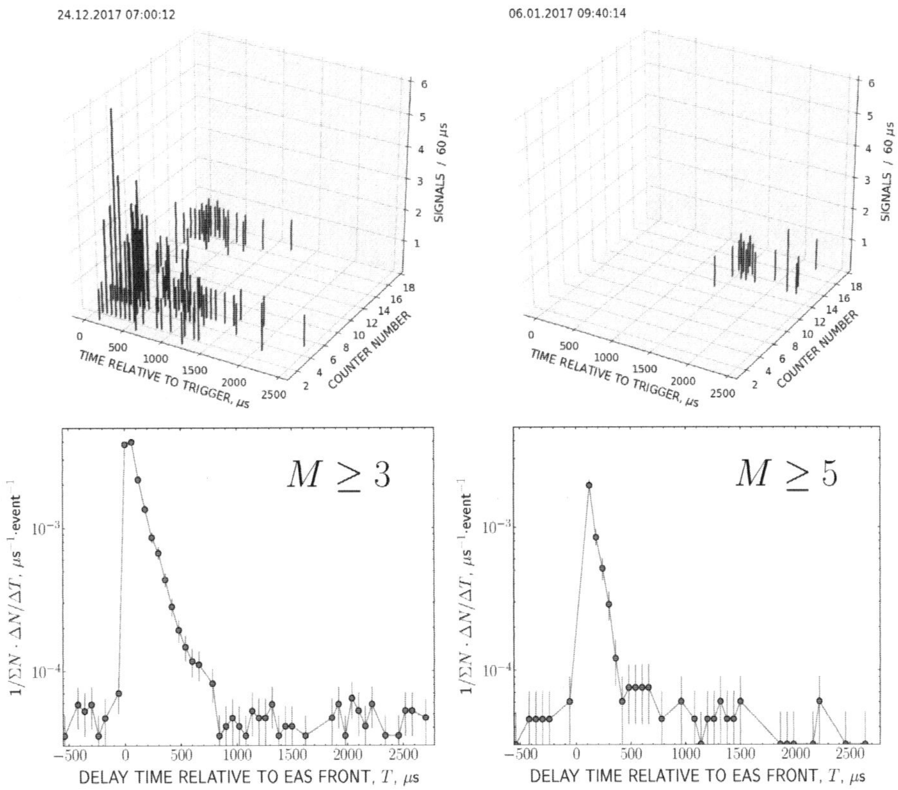

Figure 5. Top: a sample of normal (left) and delayed (right) time distributions of neutron signal in EAS connected events of underground monitor. Zero point of time axis in both plots corresponds to the moment of EAS trigger; vertical axis is graduated in the units of neutron counter pulses which have come during a single 60 μs long time interval. Neutron counters $1 \ldots 9$ correspond to the upper monitor unit, $10 \ldots 18$—to the lower one. Bottom: distribution of the time delay values between the shower front and the beginning of the neutron signal development in underground events with sum multiplicity $M \geqslant 3$ and $M \geqslant 5$.

monitor. It is natural to expect that generally this distribution must be of the exponent type which results from diffusion of the newly born evaporation neutrons within the monitor material. In fact, most part of detected events do have such exponential shape, like a sample event shown in the top left panel of the figure 5. At the same time, among all EAS connected events it was found a noticeable amount of cases when the beginning of exponential intensity decrease has been too late by a some considerable time T in relation to the moment of shower trigger, so that T can be of up to a few hundreds of microseconds order. An example of a such "delayed" event is presented in the top right panel of figure 5.

Of course, a certain part of uncommon delays seen underground can be explained by an arbitrary overlapping between the moments of EAS trigger and the background neutron events in monitor. To elucidate the role of random coincidences in the discussed effect a distribution can be built over observed values of the delay time T. Such distribution (normalized to the total amount of registered EAS cases with a neutron accompaniment underground) is presented in two bottom plots of the figure 5 which are drawn for the events with detected neutron multiplicity $M \geqslant 3$ and $M \geqslant 5$ correspondingly. As it is seen in these plots the probability to find an overtaking signal from neutron detector in the negative range of delay times $T < 0$ is about $(2-4)\cdot 10^{-4}$ $\mu s^{-1}\cdot event^{-1}$. Since obviously such pulses must be causally independent on the succeeding EAS it is this level which can be accepted as a background which owes to the random coincidences between the extensive air showers and the neutron events underground. Nevertheless, an evident and statistically reliable excess of events is seen in the plots of figure 5 in the range of delay times between $T = 0$ and $T \approx 500 - 800$ μs which means that some noticeable part of the observed delays can not be completely due to the accidental background but must have some physical reason in the EAS passage. Since the primary source of underground neutron events is the interaction of the muons same conclusion on existence of a considerable delay time relates as well to the flux of the muonic component in extensive air showers.

4. Conclusion

Presently, the investigation results of neutron generation in the underground monitor at the Tien Shan station can be summarized as the following. The comparison of the experimental and Geant4 simulated neutron multiplicity spectra has shown that the great bulk of the events observed underground can be rather convincingly explained by the interaction of the cosmic ray muons with internal material of the neutron detector. In synchronous operation of the underground neutron monitor with the Tien Shan shower installation it was found that the origination frequency of neutron events (and consequently either the average energy of EAS connected muons which are the primary source of neutron generation, or the mean muon multiplicity in EAS, or both these characteristics) starts to grow significantly around the knee of primary cosmic ray spectrum. Some peculiar EAS events were detected when the neutron signal underground reveals itself only a few hundreds of microseconds after shower trigger which circumstance means an existence of the corresponding delay of EAS connected muon flux in relation to the main front of shower particles in these events.

Acknowledgments

This work is supported by the grants #BR05236201 and #BR05236494 of IRN Program "Fundamental and applied studies in related fields of the physics of terrestrial, near-Earth and atmospheric processes and their practical application", and by the grant #0118RK00800 of the Cosmic Program of Kazakhstan Republic.

References

[1] Chubenko A P, Shepetov A L, Vildanova L I, et al 2007 *Bull. Lebedev Phys. Inst.* **34** 107–113
[2] Chubenko A P, Shepetov A L, Oscomov V V and Vildanova L I 2008 The underground neutron events at Tien-Shan *Proceedings of the 30th ICRC* vol 4 (HE part 1) (México City, México) pp 3–6
[3] Chubenko A P, Shepetov A L, Chubenko P A, et al 2008 The underground neutron calorimeter for registration of the neutron-bearing cosmic ray component at Tien Shan *Proceedings of the 30th ICRC* vol 4 (HE1) (México City, México) pp 97–100
[4] Chubenko A P, Shepetov A L, Antonova V P, et al 2016 *Nucl. Instrum. Methods A* **832** 158–178
[5] Carmichael H and Hatton C J 1964 *Can. J. Phys.* **42** 2443
[6] Chubenko A P, Shepetov A L, Antonova V P, et al 2003 Multiplicity spectrum of NM64 neutron supermonitor and hadron energy spectrum at mountain level *Proceedings of the 28th ICRC* (Tsukuba, Japan) pp 789–792
[7] Abunin A A, Pletnikov E V, Shchepetov A L and Yanke V G 2011 *Bull. Russ. Acad. Sci. Phys.* **75** 866–868
[8] Erlykin A D, Kulichenko A K, Machavariani S K and Nikolsky S I 1973 Investigation of cascades, produced by high energy muons *Proceedings of the 13th ICRC* vol 3 (Denver, Colorado, USA: NASA Conference Publication No. 2376) p 1803 (*Preprint* `http://adsbit.harvard.edu//full/1973ICRC....3.1803E/0001803.000.html`)
[9] Adamov D S, Afanasjev B N, Arabkin V V, et al 1987 Phenomenological characteristics of EAS with $N_e = 2\cdot 10^5 - 2\cdot 10^7$ obtained by the modern Tien-Shan installation "Hadron" *Proceedings of the 20th ICRC* vol 5 (Moscow, USSR) pp 460–463

IOP Conf. Series: Journal of Physics: Conf. Series **1181** (2019) 012018 doi:10.1088/1742-6596/1181/1/012018

Red Dwarfs as Sources of Cosmic Rays and First Detection of TeV Gamma-rays from these stars

V.G. Sinitysna, V.Y. Sinitsyna, Yu.I. Stozhkov

P.N. Lebedev Physical Institute, RAS, Moscow, Leninsky pr. 53

e-mail: sinits@sci.lebedev.ru, stozhkov@sci.lebedev.ru

Abstract. Conventional sources of Galactic cosmic rays are believed to be supernovae and supernova remnants that are powerful enough to accelerate particles up to 10^{17} eV. Meanwhile, an interpretation of recent experimental data obtained by PAMELA, Fermi-LAT, and AMS-02 spectrometers hints at a possibility that some cosmic-ray sources may be located in a close proximity of the solar system, at the distances less than 1 kpc. Presence of such local sources could explain the unexpected raise of the positron fraction in cosmic rays, complex shapes of the proton and helium spectra, and even anomalous low-energy cosmic rays. We consider active dwarf stars as possible sources of Galactic cosmic rays in the energy range up to 10^{14} eV. These stars are producing powerful stellar flares. Meanwhile, generation of high-energy cosmic rays should be accompanied by the high-energy γ-ray emission, which may be observed. Here we present the results of the SHALON long-term observations aimed to search for γ-ray emission above 800 GeV from the active red dwarf stars.

1. Introduction

Conventional sources of cosmic rays are believed to be supernovae and supernova remnants (SNRs) that can accelerate particles up to ~10^{17} eV. Indeed, because of their energetics SNRs could operate as powerful accelerators of cosmic-ray particles [1, 2]; this idea was prevailing from the very beginning of the cosmic-ray era. Recent observations of several SNRs in X-rays and TeV γ-rays reveal clear signatures of accelerated particles in their shells (see [3] and references within). The presence of relativistic electrons is clearly seen across the whole electromagnetic spectrum, from radio to very high-energy γ-rays, while signatures of accelerated protons can be only detected in high energy γ-rays. However, an interpretation of recent observations by PAMELA, Fermi-LAT, and AMS-02 hints at a possible presence of nearby sources of cosmic rays at the distances less than 1 kpc from the solar system. In particular, presence of such sources could explain the unexpected raise of the positron fraction in cosmic rays (Fig. 1) [4 – 7]. Such sources may also help with interpretation of the complex spectral behavior of protons, helium, and other species [8 – 10], and the appearance of the anomaly low-energy component (single ionized atoms) in cosmic rays. If the nearby cosmic-ray sources exist indeed, they could also generate high-energy γ-ray emission and may be observable with γ-ray telescopes.

The SHALON instrument [3, 11 – 15] consists of two imaging atmospheric Cherenkov telescopes located in Tien-Shan mountains at an altitude of 3340 m above the sea level. It is designed for observations of γ-ray sources in the energy range from 800 GeV to 100 TeV. Each of the telescopes has a composite mirror with an area of 11.2 m^2. The detector has the largest field of view >8° among similar instruments. It allows the background from charged cosmic-ray particles and the atmospheric transparency to be continuously monitored simultaneously with the observations of γ-ray sources. The SHALON method of selection of γ-ray showers from the background cosmic-ray showers allows the rejection of 99.93% of the background events. The minimum detectable integral flux of γ-rays at 1 TeV is 2.1×10^{-13} cm^{-2} s^{-1}. In the energy range from 1 – 50 TeV the minimum detectable flux falls to the value of 6×10^{-14} cm^{-2} s^{-1} [3, 12 – 14]. The SHALON experiment has been in operation since 1992 providing the long-term observations of many different types of sources that are of interest for many areas of astroparticle physics. One of the goals of the SHALON long-term observational program is to search for γ-ray emission above 800 GeV from the active red dwarf stars.

In this paper we study the active dwarf stars that produce powerful stellar flares and may accelerate cosmic-ray species up to 10^{14} eV. Generation of high-energy cosmic rays in their flares should be accompanied by the high-energy γ-ray emission. Here we present the results of 10 – 20 years of observations of the dwarf stars V962 Tau, V780 Tau, V388 Cas, and V1589 Cyg with the SHALON atmospheric Cherenkov telescope.

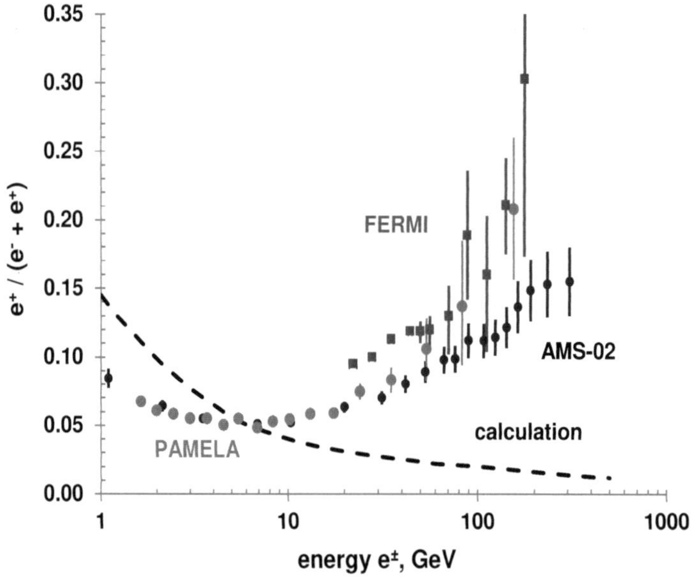

Figure 1. The positron fraction as measured by PAMELA, AMS, Fermi-LAT [4 - 6] together with theoretical predictions [16] (dashed line).

2. Observations of red dwarfs

As it was shown in the dedicated studies, the flare frequency of the active red dwarf stars is uncertain [17, 18], and their observations require a long-term monitoring program. In this study we present our results for those stars that we observed over very extended periods of time.

In accordance with the long-term program of observations of the Galactic γ-ray sources, more than ten-years long observations of the Tycho's SNR, Crab Nebula, and Cyg X-3 have been carried out by the SHALON experiment [3, 11 – 14]. During these pointed observations, some of the red dwarfs were also observed by SHALON as they are located at the angular distances less than ~ 5° from those well-known TeV γ-ray sources while the field of view of the instrument is > 8°. The observations and data processing were performed using the standard SHALON technique [3, 13].

2.1. V388 Cas

Tycho's SNR was observed by SHALON during the 123 hours from 1996 to 2010 [3, 14, 15]. All observations were made during moonless nights with zenith angles from 16° to 35°. During Tycho's SNR observations the SHALON field of view also included V388 Cas located ~4.5° south of Tycho's

SNR. V388 Cas is a UV Cet type star at a distance of ~10 pc [19]. It is also present in the ROSAT catalogue with the photon flux of ~ 1.7×10^{-12} erg cm^{-2} s^{-1} and luminosity of $10^{28.3}$ erg s^{-1} [19]. V388 Cas was observed by SHALON for a total of 93 hours from 1996 to 2010. The γ-ray source associated with the V388 Cas was detected above 1 TeV with a statistical significance of 6.8σ determined according to Li and Ma [20] method with an average γ-ray flux of $I_{V388Cas}(> 1TeV) = (0.84\pm0.21)\times10^{-12}$ cm^{-2} s^{-1}.

The signal significance of this source is much less than that of the source with a similar flux and the spectral index obtained during the same hours of observations because of the smaller collection area of the instrument relative to the standard procedure of the SHALON experiment. After processing of the Tycho's SNR observational data, first using the selection criteria associated with Tycho's SNR and then with V388 Cas, we found that less than 1% of showers were common for both sources. Recognition of the source of each of the common showers was performed through the analysis of the angular distance of the arrival directions of these showers and the source positions. This procedure does not change the average flux of Tycho's SNR. The shape of the SHALON average differential spectrum of V388 Cas (Figs. 2, 4) in the energy range from 0.8 to 25 TeV can be fitted with a soft power law: $dN/dE = (0.91\pm0.25)\times10^{-12} \times (E_\gamma/1TeV)^{-2.52\pm0.15}$ TeV^{-1} cm^{-2} s^{-1} with χ^2/Dof = 1.23, where the number of the degrees of freedom Dof = 6. During the long-term observations, V388 Cas was detected as a variable source that was visible during the flares. Its flaring spectrum, $dN/dE_{Flare} = (2.7\pm0.15)\times10^{-12}\times(E_\gamma/1\ TeV)^{-2.91\pm0.18}$ TeV^{-1} cm^{-2} s^{-1}, is shown in Fig. 2 (▲). The integral flux of V388 Cas averaged over a one-year interval is shown in Fig. 4. As it is clearly seen in Fig. 4 that V388 Cas has been detected during the flares that occurred before 2008 and that resulted in high fluxes presented in Fig. 2.

Searching for counterparts in the Fermi-LAT 8 years catalogue (FL8Y), we found an unassociated source J0106.4+5938 [21] of high energy emission that is located at a distance of ~2.5° from V388 Cas. The FL8Y source catalogue is based on the data taken during the period from 2008 August 4 to 2016 August 2. The emission from J0106.4+5938 was detected by Fermi LAT at the level of 4.2σ significance in the energy range from 100 MeV to 1 TeV. The shape of differential photon spectrum in the energy range from 100 MeV to 100 GeV is fitted with a power law $dN/dE = C(E/0.913GeV)^{-\Gamma}$, where Γ=2.78±0.16 and C= $(5.8\pm1.2)\times10^{-13}$ MeV^{-1} cm^{-2} s^{-1}.

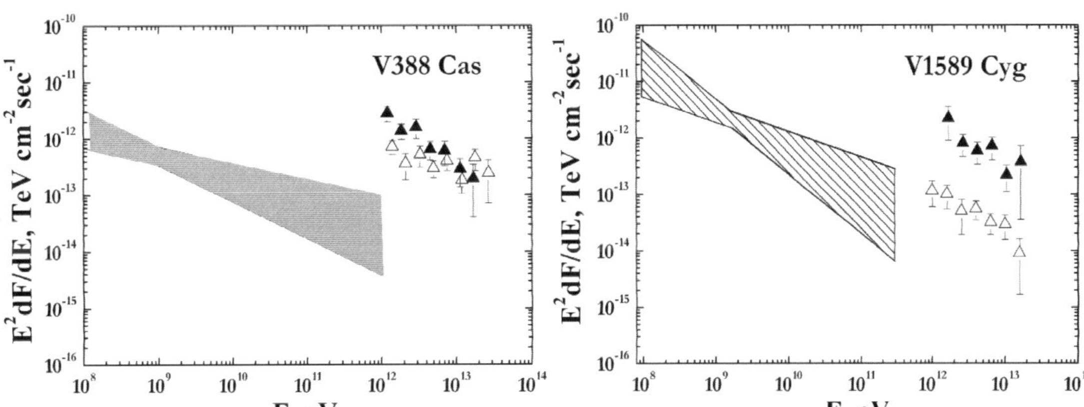

Figure 2. Spectral energy distributions of V388 Cas (left) and V1589 Cyg (right). Open and black triangles (△, ▲) show SHALON data from long-term observations (see text). The grey bow tie on V388 Cas plot indicates the minimum flux needed for a Fermi LAT 5σ detection of a soft-spectrum point source [22, 23]. The spectrum of the unassociated source J2041.2+41.76 observed from FL8Y is shown with a bow tie (right).

2.2. V1589 Cyg

The Cygnus Region was observed by SHALON in the framework of the long-term studies of microquasar Cyg X-3 at very high energies. Observations also included such sources as V1589 Cyg, TeV J2032+4130, and γCygni SNR – thanks to the large field of view of SHALON [12 – 14, 24, 25]. V1589 Cyg is an M-class flaring star located at a distance of 23-32 pc [17]. Its luminosity is significantly fluctuating with duration of flares from minutes to hours [17].

V1589 Cyg is located ~2° west of Cyg X-3. Consequently, V1589 Cyg has been systematically observed by the SHALON telescope in 1995 – 2016 during clear moonless nights at zenith angles from 5° to 34° for a total of 303.5 hours. V1589 Cyg was detected by SHALON above 0.8 TeV with the average integral flux of $I_{V1589Cyg}(> 0.8TeV) = (0.13\pm0.019)\times10^{-12}$ cm^{-2} s^{-1}. The statistical significance of the detection is 6.5σ [20]. As with observations of V388 Cas dwarf star, we found that less than 1% of Cyg X-3 showers were identified to be V1589 Cyg showers. This did not change the average flux of Cyg X-3. During the long-term observations V1589 Cyg appeared as a source with a highly variable flux. The flaring spectrum was also extracted and presented in Fig. 2 (▲). The spectral shape of V1589 Cyg in the energy range from 0.8 to 35 TeV can be fitted with a soft power law (see Fig. 2, △), $dN/dE = (0.15\pm0.05)\times10^{-12} (E_\gamma/1\ TeV)^{-2.91\pm0.18}$ TeV^{-1} cm^{-2} s^{-1} with the χ^2/Dof = 0.89, where Dof = 5. A flaring spectrum is $dN/dE_{Flare} = (1.7\pm 0.45)\times10^{-12} (E_\gamma/1TeV)^{-3.15\pm0.29}$ TeV^{-1} cm^{-2} s^{-1}. The light curve of V1589 Cyg during the observations from 1995 to 2016 is presented in Fig. 4, where the individual points are the integral fluxes averaged over each year of observations.

An unassociated object FL8Y J2041.2+41.76 [21] was detected by Fermi-LAT at a distance of ~0.45° from V1589 Cyg. The high energy emission was detected at the level of 5.6σ significance. The differential photon spectrum was obtained in the energy range from 100 MeV to 100 GeV. The spectrum was fitted with a power law $dN/dE = C(E/1.63GeV)^{-\Gamma}$, where $\Gamma = 2.82 \pm 0.26$ and $C = (7.63 \pm 1.3)\times10^{-13}$ MeV^{-1} cm^{-2} s^{-1} (see Fig. 2, right). The FL8Y spectrum of J2041.2+41.76 is consistent with SHALON observations of V1589 Cyg at TeV energies.

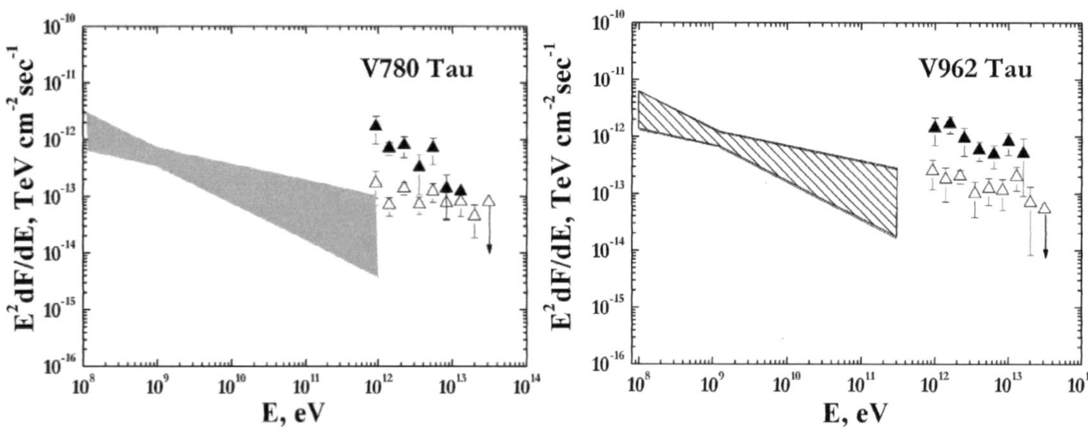

Figure 3. Spectral energy distributions of V780 Tau (left) and V962 Tau (right). Open and black triangles (△, ▲) show SHALON data from long-term observations (see text). The grey bow tie on V780 Tau plot indicates the minimum flux needed for the integral 5σ detection of point source with soft spectrum by Fermi LAT [22, 23]. The γ-ray spectrum of the unassociated source J0544.7+2239 from FL8Y is shown with a bow tie (right).

2.3. V962 Tau and V780 Tau

SHALON observations of the Crab Nebula also yielded observations of two flaring stars: V962 Tau and V780 Tau. V962 Tau and V780 Tau are the flaring type of UV Cet type with the spectrum of the red star class [19, 26] at distances of about 10 pc. V780 Tau was also observed by ROSAT with the flux of $\sim3.6\times10^{-13}$ erg cm^{-2} s^{-1} and the luminosity of $10^{27.6}$ erg s^{-1} [19].

V780 Tau is located ~3° north of Crab and V962 Tau is ~2.5° east of Crab. These stars have a separation of ~4°. Consequently, both stars were observed by the SHALON telescope from 1994 to

2014 [11, 25] for a total of 125.2 hours during clear moonless nights at zenith angles from 15° to 35°. V780 Tau and V962 Tau were detected by SHALON above 0.8 TeV with the average integral fluxes: $I_{V780Tau}(>1\ \text{TeV}) = (0.23 \pm 0.03)\times10^{-12}\ \text{cm}^{-2}\ \text{s}^{-1}$ and $I_{V962Tau}(>1\text{TeV}) = (0.39 \pm 0.04)\times10^{-12}\ \text{cm}^{-2}\ \text{s}^{-1}$, correspondingly.

V780 Tau was detected with a statistical significance 6.1σ [20]. Again, in this case we found that less than 1% of showers are common for the Crab Nebula and V780 Tau. Identification of the source of each of the common showers was performed through the analysis of the angular distances of the arrival directions of these showers and the source positions. As the result, less than ~ 0.4% of the Crab showers were identified to be V780 Tau showers. This did not change the average flux of Crab. The differential spectrum of V780 Tau in the energy range from 0.8 to 20 TeV can be fitted with a power law: $dN/dE = (0.21 \pm 0.08)\times10^{-12}\times(E\gamma/1\text{TeV})^{-2.51\pm0.15}\ \text{TeV}^{-1}\ \text{cm}^{-2}\ \text{s}^{-1}$ with the χ^2/Dof = 1.31, where Dof=5 (Fig. 3, △). Its flaring spectrum is $dN/dE_{Flare} = (2.0\pm0.15)\times10^{-12}\times(E_\gamma/1\text{TeV})^{-3.01\pm0.21}\ \text{TeV}^{-1}\ \text{cm}^{-2}\ \text{s}^{-1}$ (Fig. 3, ▲). As can be seen on Fig. 4, V780 Tau was not active since 2008.

An unassociated object FL8Y J0540.5+2305 [21] is located at a distance of $\sim1.6^o$ from V780 Tau. The emission from the unassociated source was detected by Fermi LAT at the level of 6.62σ significance in the energy range from 100 MeV to 1 TeV. The differential photon spectrum in the energy range from 100 MeV to 100 GeV was fitted with a power law $dN/dE = C(E/1.034\text{GeV})^{-\Gamma}$, where $\Gamma = 2.8\pm0.13$ and $C = (1.84 \pm 0.23) \times 10^{-12}\ \text{MeV}^{-1}\ \text{cm}^{-2}\ \text{s}^{-1}$.

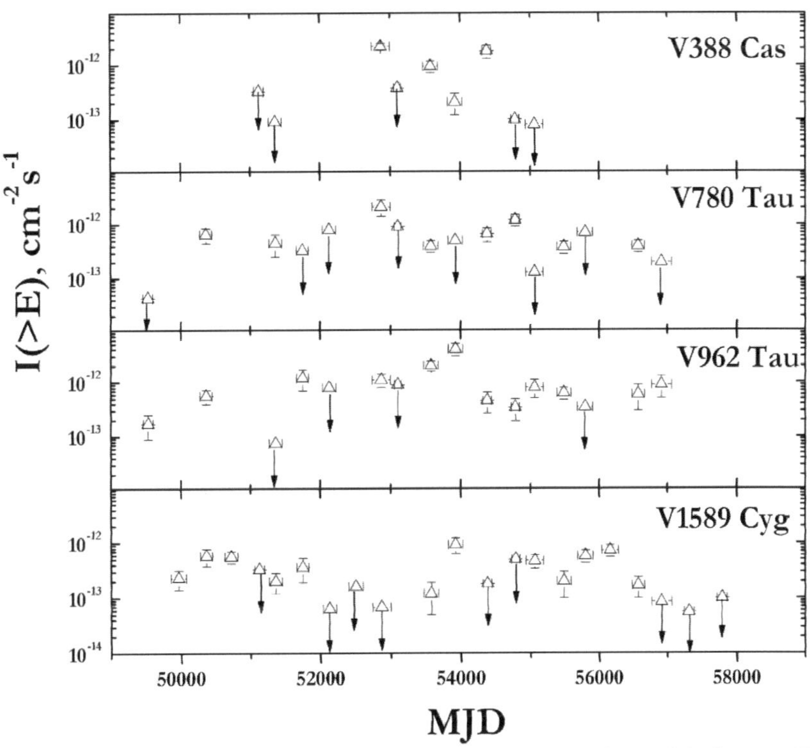

Figure 4. Light curves of V388 Cas, V780 Tau, V962 Tau and V1589 Cyg at TeV energies obtained in the long-term SHALON observations.

V962 Tau was detected by SHALON above 0.8 TeV with a statistical significance 7.7σ [20]. Here we found that there are no showers common for the Crab Nebula and V962 Tau. V962 Tau was also found to be variable. The differential spectrum of γ-ray emission from V962 Tau in the energy range from 0.8 to 20 TeV can be fitted with a power law: $dN/dE=(0.40\pm0.17)\times10^{-12}\times(E_\gamma/1\text{TeV})^{-2.54\pm0.15}\ \text{TeV}^{-1}\ \text{cm}^{-2}\ \text{s}^{-1}$ with the χ^2/Dof = 0.87, where Dof=6 (Fig. 2, △). The flaring spectrum is $dN/dE_{Flare} = (2.8\pm0.45)\times10^{-12}\ (E\gamma/1\text{TeV})^{-2.95\pm0.22}\ \text{TeV}^{-1}\ \text{cm}^{-2}\ \text{s}^{-1}$ (Fig. 3, ▲).

Both these active red dwarfs, V780 Tau and V962 Tau, are displaying variable fluxes at TeV energies. In Fig. 4 we present the light curves of V780 Tau and V962 Tau obtained during the long-term observations.

The emission from unassociated source 3FGL J0544.7+2239 [27] at the distance of $\sim 0.33^{\circ}$ from V962 Tau was detected by Fermi LAT in the energy range 100 MeV – 300 GeV with significance of 4.24σ. The Fermi-LAT spectrum was fitted with a power law $dN/dE = C(E/1.26\ GeV)^{-\Gamma}$, where $\Gamma=2.49\pm0.18$ and $C=(5.1\ \pm1.1)\times10^{-13}$ $MeV^{-1}\ cm^{-2}\ s^{-1}$ (Fig. 3). The Fermi-LAT spectrum of this unassociated source is consistent with the SHALON spectrum of V962 Tau at TeV energies.

3. Discussion

During the long-term observations, red dwarfs appear as sources with variable TeV γ-ray flux up to $\sim$10 TeV and seems are detected mostly during the flares. The observed light curves of V388 Cas, V780 Tau, V962 Tau and V1589 Cyg at TeV energies are shown in Fig. 4, where we show the integral fluxes averaged over a one-year interval. M-Dwarf stars are characterized by flaring events which occur frequently with different flare duration [17, 18]. Very high energy γ-ray fluxes presented here demonstrate that these objects produce flares with total energy of $\sim 10^{33} - 10^{35}$ ergs.

High energy emission from unassociated sources in the vicinity of V1589 Cyg and V962 Tau red dwarfs was detected by Fermi-LAT. Spectral energy distributions of these objects are shown in Figs. 2, 3. The red dwarfs are flaring stars with uncertain frequency and duration of the flares. Unfortunately, V388 Cas and V780 Tau were not active enough after 2008 to produce a significant detection by Femi-LAT launched in 2008 (see Fig. 4). In contrast, V1589 Cyg and V962 Tau were quite active during the whole period of observations and the unassociated sources nearby detected by Fermi-LAT may actually be these red dwarfs.

The so-called positron excess, or unexpected raise of the positron fraction, was discovered by the PAMELA experiment and then confirmed by Fermi-LAT and AMS-02, while this fraction is expected to decrease [16]. The observed positron fraction is starting to increase at ~5 GeV and then goes up monotonically up to $\sim$350 GeV. A possible interpretation of this effect may involve the active dwarf stars [7] as the sources of cosmic rays up to $\sim 10^{14}$ eV.

Indeed, the stellar flares of active dwarf stars could effectively accelerate cosmic rays and compete with other sources. Among the huge variety of stars in our Galaxy, in which there are $\sim 2\times10^{11}$ stars in total, more than 95% are dwarf stars of G-M spectral classes [19]. Assuming that 10% of the dwarf stars in our Galaxy produce stellar flares with frequency of flares of about 36 $year^{-1}$ and average energy released in a single flare of about 10^{35} ergs, one can find that their total energy is $\sim 1.4\times10^{54}$ erg which is comparable to the conventional estimates of the total energy of Galactic cosmic rays in the Galactic disk $W_{GCR} \sim 2\times10^{54}$ erg [7].

During powerful flares of active dwarf stars, high-energy protons are generated with a maximum energy of $E < 10^{14}$ eV. Some of these accelerated protons could escape into the interstellar medium. Others enter the photosphere of the star and undergo nuclear interactions with its matter. During these interactions, π^0 and $\pi^{\pm}$ are produced, while positrons and electrons are produced through the $\pi^{\pm}$ to $\mu^{\pm}e^{\pm}$ decay. An additional source of positrons is the decay of neutral pions into two γ-rays and the subsequent formation of (e^+e^-) pairs by these γ-rays. Such a process is observed during the large solar flares (the Sun is a yellow dwarf). Positrons and electrons from stellar flares with $E > 5$ GeV escape into the interstellar medium from a flare region with strong magnetic fields. If the energy of these particles $E < 5$ GeV they have no chance to escape from the flare region, but if their energy is >100 GeV they escape from the flare region with a high probability (~100%). In the intermediate energy interval $5< E<100$ GeV the probability of particle escape increases with energy. It is thus possible to obtain an increase in the positron fraction similar to the data plotted in Fig. 1. It should be noted that mainly electrons (and protons) are accelerated during the stellar flares, meanwhile an addition of this another component to the electron cosmic-ray flux constitutes only a small fraction of the total flux. This additional source of positrons, however, provides an essential addition to the flux of these particles up to energies of $E\sim500$ GeV. Estimates show that the mechanism suggested in [7] could explain the observed increase in the positron fraction above ~5 GeV. According to the observations, the positron flux increases from 5 GeV up to ~100 GeV and then becomes about a constant in the energy range from 200 – 350 GeV (see Fig. 1). What could we expect in the energy range above 350 GeV? According to the scenario suggested in [7] the positron fraction has to decrease to the values calculated in [16] because the highest energy of accelerated particles in the stellar flares is unlikely to exceed 10^{14} eV.

4. Conclusion

The results of long-term observations of M-Dwarf stars V388 Cas, V780 Tau, V962 Tau and V1589 Cyg by the SHALON atmospheric Cherenkov telescope are presented. Very high energy γ-ray emission from these sources is detected in the range of 1 - 10 TeV mostly during the flaring states. The emission is characterized by a soft power-low spectrum with index of ~2.6÷3.2. The high energy emission from nearby unassociated sources detected by Fermi LAT may be originated in red dwarfs. This result confirms that active dwarf stars are also the sources of high energy cosmic rays. Estimates [7] show that such sources could explain the raise of the positron fraction observed with PAMELA, Fermi-LAT, and AMS spectrometers.

References

[1] Berezhko E G and Krymskii G F 1988 Usp. Fiz. Nauk **154(1)** 49
[2] Reynolds S P 2008 Ann. Rev. Astron. Astrophys. **46** 89
[3] Sinitsyna V G, Sinitsyna V Yu et al. 2018 Advances in Space Research **62(10)** 2845
[4] Adriani O, Barbarino G C, Bazilevskaya G A, et al. 2009 Nature. **458** No 7238 607
[5] Ackermann M, et al. 2012 Phys. Rev. Lett. **108**, 011103
[6] Aguilar M, Alberti G, Alpat B, et al. 2013 Phys. Rev. Lett. **110** 141102
[7] Stozhkov Yu I 2011 Bull. of RAS, ser. Physics **75(3)** 323
[8] Aguilar M, Cavasonza L Ali, Alpat B et al, 2017 Phys. Rev. Lett. **119** 251101
[9] Aguilar M, Cavasonza L Ali, Ambrosi G et al 2018 Phys. Rev. Lett. **120** 021101
[10] Aguilar M, Cavasonza L Ali, Alpat B et al 2018 Phys. Rev. Lett. **121** 051103
[11] Sinitsyna V G 1996 Nuovo Cim. **19C** 965
[12] Sinitsyna V G 1998 Proc. of the 16th European Cosmic Ray Symposium, ed J Medina (Univ. de Alcala, Alcala de Henares, Spain) p 383
[13] Sinitsyna V G and Sinitsyna V Yu 2018 Astron. Lett. **44(3)** 162
[14] Sinitsyna V G 1997 Towards a Major Atmospheric Cherenkov Detector-V, (Kruger Park) ed O C deJager (Potchefstroom: Westprint) p 136, 190
[15] Sinitsyna V G and Sinitsyna V Yu 2011 Astron. Lett. **37(9)** 621
[16] Moskalenko I V and Strong A W, 1998 Astrophys. J., **493** 694
[17] Pettersen B R, Tsvetkov M K, Hawley S L et al., 1988 Astrophysics **29(1)** 67
[18] Ohm S and Hoischen C 2018 MNRAS **474** 1335
[19] Gershberg R E et al., 1999 Astron. Astrophys. Suppl. Ser. **139** 555
[20] Li T-P and Ma Y-Q, 1983 Astrophys. J. **272** 317
[21] https://fermi.gsfc.nasa.gov/ssc/data/access/lat/fl8y/
[22] http://www.slac.stanford.edu/exp/glast/groups/canda/lat_Performance.htm
[23] https://fermi.gsfc.nasa.gov/ssc/data/analysis/documentation/Cicerone/Cicerone_LAT_IRFs/LAT_sensitivity.html
[24] Sinitsyna V G, Sinitsyna V Y 2013 Bull. of the Lebedev Physics Institute (New York: Allerton Press, Inc.), **40(5)** 113
[25] Sinitsyna V G et al., 2016, J. Phys.: Conf. Ser. **718** 052045
[26] Hatsunzev I V 1986 Variable Stars **22(3)** 431
[27] https://fermi.gsfc.nasa.gov/ssc/data/access/lat/4yr_catalog/

Studying the Possibility of Separation of Primary Nuclei Groups in the Energy Interval 300 TeV – 10 PeV in the TAIGA-HiSCORE Experiment

L Sveshnikova[1], A Sh Elshoukrofy[1,2], E Postnikov[1], E Korosteleva[1] and H Motaweh[2]

[1] Lomonosov Moscow State University Skobeltsyn Institute of Nuclear Physics (MSU SINP), Leninskie gory, GSP-1, 119991, Moscow, Russia
[2] Faculty of Science, Damanhur University, El-Gomhouria St. 22516, Damanhur, El beheria, Egypt

E-mail: tfl10@mail.ru, evgeny.post@gmail.com, abeershehatamahmoud@yahoo.com

Abstract. We proposed a method of discrimination between light (p+He) and heavy (C+Fe) groups of primary nuclei. It's based on parametric analysis of lateral distribution of Cherenkov light in the atmosphere with approximation by the 'knee-like' fitting function, proposed and studied earlier. Two parameters most sensitive to the depth of shower maximum were revealed and used for analysis of the bulk of experimental data obtained in the TAIGA-HiSCORE experiment in the energy range 300–3000 TeV for various incident angles of particles. It is shown that the method allows estimating the contribution of light and heavy groups of primary nuclei. In the interval 300–3000 TeV we do not see a rapid decrease of the light component flux, as it was seen in the ARGO-YBJ experiment. This result will be refined after a more detailed analysis.

1. Introduction

The well-known problem of chemical composition in the range 300–3000 TeV below the knee in the cosmic-ray energy spectrum has not been solved yet due to the lack of statistics collected from direct experiment. For most extensive air shower (EAS) experiments energies lower than 1000 TeV are below threshold. In the TAIGA-HiSCORE experiment [1] due to the lower detection threshold (~100–200 TeV) we can try to distinguish showers from light and heavy groups of particles [2]. The lateral distribution functions (LDF) of Cherenkov light $Q(R)$, emitted by EAS originated from high energy primary particles (gamma rays, protons, nuclei), can be measured in details for millions of individual events with good accuracy, if the number of hit stations is sufficiently large (greater than 7) [2]. As it was well known, a steepness of the LDF is sensitive to the depth of EAS maximum and as a result to the sort of primary particles. In this work we continue the development of a parametric method of discrimination between heavy and light groups of nuclei using the 'knee-like' fitting function of the LDF, which were proposed in [3, 4]. The experimental uncertainties, caused by the procedure of measurement, were investigated and included in Monte Carlo simulations [2]. For preliminary testing of the method, we used data (several days of exposition from season 2016–2017 observation) obtained in the TAIGA-HiSCORE experiment [1].

2. Parameters of the LDF sensitive to the sort of primary particles

We developed the function, called the 'knee-like approximation', and studied its properties and applications on CORSIKA [5] simulated events in [3, 4]. It is a function of the distance to shower core *R*, *Q*(*R*), whereas earlier [6] this function was used for describing the knee in the CR spectrum, *F*(*E*). It depends on five parameters *C*, γ_1, γ_2, R_0, and α:

$$F_{\text{fitting}} = CR^{\gamma_1}(1+(R/R_0)^{\alpha})^{\gamma_2/\alpha} \tag{1}$$

where *C*, γ_1, γ_2, R_0, and α are parameters of the Cherenkov light LDF, R_0 is the knee position, γ_1 is the slope of the LDF below the knee, $\gamma_2+\gamma_1$ is the slope of the LDF above the knee, and parameter α characterizes the sharpness of the knee.

We started this work using Monte Carlo simulated data and have revealed that the most sensitive parameter to the sort of primary nuclei is R_0, but it also slightly depends on energy. For the energy 300 TeV the practically linear dependence on the distance to the shower maximum:

$$\mathrm{d}X_{\text{max}} = (X_{\text{max}} - X_{\text{obs}})/\cos\theta$$

is shown in figure 1 for four sorts of primary particles: p, He, C, Fe.

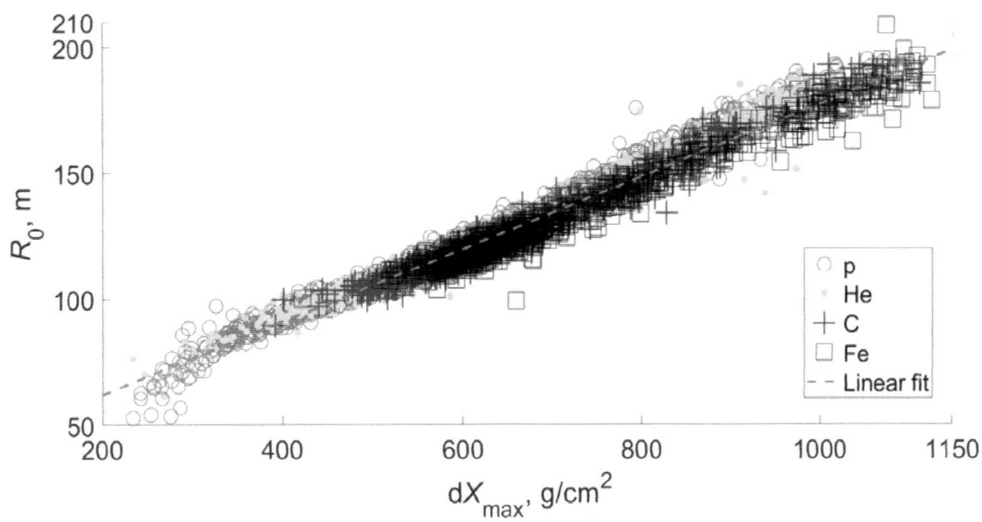

Figure 1. Dependence of parameter R_0 (from the function (1)) on the distance to the shower maximum for primary p, He, C, Fe.

We present the distribution of the parameter R_0 for two groups of showers, simulated by the CORSIKA code: the sample of light nuclei consisted of 50% protons + 50% helium nuclei, and the sample of heavy nuclei: 50% CO +50% Fe. R_0 strictly depends on X_{max}. In figure 2 it's seen that using the R_0 dependence one can try to separate light (p+He) and heavy (C+Fe) nuclei in Monte Carlo simulations.

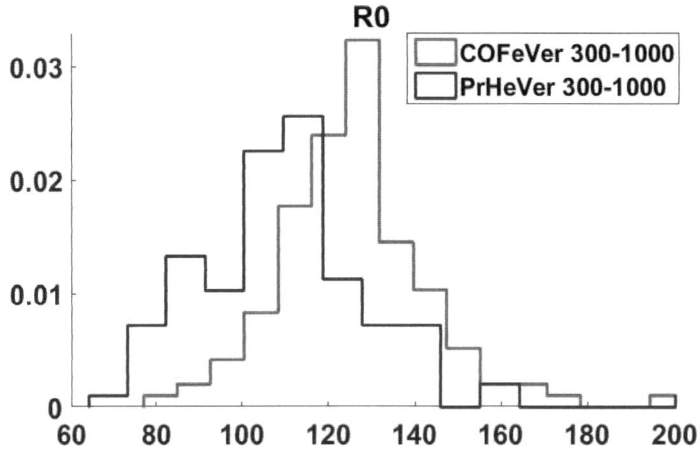

Figure 2. Distribution of the parameter R_0 for p+He (left curve) and C+Fe (right) for the energy range 300–1000 TeV and zenith angles 0–25°. MC simulations without experimental uncertainties.

However, our further study showed that the accuracy of R_0 estimation depends on the accuracy of $Q(R)$ measurement; hence, additional fluctuations of Q had to be studied and taken into account in MC simulations. We use the experimental sample of showers accumulated in the TAIGA-HiSCORE experiment during one day of 2017. The TAIGA-HiSCORE array is under development and in the season 2016–2017 it consisted of 28 stations. Each station contains 4 PMTs equipped with light-collecting Winston cones with the observation solid angle 0.6 sr and a light collection area 0.5 m^2, stations are distributed over the area of 0.25 km^2 with the step 106 m. The i-th optical station records the maximum amplitude A_i, arrival time T_i, and a total number of Cherenkov photons Q_i, obtained as the integral of the pulse. Only events with more than 12 triggered stations were considered for detailed analysis of the experimental LDF shape.

The standard procedure of the core position and arrival direction (θ, φ) reconstruction was applied for data analysis [7]. Thus, for each station we calculated the distance to the shower core, R (in the shower plane); obtained the experimental LDF, $Q_{\mathrm{exp}}(R)$; approximated it by the 'knee-like' fitting function (1) and estimated 4 parameters: R_0, γ_1, γ_2, α. In the first stage we studied the distribution of the Q_{exp} deviations from the fitting function value, Q_{fit}. Distributions of the value:

$$\mathrm{d}\lg Q = \lg Q_{\mathrm{exp}} - \lg Q_{\mathrm{fit}}$$

were analyzed for three narrow intervals of Q. Our study showed that for narrow intervals of Q a distribution $F(\mathrm{dlg}Q)$ can be successfully approximated by the Gaussian normal function. However, its standard deviation, sigma, strongly decreases with increase of Q (as it was expected). The estimated experimental distortion function was added to MC simulations.

3. Parametric approach to nuclei separation

After modelling the experimental uncertainties, the difference between R_0 values for the two groups of particles has narrowed. For that reason, one more parameter, γ_1, in fitting function (1) was added to the analysis. It reflects the steepness of the LDF at the centre of the Cherenkov light spot and it is also sensitive to the sort of primary particles.

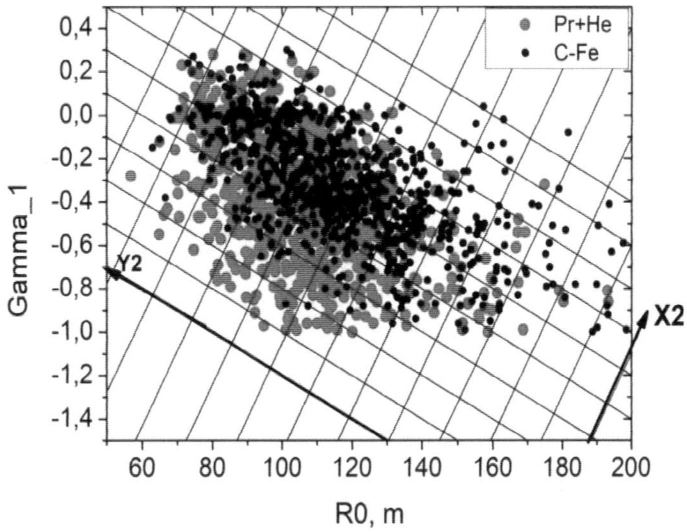

Figure 3. Correlation of the parameters γ_1 and R_0 for light (light circles) and heavy (dark circles) samples of nuclei with the energy 300–1000 TeV and zenith angles 0–25°. *X2*, *Y2* – a new coordinate system for the combined parameters.

As it is seen from the two-parametric plot γ_1–R_0 in figure 3, there are two areas with a noticeable difference between two samples, light and heavy nuclei. To stress this difference we introduced a new coordinate system *X2*, *Y2*, where *X2* is a combined parameter:

$$X2 = \gamma_1 - 0.013R_0, \tag{2}$$

and the lines correspond to different bins *k*:

$$X2(k) = -0.2 + 0.1\cdot(k-1) - 0.013R_0 \tag{3}$$

The combined parameter *X2* distribution is shown in figure 4 for three energy intervals: around 300 TeV (200–500 TeV), around 1 PeV (800–1300 TeV), and 3–10 PeV. Black curves with errors represent experimental groups of events; histogram bars demonstrate MC simulations for light (p+He) and heavy (C+Fe) nuclei; all showers having zenith angles 25–36°. For all three intervals the difference in MC simulation between the two groups is clearly seen.

The width of the experimental distribution became noticeably narrower for the energy greater than 3 PeV, whereas for the first two energy intervals the experimental distributions cover the full region from p to Fe nuclei. In this region the ARGO-YBJ experiment reported the observation of the p+He spectrum deviation from a single power law at energies lower than 1 PeV, which looks like a new sharp knee in the p+He component [8].

For the third energy interval (beyond the main knee in the cosmic ray spectrum) the experimental curve cannot cover the region occupied by proton-induced showers. As it is widely accepted, the origin of the knee is connected with a sharp cutoff of the proton spectrum at ~3 PeV and subsequent cutoff of other nuclei with the same rigidity. Hence, our method also confirms this conclusion and indicates a sensitivity to the primary nuclei content.

The accuracy of the method is not high as is well seen in figure 5, where the χ^2 value is presented depending on the ratio of primary p+He group among primary particles. The χ^2 value was calculated for three intervals of primary energy for the experimental and MC simulation curves presented in figure 4. One can also notice a sharp change of the χ^2 dependence on the ratio of light group of particles in the interval 4–10 PeV in comparison with that dependence around 300 TeV and around 1 PeV .

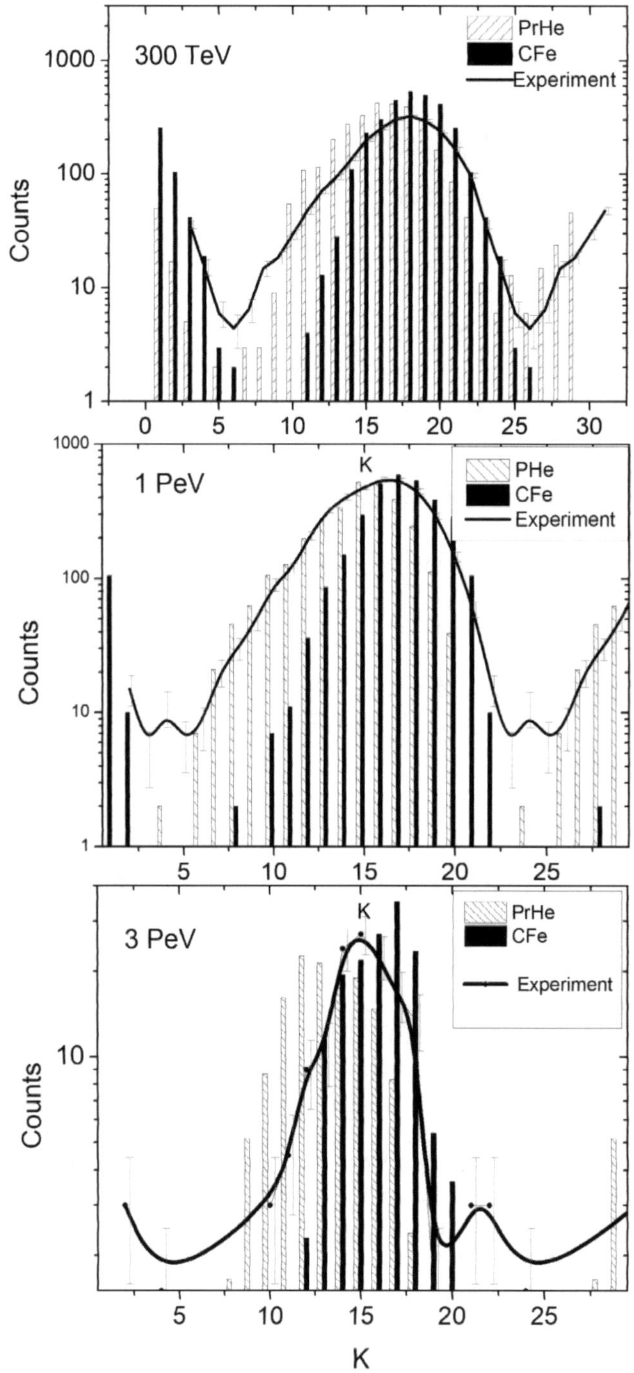

Figure 4. The combined parameter *X2* (2) distribution for three energy intervals; *k* is bin number (3). Black curves with errors – experimental groups of events. Histogram bars – MC simulations for light (p+He) and heavy (C+Fe) nuclei.

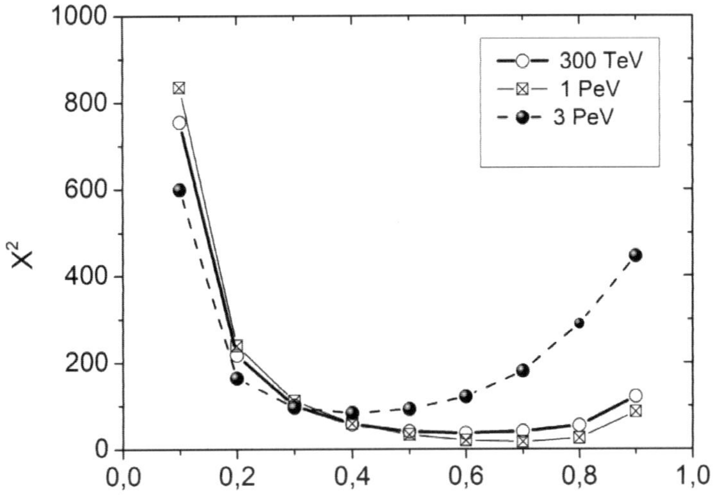

Figure 5. The χ^2 value depending on the ratio of primary p+He group among primary particles. The χ^2 was calculated for three intervals of primary energy for the experimental and MC curves presented in figure 4.

4. Conclusion

We proposed a method of discrimination between two groups of primary nuclei, light (p+He) and heavy (C+Fe), using two-parametric analysis of the lateral distribution of the Cherenkov light in the atmosphere. The LDF is approximated by the 'knee-like' fitting function (1), which was developed earlier [3, 4]. We used a small test sample of experimental events detected in the HiSCORE-TAIGA observatory in the energy range 300–10000 TeV and studied a sensitivity of the method by way of comparison of Monte-Carlo simulation and experiment. Our method does not reveal a rapid decrease of the light component flux up to 3 PeV in contrast to the ARGO–YBJ experiment [8], but indicates some decrease of p+He component's contribution beyond 3 PeV, thereby demonstrating the applicability of the method to mass groups study. To confirm this result, test sample size can be increased by several orders of magnitude; that will be done in the nearest future.

Acknowledgments. This work was supported by the RFBR, grant 16-29-13035.

References

[1] Kuzmichev L *et al.* (TAIGA collaboration) 2017 *Proc. XIX Int. Symp. Very High Energy Cosmic Ray Interactions 2016 (Moscow) EPJ Web of Conferences* **145** 01001
[2] Elshoukrofy A Sh *et al.* (TAIGA collaboration) 2018 *Kne Energy* **3(1)** 136–141
[3] Elshoukrofy A Sh M, Korosteleva E E, Postnikov E B, Sveshnikova L G and Motaweh H A 2017 *Bull. Russ. Acad. Sci. Phys.***4** 453–456
[4] Elshoukrofy A Sh M, Korosteleva E E, Postnikov E B, Sveshnikova L G and Motaweh H A 2017 *Proc. XIX Int. Symp. Very High Energy Cosmic Ray Interactions 2016 (Moscow) EPJ Web of Conferences* **145** 19006
[5] Heck D, Knapp J, Capdevielle J N, Schatz G and Thouw T 1998 Report FZKA 6019 (Forschungszentrum Karlsruhe)
[6] Hoerandel J R 2003 *Astropart. Phys.***19** 193–220
[7] Prosin V V *et al.* (TAIGA collaboration) 2014 *Nucl. Instrum. Meth.* A **756** 94–101
[8] Montini P 2016 *IL NUOVO CIMENTO* C **39** 248

IOP Conf. Series: Journal of Physics: Conf. Series **1181** (2019) 012020 doi:10.1088/1742-6596/1181/1/012020

Cosmic Ray Physics with the LOFAR Radio Telescope

T Winchen [1] and A Bonardi [2], S Buitink [1], A Corstanje [2], H Falcke [2,3,4], B M Hare [5], J R Hörandel [2,3], P Mitra [1], K Mulrey [1], A Nelles [6], J P Rachen [2], L Rossetto [2], P Schellart [2,7], O Scholten [5,8], S ter Veen [2,4], S Thoudam [2,9], T N G Trinh [5]

[1] Astrophysical Institute, Vrije Universiteit Brussel, Pleinlaan 2, 1050 Brussels, Belgium,
[2] Department of Astrophysics/IMAPP, Radboud University, P.O. Box 9010, 6500 GL Nijmegen, The Netherlands,
[3] NIKHEF, Science Park Amsterdam, 1098 XG Amsterdam, The Netherlands,
[4] Netherlands Institute of Radio Astronomy (ASTRON), Postbus 2, 7990 AA Dwingeloo, The Netherlands,
[5] KVI-CART, University Groningen, P.O. Box 72, 9700 AB Groningen,
[6] Institut für Physik, Humboldt-Universität zu Berlin, Unter den Linden 6, 10099 Berlin, Germany,
[7] Department of Astrophysical Sciences, Princeton University, Princeton, NJ 08544, USA,
[8] Interuniversity Institute for High-Energy, Vrije Universiteit Brussel, Pleinlaan 2, 1050 Brussels, Belgium,
[9] Department of Physics and Electrical Engineering, Linnéuniversitetet, 35195 Växjö, Sweden

E-mail: tobias.winchen@rwth-aachen.de

Abstract. The LOFAR radio telescope is able to measure the radio emission from cosmic ray induced air showers with hundreds of individual antennas. This allows for precision testing of the emission mechanisms for the radio signal as well as determination of the depth of shower maximum X_{max}, the shower observable most sensitive to the mass of the primary cosmic ray, to better than $20\,\mathrm{g\,cm^{-2}}$. With a densely instrumented circular area of roughly $320\,\mathrm{m}^2$, LOFAR is targeting for cosmic ray astrophysics in the energy range 10^{16}–10^{18} eV. In this contribution we give an overview of the status, recent results, and future plans of cosmic ray detection with the LOFAR radio telescope.

1. Introduction

The detection of cosmic rays via the radio technique has matured from application in single engineering projects only to usage on large scales [1]. All major experiments in the field nowadays use this technique at least complementary with more traditional methods as surface detector arrays or fluorescence telescopes. The radio technique allows for precision measurements of the depth of shower maximum of air showers X_{max} and the energy deposit in the atmosphere. Observation of cosmic rays with radio telescopes such as the Low Frequency Array (LOFAR) [2] and potentially also the future Square Kilometer Array (SKA) [3] provide observations of individual cosmic ray air showers with a high number of antennas, enabling testing details of the radio emission mechanism and shower development. High precision measurements of the cosmic ray mass composition in the energy range below the ankle may be important to understand the expected transition between galactic and extra-galactic cosmic rays [4].

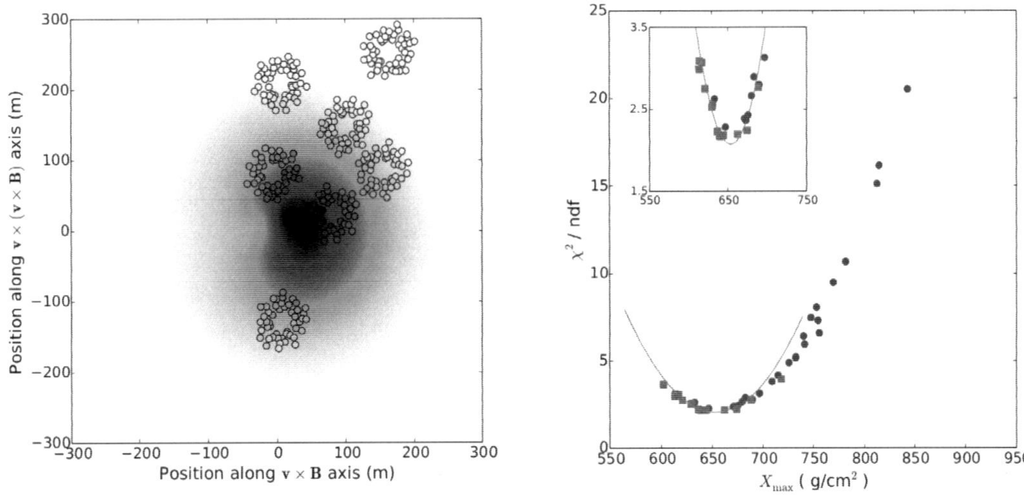

Figure 1. Event reconstruction with LOFAR. **(left)** Amplitude of the radio signal in the shower plane. Color in circles denote the observed power in the LOFAR LBA, color in the background denotes the signal from simulation of a shower with corresponding input parameters. **(right)** Reduced chi-square from the comparison of simulated iron (magenta squares) and proton (blue dots) showers with varying X_{max} with the data.

LOFAR is the first digital radio telescope with more than 50 stations distributed throughout Europe. Twenty-four of these stations are located in a dense core in the Netherlands. Five of the core stations are located within a circle of 320 m radius, the so called 'superterp'. Each core stations consists of 96 'low-band antennas' (LBA) operating from 10–90 MHz and 768 high-band antennas (HBA) operating from 110–240 MHz. Both antennas are omnidirectional V-shaped dipoles, but HBA are analog beamformed for astronomical observations which diminish their usability for cosmic ray physics. In the following we will thus discuss LBA only.

2. Cosmic Ray Detection With LOFAR

Cosmic ray detection with LOFAR currently relies on detecting the particle cascade with an array of twenty scintillators, the LOfar Radboud air shower Array (LORA) [5]. These scintillators are arranged in groups of four around the stations on the superterp. If thirteen out of the twenty stations simultaneously detect particles, indicating a high energy shower, 2 ms out of 5 s of buffered antenna data are read out for further offline analysis.

The properties of the primary cosmic-rays are reconstructed by the comparison of the data with dedicated simulations using the CORSIKA [6] and CoREAS [7] software packages. This yields a higher resolution on X_{max} then using parametrizations of the radio profile in ground plane [8] and simplifies the evaluation of the detection efficiency for the given shower parameters.

An example of a detected event and the reconstruction is shown in Figure 1. The left plot shows an example of a recorded event together with the best fitting simulated shower. The plot on the right the reduced chi-square as measure of the goodness of fit for all simulated showers. The X_{max} of the individual shower is obtained from a fit of a parabola to the simulated values.

With this procedure, LOFAR measures the cosmic ray mass composition in the energy range 10^{17}–$10^{17.5}$ eV. The reconstruction achieves an energy resolution of 32 % and a statistical uncertainty of 16 g cm^{-2} on X_{max}. The total systematic uncertainty on the energy calibration

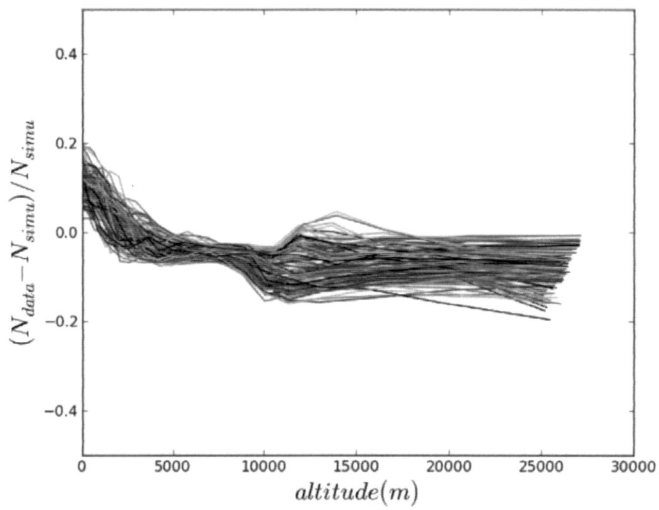

Figure 2. Variation of the refractive index along the shower profile for 100 different cosmic ray events observe with LOFAR.

is 27 % and on the depth of shower maximum $^{+14}_{-10}$ g cm^{-2}. Improvements of the accuracy of the mass-composition data in the energy range covered by LOFAR may be vital to discriminate between models for the origin of galactic and extragalactic cosmic rays.

3. Improvement of Systematic Uncertainties

3.1. Atmospheric Corrections

The observed intensity distribution of the radio signal depends on the difference in propagation time of particles and radio waves, and therefore on the refractivity profile of the atmosphere at the time of the event. The refractivity of the atmosphere depends on the density and humidity and is thus expected to vary also on short time scales [9]. In Figure 2 the difference of the refractive index at the time of 100 individual cosmic ray events to the US standard atmosphere [10] is shown. At an altitude of 5–8 km where the shower maximum is located, and thus the bulk of radiation is emitted, the variation is of 3-5%. This corresponds to an uncertainty in the inferred $X_{\max}$ of 3.5–11 g cm^{-2} [9].

So far in LOFAR and, to our knowledge, also in all other cosmic ray experiments, standardized average atmospheric models have been used for the interpretation of all showers. To account for the variation of the index of refraction, we started to use refractivity profiles obtained from atmospheric data of the Global Data Assimilation Project (GDAS) [11] in our simulations. For this, we contributed a modification to the CORSIKA simulation software to consistently use arbitrary atmospheric profiles in the simulation of the particle cascade and radio signals. The modification is included in CORSIKA version 7.6300 and following. For this purpose we fit the five layer atmospheric model to the available data for the location of the observatory and calculate the resulting index of refraction. The fit parameters describing the atmosphere layers, respectively the tabualted index of refraction as a function of height, is then fed into the particle and radio simulation. We developed a corresponding software `gdastool` that manages data download, fit, and steering file generation. This tool is also distributed alongside the CORSIKA source code [12].

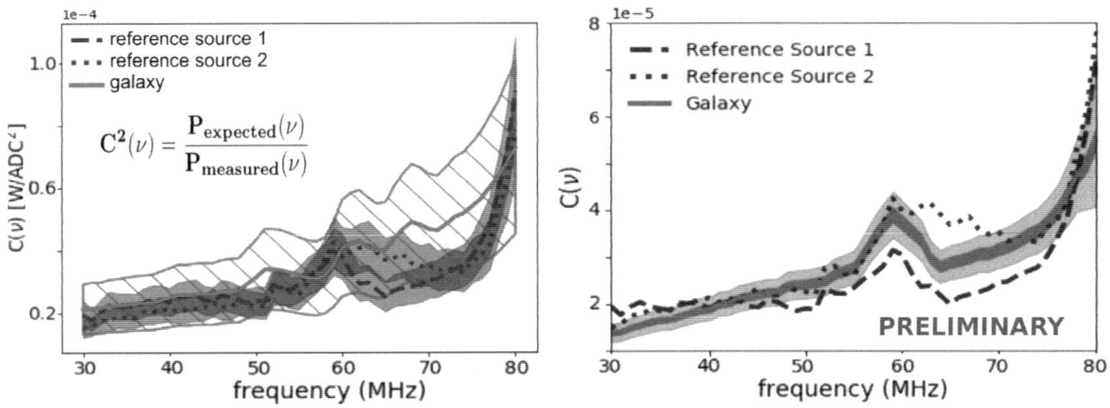

Figure 3. Antenna calibration factor as function of the frequency for galaxy calibration and reference source. **(left)** Without detailed model of the LOFAR signal chain and **(right)** with detailed model of the LOFAR signal chain.

3.2. Antenna Calibration

To reconstruct amplitude and spectral shape of radio signals an accurate calibration of the receiving system is imperative. However, at meter wavelength achieving the accuracy in calibration necessary for cosmic ray physics with antenna systems is challenging. No anechoic chambers are easily accessible that can contain the antennas to be calibrated as well as the reference sources in the far field region. Therefore even acquiring reference sources for in-situ calibrations that are well defined in the desired frequency range is difficult. Alternatively to artificial reference sources, the ubiquitous galactic background can be used as reference. Compared to artificial sources the signal emitted by the galaxy is well understood in amplitude and shape. However, the galaxy is not a point source so that calibrations using the galaxy have to be based on a model for the directivity of the antenna.

So far we used a calibration obtained from an in-situ measurement with a reference source optimized for frequencies higher than our observation band. The calibration with a reference source is consistent with a calibration using the galaxy [13]. However, by modelling of the noise contributions of the individual components of the signal processing chain we reduced the uncertainty of the galaxy calibration to a value lower than the uncertainty in calibration using the reference sources. The calibration constants obtained with and without detailed modelling of the full signal chain methods are displayed in Figure 3. In particular in the frequency range between 60–80 MHz the uncertainty on the galaxy calibration is now smaller than the systematic uncertainty arising from two different calibrations of the reference source provided by the manufacturer. This reduces a frequency dependent bias of the measurement of the slope of the spectral shape of the signal as shown in 4. This also enables detailed analysis of the frequency spectrum of the observed events that may be used as an additional observable sensitive to $X_{\max}$ [14].

4. Future Extensions

Several extensions are currently planned, respectively already under construction, that will increase the amount of data and extend the energy range for astrophysics analysis with LOFAR.

4.1. LORA Expansion

In April 2018 we installed infrastructure for twenty more particle detectors at five LOFAR stations close to the superterp. Installation of the detectors and electronics is expected to be finished

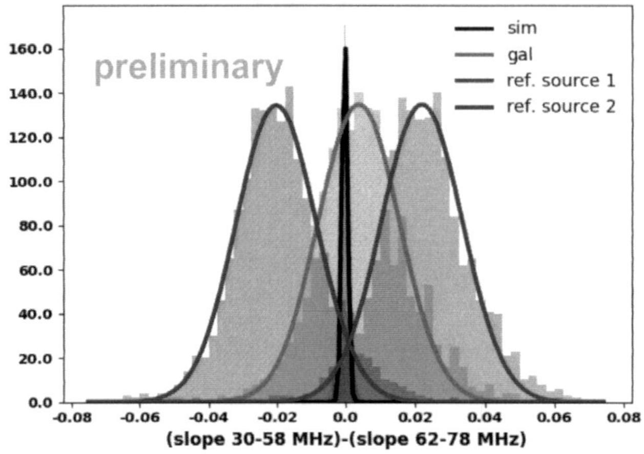

Figure 4. Difference of the slope of the frequency spectrum at high and low frequencies for LOFAR data using different calibrations and simulations.

within 2018. These additional detectors will increase the effective area for cosmic ray detection by approximately 40% and enable the development of new trigger modes for specific classes of events. Of particular interest are showers whose Cherenkov cone is located outside the superterp. While the radio signal is greatly diminished in this geometry, it may be possible to retrieve more information on the longitudinal shower development, thus improving composition measurements and eventually constraining hadronic interaction models.

4.2. Advanced Trigger

The trigger rate for cosmic ray detection is limited to approximately one trigger per hour to not disturb astronomical observations of the LOFAR telescope. This is achieved by requiring a signal over threshold in 13 out of 20 LORA particle detectors. This yields virtually 100% detection efficiency for showers above an energy of 0.1 EeV, thereby defining the lower energy threshold for the data useable for composition studies. However, as a consequence of these trigger conditions most of the recorded showers are of lower energies and approximately 80% do not even contain an observable radio pulse.

Reduction of the energy threshold for bias-free composition measurements while keeping the trigger rate constant requires to focus on showers with a detectable radio signal only. This is possible with a hybrid trigger that requires a strong signal in the antennas in coincidence with a signal in the particle array. Such an hybrid trigger will be implemented using a monitoring channel implemented in the LOFAR system that records the voltage levels in the antennas and can be used to send information to the LORA system.

Still, the current setup in LOFAR and all other radio experiments outside Antarctica requires particle detectors to trigger on cosmic ray air showers. Triggering on radio data only would instead allow for a very cost-effective increase of the instrumented area. Therefore we are developing a self-trigger aiming for a high reduction of RFI [15].

4.3. LOFAR 2.0

In the current configuration, LOFAR observations cannot overlap and data of HBA and LBA cannot be recorded at the same time. As currently roughly 50% of the observations use HBA, and also data in-between observations cannot be processed, the available time for cosmic ray observations is limited. LOFAR 2.0 is a planned extension of the technical capabilities of LOFAR allowing in particular for simultaneous observations with high-band and low-band antennas and also to perform multiple observations independently in parallel. This extension can thus significantly increase the available cosmic-ray data.

5. Conclusion

Precise measurements of the mass composition of cosmic rays are important to understand their origin and propagation. LOFAR continues to measure the cosmic ray mass composition with high precision in the expected transition region between cosmic rays of galactic and extragalactic origin. Several improvements of the analysis chain are being implemented that will allow for the reduction of the systematic uncertainties on composition and energy measurement. Additional upgrades are currently planned, respectively are already under construction, that will increase the duty cycle of LOFAR for cosmic ray measurements and reduce the energy threshold for mass composition analysis.

Acknowledgements

The LOFAR cosmic ray key science project acknowledges funding from an Advanced Grant of the European Research Council (FP/2007-2013) / ERC Grant Agreement n. 227610. The project has also received funding from the European Research Council (ERC) under the European Union's Horizon 2020 research and innovation programme (grant agreement No 640130). We furthermore acknowledge financial support from FOM, (FOM-project 12PR3041-3) and NWO (Top Grant 614- 001-454, and Spinoza Prize SPI 78-409). TW is supported by DFG grant WI 4946/1-1. LOFAR, the Low Frequency Array designed and constructed by ASTRON, has facilities in several countries, that are owned by various parties (each with their own funding sources), and that are collectively operated by the International LOFAR Telescope foundation under a joint scientific policy.

References

[1] Huege T 2016 *Phys. Rept.* **620** 1–52 (*Preprint* 1601.07426)
[2] van Haarlem M P *et al.* 2013 *Astronomy and Astrophysics* **556** A2 (*Preprint* 1305.3550)
[3] Huege T *et al.* 2015 *Proceedings, Advancing Astrophysics with the Square Kilometre Array (AASKA14): Giardini Naxos, Italy, June 9-13, 2014* vol AASKA14 p 148 (*Preprint* 1408.5288)
[4] Thoudam S, Rachen J P, van Vliet A, Achterberg A, Buitink S, Falcke H and Hörandel J R 2016 *Astron. Astrophys.* **595** A33 (*Preprint* LORA)
[5] Thoudam S *et al.* 2014 *Nucl. Instrum. Meth.* **A767** 339–346 (*Preprint* 1408.4469)
[6] Heck D *et al.* 1998 CORSIKA: a Monte Carlo code to simulate extensive air showers. Tech. Rep. FZKA 6019 Forschungszentrum Karlsruhe
[7] Huege T, Ludwig M and James C W 2013 *AIP Conf. Proc.* **1535** 128 (*Preprint* 1301.2132)
[8] Nelles A *et al.* 2015 *JCAP* **1505** 018 (*Preprint* 1411.7868)
[9] Corstanje A *et al.* 2017 *Astropart. Phys.* **89** 23–29 (*Preprint* 1701.07338)
[10] National Aeronautics and Space Administration (NASA) 1976 U.S. standard atmosphere Tech. Rep. NASA-TM-X-74335 NASA
[11] NOAA Air Resources Laboratory (ARL) 2004 Global data assimilation system (gdas1) Tech. rep. NOAA-Air URL https://ready.arl.noaa.gov/gdas1.php
[12] Mitra P *et al.* 2018 *Proceedings of the ARENA Conference (2018), Catania, Italy*
[13] Nelles A *et al.* 2015 *JINST* **10** P11005 (*Preprint* 1507.08932)
[14] Rossetto L *et al.* 2017 *Proceedings of the 35th International Cosmic Ray Conference (ICRC), Busan, Republic of Korea* (*Preprint* 1711.07749)
[15] Bonardi A *et al.* 2017 *Proceedings of the 35th International Cosmic Ray Conference (ICRC), Busan, Republic of Korea* (*Preprint* 1711.07749)

IOP Conf. Series: Journal of Physics: Conf. Series **1181** (2019) 012021 doi:10.1088/1742-6596/1181/1/012021

On the cosmic ray energy spectrum "knees"

Yu V Stenkin[a,b] and O B Shchegolev[a]

[a] - Institute for Nuclear Research, Russian Academy of Sciences, Moscow, 117312 Russia
[b] - National Research Nuclear University MEPhI (Moscow Engineering Physics Institute), Moscow, 115409 Russia

e-mail: yuri.stenkin@rambler.ru

Abstract. Primary cosmic ray energy spectrum around and above 1 PeV is of great interest due to its non-power law behavior found many years ago using the indirect EAS (Extensive Air Shower) method. The method is based on secondary particles measuring on Earth's surface under a thick atmosphere. Traditionally people use detectors sensitive to ionization produced mostly by electromagnetic component and so called "knee" was found for EAS size spectrum many years ago. Later it was assigned to a steepening of cosmic ray spectrum at 3-5 PeV. Recently some new "knees" were claimed by high altitude experiments, for primary protons and helium: at ~200-300 TeV (Tibet ASγ) and at ~700 TeV (ARGO-YBJ) thus widening the "knee" region from ~0.2 to 5 PeV and demonstrating disagreement with the existing experimental data. The natural explanation of such a strange spectrum behavior can be found in the phenomenological approach to the knee problem.

Introduction

Up to 1949 Extensive Air Shower (EAS) was considered as a pure electro-magnetic (e-m) cascade in atmosphere. Then George Zatsepin [1,2] had shown that this simplification was not true and EAS is a hadronic cascade, while e-m component is produced by π^0 and K^0 decays. This results in the two components are in equilibrium and all EAS features are defined mostly by the hadronic component being a "skeleton" of the shower. The latter means one needs to study hadronic component first of all. But, due to absence of a cheap, large and fast enough hadron detector, up to date people measure mostly e-m component, sometimes muonic one and very rarely hadronic one. Up to date people use e-m theory of cascade development (NKG - function, ages, N_e, etc.) and use N_e (number of electrons) as energy estimator when recovering primary particle energy. Up to date nobody put lower limit to primary energy when the EAS method does work properly. Probably this is a reason for appearance of 3 various "knees" for light primaries: at ~200 TeV (Tibet ASγ), at ~700 TeV (ARGO-YBJ) and at 3-5 PeV (KASCADE and many others) last years. Are all of them real (astrophysical) or "artificial" (systematic)?

Phenomenological approach to the "knee" problem proposed in 2003 [3,4] allowed us to look to the problem from another side and could explain many "puzzles" observed during decades in cosmic ray physics, including the "multiple knee" problem.

1. Phenomenology of EAS

Experimental data accumulated over a long period of ~ 70 years of EAS observations have a lot of contradictions and non-statistical dispersion. Situation in cosmic ray physics is now similar to that 70 years ago, lead Zatsepin to his discovery. By introducing hadronic component to EAS, he had explained such discrepancies as: wide particle lateral distribution, slow attenuation in atmosphere, very large fluctuations, etc. Hadronic cascade in air forms a "hadronic skeleton" defining all EAS properties, while e-m component is secondary one, being in equilibrium with cascading high energy hadrons through decays of neutral pions and kaons.

This idea was later confirmed in many experiments and nobody could doubt in its correctness. But, the problem is that not Zatsepin nor anybody else up to date put a lower energy limit to the EAS' method usage. PeV energy region is very high for e-m cascade where particle's critical energy is low (~80 MeV in air) while number of particles is very high. In hadronic cascade mean particle energy is much higher while number of cascading hadrons (N_{ch}) is low. One should keep in mind that pion production threshold put a limit to ability of hadrons to be involved in the cascading process and this results in significant reduction of the cascading hadrons N_{ch} in comparison with the total hadron number. This is a discrete value and cannot be less than 1. If the last cascading hadron lost its energy and stopped or decayed or captured by a nucleus than N_{ch}=0 and the cascade becomes hadronless or *coreless* [3] and equilibrium between EAS components becomes broken [4] (actually it is broken when N_{ch}<10). Properties of such showers differ significantly from that of normal EAS. When looking to old experimental data or making EAS simulations one could see that this crucial point in EAS development (N_{ch}=0) corresponds to $N_e \approx 10^6$ and primary energy $E_{0t} \approx 100$ TeV/nucleon (defined by the Earth's atmosphere thickness). Below this energy shower is coreless one, looking as very old electromagnetic shower with muons addition, it has shorter attenuation length, wider particle lateral distribution (not following NKG-function), etc. and cannot be processed as normal shower. Its age parameter is equal to ~2 resulting in poor core location and thus erroneous Ne estimation. The EAS method does not work properly in this region. Therefore, this energy (depending on altitude of observation) put a lower energy limit to EAS method usage. Processing of recorded showers altogether, one can obtain erroneous primary spectrum below the above-mentioned threshold E_{0t} being equal to $E_0 \approx 5$ PeV for primary iron.

2. Monte-Carlo simulations

EAS method of primary cosmic ray study is an indirect method when primary particle energy E_0 and mass A cannot be measured directly and are recovered through the experiment simulation. It is easy to show that if we have a pure power law primary spectrum $I(E_0) \sim E_0^{-\gamma}$ and a secondary component N_x also following power law dependence on E_0:

$$N_x(E_0) \sim E_0^{\alpha} \qquad (1)$$

then distribution on this component also follows power law function looking as:

$$F(N_x) \sim N_x^{-\gamma/\alpha} \qquad (2)$$

The latter means both parameters γ and α can be responsible for the above distribution behavior. α – parameter can be obtained only through calculations and therefore it depends on EAS phenomenology, on interaction models used in calculations, on Monte-Carlo statistics and on the used fitting functions, etc. Sure, the calculations should be done with the highest possible accuracy because otherwise any change of measured spectrum slope can be assigned to primary spectrum. We have made such simulations, using CORSIKA (v. 7560, QGSJET+FLUKA) and GEANT4.10, for the future PRISMA-LHAASO experiment being now under construction in China [5] at 4400 m a. s. l. Results on correlation between E_0 and N_e is shown in figure 1 for primary gamma, proton, nitrogen and iron, pure power law spectrum with γ=2.7 and PRISMA-LHAASO-64 configuration (8x8=64 en-detectors with 5 m spacing). Only EAS' producing at least 6-fold coincidences of detector hits with energy deposit above 3 m.i.p. in each, were selected for analysis. Looking to the pictures one can see that such a "comma-like" distribution cannot be fitted by a straight line below $Log(N_e) < 6 \div 6.5$ or $E_0 < 1$ PeV, especially for light primaries. Widely used simplefigure 4 scatter plot is not able to show such "comma". Such a behaviour is explained by a transition stage from coreless to normal showers, incorrect axis location and thus erroneous calculation of Ne in this region. Only above this EAS size, cascading hadrons reach observation level, the equilibrium between EAS components is achieved, hadronic axis is formed and EAS method begins working properly. Pure power law fit for heavy mass composition published by the Tibet ASγ experiment [6] is also shown along with our best power law fits for each primary. Rather good agreement with our iron curve is established in absolute values but not in the slope.

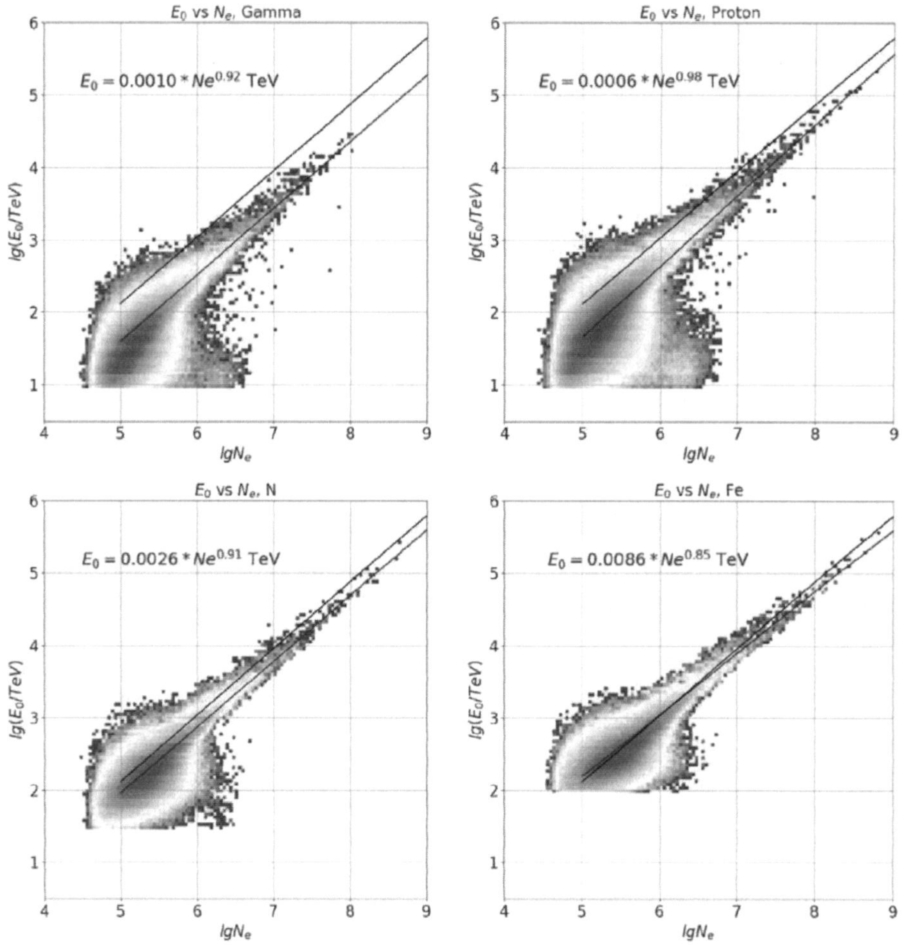

Figure 1. Correlation 3-dimension scatter plots between N_e and E_0 for different primaries: γ, p, N, Fe. Strait upper lines show best fit used by Tibet ASγ collaboration for the same altitude and our best linear fit in log-log scale (lower lines).

IOP Conf. Series: Journal of Physics: Conf. Series **1181** (2019) 012021 doi:10.1088/1742-6596/1181/1/012021

Nevertheless, the primary energy spectrum can be recovered using nonlinear fit as it is shown in figure 2, where results of simulation of PRISMA-LHAASO-64 is plotted using averaged points with errors. One can see here that power law fit similar to equation (2) cannot fit the data shown in both panels. It is interesting that $N_e(E_0)$ dependence slope changes just at the point where ARGO-YBJ experiment [7] claimed existence of a "knee" at ~700 TeV. Referring to equation 2 one could note that parameter α is highly likely responsible for the observed "knee", not primary index γ. We have to note that the fitting curve for $E_0(N_e)$ function used for primary energy recovering, depends on A (the higher the A, the higher nonlinearity) and some suppositions on c. r. mass composition should be done *a priori*, making this method of primary energy estimation (through N_e) depending on the suppositions. The latter can cause additional systematic errors, while the main systematic error appears if one uses pure power law fit for N_e to E_0 recalculation thus assigning the "knee" in measured parameter N_e to primary energy spectrum. On our opinion it would be better to use another energy estimator: N_h or N_n if the detector array can measure hadrons or neutrons as our PRISMA-LHAASO experiment will do. In this case measured spectrum on N_n starts from the point when hadrons reach observational level and all coreless (hadronless) showers are automatically skipped and dependence on EAS size in neutrons follows power law [8].

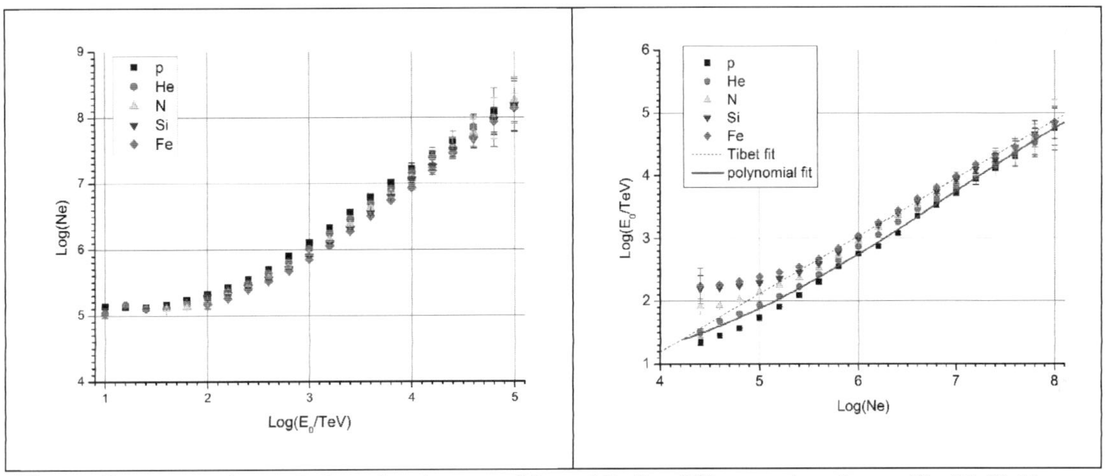

Figure 2. Averaged correlation plots between Ne and E_0 for different primaries: p, He, N, Si, Fe.
Left panel displays $N_e(E_0)$ dependence (for fixed E_0);
Right panel displays $E_0(N_e)$ dependence (for fixed Ne) where strait dashed line shows the fit used by Tibet ASγ collaboration while our best nonlinear fit for normal (light) composition is shown by a smooth curve.

Our simulations include the experimental conditions and all measurable parameters are written in the same format as experimental data and the same processing program is used both for experimental and simulated data. Results of such simulations for primary proton spectrum with the index equal to γ=2.7, is shown in figure 3.

3. Primary spectrum recovering

An interesting feature of the figure 3 is existence of a steep slope ≈1.95 below $\text{Log}(N_e)<6.5$, while above this point the spectrum becomes flatter in accordance with the primary slope. The slope difference is close to 0.45 being in a good agreement with the experimental value. We have to emphasize that the result corresponds to pure power law primary spectrum! If one applies power law fit when recalculating from N_e to E_0, then a "knee" will be "found" for pure power law primary spectrum. But, looking to figure 3

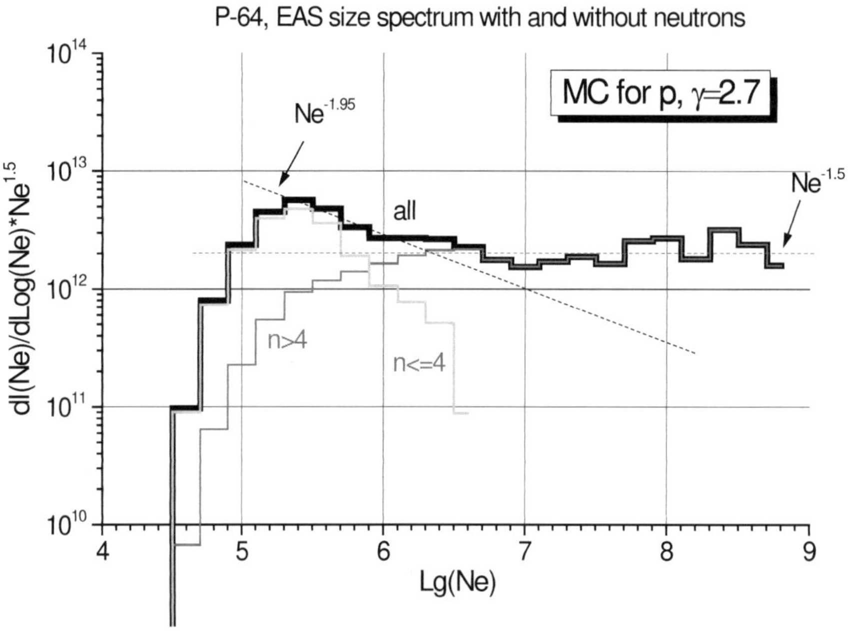

Figure 3. EAS size spectrum simulated for PRISMA-LHAASO-64.
Three groups of events are shown: with a number of recorded thermal neutrons n<=4 (neutronless), with n>4 (normal) and all events (thick histogram).

one can see that the steepening is caused by neutronless (coreless) showers marked as "n<=4" (note that neutron background included in the calculations is 1.3/event), while normal shower spectrum has no knee and goes asymptotically to the normal integral (or differential on a Log interval) slope ≈1.5. On the other hand, if one applies a polynomial fit when recalculating from N_e to E_0, then no knee is obtained [8]. The same result was obtained for primary spectrum recovered from N_n size measurements [8] being in a good agreement with direct spectrum measurements below 1 PeV.

Another interesting feature of simulated spectra can be seen in figure 4 where the same proton spectrum is compared with the proton spectrum having a knee at 1 PeV ($\gamma1=2.7$, $\gamma2=3.1$). As it is seen, the spectra below $Log(N_e)\approx6.5$ are not sensitive to the "knee" in primary spectrum and only above this size a difference can be seen. Similar spectrum behaviour (ankle-like) between 10-100 PeV can be found in results of all experiments using EAS method (see data compilation plot in [9]), thus confirming our explanation of the "knee" origin. On our opinion, only the flat part of the "ankle" at 10-100 PeV corresponds to primary spectrum, while the steep left part does not. Note again that the spectrum index obtained in different experiments above the "knee" depends on the supposition they made concerning the cosmic ray mass composition and how carefully the simulations have been done.

Conclusion

- On our opinion so-called "knees" in EAS size spectra are the features of EAS phenomenology – its specific behavior below some energy depending on altitude of observation.

- EAS method gives correct result only above Log(Ne) > ~6.5 (or E_0>1-5 PeV depending on altitude).

Only above this EAS size threshold, an equilibrium between EAS components is reached and EAS method works properly.

- There are two ways to solve the problem:
  - recording of hadronic component over full EAS area and using its number as an energy estimator;
  - make very careful simulations using traditional EAS method and take into account that recalculation from N_e to E_0 gives non power law function depending on primary mass A, altitude, trigger conditions, etc.

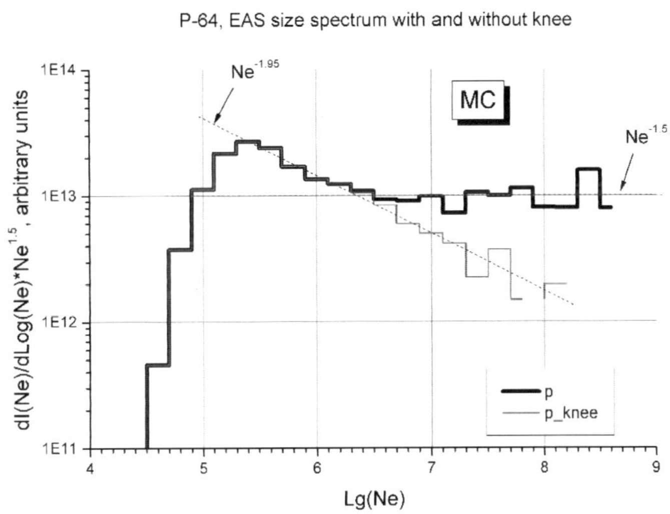

Figure 4. EAS size spectrum simulated for PRISMA-LHAASO-64 for primary proton spectrum with (thin line) and without a knee (thick line).

Acknowledgments

This work was supported by RFBR (18-02-00339, 16-29-13067 ofi_m) and "Physics of Fundamental Interactions and Nuclear Technologies" RAS Presidium Program.

References

[1] G T Zatsepin 1949 Nuclear-cascade process and its role in Extensive Air Shower development. *Doklady AN SSSR* 67 (6) 993. In Russian

[2] G T Zatsepin 1954 *PhD thesis (Doctor's dissertation).* Lebedev Phys. Inst. Moscow.

[3] Yuri V Stenkin 2003 *Mod. Phys. Lett.* A 18 (18) 1225.

[4] Yu V Stenkin 2008 *Physics Atomic Nucl.* 71 (1) 98.

[5] Victor Alekseenko, Anastasia Bagrova, Shuwang Cui, et al. 2017 The PRISMA-LHAASO project: Status and overview. *EPJ Web of Conferences* 145 07001.

[6] M Amenomori, et al., ArXiv:083.1005v1 [astro-ph] 6 Mar 2008.

[7] B Bartoli, et al., ARGO-YBJ and LHAASO Collaborations 2015 *Phys. Rev. D* 92 (9) 092005; [arXiv:1502.03164 [astro-ph.HE]] .

[8] Stenkin Yu V, Alekseenko V V, Cui S W, et al. 2017 Primary cosmic ray energy spectrum above 1 PeV as measured by the PRISMA-YBJ array. PoS(ICRC2017) 488.

[9] R U Abbasi, M Abe, T Abu-Zayyad, et al. 2018 The Cosmic-Ray Energy Spectrum between 2 PeV and 2 EeV observed with the TALE detector in monocular mode. ArXiv:1803.01288v1 [astro-ph.HE] 4 Mar 2018.

Possible explanation of results of CR investigations in the energy interval $10^{15} - 10^{17}$ eV: Nuclear-physical approach

A A Petrukhin and A G Bogdanov

National Research Nuclear University MEPhI (Moscow Engineering Physics Institute), Moscow, 115409, Russia

E-mail: AAPetrukhin@mephi.ru

Abstract. Among three possible approaches to explanation of the behaviour of CR energy spectrum and mass composition in the energy interval 10^{15} – 10^{17} eV: methodical, cosmophysical and nuclear-physical, the last of them is considered. A possibility of a new massive state of matter generation in interactions of CR with such energies is discussed. As a possible version, the production of blobs of quark-gluon plasma with a large orbital momentum is considered.

1. Introduction

Energy region ~ 10^{15} – 10^{17} eV is very interesting and important. Firstly, in this region direct measurements at satellites are impossible and only EAS investigations can be performed. Secondly, namely in this region basic characteristics of cosmic rays (CR): energy spectrum (the knee, the 2nd knee) and mass composition (becomes more heavy) are changed. But the second statement is incorrect since there are no direct experiments in which the energy spectrum and mass composition above 10^{15} eV are measured. Really we can measure EAS parameters only. The scheme of cosmic ray investigations by means of EAS observables is given in figure 1.

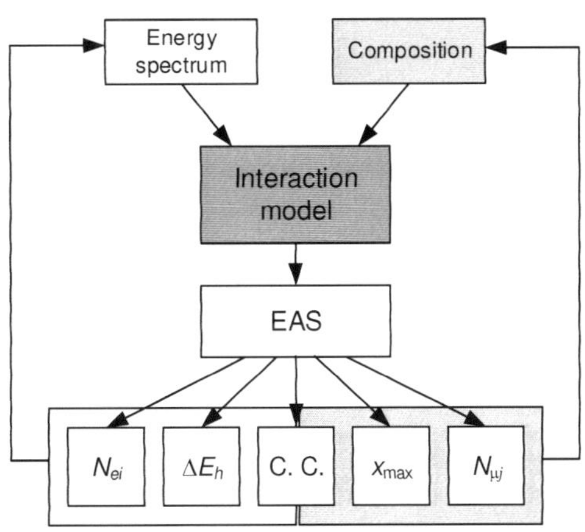

Figure 1. Existing approach to EAS data analysis [1].

What is the situation now? Today it is believed that the model of hadronic interactions is not changed significantly with increasing of energy at least up to 10^{17} eV. The reason is simple. No serious deviations from SM in pp-interactions up to energy 13 TeV (~ 10^{17} eV in laboratory system) in LHC experiments were observed. But cosmic rays consist mainly of nuclei (table 1) which interact with atomic nuclei of nitrogen or oxygen. Total amount of nuclei is about 60% at normal composition (without heaviering).

Table 1. Cosmic ray mass composition.

Particles	Z	A	Energy per nucleon	Energy per nucleus
Protons	1	1	92%	42%
α-particles	2	4	7%	21%
Light nuclei	3-5	10	0.25%	1%
Medium nuclei	6-10	15	0.5%	18%
Heavy nuclei	≥ 11	32	0.25%	18%

2. Why a new approach is required?

Now all changes in measured EAS parameters are explained by changes of the energy spectrum and mass composition. And at a first glance there are no reasons to change the interaction model.

But no exhaustive and consistent description of the knee for almost 60 years was proposed. At the same time, many unusual results in cosmic ray experiments at energies above the knee were obtained. In the last years, so-called "muon puzzle" in CR investigations appeared (excess of VHE muons and muon bundles compared with calculations). In nucleus-nucleus interactions even in LHC experiments serious deviations from existing models are observed (see figure 2).

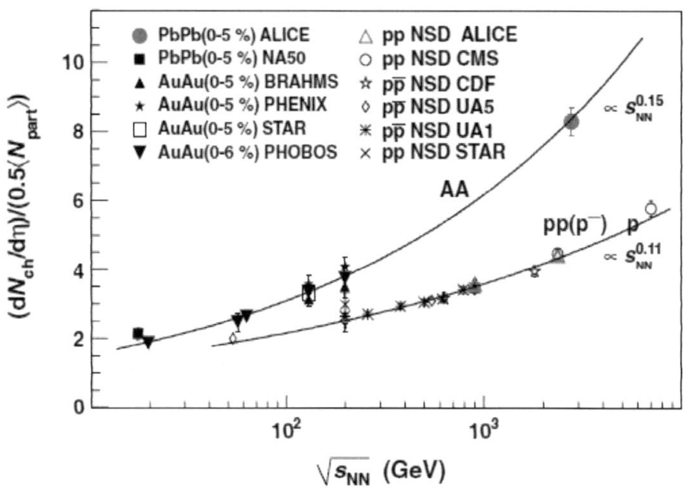

Figure 2. Multiplicities of charged particles produced in nucleus–nucleus and proton–proton collisions as a function of collision energy [2].

3. New model of nucleus-nucleus interaction

What do we need to explain all these data? Model of hadronic interactions which gives:

- threshold behaviour (all changes in CR appear at several PeV only);
- large cross section (to change EAS spectrum slope);
- increasing yield of secondary particles (excess of pions, muons and muon bundles).

Production of blobs of quark-gluon plasma (or better, quark-gluon matter, QGM) is very suitable for that.

Production of QGM provides two main conditions:

- threshold behaviour, since for that high temperature (energy) is required;
- large cross section, since the transition from quark-quark interactions to some collective interaction of many quarks and gluons occurs: $\sigma \sim \pi \lambda\!\!\!^{-2} \rightarrow \sigma \sim \pi R^2$, where R is a size of quark-gluon blob.

But for explanation of other observed phenomena a large value of orbital angular momentum is required. The question about the orbital momentum which must appear in nucleus-nucleus interactions is usually not considered. A possibility of its appearance in non-central ion-ion collisions was considered in paper [3]. Corresponding scheme is presented in figure 3. Further investigations showed [4] that the value of the orbital momentum can reach $L \sim 10^4$.

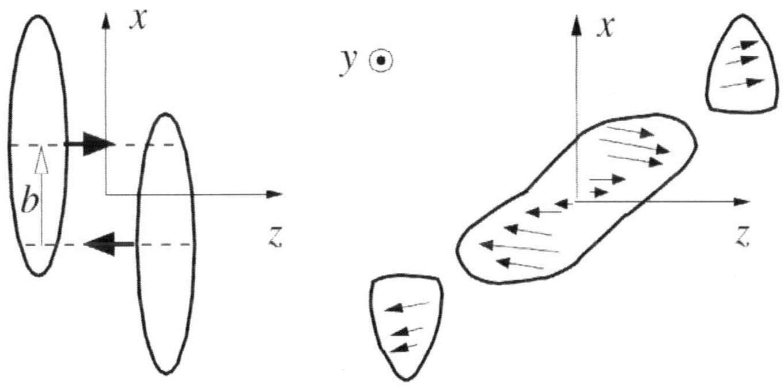

Figure 3. Production of orbital angular momentum in non-central ion-ion collisions [3].

A blob of globally polarized QGM with large orbital angular momentum can be considered as a usual resonance with a large centrifugal barrier. Centrifugal barrier $V(L) = L^2/2mR^2$ will be large for light quarks but much less for t-quarks or other heavy particles. Though in interacting nuclei t-quarks are absent, the suppression of decays into light quarks gives time for the appearance of heavy quarks.

How interaction is changed in frame of a new model? Simultaneous interactions of many quarks change the energy in the center of mass system drastically: $\sqrt{s} = \sqrt{2m_N E_1} \rightarrow \sqrt{2m_c E_1}$, where $m_c \approx n m_N$. At threshold energy, $n \approx 4$ (α-particle). Produced t-quarks take away energy $\varepsilon_t > 2m_t \approx 350$ GeV, and taking into account fly-out energy, $\varepsilon_t > 4m_t \approx 700$ GeV in the center of mass system. Top-quarks decay very quickly: $t(\bar{t}) \rightarrow W^+(W^-) + b(\bar{b})$, W-bosons decay into leptons (~30%) and hadrons (~70%); b-quarks produce jets which generate pions decaying into muons and neutrinos.

4. CR energy spectrum and mass composition in the new approach

How the energy spectrum is changed? One part of t-quark energy gives the missing energy (ν_e, ν_μ, ν_τ, μ), and another part changes EAS development due to increasing multiplicity of secondary particles. As a result, measured EAS energy E_2 will not be equal to primary particle energy E_1, and the measured spectrum will differ from the primary spectrum (figure 4). Transition of particles from energy E_1 to energy E_2 gives a bump in the energy spectrum near the threshold.

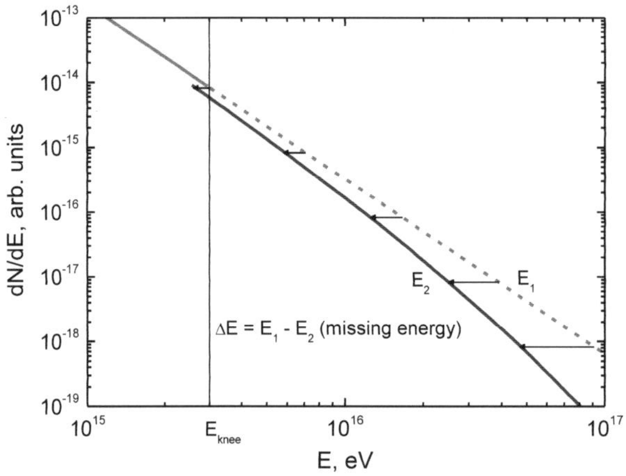

Figure 4. The change in the CR energy spectrum at the appearance of the missing energy.

How the measured composition is changed in frame of the new approach? Since for QGM production not only high temperature (energy) but also high density is required, threshold energy for production of new state of matter for heavy nuclei will be less than for light nuclei and protons. Therefore heavy nuclei (e.g., iron) spectrum is changed earlier than light nuclei and proton spectra. Measured spectra for different nuclei will not correspond to the primary composition. If to propose that CR composition is not changed with increasing of energy, the composition recalculated from results of EAS measurements begins to change due to increasing probability of heavy nuclei interactions.

5. Conclusion

To explain the behavior of CR energy spectrum and mass composition at energies $10^{15} - 10^{17}$ eV in the frame of nuclear-physical approach, the production of quark-gluon blobs with a large orbital momentum is required.

This approach can be checked as in cosmic ray so in LHC experiments. In cosmic rays some excess of very high energy muons and neutrinos (> 100 TeV) must appear. In LHC experiments some excess of *W*-bosons or/and *t*-quarks and missing energy must be detected.

Acknowledgements

Authors thank R.P. Kokoulin for fruitful discussions and help in preparation of this paper. This work was supported by Ministry of Science and Higher Education of the Russian Federation (government task and MEPhI Academic Excellence Project).

References

[1] Petrukhin A A 2013 *Acta Polytechnica* **53** (Suppl.) 707
[2] Aamodt K *et al* 2010 *Phys. Rev. Lett.* **105** 252301
[3] Zuo-Tang Liang and Xin-Nian Wang 2005 *Phys. Rev. Lett.* **94** 102301
[4] Jian-Hua Gao *et al* 2008 *Phys. Rev.* C **77** 044902

IOP Conf. Series: Journal of Physics: Conf. Series **1181** (2019) 012023 doi:10.1088/1742-6596/1181/1/012023

Origin of hardening and universality of cosmic rays spectra in GV–PV rigidity region

A A Lagutin, N V Volkov, R I Raikin and A G Tyumentsev

Altai State University, Radiophysics and Theoretical Physics Department. 61 Lenin ave., Barnaul, 656049, Russia.

E-mail: lagutin@theory.asu.ru, volkov@theory.asu.ru, altraikin@gmail.com, tyumentsev@theory.asu.ru

Abstract. Recent balloon-borne and satellite experiments have established new features in the behavior of the spectra of cosmic rays. An analysis of all the data showed that hardening of most abundant primary cosmic ray nuclei spectra with increasing rigidity is observed in > 200 GV region. At the same time, rigidity dependences of secondary cosmic rays are distinctly different. The AMS-02 data show that above 200 GV the secondary cosmic rays Li, Be, B harden more than the primary He, C, O.

In this paper we discuss a new scenario that self-consistently describes these new features of the cosmic ray spectra. We demonstrate that the measured by the AMS-02 changes in the slope of the cosmic ray spectra caused by the transition from the contribution of multiple distant Galactic sources, including the nuclear interactions of the particles accelerated by these sources with an interstellar medium during their wandering in the Galaxy, to the contribution of mainly local ones. We also found that the spectral universality is observed at rigidity $R > 10^5$ GV.

1. Introduction

In the last decade, measurements of the cosmic ray (CR) nuclei in the GV–TV rigidity region by new-generation balloon-borne and satellite instruments [1–8] have established new features in behavior of their spectra. It was found that both the spectra of most abundant primary CR nuclei proton, He, C, O, N, and the secondary cosmic rays Li, Be, B at rigidity $R > 100 - 200$ GV exhibit a hardening with increasing rigidity. As follows from figure 1, they deviate from a single power law. Above 60 GV, the three fluxes He, C, O measured by the AMS-02 have identical rigidity dependence within the measurement errors and the ratios He/O and C/O are well fit by a constant value [6].

The rigidity dependences of secondary and primary CRs are distinctly different. The AMS-02 data show that above 200 GV the secondary cosmic rays harden more than the primary He, C, O [7].

These newly discovered features are not easy to explain under standard scenario of cosmic ray origin, acceleration and their propagation in the Galaxy. Under the standard theory, the primary nuclei are thought to be produced, at least to several PV, by supernova remnant shock waves by diffusive shock acceleration mechanism that predicts power-law spectra $J \propto R^{-\gamma}$ with slope $\gamma \approx 2.0 - 2.2$. The subsequent CR transport in the turbulent Galactic magnetic fields is modeled as a diffusion process in quasi-homogeneous medium with the diffusion coefficient $D(R) = D_0(R/1\ \mathrm{GV})^{\delta}$, with $\delta \approx (0.3 - 0.8)$. Under these assumptions, the spectrum of primary

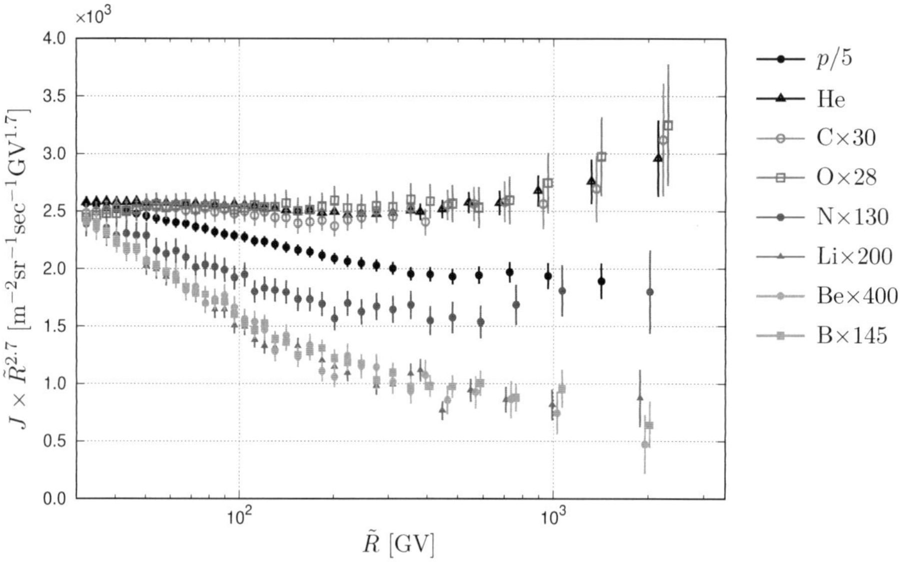

Figure 1. The AMS-02 primary and secondary cosmic ray fluxes with their total errors multiplied by $\tilde{R}^{2.7}$ as a function of rigidity above 30 GV [4–8]. The fluxes were rescaled as indicated. The He, O, Li, and B data points above 400 GV are displaced horizontally by $\sim 2\%$

nucleus i generated by the global-scale steady state distribution of sources $S(\mathbf{r})$ is described by a single power law with index $\eta = \gamma + \delta$, which is clearly at odds with the observed hardening of CR hadrons at GV–TV region.

Due to the fact that the predicted by the standard scenario observed spectrum is determined by the parameters γ, δ and the distribution of sources $S(\mathbf{r})$, the proposed solutions to the hardening problem include acceleration mechanisms modification, change of the CR diffusion properties in the different regions of the Galaxy and superposition of local and distant sources (see, e.g., [9–18]).

In this paper, we continue to investigate the spectral hardening problem observed at GV–TV rigidity region within the framework of the approach we proposed in our paper [19]. In that paper, we point to possible anomalies in the spectra of protons and helium, due to differences in the behavior of the energy spectra of distant and nearby sources in a highly inhomogeneous Galactic medium. However, the lack of experimental data for the range of $\sim 200 - 3000$ GeV at that time did not allow us to discuss these new features in [19].

In the current study, using the precision measurement of the primary and secondary cosmic ray nuclei spectra in the AMS-02 experiment [4–8], we demonstrate that the observed changes in the slope of these spectra caused by the transition from the contribution of multiple distant Galactic sources, including nuclear interactions of particles accelerated by these sources with an interstellar medium during their wandering in the Galaxy, to the contribution of mainly local ones.

2. Model

The following basic principles and key assumptions were included in our model.

(i) All particles with a rigidity $30 \lesssim R \lesssim 5 \cdot 10^7$ GV, observed in the Solar system and at Earth, are accelerated mainly by Galactic sources. The spectrum of accelerated particles in the sources is described by the power law $J \propto R^{-\gamma}$.

(ii) CR sources are divided into two groups as in our paper [19]. The first group includes the multiple old distant ($r \geq 1$ kpc) sources, the second group consists of nearby ($r < 1$ kpc) young ($t < 10^6$ yr) sources. The spatial distribution of sources also suggests the separation of the observed CR flux of the nucleus i into two components as follows:

$$J(\mathbf{r}, t, R) = J_G(\mathbf{r}, R) + J_L(\mathbf{r}, t, R).$$

Here

- J_G is the global spectrum component determined by the multiple old ($t \geq 10^6$ yr) distant ($r \geq 1$ kpc) sources, including the contribution of nuclear interactions of the particles accelerated by these steady state sources with an interstellar medium during their wandering in the Galaxy.
- J_L is the local component, i.e. the contribution of nearby ($r < 1$ kpc) young ($t < 10^6$ yr) sources; the effect of nuclear interactions of particles accelerated in these sources with an interstellar medium on the observed spectrum is negligible. It is not taken into account in this work.

(iii) The highly inhomogeneous distribution of matter and magnetic fields in the Galaxy leads to the anomalous diffusion of CRs [20]. Anomalous diffusion is manifested, in particular, by abnormally large free paths of particles (so-called "Lévy flights") with a power-law distribution $p(\mathbf{r}, R) \propto A(R, \alpha) r^{-\alpha-1}, r \to \infty, 0 < \alpha < 2$. Besides, a spatially intermittent magnetic field of the interstellar medium [21] results in a higher probability of a long stay of particles in inhomogeneities, leading to a presence of the so-called "Lévy traps". In the general case, the probability density function $q(t, R)$ of time t, during which a particle is trapped in the inhomogeneity, also has a power-law behaviour: $q(t, R) \propto B(R, \beta) t^{-\beta-1}, t \to \infty, \beta < 1$.

(iv) The equation for the density of particles with rigidity R at the location $\mathbf{r}$ and time t, generated in highly inhomogeneous medium by Galactic sources with a distribution density $S(\mathbf{r}, t, R)$ without energy losses and nuclear interactions can be written as [22, 23]

$$\frac{\partial N(\mathbf{r}, t, R)}{\partial t} = -D(R, \alpha, \beta) \mathrm{D}_{0+}^{1-\beta} (-\Delta)^{\alpha/2} N(\mathbf{r}, t, R) + S(\mathbf{r}, t, R). \quad (1)$$

Here $\mathrm{D}_{0+}^{1-\beta}$ denotes the Riemann-Liouville fractional derivative [24] and $(-\Delta)^{\alpha/2}$ is the fractional Laplacian ("Riesz operator") [24]. The anomalous diffusion coefficient is $D(R, \alpha, \beta) \sim A(R, \alpha)/B(R, \beta) = D_0(\alpha, \beta)(R/1 \text{ GV})^{\delta}$.
In case $\alpha = 2$, $\beta = 1$ we obtain Ginzburg-Syrovatsky's normal diffusion equation.

The solution of equation (1) for a point impulse source with emission time T and power-law injection spectrum $S(\mathbf{r}, t, R) = S_0 R^{-\gamma} \delta(\mathbf{r}) \Theta(T - t) \Theta(t)$ ($\Theta(\tau)$ is the step function) has the form [23]

$$N(\mathbf{r}, t, R) = \frac{S_0 R^{-\gamma}}{D(R, \alpha, \beta)^{3/\alpha}} \int\limits_{\max[0, t-T]}^{t} d\tau \tau^{-3\beta/\alpha} \Psi_3^{(\alpha, \beta)} \left(|\mathbf{r}| (D(r, \alpha, \beta) \tau^{\beta})^{-1/\alpha} \right), \quad (2)$$

where $\Psi_3^{(\alpha, \beta)}(\rho)$ is the density of the fractional stable distribution [25]

$$\Psi_3^{(\alpha, \beta)}(\rho) = \int\limits_0^{\infty} g_3^{(\alpha)}(r\tau^{\beta}) q_1^{(\beta, 1)}(\tau) \tau^{3\beta/\alpha} d\tau.$$

It should be noted that the spectrum (2) has the "knee" [19, 20, 22, 23]. The spectral index for observed particles η at the knee rigidity $R = R_k$ is equal to spectral exponent for particles generated by the source: $\eta|_{R=R_k} = \gamma$. One can also find from (2) that at $R \ll R_k$ and $R \gg R_k$ we have, respectively:

$$\eta|_{R\ll R_k} = \gamma - \delta, \quad \eta|_{R\gg R_k} = \gamma + \delta/\beta. \tag{3}$$

The spectrum of observed primary particles from a steady state source differs significantly from (2). It has no knee and the spectrum exponent turns out to be equal to the CR spectrum index above the knee [26]:

$$N_p(\mathbf{r}, R) \sim R^{-\gamma-\delta/\beta}.$$

The scenario proposed in the paper also assumes acceleration of secondary particles in Galactic sources. Some mechanisms for the acceleration of these nuclei were discussed in papers [27–31]. Since the equilibrium spectra of secondary CRs are $R^{-\delta}$ times softer than the primary ones, the spectrum of observed secondary particles from a steady state source in our model can be represented as

$$N_s(\mathbf{r}, R) \sim R^{-\gamma-\delta/\beta-\delta}.$$

3. Results

In the framework of the model described above the observed spectrum of the cosmic ray nucleus i due to all sources of the Galaxy may be presented in the form

$$J(\mathbf{r}, t, R) = \frac{v}{4\pi}\left[S_G R^{-\gamma-\delta/\beta-\varepsilon\delta} + \frac{S_L R^{-\gamma}}{D(R,\alpha,\beta)^{3/\alpha}} \times \right.$$
$$\left. \times \sum_{\substack{r_j<1\text{ kpc}\\ t_j<10^6\text{ yr}}} \int\limits_{\max[0,t_j-T]}^{t_j} d\tau \tau^{-3\beta/\alpha} \Psi_3^{(\alpha,\beta)}\left(|\mathbf{r}_j|(D(R,\alpha,\beta)\tau^\beta)^{-1/\alpha}\right)\right]. \tag{4}$$

It is clear from the physical point of view that the bulk of observed CRs with rigidity $R \sim 1 \div 10^2$ GV forms by numerous distant sources. It means that the CR fluxes in this rigidity region must be described by first term in (4). The parameter ε in this term is zero for primary nuclei and is assumed to be equal to one — for secondary nuclei. The second term defines the spectrum in the high energy region and, as discussed above, has a knee.

In the framework of our approach, we can self-consistently retrieve the main parameters of the diffusion model from experimental data. The key element in the retrieval of these model parameters is the presence of a knee in the spectrum (4). From equations (3), for example, we find

$$\gamma = \eta|_{R\ll R_k} + \delta, \quad \beta = \frac{\delta}{\eta|_{R\gg R_k} - \eta|_{R\ll R_k} - \delta}.$$

Since $\eta|_{R\ll R_k} \sim 2.56 \div 2.64$, $\eta|_{R\gg R_k} - \eta|_{R\ll R_k} \sim 0.6 \div 0.7$ [32, 33], $\delta \sim 0.27$ [7], last equations permit to estimate both spectral exponents γ and β:

$$\gamma \sim 2.8 \div 2.9, \quad \beta \sim 0.6 \div 0.8.$$

We note that a similar steep spectrum of accelerated particles has been observed from the supernova remnants W44 [34] and IC 443 [35] with the Fermi Large Area Telescope, and W49B with H.E.S.S. and Fermi-LAT [36]. The value $\gamma = 3$ has been found in [37] for RX J1713.7-3946.

At energies > 400 GeV, VERITAS and Fermi-LAT observe gamma-ray emission from Tychos SNR with power-law index ~ 2.92 [38].

To evaluate parameter α, general results for particles spectral exponent γ obtained in the framework of the diffusive shock acceleration theory extended to the case of anomalous transport with "Lévy flights" and "Lévy traps" have been used (see [Lagutin A.A., 2019, to be published]). It was shown that for $\beta = 0.6 \div 0.8$ spectral exponent $\gamma = 2.8 \div 2.9$ corresponds to nondiffusive transport with $\alpha \sim 1.7 \div 1.8$.

A set of anomalous diffusion model parameters adopted in this paper is given in Table 1. The spatial and temporal coordinates of the local sources are presented in [39]. The spherically symmetric force–field model [40] with $\varphi = 600$ MV is used to describe the solar modulation.

Table 1. The anomalous diffusion model parameters

Parameter	Value
γ	2.85
δ	0.27
$D_0(\alpha,\beta)$	$1.5 \cdot 10^{-3}$ pc$^{1.7}$y$^{-0.8}$
α	1.7
β	0.8
T	10^4 y

Figure 2 shows the model results of the spectra of nuclei measured by AMS-02 experiment [4–8]. Points on this figure are the results of AMS-02. In this and the subsequent figures, the data points are placed along the abscissa at $\tilde{R}$ calculated for a flux $f(R) = R^{-2.7}$ with the use of equation (5) from [41]. The solid green lines are the global spectrum component determined by the multiple old distant sources. Solid blue lines are the contribution of the local sources. The spectrum due to all sources of the Galaxy is shown by the solid red line.

Figures 3 and 4 show the rigidity dependence of the model spectral index of nuclei measured by AMS-02 and the flux ratios of some nuclei.

4. Conclusions

We have presented a novel scenario reproducing the features in the spectra of nuclei measured by AMS-02 experiment [4–8]. The key elements of this scenario are the following.

1. All particles with a rigidity $R \lesssim 5 \cdot 10^7$ GV, observed at Earth, are accelerated mainly by Galactic sources. The spectrum of these particles in the sources is $J \propto R^{-\gamma}$.

2. The highly inhomogeneous distribution of matter and magnetic fields in the Galaxy leads to the anomalous diffusion of CRs.

3. All galactic sources are divided into two groups — multiple old distant and nearby young ones. This separation is due to a significant difference in the spectra of distant and local sources. A much steeper spectra in GV–TV rigidity region of nuclei accelerated in distant sources, compared with the spectra from local sources, is determined by steady state mode of particle injection by an ensemble of these distant sources, as well as by nuclear interactions during their subsequent propagation in the interstellar medium.

4. Self-consistent technology to retrieve the main parameters of the model from experimental data is used. It is based on the fact that the spectrum of individual species has a knee.

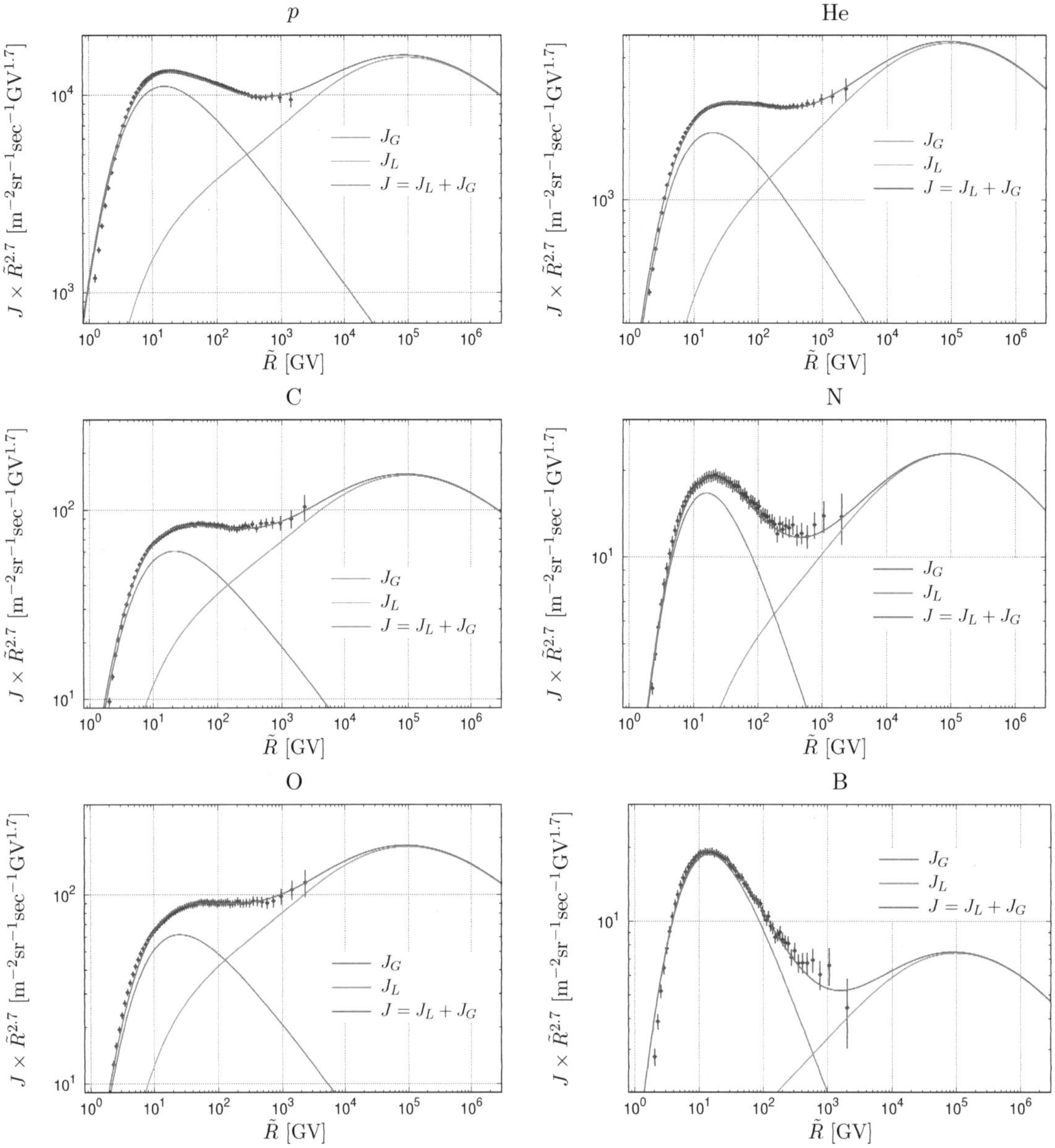

Figure 2. Spectra of p, He, C, N, O and B multiplied by $\tilde{R}^{2.7}$ as a function of rigidity, obtained in the anomalous diffusion model, compared with the AMS-02 measurements [4, 6]

We demonstrate that the observed changes in the slope of the cosmic ray spectra caused by the transition from the contribution of multiple distant Galactic sources to the contribution of mainly local ones. We also found that the spectral universality is observed at rigidity $R > 10^5$ GV.

The space missions AMS-02, DAMPE and CALET, which currently measure nuclear spectra with high precision in a wide rigidity region, could verify our model predictions in the near future.

IOP Conf. Series: Journal of Physics: Conf. Series **1181** (2019) 012023 doi:10.1088/1742-6596/1181/1/012023

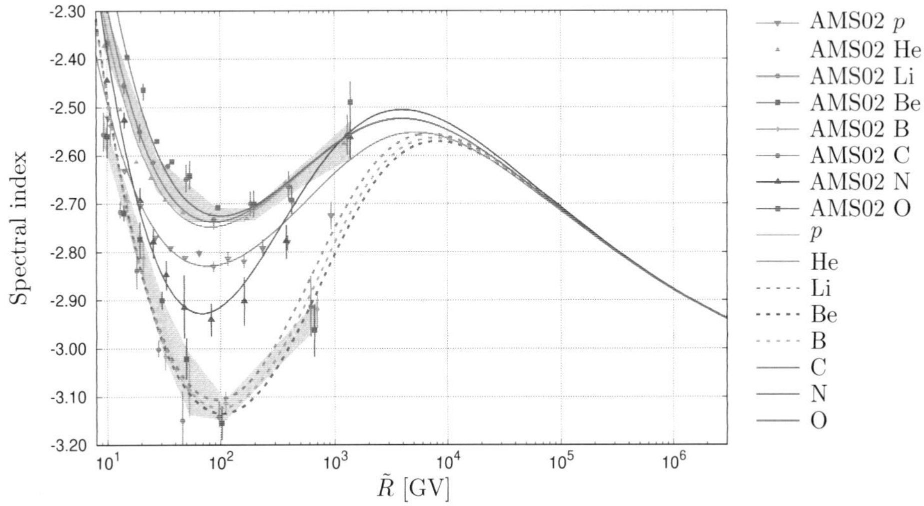

Figure 3. The dependence of the p, He, C , N, O and Li, Be, B spectral indices on rigidity in the anomalous diffusion model. Data points and the shaded region are from [5, 7]

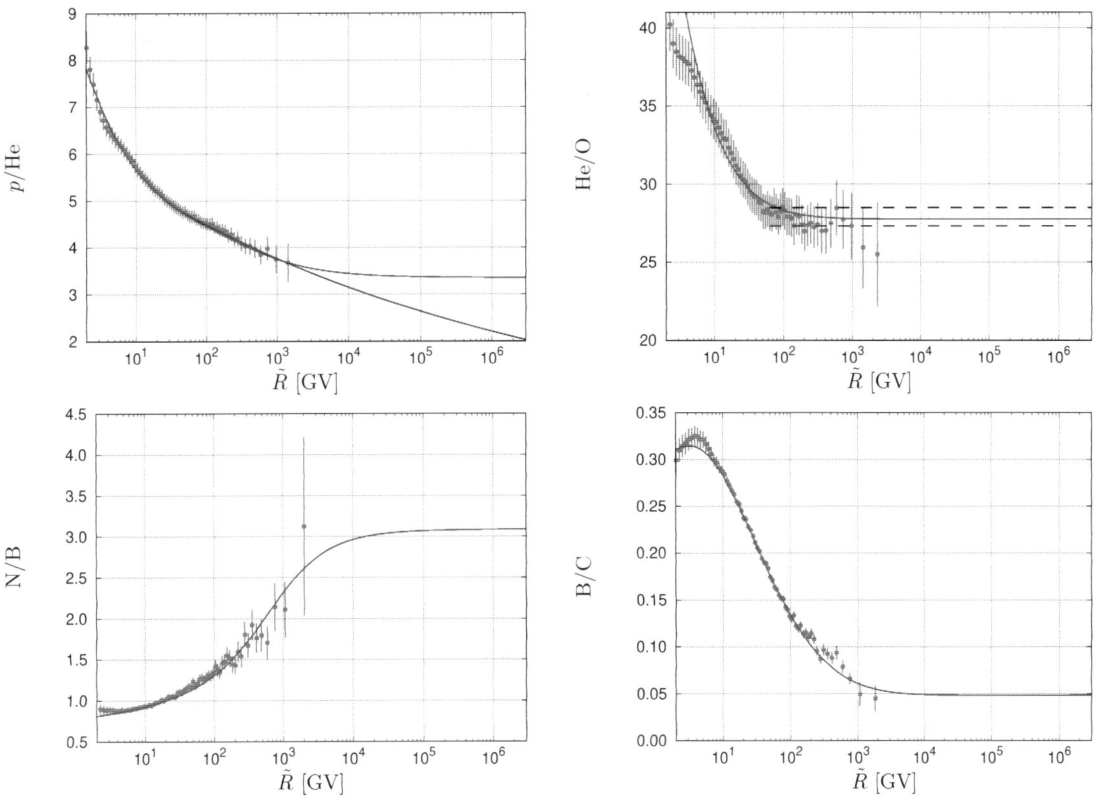

Figure 4. The p/He, He/O, N/B and B/C flux ratios as functions of rigidity, obtained in the anomalous diffusion model (solid purple lines), compared with the AMS-02 data [5,6]. The solid blue curve indicates the fit of a single power law with $\gamma_{p/\mathrm{He}} \approx -0.077$ to the AMS-02 data for p/He ratio [5]. The dashed black lines indicate the total errors of the fits by constant values of the He/O ratio [6]

IOP Conf. Series: Journal of Physics: Conf. Series **1181** (2019) 012023 doi:10.1088/1742-6596/1181/1/012023

Acknowledgments
The authors are deeply grateful to the anonymous referee for valuable comments and suggestions, which helped to improve the paper.

This work was supported in part by the Russian Foundation for Basic Research, grant no. 16-02-01103.

References
[1] Panov A and ATIC-2 Experiment 2009 *Bull. Russ. Acad. Sci.* **73** 564
[2] Ahn H *et al.* 2010 *ApJL* **714** L89
[3] Adriani O and PAMELA Collaboration 2011 *Science* **332** 69
[4] Aguilar M and AMS-02 Collaboration 2015 *Phys. Rev. Lett.* **114** 171103
[5] Aguilar M and AMS-02 Collaboration 2015 *Phys. Rev. Lett.* **115** 211101
[6] Aguilar M and AMS-02 Collaboration 2017 *Phys. Rev. Lett.* **119** 251101
[7] Aguilar M and AMS-02 Collaboration 2018 *Phys. Rev. Lett.* **120** 021101
[8] Aguilar M and AMS-02 Collaboration 2018 *Phys. Rev. Lett.* **121** 051103
[9] Biermann P *et al.* 2010 *ApJ* **725** 184
[10] Erlykin A and Wolfendale A 2012 *Astropart. Phys.* **35** 449
[11] Thoudam S and Horandel J 2012 *MNRAS* **421** 1209
[12] Tomassetti N 2012 *ApJL* **752** L13
[13] Vladimirov A *et al.* 2012 *ApJ* **752** 68
[14] Blasi P, Amato E and Serpico P 2012 *Phys. Rev. Lett.* **109** 061101
[15] Ptuskin V, Zirakashvili V and Seo E S 2013 *ApJ* **763** 47
[16] Bernard G *et al.* 2013 *A&A* **555** 48
[17] Erlykin A and Wolfendale A 2015 *J. Phys. G: Nucl. Part. Phys.* **42** 125201
[18] Zatsepin V and Sokolskaya N 2006 *A&A* **458** 5
[19] Lagutin A, Strelnikov D and Tyumentsev A 2001 *Proc. 27th ICRC* **5** 1896
[20] Lagutin A and Volkov N 2017 *EPJ Web of Conferences* **145** 19018
[21] Shukurov A *et al.* 2017 *ApJL* **839** L16
[22] Lagutin A and Uchaikin V 2001 *Proc. 27th ICRC* **5** 1900
[23] Lagutin A and Uchaikin V 2003 *Nucl. Inst. and Meth. in Phys. Res. B* **201** 212
[24] Samco S, Kilbas A and Marichev O 1993 *Fractional integrals and derivatives — Theory and Applications* (New York: Gordon and Breach) p 976
[25] Uchaikin V and Zolotarev V 1999 *Chance and Stability* (Netherlands, Utrecht: VSP) p 594
[26] Lagutin A, Makarov V and Tyumentsev A 2001 *Proc. 27th ICRC* **5** 1889
[27] Berezhko E G *et al.* 2003 *A&A* **410** 189
[28] Blasi P and Serpico P 2009 *Phys. Rev. Lett.* **103** 081103
[29] Tomassetti N and Donato F 2012 *A&A* **544** 16
[30] Berezhko E and Ksenofontov L 2014 *ApJL* **791** L22
[31] Cholis I, Hooper D and Linden T 2017 *Phys. Rev. D* **95** 123007
[32] Bartoli B and ARGO-YBJ Collaboration 2015 *Phys. Rev. D* **91** 112017
[33] Bartoli B and ARGO-YBJ and LHAASO Collaborations 2015 *Phys. Rev. D* **92** 092005
[34] Abdo A, Ackermann M, Ajello M *et al.* 2010 *Science* **327** 1103
[35] Abdo A, Ackermann M, Ajello M *et al.* 2010 *ApJ* **712** 459
[36] Abdalla H, Abramowski A, Aharonian F and HESS Collaboration 2018 *A&A* **612** A5
[37] Tanaka T, Uchiyama Y, Aharonian F *et al.* 2008 *ApJ* **685** 988
[38] Archambault S, Archer A, Benbow W *et al.* 2017 *ApJ* **836** 23
[39] Lagutin A, Volkov N, Kuzmin A and Tyumentsev A 2009 *Bull. Russ. Acad. Sci.: Physics* **73** 620
[40] Gleeson L and Axford W 1968 *ApJ* **154** 1011
[41] Lafferty G and Wyatt T 1995 *Nucl. Instr. and Meth. in Phys. Res. A* **355** 541

Seasonal and interannual changes in optical thickness of the atmosphere in the Yakutsk array region

Stanislav Knurenko[1] **and Igor Petrov**[1]

[1]Yu. G. Shafer Institute of Cosmophysical Research and Aeronomy SB RAS

E-mail: igor.petrov@ikfia.ysn.ru

Abstract. The analysis of the atmospheric optical thickness is given. Atmospheric optical thickness is obtained from spectral transparency of the atmosphere by CE318 radiometer. In the paper features of seasonal and interannual AOT of the atmosphere in the period 2004-2014 are discussed. It is noted that AOT measured at the Yakutsk sensitive to restructure of the atmosphere from winter to summer, passage of storm fronts in summer and large-scale wildfires.

1. Introduction

The formation of weather and climate in the regions is played by: microclimate, frequency of air travel (cyclones and anticyclones), including such factors as technogenic catastrophes - active volcanic activity, global forest fires and for the northern latitudes of the greenhouse effect. To monitor the changes in atmospheric parameters, a network of stations was established, located on all continents. The main goal of this project was global monitoring of the atmosphere and detection of changes that could affect the Earth's climate as a whole. Yakutsk is one of the items in the global chain of stations that supplies such information [1, 2].

At present, data on the optical thickness of the atmosphere (AOT) are analyzed in the context of long-term observations

2. Features of the atmosphere in the central Yakutia

The change in temperature conditions and the micro and macroscopic aerosol characteristics of the atmosphere affect the formation of local weather, and, with prolonged exposure to climatic conditions, as a whole. This is clearly seen from data collected over a 10-year period in the vicinity of Yakutsk. For example, an analysis of the statistical series of observations of air temperature in the surface layers of the atmosphere carried out in [3, 4], using data collected from the extensive air showers (EAS) collected over 40 years at the Yakutsk array, revealed a winter upward temperature trend with a gradient of Δt / 10 = + 0.6° (Fig. 1).

The air shower measuring array is in the valley of the Lena River 50 km south-west of Yakutsk and is located at an altitude of 100 m above sea level [5]. According to its geographical coordinates, the facility belongs to central Yakutia. The climate of central Yakutia is extreme continental: in July, air temperature often reaches 35 °C, and in winter it drops to -55 °C. In the area of the array, the wind prevails in the west and north-west directions. The average annual precipitation is 247 mm. In winter, especially during frosts below -40 °C, there is an ice fog

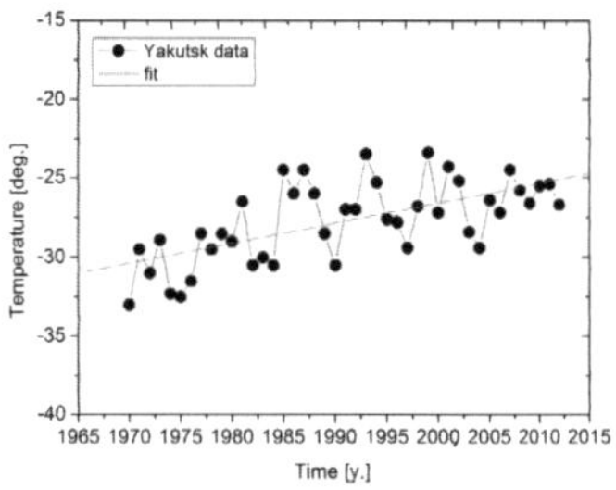

Figure 1. Interannual trend of surface air temperature near Yakutsk.

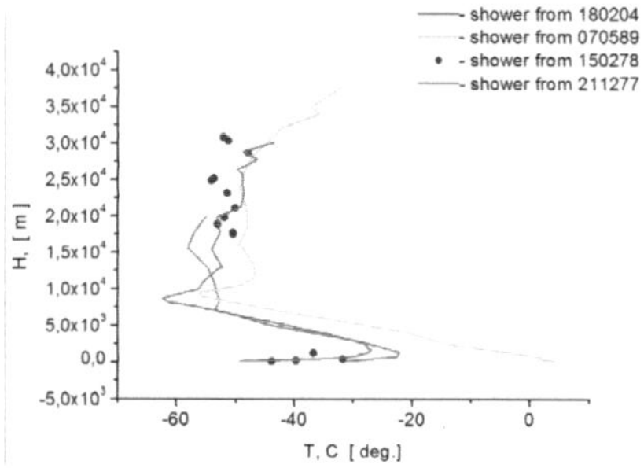

Figure 2. Vertical profile of air temperature distribution for different moments in time. The green curve (shower from 070589) was obtained in the spring period and corresponds to the standard form.

spreading through the valley. The thickness of a dense layer of the fog does not usually exceed 5-50 m from the surface of the earth. The period with low temperatures at night in the area of the test site lasts more than 6 months, beginning in October and ending in May. At winter time, there is also little precipitation, usually small needle ice crystals.

In winter, the temperature variation in altitude differs significantly from the standard distribution, and a stable temperature inversion is observed in the lower part of the atmosphere. Fig. 2 shows the distribution of air temperature in height. Individual measurements obtained with the help of atmospheric probes in the vicinity of Yakutsk are given. It can be seen from this figure that the dependence of the temperature distribution on height in the winter months is complex. In the winter, there is a gradual increase in temperature from the ground level to heights of 2000-3000 m above sea level, and above, there is a monotonous decrease to extremely low values at heights of 9000 -11000 m.

In the spring and autumn seasons, the temperature distribution in the atmosphere does not differ from the standard, i.e. the fall starts from the ground level. This difference in the temperature distribution over the height in different time periods reflects the distribution of the air density and the passage of optical waves through the troposphere.

3. Observations

Regular observations of the state of the atmosphere at the Yakutsk array began in 1970 and continuous to the present day. An automatic mini station measures the temperature, pressure and humidity of the air. Particular attention was paid to the winter period of observations, since observations of the Cherenkov radiation in the optical wavelength were conducted at the Yakutsk array, and for this reason an estimate of the state of the atmosphere was required [6]. Since 2004, a laser at a wavelength of 532 nm and a radiometer of CE 318 have been used to monitor the atmosphere [7]. On the basis of these measurements, a database on the state of the atmosphere in the area of the Yakutsk EAS installation continued to be formed.

3.1. Seasonal variations of AOT in Yakutsk region

The optical properties of the atmosphere are related to the content of aerosol particles of natural and artificial origin. Usually, observations are made between April and November in automatic

IOP Conf. Series: Journal of Physics: Conf. Series **1181** (2019) 012024 doi:10.1088/1742-6596/1181/1/012024

mode. The following characteristics were restored: AOT, albedo of single scattering of aerosol and volume distribution of particles by size.

Within the annual measurements, AOT data were averaged by monthly blocks and then summed over the entire observation period. The result is shown in Fig. 3.

It follows from the data that, in general, the AOT values in the Yakutsk region are small. As can be seen, from Fig. 3, starting from May and August, the average values of AOT in the Yakutsk region are 0.18 ± 0.02 and tend to decrease to 0.11 ± 0.02 in September-November. In winter, AOT measurement by this method is not carried out at the installation due to a short light day and low sun at the zenith ("polar night"). It can be assumed that the AOT value remains at the same level or even slightly below the average value (equal to 0.08 ± 0.02) due to the freezing of a part of the water vapor at very low temperatures. According to observations over a ten-year period of time, there are no noticeable variations of AOT in the data, except for some years, for example, 2012, when long-term and large-scale forest fires took place in the territory of central Yakutia. For this reason, the AOT values in the summer months were 1.5-2.0 times higher than in the autumn - winter and spring periods. This circumstance, as can be seen from Fig. 3, influenced the distribution of AOT in the annual block of observations. If we take into account the abnormal years (with fires), the distribution of AOT by months in the monitoring area will not have a pronounced seasonal course.

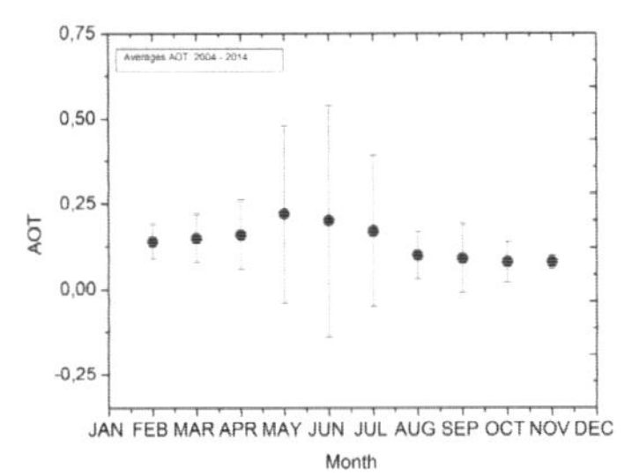

Figure 3. Seasonal change, averaged over the months, AOT values. Data are presented for the wavelength $\lambda = 500$ nm.

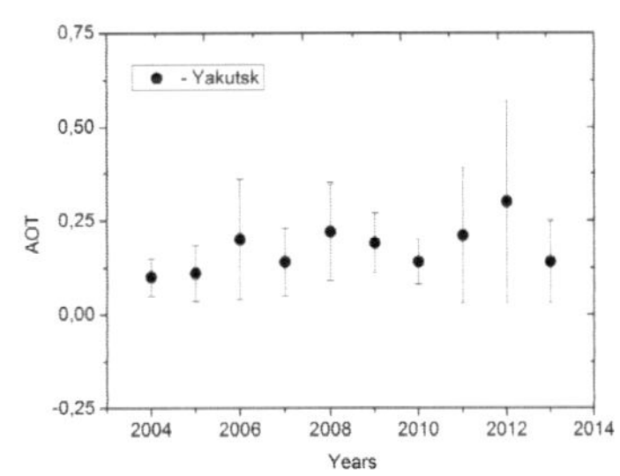

Figure 4. Interannual variability of AOT in the Yakutsk region. The averaging was carried out over the years. Observations from 2004 to 2013 are considered.

3.2. Interannual changes in AOT

Central Yakutia is located in an area with a extreme continental climate, when the air temperature in winter can reach -55°, and in summer +35 °. Such a sharp restructuring of the atmosphere is accompanied in the winter months by a temperature inversion, which leads to an increase in air density in the surface layers of the atmosphere, and this often leads to an increase in AOT and frost fogs [8, 9]. In the summer months, due to high temperatures and low rainfall, frequent forest fires are caused by dry thunderstorms. In this period of time, too, there is an increase in AOT due to the smoke of the atmosphere. Fig. 3 shows data on the parameter AOT, averaged over monthly blocks for observation periods, starting from 2004 to 2013. There is no regularity in the distribution of AOT over the years. It is only necessary to note the abnormal emissions in 2012, the cause of which is discussed below.

In the spring-summer periods with different intensity in Yakutia, forest fires are observed, which significantly affect both the average monthly and annual AOT values. This - above all, refers to the Southern and Central Yakutia, including the city of Yakutsk. As can be seen from the data shown in Fig. 4, the interannual value of AOT varies from 0.1 in quiet years, to 0.3 in years with an abnormally large number of forest fires. According to [10] large-scale forest fires were noted in 2012, which affected the annual AOT, the average value, which reached the value equal to 0.29±0.25 in 2012. 2012 is allocated, when the AOT value was three times higher than the average values for the relatively successful fire season.

4. Conclusion

Central Yakutia occupies a vast territory, has large tracts of taiga and the absence of large industrial enterprises. Therefore, the atmosphere in this area is cleaner than in regions with a developed industrial structure [11].

Nevertheless, the global warming has affected Yakutia. This is evident from Fig. 1, the data of which indicate a significant change in the mean annual temperature, and this can be traced practically throughout the entire territory of Yakutia, including the circumpolar zones. Subsequent to this, the atmosphere that responds not only to geophysical phenomena in the upper layers of the stratosphere, but also to frequent natural and anthropogenic anomalies, is gradually being rebuilt. As we can see, one of the significant factors of the interannual variability of AOT during the observation period is large-scale forest fires, as a result of which a large amount of fine-dispersed smoke aerosols enter the atmosphere. Fig. 4 shows that in the Yakutsk region for this reason, an almost threefold increase in AOT (∼0.3) relative to the multi-year rate (∼0.11) occurred.

Thus, the main role in the interannual variations of AOT in Central Yakutia, in addition to the seasonal circulation of air masses, is played by a finely dispersed aerosol from large forest fires, which significantly affects the average annual value of AOT.

Acknowledgments

The reported study was funded by RFBR according to the research project 16-29-13019

References

[1] http://aeronet.gshc.nasa.gov./
[2] Knurenko S, Petrov Z, Nikolashkin S and Sabourov A 2010 *22 ECRS (Turku)* p 263
[3] Knurenko S, Egorov Yu and Sabourov A 2011 *Int. symp. of "Atm. opt. and ocean. Physics of the atmosphere" (Tomks)* D p 9
[4] Knurenko S and Egorov A 2013 *J. Phys.: Conf. Ser.* 409 012083
[5] Artamonov V P 1994 *Bull. of RAS. Physics* vol 58 92
[6] Knurenko S P, Ivanov A A, Sleptsov I Ye and Sabourov A V 2006 *JETP Lett* vol 83 p 563
[7] Knurenko S P, Nikolashkin S V, Sabourov A V and Sleptsov I Ye 2006 *Proc. of SPIE* vol 6522
[8] Knurenko S and Petrov I 2015 *Proc. of Science* PoS(ICRC2015)584
[9] Knurenko S and Petrov I 2016 *Proc. of 25 ECRS 2016 (Torino)* Conference: C16-09-04.3
[10] Sakerin S M, Kabanov D M Panchenko M V 2005 *Optics of the atmosphere and ocean* vol 18 p 968

IOP Conf. Series: Journal of Physics: Conf. Series **1181** (2019) 012025 doi:10.1088/1742-6596/1181/1/012025

Method of EAS's Cherenkov and fluorescent light separation using silicon photomultipliers

Dmitry Chernov[1,*], Elena Bonvech[1], Timur Dzhatdoev[1], Miroslav Finger[3,4], Michael Finger[3,4] Vladimir Galkin[1,2] Gali Garipov[1], Vladimir Kozhin[1], Dmitry Podgrudkov[1,2] and Alexander Skurikhin[1]

[1] Skobeltsyn Institute of Nuclear Physics, M.V.Lomonosov Moscow State University, 1(2), Leninskie gory, GSP-1, Moscow, 119991, Russian Federation

[2] Department of Physics, M.V.Lomonosov Moscow State University, 1(2), Leninskie gory, GSP-1, Moscow, 119991, Russian Federation

[3] Faculty of Mathematics and Physics, Charles University, Ke Karlovu 3, 121 16, Prague 2, Czech Republic

[4] Joint Institute for Nuclear Research, Joliot-Curie, 6, Dubna, Moscow region, 141980, Russian Federation

E-mail: *chr@dec1.sinp.msu.ru

Abstract. Preliminary results on the development of a separation method for Cerenkov (CL) and fluorescence (FL) light from EAS are shown. The results are based on the measurement of attenuation coefficients of CL and FL for different filters. A total of six optical filters were investigated: filters from optical glass UFS-1, UFS-5, FS6 (analogue BG3) and interference filters SL 360-50, SL 280-380, FF01-375/110. The measurements were performed using silicon photomultipliers (SiPM). To improve existing fluorescent light detectors, a segment of 7 SiPM was developed, which would be able to separate both components of the light flux from EAS at the level of primary data processing.

1. Introduction

Direct and scattered Cherenkov light (CL) is one of the relevant contributions to the uncertainty of the measured flux of fluorescent light (FL) from extensive air showers (EAS). The impact of backscattered CL was noted when modelling the response of several detectors [1],[2],[3]. The problem of reliable CL and FL separation is relevant for a better detection of high energy EAS and estimation of the primary particle parameters. This study is the first step in search of both theoretical approaches to the problem and possible design of electronics for actual and future detectors.

2. The method of separation FL and CL

A separation method based on the simultaneous recording of light from one 'optical pixel by two or more pairs of silicon photomultipliers (SiPMs) is proposed. The first SiPM detects the incoming light flux in the wavelength band of its maximal sensitivity. The second SiPM detects it through an optical ultraviolet (UV) filter for fluorescent light (FL) separation. If the SiPMs' sensitivity characteristics, absorption characteristics of the filter elements and the spectra of fluorescent [4] and Cherenkov light are known, one can calculate the contribution of

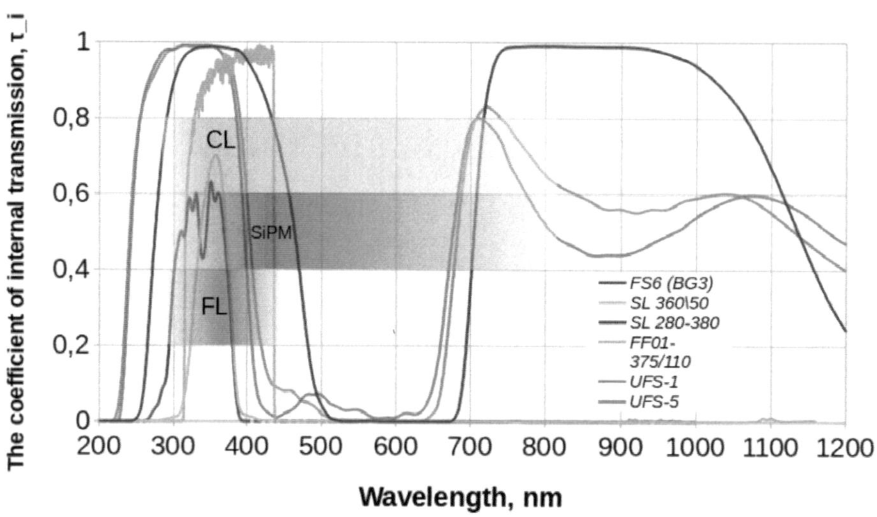

Figure 1. Spectral characteristics of light filters.

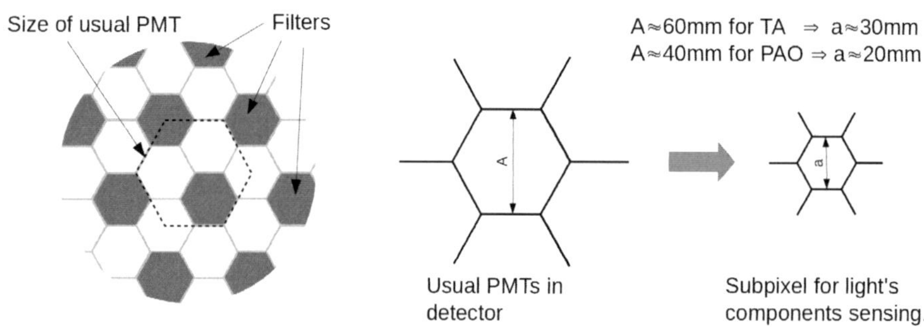

Figure 2. PMT mosaic modification example. The areas ratio (with filter)/(without filter) is equal to 1/2.

each component to the total light flux. Upon the completion of this work it will be possible to separate fluorescent and Cherenkov light at the stage of on-board primary processing of recorded data. This method will increase the methodological accuracy of fluorescent light measurements.

In this work both ordinary colored-glass filters and interference filters were studied. In theory the interference filters should allow a better FL separation due to their narrow pass-through band in the UV region. The spectral characteristics of all studied filters are show on fig. 1.

In fig. 2 the possible modification of the common photomultiplier tubes (PMT) mosaic is shown on the example of the Telescope Array (TA) [5] and Pierre Auger Observatory [6] detectors. In this case the signal loss on the filters is minimal since only one third of the mosaic area is covered by the filters. However, it is worth noting that the proposed method is effective only when the light spot from the remote point source on the mosaic is two or more times larger than the SiPM mosaic subpixel size.

3. FL and CL test stands

On the right side in fig. 3 are the general scheme and photo of the stand for the optical filters testing with CL. CL photons are generated by cosmic ray muons passing through a

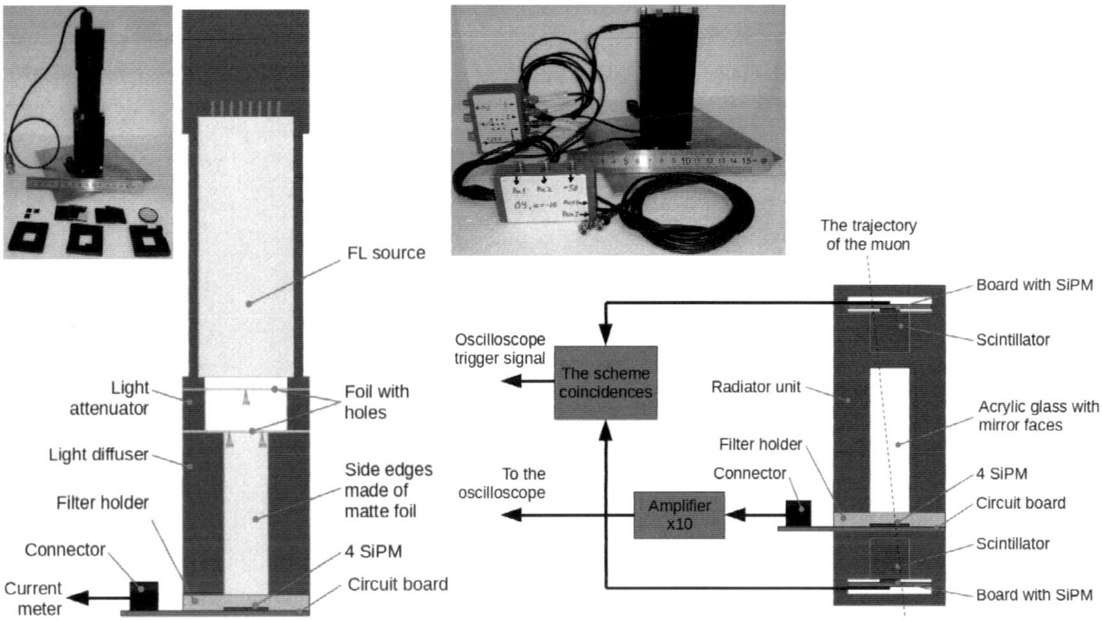

Figure 3. Schemes and photos of the stands for filter (UFS-1, UFS-5, FS6, SL 360\50, SL 280-380, FF01-375/110) attenuation coefficient measurements. The FL stand is on the left and the CL one is on the right.

$15 \times 15 \times 50$ mm acrylic radiator (acrylic glass without a UV stabilizer) with 5 reflective sides (the bottom side was transparent). The radiator's bottom side was matted to create an isotropic CL flux. The selection of vertical muons passing through the whole radiator is done using two $15 \times 15 \times 15$ mm scintillator blocks with a SiPM (shown in blue on the scheme) placed above and below the radiator. Signals from the SiPMs are routed to the coincidence scheme that in turn triggers the recording on a digital oscilloscope. The tested filters are installed between the radiator and 4 measuring SiPMs using different fittings. Each muon passing through the radiator gives around 1800 CL photons, but due to scattering, multiple reflections and bottom side absorption only about 25% of photons reach the measuring SiPMs. For more reliable measurements the SiPMs were connected in parallel in diagonal pairs. The resulting 2 signals were amplified by fast operational amplifiers AD8011 and then recorded by the oscilloscope. This scheme allows to have an average pulse amplitude of 42 mV from each channel with noise level less than 5 mV.

On the left side in fig. 3 are the general scheme and photo of the stand for testing optical filters with FL. The FL source was manufactured [7] on request and is described in detail in [8]. The stand consists of a FL source, light flux attenuator, a diffuser, filter housing and 4 SiPMs. The light attenuator consists of two parallel foils with multiple pinholes in them situated approximately 12 mm apart of each other. The pinholes are arranged so that direct light from the FL source does not reach the filters and SiPMs. The diffuser consists of a $15 \times 15 \times 50$ mm light guide made of matted foil to match the measuring geometry of the CL test stand. The optical filters, their housing and measurement SiPM arrangement and circuitry was the same as on the CL stand described above. The only difference is that FL is continuous. In this case it is more convenient to measure and compare the SiPM anode currents. For the SiPM linearity test two equal diodes with 403 nm main wavelength were used. The diodes illuminated SiPMs one by one with 5 different voltages and in combinations. The SiPMs proved to have

IOP Conf. Series: Journal of Physics: Conf. Series **1181** (2019) 012025 doi:10.1088/1742-6596/1181/1/012025

Table 1. Preliminary results on the transparency coefficient measurements for FL and CL of conventional optical filters (UFS-1, UFS-5, FS6) and interference filters (SL 360\50, SL 280-380, FF01-375/110).

Filter	FL	CL	Arbitrary separation efficiency
FF01-375/110	0.73 ± 0.07	0.39 ± 0.15	1.87
UFS-5	0.65 ± 0.01	0.36 ± 0.26	1.81
UFS-1	0.71 ± 0.01	0.48 ± 0.25	1.48
FS6 (BG3)	0.81 ± 0.01	0.55 ± 0.18	1.47
SL 280-380	0.36 ± 0.01	0.33 ± 0.11	1.09
SL 360\50	0.28 ± 0.01	0.28 ± 0.09	1.00

high linearity.

4. Preliminary results

Using the stands described above a series of tests were carried out to determine the attenuation coefficients for 6 filters (see fig. 1) for both CL and FL. The preliminary results are shown in table 1.

For filter comparison the 'Arbitrary separation efficiency' characteristic, defined as the ratio of FL transparency to CL transparency, was introduced. The best characteristics has the interference filter FF01-375/110 and colored glass filter UFS-5.The FF01-375/110 filter has a good FL transparency but as a drawback has a high cost. Moreover, the filters' angular properties need more careful analysis since in many actual experiments the light falls on the sensitive part of the detector in a wide range of angles while in this study it was collected in a narrow angle.

Since the filters were installed over SiPMs without an optical contact the transparency coefficients in table 1 also include reflection losses on the SiPM surface. Since FF01-375/110 compared to other filters has a 'perfectly' polished surface that gives visible flares the measured transparency coefficient has a higher uncertainty. In the next planned measurement series this will be corrected using a different diffuser design.

A significant uncertainty in CL measurements comes from two major sources:

(i) muon trajectory in the radiator and

(ii) low muon flux — ~3 muons per hour.

The first factor gives a difference factor of up to 2 in two measuring channels, but does not change the average value. The second factor does not allow to gather enough statistics in a reasonable time to reduce the statistical error. While the expositions can be increased by 10 times it still will not be nearly enough to reach the precision of FL measurements. To solve this issue the measurement of CL filter attenuation coefficients is planned to be carried out using CL from EAS. The measurements are planned to take place on a test site with low light pollution and aerosol levels using a 0.3 m^2 470 mm curvature radius mirror and a set of 7 SiPMs with light collectors. In comparison with the stand described above the new setup will increase the light flux per SiPM by more than 15 times and in total increase to about 150 events per hour.

5. Development of the detector

The detector is designed for ground and space based optical experiments for high and ultrahigh energy cosmic ray studies.

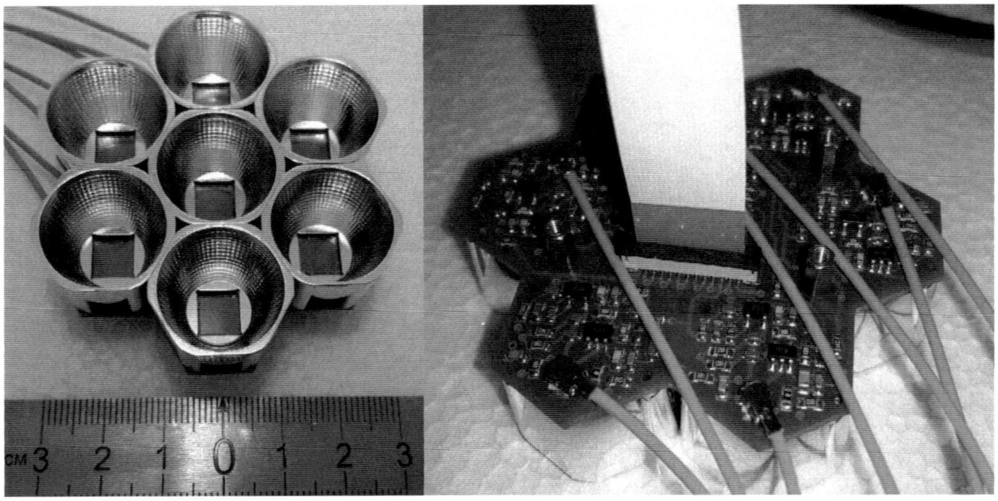

Figure 4. Matrix of seven SiPM SensL MicroFC-60035-SMT with light collectors and amplifiers.

It is planned to create a sensitive module utilizing light collectors. The module will consist of a board with 7 SiPMs [9] and a set of light collectors (fig. 4 right and left respectively), amplifiers and set of additional sensors (temperature, pressure etc). The light collectors effectively enlarge the pixel area to $\sim$3 cm^2. Unlike the commercially available ones, the new matrix allows to set the supply voltage individually for each SiPM for sensitivity equalization across the module (and between the modules). It is planned that the whole sensitive matrix will consist of 7 modules , i.e. 49 pixels in total. The effective sensitive area should be 4–5 times larger than for a single module (due to the mirror attenuation by the matrix).

On fig. 5 the 7-channel photoelectron counting board prototype is shown. The board houses a debug module with a FPGA chip and receives commands via ethernet interface. It also houses the SiPM power source ($-24 \ldots -29$ V), digital-to-analogue converters (DAC) for individual SiPM sensitivity control, comparators for photoelectron counting, DACs for the their level selection and DACs for sensor readings. The 7-SiPM module is connected to the board. The white cable is used for module powering and 7 gray cables (micro-coaxial cables) connect the SiPM amplifiers to the comparators.

6. Conclusion

A method for separating Cherenkov and the fluorescent light from EAS is being developed. Preliminary measurement results indicate good prospects from the application of the proposed method. The first elements of the detector using silicon photomultipliers for the realization of this method are produced.

Acknowledgments

This work was funded by Russian Foundation for Basic Research (RFBR) project No.16-02-00777.

References

[1] Haungs A et al. for JEM-EUSO Collaboration 2015 *J. Phys.: Conf. Series* **632** 012092
[2] The Pierre Auger Collaboration 2010 *Astropar. Phys.* **33** 108
[3] Unger M, Dawson B R, Engel R, Schussler F, Ulrich R 2008 *Nucl. Instrum. Methods* **A588** 433

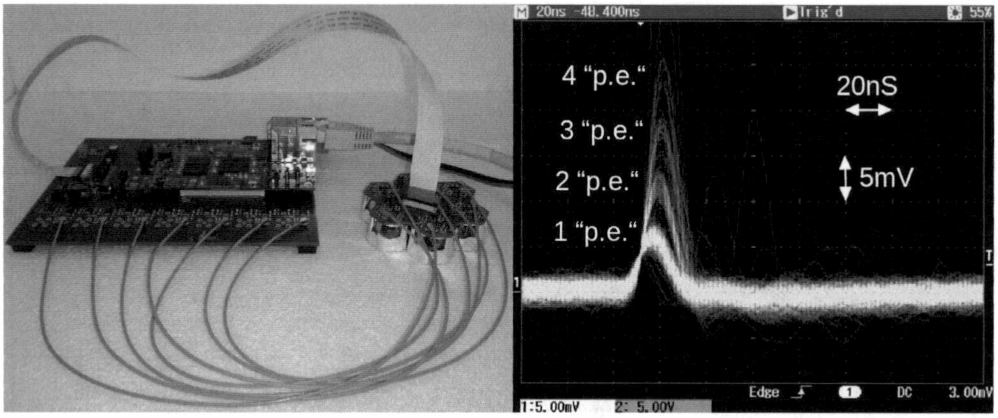

Figure 5. Left panel: an electronic board for counting photoelectrons with a Zynq-7020 FPGA debugger. Right panel: An amplified signal of SiPM's fast output.

[4] Ave M et al. for AIRFLY Collaboration 2007 Measurement of the pressure dependence of air fluorescence emission induced by electrons, *Astropart. Phys.* **28** 41
[5] The Telescope Array Collaboration 2012 *Nuclear Instruments and Methods in Physics Research A* **676** 5465
[6] The Pierre Auger Collaboration 2015 *Nuclear Instruments and Methods in Physics Research A* **798** 172-213
[7] MELZ FEU Ltd *http://www.melz-feu.ru/*
[8] Chernov D V, Bonvech E A, Dzhatdoev T A, Finger Mir., Finger Mich., Galkin V I, Garipov G K, Kozhin V A, Podgrudkov D A, Skurihin A V 2017 Position-sensitive SiPM detector for separation of Cherenkov and fluorescent light of EAS, *PoS* ICRC(2017)444
[9] SensL Technologies Ltd *http://sensl.com/*

IOP Conf. Series: Journal of Physics: Conf. Series **1181** (2019) 012026 doi:10.1088/1742-6596/1181/1/012026

Sources, spectra and composition of ultra-high energy extragalactic cosmic rays

V.N.Zirakashvili, V.S. Ptuskin, S.I.Rogovaya, E.G.Klepach

Pushkov Institute of Terrestrial Magnetism, Ionosphere and Radiowave Propagation, 108840 Moscow Troitsk, Russia

E-mail: zirak@izmiran.ru

Abstract. The inverse problem of cosmic ray transport of ultra-high energy cosmic rays is considered. The source spectrum and composition are derived based on the Auger data on energy spectrum, energy dependence of mean logarithm of atomic mass number and its variance. The regularization procedure for considered ill-posed problem and the statistical analysis of experimental data are employed.

1. Introduction

The origin of ultra-high energy cosmic rays is a key problem of cosmic ray astrophysics. The observed suppression the cosmic ray flux at energies above $\sim 5 \times 10^{19}$ eV is most probably associated with two effects: the GZK cutoff [1, 2] and acceleration limit in cosmic ray sources [3, 4]. The isotropy of the highest energy cosmic rays are indicative of their extragalactic origin. The list of potential sources which could give the observed cosmic ray flux includes active galactic nuclei, gamma-ray bursts, fast spinning newborn pulsars, interacting galaxies, large-scale structure formation shocks and some other objects, see reviews [5, 6] and references therein.

The present knowledge about the highest energy cosmic rays was mainly acquired from the Pierre Auger Observatory (Auger), Telescope Array experiment (TA), and from the Yakutsk complex EAS array, see [7, 8, 9]. The mass composition of these cosmic rays remains uncertain. The interpretation of HiRes and TA data favors predominantly proton composition at energies 10^{18} to 5×10^{19} eV, whereas the Auger data indicate that the cosmic ray composition is becoming heavier with energies changing from predominantly proton at 10^{18} eV to more heavy composition at about 5×10^{19} eV. The mass composition interpretation of the measured quantities depends on the assumed hadronic model of particle interactions which is based on not well determined extrapolation of physics from lower energies.

The energy spectrum in extragalactic sources is commonly determined by the trail-and-error method when one makes the calculations of the expected at the Earth cosmic ray intensity assuming some shape of the source energy spectrum and the source composition. The calculations follow cosmic ray propagation from the source to the observer, e. g. [10]. The standard assumption is that the source spectrum is a power law on magnetic rigidity up to some maximum rigidity.

In our work [11] we showed how to inverse the procedure and calculate the source function starting from the observed at the Earth spectrum without ad hoc assumptions about the shape of source spectrum. The only assumption done is that different nuclei follow the same spectral function on the magnetic source rigidity. Simple cases of the source composition that includes protons and Iron nuclei were considered. The case of complex source composition with a given abundances of a few elements was considered in [12]. More complete statistical analysis of cosmic ray data including the fitting of source composition is the subject of the present work.

2. Solution of inverse problem for a system of cosmic-ray transport equations

We use the following transport equation for cosmic ray protons and nuclei in the expanding Universe filled with the background electromagnetic radiation (see [13] for details):

$$-\frac{\partial}{\partial\varepsilon}\left(\varepsilon\left(\frac{H(z)}{(1+z)^3}+\frac{1}{\tau(A,\varepsilon,z)}\right)F(A,\varepsilon,z)\right)$$

$$-H(z)(1+z)\frac{\partial}{\partial z}\left(\frac{F(A,\varepsilon,z)}{(1+z)^3}\right)+\nu(A,\varepsilon,z)F(A,\varepsilon,z)$$

$$=\sum_{i=1,2...}\nu(A+i\rightarrow A,\varepsilon,z)F(A+i,\varepsilon,z)+Q(A,\varepsilon)(1+z)^m \tag{1}$$

The system of Eqs. (1) for all kinds of nuclei with different mass numbers A from Iron to Hydrogen should be solved simultaneously. The energy per nucleon $\varepsilon = E/A$ is used here because it is approximately conserved in a process of nuclear photodisintegration, $F(A,\varepsilon,z)$ is the corresponding cosmic-ray distribution function, z is the redshift, $Q(A,\varepsilon)$ is the density of cosmic-ray sources at the present epoch $z = 0$, m characterizes the source evolution (the evolution is absent for $m = 0$), $\tau(A,\varepsilon,z)$ is the characteristic time of energy loss by the production of e^-e^+ pairs and pions, $\nu(A,\varepsilon,z)$ is the frequency of nuclear photodisintegration, the sum in the right side of Eq. (1) describes the contribution of secondary nuclei produced by the photodisintegration of heavier nuclei, $H(z) = H_0((1+z)^3\Omega_m+\Omega_\Lambda)^{1/2}$ is the Hubble parameter in a flat universe with the matter density $\Omega_m(=0.3)$ and the Λ-term $\Omega_\Lambda(=0.7)$.

The numerical solution of cosmic-ray transport equations follows the finite differences method. The variables are the redshift z and $\log(E/A)$.

Let us introduce solution $G(A,\varepsilon;A_s,\varepsilon_s)$ of Eqs. (1) at $z = 0$ for a delta-source $Q(A,\varepsilon) = \delta_{AA_s}\delta(\varepsilon-\varepsilon_s)$. This source function describes the emission of nuclei with mass number A_s and energy ε_s from cosmic ray sources distributed over all z up to some $z_{\max}$. The general solution of Eqs. (1) at the observer location $z = 0$ can now be presented as

$$F(A,\varepsilon,z=0)=\sum_{A'}\int d\varepsilon' G(A,\varepsilon;A',\varepsilon')Q(A',\varepsilon'). \tag{2}$$

The observed all-particle spectrum is determined by the summation over all types of nuclei $\sum_A F(A,E/A,z=0)/A$ that is

$$N(E)=\sum_{A,A'}A^{-1}\int d\varepsilon' G(A,E/A;A',\varepsilon')Q(A',\varepsilon'). \tag{3}$$

As noted above, we shall assume that the source spectra of nuclei can be expressed in terms of one function on rigidity:

$$Q(A,\varepsilon)=k(A)q(\varepsilon A/Z) \tag{4}$$

Here $q(\varepsilon)$ is the source proton spectrum and coefficients $k(A)$ determine the source chemical composition.

IOP Conf. Series: Journal of Physics: Conf. Series **1181** (2019) 012026 doi:10.1088/1742-6596/1181/1/012026

The set of discrete values of particle energy ε_i is defined to solve the transport equation numerically. The grid with constant $\triangle\varepsilon/\varepsilon$ and with 25 energy bins per decade is used in our calculations. Eq. (3) in the discrete form is

$$N_i = \sum_j S_{ij} q_j, \tag{5}$$

$$S_{ij} = \sum_{A,A'} \frac{Z(A')k(A')}{A'A} \triangle\varepsilon_j G_{ij}(A, E_i/A; A', \varepsilon_j Z(A')/A'),$$

where the subscript indexes i and j denote the corresponding energies ε_i and ε_j.

The source term q_j can be found from this set of linear Eqs. (5) if the observed all particle spectrum $N(E)$ and chemical composition of the source are known. We considered the case of protons and iron nuclei in the source [11]. It was found that the solutions of equation (5) have a physical meaning only for a limited range of proton to iron ratio. In addition the solution can be unstable relative small deviations of the left hand side of Eq. (5) so that the inverse problem is ill-posed. Below we use the following regularization procedure [14] for the whole set of equations.

Let introduce the functional L

$$L = \sum_i \left(1 - \frac{1}{N_i}\sum_j S_{ij} q_j\right)^2 + \varepsilon_R \sum_j (q_{j-1} - 2q_j + q_{j+1})^2 \tag{6}$$

Here ε_R is the regularization parameter. The first term in this equation is simply the sum of squared relative deviations from the observable spectrum $N(E)$. For $\varepsilon_R = 0$ this functional is minimized by solutions of Eqs. (5) and and its value equals to zero.

Renormalized set of equations is found from the condition $\partial L/\partial q_j = 0$:

$$\sum_j S^R_{kj} q_j = N^R_k, \; N^R_k = \sum_i \frac{1}{N_i} S_{ik}, \; S^R_{kj} =$$

$$\sum_i \frac{1}{N_i^2} S_{ik} S_{ij} + \varepsilon_R (6\delta_{kj} - 4\delta_{k,j-1} - 4\delta_{k.j+1} + \delta_{k,j-2} + \delta_{k.j+2}), \tag{7}$$

3. Approximation of experimental data

To simplify calculations and damp the spread of data points in the measured at the Earth cosmic ray spectrum, we use its analytical approximation.

The formula

$$J(E) \propto E^{-3.23}, E < 5 \times 10^{18}\text{eV};$$

$$J(E) \propto E^{-2.63} \times [1 + \exp(\log(E/10^{19.63}\text{eV})/0.15)]^{-1}, \; E > 5 \times 10^{18}\text{eV}. \tag{8}$$

is adopted in our calculations to approximate the Auger data [15].

4. Results

The minimal value $10^{-3} - 10^{-2}$ of the parameter ε_R was adjusted to provide the smooth positive source spectrum q_j. We found that this method does not work for arbitrary chemical composition. However the diapason of the chemical composition is significantly broader in comparison with the exact solution of Eq. (5).

The results obtained for non-evolutionary sources ($m = 0$) are shown in Figures 1-5. The value of maximum redshift $z_{\max} = 3$ was used. The coefficients $k(A)$ were scanned to minimize the mean sum δ of squared deviations from the observed all-particle spectrum, mean logarithm A and variance of the logarithm A σ^2 measured by Auger collaboration [16].

$$\delta = \frac{1}{n}\sum_{i=1}^{n}\left(1 - \frac{1}{N_i}\sum_j S_{ij}q_j\right)^2 + \frac{1}{n}\sum_{i=1}^{n}\left((\ln A)_i - (\ln A)_i^{obs}\right)^2 + \frac{1}{n}\sum_{i=1}^{n}\left(\sigma_i^2 - \sigma_i^{2,obs}\right)^2 \quad (9)$$

The obtained coefficients $k(A)$ are given in the last line of Table 1. These numbers should be considered as very approximate ones because their moderate variations produce a small change of δ (cf. Figure 3).

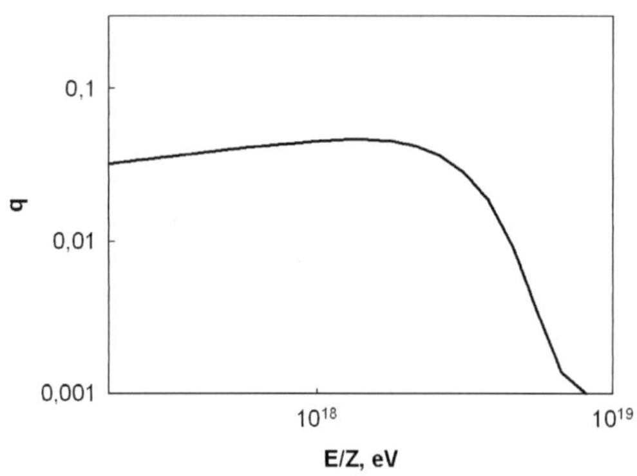

Figure 1. Calculated source spectrum $q(E/Z)$ in arbitrary units.

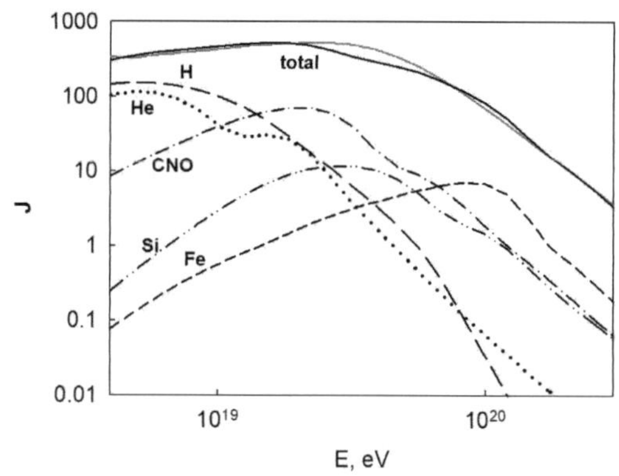

Figure 2. Calculated spectra of protons and nuclei. The all particle spectrum (thick solid) and the analytical approximation of Auger cosmic ray spectrum (gray solid line) are also shown.

It is evident that our model reproduces the observed all particle spectrum and measured mean logarithm A.

5. Discussion and Conclusion

We showed how one can find average spectrum of extragalactic sources from the cosmic ray spectrum observed at the Earth. This task was formulated as an inverse problem for the system of transport Eqs. (1) that describe the propagation of ultra-high energy cosmic rays in the expanding Universe filled with the background electromagnetic radiation.

Mathematically, the inverse problems for transport equations (1) are ill-posed in the general case that manifests itself in the instability of derived solutions. To avoid this problem we use the regularization procedure (Eqs. 6,7). The best choice of abundances for five groups of nuclei, H, He, CNO, Si and Fe, was found minimizing the value of δ. The same spectral function on the magnetic rigidity for the source spectra of different nuclei was assumed.

We found that the adjusted chemical composition permits to explain the Auger data [15, 16]. The Auger data favor the transition from a proton source composition to the heavier one as the energy is rising. With our heavy source composition, this case is most closely reproduced

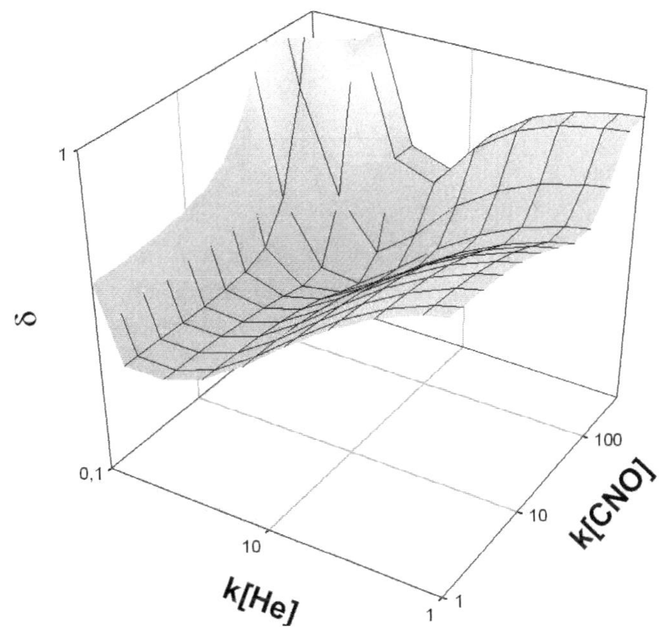

Figure 3. Dependence of the mean square of deviations δ on the Helium and CNO abundance in the source.

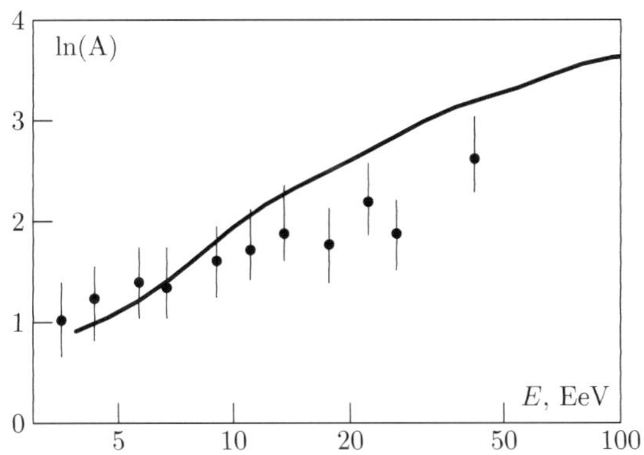

Figure 4. Calculated mean logarithm A (solid line). The experimentally measured mean logarithm A data of Auger collaboration [16] using EPOS-LHC model of particle interactions in the atmosphere (data with error-bars) are also shown.

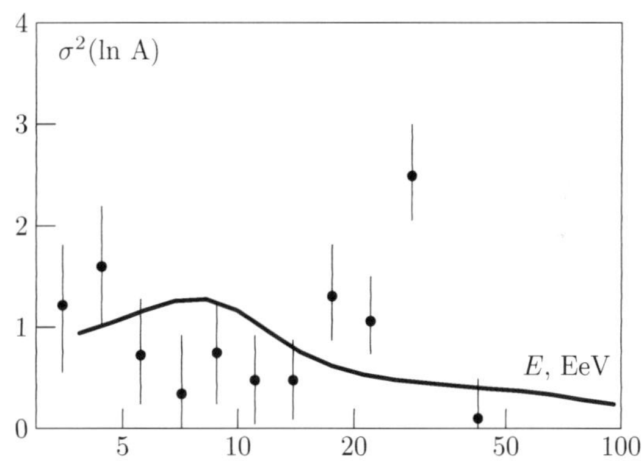

Figure 5. Calculated variance of the logarithm A (solid line). The experimentally measured variance of the logarithm A data of Auger collaboration [16] using EPOS-LHC model of particle interactions in the atmosphere (data with error-bars) are also shown.

by the calculations illustrated in Figure 2. The obtained source spectrum is hard (see Figure 1) and resemble the results [17, 18] based on the analysis of direct transport problems with a power law source spectrum. The maximum energy of accelerated particles $(3...5)Z \times 10^{18}$ eV is relatively low in this case that alleviates the problem of cosmic ray acceleration. The calculated

Table 1. Chemical composition of cosmic rays at the sources

	H, %	He, %	CNO, %	Si, %	Fe, %
A	1	4	14	28	56
galactic sources (for comparison)	92	7	0.65	0.07	0.07
extragalactic sources (this work)	0	73	24	2.5	0.8

composition of cosmic rays at the Earth shown in Figures 4,5 is also in accordance with the Auger measurements.

The study of inverse transport problem is a useful tool for the investigation of ultra high energy cosmic rays allowing the abandonment of the standard assumption of power law source spectrum with an abrupt cutoff at some maximum magnetic rigidity as it is usually assumed when the direct problem is considered.

Acknowledgments

The work was partially supported by Russian Fundamental Research Foundation grant 16-02-00255.

[1] Greisen K 1966 *Phys. Rev. Lett.* **16** 748
[2] Zatsepin G T and Kuzmin V A 1966 *JETP Lett.* **4** 78
[3] Allard D 2012 *Astropart. Phys.* **39** 33
[4] Aloisio R, Berezinsky V and Blasi P 2013 *arXiv*:1312.7459v1
[5] Kotera K and Olinto A V 2011 *Ann. Rev. Astron. Astrophys.* **49** 119
[6] Anchordoque L A 2018 *arXiv*:1807.09645v1
[7] Berat C for Pierre Auger Collaboration 2017 *EPJ Web of Conferences* **136** id.02017
[8] Sokolsky P for the TA and HiRes collaborations 2013 *EPJ Web of Conferences* **52** 06002
[9] Saburov A V. Glushkov A V, Pravdin M I *et al.* 2017 *Proc. 35th ICRC* **35** 552 POS(ICRC2017)552
[10] Allard D 2009 *arXiv*:0906.3156v1
[11] Ptuskin V S, Rogovaya S I and Zirakashvili V N 2015 *JCAP* **03** 054
[12] Zirakashvili V N, Ptuskin V S and Rogovaya S I 2017 *Radiophys. Quant. Electronics* **59** 859
[13] Ptuskin V S, Rogovaya S I and Zirakashvili V N, 2013 *Adv. Space Res.* **51** 315
[14] Tikhonov A N and Arsenin V Y 1977 *Solutions of Ill-Posed Problems* (New York: Winston)
[15] Letessier-Selvon A for the Pierre Auger Collaboration 2014 *Brazilian J. of Phys.* **44** 560 arXiv:1310.4620
[16] Aab A *et al.* (Pierre Auger Collaboration) 2014 *Phys. Rev. D* **90** 122005, arXiv:1409.4809
[17] Aab A *et al.* (Pierre Auger Collaboration) 2017 *JCAP* **04** 038
[18] Aloisio R, Berezinsky V and Blasi P 2014 *JCAP* **10** 20

IOP Conf. Series: Journal of Physics: Conf. Series **1181** (2019) 012027 doi:10.1088/1742-6596/1181/1/012027

Current Status and New Challenges of The Tunka Radio Extension

V. Lenok[1], P.A. Bezyazeekov[2], N.M. Budnev[2], D. Chernykh[2], O. Fedorov[2], O.A. Gress[2], A. Haungs[1], R. Hiller[1]*, T. Huege[1]†, Y. Kazarina[2], M. Kleifges[3], D. Kostunin[1], E.E. Korosteleva[4], L.A. Kuzmichev[4], N. Lubsandorzhiev[4], T. Marshalkina[2], R. Monkhoev[2], E. Osipova[4], A. Pakhorukov[2], L. Pankov[2], V.V. Prosin[4], F.G. Schröder[1,5], D. Shipilov[2] and A. Zagorodnikov[2] (Tunka-Rex Collaboration)

[1] Institut für Kernphysik, Karlsruhe Institute of Technology (KIT), Karlsruhe, Germany
[2] Institute of Applied Physics ISU, Irkutsk, 664020, Russia
[3] Institut für Prozessdatenverarbeitung und Elektronik, Karlsruhe Institute of Technology (KIT)
[4] Skobeltsyn Institut of Nuclear Physics MSU, Moscow, 119991, Russia
[5] Department of Physics and Astronomy, University of Delawere, Newark, DE, USA

E-mail: vladimir.lenok@kit.edu

Abstract. The Tunka Radio Extension (Tunka-Rex) is an antenna array spread over an area of about 1 km^2. The array is placed at the Tunka Advanced Instrument for cosmic rays and Gamma Astronomy (TAIGA) and detects the radio emission of air showers in the band of 30 to 80 MHz. During the last years it was shown that a sparse array such as Tunka-Rex is capable of reconstructing the parameters of the primary particle as accurate as the modern instruments. Based on these results we continue developing our data analysis. Our next goal is the reconstruction of cosmic-ray energy spectrum observed only by a radio instrument. Taking a step towards it, we develop a model of aperture of our instrument and test it against hybrid TAIGA observations and Monte-Carlo simulations. In the present work we give an overview of the current status and results for the last five years of operation of Tunka-Rex and discuss prospects of the cosmic-ray energy estimation with sparse radio arrays.

1. Introduction

Digital radio antenna arrays are a rapidly developing instrument for observation and study of ultra-high energy messengers in PeV-EeV range [1, 2]. Radio detection has evolved from understanding signal formation and early hardware development to the stage of real physical measurements.

Tunka-Rex is a radio antenna array designed for detecting cosmic rays. It is located at the site of the TAIGA detector [3] at the Tunka valley in Russia. The Tunka-Rex detector has been operating since 2012. It has been extended several times and currently consists of 57 antennas in the dense core of the detector (about 1 km^2 area) and 6 antennas in the satellite clusters

*now at the University of Zürich
†also at Vrije Universiteit Brussel, Belgium

Figure 1. *Left*: one of the clusters of Tunka-Rex. Three generations of the antennas are marked in the picture along with the Tunka-Grande and Tunka-133 detectors. The black lines schematically represent the cabling. *Right*: map of the present footprint of the detectors of TAIGA.

used as a high-energy extension. In total the Tunka-Rex radio array covers about 3 km^2. As antenna type we use SALLA [4] operating in the band of 30–80 MHz. The present configuration of Tunka-Rex detector is shown in the Fig. 1.

2. Highlights of Recent Results

Some important updates of the analysis have been done in Tunka-Rex. Recently we developed a method of template analysis of the data. Essentials of this method are the following:

- Pre-reconstruction using the standard procedure of Tunka-Rex [5]. This stage provides the shower incoming direction, the core position, and the energy estimation.
- Creating a library of CoREAS simulations [6] for each event obtained in the previous step with the goal to cover all possible depths of the shower maxima.
- Least-square fit of the simulated signal envelops to the reconstructed ones. The shower maximum and primary-particle energy are reconstructed on this stage.

Results of applying this method to the data are shown in the Fig. 2, left [7].

To better understand atmospheric impact we compared the predictions of the refractivity given by the standard CORSIKA atmosphere and the values given by GDAS [9]. The results of this comparison are displayed in the Fig. 2, right.

Also, mainly due to the worsening of the radio noise environment on the site some important updates have been introduced in the signal processing chain [10]:

- Matched filtering and background suppression with convolutional neural networks [11, 12].
- A sliding window for noise estimation was introduced to make the noise estimation procedure more robust against RFI.

Tunka-Rex has resolution of the X_{max} reconstruction about 30 g/cm^2. The resolution of the energy reconstruction is approximately 10% with systematic uncertainties of the instrument absolute calibration of about 20%.

3. Aperture and Exposure Estimation

The estimation of the instrument exposure is a key component for the correct reconstruction of the cosmic-ray flux measured by the observatory.

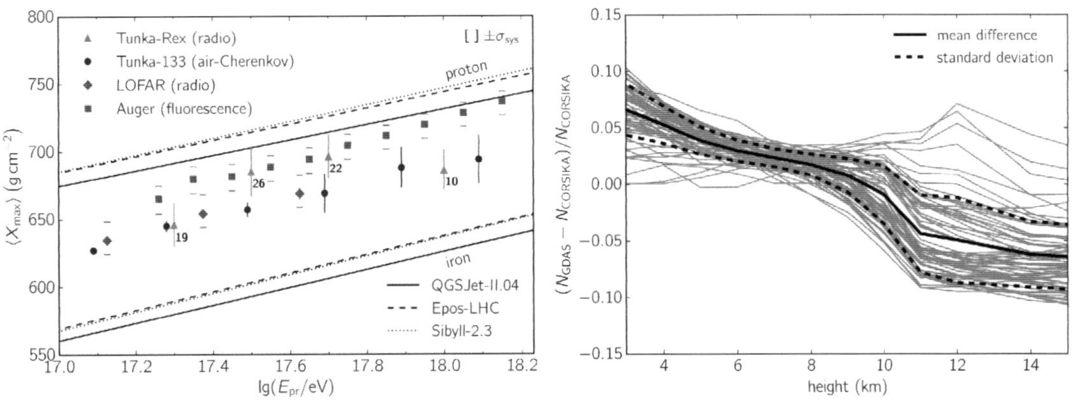

Figure 2. *Left*: Mean shower maximum as function of primary energy reconstructed by different experiments measuring the electromagnetic component of air showers [7]. *Right*: difference between the refractivity at the CORSIKA [8] standard atmosphere and the one calculated with GDAS profiles for the Tunka location.

The general way to estimate the flux J as a function of energy of the primary particle E is the following [13]:

$$J(E) = \frac{\mathrm{d}^4 N}{\mathrm{d}E\,\mathrm{d}s\,\mathrm{d}\Omega\,\mathrm{d}t} \approx \frac{\Delta N_{sel}(E)}{\Delta E}\frac{1}{\varepsilon(E)}, \quad (1)$$

where N is the number of primaries with energy between E and $E + \mathrm{d}E$ passing through a surface element $\mathrm{d}s$ within a solid angle $\mathrm{d}\Omega$ and within time $\mathrm{d}t$. $\Delta N_{sel}(E)$ is the number of events passing the selection criteria in a given analysis in the energy bin centered in E and having a width ΔE. $\varepsilon(E)$ is the energy-dependent exposure.

The exposure can be expressed in the form of the integral of the detection efficiency ξ:

$$\varepsilon(E) = \int\limits_{T_{sel}}\int\limits_{\Omega_{sel}}\int\limits_{S_{sel}} \xi\,(E, t, \vartheta, \varphi, x, y)\cos\vartheta\,\mathrm{d}s\,\mathrm{d}\Omega\,\mathrm{d}t = \int\limits_{T} A(E,t)\,\mathrm{d}t, \quad (2)$$

with $\mathrm{d}\Omega = \sin\vartheta\,\mathrm{d}\vartheta\,\mathrm{d}\varphi$ and $\mathrm{d}s = \mathrm{d}x\,\mathrm{d}y$. After partial integration one obtains the value A, which is called instant aperture. The detection efficiency depends strongly on energy, direction, and shower core location as well as on observation time due to instrumental instability over time.

The idea is to use of the full efficiency regime and fiducial selection of area and angular coverage of the sky. Using this regime of full efficiency is the most reliable way to estimate the cosmic-ray flux since instrumental effects of changing the observed mass composition near the detection threshold do not influence the final results. In this case the aperture in the equation (2) can be factorized in the following way

$$\varepsilon(E) = \int\limits_{T} S(E,t)A_\Omega(E,t)\,\mathrm{d}t, \quad (3)$$

where $S(E,t)$, $A_\Omega(E,t)$ are the fiducial area of the instrument and the angular part of the aperture with fiducial selection of the viewing angle. Full efficiency means that the detection efficiency is equal to one, and is omitted hereafter. In the simplest case of stable operation the equation (3) can be simplified

$$\varepsilon(E) = S(E)A_\Omega(E)T, \quad (4)$$

where T is the observation time at full efficiency.

In comparison to particle and non-imaging air-Cherenkov detectors, the full efficiency regime of radio measurements in the bandwidth we use has a distinguishing feature: the detection efficiency for arrival directions near the geomagnetic field is suppressed. This makes straightforward calculations of the instrumental full-efficiency angular coverage impossible.

Recently we developed a model taking into account this feature. This model uses elliptical shower footprints for calculation of detection efficiency for a given area and incoming direction. Parametrization of the ellipse parameters in this model is obtained from our experimental data.

This simple model, however, does not describe the observed detection efficiency in required detail. It works sufficiently well only in the intermediate range of zenith angles. Comparisons of the model against the hybrid observations with TAIGA detectors as well as detailed description of the model can be found in [14].

Calculations of aperture can be done with high precision by utilizing an analytical approach.

Since the main problem in the analytical calculation of the angular part of the aperture is the suppression around the magnetic field direction we can introduce a simple cut around this region. An additional cut for low zenith angles is introduced due to the low number of triggered antennas in this region.

Utilizing the considered cuts one obtains:

$$A_\Omega(E) = 2 \int\limits_{\vartheta_{min}(E)}^{\vartheta_{max}} \left[\pi - \arccos\left(\frac{\cos\alpha_{min}(E) - \cos\vartheta\cos\vartheta_m}{\sin\vartheta\sin\vartheta_m}\right)\right] \cos\vartheta\sin\vartheta\,\mathrm{d}\vartheta. \quad (5)$$

In this equation ϑ_{max} is the maximal zenith angle, ϑ_{min} — minimal zenith angle, α_{min} — minimal geomagnetic angle required (recall that geomagnetic angle is the angle between the arrival direction and the geomagnetic field), ϑ_M — zenith angle of geomagnetic field. These cuts are applied in the fiducial selection of the events. The integral is solved numerically.

Details of the derivation this formula can be found in the appendix.

The presented method allows calculation of the experimental aperture semi-analytically taking advantage of energy dependent cuts. The choice of cut parametrization for the particular mode is under investigation as well as overall performance of the method.

4. Conclusion

Tunka-Rex has been operating and successfully measuring cosmic rays since 2012. Reliable reconstruction methods and instrument calibration were developed during the last years. The main results of the Tunka-Rex can be summarized as following:

- Development of a robust reconstruction method for data containing only a few antennas with signals [7, 15].
- Experimental determination of the energy (10%) and X_{max} (30 g/cm^2) precision by comparing radio and air-Cherenkov measurements of the same events [16].
- Comparing the energy scales of the Tunka-133 and KASCADE-Grande instruments with its radio extensions [17].
- Development of an aperture and exposure model [14] and utilizing this model for the reconstruction of the mean shower maximum as function of energy [7].

Acknowledgments

The construction of Tunka-Rex was funded by the German Helmholtz association and the Russian Foundation for Basic Research (Grant No. HRJRG-303). This work has been supported by the Helmholtz Alliance for Astroparticle Physics (HAP), the Russian Federation Ministry of

Education and Science (Tunka shared core facilities, unique identifier RFMEFI59317X0005, 3.9678.2017/8.9, 3.904.2017/4.6, 3.6787.2017/7.8, 3.6790.2017/7.8), the Russian Foundation for Basic Research (Grants No. 16-02-00738, No.17-02-00905) and the Grant No. 15-12-20022 of the Russian Science Foundation (sections III and V). In preparation of this work we used calculations performed on the HPC-cluster Academician V.M. Matrosov [36] and on the computational resource ForHLR II funded by the Ministry of Science, Research and the Arts Baden-Württemberg and DFG ("Deutsche Forschungsgemeinschaft"). A part of the data analysis was performed using the radio extension of the Offline framework developed by the Pierre Auger Collaboration [18].

Appendix A. Details on Equation 5

The idea of the derivation of equation 5 can be illustrated with Fig. A1. First, we perform the integration of the cosine over area I, and then exclude the integral over area II. The coordinate system is chosen for simplification in a way that the azimuthal angle of the center of area II is zero.

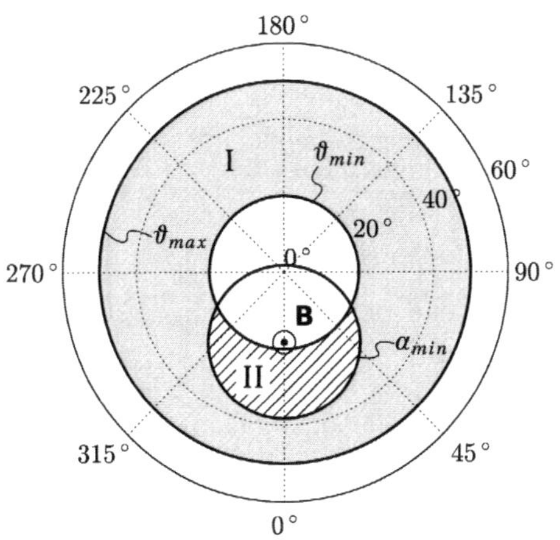

Figure A1. Cuts and areas used in the calculations (azimuthal equidistant projection of the sky). The integral is over the gray area.

Let us begin with the integral for area I. It is useful to recall that cosine in the initial integral holds information that the detector is flat.

$$(I) = \int_{\Omega'} \cos\vartheta \, \mathrm{d}\Omega = \int_0^{2\pi} \int_{\vartheta_{min}}^{\vartheta_{max}} \cos\vartheta \sin\vartheta \, \mathrm{d}\vartheta \, \mathrm{d}\varphi = 2\pi \int_{\vartheta_{min}}^{\vartheta_{max}} \cos\vartheta \sin\vartheta \, \mathrm{d}\vartheta. \tag{A.1}$$

This equation can be integrated further

$$(I) = \pi \left[\cos^2\vartheta_{min} - \cos^2\vartheta_{max}\right]. \tag{A.2}$$

This is the standard equation for the calculation of the aperture in the case of full efficiency and absence of any azimuth-dependent cuts. If $\vartheta_{min} = 0$ then $(I) = \pi\left[1 - \cos^2\vartheta_{max}\right]$.

For the area II we begin with the same equation and exploiting the symmetry for the integration:

$$(II) = \int_{\Omega'} \cos\vartheta \, \mathrm{d}\Omega = 2\int_{\tilde{\Omega}} \cos\vartheta \, \mathrm{d}\Omega = 2 \int_{\vartheta_{min}}^{\vartheta_{max}} \int_0^{f(\vartheta)} \cos\vartheta \sin\vartheta \, \mathrm{d}\varphi \, \mathrm{d}\vartheta. \tag{A.3}$$

As upper limit of the integral over φ we use dependency of φ on ϑ on the circle of radius α_{min} around the direction to the local magnetic field (ϑ_M, φ). We use equation for circle on spherical surface adopted from [19]

$$\cos\alpha_{min} = \cos\vartheta\cos\vartheta_M + \sin\vartheta\sin\vartheta_M\cos\varphi. \tag{A.4}$$

Hence it is easy to express the required dependency

$$\varphi = \arccos\left(\frac{\cos\alpha_{min} - \cos\vartheta\cos\vartheta_M}{\sin\vartheta\sin\vartheta_M}\right). \tag{A.5}$$

Using this equation as the upper limit in A.3 the integral is

$$(II) = 2\int\limits_{\vartheta_{min}}^{\vartheta'_{max}} \arccos\left(\frac{\cos\alpha_{min} - \cos\vartheta\cos\vartheta_M}{\sin\vartheta\sin\vartheta_M}\right)\cos\vartheta\sin\vartheta d\vartheta. \tag{A.6}$$

Here ϑ'_{max} is the maximal zenith angle of area II.

Thc final step is to perform the subtraction

$$(I) - (II) = \pi\left[\cos^2\vartheta_{min} - \cos^2\vartheta_{max}\right] - 2\int\limits_{\vartheta_{min}}^{\vartheta'_{max}} \arccos\left(\frac{\cos\alpha_{min} - \cos\vartheta\cos\vartheta_M}{\sin\vartheta\sin\vartheta_M}\right)\cos\vartheta\sin\vartheta d\vartheta. \tag{A.7}$$

This form is convenient for practical calculations. However, it can be can be expressed in the following shorter form

$$(I) - (II) = 2\int\limits_{\vartheta_{min}}^{\vartheta_{max}} \left[\pi - \arccos\left(\frac{\cos\alpha_{min} - \cos\vartheta\cos\vartheta_M}{\sin\vartheta\sin\vartheta_M}\right)\right]\cos\vartheta\sin\vartheta\, d\vartheta. \tag{A.8}$$

References

[1] F.G. Schröder, Prog. Part. Nucl. Phys. 93, 1 (2017), 1607.08781
[2] T. Huege, Phys. Rept. 620, 1 (2016), 1601.07426
[3] L.A. Kuzmichev et al. — TAIGA Coll., Phys.Atom.Nucl. 81 (2018) no.4, 497-507
[4] P. Abreu et al. — Pierre Auger Coll., JINST 7 (2012) P10011, 1209.3840
[5] P.A. Bezyazeekov et al. — Tunka-Rex Coll., JCAP 01 (2016) 052, doi:10.1088/1475-7516/2016/01/052
[6] T. Huege et at., AIP Conf.Proc. 1535 (2013) no.1, 128, 1301.2132
[7] P. A. Bezyazeekov et al. — Tunka-Rex Coll., Phys.Rev. D97 (2018) no.12, 122004, 1803.06862
[8] D. Heck et al., Forschungszentrum Karlsruhe Report FZKA 6019 (1998)
[9] Global Data Assimilation System (GDAS). https://www.ncdc.noaa.gov/data-access/model-data/model-datasets/global-data-assimilation-system-gdas
[10] D. Kostunin et al. — Tunka-Rex Coll., proceeding of ARENA 2018 (submitted to publication).
[11] D. Shipilov et al. — Tunka-Rex Coll., proceeding of ARENA 2018 (submitted to publication).
[12] T. Marshalkina et al. — Tunka-Rex Coll.. proceeding of ARENA 2018 (submitted to publication).
[13] P. Abreu et al. — Pierre Auger Coll., Astropart.Phys. 34 (2011) 368-381, 1010.6162
[14] O. Fedorov et al. — Tunka-Rex Coll., PoS ICRC2017 (2018) 387, 1712.00974
[15] D. Kostunin et al. — Tunka-Rex Coll., Astropart.Phys. 74 (2016) 79-86, 1504.05083
[16] P. A. Bezyazeekov et al. — Tunka-Rex Coll., JCAP 1601, 052 (2016), 1509.05652
[17] W. D. Apel et al. — Tunka-Rex and LOPES Coll., Phys. Lett. B763, 179 (2016)
[18] S. Argiro et al. — Pierre Auger Coll., Nucl.Instrum.Meth. A580 (2007) 1485-1496, 0707.1652
[19] I. N. Bronshtein, K. A. Semendyayev, G. Musiol, H. Muehlig 2007 *Handbook of Mathematics* (Berlin: Springer) chapter 3 p 175

IOP Conf. Series: Journal of Physics: Conf. Series **1181** (2019) 012028 doi:10.1088/1742-6596/1181/1/012028

Summing up Ultra-High-Energy Cosmic Rays from Radio Galaxies

Björn Eichmann

Production Editor, *Journal of Physics: Conference Series*, IOP Publishing, Dirac House, Temple Back, Bristol BS1 6BE, UK

E-mail: eiche@tp4.rub.de

Abstract. Radio galaxies are intensively discussed as the sources of cosmic rays observed above about 3 EeV, called ultra-high-energy cosmic rays (UHECRs). Here, the key issues from a recent investigation are summed up, where the individual characteristics of radio galaxies, as well as the impact by the extragalactic magnetic-field structures up to a distance of 120 Mpc has been taken into account. It is shown that the average contribution of radio galaxies taken over a very large volume cannot explain the observed features of UHECRs measured at Earth. However, we obtain excellent agreement with the spectrum, composition, and arrival-direction distribution of UHECRs measured by the Pierre Auger Observatory, if we assume that most UHECRs observed arise from only two sources: The ultra-luminous radio galaxy Cygnus A, providing a mostly light composition of nuclear species dominating up to about 60 EeV, and the nearest radio galaxy Centaurus A, providing a heavy composition dominating above 60 EeV. Here we have to assume that extragalactic magnetic fields out to 250 Mpc, which we did not include in the simulation, are able to isotropize the UHECR events at about 8 EeV arriving from Cygnus A.

1. Introduction

The origin of the Ultra-High-Energy Cosmic Rays (UHECRs) is still one of the great enigmas of modern astrophysics. Basically, there are three main observational characteristics, from observatories like the Pierre Auger Observatory (Auger) in the southern hemisphere, and the High Resolution Fly's eye (HiRes) as well as the Telescope Array (TA) experiments in the northern hemisphere, that describe our current knowledge of the UHECRs:

(1.) The energy spectrum, which changes above ~ 3 EeV — the so-called ankle — to a flatter power-law distribution with a spectral index of 2.6 and a sharp flux suppression above about $10^{19.5}$ eV [1, 2].

(2.) The chemical composition, that shows an increase of the fraction of heavier elements above $10^{18.3}$ eV [3, 4, 5].

(3.) The arrival directions, where recently Auger reported a 5σ detection of a dipole with an amplitude of $\approx 6.5\%$ in the UHECR arrival directions, while higher-order multipoles are still consistent with isotropy [6].

According to a recent study by Eichmann et al. [7] — hereafter referred to as E+18 — one of the key questions in this field (still) is: Can the origin of those extremely energetic particles be explained by radio galaxies?

IOP Conf. Series: Journal of Physics: Conf. Series **1181** (2019) 012028 doi:10.1088/1742-6596/1181/1/012028

In this paper, the extensive outcome of E+18 is summed up and discussed in the light of subsequent follow-up investigations. The UHECR propagation simulations in E+18 comprise three components: (a) a 3D structure of radio galaxies and EGMF within a radius of 120 Mpc; (b) a continuous source function derived from a luminosity function of radio galaxies for the contributions from beyond 120 Mpc; (c) a contribution of the powerful radio galaxy Cygnus A, which is near the magnetic horizon and is expected to deliver the dominant contribution to UHECRs due to its extreme radio power and brightness. All simulations are carried out with CRPropa3 [8].

2. Radio Galaxies as UHECR accelerators

Radio galaxies and in particular their jets are suitable sites for first-order Fermi acceleration at their discontinuities, as shown in numerous works in the past. Since the Fermi acceleration process is only dependent on rigidity, abundances f_i of different nuclear species with charge Z_i measured at a given rigidity R remain unchanged in the acceleration process. The abundances at injection, however, will depend on the environment or details of the injection process. In E+18, we suggest a simple relation where particles are injected at some minimal rigidity $\check{R}$ with abundances $f_i = f_\odot Z_i^q$, where $f_\odot$ denotes solar abundances, that enables to increase the heaviness of the composition based on a single parameter q.

As shown in great detail in E+18, the established correlation between the jet power $Q_{\rm jet}$ and the extended radio luminosity $L_{1.1}$ from Willott et al. [9] gives for the minimal energy condition [10] a good estimate of the cosmic ray luminosity

$$Q_{\rm cr} \simeq \frac{4}{7} Q_{\rm jet} = 1.3 \times 10^{42}\, g_{\rm cr} \left(\frac{L_{1.1}}{L_\star}\right)^{6/7} \frac{\rm erg}{\rm s}\,, \tag{1}$$

and the maximal rigidity

$$\hat{R}(Q_{\rm cr}) = g_{\rm acc}\sqrt{Q_{\rm cr}/c}\,, \tag{2}$$

where $L_\star \approx 4.9 \times 10^{40}$ erg/s denotes a characteristic luminosity according to Mauch and Sadler [11]. In doing so, we suppose that the total power in CRs is significantly higher than in relativistic electrons. From the uncertain efficiency of converting jet internal energy into observable radio luminosity, as well as the uncertain details of the acceleration process we obtain the dimensionless coefficients

$$1 \lesssim g_{\rm cr} \lesssim 50 \quad \text{and} \quad 0.01 \lesssim g_{\rm acc} \lesssim 0.5\,. \tag{3}$$

Introducing different nuclei species i with charge number Z_i and an abundance f_i, the total cosmic ray power per charge number yields

$$Q_{{\rm cr},i} \equiv Q_{\rm cr}(Z_i) = f_i\, Z_i\, Q_{\rm cr}/\bar{Z}\,. \tag{4}$$

3. Simulation setup

While the previously introduced physics are used to characterize the individual CR sources, the Monte Carlo based algorithm of CRPropa3 [8] is used to describe the propagation of UHECR nuclei in extragalactic background light, i.e. the cosmic microwave background (CMB) and the UV/optical/IR background (IRB), and cosmic magnetic fields. For the latter, the local EGMF structure from Dolag et al. [12] — hereafter referred to as D05 — which extends up to a distance of 120 Mpc as well as the GMF model of Jansson & Farrar [13] is adopted. However, an impact on the energy distribution by the GMF, i.e. energy losses or rigidity dependent arrival probabilities, is not taken into account. Due to rather technical issues like the limited EGMF structure and a limited CPU time, there is a need to distinguish the source contributions in the simulation setup.

3.1. Local sources

Sources up to a distance of 120 Mpc from Earth, called *local* RGs, are treated as individual sources in a 3D simulations, where the impact of cosmic magnetic fields is taken into account. To obtain a complete sample of radio galaxies powerful enough to contribute to the UHECR spectrum we use the catalog of van Velzen et al. [14] — hereafter referred to as *vV12 catalog.* After excluding 52 starforming galaxies, we are finally left with 121 radio galaxies within the considered volume, predominantly of FR-I type. Due to the flux limitation of the catalog as well as the limited sky coverage (88%), the catalog misses a multiplicity of predominantly low luminous, rather distant RGs. In E+18 it is described in great detail, that the radio luminosity function (RLF) can be used to complete the sample, however, the resulting contribution of these missing sources to the total UHECR flux is shown to be negligible. Thus, each source isotropically emits an individual UHECR flux that is *absolutely normalized* based on its CR power (1) and provides a maximal rigidity according to Eq. (2).

3.2. Average non-local sources

Sources beyond a distance of 120 Mpc from Earth, called *non-local* RGs, are treated by a continuous source function (CSF) in a 1D simulation, where only the impact of cosmic photon targets is taken into account. In doing so, the local CSF, $\Psi_{i,0}(R)$, is derived as

$$\Psi_{i,0}(R) \equiv \frac{\mathrm{d}N_{\mathrm{cr}}(Z_i)}{\mathrm{d}V\mathrm{d}R\,\mathrm{d}t} = \int_{\check{Q}_{\mathrm{cr}}}^{\hat{Q}_{\mathrm{cr}}} S_i\big(R, \hat{R}(Q_{\mathrm{cr}})\big)\,\frac{\mathrm{d}N_{\mathrm{RG}}}{\mathrm{d}V\,\mathrm{d}Q_{\mathrm{cr}}}\,\mathrm{d}Q_{\mathrm{cr}}\,, \tag{5}$$

using the local RLF, $\mathrm{d}N_{\mathrm{RG}}/(\mathrm{d}V\,\mathrm{d}Q_{\mathrm{cr}})$, and the CR spectrum, $S_i\big(R, \hat{R}(Q_{\mathrm{cr}})\big)$, of element species i with charge number Z_i, emitted by a source with total cosmic ray power per charge number, $Q_{\mathrm{cr},i}$, up to a maximal rigidity $\hat{R}(Q_{\mathrm{cr}})$. The limits of integration are the smallest, $\check{Q}_{\mathrm{cr}}$, respectively largest, $\hat{Q}_{\mathrm{cr}}$, CR powers we need to consider. To solve the integral analytically, as shown in E+18, a sharp cutoff of the individual source spectra at $\hat{R}(Q_{\mathrm{cr}})$ is supposed. The function $\Psi_{i,0}(R)$ is the local continuous source function as it is derived from a radio luminosity function determined in the local universe ($z < 0.3$). To extend it to larger redshifts, we use the approximation

$$\Psi_i(R, z) \approx \Psi_{i,0}(R)\,(1+z)^{m-1}\,, \tag{6}$$

where $m \in [2, 4]$ is called the source evolution index. The analytical solution of the CSF (6) shows a spectral break at a critical rigidity $R_* = g_{\mathrm{acc}}\sqrt{g_{\mathrm{cr}}\,Q_*/c} \approx 2 \times 10^{18}\,g_{\mathrm{acc}}\,\sqrt{g_{\mathrm{cr}}}\,\mathrm{V}$, with $\Psi_i(R < R_*, z) \propto R^{-a}$, where a denotes the spectral index of the individual sources, and $\Psi_i(R > R_*, z) \propto R^{-3}$ or even steeper. This means that for light nuclei the spectral index in the UHECR regime is too steep to explain the observed UHECR spectrum (as propagation will make the spectrum even steeper). For heavy nuclei, $eZR_\star$ may be large enough to allow UHECR spectra consistent with the data, but this would mean that they have to dominate in the entire UHECR regime, which is inconsistent with the measured composition. We can therefore already get to the preliminary conclusion that the *cosmic-ray spectrum averaged over the radio luminosity function cannot explain the known UHECR data.*

3.3. Cygnus A

The UHECR contribution by non-local, ultra-luminous RGs like the FR II source Cygnus A is not contained in the previously introduced average non-local source contribution (6). Although, Cygnus A with a CR luminosity $Q_{\mathrm{cr}}(L) = 1.3 \times 10^{45}\,g_{\mathrm{cr}}\,\mathrm{erg/s}$ and a corresponding maximal rigidity $\hat{R}(L) = 62.6\,g_{\mathrm{acc}}\,\sqrt{g_{\mathrm{cr}}}\,\mathrm{EV}$ at a distance of 255 Mpc [14], represents an ideal UHECR accelerator. Due to the lack of a reliable EGMF structure out to its distance as well as other technical limitations, we use a simplified approach where deflections in the EGMF are neglected.

Nevertheless, we are still able to normalize the flux in the same manner as we do it for the other sources. Deflections by the EGMF are then estimated using a homogeneous random field with a given mean field strength and coherence length. Deflections inside the Galaxy can still be applied in the usual way.

4. Best-fit results

Finally, the previously introduced parameters within their plausibility ranges, i.e. the spectral index $a \in [1.7, 2.2]$, the cosmic-ray load $g_{\rm cr} \in [1, 50]$, the acceleration efficiency $g_{\rm acc} \in [0.01, 0.5]$, and the abundance parameter $q \in [0, 2]$, are used to fit the observational data of the flux and mean logarithm of the mass number, $\langle \ln(A) \rangle$, above 5 EeV from Auger. The source-evolution parameter was fixed to $m = 3$ as its influence is mainly important below the ankle [15].

The normalization of the observed $\langle \ln(A) \rangle$ data depends strongly on the used hadronic interaction model, therefore the fitting is performed for the three different contemporary models, QGSJetII-04, Epos-LHC, and Sibyll2.1, that have been used by Auger [4].

The first fit approaches showed that the UHECR data cannot be explained, if all sources provide the same parameter values.

So, we enabled exceptional values for certain parameters of the dominant sources (as summarized in table 1) and obtained an accurate explanation of the observational data as shown in Fig. 1. As indicated by the reduced chi-squared value χ^2, the accuracy of the fit does not significantly differ with respect to the used hadronic interaction model. The results show that the UHECR data is accurately explained by the CR contributions from Centaurus A and Cygnus A with an isotropically ejected CR power above the ankle of $Q_{\rm uhecr}^{\rm CenA} \simeq 2 \times 10^{43}$ erg/s and $Q_{\rm uhecr}^{\rm CygA} \simeq 2 \times 10^{47}$ erg/s, weakly dependent on the chosen scenario or the hadronic interaction model, while the contribution of all other RGs is irrelevant. In particular the average non-local sources are kept negligible by a significantly smaller parameter value of $\bar{g}_{\rm cr}$ compared to the values of Centaurus A or Cygnus A due to its steep spectral behavior. However, an inclusion of some data below the ankle might also increase its significance, and the value of $\bar{g}_{\rm cr}$ could become comparable to the one from Centaurus A and Cygnus A.

Table 1: Best-fit parameters, with fixed $\bar{q}^{\rm CygA} = 0$.

	a	$\bar{g}_{\rm cr}$	$g_{\rm cr}^{\rm CenA}$	$g_{\rm cr}^{\rm CygA}$	$\bar{g}_{\rm acc}^{\rm CenA}$	$g_{\rm acc}^{\rm CygA}$	$q^{\rm CenA}$	χ^2
EPOS-LHC	1.85	7.73	41.54	43.94	0.127	0.059	2	1.1
QGSJetII-04	1.82	6.31	21.20	48.84	0.220	0.056	2	1.4
Sibyll2.1	1.83	6.67	24.77	47.90	0.19	0.056	2	1.3

Upper index of the parameter indicates the corresponding source (Centaurus A or Cygnus A), bar on top of the parameter corresponds to all sources except Centaurus A and Cygnus A

To estimate EGMF deflections for Cygnus A, we derived the rms deflection angle for a distance $d = 255$ Mpc analytically [16] by

$$\theta_k \simeq 0.8^\circ \, \bar{Z}_k \left(\frac{\bar{E}_k}{100\,\rm EeV}\right)^{-1} \left(\frac{d}{10\,\rm Mpc}\right)^{1/2} \left(\frac{\lambda_{\rm c}}{1\,\rm Mpc}\right)^{1/2} \left(\frac{B_{\rm rms}}{1\,\rm nG}\right) , \qquad (7)$$

where we used an approximated mean candidate energy $\bar{E}_k = (E_k + E_k^*)/2$, an approximated mean candidate charge number $\bar{Z}_k = (Z_k + Z_k^*)/2$, and used as the rms strength $B_{\rm rms} = 1.18$ nG of the D05 field. Further, a correlation length $0.1\,{\rm Mpc} \leq \lambda_{\rm c} \leq 10\,{\rm Mpc}$ is assumed, as for $\lambda_{\rm c} \ll 0.1$ Mpc we obtained negligible deflections, whereas in the case of $\lambda_{\rm c} \gg 10$ Mpc the UHECRs with $E_i \lesssim 10$ EeV from Cygnus A were completely isotropized. Using the previously

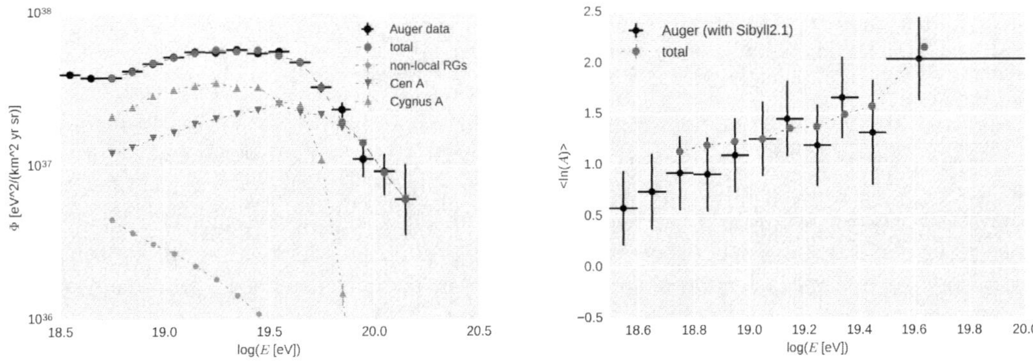

Figure 1: Best-fit results to energy spectrum (left) and chemical composition (right) using Sibyll2.1.

derived best-fit parameters the Fig. 2 shows the resulting angular power spectra of the arrival directions for different deflection scenarios: Three different coherence lengths λ_c of a turbulent field with $B_{rms} = 1.18\,\mathrm{nG}$, as well as the case of the complete isotropization (i.e. $\lambda_c \gg 10\,\mathrm{Mpc}$), and the complete isotropization where initial H and He candidates from Centaurus A are excluded. The latter assumption does not affect the total energy spectrum or the chemical composition, but lowers the level of anisotropy and improves consistency with the data set at 4–8 EeV. The motivation for this is that models for a particular heavy enhancement of Centaurus A would expect a pure heavy composition [17]. We also note that at 4–8 EeV an impact of cosmic rays from the Galaxy or other extragalactic components cannot be excluded, and that at $E > 8\,\mathrm{EeV}$ our results are in good agreement with the data, for the case that UHECR from Cygnus A are completely isotropized. Further, an agreement with the observed arrival directions at energies $E > 8\,\mathrm{EeV}$ is obtained for

$$\lambda_c^{1/2}\, B_{rms} \geq 6\,\mathrm{Mpc}^{1/2}\,\mathrm{nG}, \tag{8}$$

i.e. a global coherence length $\lambda_c = 26\,\mathrm{Mpc}$ in the case of a rms field strength of $B_{rms} \geq 1.18\,\mathrm{nG}$.

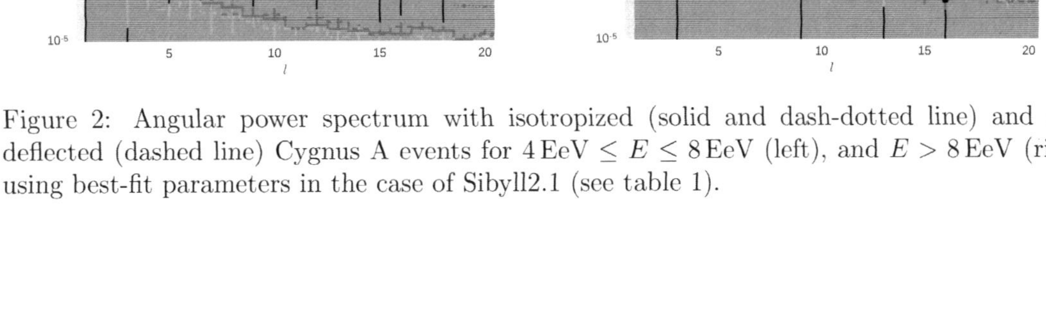

Figure 2: Angular power spectrum with isotropized (solid and dash-dotted line) and non-deflected (dashed line) Cygnus A events for $4\,\mathrm{EeV} \leq E \leq 8\,\mathrm{EeV}$ (left), and $E > 8\,\mathrm{EeV}$ (right) using best-fit parameters in the case of Sibyll2.1 (see table 1).

5. Discussion

The most solid conclusion that can be drawn is that an average contribution of all radio galaxies cannot explain both spectrum and composition of UHECR.

The second conclusion is, that we need some outstanding, individual sources which explain the cosmic ray flux above the ankle. Here, two sources recommend themselves: Cygnus A, the by far brightest radio galaxy in the sky, and Centaurus A, the second-brightest radio galaxy and also the one nearest to Earth. In doing so, either both sources have a cosmic ray load significantly above the average of the bulk of radio galaxies, or the average non-local source contribution explains the CRs between $\sim 1\,\mathrm{EeV}$ and the ankle, which seems likely due to its spectral behavior. In addition, it cannot be ruled out that there is also a significant contribution by the other two exceptionally bright local radio galaxies, M87 and Fornax A, but as showed in E+18 they cannot solve the issue alone — Centaurus A is needed.

The third conclusion is that Cygnus A needs to provide a solar-like composition, at least if its contribution dominates over the average of the bulk of radio galaxies at the ankle, while Centaurus A needs to have a heavy composition, with an iron fraction comparable to protons at a given cosmic-ray energy.

A necessary condition for all this to work is that UHECRs from Cygnus A are significantly deflected by the EGMF. We estimate that for $\lambda_c^{1/2}\,B_{\mathrm{rms}} = 6\,\mathrm{Mpc}^{1/2}\,\mathrm{nG}$ all anisotropy constraints at energies $> 8\,\mathrm{EeV}$ are satisfied.

Finally, the dominance of a few single sources, whereof Centaurus A is only visible by Auger and Cygnus A is predominantly visible by TA, indicates the need for future investigations that include the impact of the GMF as well as the geometrical exposure of the experiments on the UHECRs from these sources. Further, the influence of the EGMF on the energy spectrum and composition of Cygnus A needs to be taken into account and a possible average contribution by the bulk of radio galaxies between the ankle and the second knee needs to be investigated.

References

[1] Abbasi R U *et al.* (HiRes) 2008 *Phys. Rev. Lett.* **100** 101101 (*Preprint* astro-ph/0703099)
[2] Abraham J *et al.* (Pierre Auger) 2010 *Phys. Lett. B* **685** 239–246 (*Preprint* 1002.1975)
[3] Abraham J *et al.* (Pierre Auger) 2010 *Phys. Rev. Lett.* **104** 091101 (*Preprint* 1002.0699)
[4] Aab A *et al.* (Pierre Auger) 2014 *Phys. Rev.* **D90** 122006 (*Preprint* 1409.5083)
[5] Unger for the Pierre Auger Collaboration M (Pierre Auger) 2017 *PoS* **ICRC2017** 1102 (*Preprint* 1710.09478)
[6] Aab A *et al.* (Pierre Auger) 2017 *Science* **357** 1266–1270 (*Preprint* 1709.07321)
[7] Eichmann B, Rachen J, Merten L, van Vliet A and Tjus J B 2018 *Journal of Cosmology and Astroparticle Physics* **2018** 036
[8] Alves Batista R *et al.* 2016 *JCAP* **1605** 038 (*Preprint* 1603.07142)
[9] Willott C J, Rawlings S, Blundell K M and Lacy M 1999 *Mon. Not. Roy. Astron. Soc.* **309** 1017–1033 (*Preprint* astro-ph/9905388)
[10] Pacholczyk A G 1970 *Radio Astrophysics. Nonthermal Processes in Galactic and Extragalactic Sources* (W. H. Freeman & Co Ltd, San Francisco)
[11] Mauch T and Sadler E M 2007 *Mon. Not. Roy. Astron. Soc.* **375** 931–950 (*Preprint* astro-ph/0612018)
[12] Dolag K, Grasso D, Springel V and Tkachev I 2005 *JCAP* **1** 009 (*Preprint* astro-ph/0410419)
[13] Jansson R and Farrar G R 2012 *Astrophys. J.* **757** 14 (*Preprint* 1204.3662)
[14] van Velzen S, Falcke H, Schellart P, Nierstenhöfer N and Kampert K H 2012 *Astron. Astrophys.* **544** A18 (*Preprint* 1206.0031)
[15] Berezinskii V S and Grigor'eva S I 1988 *Astron. Astrophys.* **199** 1–12
[16] Sigl G, Miniati F and Enßlin T A 2004 *Phys. Rev. D* **70** 043007 (*Preprint* astro-ph/0401084)
[17] Gopal-Krishna, Biermann P L, de Souza V and Wiita P J 2010 *Astrophys. J. Lett.* **720** L155–L158 (*Preprint* 1006.5022)

IOP Conf. Series: Journal of Physics: Conf. Series **1181** (2019) 012029 doi:10.1088/1742-6596/1181/1/012029

Current status of the CODALEMA/EXTASIS experiments

B. Revenu[1,3], D. Charrier[1,3], R. Dallier[1,3], A. Escudie[1], D.García-Fernández[1], A. Lecacheux[2], and L. Martin[1,3]

[1] Subatech, Institut Mines-Telecom Atlantique, CNRS, Université de Nantes, Nantes, France

[2] CNRS-Observatoire de Paris, Meudon, France

[3] Unité Scientifique de Nançay, Observatoire de Paris, CNRS, PSL, UO/OSUC, Nançay, France

E-mail: `revenu@in2p3.fr`

Abstract. Secondary electrons and positrons of air showers emit a coherent radio electric field in a wide frequency band. The CODALEMA experiment installed at the Nançay radioastronomy observatory since 2002 detects air showers and the associated electric field in [20;200] MHz. EXTASIS, triggered by the CODALEMA scintillators, detects since 2016 the air shower electric field in [1;6] MHz. We also expect an additional signal at low frequency: the sudden death pulse, corresponding to the sudden disappearance of the shower front particles when they reach the ground level. We present the instrumental setups, their performances and the first results of EXTASIS.

1. The radio signal from extensive air showers

The air showers created following the interaction of ultra-high energy cosmic ray in our atmosphere contain many information on the primary particle. The secondary particles, in particular the $e^{\pm}$, emit a detectable electric field. The emission mechanisms are understood in great details [1, 2, 3]. The principal mechanism is the geomagnetic effect that produces an electric field linearly polarized in the direction $\mathbf{v} \times \mathbf{B}$ where $\mathbf{v}$ is the shower axis direction and $\mathbf{B}$ is the geomagnetic field at the location of the experiment. The second effect is the field emitted by the excess of e^- with respect to e^+ [4] which produces a field with a radial polarization with respect to the shower axis. The total electric field is the superposition of both mechanisms with a possible Cherenkov-like amplification for adequate observer positions with respect to the shower axis. Various radio experiments are currently taking data all over the world [5, 6]. By the comparison of the measured electric field distribution at ground level to the expected distribution, it is possible to estimate the atmospheric depth $X_{\max}$ of maximum development of the shower which is an excellent way to constrain the nature of the primary cosmic ray. The cosmic ray composition is the key to a much better understanding of their sources that are still unidentified.

2. The experimental setup

The CODALEMA/EXTASIS experiment aims, among other topics, at estimating the cosmic ray composition in the energy range $10^{16} - 10^{18}$ eV. It covers the full area of 1 km^2 of the

Nançay radio observatory. We have various instruments allowing a precise reconstruction of the air shower characteristics: axis direction θ, ϕ, core position x_c, y_c and the atmospheric depth of maximum development $X_{\max}$. From these informations, we can estimate the primary cosmic ray properties: energy and nature.

The shower observation is made through:

- the direct detection of the secondary particles by the scintillator array (13 detectors spread over a square of 0.16 km^2) with an energy threshold around 10^{15} eV;
- the detection of the electric field emitted by the secondary charged particles
  - with the 57 autonomous (self-trigger) stations covering the full area of 1 km^2, measuring the electric field in both north/south and east/west polarizations with a Butterfly antenna (in 20-200 MHz).
  - with the 10 cross-polarized antennas (in 10-200 MHz) covering a very small area to study the electric-field at short distances.
  - a single 3D antenna, triggered by the scintillator array, in order to measure the three components of the electric-field vector, allowing to check the validity of the far-field approximation.
- the detection of the electric field at low frequency (1 − 6 MHz), with the 7 radio stations spread over the area of 1 km^2.

The setup is presented in Fig. 1. The scintillator array has a very low background as

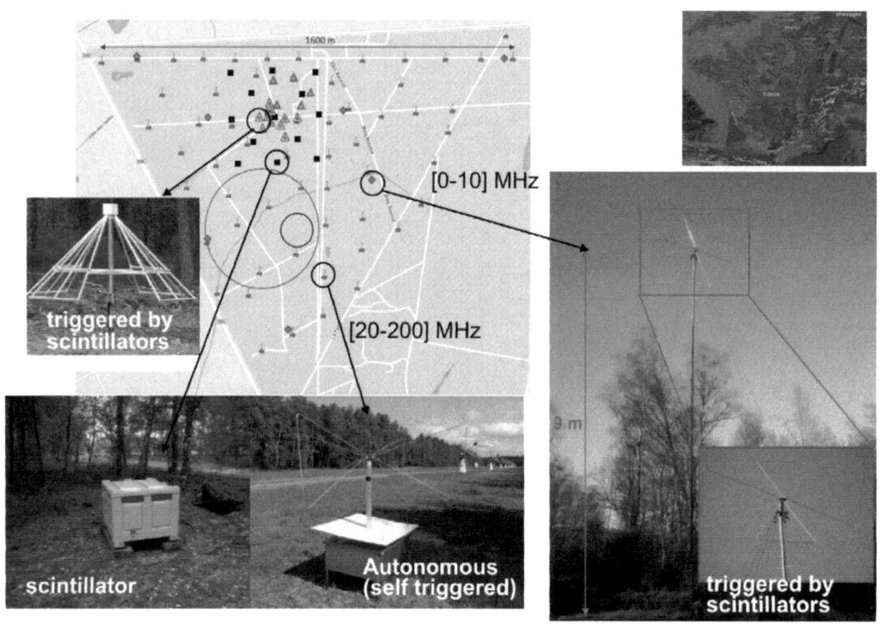

Figure 1. Setup of the CODALEMA/EXTASIS experiment in Nançay, France. The black squares represent the scintillators. The red symbols are the 57 autonomous radio stations. The white triangles are the antennas of the compact array and the blue diamonds are the low-frequency antennas of EXTASIS.

these detectors are sensitive to high energy particles of the shower only. The scintillator DAQ triggers in real time the radio compact array, the 3D antenna and the low frequency antennas of EXTASIS.

3. Analysis of the radio signal

The expected electric field is obtained with the code SELFAS3, developed at Subatech. It is based on the SELFAS2 version of the code [7, 8] but we significantly improved the computation in 2018. First, we are not using anymore the common US Standard model for the description of the atmosphere. This model is static and does not include day/night effects nor seasonal effects. We are now using the data from the GDAS [9] to estimate the precise atmospheric status at the time of the detection of our events at the location of Nançay. These data allow us to compute the air density and air refractive index which are key ingredients to obtain with high precision the electric field emitted by the secondary particles [10]. We also use the exact formula for the electric field estimation, valid at all frequencies [11]. We don't rely anymore on the far-field hypothesis ($kR \gg 1$) which is wrong in particular for the wavelengths of interest for EXTASIS (1-6 MHz, ie some hundreds of meters). For instance, at 100 m from the source and at a frequency of 1 MHz, we have $kR \sim 2$ which is clearly not satisfying the far-field condition.

For each shower detected in CODALEMA/EXTASIS, we compute the arrival direction with a resolution of half a degree or so. Using SELFAS3, we simulate 40 protons showers and 10 iron showers at 10^{17} eV with the same arrival direction, on a virtual array of 300 antennas. Then we compare the measured values of the electric field in 20-80 MHz to the simulated ones and the best agreement is the estimated shower core; the procedure also provides the scaling factor we must apply to correct from the arbitrary choice of the primary energy of 10^{17} eV [12]. The typical accuracy we obtain at the end of the analysis chain is 15 m on the core position, 20% on the primary energy and 20 g/cm^2 on the $X_{\max}$. FIG. 2 shows the reconstruction performances on an event detected by CODALEMA on March 9th, 2017. As the CODALEMA radio stations are

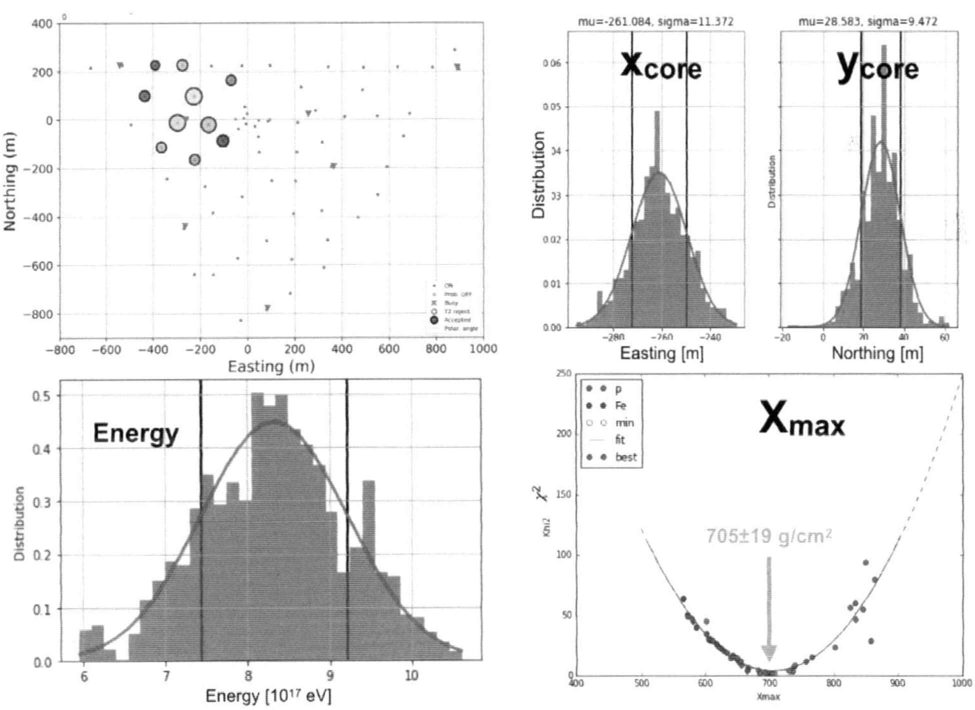

Figure 2. Reconstruction of an event detected by CODALEMA. The ground footprint is on the top left. The color scale indicates the arrival time of the transient signal in each antenna (blue: early, red: late). The other plots indicates the accuracy on the shower parameters.

able to measure the electric field up to 200 MHz, we can improve the 20-80 MHz reconstruction by taking into account the high frequency data. For geometrical reasons, the electric field emitted by inclined showers has an important high frequency component. The electric field amplification due to Cherenkov-like effects is clearly visible on inclined events. FIG. 3 shows the expected lateral distribution functions for the frequency band 20-80 MHz (left) and 120-200 MHz (right). For inclined showers, all stations are roughly at the same axis distance and

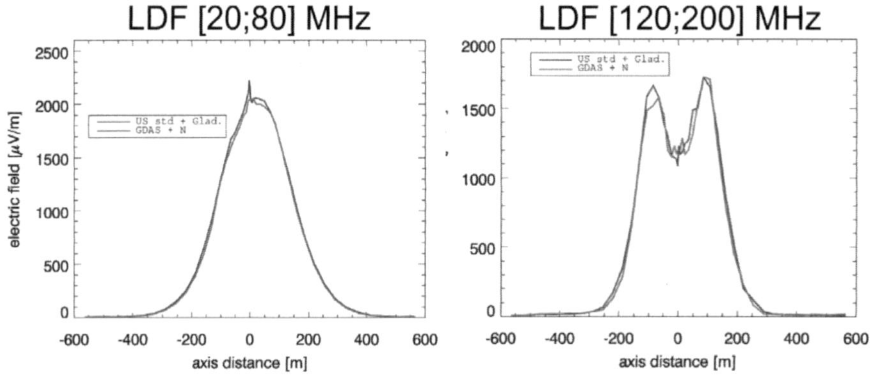

Figure 3. Expected lateral distribution functions (LDF) for an inclined shower, between 20-80 MHz (left) and 120-200 MHz (right) using SELFAS3. The LDF looks like a Gaussian, as a function of axis distance, in 20-80 MHz. The shape is more complicated at high frequency as the Cherenkov-like amplification of the electric field is visible at 140 m of the axis. The two curves correspond to the profiles obtained using two different atmospheric models.

the electric field gradient in 20-80 MHz is very small: the shower parameters reconstruction is not well constrained by these data alone. At higher frequency (120-200 MHz), the electric field variations are much stronger due to the Cherenkov-like amplification at $\sim$ 140 m of the shower axis. This permits a much better shower reconstruction.

4. EXTASIS: air showers observed at low frequency

There are several motivations for the study of the low frequency radio emission of air showers. Air showers develop within some μs so that we expect some contribution at some hundreds of kHz. The simulation also predicts an emission at some MHz comparable to that of the 20-80 MHz band but the range is expected to be larger. We also aim to detect the emission from the air shower sudden death: it is the coherent emission from the sudden deceleration of $e^{\pm}$ when the shower front reaches the ground level [13, 11]. This emission is directly related to the shower front lateral dimensions, implying a contribution at the MHz level. We installed in Nançay 7 Buttefly antennas, 9 m above the ground level, as suggested by NEC2 simulations for the antenna response at these frequencies. These low frequency antennas are triggered by the scintillator array and data are produced by a scope with a sampling rate of 500 MHz on 8 bits, in a time window of 2 ms. The associated amplifier is a modified version of the LONAMOS used in the 57 radio stations of CODALEMA.

In 18 months, we detected 18 showers. We observed that the detection range is higher than in the 20-80 MHz band as predicted by the SELFAS3 simulation [14]: in these events, the maximum distance between any pair of low frequency antennas is always larger than for the CODALEMA antennas. We did not detect the sudden death signal so far. This could be due to the fact that the Nançay observatory has an altitude of 180 m: the atmospheric depth at this altitude is of the order of 1000 g/cm^2 for vertical showers and the number of

$e^{\pm}$ reaching the ground level is not sufficient to emit a measurable sudden death electric field. The detection threshold for EXTASIS is around 5 μV/m which corresponds in the best case to vertical showers of 2×10^{18} eV, making candidate showers quite rare. The situation would be much more favorable for a site at higher altitude. For instance, at 1400 m above sea level (the Auger site for instance) candidate showers have an energy of some 10^{17} eV, the flux would be multiplied by a factor of ~130. We also observed that the probability to detect showers at low frequency is significantly improved when there is a high atmospheric electric field, as it is the case in thunderstorm conditions. In FIG. 4, we present an event detected on January 1st, 2018. The shower is coming from the south-west, as indicated by the blue arrow. The

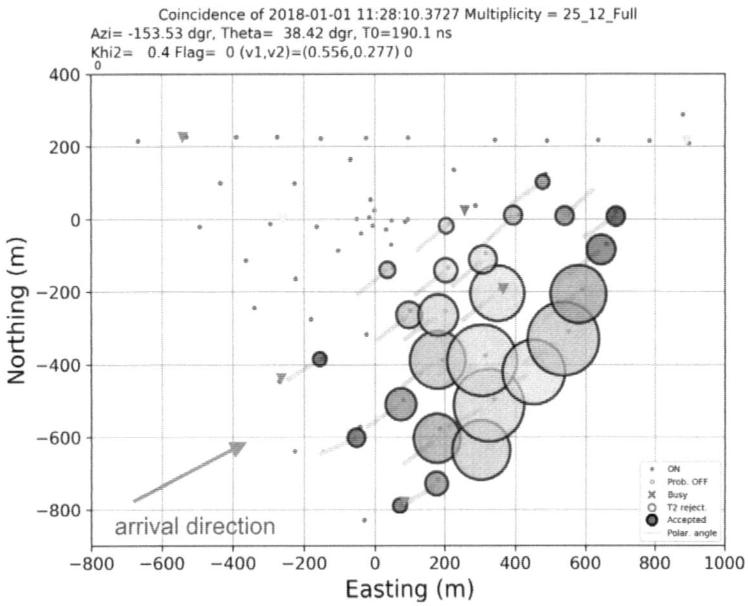

Figure 4. Ground footprint of an air shower detected by 6 EXTASIS antennas (trinagles) and 25 CODALEMA antennas (circles). The shower arrival direction projected in the horizontal plane is indicated by the blue arrow. The electric field polarization measured by the CODALEMA antennas is indicated by the green lines. See text for details.

radio stations of CODALEMA that detected the shower are indicated by the colored circles, with the same color code than in FIG. 2. The 6 EXTASIS antennas involved in the event are indicated by the colored triangles; the arrival direction estimated using the low-frequency data only is in very good agreement with the direction obtained by the CODALEMA data. We see that the ground pattern of the EXTASIS antennas has a much larger area than the pattern of the CODALEMA stations: this confirms the more extended lateral width of the signal at low-frequency compared to the signal in the 20-80 MHz band. It is very interesting to focus on the polarization angle of the electric in all triggered CODALEMA stations. This linear polarization is indicated with the green lines. We see that these lines are roughly parallel to the incoming direction (in the horizontal plane) which is not what we expect from the geomagnetic effect. We should have observed an electric field orthogonal to the incoming direction as it is polarized at first order along $\mathbf{v} \times \mathbf{B}$. The discrepancy is due to the presence of a thunderstorm: the strong atmospheric electric field overtakes the geomagnetic effect and we observe a polarization that could be parallel to $\mathbf{E}_{\text{storm}} + \mathbf{v} \times \mathbf{B}$ with $||\mathbf{E}_{\text{storm}}|| \gg ||\mathbf{v} \times \mathbf{B}||$. The amplitude of the signal is also strongly amplified by the presence of $\mathbf{E}_{\text{storm}}$. This storm is not a regular one: it is the storm Carmen; at the time of the detection of the cosmic ray, the atmospheric electric field had a value at 20σ of its value in quiet conditions.

5. Conclusion

CODALEMA/EXTASIS is a multi-wavelength experiment that observes cosmic-ray induced air showers in 1-6 MHz, 20-80 MHz and 120-200 MHz. This allows us to use the information at all frequencies allowing improved reconstruction of the primary cosmic ray characteristics. We observe showers in 1-6 MHz: the detected signal is the counterpart at low frequency of the regular emission when the shower develops in the atmosphere. But the number of events detected is small (18 only in 1.5 years) and the detection is favored when the atmospheric electric field is high (thunderstorm conditions). We did not detect yet the sudden death signal, corresponding to the electric field emitted when the shower front hits the ground. We believe the reason for this is the low altitude of the site of Nançay, making the number of secondary particles reaching the ground level too small to produce an electric field that would be above the detection threshold. It would be very interesting to try the sudden death pulse detection at higher altitudes. Our analysis relies on the code SELFAS3 which uses a formula of the electric field valid at all frequencies.

Acknowledgments

We thank the Région Pays de la Loire for its financial support of the Astroparticle group of Subatech and in particular for its contribution to the EXTASIS experiment.

References

[1] D. Ardouin, A. Belletoile, C. Berat, D. Breton, D. Charrier, et al. Geomagnetic origin of the radio emission from cosmic ray induced air showers observed by CODALEMA. *Astropart.Phys.*, 31(3):192–200, 2009.

[2] A. Belletoile, R. Dallier, A. Lecacheux, V. Marin, L. Martin, B. Revenu, and D. Torres Machado. Evidence for the charge-excess contribution in air shower radio emission observed by the CODALEMA experiment. *Astroparticle Physics*, 69:50–60, April 2015.

[3] The Pierre Auger Collaboration. Probing the radio emission from cosmic-ray-induced air showers by polarization measurements. *Phys. Rev. D*, 89(052002), March 2014.

[4] G.A. Askaryan. *J. Exp. Theor. Phy.*, 21:658, 1962.

[5] E. Holt for the for the Pierre Auger Collaboration. Recent Results of the Auger Engineering RadioArray (AERA). In *ICRC2017*, volume 301, 2017.

[6] F. G. Schröder for the Tunka-Rex collaboration. Overview on the Tunka-Rex antenna array forcosmic-ray air showers. *PoS*, ICRC2017:459, 2018.

[7] Vincent Marin and Benoît Revenu. Simulation of radio emission from cosmic ray air shower with SELFAS2. *Astropart.Phys.*, 35:733–741, 2012.

[8] V. Marin. SELFAS2 : radio emission from cosmic ray air showers. Effect of realistic air refractive index. In *Proceedings of the ARENA 2012 workshop (Erlangen, Germany)*, AIP Conference Proceedings, 2012.

[9] GDAS Archive Information, http://ready.arl.noaa.gov/gdas1.php. Technical report, NOAA.

[10] F. Gaté, B. Revenu, D. García-Fernández, V. Marin, R. Dallier, A. Escudié, and L. Martin. Computing the electric field from extensive air showers using a realistic description of the atmosphere. *Astroparticle Physics*, 98:38 – 51, 2018.

[11] Daniel García-Fernández, Benoît Revenu, Didier Charrier, Richard Dallier, Antony Escudie, and Lilian Martin. Calculations of low-frequency radio emission by cosmic ray-induced particle showers. *Phys. Rev.*, D97(10):103010, 2018.

[12] F. Gaté and The Pierre Auger Collaboration. $X_{\max}$ reconstruction from amplitude information with AERA. In *Proceedings of the ARENA 2016 workshop (Groningen, The Netherlands)*. ARENA, 2016.

[13] Benoît Revenu and Vincent Marin. Coherent radio emission from the cosmic ray air shower sudden death. In *proceedings of the 33rd ICRC, Rio de Janeiro, Brasil*, number 398, July 2013.

[14] Antony Escudie, Didier Charrier, Richard Dallier, Daniel García-Fernández, Alain Lecacheux, Lilian Martin, and Benoît Revenu. Low frequency observations of cosmic ray air shower radio emission by CODALEMA/EXTASIS. *PoS*, ICRC2017:419, 2018.

IOP Conf. Series: Journal of Physics: Conf. Series **1181** (2019) 012030 doi:10.1088/1742-6596/1181/1/012030

Estimating the air shower $X_{\rm max}$ from radio measurements

B. Revenu[1,3], R. Dallier[1,3], A. Escudie[1], D.García-Fernández[1], L. Martin[1,3]

[1] Subatech, Institut Mines-Telecom Atlantique, CNRS, Université de Nantes, Nantes, France
[2] Nantes, France
[3] Unité Scientifique de Nançay, Observatoire de Paris, CNRS, PSL, UO/OSUC, Nançay, France

E-mail: revenu@in2p3.fr

Abstract. It is possible to use the radio signal emitted by air showers to estimate the primary cosmic ray characteristics, in particular its nature through the shower $X_{\rm max}$. The electric field emitted by air showers is computed assuming an atmospheric model; we need to know the air density and refractive index since these parameters have and influence on the development of the shower and the emitted electric field. We can get an unbiased estimation using the Global Data Assimilation System (GDAS) to compute the precise atmospheric conditions at the time and location an event is detected. We show how to use these data.

1. Introduction

Many experiments are using the radio technique in addition to older technologies like a fluorescence telescope or a surface detector [1, 2, 3]. The radio signal has a duty cycle close to 100% and it permits to estimate directly the shower $X_{\rm max}$ with an uncertainty similar to what is obtained with a fluorescence telescope. We use the signal in the band 20-80 MHz where the electric field is coherent (the exact coherence band depends on the shower geometry). The estimation of the primary cosmic ray characteristics is performed by a comparison between the electric field measurements (amplitude, polarization, spectrum) with the theoretical expectation from simulation codes [4, 5, 6]. These codes are able to compute the electric field emitted by a shower as a function of time for any observer position. They all need as input the shower geometry (zenith θ, azimuth ϕ), the nature of the primary cosmic ray and its energy. After the first interaction at the atmospheric depth X_1, the secondary particles of the shower are created and tracked along their trajectory in the atmosphere. The electric field is computed for each of these particles and summed up to get the total electric field at the observer location. The propagation is driven by the medium and the first key ingredient is the air density $\rho_{\rm air}(z)$ at every position of the secondary particles, i.e. for each altitude z, if we assume the atmosphere density only depends on z at fixed time. The second key ingredient is the air refractive index η

which appears explicitely in the electric field formula (see [7] for the full derivation):

$$\mathbf{E}(\mathbf{x},t) = \frac{1}{4\pi\epsilon}\int \mathrm{d}^3x' \left\{ \left[\frac{\rho(\mathbf{x}',t')\mathbf{r}}{R^2(1-\eta\beta\cdot\mathbf{r})}\right]_{\mathrm{ret}} + \frac{\eta}{c}\frac{\partial}{\partial t}\left[\frac{\rho(\mathbf{x}',t')\mathbf{r}}{R(1-\eta\beta\cdot\mathbf{r})}\right]_{\mathrm{ret}} - \frac{\eta^2}{c^2}\frac{\partial}{\partial t}\left[\frac{\mathbf{J}(\mathbf{x}',t')}{R(1-\eta\beta\cdot\mathbf{r})}\right]_{\mathrm{ret}}\right\}. \quad (1)$$

This is the expected value at the observers coordinates $\mathbf{x}, t$ from the charge density ρ, current $\mathbf{J}$, $\mathbf{r}$ being the normalized vector particle-observer and $\beta = \mathbf{v}/c$. R is the distance particle-observer. Integration is performed over x' at retarded time $t' = t - \eta R/c$. The dependence in the air density ρ_{air} appears implicitely through the charge density ρ. The air refractive index drives the amplitude and arrival time of the electric field (it appears in the denominators of EQ. 1) i.e. the pulse shape, in particular close to the Cherenkov angle. It is therefore of prime importance to provide a realistic atmospheric model, which is not possible using the commonly-used US Standard description [8]. The details of the method are in [9].

2. Atmospheric model

As we experience daily, the atmosphere is a quickly varying medium with day/night effects and seasonal variations (winter/summer). The weather conditions also depend on the considered location on Earth. It is natural to use an atmospheric model that follows these variations.

2.1. The US Standard atmospheric model

It has been proposed in 1976 and it is an idealized modelisation of the atmosphere from an altitude of 0 up to 1000 km. The atmosphere characteristics are obtained from yearly averages assuming a hydrostatic equilibrium of the fluid. In this model, air is considered as a homogeneous mixture of various gases. The full height of the atmosphere is divided in 5 layers with some continuity conditions. Air density $\rho_{\mathrm{air}}(z)$ is a static average, valid for all locations on Earth, at all times (no day/night nor winter/summer influence). This very rough description of the atmosphere can lead to large deviations compared to actual weather conditions; and we can also expect some deviations in the resulting electric field from air showers. The code SELFAS used this parameterization until December 2016; then we switched to a more refined model, using the GDAS data.

2.2. The GDAS atmospheric model

GDAS [10] is the Global Data Assimilation System. It is the system used by "the Global Forecast System (GFS) model to place observations into a gridded model space for the purpose of starting, or initializing, weather forecasts with observed data." They produce a 3D model space that is based on ground observations, balloon measurements, satellite data, aircraft reports etc. Data are available from year 2001 on grids of various sizes ($1° \times 1°$, $0.5° \times 0.5°$ up to $0.25° \times 0.25°$) in longitude and latitude. The data is updated every 3 hours between the sea level up to an altitude of $z_{\mathrm{max}}^{\mathrm{GDAS}} = 26$ km. Many physico-chemical data are available but we need a small fraction of them to compute the air density and air refractive index above the observation site. These mandatory data are: the relative humidity R_h, the temperature T and the pressure P. As an illustration of the dynamics of the atmospheric conditions, we show in FIG. 1 the variation in relative humidity with time during a single day (March 18, 2014). We see very large variations that will affect in particular the air refractive index value: the pulse shape must be estimated using the air refractive index computed from the actual relative humidity and not an average value.

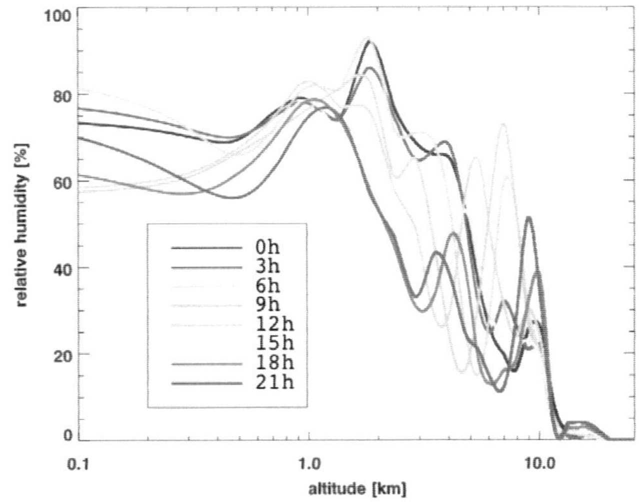

Figure 1. Relative humidity as a function of altitude above Nançay on March 18, 2014, for different time in the day.

2.3. Air density computation

The air density can be estimated using the ideal gas law:

$$\rho_{\rm air}(z) = \frac{p_d(z(Z_g,\phi))M_d + p_v(z(Z_g,\phi))M_v}{RT(z(Z_g,\phi))},$$

where $z(Z_g,\phi)$ is the altitude above sea level, depending on the geopotential height Z_g at a latitude ϕ; p_d and p_v are the partial pressures of dry air and water vapor respectively; M_d and M_v are associated the molar masses. GDAS data are given as a function of Z_g that we can convert in altitude above sea level using:

$$z(Z_g,\phi) = \frac{g_0}{C}\left(\Lambda(\phi) - \sqrt{\Lambda^2(\phi) - \frac{2C\,Z_g}{g_0}}\right)$$

with:

$$\Lambda(\phi) = 1 + A\sin^2(\phi) - B\sin^2(2\phi)$$
$$\text{where } A = 0.0053024,\ B = 5.8\times10^{-6},\ C = 3.086\times10^{-6}\ \text{s}^{-2}.$$

We compute the water vapor partial pressure from $p_v = R_h\,p_{\rm sat}$ where $p_{\rm sat}$ is given by (see [11, 12]):

$$p_{\rm sat} = 6.1121\exp\left[\left(18.678 - \frac{T}{234.5}\right)\left(\frac{T}{257.14+T}\right)\right] \quad (T \text{ in } {}^\circ\text{C}).$$

The dry air partial pressure is $p_d = P - p_v$. We can use these results in the range $[-80; +50]$°C, which is precisely our range of interest.

At this stage, we know the air density $\rho_{\rm air}$ between sea level and $z_{\rm max}^{\rm GDAS} = 26$ km. Since we deal with air showers that can start to develop at much higher altitude, we need to extend the

$\rho_{\rm air}$ estimation up to the atmosphere limit of $\sim$ 110 km. For altitude above $z_{\rm max}^{\rm GDAS}$, we use the US Standard model with a scaling factor to ensure continuity with our GDAS estimation of $\rho_{\rm air}$. This is possible as the atmosphere, above 26 km of altitude, is much more stable than at lower altitudes and the US Standard model is relevant. So in the end, we have the value of air density for all altitudes between sea level up to 110 km which can be seen as the limit of the atmosphere (i.e. where no shower can develop yet). Note that the relative difference $(\rho_{\rm air}^{\rm GDAS} - \rho_{\rm air}^{\rm US})/\rho_{\rm air}^{\rm US}$ when comparing the air density obtained the GDAS data or only the US Standard model can reach some tens of %. For instance, for the sample day of March 18, 2014, the maximum difference is of 15% for an altitude around 15 km. We can estimate the influence of such a difference on the electric field at ground level. For this, we simulate with SELFAS3 a shower with zenith angle $\theta = 30°$ and azimuth $\phi = 90°$ (i.e. coming from the North, azimuth are counted anti clockwise). The primary particle is a proton of 10^{18} eV (1 EeV) and the first interaction occurs at an atmospheric depth of $X_1 = 100$ g/cm^2. We ran the code with the very same shower once with the US Standard model, and another time with the actual atmospheric conditions of March 18, 2014 using the GDAS data. The difference in the total electric field amplitude is shown in FIG. 2. It is the amplitude in the horizontal direction $\mathbf{v} \times \mathbf{B}$ projected onto the ground frame, $\mathbf{v}$ being the shower axis direction and $\mathbf{B}$ the geomagnetic field at the observation site. In this

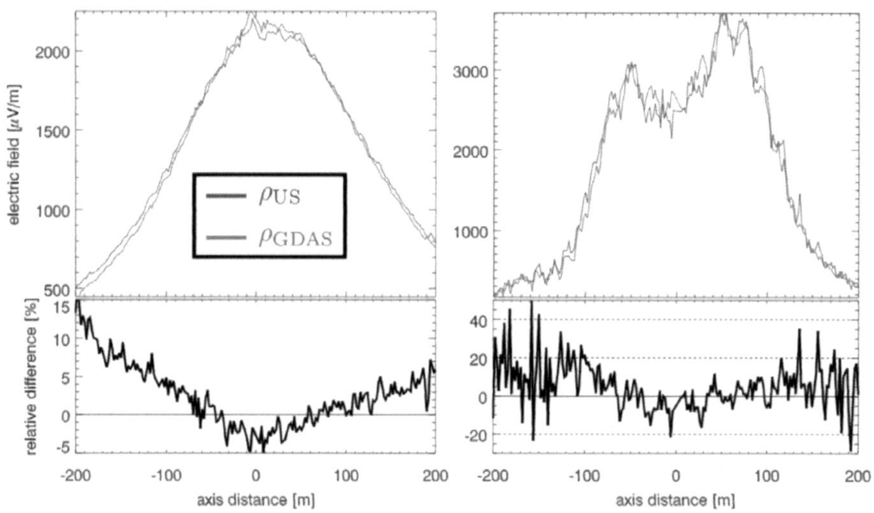

Figure 2. Electric field amplitude as a function of shower axis distance, in [20; 80] MHz (left) and [120; 250] MHz (right). The blue curve corresponds to the US Standard model ($\rho_{\rm air}^{\rm US}$) and the red curve to the refined model based on the GDAS data on March 18, 2014 at noon ($\rho_{\rm air}^{\rm GDAS}$). We show the relative difference between both models.

example, we see that the electric field profile is wider when using the US Standard model that will lead to a larger estimation of the shower $X_{\rm max}$. Using the US Standard model is a source of systematic uncertainties.

2.4. Air refractive index computation

The simulation codes usually take the Gladstone and Dale law to compute the air refractive index:

$$\eta(z(l)) = 1 + \kappa\, \rho_{\rm air}(z(\ell)) \text{ with } \kappa = 0.226 \text{ cm}^3/\text{g}.$$

We need to know the air refractive index value at the emission point. We also need its average value along the line of sight between the observer position and the emission point:

$$< \eta(z(\ell)) >= 1 + \frac{\kappa}{\ell} \int_0^\ell \rho_{\mathrm{air}}(z(\ell'))\, \mathrm{d}\ell'$$

We see that the air density also appears in this computation. The problem with the Gladstone and Dale law is that it is valid for optical wavelengths [13] and is not valid for our much larger wavelengths ($\lambda = 7.5$ m at 40 MHz). We use instead the relation proposed in [14], valid at high frequency (HF):

$$\eta = 1 + 10^{-6} N \quad \text{with} \quad N_{\mathrm{HF}} = \frac{77.6}{T}\left(P + 4810\,\frac{p_v}{T}\right) \qquad T \text{ in K},$$

where N is the refractivity. The relative humidity is taken into account through p_v so that we can estimate the refractivity $N_{\mathrm{HF}}^{\mathrm{GDAS}}$. At higher altitude, the relative humidity can be considered as null, as no clouds are usually observed above 12 km and the refractivity reduces to:

$$N_{\mathrm{HF}}^{\mathrm{GDAS}} = 77.6\,\frac{R\,\rho_{\mathrm{air}}}{M_d}$$

where ρ_{air} is the air density above $z_{\mathrm{max}}^{\mathrm{GDAS}}$ i.e. is given by $\rho_{\mathrm{air}}^{\mathrm{US}}$ with the scaling factor. The refractivity can be compared with different combinations: $N_{\mathrm{GD}}^{\mathrm{GDAS}}$ based on $\rho_{\mathrm{air}}^{\mathrm{GDAS}}$ and the Gladstone-Dale law, $N_{\mathrm{GD}}^{\mathrm{US}}$ based on $\rho_{\mathrm{air}}^{\mathrm{GDAS}}$ and the Gladstone-Dale law and $N_{\mathrm{HF}}^{\mathrm{GDAS}}$ based on $\rho_{\mathrm{air}}^{\mathrm{GDAS}}$ and the HF law. Tab. 1 presents the relative differences between all estimations. The

altitude (km)	$(N_{\mathrm{GD}}^{\mathrm{GDAS}} - N_{\mathrm{GD}}^{\mathrm{US}})/N_{\mathrm{GD}}^{\mathrm{US}}$	$(N_{\mathrm{HF}}^{\mathrm{GDAS}} - N_{\mathrm{GD}}^{\mathrm{US}})/N_{\mathrm{GD}}^{\mathrm{US}}$	$(N_{\mathrm{HF}}^{\mathrm{GDAS}} - N_{\mathrm{GD}}^{\mathrm{GDAS}})/N_{\mathrm{GD}}^{\mathrm{GDAS}}$	$(\eta_{\mathrm{HF}}^{\mathrm{GDAS}} - \eta_{\mathrm{GD}}^{\mathrm{US}})/\eta_{\mathrm{GD}}^{\mathrm{US}}$
0	$(1.3 \pm 4.35)\%$	$(18.1 \pm 24.0)\%$	$(16.5 \pm 5.0)\%$	$(5.0 \pm 6.6) \times 10^{-5}$
2.5	$(-0.1 \pm 1.6)\%$	$(6.5 \pm 11.0)\%$	$(6.6 \pm 4.3)\%$	$(1.4 \pm 2.4) \times 10^{-5}$
5	$(-0.4 \pm 1.0)\%$	$(1.4 \pm 3.5)\%$	$(1.76 \pm 2.2)\%$	$(2.3 \pm 5.8) \times 10^{-6}$
7.5	$(0.3 \pm 1.4)\%$	$(0.2 \pm 1.3)\%$	$(-0.17 \pm 1.2)\%$	$(1.2 \pm 16.7) \times 10^{-7}$
10	$(0.3 \pm 3.2)\%$	$(-0.8 \pm 2.2)\%$	$(-1.1 \pm 3.0)\%$	$(-7.2 \pm 20.3) \times 10^{-7}$
12.5	$(0.6 \pm 6.1)\%$	$(-0.8 \pm 4.7)\%$	$(1.35 \pm 5.6)\%$	$(-6.0 \pm 29.8) \times 10^{-7}$
15	$(0.6 \pm 5.3)\%$	$(-0.8 \pm 3.9)\%$	$(-1.4 \pm 4.8)\%$	$(-3.6 \pm 16.7) \times 10^{-7}$
17.5	$(0.6 \pm 4.8)\%$	$(-0.8 \pm 3.3)\%$	$(-1.4 \pm 4.2)\%$	$(-2.4 \pm 9.5) \times 10^{-7}$
20	$(0.6 \pm 4.0)\%$	$(-0.8 \pm 2.5)\%$	$(-1.4 \pm 3.4)\%$	$(-2.4 \pm 4.8) \times 10^{-7}$

Table 1. Relative differences of the refractivity (N) and the refractive index (η) between several GDAS-based and US Standard-based models for several altitudes of interest for air showers physics. For GDAS-based models all the data of the year 2014 at Nançay were used to compute the mean differences along the year. The errors show the standard deviation at each altitude.

relative difference with the canonical estimation $N_{\mathrm{GD}}^{\mathrm{US}}$ (i.e. used in most simulation codes) can reach some tens of %. For the day of March 18, 2014, the relative difference reaches 35% at sea level and 15% at altitudes of interest for shower development, around 20 km. The relative difference on the air refractive index is smaller because of the 10^{-6} factor and the effect on the electric field is of the order of some % in the amplitude in 20-80 MHz and can reach 40% in 120-250 MHz. In conclusion, one should be cautious with the model for the refractive index when considering frequencies above ~ 100 MHz.

3. X_{max} reconstruction

We can study the influence of the models chosen for both air density ρ_{air} and refractivity N on simulated showers. We ran simulations to test 6 differents configurations of (ρ_{air}, N). For

data set	air density	air refractive index	ΔX_{30° [g/cm^2]
#1	US Std.	US Std. + GD	34.1 ± 8.9
#2	GDAS	US Std. + GD	5.7 ± 5.4
#3	GDAS	GDAS + GD	4.6 ± 3.6
#4	GDAS	GDAS + HF	0.1± 2.4
#5	GDAS	GDAS + HF (N+10%)	2.9 ± 4.8
#6	GDAS	GDAS + HF (N+20%)	9.3 ± 16.4

Table 2. Quality of the $X_{\max}$ reconstruction for 6 different configurations of air density and air refractive index. The ΔX column presents the mean difference with the true value and the 1σ deviation. GD stands for Gladstone-Dale.

each of these configurations, we used 10 iron showers and 40 proton showers with random $X_{\max}$ coming from $\theta = 30°$, $\phi = 90°$. In TAB. 2, we show the result of the comparison. The reference configuration is #4: $\rho_{\rm air}^{\rm GDAS}$ and $N_{\rm HF}^{\rm GDAS}$, which uses the most refined information to compute these quantities. We consider the electric field computed by each of the simulated showers of #4. Using this electric field pattern at ground level, we apply to procedure of [15] to reconstruct the $X_{\max}$, from the 50 showers simulated using the air density and refractivity model of the other configurations #1, #2, #3, #5, #6 and the 49 remaining independent showers of #4. We then get the average and standard deviation of the obtained $X_{\max}$ values and this is shown in the last column of TAB. 2.

This shows that using the most refined model gives an error on $X_{\max}$ compatible with zero (0.1± 2.4 g/cm^2). We conclude by saying that the choice of the atmospherical model for $\rho_{\rm air}$ is critical for the $X_{\max}$ estimation. The choice of the refractivity model has a smaller impact on $X_{\max}$.

4. Conclusion

The source of ultra-high energy cosmic rays is still unknown. We hope that a much more precise measurement of the composition will help into understanding this old problem in astrophysics. Both the Pierre Auger Observatory and the Telescope Array will proceed to a significant upgrade of their instruments in this direction. The radio signal will also play a key role in the determination of the nature of the cosmic rays, since it can provide the shower $X_{\max}$ with competitive uncertainties, of the same order of what a fluorescence telescope can do ($X_{\max}$ resolution of $\sim$ 19 g/cm^2 at 1 EeV, with systematic uncertainties of $+8/-13$ g/cm^2, see [16]) but with a $\sim$ 100% duty cycle. To reach its best performances, the simulation codes that compute the expected electric field emitted by air showers should use a precise atmospheric model for both the air density and air refractive index. This can be achieved using the GDAS data and using a refractive index valid at the frequencies of interest (i.e. some tens of MHz). With these detailed models, the shower $X_{\max}$ can be retrieved with no systematic bias, as it is the case when using basic models. The code SELFAS3 uses the GDAS data and the correct law for the refractive index.

Acknowledgments

We thank the Région Pays de la Loire for its financial support of the Astroparticle group of Subatech and in particular for its contribution to the EXTASIS experiment.

References

[1] B. Revenu for the CODALEMA/EXTASIS collaboration. Current status of the CODALEMA/EXTASIS experiments. In *Proceedings of the 26th E+CRS/35th RCRC*, number 268, July 2018.

[2] E. Holt for the for the Pierre Auger Collaboration. Recent Results of the Auger Engineering RadioArray (AERA). In *ICRC2017*, volume 301, 2017.

[3] F. G. Schröder for the Tunka-Rex collaboration. Overview on the Tunka-Rex antenna array forcosmic-ray air showers. *PoS*, ICRC2017:459, 2018.

[4] Vincent Marin and Benoît Revenu. Simulation of radio emission from cosmic ray air shower with SELFAS2. *Astropart.Phys.*, 35:733–741, 2012.

[5] J. Alvarez Muñiz, W. R. Carvalho Jr, and E. Zas. Monte carlo simulations of radio pulses in atmospheric showers using ZHAireS. *Astropart. Phys.*, 35:325–341, 2012.

[6] T. Huege and C. James. Full Monte Carlo simulations of radio emission from extensive air showers with CoREAS. In *proceedings of the 33rd ICRC, Rio de Janeiro, Brasil*, number id 548. arXiv:1307.7566, July 2013.

[7] Daniel García-Fernández, Benoît Revenu, Didier Charrier, Richard Dallier, Antony Escudie, and Lilian Martin. Calculations of low-frequency radio emission by cosmic ray-induced particle showers. *Phys. Rev.*, D97(10):103010, 2018.

[8] NASA NOAA. https://ntrs.nasa.gov/archive/nasa/casi.ntrs.nasa.gov/19770009539.pdf. Technical report, 1976.

[9] F. Gaté, B. Revenu, D. García-Fernández, V. Marin, R. Dallier, A. Escudié, and L. Martin. Computing the electric field from extensive air showers using a realistic description of the atmosphere. *Astroparticle Physics*, 98:38 – 51, 2018.

[10] GDAS Archive Information, http://ready.arl.noaa.gov/gdas1.php. Technical report, NOAA.

[11] A. L. Buck. New equations for computing vapor pressure and enhancement factor. *J. Appl. Meteorol.*, (20):1527:1532, 1981.

[12] A. L. Buck. Buck research cr-1a user's manual. Technical report, 1996.

[13] W. Merzkirch. *Flow Visualization (Second Edition)*. Number ISBN: 978-0-12-491351-6. Elsevier, 1987.

[14] R. L. Freeman. *Radio System Design for Telecommunications, Third Edition*. 2006.

[15] Lilian Martin, Richard Dallier, Antony Escudie, Daniel García-Fernández, Florian Gaté, Alain Lecacheux, and Benoît Revenu. Main features of cosmic ray induced air showers measured by the CODALEMA experiment. *PoS*, ICRC2017:414, 2018.

[16] Pierre Auger Collaboration and Alessio Porcelli. Measurements of xmax above 1017 ev with the fluorescence detector of the pierre auger observatory. In *34th International Cosmic Ray Conference, ICRC 2015, The Hague, NL, July 30 - August 6, 2015*, Proceedings of Science, page Art.Nr.: 420. SISSA (International School for Advanced Studies), Triest, 2016. 51.03.03; LK 01.

Lower limit on the ultra-high-energy proton-to-helium ratio from the measurements of the tail of $X_{\rm max}$ distribution

Yana Zhezher

Institute for Nuclear Research of the Russian Academy of Sciences, 117312, Moscow, Russia, Faculty of Physics, M.V. Lomonosov Moscow State University, 119991, Moscow, Russia

E-mail: zhezher.yana@physics.msu.ru

Abstract. There are multiple techniques to determine the chemical composition of the ultra-high-energy cosmic rays. While most of the methods are primarily sensitive to the average atomic mass, it is challenging to discriminate between the two lightest elements: proton and helium. In this paper, the proton-to-helium ratio in the energy range from $10^{18.3}$eV to $10^{19.3}$eV is estimated using the tail of the distribution of the depth of the shower maximum $X_{\rm max}$. Using the exponential decay scale Λ measured by the Telescope Array experiment we derive the constraint on the proton-to-helium ratio p/He > 0.43 (68% CL). The result is conservative with respect to the admixture of heavier elements. We evaluate the impact of the hadronic interaction model uncertainty. The implications for the astrophysical models of the origin of cosmic rays and the safety of the future colliders are discussed.

1. Introduction

The mass composition of the ultra-high-energy cosmic rays lies among the key tasks of major present-day and upcoming experiments. The precise knowledge of the composition is important for understanding the cosmic-ray production mechanism in the sources and its population [1]. Moreover, composition at the highest energies is the decisive factor for the observable flux of cosmogenic photons [2, 3] and neutrinos [4, 5], see [6] for a review. The photons and neutrinos are most efficiently produced by the primary protons due to the highest energy per nucleon compared to heavier elements. The diversity of the models may be illustrated with the two antipodal examples namely the dip model [7, 8, 9] and the disappointing model [10]. The dip model has purely proton composition and as a consequence predicts observable fluxes of the cosmogenic photons and neutrino. The model is named after the dip spectral feature which is naturally explained with the electron-positron pair production by protons. The disappointing model includes both protons and nuclei in the source and assumes that the acceleration of primary nuclei in the sources is rigidity-dependent with relatively low maximum energy of acceleration. It was named "disappointing", because in this case there are no pion photo-production on CMB photons in extragalactic space and consequently no high-energy cosmogenic neutrino fluxes. Disappointing model predicts no GZK-cutoff [11, 12] in the spectrum and shows no correlation with nearby sources due to nuclei deflection in the galactic magnetic fields up to highest energies.

Another implication of the mass composition at ultra-high energies is the investigation of safety of the future colliders. In certain theoretical models characterized by additional spatial dimensions, the production of non-evaporating microscopic black holes becomes possible. This phenomenon was taken into consideration in the framework of the Large Hadron Collider (LHC) safety analysis [13, 14]. The proof of the LHC safety is based on the constraints on the black hole production derived from the stability of dense astrophysical objects, such as white dwarfs and neutron stars. The latter interact with the ultra-high-energy cosmic rays with the center of mass energies larger than ones achieved at LHC. One may ascertain the safety of the future 100 TeV colliders by studying the interaction of the cosmic rays of the highest energies. The primary protons again play an important role as the production of the black holes is determined by the energy per nucleon. It was shown that the charged stable microscopic black hole production may be excluded already, while the exclusion of the neutral black holes would require a precise knowledge of the proton fraction at the ultra-high energy [15].

One of the most common approaches is the measurement of the longitudinal shape of the extensive air showers (EAS). The depth of a shower maximum, or $X_{\max}$, is used as a composition-sensitive variable [16]. The measurements of the mean $X_{\max}$ gives the estimate of the average atomic mass, while the study of the $X_{\max}$ distribution and its moments may resolve the multicomponent composition, see [17, 18, 19] for review.

The tail of the $X_{\max}$ distribution may be studied independently on the main part of the distribution. It may be fit with an exponential function $\exp(-X_{\max}/\Lambda)$, where Λ is called the attenuation length. The attenuation length is found to be sensitive to the proton-air interaction cross-section. The first results by this method were obtained by the Fly's Eye Collaboration [20, 21] followed by the results of the Pierre Auger and Telescope Array Collaborations [23, 22, 24]. It was shown in [25] that the attenuation length may be used to estimate the proton-to-helium ratio p/He. The latter estimate has only minor dependence on the hadronic interaction models and $X_{\max}$ experimental systematic uncertainties. The proton-to-helium ratio is directly measured below the knee and it allows to constrain different astrophysical models of the origin of cosmic rays [26, 27]. The measurements of the proton-to-helium ratio at the ultra-high energies may be used similarly to discriminate between different source models. As a recent example, a modified dip model [28] agrees with the measured spectrum of the ultra-high-energy cosmic rays for the value of proton-to-helium ratio p/He = 5. Furthermore, the value of p/He used jointly with the other composition studies will allow to pinpoint the flux of the primary protons. The latter is an important quantity as discussed above.

The present work is dedicated to the determination of proton-to-helium ratio of ultra-high-energy cosmic rays in the energy range from $10^{18.3}$eV to $10^{19.3}$eV based on the Telescope Array measurements of the attenuation length [22]. The data are compared to the Monte-Carlo simulations using the CORSIKA package [29] along with the QGSJET II-04 [30, 31] and EPOS-LHC [32] hadronic interaction models.

2. Method and Monte-Carlo simulations

The method generally follows the work of Yushkov et. al [25] to derive the proton-to-helium ratio using the measurements of the attenuation length by the Telescope Array collaboration [22].

At first, the simulated sets of extensive air showers initiated by primary protons, helium and carbon are produced with the use of the CORSIKA package [29]. Simulations are performed separately with QGSJET II-04 [31] and EPOS-LHC [32] hadronic interaction models. 17 354 events are simulated for each species in the energy range from $10^{18.3}$eV to $10^{19.3}$eV with the spectrum obtained by the Telescope Array collaboration defined by the spectral index -3.226 for $E < 10^{18.72}$ eV and -2.66 for $E > 10^{18.72}$ eV [33].

At the second step, the simulated sets are "mixed" in different proportions from p/He = 0.01 to p/He = 100.0. For each mixture the $X_{\max}$ distribution slope is fitted exponentially to derive the attenuation length for a mixed composition model.

An important constituent of this method is the choice of the starting point of the fit: it can

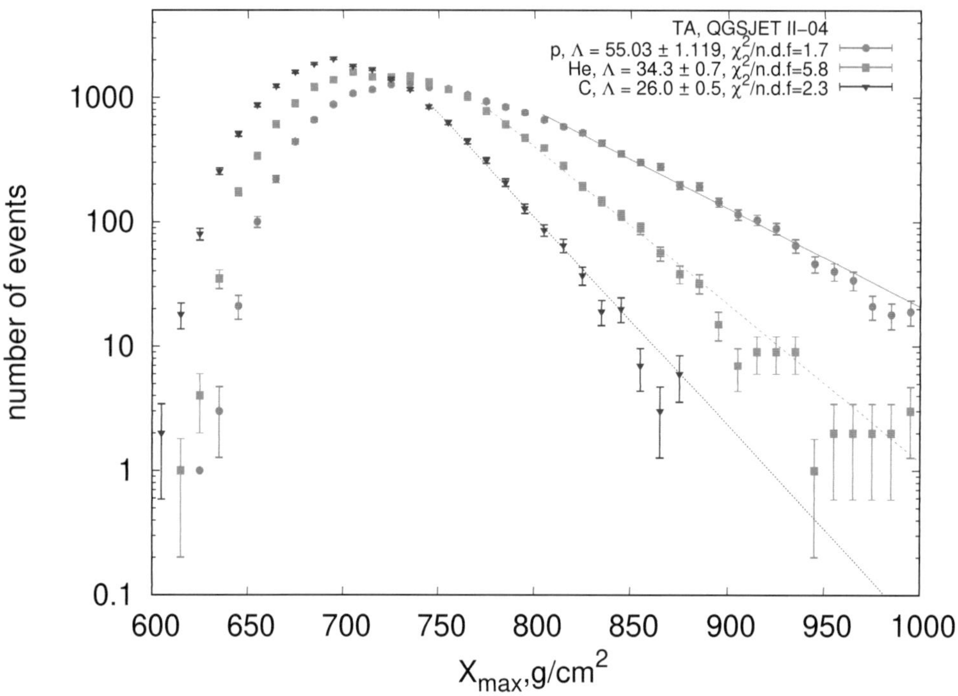

Figure 1. $X_{\rm max}$ distributions for $10^{18.3}$eV $< E <$ $10^{19.3}$ eV for the Telescope Array for proton (magenta), helium (orange) and carbon (blue) Monte-Carlo distributions simulated with QGSJET II-04. $X_{\rm max}$ distribution's tail exponential fit $\exp(-X_{\rm max}/\Lambda)$ is shown for each Monte-Carlo with a line of the corresponding color.

be defined in many different ways. In the initial papers [20, 21] the lower range of $X_{\rm max}$ fit was fixed at the constant values $X_{max} = 760$ g/cm^2 and $X_{max} = 830$ g/cm^2, respectively. Yushkov et. al [25] have proposed another determination of lower fit range, which involves carbon $X_{\rm max}$ distribution: the lower limit is defined as a value at which only $\approx 0.5\%$ of the carbon-initiated showers get into the fitting range. In the present work we follow the method implemented by Abbasi et al. [22], where the lower limit is defined as the $X_i = \langle X_{max}\rangle + 40$g/cm^2, where $\langle X_{max}\rangle$ is the average value of a given distribution.

In the present analysis, the implemented method leads to the following starting points for protons, helium and carbon nuclei: $X^p_{max,start} = 804$ g/cm^2, $X^{He}_{max,start} = 772$ g/cm^2 and $X^C_{max,start} = 745$ g/cm^2. For each MC-set, ending point for X_{max} fit is 1000 g/cm^2.

Finally, after performing the fit of each mixture's $X_{\rm max}$ distribution, Λ for the Telescope Array values are obtained as a function of p/He ratio. The constraints on the proton-to-helium ration are then obtained by comparing these values with the experimental Λ values [22]. Lower limit on the proton to helium ratio at 68% CL corresponds to the lower limit of the experimentally measured Λ value derived with the use of measured experimental uncertainties. Expansion of the experimental Λ range to 2σ than leads to the derivation of proton-to-helium ratio with the 95% CL.

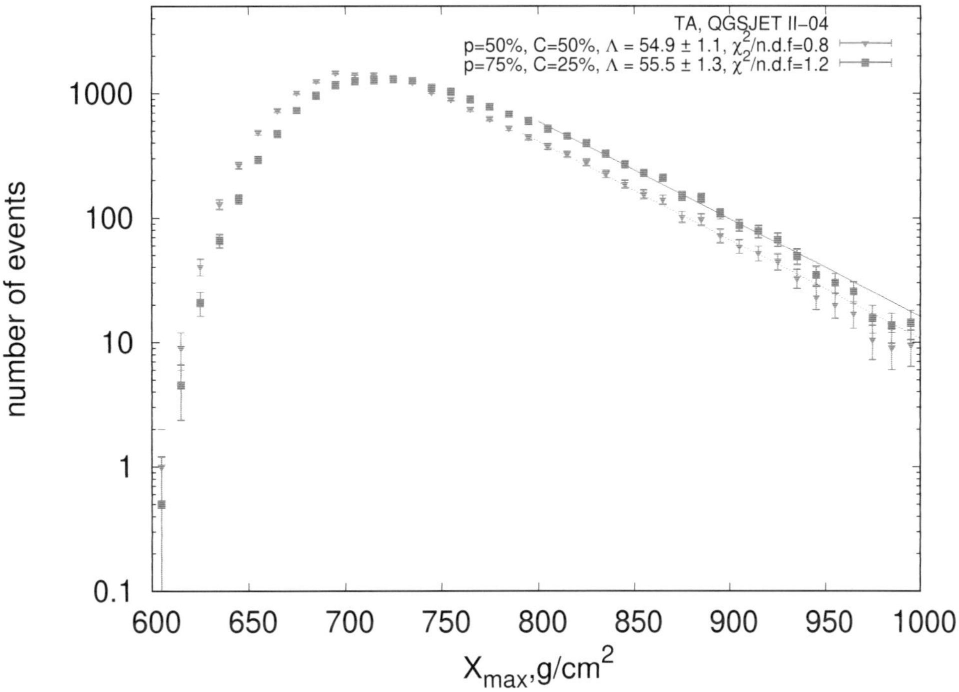

Figure 2. $X_{\max}$ distributions for $10^{18.3}$eV $< E < 10^{19.3}$ eV for the Telescope Array for mixtures $p : C = 3 : 1$ (magenta) and $p : C = 1 : 1$ (orange). $X_{\max}$ distribution's tail exponential fit $\exp(-X_{\max}/\Lambda)$ is shown for each mixture with a line of the corresponding color.

3. Results

We present the $X_{\max}$ distributions and corresponding fits of exponential tails for proton, helium and carbon Monte-Carlo simulated sets in Figure 1. $X_{\max}$ distributions and corresponding fits of exponential tails for mixtures of proton and carbon with proportions $p : C = 3 : 1$ and $p : C = 1 : 1$ are shown in Figure 2.

Λ as a function of proton-to-helium ratio in QGSJET II-04 and EPOS-LHC models is shown in Figure 3 with a black line. The plot includes the proton-to-helium ratio range from p/He = 0.01 to p/He = 100 with a step $\Delta = 10^{0.2}$. Comparing the Monte-Carlo function with the experimental value $\Lambda = 50.47 \pm 6.26$ g/cm^2 obtained by the Telescope Array collaboration [22] we arrive at the following lower limits on the proton-to-helium ratio for the $\langle E \rangle = 3.6 \times 10^{18}$ eV:

$$\text{p/He} > 0.43 \ (68\%\,\text{CL}) \quad \text{QGSJET II-04}, \tag{1}$$

$$\text{p/He} > 0.63 \ (68\%\,\text{CL}) \quad \text{EPOS-LHC}. \tag{2}$$

We note that the pure proton composition is well compatible with the measured attenuation length.

The stability of the method in respect to the admixture of the heavier elements is studied. For this reason, the analysis is repeated for three-component mixtures containing 25%, 50% and 75% of carbon and corresponding Λ is shown in Figure 3 by red triangles and green squares respectively. One may see that the constraints 1 and 2 are conservative to addition of the heavier elements as expected in [25].

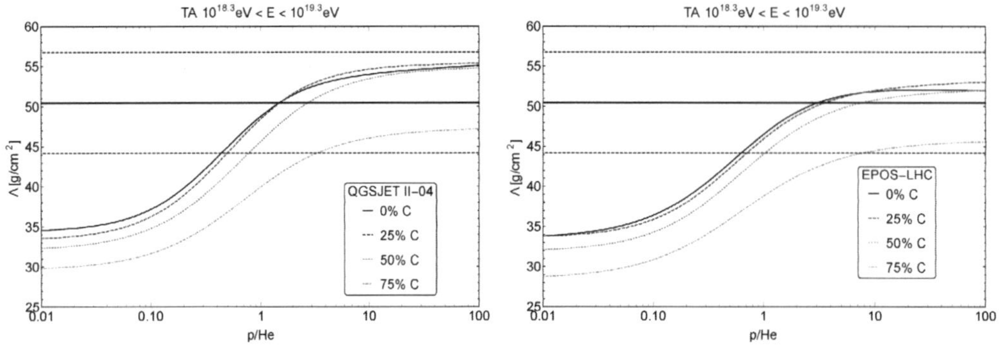

Figure 3. Λ parameter as a function of proton-to-helium ratio for two-component mixture (p and He, black line) and three component mixtures (p, He and 25 % C – blue line; p, He and 50 % C – red line; p, He and 75 % C – orange line) of Monte-Carlo events simulated with QGSJET II-04 (left) and EPOS-LHC (right). Black solid and dashed lines correspond to the experimental value $\Lambda = 50.47 \pm 6.26$ g/cm^2 obtained by the Telescope Array collaboration [22].

It is also important to mention that the detector effects, such as acceptance of different types of primary particles and detector resolution may affect the measured Λ value. It was shown in [22], that the detector effects are negligible for the Λ value obtained by the Telescope Array.

One may further study the three-component mixture of protons, helium and carbon. By calculating Λ for all possible combinations we arrive to the following lower limits on the fraction of protons in the three-component mixture for $\langle E\rangle = 3.6 \times 10^{18}$ eV:

$$\mathrm{p/(p + He + C)} > 0.20 \ (68\%\,\mathrm{CL}) \ \ \mathrm{QGSJET\ II\text{-}04}, \tag{3}$$
$$\mathrm{p/(p + He + C)} > 0.23 \ (68\%\,\mathrm{CL}) \ \ \mathrm{EPOS\text{-}LHC}. \tag{4}$$

One may note that the possible systematic error of the method is mainly caused by the uncertainties in the hadronic models used for the Monte-Carlo simulations, all of which are known to be inconsistent with the muonic component of the EAS [34, 35].

4. Conclusion

Let us finally discuss the possible applications of the obtained lower limit on the proton-to-helium ratio at the energies $10^{18.3}\mathrm{eV} < E < 10^{19.3}\mathrm{eV}$. First of all, we consider the properties of the sources of the ultra-high-energy cosmic rays in the view of the constraints 1 and 2. The present limits constrain the models with helium domination in the energy range under study, e.g. the helium version of the disappointing model [10]. These models generally include the preferential acceleration of helium or an excessive helium abundance at the acceleration region. At the same time, the result is fully compatible with the original pure proton dip model [7, 8, 9] and with the modification of the dip model with p/He = 5 [28] as long as with the standard disappointing model [10] with p/He $\sim$ 1.

Secondly, let us discuss the safety of the future colliders. The proof of the safety relies largely on the existence of non-zero flux of the ultra-high-energy protons [15]. One may see that this work confirmed the existence of the primary protons at the 2σ level. Indeed, the following 95% CL limits may be obtained repeating the analysis done for Equations 3 and 4 for $\langle E\rangle = 3.6 \times 10^{18}$ eV:

$$\mathrm{p/(p + He + C)} > 0.09 \ (95\%\,\mathrm{CL}) \ \ \mathrm{QGSJET\ II\text{-}04}, \tag{5}$$

$$\mathrm{p/(p + He + C)} > 0.11 \ (95\%\,\mathrm{CL}) \ \ \mathrm{EPOS\text{-}LHC}. \tag{6}$$

The results are in favor of the safe operation of the future colliders. Still the importance of issue demand higher confidence which may be achieved with the future precision measurements of the attenuation length.

References

[1] R. Aloisio, V. Berezinsky and P. Blasi, JCAP **1410**, no. 10, 020 (2014).
[2] G. Gelmini, O. E. Kalashev and D. V. Semikoz, J. Exp. Theor. Phys. **106**, 1061 (2008).
[3] D. Hooper, A. M. Taylor and S. Sarkar, Astropart. Phys. **34**, (2011) 340.
[4] V. S. Berezinsky and G. T. Zatsepin, Phys. Lett. **28B**, 423 (1969).
[5] K. Kotera, D. Allard and A. V. Olinto, JCAP **1010**, 013 (2010).
[6] T. Karg *et al.* [IceCube and Pierre Auger and Telescope Array Collaborations], JPS Conf. Proc. **9**, 010021 (2016).
[7] V. Berezinsky, A. Z. Gazizov and S. I. Grigorieva, Phys. Rev. D **74**, 043005 (2006).
[8] V. Berezinsky, A. Z. Gazizov and S. I. Grigorieva, Phys. Lett. B **612**, 147 (2005).
[9] R. Aloisio, V. Berezinsky, P. Blasi, A. Gazizov, S. Grigorieva and B. Hnatyk, Astropart. Phys. **27**, 76 (2007).
[10] R. Aloisio, V. Berezinsky and A. Gazizov, Astropart. Phys. **34**, 620 (2011).
[11] K. Greisen, Phys. Rev. Lett. **16**, 748 (1966)
[12] Z. T. Zatsepin and V. A. Kuz'min, Zh. Eksp. Teor. Fiz. Pis'ma Red. **4**, 144 (1966)
[13] J. R. Ellis, G. Giudice, M. L. Mangano, I. Tkachev and U. Wiedemann, J. Phys. G **35**, 115004 (2008), [arXiv:0806.3414 [hep-ph]].
[14] S. B. Giddings and M. L. Mangano, Phys. Rev. D **78**, 035009 (2008), [arXiv:0806.3381 [hep-ph]].
[15] A. V. Sokolov and M. S. Pshirkov, Eur. Phys. J. C **77**, no. 12, 908 (2017), [arXiv:1611.04949 [hep-ph]].
[16] T. K. Gaisser *et al.* [HiRes Collaboration], Phys. Rev. D **47**, 1919 (1993).
[17] K. H. Kampert and M. Unger, Astropart. Phys. **35**, 660 (2012).
[18] E. Barcikowski *et al.* [Pierre Auger and Yakutsk Collaborations], EPJ Web Conf. **53**, 01006 (2013).
[19] W. Hanlon *et al.*, JPS Conf. Proc. **19**, 011013 (2018).
[20] R. Ellsworth *et al.* [Fly's Eye Collaboration], Phys. Rev. D **26**, 336 (1982).
[21] R. Baltrusaitis *et al.* [Fly's Eye Collaboration], Phys. Rev. Lett. **52**, 1380 (1984).
[22] R. U. Abbasi *et al.* [Telescope Array Collaboration], Phys. Rev. D **92**, no. 3, 032007 (2015).
[23] P. Abreu *et al.* [Pierre Auger Collaboration], Phys. Rev. Lett. **109**, 062002 (2012).
[24] R. Ulrich, PoS ICRC **2015**, 401 (2016).
[25] A. Yushkov, M. Risse, M. Werner and J. Krieg, Astropart. Phys. **85**, 29 (2016), [arXiv:1609.08586 [astro-ph.HE]].
[26] P. L. Biermann, T. K. Gaisser and T. Stanev, Phys. Rev. D **51**, 3450 (1995), [astro-ph/9501001].
[27] Y. Ohira, K. Ioka, Astrophys. J Lett. **729**, L13 (2011).
[28] R. Aloisio and V. Berezinsky, arXiv:1703.08671 [astro-ph.HE].
[29] D. Heck *et al.*, Report FZKA-6019 (1998), Forschungszentrum Karlsruhe.
[30] S. Ostapchenko, Nucl. Phys. Proc. Suppl. **151**, 143 (2006) [hep-ph/0412332].
[31] S. Ostapchenko, Phys. Rev. D **83**, 014018 (2011)
[32] T. Pierog, I. Karpenko, J. M. Katzy, E. Yatsenko and K. Werner, Phys. Rev. C **92**, no. 3, 034906 (2015) [arXiv:1306.0121 [hep-ph]].
[33] D. Ivanov, PoS ICRC **2015**, 349 (2016).
[34] A. Aab *et al.* [Pierre Auger Collaboration], Phys. Rev. Lett. **117** (2016) no.19, 192001, [arXiv:1610.08509 [hep-ex]].
[35] R. U. Abbasi *et al.* [Telescope Array Collaboration], Phys. Rev. D **98**, no. 2, 022002 (2018) [arXiv:1804.03877 [astro-ph.HE]].

On the efficiency of the evaluation of the primary cosmic ray composition using lateral distributions of air shower electromagnetic component

Roman Raikin, Tatyana Serebryakova, Nikolay Volkov and Anatoly Lagutin

Altai State University. 61 Lenin Ave., Barnaul, 656049, Russia

E-mail: raikin@theory.asu.ru

Abstract. New analysis of lateral distributions (LD) of electrons, muons and charged particles in extensive air showers generated by $10^{15} - 10^{19}$ eV cosmic rays has been made with respect to the scaling formalism for LD and air shower universality paradigm. It is found that lateral distributions of electrons and muons, when considered separately, are well described by one-parametric scaling functions, while for the charged particles mixture distributions the scaling description is not accurate in radial distance range where electrons and muons give a comparable contribution to the local particle densities. The use of scaling formalism enables enhancing the electron/muon separation capabilities and, when combined with theoretically motivated air shower universality concept, provides robust mass composition estimates. The proposed approach could be implemented for the present and future (multi-) hybrid air shower observations, especially in realizing the potential of Auger and Telescope Array observatories upgrades, as well as for re-analysis and cross-calibration of the data collected from different air shower arrays in a broad primary energy range.

1. Introduction

The evaluation of primary cosmic ray mass composition is at present stage the crucial issue for the interpretation of the experimental data in terms of sources and propagation models [1,2]. The mass composition is inferred from the extensive air shower (EAS) observations by comparisons with simulations results, which rely on hadronic interaction models fundamentally uncertain in the relevant energy range. Numerous methods and techniques are implemented, including analysis of mean values, fluctuations, correlations or even particular features of distributions of different observables characterized both longitudinal and spatial shower development (depth of maximum, muon production depth, total number of electrons and muons at the observation level and their local densities at various distances from the shower axis, particles arrival time profiles, spatial distribution of radio emission, Cherenkov light) (see [3–12]).

Large efforts have been made recently in both gaining experimental data with increased resolution in detection of various EAS components and developing improved methods for physical interpretation of the data along with evolving hadronic interaction models after the LHC results. Nevertheless, the composition results remain ambiguous in the entire energy range available for EAS studies.

The discrepancies between estimates of composition derived by various methods from the data of different experiments are apparently caused by a complex of instrumental and methodical systematic biases of different nature as well as by strong model dependence of observables, mostly in case of muon component characteristics, the so-called *muon problem* (for the up-to-date state one can refer to [13, 14]), disadvantages in taking into account meteorological effects etc.

In this paper we report on the further development of the approach based on the air shower universality and scaling formalism for lateral distribution (LD) that exhibits the potential in improved primary composition evaluation from the 100% duty cycle surface detectors data [15–22].

The paper is organized as follows. In Section 2 the universality in air shower development and scaling formalism for LD are explained. The Monte-Carlo simulations and results of analysis are presented in Section 3. Section 4 contains brief discussion and concluding remarks.

2. Air shower universality and scaling formalism

The concept of air shower universality is inspired by the classical results of the electromagnetic cascade theory and in a broad sense expressed in the similarity of the spectra of low-energy secondary particles with respect to the cascade development stage. As a consequence, the relations between various shower characteristics represented using the cascade age parameter demonstrate to one degree or another the invariance with respect to the properties of a primary particle. Different aspects of the universality (phenomenology, possible generalisations, deeper understanding of the origin and limitations, sensitivity to the hadronic interaction model, applicability under conditions of real experiments) are subjected to remarkable interest [23–39].

In our papers [15–22, 40] it was shown that on the basis of scaling formalism for lateral distribution of electrons in air showers one can observe the generalized manifestation of the universality very promising for model-independent EAS reconstruction in a wide primary energy range.

The basic idea of the one-valued relation between the shape of electron LDF and shower age dates back to the classical Nishimura-Kamata-Greizen (NKG) function [41, 42]:

$$\rho(r; E, s) = N(E, s)\frac{C}{r_0^2}\left(\frac{r}{r_0}\right)^{s-2}\left(1+\frac{r}{r_0}\right)^{s-4.5}. \tag{1}$$

Here $\rho(r; E, s)$ is local particle density at radial distance r from the core position in shower with primary energy E and the longitudinal age parameter $s = 3X/(X + 2X_{\max})$, where X is the observation depth in the atmosphere, $X_{\max}$ is the depth of shower maximum, $N(E, s)$ – total number of particles at the observation depth (shower size), r_0 – constant shower scale radius, which does not depend on primary particle type and energy (originally – the Moliere unit r_M).

The NKG function is insufficient to describe electron LDFs accurately at large distances (see e.g. [43]). This led to various modifications of the NKG form, such as changing the values of exponents, introducing different constant or variable scale factors, lateral ($s_\perp$) or local ($s(r)$) age parameters and also generalizations of the function by using a third power-law term. A comprehensive review is beyond the scope of this paper (some discussions can be found in [33, 34, 36, 38, 44]).

According to the scaling approach introduced in [40,45] the lateral distribution of electrons in both gamma- and hadron-induced air showers can be properly described by the scale-invariant function:

$$\rho(r; E, X) = \frac{N(E, X)}{R_0^2(E, X)}F\left(\frac{r}{R_0(E, X)}\right). \tag{2}$$

The principal difference of function (2) from the original NKG form (1) and its modifications mentioned above is that the scale factor R_0 is variable, while function $F(x)$ does not depend explicitly on primary particle properties and observation depth. Consequently, variations of the shape of lateral distribution are totally described by rescaling the lateral distance.

According to our calculations based on different methods and codes [19, 20, 40, 45, 46], the scaling approach is valid for the lateral distribution of electrons over the interval of scaling variable $x = r/R_0 = (0.05-20)$, corresponding to the region of radial distances from $r \sim 10$ m to $r \sim (2-4)$ km depending on shower age. The energy range where scaling formalism can be used extends up to 10^{22} eV for electromagnetic showers taking into account Landau-Pomeranchuk-Migdal effect and interactions with geomagnetic field [19, 20]. For nuclei-initiated extensive air showers the energy region verified by simulations is $(10^{15}-10^{19})$ eV. It is also worth mentioning that for lateral distribution of electrons the scale factor R_0 is equal to the root mean square radius of electron component $R_{\rm ms}$, which is defined in a standard way as

$$R_{\rm ms}(E,X) = \left(\frac{2\pi}{N(E,X)}\int_0^\infty r^2\rho(r;E,X)rdr\right)^{1/2}, \tag{3}$$

and the following expression suggested in [46] for $F(x)$ could be used

$$F(x) = Cx^{-\alpha}(1+x)^{-(\beta-\alpha)}(1+(x/10)^\gamma)^{-\delta}, \tag{4}$$

with the set of constant parameters $C = 0.28$, $\alpha = 1.2$, $\beta = 4.53$, $\gamma = 2.0$, $\delta = 0.6$.

The relation between scale factor R_0 and cascade age parameter s can be well reproduced by the expression $R_0 = \eta f(s)$, where $f(s)$ – the universal function, $\eta(E, X, \rho(X))$ - correction factor, describing the influence of atmospheric density profile. Thus, for constant atmospheric conditions at fixed observation depth we have one-valued mapping between lateral distribution shape and shower age s as well as depth of maximum $X_{\rm max}$, that could be used as a theoretical basis for primary composition deduction from surface detectors data as well as from hybrid measurements.

The function of the form (4) was implemented by several groups for the description of LD of electrons, muons and charged particles in different energy and radial distance ranges (see e.g. [38, 44, 47–56]). Nevertheless, the use of the shape of the lateral distribution to infer the mass composition is not straightforward due to the impossibility of high-precision measurements of the lateral distribution in a wide range of radial distances, the insufficient capabilities for effective discrimination of the electron contribution to the local densities of charged particles far from the shower core and also the limitations of the scaling approach itself.

3. Simulations and results

In this paper we present the results of the analysis of Monte-Carlo simulations of EAS initiated by protons and iron nuclei in the energy range of $10^{15}-10^{19}$ eV performed using CORSIKA v.7.4100 [57] with EPOS LHC v.3400 and QGSJet-II-04 (FLUKA 2011.2c.2) hadronic interaction models. For the first time electron, muon and charged particles lateral distributions were considered separately in the framework of scaling and universality approach described above. In order to get reliable data on local particle densities at very large distances from the shower core the thinning level and particle weight limit were set as $\varepsilon_{\rm th} = 10^{-8}$ and $\omega = 10^2$ respectively.

Radial scale factors were evaluated for averaged and individual showers by fitting simulated LDs. Note, that to obtain the unbiased estimates of R_0 an iterative procedure was implemented for discrimination of data at distances where the scaling formalism is not valid and also for using fixed distances ranges with respect to the scaling variable $x = r/R_0$ for showers of a specific energy and primary particle type. For $F(x)$, different expressions of function $F(x)$ were used. We have checked out function (4) with a refined set of parameters α, β, γ, δ and other representations, including polynomial approximation for $x^2F(x)$ giving a considerably better

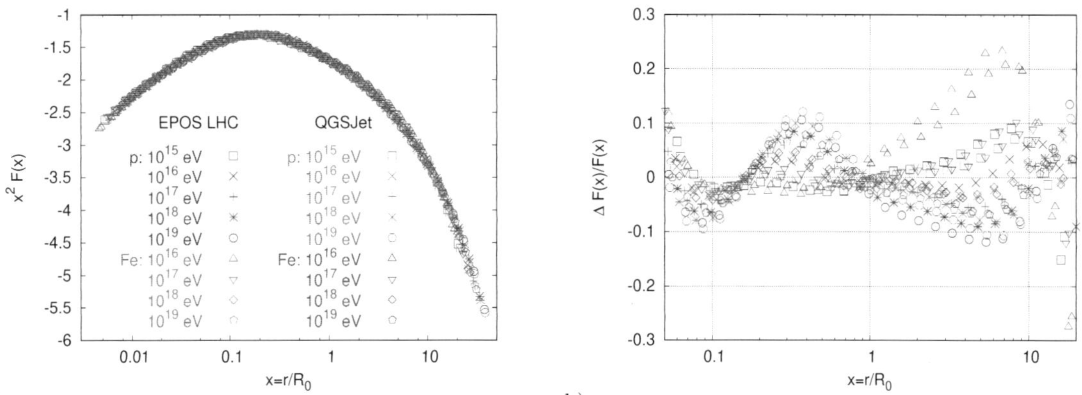

a) b)

Figure 1. Invariant part (a) and relative uncertainties (b) of the scaling description of average lateral distribution of electrons at sea level for vertical simulated EAS. Results obtained for proton and iron primaries in the energy range of $(10^{15} - 10^{19})$ eV using EPOS LHC and QGSJet hadronic interaction models are shown together by coloured symbols described in the Figure. See text for details.

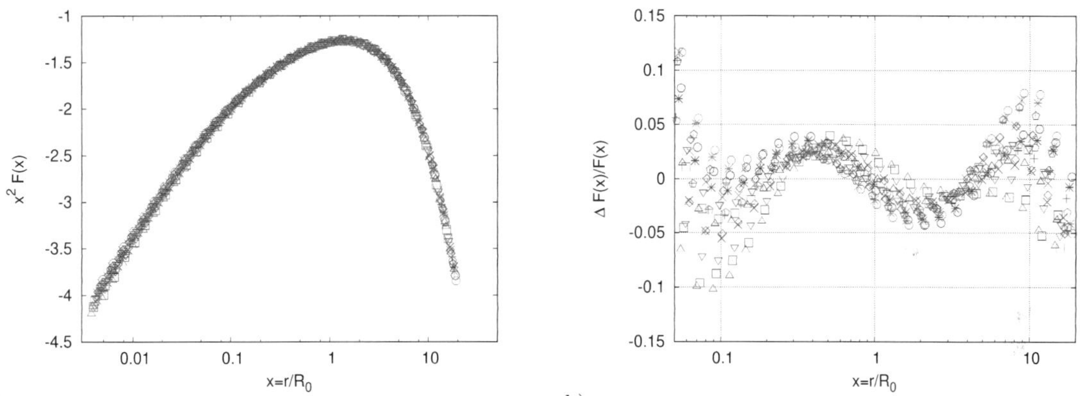

a) b)

Figure 2. Invariant part (a) and relative uncertainties (b) of the scaling description of average lateral distribution of muons at sea level for vertical simulated EAS. All the parameters and designations are same as in Figure 1.

overall fit of simulated data. It was found that the resulting R_0 values demonstrate only a weak dependence on the explicit form of $F(x)$ choosing for fitting.

In Figure 1 the invariant part (a) and relative uncertainties (b) of the scaling description (2) with polynomial $x^2F(x)$ fit of average lateral distribution of electrons with energy > 1 MeV at sea level for vertical simulated EAS are shown. The invariant part of LD is given in more illustrative form $x^2F(x)$. In Figure 2 the same results but for lateral distributions of muons with energy > 10 MeV are presented.

It is seen from the figures, that lateral densities of electrons and muons, when considered separately, are well described by one-parametric scaling functions in the range of scaling variable $x = (0.05 - 20)$. The relative uncertainties do not exceed 15% for electron LDF except iron induced showers with lowest energy 10^{16} eV, where the relative uncertainties are within 30%. For muon LD the accuracy of scaling descriptions is even better (mostly within 10%). For

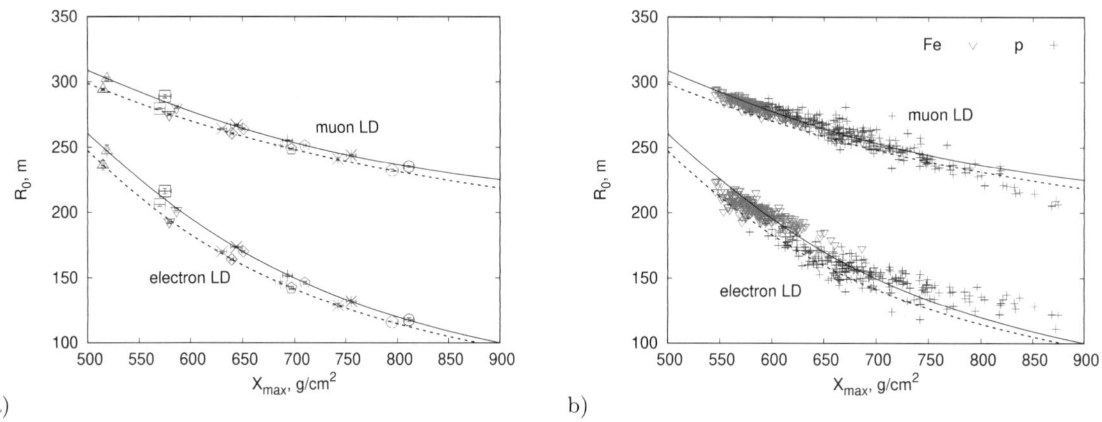

Figure 3. Scaling factors R_0 of lateral distributions of electrons (bottom) and muons (top) at sea level vs depth of shower maximum $X_{\max}$. a) Data points represent parameters of average vertical showers generated by protons and iron nuclei in the energy range of $(10^{15} - 10^{19})$ eV calculated using EPOS LHC and QGSJet hadronic interaction models. Designations are same as in Figure 1. $R_0(X_{\max})$ approximation curves for EPOS LHC (solid) and QGSJet (dotted) are shown to guide the eye. b) Parameters of individual showers generated by protons (blue crosses) and iron nuclei (red triangles) of 10^{17} eV simulated using EPOS LHC model are shown in a scatter plot. 200 events are included in each data set. The curves are average $R_0(X_{\max})$ approximations same as in Figure 3a).

charged particles distributions it was found that scaling description with single $F(x)$ is violated considerably in radial distance range where electrons and muons give a comparable contribution to the local particle densities. The relative uncertainties reach 50%.

Examining the universality in relation between radial scale factors and depth of shower maximum we plot in Figure 3a) the $R_0(X_{\max})$ dependencies for electron (bottom) and muon (top) average LDs. Functional dependence $R_0(X_{\max})$ is observed for both electrons and muons. It is remarkable that the rates of change of scaling factors with increasing depth of maximum practically do not differ for two hadronic interaction models.

In Figure 3b) the R_0 vs $X_{\max}$ scatter plot for individual showers induced by protons and iron nuclei of 10^{17} eV (EPOS LHC model) is demonstrated. 200 showers are included in each data set. Strong anticorrelation between the parameters allowing primary mass discrimination on the event-by-event basis should be noted.

Finally, it should be mentioned that due to its high accuracy for pure electromagnetic cascades in a wide range of radial distances, the scaling formalism can be successfully applied for fast semi-analytical EAS simulations. In Figure 4 the results of CORSIKA simulations with NKG-like option (scaling function with the appropriate values of R_0 is used instead of NKG to describe electromagnetic subshowers analytically) are shown in comparison with full Monte-Carlo treatment of EAS for primary protons of $(10^{17} - 10^{20})$ eV (EPOS LHC hadronic interaction model). Despite fluctuations suppression such a fast simulations could be very useful for the analysis of sensitivity to hadronic interaction models and meteorological effects.

4. Summary

Solving the mass composition problem is a key factor for understanding the origin of cosmic rays. A possible solution might be achieved with refined (multi-)hybrid measurements together with generalizations of the analysis by revealing universal features, based on the intrinsic physical

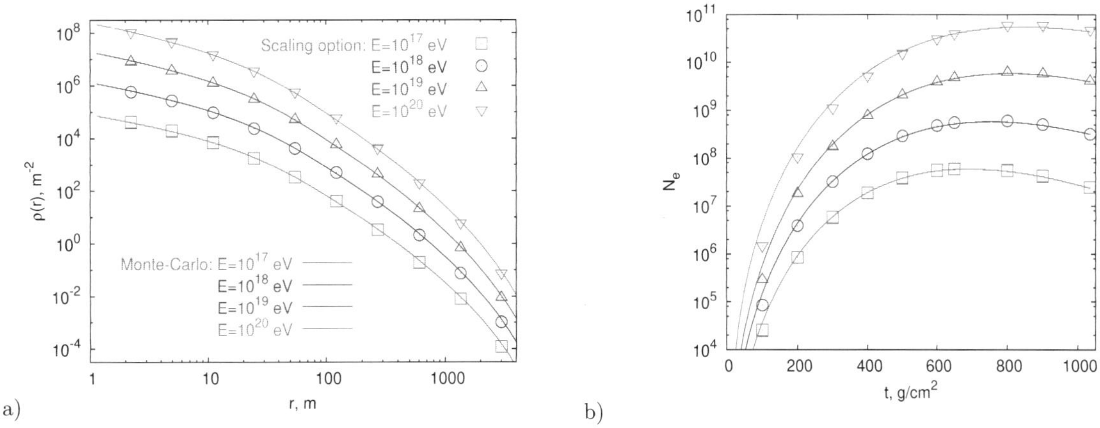

Figure 4. Comparison of Monte-Carlo simulation results (lines) with fast semi-analytical calculations based on the scaling formalism for lateral distributions (points). Primary protons of $(10^{17} - 10^{20})$ eV, EPOS LHC hadronic interaction model. Different symbols and colors correspond to different primary energies as described in the Figure. a) Average lateral distributions of electrons in the radial distance range $(2 - 3000)$ m. b) Average cascade curves.

properties of air showers, evaluation of new parameters and functionals, which are weakly sensitive to the hadronic interaction model being good primary mass indicators.

In this paper we have demonstrated the extended scaling formalism, which enables accurate description of lateral distributions of electrons and muons by one-parametric scale-invariant functions in wide primary energy and radial distance ranges. The scale-invariance of LD and air shower universality manifesting through the functional dependence between radial scale factors and longitudinal shower age are both insensitive to hadronic interactions. The proposed analysis relies on general features of air showers and is thus less affected by sophisticated mixture of systematics in the measurement of the cascade curve and lateral profile. Consequently, the proposed approach could be implemented for the present and future (multi-)hybrid air shower observations, especially in realizing the potential of Auger and Telescope Array observatories upgrades, as well as for re-analysis and cross-calibration of the data collected from different air shower arrays within the single method in a broad primary energy range.

Acknowledgments

This work was supported by Russian Foundation for Basic Research (grant #16-02-01103 a)

References

[1] Aloisio R, Berezinsky V and Blasi P 2014 *Journal of Cosmology and Astroparticle Physics* **2014** 020
[2] Aloisio R 2017 *Progress of Theoretical and Experimental Physics* **2017**
[3] Kampert K H and Unger M 2012 *Astropart. Phys.* **35** 660–678
[4] Haungs A 2015 *Physics Procedia* **61** 425–434
[5] Abbasi R *et al.* 2015 *Astropart. Phys.* **64** 49 – 62
[6] Yushkov A *et al.* 2016 *Astropart. Phys.* **85** 29 – 34
[7] Aab A *et al.* 2016 *Physics Letters B* **762** 288 – 295
[8] Yushkov A 2017 *EPJ Web of Conferences* **145** 05002
[9] Horandel J 2017 *EPJ Web of Conferences* **136** 02001
[10] Chiavassa A 2018 *EPJ Web of Conferences* **172** 07001
[11] Hanlon W *et al.* 2018 *JPS Conf. Proc.* **19** 011013
[12] Karpikov I S, Rubtsov G I and Zhezher Y V 2018 *Phys. Rev. D* **98**(10) 103002
[13] Aab A *et al.* (Pierre Auger Collaboration) 2016 *Phys. Rev. Lett.* **117**(19) 192001

[14] Abbasi R U *et al.* (Telescope Array Collaboration) 2018 *Phys. Rev. D* **98**(2) 022002
[15] Raikin R I, Lagutin A A and Yushkov A V 2008 *Nucl. Phys. Proc. Suppl.* **175-176** 559–562
[16] Raikin R I, Lagutin A A and Tyumentsev A G 2009 *Nucl. Phys. Proc. Suppl.* **196** 383–386
[17] Raikin R I and Lagutin A A 2011 *Proc. 32nd ICRC 2011: Beijing, China* vol 1 pp 299–302
[18] Raikin R I and Lagutin A A 2011 *Bull. Russ. Acad. Sci. Phys.* **75** 305
[19] Lagutin A A, Raikin R I and Serebryakova T L 2013 *Bull. Russ. Acad. Sci. Phys.* **77** 623–625
[20] Lagutin A A, Raikin R I and Serebryakova T L 2013 *J. Phys. Conf Ser.* **409** 012092
[21] Raikin R *et al.* 2017 *EPJ Web Conf.* **145** 19014
[22] Raikin R I *et al.* 2017 *Bull. Russ. Acad. Sci. Phys.* **81** 450–452
[23] Giller M *et al.* 2005 *J. Phys. G* **31** 947
[24] Nerling F *et al.* 2006 *Astropart. Phys.* **24** 421
[25] Capdevielle J N and Cohen F 2005 *J. Phys. G* **31** 507
[26] Gora D *et al.* 2006 *Astropart. Phys.* **24** 484
[27] Apel W *et al.* 2008 *Astropart. Phys.* **29** 412
[28] Lipari P 2009 *Phys. Rev. D* **79** 063001
[29] Lafebre S *et al.* 2009 *Astropart. Phys.* **31** 243
[30] Matthews J A J *et al.* 2010 *J. Phys. G* **37** 025202
[31] Yushkov A *et al.* 2010 *Phys. Rev. D* **81** 123004
[32] Ivanov A A, Pravdin M I and Sabourov A V 2001 *Proc. 32nd ICRC: Beijing, China* vol 2 p 35
[33] Tapia A *et al.* 2013 *Proc. 33rd ICRC: Rio de Janeiro, Brazil* p 0251
[34] Dey R K, Bhadra A and Capdevielle J N 2012 *J. Phys. G* **39** 085201
[35] Giller M, Smialkowski A and Wieczorek G 2015 *Astropart. Phys.* **60** 92–104
[36] Dey R K and Dam S 2016 *Eur. Phys. J. Plus* **131** 402
[37] Bridgeman A (Pierre Auger) 2018 *PoS* **ICRC2017** 323
[38] Bartoli B *et al.* (ARGO-YBJ) 2017 *Astropart. Phys.* **93** 46–55
[39] Smialkowski A and Giller M 2018 *ApJ* **854** 48
[40] Lagutin A A *et al.* 2002 *J. Phys. G* **28** 1259–1274
[41] Nishimura J and Kamata K *Progr. Theor. Phys.* **6** 93
[42] Greisen K *Ann. Rev. Nucl. Part. Sci.* **10** 63
[43] Lagutin A A, Pljasheshnikov A V and Uchaikin V V 1979 *Proc. 16th ICRC: Kyoto, Japan* vol 7 p 18
[44] Apel W *et al.* 2006 *Astropart. Phys.* **24** 467
[45] Lagutin A A *et al.* 1997 *Proc. 25th ICRC: Durban, South Africa* vol 6 p 285
[46] Lagutin A A and Raikin R I 2001 *Nucl. Phys. Proc. Suppl.* **97** 274–277
[47] Antoni T *et al.* 2001 *Astropart. Phys.* **14** 245 – 260
[48] Cotzomi J *et al.* 2007 *Proc. 30th ICRC: Merida, Mexico* vol 4 pp 183–186
[49] Ivanov A A and Knurenko S Pand Sleptsov I E 2007 *JETP* **104** 872–886
[50] Kalmykov N N *et al.* 2007 *Bull. Russ. Acad. Sci. Phys.* **71** 522–524
[51] Fomin Y *et al.* 2008 *Nucl. Phys. Proc. Suppl.* **175-176** 334–337
[52] Knurenko S *et al.* 2008 *Nucl. Phys. Proc. Suppl.* **175-176** 201 – 206
[53] Apel W *et al.* 2010 *Nucl. Instr. Meth. Phys. Res. A* **620** 202 – 216
[54] Sabourov A V *et al.* 2007 *Proc. 30th ICRC: Merida, Mexico, 2008* vol 4 p 291
[55] Arteaga-Velazquez J *et al.* 2013 *EPJ Web of Conferences* **52** 07002
[56] Apel W *et al.* 2017 *Astropart. Phys.* **95** 25 – 43
[57] Heck D *et al.* (Report FZKA 6019. Forschungszentrum Karlsruhe, 1998)

Cosmic ray acceleration in accretion flows of galaxy clusters

V.N.Zirakashvili, V.S.Ptuskin

Pushkov Institute of Terrestrial Magnetism, Ionosphere and Radiowave Propagation, 108840 Moscow Troitsk, Russia

E-mail: zirak@izmiran.ru

Abstract. We investigate acceleration of cosmic rays by shocks and accretion flows in galaxy clusters. Numerical results for spectra of accelerated particles and nonthermal emission are presented. It is shown that the acceleration of protons and nuclei in the nearby galaxy cluster Virgo can explain the observed spectra of ultra high energy cosmic rays.

1. Introduction

Clusters of galaxies are considered as a possible candidate for the origin of ultra-high energy cosmic rays (UHECRs) (see e.g. [1] for a review). Primordial density fluctuations are amplified via gravitational instability in the expanding Universe and result in the appearance of different coherent density structures at the present epoch. The galaxy clusters are formed at the latest times and continue to grow presently due to accretion of the cicumcluster gas and dark matter. The inflow of matter is accompanied by an accretion shock at a distance of several Mpc from the cluster center.

Virgo cluster of galaxies with the total mass of $\sim 10^{15}M_{\odot}$ at the distance $d \sim 15-20$ Mpc [2] from the Milky Way is the nearest large galaxy cluster. It is located at the center of the Local super cluster of galaxies. Because of its proximity Virgo has been proposed as the source of observed UHECRs within a phenomenological diffusive model [3].

The diffusive shock acceleration (DSA) process [4, 5, 6, 7] is believed to be the principal mechanism for the production of galactic cosmic rays (CR) in supernova remnants (SNRs). Over the last decade the excellent results of X-ray and gamma-ray astronomy have been providing evidence of the presence of multi-TeV energetic particles in these objects (see e.g. [8]).

Since the accretion shocks are very large they can accelerate particles to significantly higher energies compared to SNRs and produce a significant fraction of observable UHECRs [9, 10, 11]. In this paper we describe the modifications of our non-linear DSA model [12] designed for the investigation of DSA in SNRs and apply it for the investigation of particle acceleration in clusters of galaxies. In particular we apply the model for the Virgo cluster of galaxies.

2. Nonlinear diffusive shock acceleration model

Details of our model of nonlinear DSA can be found in [12]. The model contains coupled spherically symmetric hydrodynamic equations and the transport equations for energetic protons, ions and electrons.

The hydrodynamical equations for the gas density $\rho(r,t)$, gas velocity $u(r,t)$, gas pressure $P_g(r,t)$, magnetic pressure $P_m(r,t)$, and the equation for isotropic part of the cosmic ray proton momentum distribution $N(r,t,p)$, the momentum per nucleon distribution $N_i(A,r,t,p)$ of nuclei with atomic number A and dark matter velocity distribution $F(r,t,w)$ in the spherically symmetrical case are given by

$$\frac{\partial \rho}{\partial t} = -\frac{1}{r^2}\frac{\partial}{\partial r} r^2 u\rho \tag{1}$$

$$\frac{\partial u}{\partial t} = g(r) - u\frac{\partial u}{\partial r} - \frac{1}{\rho}\left(\frac{\partial P_g}{\partial r} + \frac{\partial P_c}{\partial r} + \frac{\partial P_m}{\partial r}\right) \tag{2}$$

$$\frac{\partial P_g}{\partial t} + u\frac{\partial P_g}{\partial r} + \frac{\gamma_g P_g}{r^2}\frac{\partial r^2 u}{\partial r} = -(\gamma_g - 1)\Lambda_c(T_e)n^2 \tag{3}$$

$$\frac{\partial P_m}{\partial t} + u\frac{\partial P_m}{\partial r} + \frac{\gamma_m P_m}{r^2}\frac{\partial r^2 u}{\partial r} = 0 \tag{4}$$

$$\frac{\partial N}{\partial t} = \frac{1}{r^2}\frac{\partial}{\partial r} r^2 D(p,r,t)\frac{\partial N}{\partial r} - u\frac{\partial N}{\partial r} + \frac{\partial N}{\partial p}\frac{p}{3r^2}\frac{\partial r^2 u}{\partial r} + \frac{1}{p^2}\frac{\partial}{\partial p}p^2 b(p)N + 4\nu_{ph}(4)N_i(4) + \sum_{A=5}^{56}\nu_{ph}(A)N_i(A) \tag{5}$$

$$\frac{\partial N_i(A)}{\partial t} = \frac{1}{r^2}\frac{\partial}{\partial r} r^2 D_i(p,r,t)\frac{\partial N_i(A)}{\partial r} - u\frac{\partial N_i(A)}{\partial r} + \frac{\partial N_i(A)}{\partial p}\frac{p}{3r^2}\frac{\partial r^2 u}{\partial r} + \frac{1}{p^2}\frac{\partial}{\partial p}p^2 b(p)N_i(A) - \nu_{ph}(A)N_i(A) + \nu_{ph}(A+1)N_i(A+1) \tag{6}$$

$$\frac{\partial F}{\partial t} = -\frac{1}{r^2}\frac{\partial}{\partial r} r^2 wF - g(r)\frac{\partial F}{\partial w} \tag{7}$$

Here $P_c = 4\pi\int dpp^3 v(N + \sum_A AN_i(A))/3$ is the cosmic ray pressure, T_e, γ_g and n are the gas temperature, adiabatic index and number density respectively, γ_m is the magnetic adiabatic index, $D(r,t,p)$ is the cosmic ray diffusion coefficient. The radiative cooling of gas is described by the cooling function $\Lambda_c(T_e)$.

The gravitational acceleration $g(r)$ is given by the following expression

$$g(r) = \Lambda r - \frac{4\pi G}{r^2}\int_0^r r_1^2 dr_1(\rho(r_1) + \rho_{DM}(r_1)) \tag{8}$$

where G is the gravitational constant, Λ is the lambda term and $\rho_{DM}(r) = \int dwF(r,w)$ is the dark matter density.

The function $b(p)$ describes the energy losses of particles. In particular the losses of multi EeV protons and nuclei via electron-positron pair production on the microwave background radiation (MWBR) are important in galaxy clusters. Another losses of energy of the nuclei occur in photonuclear reactions with infrared photons. They are described by two last terms in Eq. (6). We use a simplified approach and assume that the nuclei lose one nucleon in every

IOP Conf. Series: Journal of Physics: Conf. Series **1181** (2019) 012033 doi:10.1088/1742-6596/1181/1/012033

photonuclear reaction with the rate $\nu_{ph}(A)$. The corresponding proton source term is added in Eq. (5). For the calculation of photonuclear cross-sections we use the background spectrum of infrared photons $n(\epsilon) = 0.87 \cdot 10^{-3} a^{-4}(t)\epsilon^{-2}$ cm^{-3}eV^{-1}, where ϵ is the photon energy expressed in eV and $a(t)$ is the expansion factor of the Universe. At the present epoch ($a = 1$) the spectrum is three times lower than LIR spectrum of Puget et al. [13] and is close to the lower limit of Malkan & Stecker [14].

Cosmic ray diffusion is determined by particle scattering on magnetic inhomogeneities. Here we consider the simplest model with strong background random magnetic fields in the circumcluster medium. The properties of the gas and magnetic field in the circumcluster medium are determined by galactic winds of individual galaxies. These winds were strong in past when the star formation rate was significantly higher than at the present epoch. The winds were driven by cosmic rays and hot gas produced by supernova explosions. The gas was heated, galactic cosmic rays were reaccelerated and magnetic fields were amplified at termination shocks of the galactic winds. That is why we expect that these three components have comparable energy densities in the circumcluster medium. Additionally the galactic winds transport metals and therefore the metallicity of the circumcluster medium is different from the primordial one that is indeed observed.

Bearing in mind that the magnetic field is strong in the circumcluster medium we neglect its generation via cosmic ray streaming instability in Eq. (4). We also neglect the gas heating via magnetic dissipation in Eq. (3). So the gas and magnetic pressures P_g and P_m evolve in accordance with adiabatic compression of the gas. Below we use the magnetic adiabatic index $\gamma_m = 3/2$.

We use Bohm-like dependence of the diffusion coefficient D of energetic particles with the charge q, momentum p and speed v

$$D = \eta_B D_B,\ D_B = \frac{cpv}{3qB},\ B = \sqrt{8\pi P_m} \tag{9}$$

where B is the total magnetic field strength, while η_B is a free parameter characterizing speed of diffusion in the terms of Bohm diffusion. Since the Bohm diffusion coefficient D_B is the minimal possible value we shall use the value of $\eta_B = 2$ below.

In the shock transition region the magnetic pressure is increased by a factor of σ^{γ_m}, where σ is the shock compression ratio. Its impact on the shock dynamics is taken into account via the Hugoniot conditions.

We do not consider the injection of thermal ions at the shock front in the present study. As was mentioned before there are background cosmic rays in the circumcluster medium. We assumed that the spectrum of these particles is given by

$$N_{IG}(p) = N_0 \left(\frac{a(t)p}{p_{IG}}\right)^{-\gamma-2} \exp\left(-\frac{a(t)p}{p_{IG}}\right) \tag{10}$$

Here p_{IG} is the maximum momentum of background cosmic rays at the present epoch and γ is the spectral index of differential energy distribution of background cosmic rays.

The similar expression was used for the background spectra of nuclei. Helium nuclei were taken 10 times less abundant than protons. For heavier nuclei we assumed an enrichment proportional to the atomic number A as observed in Galactic cosmic rays.

The following expressions were used for the circumcluster magnetic and gas pressures

$$P_{gIG} = a(t)^{-3\gamma_g} P_{g0},\ P_{mIG} = a(t)^{-3\gamma_m} P_{m0}, \tag{11}$$

where P_{g0} and P_{m0} are the gas and magnetic circumcluster pressures at the present epoch respectively.

3. Modeling of DSA in the Virgo cluster

We model the Virgo cluster formation and particle acceleration for the cosmological parameters $\Omega_b = 0.05$, $\Omega_{DM} = 0.25$, $\Omega_\Lambda = 0.7$ and the Hubble constant $H = 70$ km s^{-1} Mpc^{-1}. The corresponding mean barion density ρ_b of the Universe is $\rho_b = 3\Omega_b H^2/8\pi G = 4.6 \cdot 10^{-31}$g cm^{-3}. The lambda term in Eq. (8) is given by $\Lambda = \Omega_\Lambda H^2$. The initial instant of time is $t_0 = 4$ Gyr from the Big Bang.

The parameters of the background cosmic ray spectrum given by Eq. (10) were adjusted to reproduce observations of Auger Collaboration. We used the values $\gamma = 1.8$ and $p_{IG} = 8$ PeV/c. The spectra of background cosmic rays are somewhat harder than the expected spectra of galactic sources. However the additional reacceleration at galactic wind termination shocks indeed will make galactic cosmic ray spectra harder.

We use the initial condition corresponding to the self-similar solution of Bertschinger [15]. At latest times the accelerated expansion of the Universe results in the higher accretion shock speed and radius in comparison with this solution.

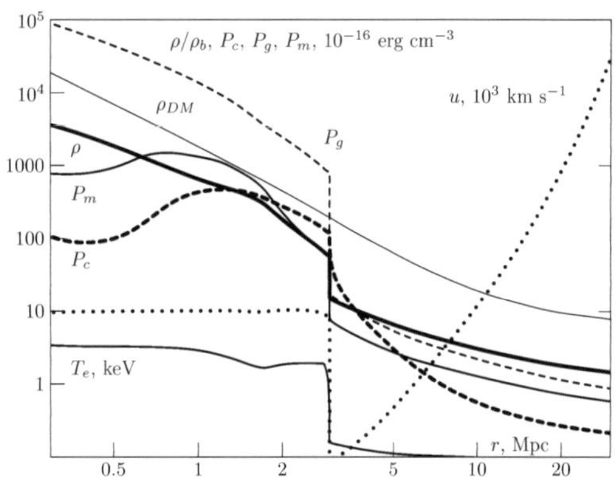

Figure 1. The radial dependence of the gas density (thick solid line), dark matter density (thin dotted line), the gas velocity (dotted line), cosmic ray pressure (thick dashed line), the magnetic pressure $B^2/8\pi = P_m$ (solid line), the gas temperature T_e (thin solid line) and the gas pressure P_g (dashed line) at $t =$ 13.5 Gyr in the Virgo cluster.

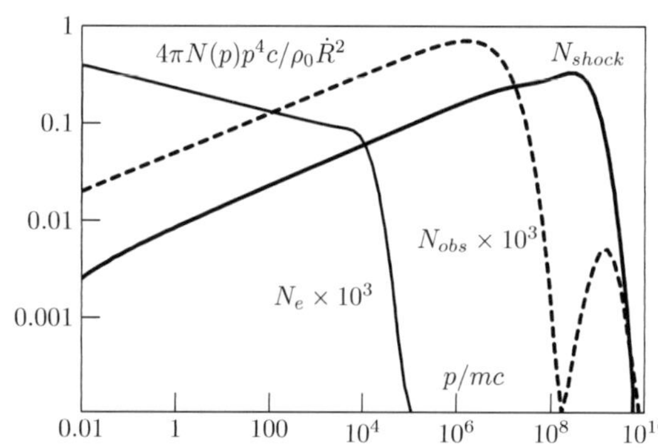

Figure 2. Spectra of accelerated particles at the accretion shock of Virgo cluster. The spectrum of protons (thick solid line) and electrons (thin solid line) are shown. We also show the proton spectrum at the Milky Way position at $r = 20$ Mpc (dashed line).

Figures (1)-(5) illustrate the results of our numerical calculations.

The radial dependence of physical quantities in the Virgo cluster at the present epoch ($t = 13.5$ Gyr) are shown in Fig.1. At large distances we have the uniform expansion of the Universe. Closer to the cluster deviations from the Hubble law due to the initial local overdensity occur. We observe an inflow of matter inside a so-called "turnaround' point at $r \approx$10 Mpc. The accretion shock is situated at $r = R = 3$ Mpc. The gas temperature is several keV inside the shock. The calculated gas density profile is in agreement with observations of the Virgo cluster [2]. The total mass of the gas inside the shock $1.5{\cdot}10^{14} M_\odot$ is also close to the observed value [2].

The cosmic ray, magnetic and gas pressures at large distances are of the order of 10^{-16} erg cm^{-3} that is considered as a characteristic value for the extragalactic pressure around the Milky Way. The corresponding strength of the background magnetic field is $B_{IG} = 3 \cdot 10^{-8}$G that is a reasonable value for the Local supercluster.

It is interesting that the adjusted level of background cosmic rays results in the pressure of reaccelerated particles close to 10% of the ram pressure of the accretion shock. This number is similar to the one found in the hybrid modeling of collisionless quasiparallel shocks [16] where the ions are injected at the shock front.

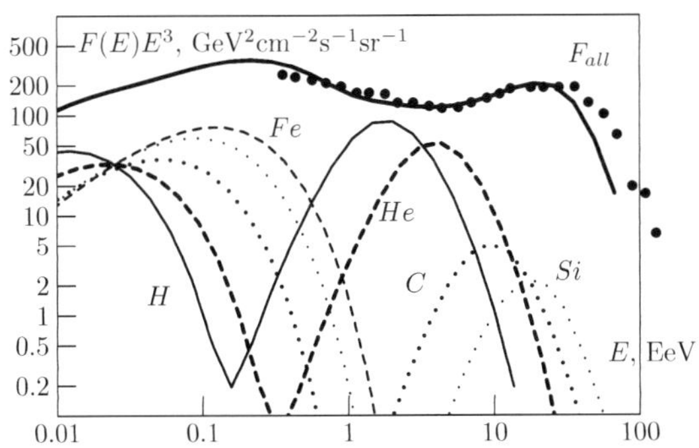

Figure 3. Calculated all-particle spectrum (thick solid curve), the spectrum of protons (thin solid line) and the spectra of other nuclei at the Milky Way position. Observational data of Auger Collaboration [17] (circles) are also shown.

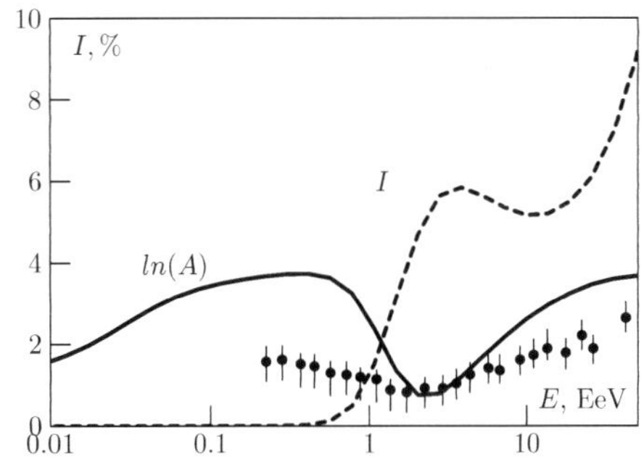

Figure 4. Calculated mean logarithm (solid line) and cosmic ray anisotropy (dashed line) at the Milky Way position. Experimentally measured mean logarithm data of Auger collaboration [17] using EPOS-LHC model of particle interactions in the atmosphere (data with error-bars) are also shown.

Spectra of accelerated in Virgo protons and electrons at $t = 13.5$ Gyr are shown in Fig.2. At this point the maximum energy of accelerated protons is close to 10^{18} eV. Only highest energy protons and nuclei from the cut-off region reach the Milky Way. Lower energy particles simply have not time to propagate from the shock surface to the Milky Way. We can observe only exponentially small fraction of protons (10^{-4}) reaccelerated at the accretion shock. So the particles are in a rather unusual propagation mode (see also [10]). At lower energies we observe the background circumcluster cosmic rays.

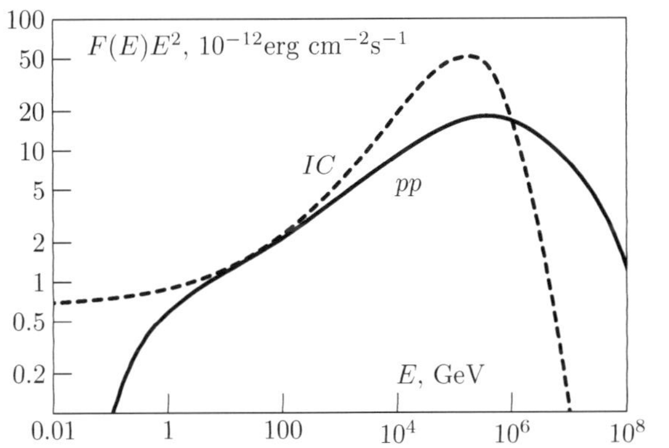

Figure 5. The results of modeling of gamma radiation of Virgo. We show IC emission (dashed line) and gamma-ray emission from pion decay (solid line). The absorption of multi-TeV photons due to pair production is not taken into account.

All-particle spectrum and spectra of individual nuclei calculated for the Milky Way position ($r = 20$ Mpc) are shown in Fig.3. The ankle at 5 EeV corresponds to the transition from the proton-Helium composition to the heavier nuclei dominated composition in our model. The heaviest nuclei do not present in the spectrum because of the photonuclear disintegration. The cut-off energy of the spectrum is mainly determined by the photonuclear reactions of heavy nuclei during acceleration and propagation. If we neglect the photonuclear reactions the maximum energy will be several times higher.

IOP Conf. Series: Journal of Physics: Conf. Series **1181** (2019) 012033 doi:10.1088/1742-6596/1181/1/012033

Calculated mean logarithm of atomic number and anisotropy are shown in Fig.4. The model reproduces light composition at EeV energies and the dipole anisotropy below the value of 6.5 percent measured by Auger Collaboration [18].

The spectra of unabsorbed gamma rays expected from the Virgo cluster are shown in Fig. 5. The gamma emission shows two components. One is produced by the nuclear collisions of protons and nuclei with gas in the cluster interior while another one is the inverse Compton component that at TeV energies is produced by electron positron pairs generated via interaction of accelerated particles with background photons of MWBR. The last process is rather efficient in galaxy clusters [19, 20]. In spite of the high expected flux the large angular size of the source (several degrees) makes the gamma detection of the Virgo cluster very difficult.

4. Discussion

We show that almost all UHECRs can be accelerated in the nearby Virgo cluster of galaxies. Our consideration is close to the earlier pure proton model of Kang et al. [10]. Note that we used more realistic value of the magnetic field strength that is 10 times lower. The corresponding decrease of the maximum energy is compensated by the presence of heavier nuclei in UHECR spectrum.

The drawback of the model is its sensitivity to adjusted parameters to fit UHECR data. As mentioned before we observe an exponentially small part (10^{-4}) of the highest energy particles accelerated at the accretion shock. That is why modest variations of the magnetic field strength result in the significant change of UHECR intensity.

Probably more massive and more distant clusters like Coma or another astrophysical sources (e.g. active galactic nuclei) also might give some contribution at highest energies. The particles accelerated in distant astrophysical objects can reach the Local supercluster propagating in the cosmological voids where the magnetic fields are very weak.

The adjusted maximum energy of background cosmic rays $E_{IG} = cp_{IG} = 8\ Z$ Pev is not far from the "knee" energy. It is not excluded that this is not a coincidence. The value of E_{IG} is rather well constrained by available UHECR data. For lower values of E_{IG} we would observe a large dip in the spectrum between 0.1 and 1 EeV. For higher values of E_{IG} we would observe heavy UHECR composition at 1 EeV.

The work was supported by Russian Foundation of Fundamental Research grant 16-02-00255.

[1] Brunetti G and Jones T 2014 *Intern. Journal of Modern Physics D* **29** 3007
[2] Ade P A R, Aghanim N, Arnaud M. *et al* 2016, *Astron. Astrophys.* **596** 101
[3] Giler M, Wdowczyk W and Wolfendale A W 1980 *J. Phys. G* **6** 1561
[4] Krymsky G F 1977 *Soviet Physics-Doklady* **22** 327
[5] Bell A R 1978 *MNRAS* **182** 147
[6] Axford W I, Leer E and Skadron G 1977 *Proc. 15th ICRC, Plovdiv* **90** 937
[7] Blandford R D and Ostriker J P 1978 *Astrophys. J.* **221** L29
[8] Lemoine-Goumard M 2004 *Proc. IAU Symp.* **296** 287
[9] Norman C A, Melrose D B and Achterberg A, 1995 *Astrophys. J.* **454** 60
[10] Kang H, Ryu D and Jones T W 1996 *Astrophys. J.* **456** 422
[11] Kang H, Rachen J P and Biermann P L, 1997 *MNRAS* **286** 257
[12] Zirakashvili V N and Ptuskin V S 2012 *Astropart. Phys.* **39** 12
[13] Puget J L, Stecker F W and Bredekamp J H 1975 *Astrophys. J.* **205** 638
[14] Malkan M A and Stecker F W 1998 *Astrophys. J.* **496** 13
[15] Bertschinger M 1985 *Astrophys. J. Suppl.* **58** 39
[16] Caprioli D 2014 *Nuclear Physics B (Proc. Suppl.)* **256** 48
[17] Letessier-Selvon A for the Pierre Auger Collaboration 2014 *Brazilian J. of Phys.* **44** 560 arXiv:1310.4620
[18] Pierre Auger Collaboration 2017 arXiv:1709.0732
[19] Vannoni G, Aharonian F A, Gabici S, Kelner S R and Prosekin A 2011 *Astron. Astrophys.* **536** 56
[20] Inoue S, Sigle G, Miniati F and Armengaud E 2008 *Proc. 30th ICRC* **4** 555

IOP Conf. Series: Journal of Physics: Conf. Series **1181** (2019) 012034 doi:10.1088/1742-6596/1181/1/012034

CRPropa - A Toolbox for Cosmic Ray Simulations

R. Alves Batista[1], J. Becker Tjus[2], A. Dundovic[3], M. Erdmann[4], C. Heiter[4,5], K.-H. Kampert[6], D. Kuempel[4], L. Merten[2], G. Müller[4], G. Sigl[3], A. v. Vliet[3,7,8], D. Walz[4], T. Winchen[9], M. Wirtz[4]

[1] Universidade de São Paulo, Instituto de Astronomia, Geofísica e Ciências Atmosféricas; Rua do Matão, 1226, 05508-090, São Paulo-SP, Brazil,
[2] Ruhr-Universität Bochum, Theoretische Physik IV: Plasma-Astroteilchenphysik, Universitätsstrasse 150, 44801 Bochum, Germany,
[3] Universität Hamburg, II Institut für Theoretische Physik, Luruper Chaussee 149, 22761 Hamburg, Germany,
[4] RWTH Aachen University, III. Physikalisches Institut A Otto-Blumenthal-Str., 52056 Aachen, Germany,
[5] Max Planck Institut für Radioastronomie, Auf dem Hügel 69, 53121 Bonn, Germany,
[6] Bergische Universität Wuppertal, Department of Physics, Gaußstr. 20, 42097 Wuppertal, Germany,
[7] Department of Astrophysics/IMAPP, Radboud University, P.O. Box 9010, 6500 GL Nijmegen, The Netherlands,
[8] DESY Zeuthen, Platanenallee 6, 15738 Zeuthen, Germany,
[9] Vrije Universiteit Brussel, Astrophysical Institute, Pleinlaan 2, 1050 Brussels, Belgium

E-mail: `tobias.winchen@rwth-aachen.de`

Abstract. The astrophysical interpretation of recent experimental observations of cosmic rays relies increasingly on Monte Carlo simulations of cosmic ray propagation and acceleration. Depending on the energy range of interest, several different propagation effects inside the Milky Way as well as in extragalactic space have to be taken into account when interpreting the data. With the CRPropa framework we aim to provide a toolbox for according simulations. In recent versions of CRPropa, the ballistic single particle propagation mode aiming primarily at extragalactic cosmic rays has been complemented by a solver for the differential transport equation to address propagation of galactic cosmic rays. Additionally, modules have been developed to address cosmic ray acceleration and many improvements have been added for simulations of electromagnetic secondaries. In this contribution we will give an overview of the CRPropa simulation framework with a focus on the latest improvements and highlight selected features by example applications.

1. Introduction

Current and future detectors for ultra-high energy cosmic rays (UHECR), neutrinos, and gamma-rays provide us with an increasing amount of precision data. The Pierre Auger Observatory [1] and Telescope Array [2] measured the spectrum [3, 4] and composition [5, 4] of the UHECR flux at the highest energies, and also observed anisotropy in the arrival direction of UHECR [6, 7]. The IceCube neutrino detector measured a flux of extraterrestrial neutrinos [8], and, recently, observed a first coincident event with a flaring gamma ray blazar [9].

However, understanding this wealth of data is no easy task. The place and mechanism of acceleration of cosmic rays remains unknown and the newest data challenges the so far prevailing models. Accounting for propagation effects, the observations point to a surprisingly hard spectral index and heavy composition at the accelerators [10], requiring at least extensions to the simplest models for the acceleration process. Observation of mass composition at the ankle [11] indicate a surprisingly light composition that is at odds with simple models for the transition between galactic and extragalactic cosmic rays, and even at the lowest energies the observed particle spectra challenge simple models for acceleration in supernovas [12]. To develop – and test – the increasingly complex models required to understand our data, detailed simulations of the propagation and acceleration mechanisms are needed.

2. The CRPropa Software

The CRPropa software in its third major version [13] has been developed as a versatile simulation tool to efficiently develop predictions from astrophysical models. The code includes a variety of components to set up simulations for the propagation of particles, including models for energy losses in interactions and creation of secondary particles. In particular, particles can be propagated through arbitrary three-dimensional magnetic fields and be subjected to energy losses due to interactions in photon fields by pair-production, photo-meson production, photo-disintegration, due to nuclear decay, and due to the adiabatic expansion of the universe. The full propagation can be performed including the combination of both, cosmological effects and three-dimensional propagation. The simulation of the physical processes is limited to the highly relativistic propagation regime where the rest-mass of the particles can be neglected. In addition to modules subjecting the propagated particles to physical effects, a large collection of tools to inject particles with desired properties into the simulation respectively record the requested output are included. Execution of the simulation utilizes shared-memory parallel architectures using OpenMP. Parallelization on distributed-memory architectures can be implemented by users e.g. using MPI.

The individual components are independent and can be combined by the user to a customized simulation. Additional components not yet included in CRPropa can be written by the users in Python or C++ and included in the simulations, usually without requiring a new compilation of the CRPropa code. We strive to collect the contributions of users to the code and either include them into CRPropa or facilitate sharing of code amongst user by referencing them on the CRPropa webpage.

Code development follows an open procedure. All changes to the code are tracked using a public git repository. Discussions on bugs and future enhancements as well as support questions are transparent in an online issue tracker. Contributions to the code by users as well as any larger modification is subjected to peer review before inclusion to not only minimize wrong computations, but also enforce technical standards and code quality to ensure the future maintainability of the project. Guidelines for contributions, and thus also for the review, are public and distributed alongside the CRPropa code. In particular, unit-tests, inline documentation, and also usage examples published on the CRPropa website are required. CRPropa is licensed under the GPLv3 [14].

3. Recent Developments and Example Applications

3.1. Galactic Cosmic Rays

CRPropa has been recently extended to simulate propagation of galactic cosmic rays [15]. For low energy cosmic rays the calculation of individual particle trajectories is not feasible, as integrating the equation of motion becomes too computational intensive. Instead, cosmic ray propagation is described stochastically via the modified Parker transport equation

$$\frac{\partial n}{\partial t} + \vec{u} \cdot \nabla n = \nabla \cdot (\hat{\kappa} \nabla n) + \frac{1}{p^2} \frac{\partial}{\partial p} \left(p^2 \kappa_{pp} \frac{\partial n}{\partial p} \right) + \frac{1}{3} (\nabla \vec{u}) \frac{\partial n}{\partial \ln p} + S(\vec{x}, p, t) \tag{1}$$

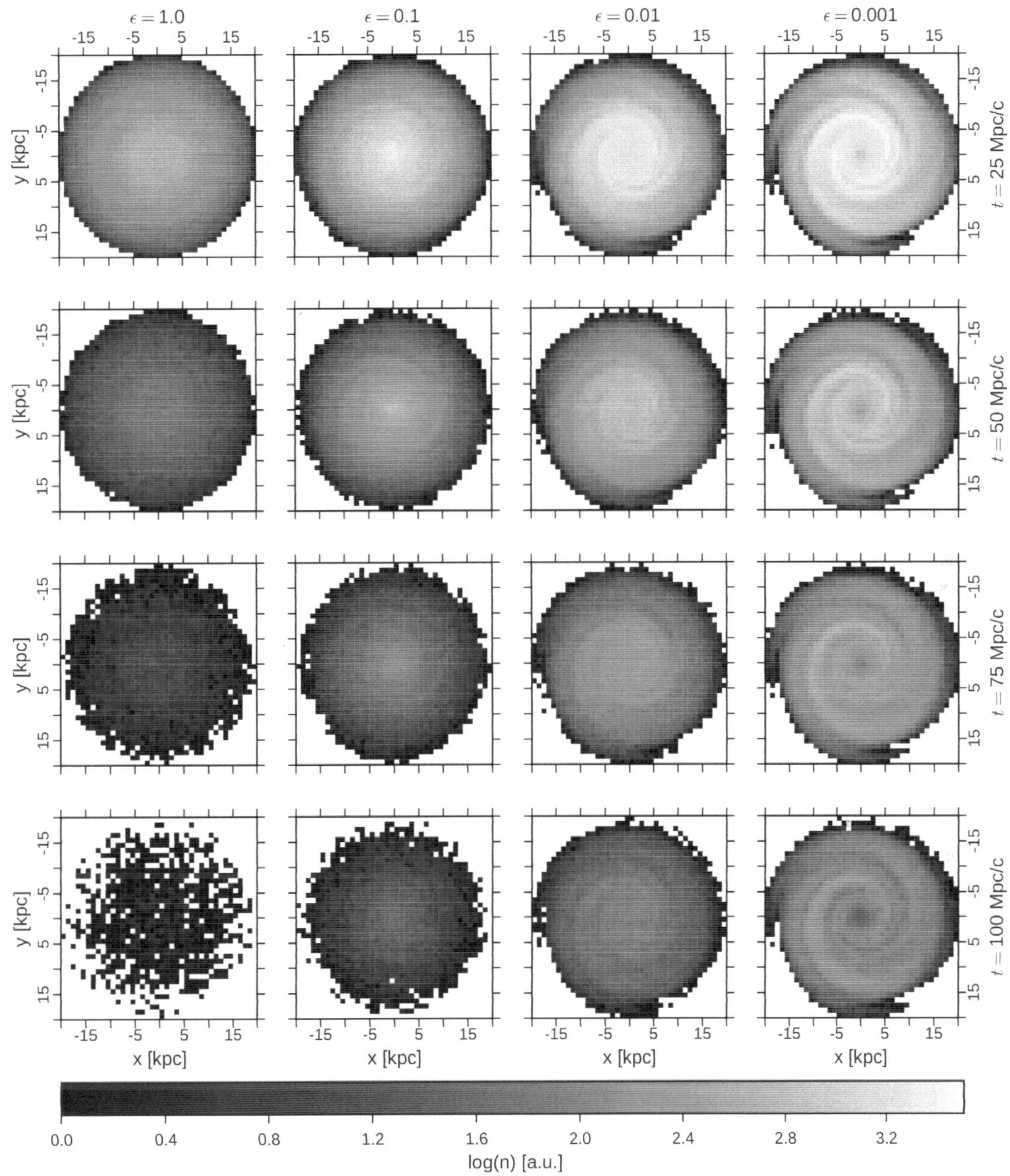

Figure 1. Time evolution of the density of cosmic rays in the Galactic plane resulting from homogeneous injection of cosmic rays in a JF 12 galactic magnetic field model for several choices of the diffusion parameter ϵ (From [15]).

which relates the particle density n at point $\vec{x}$, momentum p, and time t of a system with the local spatial diffusion tensor $\hat{\kappa}$, momentum diffusion coefficient κ_{pp}, advection speed $\vec{u}$ and particle sources $S(\vec{x}, p, t)$. Additional terms can be added to eq. 1 to account for particle interactions and decays.

The stochastic description can be reconciled with the single particle description in CRPropa by treating the corresponding stochastic partial-differential equation instead of eq. 1. Here, individual 'pseudo' particles make in every step of the simulation effective steps in space and momentum according to the diffusion coefficients. In the limit of many particles the resulting distributions of the particles are identical to the direct solution of eq. 1, respectively the simulation of ensembles of individual trajectories.

While this has the advantage that the propagation of ensembles of low energy particles can be calculated very efficiently and interactions can be accounted for at the same time with the regular CRPropa modules, additional assumptions on the diffusion coefficients have to be made in the simulations. As an example application, in figure 1 the time evolution of the cosmic ray density in the Galactic disc is displayed for several choices of the diffusion coefficient. In case of large ratios between parallel and perpendicular diffusion $\kappa_{\perp}/\kappa_{\parallel} = \epsilon$, particles leak quickly out of the galaxy; For small ϵ particles are bound to the galactic field and follow the structure of the magnetic field models as expected.

3.2. Particle Acceleration

In the majority of models for cosmic ray sources, particles are believed to be accelerated in shocked plasmas via the first-order Fermi mechanism (e.g. [17]) which predicts a power-law emission spectrum $dN/dE \propto E^{\gamma}$ with $\gamma \leq -2$. However, evaluation of the data of the Pierre Auger Observatory indicates [10, 18, 19] harder injection spectra that require new or at least modified acceleration models (e.g. [20, 16]).

To simulate the processes in the sources, we created modules to calculate particle acceleration via scatter processes, e.g. by plasma waves. Input to the simulation is a user defined step length corresponding to the frequency of scatter events of the particles. With appropriate choices of these parameters, the simulation of first-order Fermi acceleration [21] and second-order Fermi acceleration is possible. Using this new feature it was demonstrated that acceleration via the second-order Fermi mechanism after pre-acceleration, geometrical effects of the acceleration region can modify the spectrum from a power-law to a peaking distribution as shown in figure 2. With the modified spectrum, the observations of the Pierre Auger Observatory can be explained without implying an unexpected high abundance of heavy elements at the sources [16].

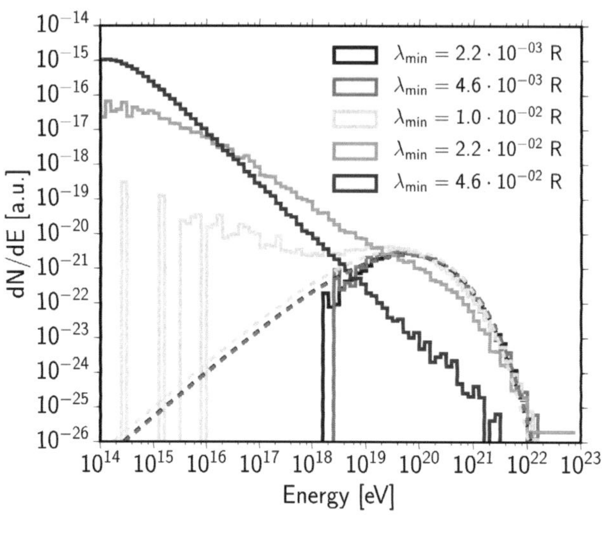

Figure 2. Acceleration spectra from second order Fermi acceleration with a fixed minimum step length after injection in center of spherical acceleration region with radius R (From [16]).

3.3. Photon Propagation and Production

Photon propagation in CRPropa relied so far on external codes such as DINT [22] or EleCa [23]. Consequently, the simulations were not easily extendable and full consistency between both components were hard to achieve, as for example some photon models were not available in the independent codes and effects of the local magnetic field were not easy to account for.

To increase the precision of photon propagation as important secondary messenger, new photon propagation modules have been created inside the CRPropa core structure that allow a consistent simulations setup. For this, modules for all energy loss process present in DINT and ELeCa, have been implemented, and also additional photon production channels are now considered [24]. In particular, photon creation during photo-disintegration of nuclei, elastic scattering and radiative decay have been included and the simulation accounts for deflections in the local magnetic field. To also speed up the computation, which, in some circumstances, can be otherwise prohibitively time consuming, a new thinning procedure is currently under development.

4. Conclusion

The wealth of data in astroparticle physics provides new detailed insights in the highest energetic processes of the universe. The CRPropa software framework provides the necessary means to develop our understanding and compare the predictions of the models with data. The latest development efforts extend the scope of the project from propagation of the highest energetic particles only to include also lower energy and acceleration processes, thus enabling a consistent view on high energy astroparticle physics.

Acknowledgements

Tobias Winchen is supported by the German Research Foundation (DFG), grant WI 4946/1-1.

References

[1] Aab A *et al.* (The Pierre Auger Collaboration) 2015 *Nucl. Instrum. Meth.* **A798** 172–213 (*Preprint* 1502.01323)
[2] Kawai H *et al.* (The Telescope Array Collaboration) 2008 *Nuclear Physics B - Proceedings Supplements* **175** 221 – 226
[3] Fenu F for the Pierre Auger Collaboration 2017 *Proceedings of the 35th International Cosmic Ray Conference (ICRC)* (*Preprint* 1708.06592)
[4] Matthews J (The Telescope Array Collaboration) 2017 *Proceedings of the 35th International Cosmic Ray Conference (ICRC)*
[5] Bellido J for the Pierre Auger Collaboration 2017 *Proceedings of the 35th International Cosmic Ray Conference (ICRC)* (*Preprint* 1708.06592)
[6] Abbasi R *et al.* (The Telescope Array Collaboration) 2014 *Astrophys. J.* **790** L21 (*Preprint* 1404.5890)
[7] Aab A *et al.* (The Pierre Auger Collaboration) 2017 *Science* **357** 1266–1270 (*Preprint* 1709.07321)
[8] Aartsen M G *et al.* (The IceCube Collaboration) 2013 *Science* **342** 1242856 (*Preprint* 1311.5238)
[9] Aartsen M G *et al.* (The Liverpool Telescope, MAGIC, H.E.S.S., AGILE, Kiso, VLA/17B-403, INTEGRAL, Kapteyn, Subaru, HAWC, Fermi-LAT, ASAS-SN, VERITAS, Kanata, IceCube, Swift, and NuSTAR Collaborations) 2018 *Science* **361** eaat1378 (*Preprint* 1807.08816)
[10] Aab A *et al.* (The Pierre Auger Collaboration) 2017 *JCAP* **1704** 038 (*Preprint* 1612.07155)
[11] Buitink S *et al.* 2016 *Nature* **531** 70–73
[12] Marrocchesi P S 2017 *25th European Cosmic Ray Symposium (ECRS 2016)* (*Preprint* 1704.00304)
[13] Alves Batista R *et al.* 2016 *Journal of Cosmology and Astroparticle Physics* **05** 038 (*Preprint* 1603.07142)
[14] Free Software Foundation 2007 GNU general public license URL https://www.gnu.org/licenses/gpl.html
[15] Merten L, Becker Tjus J, Fichtner H, Eichmann B and Sigl G 2017 *JCAP* **1706** 046 (*Preprint* 1704.07484)
[16] Winchen T and Buitink S 2018 *Astroparticle Physics* **102** 25
[17] Kotera K and Olinto A V 2011 *Annual Review of Astronomy and Astrophysics* **49** 119–153 (*Preprint* 1101.4256)
[18] Allard D, Busca N, Decerprit G, Olinto A V and Parizot E 2008 *Journal of Cosmology and Astroparticle Physics* **10** 033 (*Preprint* 0805.4779)

[19] Aloisio R, Berezinsky V and Blasi P 2014 *Journal of Cosmology and Astroparticle Physics* **2014** 020
[20] Unger M, Farrar G R and Anchordoqui L A 2015 *Phys. Rev.* **D92** 123001 (*Preprint* 1505.02153)
[21] Ghosh A *et al.* 2017 *Proceedings of the 35th International Cosmic Ray Conference (ICRC)*
[22] Lee S 1998 *Phys.Rev.* **D58** 043004 (*Preprint* astro-ph/9604098)
[23] Settimo M and De Domenico M 2015 *Astropart.Phys.* **62** 92–99 (*Preprint* 1311.6140)
[24] Heiter C, Kuempel D, Walz D and Erdmann M 2018 *Astropart. Phys.* **102** 39–50 (*Preprint* 1710.11406)

IOP Conf. Series: Journal of Physics: Conf. Series **1181** (2019) 012035 doi:10.1088/1742-6596/1181/1/012035

TeV-PeV cosmic-ray anisotropy and local interstellar turbulence

Gwenael Giacinti and John G. Kirk
Max-Planck-Institut für Kernphysik, Postfach 103980, D-69029 Heidelberg, Germany
E-mail: Gwenael.Giacinti@mpi-hd.mpg.de

Abstract. We calculate the shape of the large-scale anisotropy of TeV–PeV cosmic-rays (CR) in different models of the interstellar turbulence. In general, the large-scale CR anisotropy (CRA) is not a dipole, and its shape can be used as a new probe of the turbulence. The 400 TeV and 2 PeV data sets of IceTop can be fitted with Goldreich-Sridhar turbulence and a broad resonance function, but other possibilities are not excluded. We then present our first numerical calculations of the CRA down to 3 TeV energies in 3D isotropic Kolmogorov turbulence. At these low energies, the large-scale CRA aligns well with the direction of local magnetic field lines around the observer. In this type of turbulence, the CR intensity is flat in a broad region perpendicular to field lines. Even though the CRA is quite gyrotropic, we show that the local configuration of the turbulence around the observer does result in the appearance of weak, "non-gyrotropic" small-scale anisotropies, which contain information on the local turbulence level.

1. Introduction

The flux of TeV–PeV cosmic-rays (CR) at Earth is anisotropic at a level of $\sim 10^{-3}$. See e.g. [1, 2] for reviews. The direction of the large-scale CR anisotropy (CRA) is broadly consistent with that of the local interstellar magnetic field [3], and its shape cannot be described by a pure dipole [4]. For instance, IceCube and IceTop data at ≥ 100 TeV energies hint at a flattening of the CR intensity in directions perpendicular to local magnetic field lines [5]. In the present work, we first report on how the shape of the large-scale CRA depends on the properties of the local interstellar turbulence within about a CR mean free path from Earth. We then illustrate and test the main assumptions of our theory [5] by extending the numerical simulations of Ref. [6] down to TeV energies: we calculate the CRA in different realizations of 3D isotropic Kolmogorov turbulence for a realistic ratio of the CR gyroradius to the turbulence coherence length.

2. Large-scale CRA and local interstellar turbulence

The gyroradius of TeV–PeV CR ($\sim 10^{-4} - 10^{-1}$ pc) is substantially smaller than the typical coherence length of the interstellar turbulence $l_c \sim 1 - 10$ pc [7], and CR are expected to diffuse preferentially along magnetic field lines. The fact that the observed CRA points in the direction of the local magnetic field [3] corroborates this expectation. Assuming that CR undergo pitch-angle diffusion in a 1D magnetic flux tube of length $d \leq l_c$, which contains the Earth, and assuming that the problem is stationary, we have shown in Ref. [5] that the CRA at Earth is proportional to:

$$g(\mu) = \frac{\int_0^\mu \mathrm{d}\mu' \left(1-\mu'^2\right)/D_{\mu'\mu'}}{\int_0^1 \mathrm{d}\mu' \left(1-\mu'^2\right)/D_{\mu'\mu'}}, \tag{1}$$

where $\mu = \cos\theta$, θ is the pitch-angle (angle between the direction of the ordered magnetic field and the CR momentum), and $D_{\mu\mu}$ is the pitch-angle diffusion coefficient, assumed to be homogeneous in the flux tube. Eq. (1) is valid as long as the CR mean free path is smaller than a few times d, see [5] for more details. The amplitude of the CRA depends on the *a priori* unknown value of the CR flux in the magnetic flux tube, but its shape does not. Therefore, we work with $g(\mu)$ which corresponds to the CRA with its amplitude renormalized to 1. Eq. (20) of Ref. [5] expresses $D_{\mu\mu}$ as a function of a resonance function R_n, for which we try the two following functions. One with a narrow, "N", and one with a broad, "B", resonance:

$$R_n^{\mathrm{N}} = \frac{\tau^{-1}}{(k_\parallel v_\parallel - \omega + n\Omega)^2 + \tau^{-2}}, \quad \text{and} \quad R_n^{\mathrm{B}} = \frac{\sqrt{\pi}}{\left|k_\parallel\right| v_\perp \delta\mathcal{M}_{\mathrm{A}}^{1/2}} \exp\left(-\frac{(k_\parallel v_\parallel - \omega + n\Omega)^2}{k_\parallel^2 v_\perp^2 \delta\mathcal{M}_{\mathrm{A}}}\right), \tag{2}$$

where $k_\parallel$ is the parallel component of the wavevector, $v_\parallel = c\mu$, $v_\perp = c\sqrt{1-\mu^2}$, ω is the angular frequency of the waves, Ω the CR gyrofrequency, and $n = 0, \pm1$ in our calculations. For R_n^{N}, the broadening of the resonance is assumed to be dominated by the Lagrangian correlation time of the turbulence, τ, and for R_n^{B}, by fluctuations of the parallel magnetic field strength, which is encapsulated in the parameter $\delta\mathcal{M}_{\mathrm{A}} < 1$. We test two models of turbulence: Fast magnetosonic mode turbulence with an isotropic power spectrum $I_{\mathrm{F}}(k) \propto k^{-3/2}$ [8] and Goldreich-Sridhar turbulence whose Alfvén and pseudo-Alfvén modes have the power spectrum $I_{\mathrm{A,S}}(\mathbf{k}) \propto k_\perp^{-10/3} \exp(-k_\parallel l^{1/3}/k_\perp^{2/3})$ [9], where l denotes the outer scale of the turbulence.

In Fig. 1, we present our calculations of the dimensionless pitch-angle scattering rate $\nu(\mu) = 2D_{\mu\mu}/(1-\mu^2) \times (l/c)$ (upper left panel), and normalized large-scale CRA, $g(\mu)$ (lower left panel), for these models. The dashed lines are for fast mode turbulence, and the solid ones for Goldreich-Sridhar turbulence. In the middle column of Fig. 1, we plot the relative CR intensity versus right ascension in the field of view of IceTop experiment, which observes part of the Southern hemisphere, and compare with its 2 PeV (top, green boxes) and 400 TeV (bottom, blue boxes) data [4]. The dashed purple lines correspond to fast mode turbulence with the narrow resonance function R_n^{N}, and with τ calculated for an Alfvén velocity equal to $u_{\mathrm{A}} = 10\,\mathrm{km/s}$ — see [5] for more details. We set here the dimensionless CR rigidity to $\epsilon = c/(l\Omega) = 10^{-3}$. For this turbulence, ϵ only changes the normalization of $\nu(\mu)$, and not the shape of the CRA, $g(\mu)$. As can be seen in the upper left panel, ν has a very narrow peak around $\mu = 0$, which means that pitch-angle scattering is strongly enhanced for CR whose momenta are almost perpendicular to the local coherent field. This peak is due to the $n = 0$ term in the expression for $D_{\mu\mu}$. At small $|\mu|$, and on both sides of the peak, ν goes through a minimum and recovers at $\mu \to \pm1$. Since the derivative of $g(\mu)$ is proportional to $1/\nu$ (cf. Eq. (1)), this results in a CRA that strongly varies at small values of $|\mu|$, and has broad, almost flat minima and maxima at $|\mu| \geq 0.5$, see the lower left panel. For comparison, the thin black dotted line shows $g(\mu) = \mu$ which corresponds to a dipole anisotropy. The anisotropy for the purple line is clearly not a dipole. For a better visualisation, we plot it in equatorial coordinates in the upper right panel of Fig. 1, assuming that its direction is given by that of the local coherent field measured by [10, 11]. By comparing with Fig. 6 of Ref. [4], one can see by eye that the shape of the CRA in the Southern hemisphere is very different from that measured by IceTop. This can also be seen in the upper middle panel, where we compare the resulting CR intensity with IceTop 2 PeV data. In general, we find that models with R_n^{N} provide a bad match to the observations, which apparently require a flattening of g in a moderately broad region around $\mu = 0$, and hence a broad "bump" in ν around $\mu = 0$. R_n^{B} gives a better fit to the data: we show with dashed blue lines the results for fast modes

IOP Conf. Series: Journal of Physics: Conf. Series **1181** (2019) 012035 doi:10.1088/1742-6596/1181/1/012035

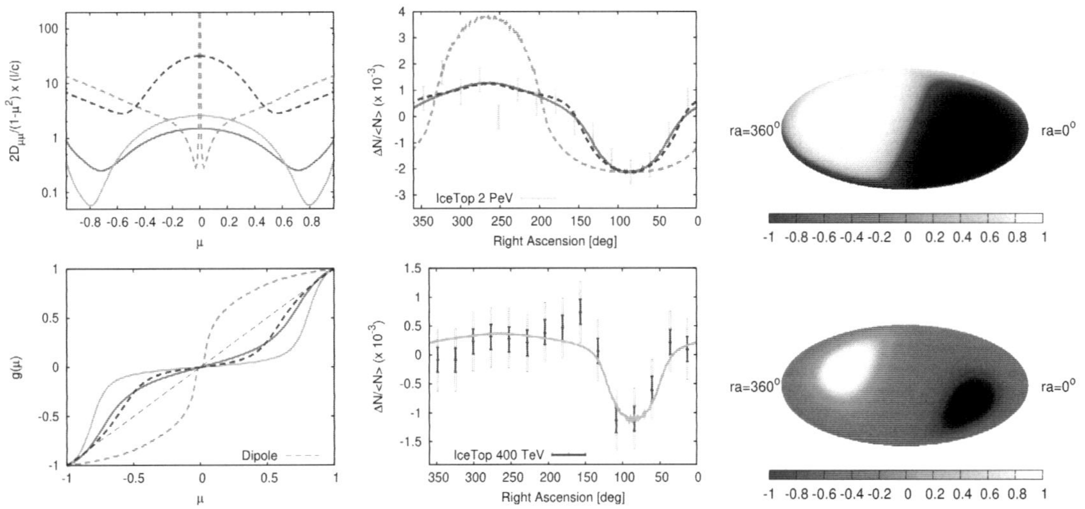

Figure 1. Pitch-angle scattering rate $\nu(\mu)$ (upper left), anisotropy $g(\mu)$ (lower left), and relative CR intensity versus right ascension in the field of view of IceTop (middle column), for three models of the interstellar turbulence: Fast mode turbulence with $\epsilon = 10^{-3}$ and with either $R_n^{\rm N}$ and $u_{\rm A} = 10\,{\rm km/s}$ (dashed purple lines) or $R_n^{\rm B}$ and $\delta M_{\rm A} = 0.1$ (dashed blue lines); Goldreich-Sridhar turbulence with $R_n^{\rm B}$, $\delta M_{\rm A} = 0.33$, and $\epsilon = 10^{-2}$ (solid red lines) or $\epsilon = 10^{-3}$ (solid orange lines). Also shown are the 2 PeV (upper middle) and 400 TeV (lower middle) data taken from Ref. [4]. Right column: CRA in equatorial coordinates for the dashed purple lines (upper panel), and the solid orange ones (lower panel).

with $R_n^{\rm B}$ and $\delta M_{\rm A} = 0.1$. ν has a broad bump at $|\mu| < 0.5$ and minima at $\mu \simeq \pm 0.5$. This results in a CRA with a broad flat region at $|\mu| < 0.5$ and with two extrema whose angular half-widths are smaller than those of a dipole, see the lower left panel. This model provides a good fit to the 2 PeV IceTop data, with a flat CR intensity at RA $\geq 160°$, and a small cold spot around RA $\simeq 80 - 90°$. However, the smaller angular size of the cold spot in the 400 TeV data set cannot be well fitted with this model. If this change in the size of the cold spot with CR energy is real, this may point at an anisotropy in the power spectrum of the turbulence: If the anisotropy in Fourier space varies with $|\mathbf{k}|$, then CR with different energies interact with modes whose level of anisotropy is different. Goldreich-Sridhar turbulence is anisotropic, and the level of anisotropy of its modes in $\mathbf{k}$-space varies with $|\mathbf{k}|$. Solid lines show results for this turbulence with $R_n^{\rm B}$ and $\delta M_{\rm A} = 0.33$. The red lines are for $\epsilon = 10^{-2}$, and the orange ones are for $\epsilon = 10^{-3}$. The ratio of these two values of ϵ is similar to the ratio of the energies of the two IceTop data sets — the factor 2 difference does not affect our conclusions and is only due to the binninig in energy we chose for our scan of the parameter space. The scattering rate (upper left) is lower than for fast mode turbulence, but also presents broad maxima around $\mu = 0$. The maximum is broader for $\epsilon = 10^{-3}$ than for $\epsilon = 10^{-2}$, and the CRA (lower left) flattens at $|\mu| < 0.7$ for $\epsilon = 10^{-3}$ and at $|\mu| < 0.5$ for $\epsilon = 10^{-2}$. The lower right panel shows the CRA for $\epsilon = 10^{-3}$ in equatorial coordinates. A small hot spot and a small cold spot are visible in the direction of the coherent magnetic field ($\mu = \pm 1$). They are separated by a wide magenta region, where the CR intensity is very flat. The two data sets from IceTop can be well fitted with this model of turbulence: The calculations for $\epsilon = 10^{-2}$ are compatible with the shape of the measured CR intensity at 2 PeV (see the red line in the upper middle panel), and $\epsilon = 10^{-3}$ fits well the 400 TeV data (see the orange line in the lower middle panel).

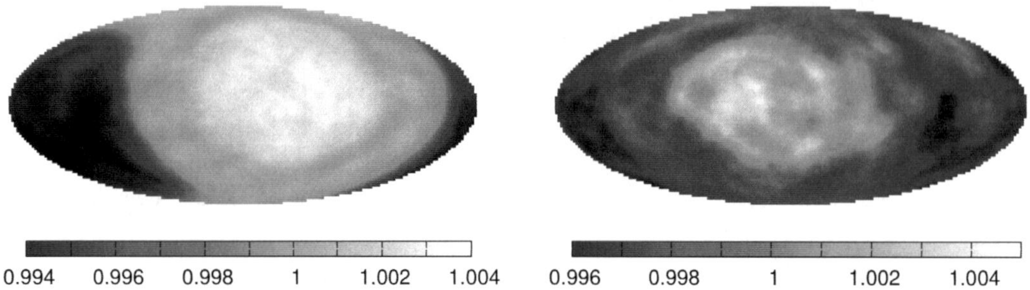

Figure 2. Numerical calculations of the relative CR flux at 3 TeV in two different realizations of 3D isotropic Kolmogorov turbulence with $B_{\rm rms} = 4\,\mu$G and $l = 150\,$pc. The CR intensity in these maps is averaged over 10°-radius circles.

3. Numerical simulations

In the above analytical calculations, we made the assumption of gyrotropy. In this section, we investigate numerically the validity of this assumption. We propagate individual CR in different realizations of 3D isotropic Kolmogorov turbulence and calculate the CRA, using the backtracking method described in Ref. [6]. We consider here, for the first time, CR with energies as low as 3 TeV. This is more than 3 orders of magnitude smaller than the energies probed in Ref. [6], and, therefore, requires substantially longer calculation times. This value of 3 TeV is in the relevant range for a direct comparison with the low energy measurements of the CRA at TeV energies. We note that, at even lower CR energies, time variations of the local turbulence would start to make small-scale features of the CRA vary on time scales smaller than the typical lifetime of an experiment, assuming local fluid and Alfén velocities of a few tens of km/s. We do not investigate this regime here. We use turbulence with root-mean-square strength $B_{\rm rms} = 4\,\mu$G, and outer scale $l = 150\,$pc, which corresponds to $l_c = 30\,$pc. The only difference between our numerical technique here and that of Ref. [6] is that we propagate here CR on finite distances ct (as in Ref. [12]), instead of stopping the trajectories on a sphere with a fixed radius. This difference is unimportant for the purpose of the present study. We propagate CR on distances of a few hundreds of pc, until the CRA converges. The angular shape of the CRA converges quickly, after about a CR mean free path. However, its absolute amplitude converges only after the CR have probed distances greater than a few coherence lengths of the turbulence. For shorter distances, the centre of mass of a set of initially nearby trajectories still strongly depends on the bending of local magnetic field lines around the observer.

In Fig. 2, we show two sky maps of the relative CR flux (normalized to the angle-averaged flux) in two different realizations of the turbulence, and averaged over 10°-radius circles. As expected, the direction of the large-scale anisotropy is found to point in the direction of local magnetic field lines around the observer in both simulations. We do not add any regular field to the turbulence: the local coherent field is provided by the modes whose wavelengths are much larger than the CR gyroradius. The two panels in Fig. 2 show two limiting cases. In the left panel, the local turbulence level on the scale of a 3 TeV CR gyroradius, and within a CR mean free path from the observer, is quite small in this realization. The anisotropy looks very smooth and almost perfectly gyrotropic. In the right panel, the local turbulence level happens to be higher in that realization. The anisotropy looks less regular, with non-gyrotropic small-scale fluctuations that are well above the numerical noise level. This confirms that the mechanism suggested in [6] for creating small-scale anisotropies also works at energies of a few TeV. In all these simulations, the amplitude of "non-gyrotropic" small-scale anisotropies is substantially smaller than that of the large-scale anisotropy. This is in line with observations, which suggest that the amplitude of small-scale anisotropies is about an order of magnitude smaller than that of the large-scale anisotropy. This also justifies the assumption of gyrotropy in the calculations of $g(\mu)$ in the previous section.

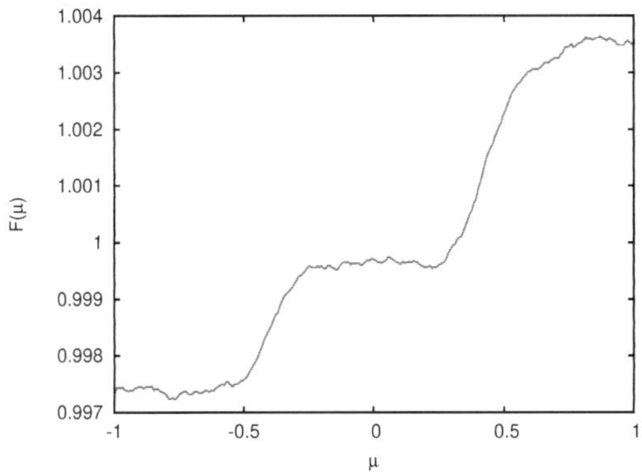

Figure 3. Numerical calculation of the gyrophase-averaged relative CR flux $F(\mu)$, at 3 TeV, for an observer in a given realization of 3D isotropic Kolmogorov turbulence with $B_{\rm rms} = 4\,\mu$G and $l = 150$ pc.

In Fig. 3, we plot our calculation of the gyrophase-averaged relative CR flux, F, versus μ for one of our simulations. The shape of $g(\mu)$ ($\propto F(\mu) - 1$) shows some variations from one realization of the turbulence to another, but all cases we tested show a flattening of $g(\mu)$ at $|\mu| < 0.25 - 0.5$. This tends to suggest that the CR pitch-angle scattering rate in 3D isotropic Kolmogorov turbulence may too have a broad bump around $\mu = 0$.

4. Discussion

Both IceCube [13] and HAWC [14] have provided measurements of the angular power spectrum of the CRA. We have shown in Section 2 that the large-scale CRA $g(\mu)$ is not a pure dipole, except in the unphysical case of isotropic CR pitch-angle scattering where $D_{\mu\mu} \propto 1 - \mu^2$. Therefore, a fraction of the power in the C_ℓ's, including those with $\ell \geq 2$, should be due to multipoles pointing roughly in the direction of local magnetic field lines. Among the several small-scale anisotropies detected by HAWC, one of them does not seem to vary substantially with CR energy: A small hot spot, located roughly in the direction of RA $\sim 60^\circ$ and slightly below dec $= 0^\circ$, remains at almost the same place on the sky from 2.0 TeV to 72.8 TeV, see Figs. 10 and 9 in Ref. [14]. When comparing with the Figs. 4 and 5 of Ref. [14], it seems that the direction of this hot spot is quite roughly compatible with that of the large-scale anisotropy. We suggest here that this hot spot may be an additional sign that the "large-scale" anisotropy pointing in the direction of magnetic field lines deviates from a dipole and contains high-order multipoles. This would naturally account for the apparent stability of this spot with energy. From a conceptual point of view, this suggestion and our study in Section 2 contain similarities with the suggestion of Ref. [15] for the origin of one of Milagro's hotspots, "Region A". However, our study differs in particular by the fact that we have used here the full general solution of the CR transport equation. We also note that the maximum significance of Milagro's Region A is around dec $\simeq 10^\circ - 20^\circ$ (see Fig. 1 of Ref. [16]), which is $\simeq 20^\circ$ away from the centre of HAWC's hot spot, and that Region A does not coincide with the direction of the maximum of the large-scale anisotropy as is visible in Fig. 2 of Ref. [16]. Some of these apparent differences may just be due to the different exposures of these experiments and data analysis techniques. A deeper study of these points would nonetheless be necessary in order to reach a firm conclusion on the origin of this particular hot spot in HAWC data.

The other small-scale anisotropies detected by HAWC, as well as those detected by IceCube, are not aligned with the large-scale anisotropy and require another explanation. They imply that the CR distribution at Earth is not perfectly gyrotropic. The mechanism discussed in Section 3, and initially presented in Ref. [6], provides a natural explanation for them. Below a few TeV, heliospheric magnetic fields may too generate small-scale anisotropies for similar reasons, see e.g. Ref. [17]. In any case, the fact that "non-gyrotropic" small-scale anisotropies have been detected at 20 TeV energy by IceCube [13] suggests that the mechanism of Ref. [6] should be at work, at least above several TeV.

5. Conclusions

Assuming pitch-angle diffusion of TeV–PeV CR in our local interstellar medium, we show that their large-scale anisotropy at Earth, $g(\mu)$, is in general not a dipole, see Section 2. $g(\mu)$ contains information on the local interstellar turbulence and CR propagation properties. Moderately broad resonance functions are favoured, and the 2 PeV data set of IceTop can be fitted with isotropic fast modes or Goldreich-Sridhar turbulence. Thanks to its $|\mathbf{k}|$-dependent power spectrum, the latter type of turbulence can also explain the change in shape of the CRA between the 400 TeV and the 2 PeV data sets of IceTop. In Section 3, we present our first numerical calculations of the CRA in 3D isotropic Kolmogorov turbulence down to 3 TeV. We find that $g(\mu)$ has a flattening in directions around $\mu = 0$ in this type of turbulence too. At these low energies, the CRA aligns well with the direction of local magnetic field lines around the observer and is quite gyrotropic. Weak, "non-gyrotropic" small-scale anisotropies do nonetheless appear due to the local configuration of the turbulence around the observer at the time of the observations, as initially suggested in Ref. [6]. Their amplitude, which is always much smaller than that of the large-scale anisotropy $g(\mu)$, is connected to the turbulence level on resonant scales in our local magnetic flux tube.

Acknowledgments

This research was supported by a Grant from the GIF, the German-Israeli Foundation for Scientific Research and Development. GG also acknowledges useful discussions at the team meeting "The Physics of the Very Local Interstellar Medium and Its Interaction with the Heliosphere", led by J. Giacalone and J. R. Jokipii, and supported by the International Space Science Institute (ISSI) in Bern, Switzerland.

References

[1] Di Sciascio G and Iuppa R 2014 *ArXiv e-prints* (*Preprint* arXiv:1407.2144)
[2] Ahlers M and Mertsch P 2017 *Prog. Part. Nucl. Phys.* **94** 184–216 (*Preprint* 1612.01873)
[3] Schwadron N A, Adams F C, Christian E R, Desiati P, Frisch P, Funsten H O, Jokipii J R, McComas D J, Moebius E and Zank G P 2014 *Science* **343** 988–990
[4] Aartsen M G *et al* 2013 *Astrophys. J.* **765** 55 (*Preprint* 1210.5278)
[5] Giacinti G and Kirk J G 2017 *Astrophys. J.* **835** 258
[6] Giacinti G and Sigl G 2012 *Phys. Rev. Lett.* **109** 071101 (*Preprint* 1111.2536)
[7] Haverkorn M, Brown J C, Gaensler B M and McClure-Griffiths N M 2008 *Astrophys. J.* **680** 362-370 (*Preprint* arXiv:0802.2740)
[8] Cho J and Lazarian A 2002 *Phys. Rev. Lett.* **88** 245001 (*Preprint* astro-ph/0205282)
[9] Cho J, Lazarian A and Vishniac E T 2002 *Astrophys. J.* **564** 291–301 (*Preprint* astro-ph/0105235)
[10] Frisch P C *et al* 2012 *Astrophys. J.* **760** 106 (*Preprint* 1206.1273)
[11] Frisch P C *et al* 2015 *Astrophys. J.* **814** 112 (*Preprint* 1510.04679)
[12] Ahlers M and Mertsch P 2015 *Astrophys. J. Lett.* **815** L2 (*Preprint* 1506.05488)
[13] Aartsen M G *et al* 2016 *Astrophys. J.* **826** 220
[14] Abeysekara A U *et al* 2018 *Astrophys. J.* **865** 57 (*Preprint* 1805.01847)
[15] Malkov M A, Diamond P H, O'C Drury L and Sagdeev R Z 2010 *Astrophys. J.* **721** 750–761 (*Preprint* 1005.1312)
[16] Abdo A A *et al* 2008 *Phys. Rev. Lett.* **101** 221101 (*Preprint* 0801.3827)
[17] Zhang M, Zuo P and Pogorelov N 2014 *Astrophys. J.* **790** 5

IOP Conf. Series: Journal of Physics: Conf. Series **1181** (2019) 012036 doi:10.1088/1742-6596/1181/1/012036

Observation of cosmic ray anisotropy with GRAPES-3 Experiment

P K Mohanty[1], S Ahmad[5], A Chandra[5], S R Dugad[1], S K Gupta[1], B Hariharan[1], Y Hayashi[2], P Jagadeesan[1], A Jain[1], V B Jhansi[1], S Kawakami[2], H Kojima[4], S D Morris[1], P K Nayak[1], A Oshima[3], P Rakshe[1], K Ramesh[1], B S Rao[1], L V Reddy[1], S Shibata[3], M Zuberi[1]

[1] Tata Institute of Fundamental Research, Homi Bhabha Road, Mumbai 400005, India
[2] Graduate School of Science, Osaka City University, Osaka 558-8585, Japan
[3] College of Engineering, Chubu University, Kasugai, Aichi 487-8501, Japan
[4] Faculty of Engineering, Aichi Institute of Technology, Toyota City, Aichi 470-0392, Japan
[5] Aligarh Muslim University, Aligarh 202002, India

E-mail: pkm@tifr.res.in

Abstract.
Large scale anisotropy of cosmic rays (CRs) in the TeV-PeV energy region has been reported by several experiments located in the Northern Hemisphere as well as by the IceCube, and its surface array IceTop at the South Pole. The GRAPES-3 experiment in Ooty, India is designed to observe CRs in the energy range from a few TeV to over 10 PeV. Its near-equatorial location (11.4°N) provides a unique advantage for observations of CR anisotropy, overlapping with experiments in both the hemispheres. The GRAPES-3 consists of a dense array of ~400 plastic scintillator detectors spread over an area of 25000 m^2. It also contains a large area (560 m^2) tracking muon detector which provides an excellent capability to discriminate γ-rays against CRs. It has recorded over 10 billion showers since its operation began in 2000. Measurement of CR anisotropy with a subset of this data is presented here.

1. Introduction

The long standing issue of CR origin may be attributed to the fact that the CRs are charged particles because of which their paths undergo deflection and directions are randomized by the magnetic field in the interstellar medium through which they propagate. This leads to an isotropic distribution of the arrival directions of CRs observed on Earth making it impossible to trace their sources. CRs below 10^{17} eV are considered to be Galactic origin and are mostly believed to be produced and accelerated by shocks in supernova remnants [1]. The galactic magnetic field is $\sim 3\,\mu$G which contains a regular and a turbulent component where the latter is believed to have a scale-size much larger than the Larmor radii for CRs below 10^{17} eV. However, during the past one decade, several ground based air shower experiments have observed a small anisotropy in the arrival directions of CRs at the level of 10^{-4} to 10^{-3} in the energy range of TeV - PeV. The excess in CR flux is observed in a region 40° - 90° in right ascension and deficit in a region 150° - 240° in right ascension [2]. Thanks to large statistics collected by these experiments with duty cycle close to 100% and a field of view covering the full sky which has made it possible to detect this tiny effect. The cause of anisotropy may be understood to be due to the presence of an ordered magnetic field in our local environment with a strength at the level of 3μG which induces a circular motion of CRs around magnetic field lines with a Larmor radius $r_L \simeq 0.4E_{PeV}/(ZB_{3\mu G})$ pc. This length scale is much smaller than the typical scattering length for CRs in the TeV - PeV energies predicted by isotropic diffusion models which may indicate a strong anisotropic diffusion in our local environment [3]. Thus, the study of the CR anisotropy may provide important information on their source distribution as well as the structure of the magnetic field.

Large scale CR anisotropy in TeV - PeV energies has been measured by several air shower experiments including Tibet ASγ [4], ARGO-YBJ [5], Milagro [6], HAWC [7], EAS-TOP [8], IceCube/IceTop [9, 10]. Each of these experiments is sensitive in a different energy range of CR spectrum. A recent compilation of CR anisotropy results shows that the amplitude is energy dependent and phase flip occurs around 100 TeV [7]. Two dimensional mapping of CR anisotropy in right ascension and declination coordinates has been successfully carried out by several of these experiments due to the high statistics data collected by them. Being located at the south pole, the IceCube and IceTop are sensitive to CR anisotropy in the southern hemisphere while the rest are located in the northern hemisphere. The fortuitous location of GRAPES-3 air shower array at Ooty in India close to the equator (11.4°N) has the unique advantage of observing CR anisotropy with overlapping fields of view of detectors with both the hemispheres. The GRAPES-3 also could provide a good overlap of CR anisotropy measurement in energy as it is sensitive from a few TeV to well above PeV.

2. The GRAPES-3 experiment

The GRAPES-3 experiment comprises a dense array of scintillator detectors with a coverage area of 25,000 m^2. It is currently operating with 400 scintillator detectors of 1 m^2 area each with 8 m inter-detector separation arranged in a hexagonal configuration [11, 12]. Each detector is designed to measure densities and arrival times of shower particles to obtain energy and incident direction of primary CRs. It began operating in year 2000 with a modest array of 217 plastic scintillator detectors of a traditional type [11]. It was subsequently upgraded and expanded with a vastly superior design based on in-house produced plastic scintillator and wavelength shifting fiber obtained from Kuraray, Japan [12]. The trigger rate of the array is currently 35 Hz which is three times higher compared to the rate prior to the upgrade. The PMT signal from each detector is digitized using a charge integrating CAEN ADC and the relative arrival times are digitized using an in-house developed 32-channel TDC module based on a CERN HPTDC chip [13]. The other major component of the GRAPES-3 experiment is a large area (560 m^2) tracking muon detector that consists of 16 separate modules of 35 m^2 area each [14]. Each module consists of 4 layers of proportional counters (PRC) made of 6 m long and 0.1 m$\times$0.1 m square cross section steel tubes to track individual muons that survive after passing through a concrete absorber of 550 gm cm^{-2} which filters out the electromagnetic components and sets a threshold of 1 GeV for vertically incident muons. The muon detector provides an excellent discrimination of γ-rays against charged CRs enabling the study of multi TeV γ-ray sources. The muon component provides a sensitive measurement of CR composition [15]. The GRAPES-3 has recorded over 10 billion showers since its operation began in 2000. A recent detail simulation study has shown that 50% of showers recorded by the array are below 10 TeV.

3. Analysis and discussion

In this report, the analysis of data carried out for the period 2000-2006 comprising 1.9 billion showers is presented. The shower direction (zenith angle θ and azimuth angle ϕ) was determined using a plane fit to the relative arrival times of shower front at various detectors after correcting for the shower front curvature with a slope of 215 ps m^{-1}. Details of the direction reconstruction and angular resolution study are discussed elsewhere [16]. The study has found that the angular resolution of the GRAPES-3 array is 1.4° at 10 TeV which improves to 0.4° at 80 TeV. The angular resolution obtained using three different methods namely left-right, even-odd and moon-shadow have provided similar results. Various other shower parameters such as core location (X_c, Y_c), age 's' and size 'N_e' were obtained by fitting NKG function to the lateral distribution of particle densities through a minimization algorithm using MINUIT which is discussed elsewhere [15].

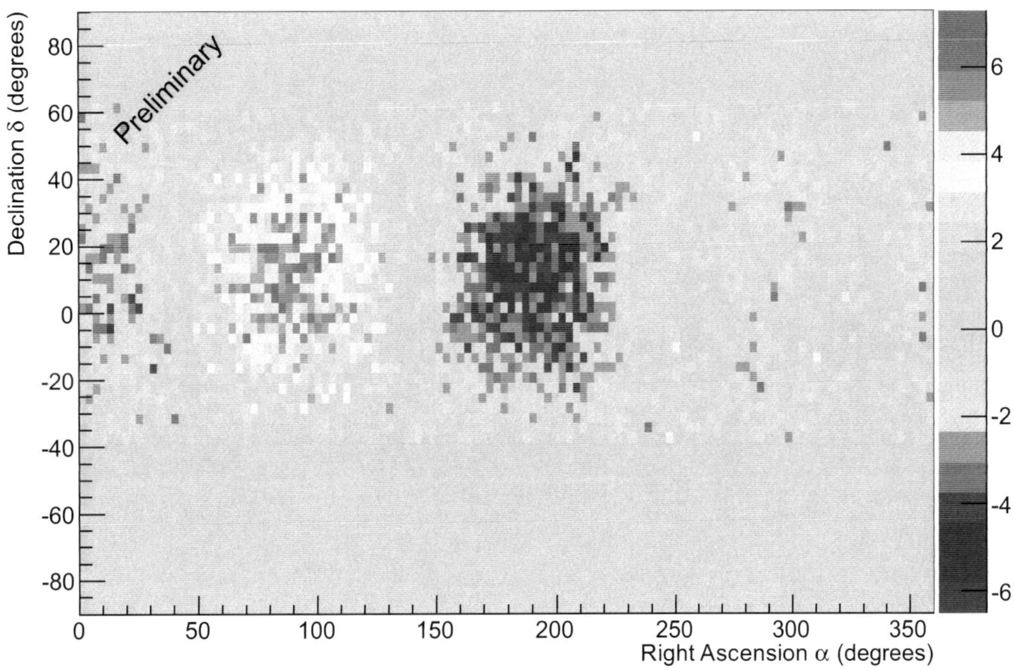

Figure 1. Statistical significance map of anisotropy measured by GRAPES-3

The arrival direction of showers (θ, ϕ) were converted into equatorial coordinates, right ascension (α) and declination (δ) and then binned into cells of size 3° ×3° in α and δ where α is from 0° to 360° and δ is from -90° to 90°. Events in each cell were normalized with the average value for the corresponding declination band. The anisotropy was calculated by subtracting a dummy cosmic ray distribution from the real cosmic ray distribution where the dummy distribution was produced using the real data by shuffling event time. The statistical significance value for each bin was calculated using Li Ma formula [17]. A 2-D sky map of the significance without applying any smoothing is shown in Fig.1. The significance of same data in three different declination bands for δ between +40° and -20° is shown in Fig.2.

The significance map in Fig.1 indicates an excess in a broad region of α=50°-130° and a deficit in α=150°-230°. The anisotropy is clearly visible over δ=+40° in northern hemisphere and δ=-30° in southern hemisphere. The anisotropy shown in Fig.2 in three different declination bands shows the excess about α=90° and deficit about α=190°. It is to be noted that the GRAPES-3 is located at latitude of 11.4°N and beyond δ=+40° and δ=-40°, the anisotropy is not significant due to poor statistics with the subset of the data used in this analysis.

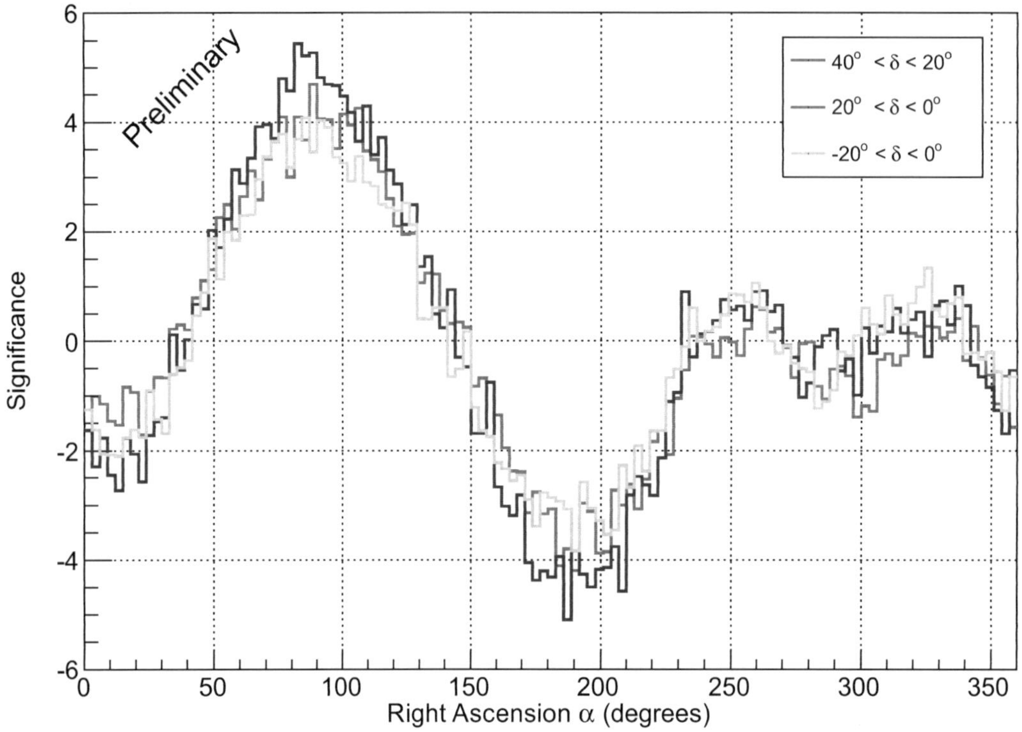

Figure 2. Statistical significance of anisotropy in different declination bands

4. Summary

Preliminary results of anisotropy measured with a subset of the GRAPES-3 data from 2000 to 2006 are presented in this paper. The results indicate that the GRAPES-3 could observe the anisotropy extending to both the hemispheres due to its near equatorial location. An on-going analysis effort with higher statistics and a better understanding of various systematics is expected to complement the observations from both the hemisphere experiments.

4.1. Acknowledgments

We thank D.B. Arjunan, G.P. Francis, V. Jeyakumar, S. Kingston, K. Manjunath, S. Murugapandian, S. Pandurangan, B. Rajesh, K. Ramadass, V. Santoshkumar, M.S. Shareef, C. Shobana, R. Sureshkumar and other colleagues for their help in running and maintenance of the GRAPES-3 experiment.

References

[1] Blasi P 2013 *Astron. Astrophys. Rev.* **21** 70
[2] Sciascio G Di and Iuppa R *arXiv:1407.2144*
[3] Ahlers M 2016 *Phys. Rev. Lett.* **117** 151103
[4] Amenomori M et al. 2017 *Astrophys. J.* **836** 153
[5] Bartoli B et al. 2015 *Astrophys. J.* **809** 90
[6] Abdo A A et al. 2008 *Phys. Rev. Lett.* **101** 221101
[7] Abeysekara A U et al. 2018 *Astrophys. J.* **865** 57
[8] Aglietta M et al. 2009 *Astrophys. J. Lett.* **692** L130

[9] Aartsen M et al. 2013 *Astrophys. J.* **765** 55
[10] Aartsen M et al. 2016 *Astrophys. J.* **826** 220
[11] Gupta S K et al. 2005 *Nucl. Instrum. Meth. A* **540** 311-23
[12] Mohanty P K et al. 2009 *Astropart. Phys.* **31** 24-36
[13] Gupta S K et al. 2012 *Expt. Astron.* **35** 507-26
[14] Hayashi Y et al. 2005 *Nucl. Instrum. Meth. A* **545** 643-57
[15] Tanaka H et al. 2012 *J. Phys. G: Nucl. Part. Phys.* **39** 025201
[16] Oshima A et al. 2010 *Astropart. Phys.* **33** 97-107
[17] LI T & MA Y 1983 *Astrophys. J.* **272** 317-324

Puzzles of the Cosmic Ray Anisotropy

A D Erlykin[1], S K Machavarian[1] and A W Wolfendale[2]
[1]P.N.Lebedev Physical Institute, Leninsky prosp. 53, Moscow, Russia
[2]Physics Dept., Durham Univ., Durham DH1 3LE, UK

E-mail: erlykin@sci.lebedev.ru

Abstract. We discuss three of the known puzzles of the cosmic ray anisotropy in the PeV and sub-PeV energy region. They are 1) the so called inverse anisotropy, 2) the irregularity in the energy dependence of the amplitude and phase of the first harmonic and 3) the contribution of the single source.

1. Introduction

One of the most important and simultaneously most difficult studies of the origin of cosmic rays (CR) is the study of their anisotropy. The difficulty is due to the extremely low level of the anisotropy and the steeply falling energy spectrum of CR. In their seminal book 'Origin of Cosmic Rays' issued in 1964 V.L.Ginzburg and S.I.Syrovatskii could only give an upper limit of the anisotropy as $< 1\%$ [1]. More than 20 years later V.L.Ginzburg and his co-authors on the basis of data obtained by Linsley J. could show that the amplitude and phase of the anisotropy are not constant, but vary with the energy [2]. This variability is the consequence of the non-uniformity of the spatial distribution of CR sources and properties of the interstellar medium (ISM). Sources of different locations and ages contribute in different energy regions and magnetic fields of different strength and orientations tend to isotropise the arrival directions of CR particles.

One of the most interesting regions of the CR energy spectrum is sub-PeV and PeV region where the well-known 'knee' is observed. The study of this region is difficult because of the very small intensity of particles and low anisotropy level. That is why there are many unsolved problems and puzzles here which still wait for their solution. Below we describe only three of them.

2. Puzzle 1: an inverse anisotropy

Due to CR scattering on the chaotic magnetic fields their motion in the Galaxy is like a diffusion from regions of higher to lower CR density. Since the solar system is located at about 8 kpc from the galactic center then the main source of CR - supernova remnants (SNR) and pulsars are concentrated in the Inner Galaxy. The CR density is higher towards the Galactic Center and their gradient is directed from the Inner to Outer Galaxy. Hence we have to expect some anisotropy with the maximum flux from the Inner Galaxy.

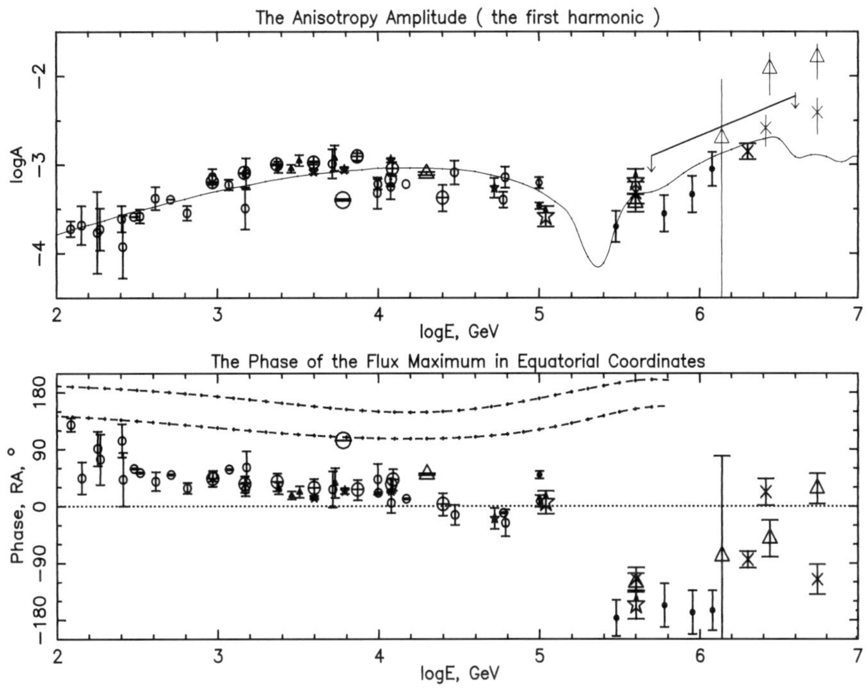

Figure 1. The observed amplitude (upper panel) and equatorial phase (lower panel) of the first harmonic of the CR anisotropy. Data denoted as ○ are taken from the survey presented in [3], ⊙ - Super-Kamiokande-I [3], ⋆ - Tibet III [4], △ - Baksan [5], ◇ - Andyrchi [5], • - EAS-TOP [6]. Thick line above $logE = 5.7$ with an arrow - upper limits of the amplitude, given by KASCADE [7]. Here and in the text below the energy E is in GeV. The full line in the upper panel relates to calculations with our model described in Section 3 [8]. The area between thin dashed lines in the lower panel denotes the region in *galactic* coordinates corresponding to the RA region occupied by experimental points in equatorial coordinates assuming that the observations were made at declinations about 30°-60°.

The expected and observed situation with the first harmonic of the anisotropy in the sub-TeV region is shown in Figure 1. Ihe upper panel shows the amplitudes and it is seen that they do not exceed the value of 10^{-3}. The lower panel presents measurements of the phase in equatorial coordinates. If transferred to galactic coordinates they occupy the region delimited by two dashed lines. It is seen that when the expected phase of the maximum CR flux is from the Inner Galaxy with $\ell \approx 0°$ [8], the observed phase occupies the longitude ℓ-region between 90° and −90°, i.e. CR come preferentially from the Outer Galaxy. We call this phenomenon an 'inverse anisotropy'.

One of the possible explanations of an inverse anisotropy is to assume that it is a local phenomenon caused by a spatial orientation of the magnetic field or the ISM density fluctuation. The Sun is located in the Local Bubble at the inner edge of the Orion spiral arm. The strength of the magnetic field and of the ISM density in the arm is higher than in the interarm region where the Local Bubble is located. The diffusion in the nearby Outer side of the Galaxy is slower than locally. CR moving from the Inner Galaxy meet like a wall, a part of them reflect from it, accumulate in a number and create an inverse gradient [9].

There might be an alternative explanation [10]. Recent measurements of CR at sub-TeV - sub-PeV energies demonstrated the irregular non-power law shape of the energy spectrum.

IOP Conf. Series: Journal of Physics: Conf. Series **1181** (2019) 012037 doi:10.1088/1742-6596/1181/1/012037

As an example the spectral hardening at 200-300 GeVnoticed in CREAM, PAMELA, AMS-2 experiments can be mentioned, the steepening at the magnetic rigidity of 10TV in the NUCLEON experiment and others. The observed irregularities point to the possible role of local sources. However, to be responsible for an observed inverse anisotropy these local sources should be located mostly in the Outer Galaxy.

We should underline that these explanations are only a possible assumption with many internal uncertainties. However, we should emphasize that the inverse anisotropy is most likely a local phenomenon, which is caused by the reflection of CR from a nearby region of higher ISM density or the dominance of some local sources in the Outer Galaxy.

3. Puzzle 2: peculiarity of the amplitude and phase

In general there are several features noticable in Figure 1:
(i) a good consistency of the results at energies up to a few PeV;
(ii) the extremely small $\sim 10^{-4} \div 10^{-3}$ amplitude of the anisotropy;
(iii) the visible rise of the amplitude A with energy E up to $logE \approx 4$;
(iv) a moderate fall of the amplitude above $logE \approx 4$ up to a minimum at $logE \approx 5.3 \div 5.5$;
(v) the rise of the amplitude beyond this minimum up to a few PeV;
(vi) the approximately constant phase at low energies which suddenly changes its direction at about the same energy of $logE \approx 5.3 \div 5.5$ where the amplitude has a minimum;
(vii) in the PeV region, where the rise of the amplitude is observed, the phase has an apparent trend to recover up to its previous direction close to $RA \sim 0$.

In what follows we shall endeavour to build a model which can reproduce these features with the minimum number of assumptions. This model contains three basic ingredients: the Galactic Disk, the Halo and the Single Source (SS). Although we separate here the role of the Single Source, we understand that, in fact, it is just part of CR in the Disk.

The details of the model suggested to explain all these peculiar features are given in [8]. Here we give just its main features and they are illustrated in Figure 2. We assume that the anisotropy appears only in the vicinity of the sources, i.e. in the Disk. The amplitude of the anisotropy A is connected with the CR intensity I, its gradient, $gradI$, and the diffusion coefficient D as $A = \frac{3DgradI}{cI}$. If the relative gradient $\frac{gradI}{I} \neq 0$ and $D \propto E^{\delta_d}$ then A in the Disk rises with energy as $A_d \propto E^{\delta_d}$. The anisotropy of CR in the Halo is postulated as being $A_h = 0$ and the isotropy of CR re-entrant from the Halo back into the Disk dilutes the anisotropy of CR produced and trapped in the Disk.

In this treatment we just consider the first harmonic. Later work will deal with higher multipoles. We calculate the amplitude of the first harmonic for the case where only Disk and Halo contribute to CR as

$$A = \frac{A_d I_d}{I_d + I_h} \tag{1}$$

The result is shown in Figure 2b by the dotted line. The rise of the amplitude at energies above 100 GeV is due to the rise of the diffusion coefficient in the expression for A_d mentioned above. The slow decrease of A above ~10 TeV is explained by the rising fraction of isotropic CR from the Halo, which overcomes the rise of A_d. However, this scenario does not reproduce the remarkable dip in the amplitude visible in the experimental data at $logE = 5.3 - 5.5$ and the subsequent rise of the amplitude above this dip (Figure 1a).

We think that these features, if they are real, are connected with the existence of the Single Source, from which the CR energy spectrum is schematically shown by the dotted line in Figure 2a and denoted as 'SS'. However, this idea alone is not enough to reproduce the experimental data and here the examination of the phase of the first harmonic could help. In Figure 2b it is seen that after the moderate decrease in the energy interval 0.1 - 100 TeV the phase suddenly changes to its opposite. We consider this change seriously and propose that CR from the Single

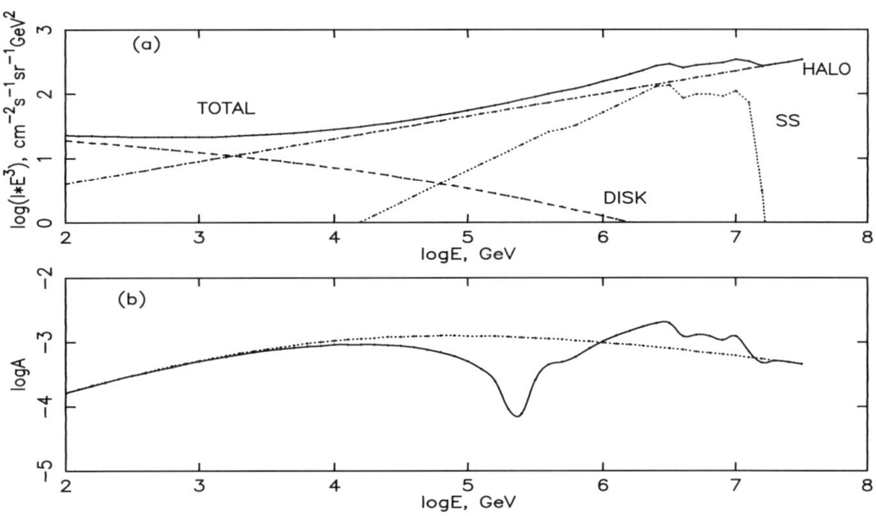

Figure 2. (a) Schematic formation of the CR energy spectrum from the Disk, the Halo and the Single Source. (b) The amplitude of the first harmonic of the CR anisotropy obtained with contributions from Disk + Halo + Single Source (full line) and the same with contributions only from Disk + Halo without Single Source (dotted line).

Source have a phase opposite to that of the background at lower energies. This is a necessary complementary requirement in our model. The rising part of CR coming from the opposite direction would reduce the anisotropy of the background from the Disk and Halo as

$$A = |\frac{A_d I_d - A_{ss} I_{ss}}{I_d + I_h + I_{ss}}| \tag{2}$$

The result of the calculations with the contribution of the Single Source is shown in Figure 2b by the full line. A comparison with the experimental data is shown also in Figure 1a by the full line. It is seen that after minimum in the dip the amplitude of the anisotropy starts rising again and it is caused by the rising contribution of the Single Source, which has the opposite phase.

We think that the described model with three basic ingredients: Disk, Halo and Single Source is reasonable and is worthy of discussion. The new features advocated here are:

(a) The dominance of the Halo component in the sub-PeV region. It means that the CR which we observe and study in spite of being ourselves inside the Disk come mostly from the Halo. It is disputable but helps to understand the low anisotropy, small radial gradient of CR intensity and small level of irregularities in the regular power law energy spectrum.

(b) The idea about the Single Source, which has to be nearby and young and creates the knee, usually raises questions: 'if it is nearby why do'nt we see it in the anisotropy ?'. It is a very reasonable question and this work gives the answer. The Single Source causes the stronger decrease and the dip in the amplitude of the dipole anisotropy at sub-PeV energies. It is also seen in the change of the phase of the anisotropy at the same energies. It means that the Single Source should deliver CR from the direction opposite to the direction of CR from the background and it is a new assumption in the Single Source scenario. Above the dip energy the amplitude starts to rise again with the opposite phase, as expected.

The dip in this model appears as the result of subtraction of two bigger values. Its position and shape are extremely sensitive to the choice of parameters participating in expression (2).

The relatively good agreement with the experimental data is the result of the fitting procedure, but, nevertheless:
(i) it demonstrates the possibility of achieving agreement within the framework of our simplistic model and
(ii) the high sensitivity of the dip to the input parameters of the expression (2) gives the possibility of investigating these parameters when precise results in this energy region are obtained.

(iii) The phase of the first harmonic in the PeV region, where the contribution of the Single Source is big enough, could help to locate it on the sky. The present experimental data have a too big spread to make a conclusion.

We understand that this scenario raises more questions than gives answers. For instance, the main questions are:

(a) Do the Halo and the Single Source really exist ? Arguments for positive answers are given in the §3, but more supportive arguments are needed.

(b) Why numerical estimates for CR intensity in the Disk and the Halo, calculated with the expression (2), are smaller than in the observations ?

(c) Why is the energy spectrum of CR in the Disk steeper than the spectrum in the Halo ? According to our conception developed in [8] the spectrum in the Disk, with its higher turbulence in the interstellar medium due to SN explosions, should be flatter than in the Halo where there are no such powerful sources of turbulence as SN.

(d) To what extent are the simplified assumptions about the shapes and normalisation of the Disk, Halo and Single Source spectra as well as other parameters: A_d, A_h and A_{ss}, reasonable and what will the more sophisticated approach do for the result ?

The answers to these and other puzzling questions are the subject of further work.

4. Puzzle 3: the nature of the Single Source

The existence of the Single Source has been proposed by us to explain the puzzling sharpness of the knee in the size spectrum of extensive air showers (EAS) (see [11] and later publications). The physical basis of this proposal is the evident non-uniformity of the spatial and temporal distributions of SN explosions and subsequent SNR. As a result one SN could explode not very long ago and close to the solar system.. Its contribution to the CR intensity is rather high and it gives rise to a small peak (knee) above the background from other SNR - it is our Single Source.

To search for the Single Source we used the so called difference metod [12]. It is different from the traditional study of the CR intensity. Its main idea is that the difference between properties of EAS coming from the direction of the Single Source and from the opposite direction should be maximum. The method is stable against random experimental errors and allows to separate anomalies connected with the laboratory coordinate system from anomalies in the celestial coordinate system. The method allows to study the whole celestial sphere including regions outside of the line of sight of the experimental installation.

To search for an anisotropy we used experimental data obtained with the GAMMA experimental EAS array and have taken the EAS age parameter as a characteristic of the EAS property. The difference between EAS age distributions in two opposite directions was quantified by the reduced χ^2 value [12]. The result of the search is given in Figure 3. The maximum difference in galactic coordinates has been found at $\ell = 277 \pm 3°, b = -5 \pm 3°$. The closest source to this location is cluster Vela and therefore it is a good candidate for the role of the Single Source, which is responsible for the knee in the CR spectrum and the minimum in the amplitude of the anisotropy. However, a firm conclusion can be drawn only if it will be confirmed by the analysis of data from other EAS arrays.

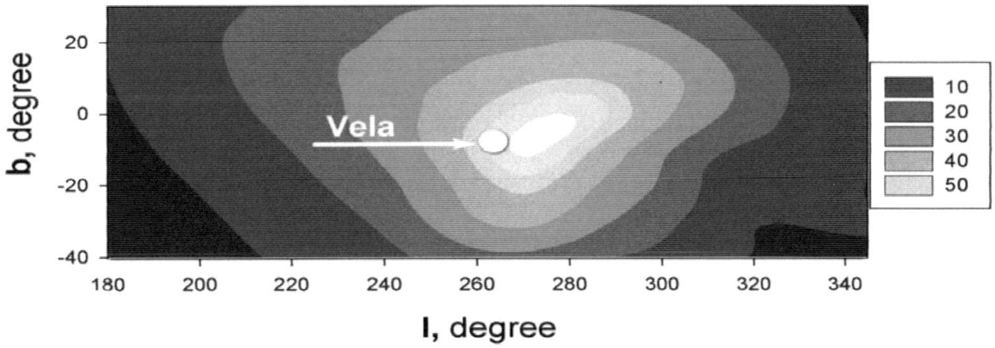

Figure 3. Two-dimensional profile of the reduced χ^2 of the difference between age distributions of EAS coming from opposite directions in galactic coordinates.

An alternative explanation is the influence of the regular magnetic field in the area surrounding the Earth especially taking into account the nearby spiral arm (see Puzzle 1 of this paper). The only point we should like to stress is that the effect of the regular magnetic field in PeV and sub-PeV energies should be small compared with the effect of the single source in order not to destroy the diffusive character of CR propagation. Additional asguments in favour of the dominance of the single source in the formation of the maximum difference in age distributions of EAS coming from the opposite directions could be the vicinity of Vela source and the 'younger' age of EAS coming from the direction of the maximum since magnetic fields do not change the energy spectrum of EAS and their ages.

In the section 6 of our paper [12] we speculated that a minor shift of the maximum difference from Vela source seen in Figure 3 could be due to the influence of this small regular magnetic field.

5. Conclusion

We mentioned only 3 puzzles existing in the PeV and sub-PeV energy region. However, there are many other there and after Prince Hamlet we can conclude that 'there are many things in heaven and earth, Horatio, than are dreamt of in your philosophy'.

References

[1] Ginzburg V L and Syrovatskii S I 1964 *The Origin of Cosmic Rays* Pergamon Press Oxford
[2] Berezinsky V S, Bulanov S V, Ginzburg V L, Dogiel V A and Ptuskin V S 1990 *Astrophysics of Cosmic Rays* North Holland Amsterdam
[3] Guillian G for the SK coll 2007 *Phys Rev* **D75**:062003
[4] Amenomori M et al, Tibet AS-gamma coll 2017 *Astrophys J* **836** 153
[5] Kozyarivsky V A et al 2004 *Preprint* astro-ph/0406059
[6] Aglietta M et al 2009 *Astrophys J Lett* **692** L130
[7] Antoni T et al for KASCADE coll 2003 *Astrophys J* **604** 687
[8] Erlykin A D and Wolfendale A W 2006 *Astropart Phys* **25** 183
[9] Erlykin A D, Sibatov R T, Uchaikin V V and Wolfendale A W 2015 *Proc 34 Int Cosm Ray Conf* Hague POS(ICRC2015) **463**
[10] Ahlers M 2016 *Phys Rev Lett* **117** 151103
[11] Erlykin A D and Wolfendale A W 1997 *J Phys D Nucl Part Phys* **23**, 979
[12] Pavlyuchenko V P, Martirosov R M, Nikolskaya N M and Erlykin A D 2018 *J Phys G Nucl Part Phys* **45**, 015202 (16pp)

Observation of the cosmic ray large-scale anisotropy with the ARGO-YBJ experiment

Songzhan Chen[1*], Wei Gao[2], on behalf of the ARGO-YBJ collaboration

[1]Key Laboratory of Particle Astrophysics, Institute of High Energy Physics, CAS, Beijing 100049, China.

[2]Institute of Frontier and Interdisciplinary Science and Key Laboratory of Particle Physics and Particle Irradiation (MOE), Shandong University, Qingdao, Shandong 266237, China.

E-mail: *chensz@ihep.ac.cn

Abstract. This paper reports on the observation of the sidereal large-scale anisotropy of cosmic rays using the ARGO-YBJ data collected from 2008 January to 2012 December. The anisotropy is investigated within a wide energy range from 4 to 520 TeV. Below 100 TeV, the anisotropy is dominated by two wide regions, the so-called "tail-in" and "loss-cone" features. The maximum value of the anisotropy is around 7 TeV. No noticeable variation of cosmic-ray anisotropy with solar activity is observed during the 5 years even though the solar activity varied from near minimum to maximum. Above 100 TeV, a dramatic change of the morphology is confirmed. The physical implications of these results are also discussed.

1. Introduction

Cosmic rays are high energy particles, mainly originating outside the solar system. The primary energy spectrum extends over many orders of magnitude from 10^9 eV up to 10^{20} eV, generally following a steeply falling power-law spectrum. A prominent feature is the change in the spectral index around 4 PeV, known as the cosmic ray knee, which was discovered 60 years ago [1]. Recent measurement has shown that the knee may be attributed to the elements heavier than hydrogen and helium [2]. It is believed that the cosmic rays with energies around and below the knee region are the products of energetic process within Galactic objects. The knee feature may be caused by the acceleration limit within the acceleration sites. The shocks of supernova remnants have been suspected to be the main acceleration sites for a long time. The population of young massive stars has been also suggested by [3] as the possible main origin.

Recently, important progresses have been achieved by several spaced-based detectors. A nearly universal hardening of the spectra near the magnetic rigidity of 200 GV was observed in the spectra of all abundant cosmic ray nuclei [4, 5, 6]. This feature is consistent with the new universal cosmic ray knee around 10 TV revealed by the NUCLEON detector [7]. These results may indicate three source classes for the Galactic cosmic rays as suggested in [8]. To further investigate the origins of cosmic rays, directional information is needed. For such a scenario, the cosmic ray anisotropy will be helpful.

In the past decades, the anisotropy of cosmic ray intensity was observed across a wide energy range from 60 GeV to 8 EeV. The morphology and amplitude are energy dependent. The large-scale anisotropy at energy from sub-TeV to tens of TeV has been detected by many experiments

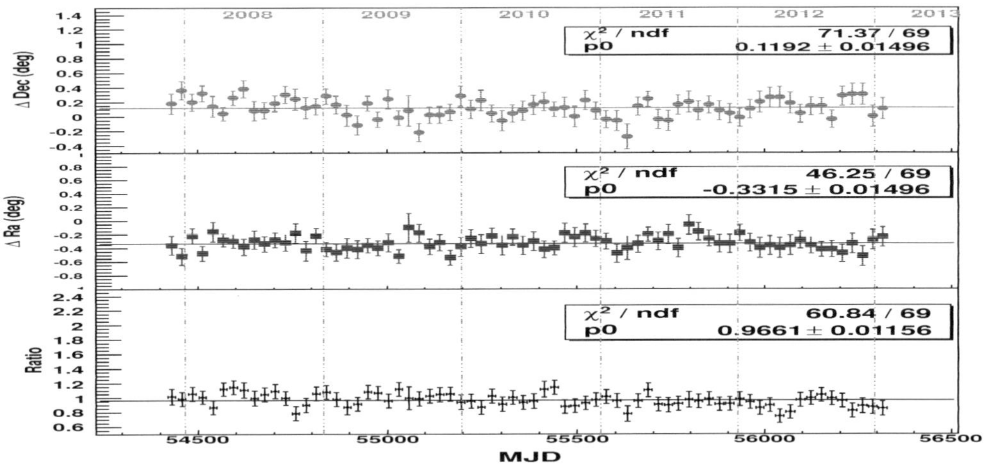

Figure 1. Displacement of the Moon shadow from the apparent center in the North-South (upper) and East-West (middle) direction. Lower panel is the ratio of the observed deficit count to the expected one. The solid line in each panel is the fitting result with a constant parameter.

[9]. The observed anisotropy map is dominated by two wide regions of excess and deficit, recognized as "tail-in" and "loss-cone" features, respectively. At energy above 100 TeV, a major change in the morphology was observed by EAS-TOP [10], IceCube [11], and Tibet ASγ [12].

The ARGO-YBJ experiment is a full-coverage extensive air shower array, sensitive to cosmic rays with the energy threshold of a few hundred GeV. A specific analysis on medium-scale anisotropy using ARGO-YBJ data has been reported in [13]. Several hot spots were observed. Based on the first 2 years of ARGO-YBJ data (2008 and 2009), an energy dependence of the large-scale anisotropy at energies from 1 to 30 TeV was reported in [14]. Very recently, a new analysis with the 5 years of ARGO-YBJ data collected from 2008 to 2012 was implemented [9]. A new method was used to improve the energy reconstruction, allowing to cover a much wider energy range, from 4 to 520 TeV. In this paper, we will review the data analysis and the results.

2. The ARGO-YBJ experiment and the data

The ARGO-YBJ experiment, a collaboration among Chinese and Italian institutions, is a full coverage extensive air shower detector with RPCs at Yangbajing (Tibet, China, 4300 m a.s.l.). It is mainly devoted to γ-ray astronomy [15, 16, 17] and cosmic ray physics [18, 2]. The full detector was in stable data taking from November 2007 to January 2013. The trigger rate is $\sim$3.5 kHz with a dead time of 4% and the average duty cycle is higher than 86%.

To monitor the long-term stability of the detector performance as far as the pointing accuracy and the angular resolution, both the position of the Moon shadow, separately along R.A. and DEC. projections, and the amount of shadow deficit events were monitored monthly as shown in Figure 1. The position of the Moon shadow turned out to be stable at level of 0.1° and the angular resolution was stable at level of 10% [19].

Figure 2 shows the daily average temperature $< T_{in} >$ inside the ARGO Hall during the period from 2008 to 2012. The average temperature was lower than 10°C in the winters of 2010, 2011 and 2012, while this feature was absent in 2009. To further check the data, the stability of the azimuth distribution within one day during the winter periods when $< T_{in} >< 10$°C was investigated. A tiny variation ($< 0.5\%$) of the azimuth distribution was observed, which was correlated with the variation of temperature. Such a variation was not observed during the summer periods when $< T_{in} >> 10$°C. According to Figure 1, the influence on the Moon shadow, as well as on the γ-ray data, can be ignored. However, the analysis of large-scale anisotropy is much sensitive to the stability of the azimuth distribution. Therefore, only the data collected in days with $< T_{in} >> 10$°C were used in this work. This criterion removed 35, 78, 130, and 87

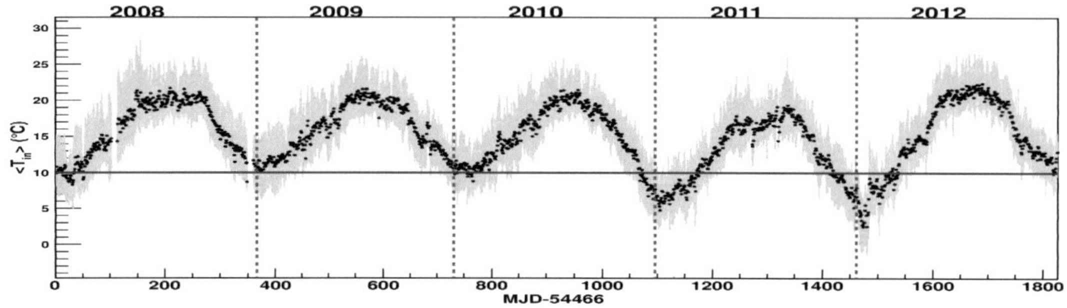

Figure 2. The daily average temperature inside the ARGO Hall during the period from 2008 to 2012. The error bar shows the variation range of the temperature within one day.

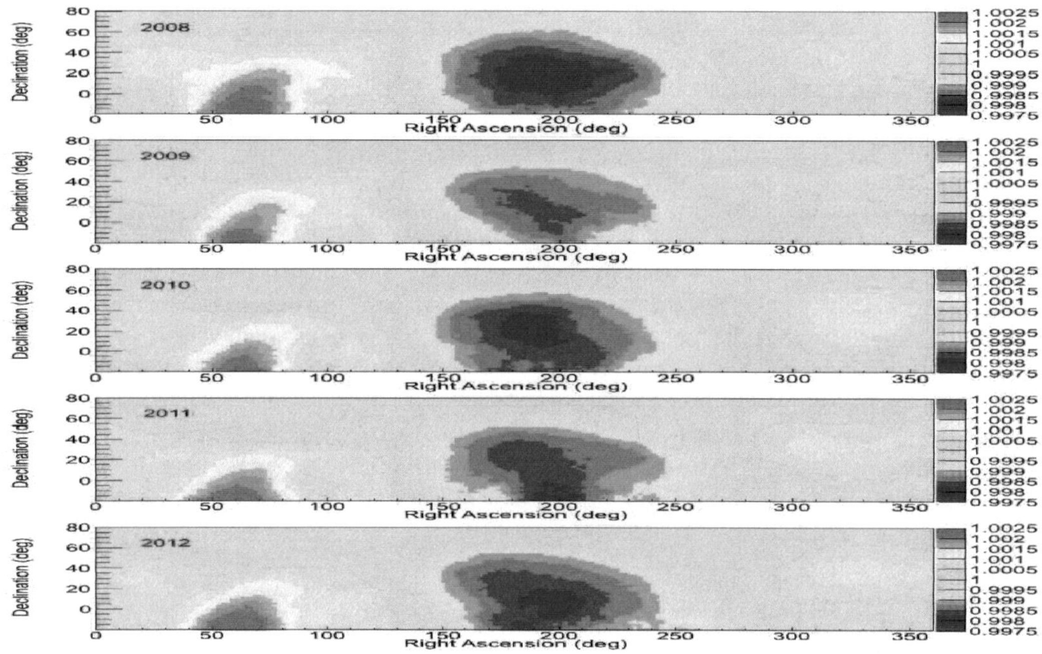

Figure 3. Relative intensity of LSA from 2008 to 2012. The median energy is 7 TeV. The maps are smoothed with 15° angular radius.

days of data during the years of 2008, 2010, 2011, and 2012, respectively.

Generally, the primary energy of an event is positively correlated with the number of fired pick-up electrodes of the detector. The number of fired electrodes was solely adopted to infer the primary energy in our previous analysis [14]. In this work, a new method was used to improve the energy reconstruction. Both the number of fired electrodes and the incident zenith angle were used to reconstruct the primary energy. More details about this reconstruction can be found in [9]. The events with reconstructed energy $\geq 10^{3.5}$ GeV, about 3.03×10^{10} events, were used in this analysis. The median energy is 7 TeV.

3. Results

In this work, the same analysis method as that used in [14] was adopted. The background map was estimated via the equi-zenith angle method based on an iterative procedure. With this approach, the 2D large-scale anisotropy can be determined.

Figure 3 shows the yearly relative intensity of the large-scale anisotropy from 2008 to 2012. The "tail-in" and "loss-cone" structures are distinct and almost stable during these

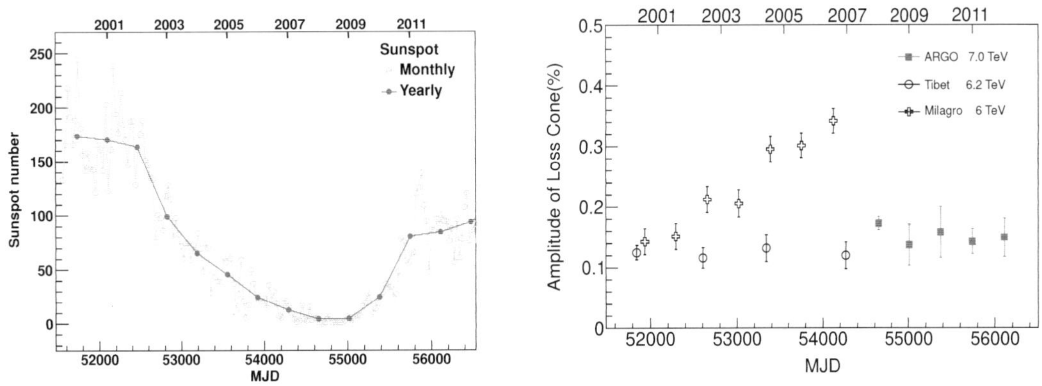

Figure 4. Left: The yearly and monthly variation of sunspot number. Right: Temporal variation of the "loss-cone" amplitude obtained by ARGO-YBJ, Milagro [20] and Tibet ASγ [21].

years. It is worth to note that the Milagro collaboration reported a steady increase in the amplitude of the "loss-cone" which was coincident with the decreasing of the solar activity [20], however, this result conflicted with the Tibet ASγ result [21]. Figure 4 shows the yearly "loss-cone" amplitude observed by ARGO-YBJ. The error bars represent the sum of statistical and systematic errors. The systematic error was estimated using the anti-sidereal anisotropy. The "loss-cone" amplitude was constant within errors. No significant time dependence was observed even though the solar activity varied from near minimum to maximum as shown in Figure 4. For comparison, Figure 4 also shows the results of Milagro and Tibet ASγ. The ARGO-YBJ result is mostly consistent with the same constant value as Tibet ASγ.

Figure 5 shows the anisotropy at eight energy intervals from 4.0 TeV to 520 TeV. The "tail-in" and "loss-cone" features are significant at energies from 4 TeV to 22 TeV. At energies from 39 TeV to 71 TeV, the "tail-in" and "loss-cone" features gradually fade away. At the same time, a new excess feature around the right ascension of 250°-300° gradually appears, replacing the structure of the "loss-cone". At energies above 160 TeV, the "tail-in" and "loss-cone" features completely disappear, and the map is dominated by a new pattern with an excess around the right ascension of 200°-310° and a deficit around 0°-100°. To quantitatively estimate the evolution of the anisotropy, the 1D profiles of the anisotropy were fitted by a first-order harmonic function, as shown in Figure 5. The fitted parameters, i.e., the amplitudes and phases, as a function of energy are shown in Figure 6. Combining the current result and the previous one [14], the energy dependence of the anisotropy is more evdent. The amplitude increases with energy, with a maximum intensity around 7 TeV, above which the amplitude begins to decrease with the phase gradually shifting. At energies above 100 TeV, a sudden change of the phase is observed and the amplitude also begins to increase. For comparison, the results reported by other detectors are also shown in Figure 6. The results obtained in this work generally agree with others.

4. Discussion and summary

According to current results, the morphology and amplitude variation trend of anisotropy below 100 TeV is very different from that above 100 TeV, implying different origins. This may be consistent with recent progress on the cosmic ray spectrum, which also indicates that the cosmic ray at energies from 1 TeV to 1 PeV may be a mixture from two different source classes [7, 8]. For the anisotropy below 100 TeV, different models have been proposed to explain the origin, concerning different aspects of cosmic ray physics, from the sources of cosmic rays to the

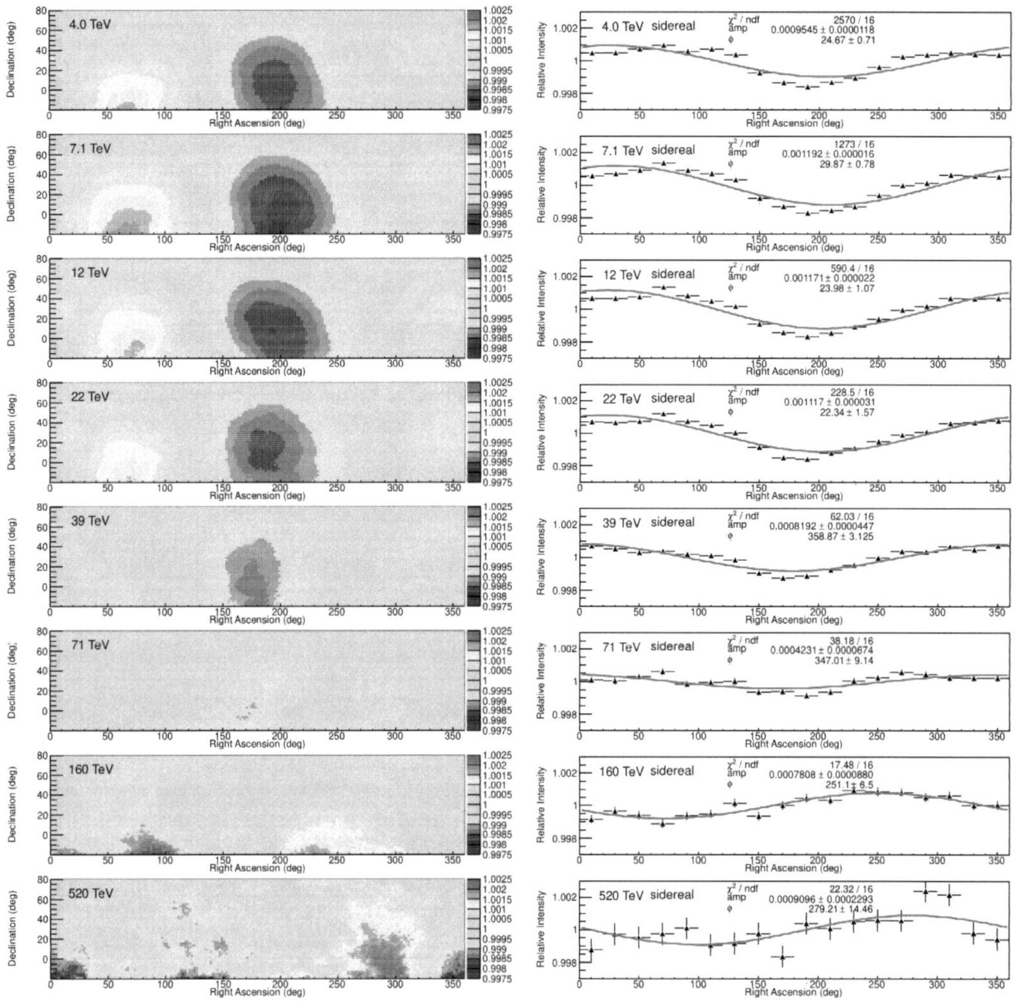

Figure 5. The 2D maps of relative intensity of large-scale anisotropy (left panels) and 1D projections on R.A. (right panels) for eight energy bins. Each 2D plot is smoothed with a 30° angular radius.

propagation to the Earth. Some models consider the anisotropy due to the spatial distribution of cosmic ray sources, as the presence of a nearby strong source [25]. Other interpretations concern the local interstellar magnetic fields [22] and interplanetary magnetic fields [23]. According to the time dependence of the anisotropy achieved in this work, the anisotropy above 3 TeV is not correlated with the solar activity. This indicates that the influence of the magnetic field within the heliosphere is not important, which is consistent with the estimation in [24]. If the anisotropy is caused by the local interstellar magnetic fields [22], the amplitude should monotonously decrease with energy increasing. This may be in principle not consistent with observations that the amplitude increases with energy below 7 TeV and decreases above 7 TeV. One plausible reasonable interpretation may be due to a single nearby strong source [25]. The new universal knee of cosmic ray spectrum around 10 TV [7] is close to the peak energy of the anisotropy. This may indicate that the new spectral feature and the anisotropy are correlated for their origin. For the anisotropy above 100 TeV, the excess region is near the direction of the Galactic center, suggesting this region as a possible source of cosmic rays. The variation trend of amplitude increasing as energy increasing is also consistent with scenario of cosmic ray diffusion.

In summary, this paper reports on the measurement of the large-scale cosmic ray anisotropy

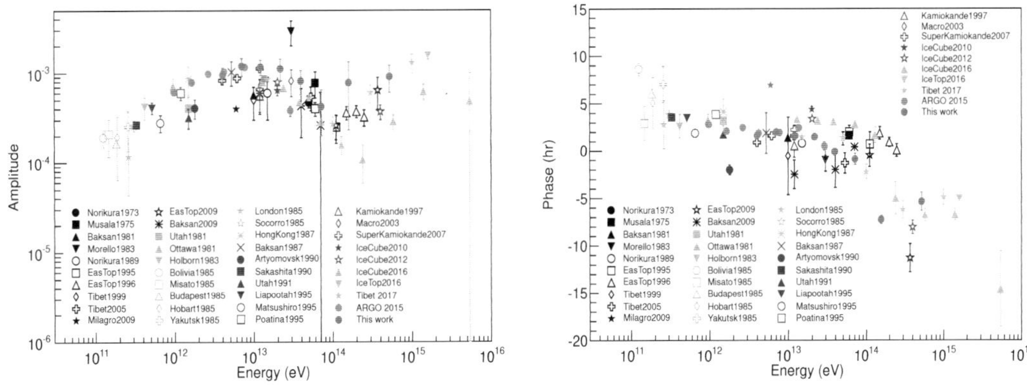

Figure 6. The amplitude (left panel) and phase (right panel) of the first harmonic of the sidereal anisotropy as a function of the cosmic-ray energy measured by ARGO-YBJ and other experiments.

by the ARGO-YBJ experiment with data collected from 2008 January to 2012 December. In contrast with a previous report by the Milagro experiment, no significant correlation with solar activity was detected. With an improved energy reconstruction procedure, we extended the energy range investigated by ARGO-YBJ up to 520 TeV. A dramatic change of the morphology, consistent with the observations reported by IceCube in the southern hemisphere and Tibet ASγ in the northern hemisphere, was clearly observed starting from about 50 TeV.

Acknowledgments
This work is supported in China by the National Natural Science Foundation of China (NSFC) under the grant No.11575203.

References
[1] Kulikov G V, et al. 1958 J. Exp. Theor. Phys. 35 635
[2] Bartoli B, et al. 2015 PHYSICAL REVIEW D 92 092005
[3] Aharonian F, Yang R Z, Wilhelmi E O 2018 arXiv:1804.02331
[4] Yoon Y S, et al. 2011 ApJ 728 122
[5] Aguilar M, et al. 2015 PRL 114 171103
[6] Aguilar M, et al. 2015 PRL 115 211101
[7] Atkina E, et al. 2018 JETP Letters 108 5
[8] Zatsepin V I, Sokolskaya N V 2006 Astronomy & Astrophysics 458 1
[9] Bartoli B, et al. 2018 ApJ 861 93
[10] Aglietta M, et al. 2009 ApJ 692 L130
[11] Aartsen M G, et al. 2016 ApJ 826, 220
[12] Amenomori M, et al. 2017 ApJ 836 153
[13] Bartoli B, et al. 2013 Phys. Rev. D 88 082001
[14] Bartoli B, et al. 2015 ApJ 809 90
[15] Bartoli B, et al. 2013 ApJ, 779 27
[16] Bartoli B, et al. 2014 ApJ, 790 152
[17] Bartoli B, et al. 2016 ApJS, 222 6
[18] Bartoli B, et al. 2012 Phys. Rev. D 85, 092005
[19] Bartoli B, et al. 2011 Phys. Rev. D 84, 022003
[20] Abdo A A, et al. 2009 ApJ 698, 2121
[21] Amenomori M, et al. 2012 APh 36 237
[22] Schwadron N A, et al. 2014 Science 343 988
[23] Nagashima K, Fujimoto K, Jacklyn R M 1998 J.Geophys.Res. 17 429
[24] Zhang M, Zuo P B, Pogorelov N 2014 ApJ 790 5
[25] Liu W, et al. 2017 Physical Review D 96 023006

Anisotropic diffusion and the cosmic ray anisotropy

M. Kachelrieß
Institutt for fysikk, NTNU, Trondheim, Norway

Abstract. We argue that the diffusion of cosmic rays in the Galactic magnetic field has to be strongly anisotropic. As a result, the number of CR sources contributing to the local CR flux is strongly reduced. The CR density is therefore less smooth, and the contribution of individual sources to the CR dipole anisotropy becomes more prominent. In the case of anisotropic diffusion, the observed plateau in the CR dipole anisotropy around 2–20 TeV can be explained by a 2–3 Myr old CR source which dominates the local CR flux in this energy range.

1. Introduction

The observed distribution of cosmic ray (CR) arrival directions is highly isotropic. Since Galactic CR sources are strongly concentrated in the Galactic disc, an efficient mechanism for the isotropisation of the CR momenta exists. Agent of this isotropisation are turbulent magnetic fields, since charged CRs scatter efficiently with resonant field modes which wavelength matches their Larmor radius. As a result, CRs perform on scales larger than the coherence length of the turbulent field a random walk, and the memory of the initial source location is mostly erased. Residual anisotropies are connected to the structure of the local magnetic field and, e.g., to a remaining net flux of CRs.

Since large wavelengths of the turbulent field modes are less abundant, CRs with higher energy are scattered less efficiently. Therefore, the diffusion picture predicts that the CR anisotropy should increase monotonically with energy. More precisely, if the turbulent field follows a Kolmogorov power law as suggested by the observed B/C ratio, the dipole anisotropy δ should increase with energy as $\delta \propto E^{1/3}$. Both the energy-dependence and the absolute value of the dipole anisotropy predicted in simple isotropic diffusion models do not agree with observations. This discrepancy was dubbed the "CR anisotropy problem" by Hillas [1].

In this short review based on the results of Refs. [2, 3], we will first argue that the diffusion of CRs in the Galactic magnetic field (GMF) has to be strongly anisotropic. As a result, the number of CR sources contributing to the local CR flux is strongly reduced. Therefore, the CR density is less smooth, and the contribution of individual sources to the CR dipole anisotropy becomes more prominent than in the standard picture. Then we argue that the observed plateau in the CR dipole anisotropy around 2–20 TeV is connected to a 2–3 Myr old CR source which dominates in this energy range the local CR flux. Finally, we comment on the alternative that a young source like Vela is responsible for the observed plateau in the dipole anisotropy.

2. Galactic magnetic field and anisotropic diffusion

In the diffusion approach to CR propagation one considers typically the CR density in the stationary limit. The measured ratios of CR isotopes like Be^{10}/Be^{9} and of secondary/primary ratios like B/C indicate a residence time of CRs with rigidity $\mathcal{R}$ of order $\tau_{\rm esc} \simeq$ few $\times$

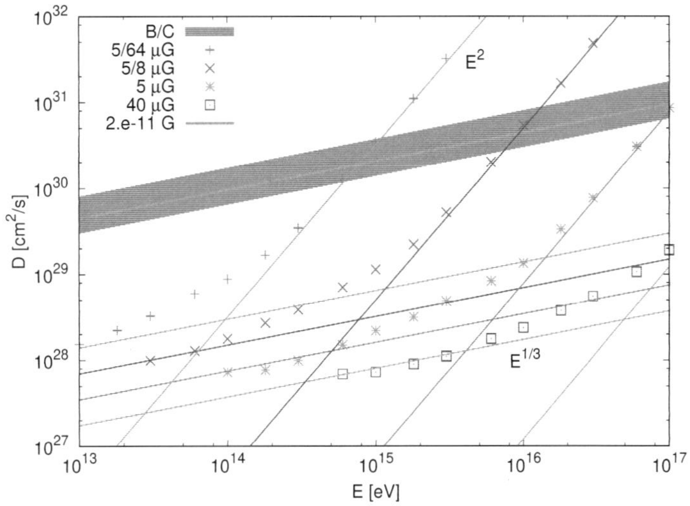

Figure 1. CR diffusion coefficient $D(E)$ in pure isotropic Kolmogorov turbulence with $L_{\max} =$ 25 pc for four values of the strength $B_{\rm rms}$ of the turbulent field. The green band shows the range of magnetic field strengths for which the diffusion coefficient satisfies $D_0 = (3-8) \times 10^{28}{\rm cm}^2/{\rm s}$ at $E_0 = 10\,{\rm GeV}$; from Ref. [3].

$10^7{\rm yr}\,(\mathcal{R}/5\,{\rm GV})^{-\beta}$ with $\beta \simeq 1/3$. Then the flux from some 10^4 sources accumulates at low rigidities, forming a "sea" of Galactic CRs, if one assumes that the main CR sources are supernovae (SN) injecting $\simeq 10^{50}$ erg every $\simeq 30$ yr in the form of CRs. Since many sources contribute, the discrete nature of the CR sources can be neglected. Assuming additionally that the turbulent magnetic field dominates relative to the regular field, one often replaces the diffusion tensor D_{ij} by a scalar diffusion coefficient D.

The diffusion approach based on the approximations described above has been sufficient to describe the bulk of experimental data obtained until $\simeq$ 2005. With the increased precision of newer experiments, several discrepancies like the "positron excess" or breaks in the CR spectra of nuclei have been emerged. Here we will discuss a more theoretical challenge for the approximations employed in the standard diffusion approach, which has also the potential to solve other observational anomalies.

In Fig. 1, we show the diffusion coefficient

$$D_{ij} = \lim_{t\to\infty} \frac{1}{2Nt} \sum_{a=1}^{N} (x_i^{(a)} - x_{i,0})(x_j^{(a)} - x_{j,0}) \qquad (1)$$

calculated numerically following the trajectories $x_i^{(a)}(t)$ of N CRs injected into a pure random field with a Kolmogorov power spectrum with $L_{\max} = 25$ pc for various field strengths. The transition at $R_L(E_{\rm cr}) = L_{\max}$ between the asymptotic low-energy ($D \propto E^{1/3}$, large-angle scattering) and high-energy ($D \propto E^2$, small-angle scattering) behaviour is clearly visible. However, for all used field strengths the diffusion coefficients are much smaller than those extracted using e.g. Galprop [4] or DRAGON [5]. Therefore CR propagation cannot be isotropic, because otherwise CRs overproduce secondary nuclei like boron for any reasonable values of the strength and the coherence scale of the turbulent field, cf. with Fig. 2. Such an anisotropy

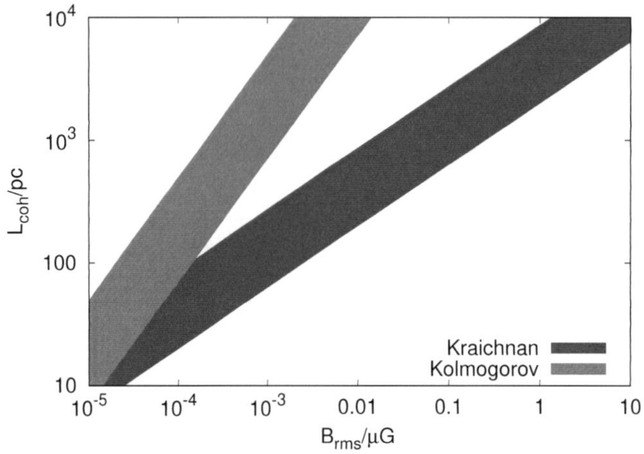

Figure 2. Allowed ranges of $B_{\rm rms}$ and $L_{\rm coh}$ compatible with $D_0 = (3-8)\times 10^{28}{\rm cm}^2/{\rm s}$ at $E_0 = 10\,{\rm GeV}$ for Kolmogorov and Kraichnan turbulence. These ranges should be compared with the typical order-of-magnitude values that are relevant for the Galactic magnetic field: $B_{\rm rms} \sim (1-10)\,\mu{\rm G}$ and $L_{\rm coh} \lesssim$ a few tens of pc, from Ref. [3].

may appear if the turbulent field at the considered scale does not dominate over the ordered component, or if the turbulent field itself is anisotropic.

One can estimate the level of anisotropy required considering the following toy model: Let us adopt a thin matter disc with density $\rho/m_{\rm p} \simeq 1/{\rm cm}^3$ and height $h = 150\,{\rm pc}$ around the Galactic plane, while CRs propagate inside a larger halo of height $H = 5\,{\rm kpc}$. We assume that the regular magnetic field inside this disc and halo has a tilt angle ϑ with the Galactic plane, so that the component of the diffusion tensor relevant for CR escape is given by

$$D_z = D_\perp \cos^2\vartheta + D_\parallel \sin^2\vartheta\,. \tag{2}$$

Applying a simple leaky-box approach, the grammage follows as $X = c\rho hH/D_z$. Using now as allowed region for the grammage $5 \leq X \leq 15\,{\rm g/cm}^2$, the permitted region in the ϑ–η plane shown in the left panel of Fig. 3 follows. For not too large values of the tilt angle, $\vartheta \lesssim 30°$, the regular field should strongly dominate, $\eta \lesssim 0.35$. This results in a strongly anisotropic propagation of CRs, where the diffusion coefficient perpendicular to the ordered field can be between two and three orders of magnitude smaller than the parallel one, $D_\perp \ll D_\parallel$. As a result, the z component of the regular magnetic field can drive CRs efficiently out of the Galactic disk. For instance, the "X-field" in the Jansson-Farrar model [6] for the GMF leads to the correct CR escape time, if one chooses as turbulence level $\eta \equiv B_{\rm rms}/B_0 \simeq 0.25$ [7].

For this choice, the diffusion coefficients satisfy $D_\parallel \simeq 5D_{\rm iso}$ and $D_\perp \simeq D_{\rm iso}/500$, where $D_{\rm iso}$ denotes the isotropic diffusion coefficient $D_{\rm iso}$ satisfying the B/C constraints. In the regime, where the CRs emitted by a single source fill a Gaussian with volume $V(t) = \pi^{3/2} D_\perp D_\parallel^{1/2} t^{3/2}$, the CR density is increased by a factor $500/\sqrt{5} \simeq 200$ compared to the case of isotropic diffusion. The smaller volume occupied by CRs from each single source leads to a smaller number of sources contributing substantially to the local flux, with only $\sim 10^2$ sources at $\mathcal{R} \sim 10\,{\rm GV}$ and about ~ 10 most recent SNe in the TeV range. This reduction of the effective number of sources may invalidate the assumption of a continuous CR injection and a stationary CR flux.

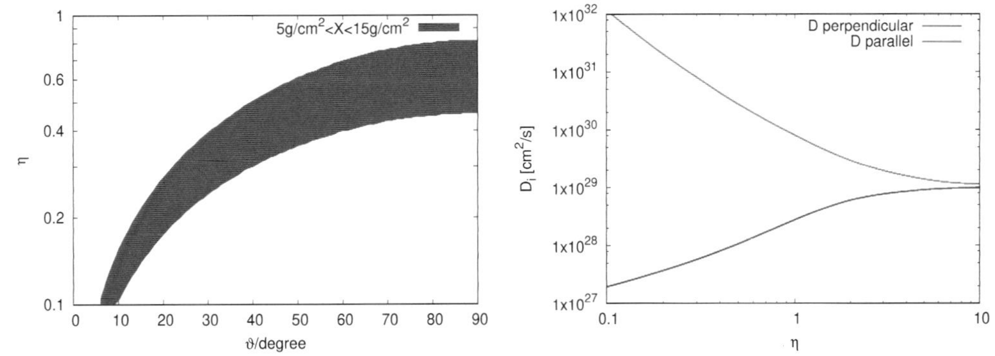

Figure 3. Left: Grammage X crossed by CRs in a "disc and halo" model as a function of the tilt angle ϑ between the regular magnetic field and the Galactic plane, and of the turbulence level η. Right: Fit of the diffusion coefficients $D_{\parallel}$ and $D_{\perp}$ at $E = 10^{15}\,\mathrm{eV}$ as a function of η, from Ref. [3].

3. A local source and the cosmic ray anisotropy

In the diffusion approximation, Fick's law is valid and the net CR current $\boldsymbol{j}(E)$ is determined by the gradient of the CR number density $n(E) = \mathrm{d}N/(\mathrm{d}E\mathrm{d}V)$ and the diffusion tensor $D_{ab}(E)$ as $j_a = -D_{ab}\nabla_b n$. The dipole vector $\boldsymbol{\delta}$ of the CR intensity $I = c/(4\pi)n$ follows then as

$$\delta_a = \frac{3}{c}\frac{j_a}{n} = -\frac{3D_{ab}}{c}\frac{\nabla_b n}{n}\,. \tag{3}$$

In the case of a strong ordered magnetic field $\boldsymbol{B}$, the tensor structure of the diffusion tensor simplifies to $D_{ab} \propto B_a B_b$. This corresponds to a projection of the CR gradient onto the magnetic field direction [8]. Hence, anisotropic diffusion predicts that the dipole anisotropy should align with the local ordered magnetic field instead of pointing to the source [8, 2]. Note that the ordered magnetic field corresponds to the sum of the regular magnetic field and the sum of turbulent field modes with wavelengths larger than the Larmor radius at the corresponding CR energy.

In the case of a (three-dimensional) Gaussian CR density n, the formula (3) can be evaluated analytically. The result $\delta = 3R/(2cT)$ for a single source with age T and distance R is independent of the regular and turbulent magnetic field. In Ref. [2], it was shown that the CR density of a single source is quasi-Gaussian, if CRs propagate over length scales $l \gg L_{\mathrm{coh}}$. Numerically, the dipole anisotropy δ of a source contributing the fraction f_i to the total observed CR flux is thus

$$\delta_i = f_i\,\frac{3R}{2cT} \simeq 5.0\times 10^{-4}\,f_i\left(\frac{R}{200\,\mathrm{pc}}\right)\left(\frac{T}{2\,\mathrm{Myr}}\right)^{-1}\,. \tag{4}$$

In Fig. 4, we show experimental data for the dipole anisotropy from Refs. [9, 10] as a green band: The anisotropy grows as function of energy until $E \simeq 2\,\mathrm{TeV}$, remains approximately constant in the range 2–20 TeV, before it decreases again. The plateau in the range 2–20 TeV is naturally explained by the energy-independent contribution to the dipole anisotropy of a single source. This is supported by the fact that the dipole phase remains approximately constant in this range too, before it flips by $\sim 180^\circ$. Such a flip is naturally explained by the projection effect on the magnetic field line, if above 20 TeV another source, which is located in the opposite hemisphere, dominates the CR dipole anisotropy.

More specifically, it was suggested in Ref. [2] that a 2–3 Myr old source at the distance 200–300 pc dominates the dipole anisotropy in the range 2–20 TeV. Previously, it was shown in

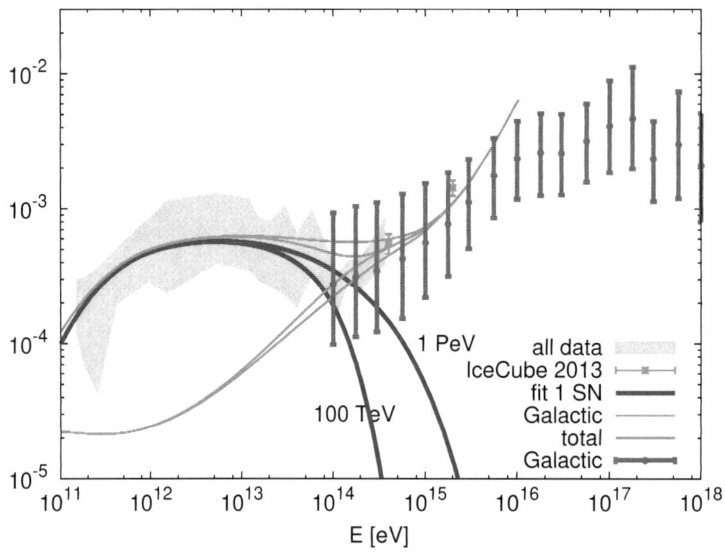

Figure 4. Lower and upper limit (green band) on the dipole anisotropy and data from IceCube (blue error-bars) compared to the contribution from the local source (red, for two values of $E_{\max}$) and from the average CR sea (magenta) as function of energy; adapted from Ref. [2].

Ref. [11] that the same source can explain the "positron excess", as well as the breaks of the nuclei spectra and the different slope of the proton spectrum [12]. The contribution of this local source is shown by two magenta lines for two different high-energy cutoffs: In one case, it was assumed that the source can accelerate up 100 TeV, in the other that it is a PeVatron. In both cases, the CR flux was calculated following the trajectories of individual CRs, as discussed in Refs. [7, 13, 11, 12]. Additionally, the total anisotropy beyond 10^{14} eV of all Galactic SNe is shown by red error-bars which is calculated in the escape model which uses the same magnetic field configuration as the one used for the loal source [7, 13].

A characteristic feature of this proposal is that a relatively old source dominates the observed CR flux. This is only possible in the case of anisotropic diffusion, and requires additionally that the perpendicular distance $d_\perp$ of the Sun to the magnetic field line connecting it to the source is not too large. Even for small $d_\perp$, the CR flux from the single source is suppressed at low-energies, because of the slower perpendicular diffusion. In Refs. [12], the value $d_\perp \simeq 70$ pc was estimated requiring that the low-energy break in the source spectrum explains the breaks in the energy spectra of CR nuclei. For this choice of $d_\perp$, the flux of the local source is suppressed below $\simeq 1$ TeV (cf. with Fig. 2 of Ref. [12]), leading to a decreasing f_i and the transition to the standard $\delta \propto E^{1/3}$ behavior below this energy.

Another choice for the age of the source was suggested in Refs. [14, 15]. Here, Vela with the age around 11,000 yr and distance 300 pc was proposed as the single source responsible for the plateau in the dipole anisotropy in the energy range 2–20 TeV. In this case, the contribution of Vela to the dipole amplitude has to be suppressed by a factor $\simeq 200$. Three mechanisms for such a suppression may be operating: First, if the regular magnetic field and the CR gradient are not parallel, the projection effect in $D_{ab}\nabla_b n$ can reduce the dipole [16, 15]. Second, the measured CR dipole is a projection into the equatorial plane and is thus reduced compared to the true one. Finally, the CR flux contributed by Vela may be small. Calculating the CR fluxes from nearby young sources using the standard isotropic diffusion coefficient and taking into account these effects, [15] argued that Vela leads to correct level of anisotropy. There is however a caveat in this conclusion: While Ref. [15] calculates the CR fluxes from individual sources for *isotropic* diffusion,the remaining analysis is based on strongly anisotropic diffusion. In the latter case, the CR flux depends however crucially on the perpendicular distance $d_\perp$ of the source to the magnetic field line connecting it with the Sun, and a calculation of the CR flux following the lines of Refs. [2, 11, 12] is required. Moreover, the number of sources is strongly reduced and correspondingly the flux of nearby sources with small perpendicular distance strongly enhanced.

4. Conclusions

We have argued that the diffusion in the GMF has to be strongly anisotropic, because otherwise CRs overproduce secondary nuclei like boron for any reasonable values of the strength and the coherence scale of the turbulent field. Therefore the number of CRs contributing to the local CR flux is strongly reduced compared to the "standard picture". As a result, the CR density is less smooth, and the contribution of individual sources to the CR dipole anisotropy becomes more prominent. In this picture, the observed plateau in the CR dipole anisotropy around 1–20 TeV can be explained by a 2–3 Myr old CR source which dominates the local CR flux in this energy range. Such a source can explain also several other CR puzzles such as the "positron excess" , the difference in the slope of the proton and nuclei spectra as well as their breaks [12].

Acknowledgments
It is a pleasure to thank Gwenal Giacinti, Andrii Neronov, Volodymyr Savchenko and Dimitri Semikoz for fruitful collaborations.

References

[1] Hillas A M 2005 *J. Phys.* **G31** R95–R131
[2] Savchenko V, Kachelrieß M and Semikoz D V 2015 *Astrophys. J.* **809** L23 (*Preprint* 1505.02720)
[3] Giacinti G, Kachelrieß M and Semikoz D V 2018 *JCAP* **1807** 051 (*Preprint* 1710.08205)
[4] Jóhannesson G *et al.* 2016 *Astrophys. J.* **824** 16 (*Preprint* 1602.02243)
[5] Evoli C, Gaggero D, Grasso D and Maccione L 2008 *JCAP* **0810** 018 [Erratum: JCAP1604,no.04,E01(2016)] (*Preprint* 0807.4730)
[6] Jansson R and Farrar G R 2012 *Astrophys.J.* **761** L11 (*Preprint* 1210.7820)
[7] Giacinti G, Kachelrieß M and Semikoz D V 2014 *Phys. Rev.* **D90** 041302 (*Preprint* 1403.3380)
[8] Jones F C 1990 *Astrophys. J.* **361** 162–172
[9] Di Sciascio G and Iuppa R 2014 (*Preprint* 1407.2144)
[10] Ahlers M and Mertsch P 2017 *Prog. Part. Nucl. Phys.* **94** 184–216 (*Preprint* 1612.01873)
[11] Kachelrieß M, Neronov A and Semikoz D V 2015 *Phys. Rev. Lett.* **115** 181103 (*Preprint* 1504.06472)
[12] Kachelrieß M, Neronov A and Semikoz D V 2018 *Phys. Rev.* **D97** 063011 (*Preprint* 1710.02321)
[13] Giacinti G, Kachelrieß M and Semikoz D V 2015 *Phys. Rev.* **D91** 083009 (*Preprint* 1502.01608)
[14] Sveshnikova L G, Strelnikova O N and Ptuskin V S 2013 *Astroparticle Physics* **50** 33–46 (*Preprint* 1301.2028)
[15] Ahlers M 2016 *Phys. Rev. Lett.* **117** 151103 (*Preprint* 1605.06446)
[16] Mertsch P and Funk S 2015 *Phys. Rev. Lett.* **114** 021101 (*Preprint* 1408.3630)

IOP Conf. Series: Journal of Physics: Conf. Series **1181** (2019) 012040 doi:10.1088/1742-6596/1181/1/012040

Measurement of Low-Energy Cosmic Rays via the Neutral Iron Line

Kumiko K. Nobukawa[1], Shigetaka Saji[2], Arisa Hirayama[1], Masayoshi Nobukawa[3], Shigeo Yamauchi[1], Hironori Matsumoto[4], and Katsuji Koyama[5]

[1] Department of Physics, Nara Women's University, Kitauoyanishi-machi, Nara, Nara 630-8506, Japan

[2] Division of Particle and Astrophysical Science, Graduate School of Science, Nagoya University, Furo-cho, Chikusa-ku, Nagoya, Aichi 464-8602, Japan

[3] Faculty of Education, Nara University of Education, Takabatake-cho, Nara, Nara 630-8528, Japan

[4] Department of Earth and Space Science, Graduate School of Science, Osaka University, 1-1 Machikaneyama-cho, Toyonaka, Osaka 560-0043, Japan

[5] Department of Physics, Graduate School of Science, Kyoto University, Kitashirakawa-oiwake-cho, Sakyo-ku, Kyoto, Kyoto 606-8502, Japan

E-mail: kumiko@cc.nara-wu.ac.jp

Abstract. There has been little information of Galactic low-energy cosmic rays (LECRs) so far because observations of LECRs in the solar system are affected by solar modulation and there is no effective way of indirect measurement. When LECRs in the MeV energy band collide with neutral iron atoms in the interstellar medium, fluorescent X-rays at 6.4 keV are produced via inner-shell ionization of neutral iron atoms. We investigate the 6.4 keV line in supernova remnants (SNRs) as a prime candidate for the Galactic cosmic-ray origin. The line emission is discovered from 11 SNRs. The spectra and morphologies suggest that the 6.4 keV line is produced by LECRs interacting with cold gas. The proton energy density is estimated to be 10–100 eV cm^{-3}. Furthermore, we measure the distribution of the 6.4 keV line emission near the Galactic center. The intensity profile is very similar to that of molecular clouds. The most plausible origin of the enhancement is the LECR proton bombardment. The energy density of MeV protons is estimated to be $\sim$ 80 eV cm^{-3}. Since the diffusion length of MeV protons is short, they should be produced *in situ*. Surprisingly, there is no SNR in the vicinity, and thus another mechanism such as the stochastic acceleration would possibly work.

1. Introduction

Cosmic rays (mainly protons) below the knee energy are thought to be accelerated in the Galaxy. Supernova remnants (SNRs) are the prime candidate for the Galactic cosmic-ray origin. Gamma-ray observations have obtained evidence for acceleration of high-energy cosmic rays (HECRs; in the GeV–TeV band) and their spectrum in SNRs (e.g., [1],[2]). Furthermore they provide the implication for the distribution of the HECRs in the Galaxy [3]. On the other hand, to date there are very few observations that give us the information of the Galactic low-energy cosmic rays (LECRs). Since solar modulation affects LECR measurements inside the heliosphere, a direct and convincing observation is performed only by Voyager-I [4]. The indirect measurements of

IOP Conf. Series: Journal of Physics: Conf. Series **1181** (2019) 012040 doi:10.1088/1742-6596/1181/1/012040

LECRs are also limited. In principle, observations of gamma-rays from π^0-decays provide the information of high-energy protons above 280 MeV. Although the infrared and radio observations can measure the ionization rate via the absorption line of H_3^+ [5] and the [DCO^+]/[HCO^+] abundance ratio [6], respectively, there still remain large uncertainties of CR density and spectral shape.

The neutral iron line can be an effective prove to investigate the LECRs. When LECR protons or electrons collide with neutral iron atoms in cold gas, the 6.4 keV line is produced via inner-shell ionization. Iron has high abundance in the interstellar medium and has high fluorescence yield, and hence the 6.4 keV line is the best among fluorescent X-rays from neutral heavy atoms. Furthermore, the X-ray continuum is also produced via bremsstrahlung (by electrons) or inverse bremsstrahlung (by protons). Since the continuum intensity is different between the proton and electron cases, the intensity ratio of the 6.4 keV line and continuum, or an equivalent width, is used to distinguish the irradiating particles are protons or electrons.

The purpose of this study is to investigate LECRs by measuring the particle-induced 6.4 keV line. We first focus on SNRs because they are the prime candidate for the Galactic CR origin, and then move to the Galactic ridge in order to examine whether there are some CR accelerators other than SNRs. We utilize the Suzaku data because the X-ray CCD camera (X-ray Imaging Spectrometer: XIS) aboard the satellite has advantages of a large effective area and a low and stable background especially in the iron K-shell band [7, 8].

2. Results

2.1. Supernova remnants

We started from the five SNRs: W28, Kes 67, Kes 69, Kes 78, and W44. They all are middle-aged or old SNRs ($\sim 10^4$–10^5 yr; see table 1), and three out of them, W28, Kes 69, and W44, are classified as mixed-morphology SNRs [9]. Furthermore, they all interact with molecular clouds (MCs) [10]. LECRs accelerated in individual SNRs can collide the MCs to produce the 6.4 keV line. Detailed analysis procedure and results are described in Nobukawa *et al.* (2018) [11]. For each SNR, we discovered that the 6.4 keV line emission is locally enhanced. The 6.4 keV line emission would associate with the cold gas, such as MCs and HI clouds for the five SNRs. This fact indicates that the 6.4 keV line is generated by cosmic rays interacting with the interstellar medium.

We extracted a spectrum from the region with a high line intensity in each SNR, and then stacked all the spectra. A clear enhancement of the 6.4 keV line in the stacked spectrum compared to the background one was found with more than 4.3σ confidence. We, then, obtained the equivalent width of the 6.4 keV line of ≥ 1 keV. This value favors the MeV protons origin [12]. We estimated the energy density to be more than 10–100 eV cm^{-3} based on the 6.4 keV line intensity and the density of the interacting clouds. Since the diffusion length of LECRs is short, LECRs should be produced quite near the region with the 6.4 keV line emission. Detection of the 6.4 keV line suggests that acceleration of LECRs would indeed occur in SNRs.

Including the five SNRs, the 6.4 keV line that is most likely due to LECRs has been detected from more than 10 SNRs. We summarize those results in table 1. Most part of the listed SNRs are known to be middle-aged or old. This might be caused by bias due to the sensitivity at the 6.4 keV line band. In young SNRs, acceleration is highly activated, and a considerable number of high-energy CR electrons are produced, which entails a high-intensity continuum via synchrotron radiation (e.g., SN1006 [13], RX J1713.7-3946 [14]). Also high bremsstrahlung emission comes from high temperature plasma of kT = several keV. Furthermore, the ionization state of iron in the plasma is often lower than helium-like iron, and thus these SNRs show a iron line of $\sim$6.4 keV (e.g., RCW 86 [15], SN 1006 [16]). Even if the LECRs produce the 6.4 keV line in young SNRs, it can be buried in the continuum or the plasma emission. In middle-aged and old SNRs, on the other hand, it is plausible that the particle-induced 6.4 keV line is more detectable because the nonthermal continuum is suppressed, and in many cases plasma temperature have decreased so that the iron K-shell line is emitted with the intensity under the detection limit of the Suzaku/XIS.

More than a half of the SNRs with the 6.4 keV line are accompanied by gamma-ray emission (table 1). LECRs can be a seed particle of HECRs, and therefore one may guess that the distribution of the 6.4 keV line is associated with those of the GeV and very-high-energy (VHE; > 0.1 TeV) gamma-rays. Observations indicate that the distributions of the 6.4 keV line and gamma-ray emission are not always associated with each other. In case of W44, where the gamma-rays are supposed to be originated from the interactions between high-energy protons and molecular clouds [1], the GeV gamma-ray and 6.4 keV line regions well overlap [11]. On the other hand, the peak of the VHE gamma-ray emission in G323.7−1.0 is about $20'$ away from the 6.4 keV clumps (figure 1). The difference of the diffusion lengths between LECRs and HECRs can cause the discrepancy of the morphologies. It is possible that the escape effect from the acceleration site contribute to the morphology difference: HECRs have larger gyro radius and can escape faster (e.g., [17, 18]). Another possibility is that the particle acceleration is not effective and thus LECRs are produced but HECRs are not in the 6.4 keV region. Sensitive and detailed observations of gamma-rays with angular scales of at least a few arcmin are required in order to discuss their association with the 6.4 keV line.

Table 1. List of the SNRs from which the 6.4 keV line is detected. Errors are quoted at 68% confidence levels.

Object	6.4 keV intensity (10^{-8} ph s^{-1} cm^{-2} arcmin^{-2})	SNR age (yr)	Gamma-ray detection (GeV/VHE)
W28	0.8 ± 0.5 [11]	$\sim 36,000$ [19]	GeV[20], VHE[21]
Kes 67	3.6 ± 1.4 [11]	$\sim 100,000$ [22]	–
Kes 69	2.8 ± 1.0 [11]	$\sim$ 5,000–10,000 [23, 24]	–
Kes 78	1.8 ± 0.9 [11]	13,000–63,000 [25]	GeV[20], VHE[26]
Kes 79	6.0 ± 1.3 [27]	$\sim 27,000$ [27]	–
W44	1.3 ± 0.6 [11]	$\sim 20,000$ [28]	GeV[1]
G323.7−0.1	4.1 ± 0.8 [29]	$\sim 100,000$ [29]	GeV[30], VHE[31]
G346.6−0.2	$4.9^{+1.4}_{-1.6}$ [32]	14,000–16,000 [33]	–
Kes 17	$3.8^{+1.0}_{-1.1}$ [32]	28,000–64,000 [34]	GeV [20]
N132D	$5.3^{+2.0}_{-2.2}$ [35]	$\sim 2,500$ [36]	GeV[37], VHE[38]
IC 443	0.54 ± 0.17 [39]	3,000–30,000 [40, 41]	GeV[1]

2.2. Near the Galactic center region

We also discovered the particle-induced 6.4 keV line in the Galactic ridge near the Galactic center [42]. We measured an intensity profile of the 6.4 keV line at $l \sim 3^\circ$, where the giant molecular clouds called the Bania's Clump 2 is located (figure 2; [43]). The profile of the 6.4 keV line is very similar to that of the ^{12}CO distribution. This indicates that the origin of the 6.4 keV line would be due to the molecular clouds bombarded by LECRs or irradiated by X-rays. Since there is no X-ray source that has enough luminosity to explain the line intensity in the vicinity, the cosmic-ray origin is the most plausible also in this case. The equivalent width is estimated to be 1.3 keV and the proton origin is favored.

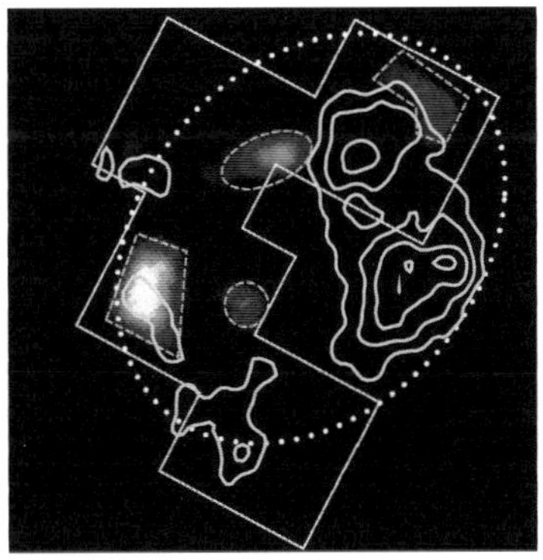

Figure 1. X-ray image of the 6.3–6.5 keV line band [29] obtained by the Suzaku/XIS, overlaid by the contours of the VHE gamma-ray emission [31]. The white solid line shows the fileds of view of the XIS. The white dashed ellipse roughly represents the shape of the radio shell.

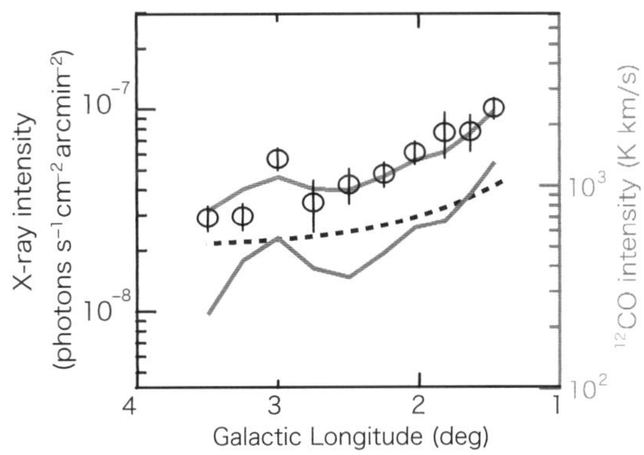

Figure 2. Intensity profile of the 6.4 keV line on the Galactic plane ($b = -0°.046$). Errors are quoted at 68% confidence levels. The dashed line shows a model that assumes the distribution symmetric with respect to Sgr A*. The blue line is the ^{12}CO intensity profile (the unit is on the right-side axis). The red line is the sum of the symmetric distribution model and the ^{12}CO intensity.

We estimated the energy density of LECR protons (0.1–1000 MeV) near the Galactic center to be 80 eV cm^{-3}. This is about one or two orders of magnitude higher than the energy density of CRs (> 3 MeV) in the local interstellar medium determined by the Voyager-I observation, $\sim$ 1 eV cm^{-3} [44]. Whereas the LECRs should be generated *in situ*, there is no SNR in the vicinity. The LECRs should be produced by another mechanism (e.g. stochastic acceleration) in the Galactic disk.

Yamauchi *et al.* (2016) [45] and Nobukawa *et al.* (2016) [46] measured the intensity profile of the iron K-shell lines and global X-ray spectrum in the inner Galaxy by utilizing Suzaku, and argued that a significant fraction of the 6.4 keV line emission from the Galactic ridge can be due to local bombardment of LECRs to the molecular clouds. H_3^+ observations revealed that the ionization rate in the Galactic interstellar medium would be controlled by the proximity to local sites of particle acceleration [5]. Acceleration of LECRs can occur through the interstellar space by not only SNRs but also other mechanisms. It is our future work to compare between the intensity profile of the 6.4 keV line and those of GeV and VHE gamma-rays, which will provide a key to the CR production and propagation mechanism in the Galactic plane.

3. Conclusion

The neutral iron line at 6.4 keV can be an effective probe to investigate LECRs. We searched for the particle-induced 6.4 keV line in SNRs, and found it from 11 SNRs, most of which are middle-aged or old. Therefore, LECRs would be indeed accelerated in SNRs. Also the diffuse 6.4 keV line emission was discovered at $l = 1^\circ.5$–$3^\circ.5$. Since there is no SNR in the vicinity, other acceleration mechanisms other than SNR shocks seem to work in the Galactic ridge.

Acknowledgments

We thank the organizing committee of the 26th E+CRS and 35th RCRC for kind hospitality. We are grateful to all the members of the Suzaku team. KKN is supported by Research Fellowships of JSPS for Young Scientists. This work is supported by JSPS Numbers JP16J00548 (KKN), JP15H02090, JP17K14289 (MN).

References

[1] Ackermann M *et al.* 2013 *Sci* **339** 807
[2] Aharonian F *et al.* 2007 *Astron. Astrophys.* **464** 235
[3] Ackermann M *et al.* 2011 *Astrophysical J.* **726** 81
[4] Stone E C, Cummings A C, McDonald F B, Heikkila B C, Lal N, and Webber W R 2013 *Sci* **341** 150
[5] Indriolo N and McCall B J 2012 *Astrophysical J.* **745** 91
[6] Ceccarelli C, Hily-Blant P, Montmerle T, Dubus G, Gallant Y, and Fiasson A 2011 *Astrophysical J.* **740** L4
[7] Koyama K *et al.* 2007 *Publ. Astron. Soc. Jpn.* **59** S23
[8] Mitsuda K *et al.* 2007 *Publ. Astron. Soc. Jpn.* **59** S1
[9] Rho J and Petre R 1998 *Astrophysical J.* **503** L167
[10] Jiang B, Chen Y, Wang J, Su Y, Zhou X, Safi-Harb S, and DeLaney T 2010 *Astrophysical J.* **712** 1147
[11] Nobukawa K K, Nobukawa M, Koyama K, Yamauchi S, Uchiyama H, Okon H, Tanaka T, Uchida H and Tsuru T G 2018 *Astrophysical J.* **854** 87
[12] Dogiel V, Chernyshov D, Koyama K, Nobukawa M, and Cheng K -S 2011 *Publ. Astron. Soc. Jpn.* **63** 535
[13] Koyama K, Petre R, Gotthelf E V, Hwang U, Matsuura M, Ozaki M, and Holt S S 1995 *Nature* **378** 255
[14] Koyama K, Kinugasa K, Matsuzaki K, Nishiuchi M, Sugizaki M, Torii K, Yamauchi S, Aschenbach B 1997 *Publ. Astron. Soc. Jpn.* **49** L7
[15] Yamaguchi H, Koyama K, Nakajima H, Bamba A, Yamazaki R, Vink J, and Kawachi A 2008 *Publ. Astron. Soc. Jpn.* **60** S123
[16] Yamaguchi H, Koyama K, Katsuda S, Nakajima H, Hughes J P, Bamba A, Hiraga J S, Mori K, Osaki M, and Tsuru T G 2008 *Publ. Astron. Soc. Jpn.* **60** S141
[17] Ohira Y, Murase K, and Yamazaki R 2010 *Astron. Astrophys.* **513** A17
[18] Ohira Y, Murase K, and Yamazaki R 2011 *Mon. Not. R. Astron. Soc.* **410** 1577
[19] Rho J and Borkowski K J 2002 *Astrophysical J.* **575** 201
[20] Acero F *et al.* 2016 *Astrophys. J. Suppl. Series,* **224** 8
[21] Aharonian F *et al.* 2008 *Astron. Astrophys.* **481** 401
[22] Dubner G, Giacani E, Reynoso E, and Paron S 2004 *Astron. Astrophys.* **426** 201
[23] Tian W W and Leahy D A 2008 *Mon. Not. R. Astron. Soc.* **391** L54
[24] Bocchino F, Bykov A M, Chen Y, Krassilchtchikov A M, Levenfish K P, Miceli M, Pavlov G G, Uvarov Y A, and Zhou X 2012 *Astron. Astrophys.* **541** A152
[25] Bamba A, Terada Y, Hewitt J, Petre R, Angelini L, Safi-Harb S, Zhou P, Bocchino F, Sawada M 2016 *Astrophysical J.* **818** 63
[26] Kosack K, Chaves R C G, Acero F 2011 *Proc. of 32nd Int. Cosmic Ray Conf.* vol 7 76
[27] Sato T, Koyama K, LEE S H and Takahashi T 2016 *Publ. Astron. Soc. Jpn.* **68** S8
[28] Wolszczan A, Cordes J M and Dewey R J 1991 *Astrophysical J.* **372** L99
[29] Saji S, Matsumoto, H, Nobukawa, M, Nobukawa K K, Uchiyama H, Yamauchi S, and Koyama K 2018 *Publ. Astron. Soc. Jpn.* **70** 23
[30] Araya M 2017 *Astrophysical J.* **843** 12
[31] H. E. S. S. Collaboration 2018 *Astron. Astrophys.* **612** A8
[32] Saji S 2018 Ph.D. dissertation (Nagoya University)

[33] Yamauchi S, Nobukawa M, Koyama K, and Yonemori M 2013 *Publ. Astron. Soc. Jpn.* **65** 6
[34] Combi J A, Albacete Colombo J F, Sánchez-Ayaso E, Romero G E, Martí J, Luque-Escamilla P L, Muñoz-Arjonilla A J, Sánchez-Sutil J R, and López-Santiago J 2010 *Astron. Astrophys.* **523** A76
[35] Bamba A *et al.* 2018 *Astrophysical J.* **854** 71
[36] Vogt F and Dopita M A 2011 *Astrophys. Spac. Sci.* **331** 521
[37] Ackermann M *et al* 2016 *Astron. Astrophys.* **586** A71
[38] H.E.S.S. Collaboration 2015 *Sci* **347** 406
[39] Hirayama A, Yamauchi S, Nobukawa K K, Nobukawa M, and Koyama K 2018 *Publ. Astron. Soc. Jpn.* submitted
[40] Petre R, Szymkowiak A E, Seward F D and Willingale R 1988 *Astrophysical J.* **335** 215
[41] Olbert C M, Clearfield C R, Williams N E, Keohane J W, and Frail D A 2001 *Astrophysical J.* **554** L205
[42] Nobukawa K K, Nobukawa M, Uchiyama H, Tsuru T G, Torii K, Tanaka T, Chernyshov D O, Fukui Y, Dogiel V A, and Koyama K 2015 *Astrophysical J.* **807** L10
[43] Bania T M 1977 *Astrophysical J.* **216** 381
[44] Cummings A C, Stone E C, Heikkila B C, Lal N, Webber W R, Jóhannesson G, Moskalenko I V, Orlando E, and Porter T A 2016 *Astrophysical J.* **831** 18
[45] Yamauchi S, Nobukawa K K, Nobukawa M, Uchiyama H, and Koyama K 2016 *Publ. Astron. Soc. Jpn.* **68** 59
[46] Nobukawa M, Uchiyama H, Nobukawa K K, Yamauchi S, and Koyama K 2016 *Astrophysical J.* **833** 268

IOP Conf. Series: Journal of Physics: Conf. Series **1181** (2019) 012041 doi:10.1088/1742-6596/1181/1/012041

Space-based GAMMA-400 mission for direct gamma- and cosmic-ray observations

N P Topchiev[1,*], A M Galper[1,2], I V Arkhangelskaja[2], A I Arkhangelskiy[1,2], A V Bakaldin[1,3], I V Chernysheva[1,2], O D Dalkarov[1], A E Egorov[1], Yu V Gusakov[1], M D Kheymits[2], A A Leonov[1,2], P Yu Naumov[2], N Yu Pappe[1], M F Runtso[2], Yu I Stozhkov[1], S I Suchkov[1], Yu T Yurkin[2] and V G Zverev[1]

[1] Lebedev Physical Institute, RU-119991 Moscow, Russia
[2] National Research Nuclear University MEPhI (Moscow Engineering Physics Institute), RU-115409 Moscow Russia
[3] Scientific Research Institute for System Analysis, RU-117218 Moscow, Russia

E-mail: *tnp51@yandex.ru

Abstract. The future space-based GAMMA-400 mission is intended for direct gamma- and cosmic-ray observations in the highly elliptic orbit during 7-10 years. GAMMA-400, currently developing gamma-ray telescope, will observe in the energy range from ~20 MeV to several TeV some regions of the Universe (such as Galactic Center, Fermi Bubbles, Crab, Cygnus, etc.) with the unprecedented angular (~0.01° at E_γ = 100 GeV) and energy (~1% at E_γ = 100 GeV) resolutions better than the Fermi-LAT, as well as ground gamma-ray telescopes, by a factor of 5-10. GAMMA-400 will also study cosmic rays in the energy range of up to ~20 TeV due to deep calorimeter (22 r.l. and 53 r.l. for vertical and lateral events, respectively). GAMMA-400 will permit to resolve gamma rays from dark matter particles, identify many discrete sources (many of which are variable), to clarify the structure of extended sources, to specify the data on the diffuse emission. GAMMA-400 will also specify the sources and the spectra of cosmic-ray electrons + positrons.

1. Introduction

At present, AGILE, Fermi-LAT, CALET, and DAMPE perform observations of discrete gamma-ray sources in space. The third Fermi-LAT catalog (3FGL) contains 3033 sources for the energy range from 100 MeV to 300 GeV, but 33% of gamma-ray sources are unidentified [1]. The ground-based facilities VERITAS, MAGIC, H.E.S.S., HAWC and others observe only 215 gamma-ray sources in the energy range more than 100 GeV (http://tevcat.uchicago.edu/). It is important to note that the observational data were mainly obtained for the energy ranges < 100 GeV for Fermi-LAT and > 100 GeV for ground-based facilities and these data overlap poorly for many gamma-ray sources. The frontier range around 100 GeV is still very interesting for investigations.

Another very interesting and important goal when studying gamma-ray sky is indirect searches of dark matter (DM). WIMPs with mass between several GeV and several TeV are still considered as the most probable candidate. WIMPs can annihilate or decay with the production of gamma rays. This emission can have both continuous energy spectrum or monoenergetic narrow lines. Up to now, there are no data on DM gamma-ray lines from space- and ground-based instruments.

Fermi-LAT, PAMELA, AMS-2, CALET, and DAMPE obtained energy spectra for primary cosmic-ray (CR) electrons + positrons, but their fluxes practically do not match in the energy range more than 50 GeV [2].

Therefore new direct observations of gamma-ray emission in the energy range from ~20 MeV up to several TeV and electron + positron fluxes and up to 20 TeV are required using the space-based telescope with much better separation from background, angular and energy resolutions in order to identify many gamma-ray sources, resolve DM gamma-ray lines, clarify energy spectra of CR electrons + positrons.

2. The GAMMA-400 gamma-ray telescope

The GAMMA-400 gamma-ray telescope will be installed onboard the Russian space astrophysical observatory [3-5]. The GAMMA-400 main scientific goals are: precise uninterrupted up to hundred days measurements of Galactic Center, Fermi Bubbles, Crab, Vela, Cygnus, Geminga, Sun, and other regions, extended and point gamma-ray sources, diffuse gamma rays, dark matter searching, measuring CR electron + positron fluxes with unprecedented angular (~0.01° at E_γ = 100 GeV) and energy (~1% at E_γ = 100 GeV) resolutions.

2.1. The GAMMA-400 physical scheme

The physical scheme of the GAMMA-400 gamma-ray telescope is shown in figure 1. GAMMA-400 includes:

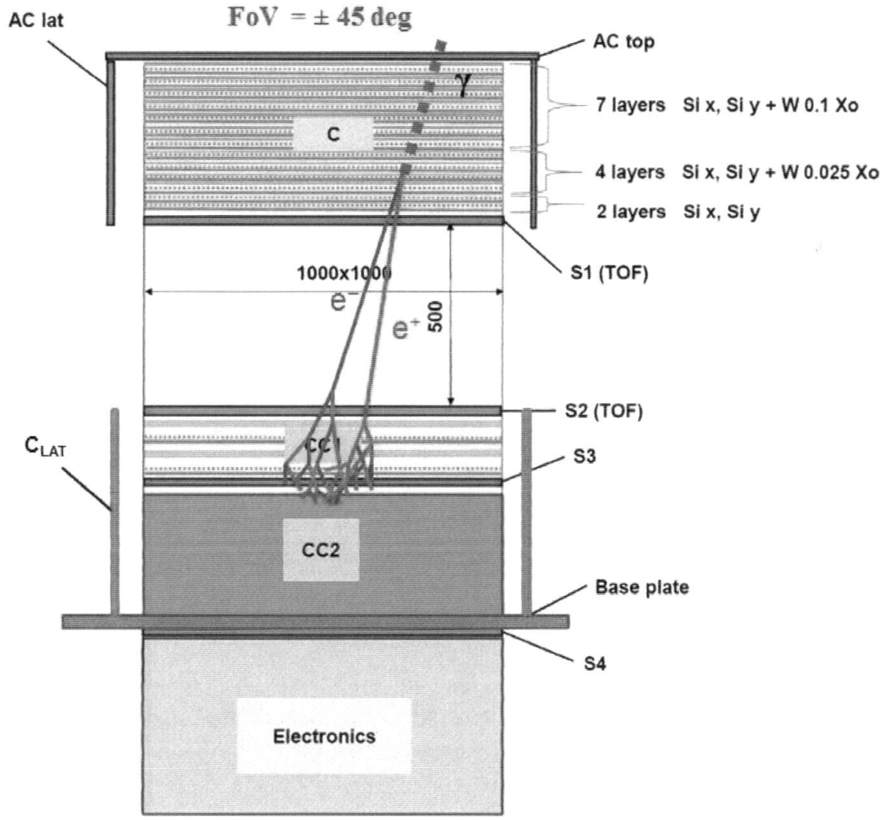

Figure 1. The GAMMA-400 physical scheme.

- plastic scintillation anticoincidence top ACtop (1280×1280 mm) and four lateral AClat (1280×600 mm) detectors with efficiency of 0.99995 and time resolution of 300 ps;

- converter-tracker (C), consisting of 13 pairs of planes of silicon strip detectors (X and Y coordinates, 1000×1000 mm) with pitch of 0.08 mm and analog readout. First 7 and 4 pairs have W converter foils of 0.1 X_0 and 0.025 X_0 thick each, respectively. Last 2 pairs do not have W;
- time of flight system (ToF) from plastic scintillation detectors (S1 and S2, 1000×1000 mm) spaced by 500 mm with coefficient of separation for top/down moving particles of 1000 and time resolution of 300 ps;
- calorimeter (CC1 and CC2). CC1 (1000×1000 mm) consists of two layers of CsI(Tl) scintillation crystals and silicon strip detectors (X and Y coordinates) with pitch of 0.08 mm. CC2 (1000×1000 mm) consists of 28×28 CsI(Tl) crystals. Total thickness of calorimeter is ~22 X_0 (~1.0 λ_0) and ~54 X_0 (~2.5 λ_0) for vertical and lateral particle detection, respectively;
- plastic scintillation detectors (S3 and S4, 1000×1000 mm) for improving hadron and electromagnetic shower separation;
- four lateral scintillation detectors of calorimeter (Clat) for detecting lateral electrons + positrons.

After interaction of incident gammas with GAMMA-400 matter the backscattering omnidirectional particles (mainly, 1-MeV photons and electrons) are arisen. In order to exclude backscattering particles, all scintillation detectors consist from two independent 10-mm layers of 100-mm strips and fast timing methods are used. Timing methods use the difference in time in AC and S1 ToF to separate incoming and backscattering particles.

The GAMMA-400 energy range for gamma-ray studies is from ~20 MeV to several TeV and up to ~20 TeV for electrons + positrons. The field of view (FoV) for detecting particles from top is ±45°. The geometrical factor for detecting electrons + positrons from vertical directions is ~1.3 m^2sr and from four lateral directions using Clat is ~2 m^2sr.

2.2. The GAMMA-400 performance

2.2.1. Calculation results

Model calculations of the GAMMA-400 gamma-ray telescope performance were carried out using the "GEANT4.10.01.p02" software package. Examples of interaction of gammas with the GAMMA-400 matter for E_γ = 0.1, 1, 10 TeV are shown in figures 2a-2c.

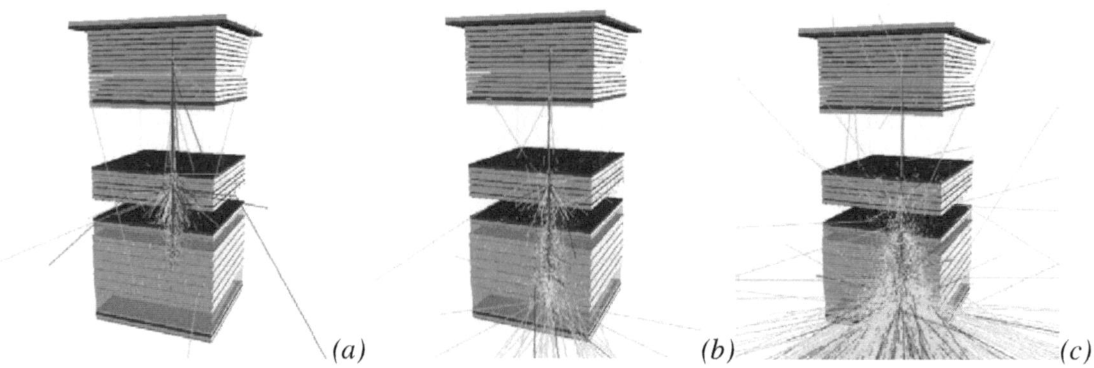

Figure 2. The development of electromagnetic shower in the GAMMA-400 gamma-ray telescope for: (a) E_γ = 0.1 TeV, (b) E_γ = 1 TeV, (c) E_γ = 10 TeV. Electrons and positrons are marked by blue and violet colors, respectively, secondary gammas are not shown for better visualization.

As a result of calculations, we obtained the following dependences:
- the effective area vs the energy (figure 3a). The effective area is ~5000 cm^2 for the energy more than 10 GeV;
- the effective area vs the angle of incidence of particles for E_γ = 1, 10, 100 GeV (figure 3b);
- the energy resolution vs the energy (figure 3c). The energy resolution for E_γ = 100 GeV is ~1%;

- the angular resolution vs the energy (figure 3d). The angular resolution for E_γ = 100 GeV is ~0.01°.

Using the combined information from all GAMMA-400 detector systems, it is possible to reach an effective rejection of protons from electrons. The methods to separate electron from protons presented in [6] are based on the difference of the development of hadronic and electromagnetic showers inside the instrument. For the current physical scheme the rejection factor for vertical protons is about 3×10^5.

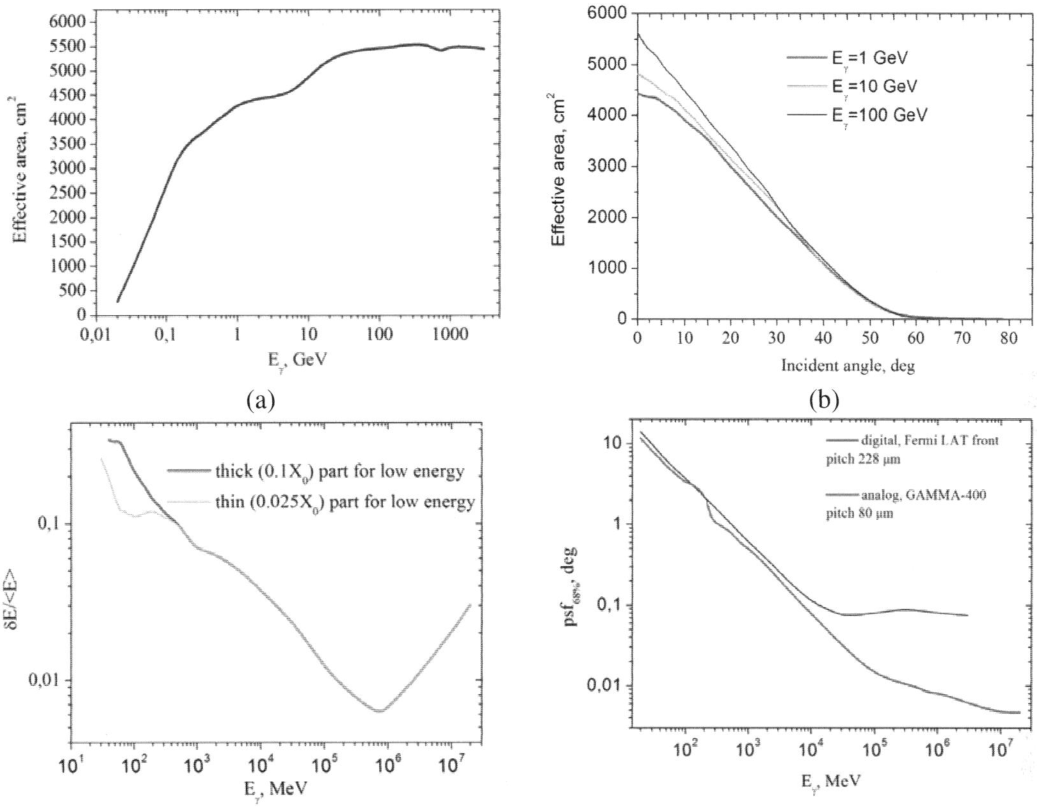

(c) (d)

Figure 3. The GAMMA-400 dependences: (a) the effective area vs the energy for vertically incident particles; (b) the effective area vs the angle of incidence of particles for E_γ = 1, 10, 100 GeV; (c) the energy resolution vs the energy for two parts of converter: 4 panels with W of 0.025 X_0 and 7 panels with W of 0.1 X_0; (d) the angular resolution vs the energy for: GAMMA-400 (80-μm pitch, analog readout) and Fermi-LAT (228-μm pitch, digital readout).

2.2.2. Calibration results

Some laboratory prototypes of the GAMMA-400 detector systems were calibrated at the synchrotron of Lebedev Physical Institute on positron beam up to the energy of 300 MeV.

The CC2 prototype consisted of 4 scintillator crystals CsI(Tl) with dimensions of 36×36×372 mm and silicon photomultipliers SensL MicroFC-60035 mounted on the end of each crystal. Positron beam with a diameter of 10 mm was directed to the center of crystal along its axis. The prototype energy resolution of 10% for the positron energy of 300 MeV was obtained. This result coincides with calculations.

Studies of time resolution of the AC prototype were performed. The prototype consists of two strips of scintillator BC-408 with dimensions of 1280×100×10 mm each viewed by four silicon photomultipliers SensL MicroFC-60035-SMT and front-end electronics. The time resolution for scintillator strips was about 300 ps.

3. The preliminary GAMMA-400 scientific program

3.1. The GAMMA-400 astrophysical observatory

The GAMMA-400 astrophysical observatory will be installed onboard of the Navigator space platform, which is designed and manufactured by the Lavochkin Association.

The GAMMA-400 experiment will be initially launched into a highly elliptical orbit (with an apogee of 300 000 km and a perigee of 500 km, with an inclination of 51.4°), with 7 days orbital period. Under the influence of gravitational disturbances of the Sun, Moon, and the Earth after ~6 months the orbit will transform to about an approximately circular one with a radius of ~200 000 km and will not suffer from the Earth's occultation and be outside the radiation belts. A great advantage of such an orbit is the fact that the full sky coverage will always be available for gamma-ray astronomy, since the Earth will not cover a significant fraction of the sky, as is usually the case for low-Earth orbit. Therefore, the GAMMA-400 source pointing strategy will hence be properly defined to maximize the physics outcome of the experiment. The launch of the GAMMA-400 space observatory is planned for the middle of the 2020s.

3.2. Observation of the Galactic plane

GAMMA-400 will study continuously over a long period of time different regions of Galatic plane, for example, Galactic center, Fermi Bubbles, Crab, Vela, Cygnus, Geminga with FoV of ±45° (figure 4). Using the gamma-ray fluxes from 3FGL, we can expect the number of sources (N_s) with gammas more than 10 and gammas (N_γ) for different energy ranges, when GAMMA-400 will observe several regions during 100 days (table 1).

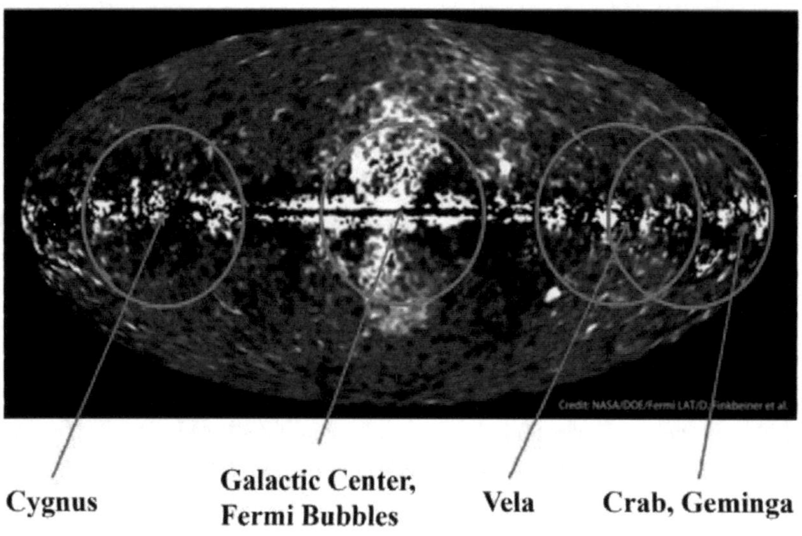

Figure 4. The Fermi-LAT all-sky gamma-ray data (in galactic coordinates), where Galactic center, Fermi Bubbles, Cygnus, Vela, Crab,and Geminga are marked by red circles, corresponding to the GAMMA-400 FoV of ±45°. These regions will be observed with GAMMA-400.

3.3. Dark matter searching

Main targets to search for gamma rays from dark matter are:

Milky Way. The center of Milky Way is, apparently, the best potential source of dark matter emission possessing the largest J-factor. Moreover, recently, the anomaly excess of gamma-ray emission in the GeV energy range was revealed near the Galactic center (the region of about one degree) [7], which can be well described by dark matter with mass of several tens of GeV and annihilation cross section of about standard thermal 10^{-26} cm^3/s. However, this observed excess can have another interpretation - the presence of population of millisecond pulsars [8]. Therefore, the new GAMMA-400 observational data can help to solve this problem, using the best angular and time resolutions.

Milky Way satellites are considered for a long time as the strongest sources of constraints for dark matter, because they have sufficiently large J-factors and at the same time have considerably less gamma-ray background in comparison with the Galactic center. GAMMA-400 will be able to specify the constrain area.

Other objects. Another potentially interesting objects are other galaxies and their clusters, where dark matter presents and can emit gamma rays. GAMMA-400 with the highest energy resolution of ~1% will have unique sensitivity for detecting dark matter.

Table 1. Expected number of sources (N_s) with gammas more than 10 and gammas (N_γ) for different energy ranges, when GAMMA-400 will observe several regions during 100 days.

Energy range	100 MeV-100 GeV		1 GeV-100 GeV		10 GeV-100 GeV	
	N_s	N_γ	N_s	N_γ	N_s	N_γ
Galactic center b=0°, l=0°	723	$5.2x10^5$	422	$4.8x10^4$	21	1365
Crab + Geminga b=0°, l=190°	495	$3.1x10^5$	175	$3.9x10^4$	11	1020
Vela b=0°, l=265°	649	$5.2x10^5$	280	$6.3x10^4$	9	1165
Cygnus b=0°, l=75°	604	$3.2x10^5$	269	$3.1x10^4$	12	1010

References

[1] Acero F et al. 2015 *ApJS* 218 23
[2] Adriani O et al. 2018 *Phys. Rev. Lett.* 120 261102, *arXiv*:1806.09728
[3] Galper A et al. 2013 *ASR* 51 297
[4] Topchiev N et al. 2016 *J. of Phys.: Conf. Ser.* 675 032009
[5] Galper A et al. 2017 *Phys. Atomic Nucl.* 80 1141
[6] Leonov A et al. 2015 *ASR* 56 1538
[7] Abazajian K and Kaplinghat M 2012 *Phys. Rev*. D86 083511
[8] Bartels R, Krishnamurthy S and Weniger C 2016 Phys. Rev. Lett. 116 051102, *arXiv*:1506.05104

IOP Conf. Series: Journal of Physics: Conf. Series **1181** (2019) 012042 doi:10.1088/1742-6596/1181/1/012042

Measurements of Heavy Cosmic-Ray Nuclei Spectra with CALET on the ISS

Yosui Akaike for the CALET Collaboration

CRESST/NASA Goddard Space Flight Center and University of Maryland Baltimore County, 8800 Greenbelt Road, Greenbelt, MD 20770, USA

E-mail: yosui.akaike@nasa.gov

Abstract. CALorimetric Electron Telescope (CALET) has been accumulating data of high-energy cosmic rays on the International Space Station since August 2015. In addition to the primary observation of the all-electron spectra, CALET also measures the spectra of nuclei, their relative abundances and secondary-to-primary ratios to the highest energy region ever directly observed in order to investigate details of their origin and propagation in the galaxy. The CALET instrument consists of two layers of segmented plastic scintillators to identify the individual elements from $Z = 1$ to 40, a 3 radiation length thick tungsten-scintillating fiber imaging calorimeter to obtain complementary charge and tracking information, and a 27 radiation length thick segmented PWO calorimeter to measure the energy. In this paper, the capability of CALET to perform nuclei measurements and preliminary energy spectra of heavy nuclei components using 962 days of data is presented.

1. Introduction

Precision measurements of cosmic-ray spectra and their flux ratios are important to understand their origin and propagation in our galaxy. While the acceleration via supernova shock waves and diffusive propagation in the galactic magnetic fields are commonly accepted as a plausible scenario for cosmic-ray origins, there are many outstanding questions about the details. In particular, recent observations such as CREAM [1–3], PAMELA [4] and AMS-02 [5–8] reported that a deviation from a single power-law has been observed for proton, helium and light nuclei at a few hundred GeV/n which indicates an unexpected hardening of the spectra.

CALET is a cosmic ray experiment on the International Space Station [9, 10], and has been collecting science data since mid-October 2015 [11]. The detector is optimized to measure electron spectrum in the multi-TeV region and the results of electron spectrum measurement have been reported [12, 13]. CALET also measures the energy spectra and elemental composition of cosmic-ray nuclei from proton to iron in the range from a few tens of GeV to the PeV scale. The features of the CALET instrument include very good energy resolution provided by its thick calorimeter and excellent charge resolution and robust track identification based on the segmented scintillator paddles and scintillating fibers. Also the dynamic range of CALET covers 6 orders of magnitude in energy range from 1 MIP to 1 PeV shower energy. In this paper, we present the analysis procedure of nuclei measurements and preliminary energy spectra of heavy nuclei components with the data obtained in the period from October 13, 2015 to May 31, 2018.

2. CALET instrument

The CALET detector consists of three subsystems; CHarge Detector (CHD), IMaging Calorimeter (IMC) and Total AbSorption Calorimeter (TASC). CHD located at the top part is composed of two layers of 14 plastic scintillator paddles for measurement of the primary particle charge. Each scintillator has dimensions of 32 mm × 450 mm × 10 mm. IMC is a sampling calorimeter composed of $(X,\ Y) \times 8$ layers of scintillating fiber (SciFi) belts and 7 tungsten plates. Each layer of SciFi belts is made of 448 SciFis with a 1 mm square cross section and 448 mm in length. The tungsten plates interleaved between the SciFi layers have thickness of $0.2\ X_0 \times 5$ layers and $1.0\ X_0 \times 2$ layers from top to bottom. The IMC is used for track reconstruction and charge identification. TASC is a total absorption calorimeter made of 12 layers of PWO scintillator logs for energy measurement and discrimination of electromagnetic shower and hadronic shower. Each layer has 16 PWO logs and each log has dimensions of 19 mm × 326 mm × 20 mm. The total thickness of the calorimeter is 30 X_0 for electromagnetic particles or 1.3 λ_I for protons. Figure 1 and 2 show examples of schematic view of the detector with shower images of carbon and iron nuclei from the flight data.

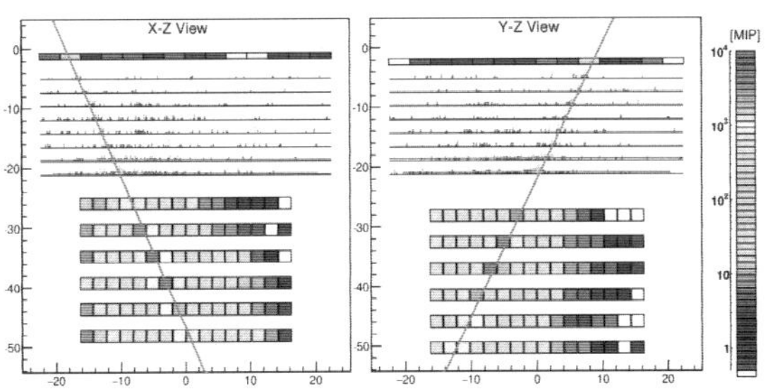

Figure 1. An example of carbon event with $\Delta E_{\mathrm{TASC}} = 1.4$ TeV.

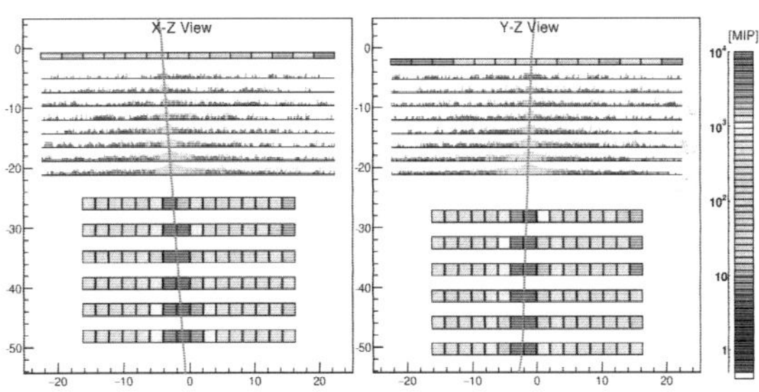

Figure 2. An example of iron event with $\Delta E_{\mathrm{TASC}} = 9.2$ TeV.

3. Data analysis

Event trigger The events detected by the onboard high energy trigger (HET) are used for the flux analysis. HET requires a coincidence of the two bottom layers of IMC and the top layer of TASC with thresholds to detect electrons above 10 GeV. Therefore, for light nuclei only events with a shower induced in the detector are triggered by HET, while for heavier nuclei with $Z \geq 10$ also penetrating particles are triggered because their dE/dx is large enough to exceed the trigger threshold. HET efficiency for light nuclei is evaluated using a data set taken with the same trigger logic, but with a lower threshold that allows to detect penetrating particles. Figure 3 shows the HET efficiency compared with MC simulation. Data and MC are in good agreement.

Detector calibration Energy calibration of each channel has been carried out by using penetrating cosmic-ray protons and helium [14]. The detector response includes position, temperature and time dependence of the plastic scintillators and PWO logs have been well studied and equalized with the MIP calibration. For charge identification, the non-linearity of the CHD and IMC between detector response and deposit energy which is caused by the scintillation quenching effects is obtained from the flight data. Figure 4 shows the peak values of the CHD signal distributions of some charge derived from the flight data as a function of Z^2.

IOP Conf. Series: Journal of Physics: Conf. Series **1181** (2019) 012042 doi:10.1088/1742-6596/1181/1/012042

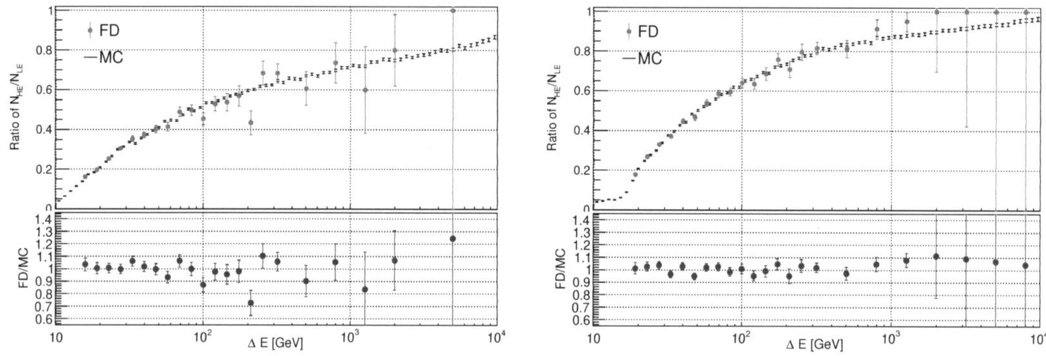

(a) Carbon (b) Oxygen

Figure 3. High energy trigger efficiency of flight data (FD) compared with MC simulations.

It is fitted by a function based on the halo model [15] described as:

$$f(Z^2) = A\left(\frac{1 - f_h}{1 + B_S(1 - f_h)\alpha Z^2} + f_h\right)\alpha Z^2, \tag{1}$$

where f_h is the ratio of halo region, B_S is the quenching parameter, A is the normalization factor and α is a constant (2 MeV). Since the signal of CHD-X and Y are affected the δ-ray from upstream material, the parameters of CHD-X and Y are obtained independently as shown in Fig. 4. Figure 5 shows the correlation of CHD-X and CHD-Y from proton to nickel. Non-linearity of IMC has also been corrected with the same procedure used for the charge identification.

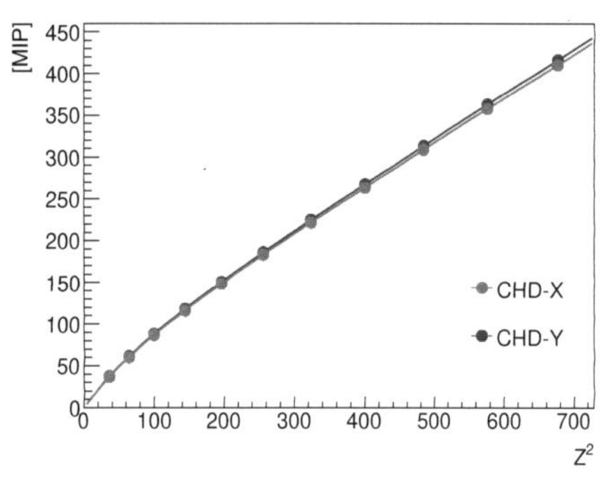

Figure 4. Peak value of CHD signals of some charge as a function of Z^2.

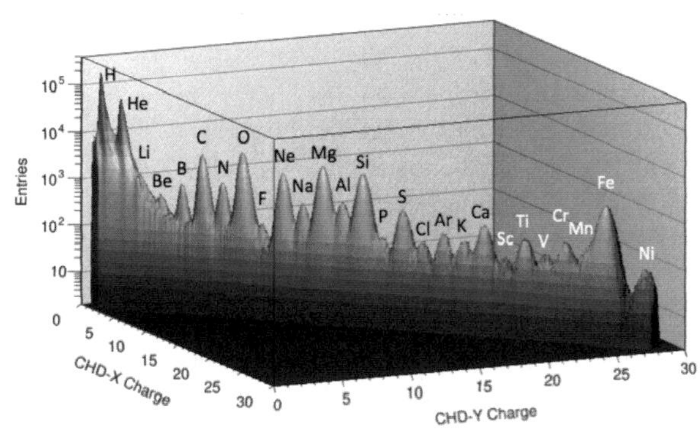

Figure 5. Correlation between CHD-X and CHD-Y.

Track reconstruction The accurate reconstruction of shower axis and estimation of the impact point at the CHD are essential for the charge identification. The shower axis is reconstructed by using the IMC signals. Although heavy ions create many shower particles in the IMC, which could be a large background for the track reconstruction, the signals from the ionizing primary particle are commonly larger than the signals of the shower particles thanks to their large dE/dx

which is proportional to Z^2 disregarding the quenching effect. The track is reconstructed by a least-square fit using the channels with maximum deposit in the upper four IMC layers. The impact point at the CHD can be estimated with an accuracy of 330 μm for CHD-X and 300 μm for CHD-Y with minor charge and energy dependence. Events whose reconstructed track is fully contained in the detector are selected for the nuclei analysis in this paper.

Charge identification To perform precise charge measurements and maintain good charge resolution, charge consistency of CHD-X, CHD-Y and upper IMC layers are required. Events are selected if the difference of the charge with CHD-X and CHD-Y is less than 10%. For light nuclei with $Z \leq 8$, the charge consistency of IMC with 15% is also required. In addition, the track width defined as the difference of the sum of 7 SciFis and the sum of 3 SciFis associated with the reconstructed track and normalized to the reconstructed charge squared is used for the selection because the events interacting in the CHD make a wider track due to the secondary particles than that of penetrating events as shown in Fig. 6. Applying these cuts rejects events with poor charge resolution due to interactions in the CHD or upper IMC layers. The distribution of estimated charge in the range of $Z = 5 - 28$ is shown in Fig. 7. Events within ± 0.4 charge units around the Z-peaks are identified as particular elements in the flux analysis presented in this paper. The total efficiency is ~30% for carbon, ~40% for oxygen and more than ~60% for heavier particles. The contamination from the other particles is estimated by means of MC simulations. MC events in each energy bin for all species with $Z = 1 - 28$ are normalized to the flight data based on the charge distribution with CHD. The number of contaminant events is calculated by the integration of all contaminant MC events within ± 0.4 charge unit. Figure 8 shows the energy distribution of carbon, the estimated background and the ratio. The total background ratio is less than 1% for carbon, oxygen and iron, less than 5% for neon, magnesium and silicon.

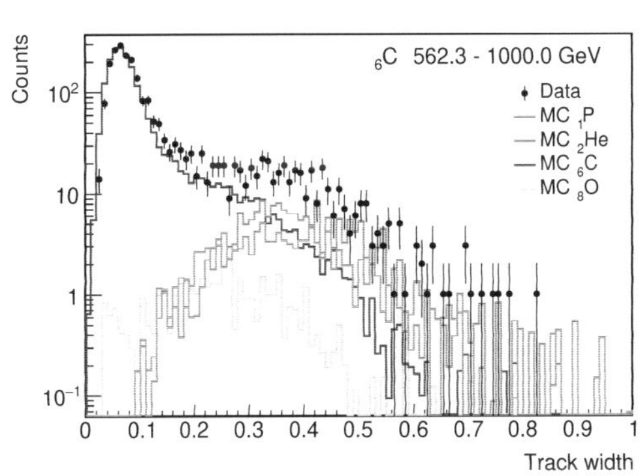

Figure 6. The distribution of track width for carbon before charge consistency cuts.

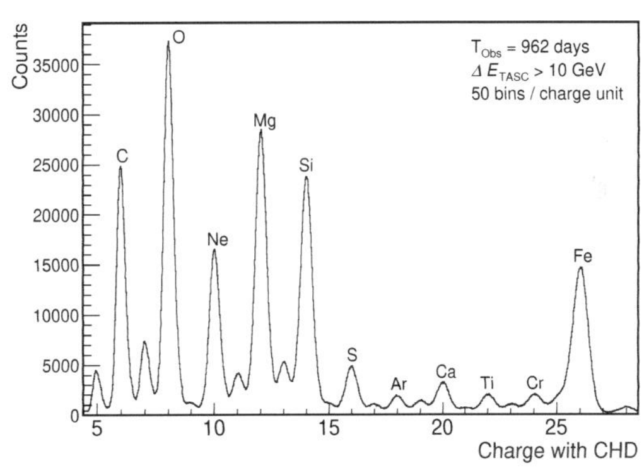

Figure 7. Charge distribution of CHD with $\Delta E_{\mathrm{TASC}} > 10$ GeV.

Energy measurements and unfolding The shower energy for an event is determined by the sum of energy deposits in the TASC. Since the energy leakage for nuclei from the calorimeter is unavoidable due to the characteristics of hadron induced showers and the finite detector resolution, the energy unfolding procedure for the derivation of the primary energy spectrum to correct bin-to-bin migration is necessary. Iterative procedure based on the Bayes's theorem [16] with the RooUnfold package [17] is applied. The response function for the energy unfolding is made from detailed MC simulations including the detector response. MC events are selected with the same cuts used for flight data. Figure 9 shows the distribution of observed energy and unfolded energy for carbon.

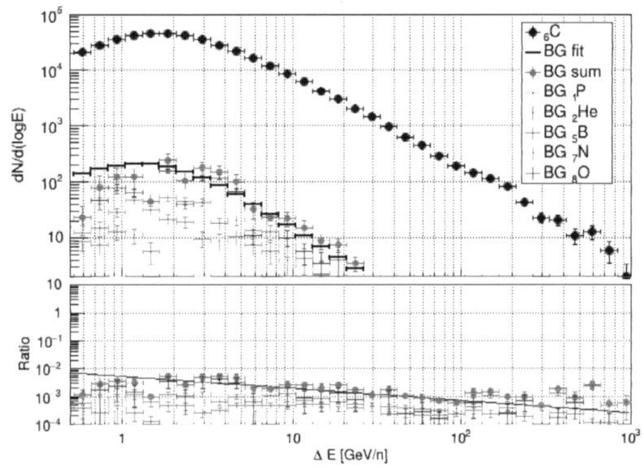

Figure 8. dN/dE, estimated background and the ratio of background to signal for carbon.

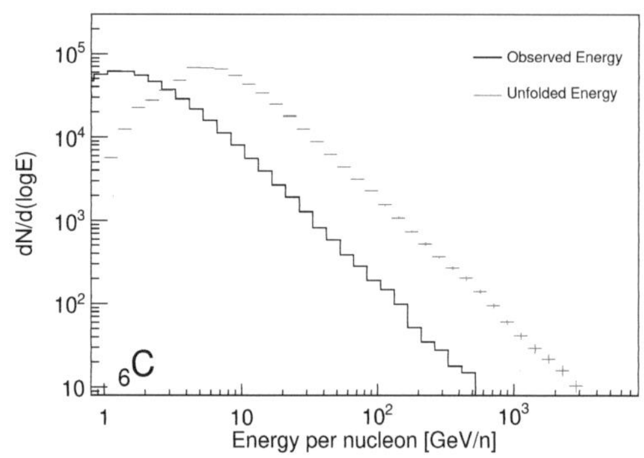

Figure 9. Energy distributions before and after the energy unfolding for carbon.

4. Preliminary Results

Energy flux, $\Phi(E)$, is calculated as follows,

$$\Phi(E) = \frac{N(E)}{\varepsilon S\Omega T \Delta E} \tag{2}$$

where $N(E)$ is the number of events in the unfolded energy bin, $S\Omega$ is the geometrical acceptance, T is live time, ΔE is the width of energy bin and ε is the total efficiency. Figure 10 shows the preliminary energy spectra of carbon, oxygen, neon, magnesium, silicon and iron as a function of kinetic energy per particle with 962 days of operations. These spectra are comparable with previous observations [7, 8, 18–24]. We note that the event selection we have applied here is based on a preliminary analysis as compared to what can eventually be achieved with CALET. Further studies are now underway to provide a more detailed analysis of these spectra.

5. Summary

The ability of CALET to measure heavy cosmic-ray nuclei has been successfully demonstrated and preliminary energy spectra have been derived for the primary cosmic ray elements up to 100 TeV using 5.57×10^6 events in the range of C–Fe with $\Delta E > 10$ GeV from the 962 days of operations. The derived spectra illustrate the excellent capability of CALET to measure heavy ions with high statistics in a wide energy range. Further studies on an increased data set and detailed systematic study will increase the sensitivity to detailed spectral features, which may be a key to solve questions about galactic cosmic-ray acceleration and propagation.

References

[1] Ahn H S *et al.* 2010 *Astrophys. J. Lett.* **714** 89
[2] Yoon Y S *et al.* 2011 *Astrophys. J.* **728** 122
[3] Yoon Y S *et al.* 2017 *Astrophys. J.* **839** 5
[4] Adriani O *et al.* 2011 *Science* **332** 69
[5] Aguilar M *et al.* 2015 *Phys. Rev. Lett.* **114** 171103
[6] Aguilar M *et al.* 2015 *Phys. Rev. Lett.* **115** 211101

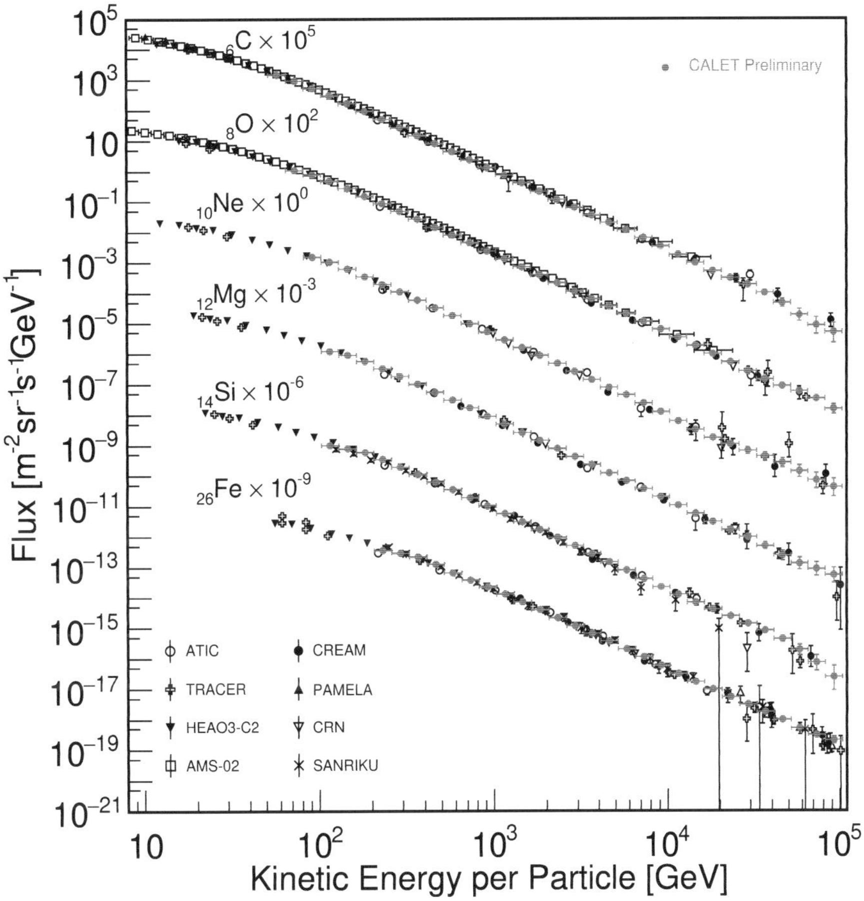

Figure 10. Preliminary energy spectra of carbon, oxygen, neon, magnesium, silicon and iron as a function of kinetic energy per particle after 962 days of CALET operation compared with previous observations [7, 8, 18–24]. Only statistical errors are shown.

[7] Aguilar M *et al.* 2017 *Phys. Rev. Lett.* **119** 251101
[8] Aguilar M *et al.* 2018 *Phys. Rev. Lett.* **120** 021101
[9] Torii S *et al.* 2017 *Proceedings of Science (ICRC2017)* 1092
[10] Asaoka Y *et al.* 2018 *this volume*
[11] Asaoka Y, Ozawa S, Torii S *et al.* 2018 *Astropart. Phys.* **100** 29
[12] Adriani O *et al.* 2017 *Phys. Rev. Lett.* **119** 181101
[13] Adriani O *et al.* 2018 *Phys. Rev. Lett.* **120** 261102
[14] Asaoka Y, Akaike Y, Komiya Y, Miyata R, Torii S *et al.* 2017 *Astropart. Phys.* **91** 1
[15] Tarlé G, Ahlen S and Cartwright B 1979 *Nucl. Instrum. Neth. A* **230** 607
[16] D'Agostini G 1995 *Nucl. Instrum. Neth. A* **362** 487
[17] Adye T 2011 *arXiv* 1105.1160
[18] Panov A *et al.* 2009 *Bull. Russian Acad. Sci.* **73** 564
[19] Ahn H *et al.* 2009 *Astrophys. J.* **707** 593
[20] Obermeier A *et al.* 2011 *Astrophys. J.* **724** 14
[21] Adriani O *et al.* 2014 *Astrophys. J.* **791** 93
[22] Engelmann J *et al.* 1990 *Astron. Astrophys.* **233** 96
[23] Muller D *et al.* 1991 *Astrophys. J.* **374** 356
[24] Ichimura M *et al.* 1993 *Phys. Rev. D.* **48** 1949

IOP Conf. Series: Journal of Physics: Conf. Series **1181** (2019) 012043 doi:10.1088/1742-6596/1181/1/012043

Main scientific results of the DAMPE mission

Paolo Bernardini, on behalf of the DAMPE Collaboration

Dipartimento di Matematica e Fisica, Università del Salento, Lecce, Italy
Istituto Nazionale Fisica Nucleare, Sezione di Lecce, Italy

E-mail: paolo.bernardini@le.infn.it

Abstract. The DArk Matter Particle Explorer (DAMPE) is a space mission, resulting from the collaboration of Chinese, Italian, and Swiss institutions. Since December 2015, DAMPE orbits at the altitude of 500 km and collects data regularly. The detector is made of four sub-detectors: top layers of plastic scintillators, a silicon-tungsten tracker, a deep BGO calorimeter (32 radiation lengths), and a bottom boron-doped scintillator to detect delayed neutrons. The main goal of the mission is the search for indirect signals of Dark Matter in the electron and photon spectra with energies up to 10 TeV. Furthermore DAMPE is studying cosmic charged and gamma radiation. The calorimeter depth and the large effective acceptance allow to measure cosmic ray fluxes in the range from 20 GeV up to hundreds of TeV. An overview of the latest results about electron and positron flux, light component (p+He) of charged cosmic rays, and gamma rays will be presented.

1. Introduction

The DArk Matter Particle Explorer (DAMPE) is a space mission supported by the strategic space projects of the Chinese Academy of Sciences with the contribution of Swiss and Italian institutions [1, 2].

The rocket has been successfully launched on December 17, 2015. After a week the HV system was switched on and the data acquistion began very soon. Presently DAMPE flies regularly on a Sun-synchronous orbit at the altitude of 500 km. Four different detectors are arranged on the satellite: a plastic scintillator array, a silicon-tungsten tracker, a BGO calorimeter, and a neutron detector. They are devoted to measure the fluxes of charged cosmic rays (electrons, protons, and heavier nuclei), to study the high energy gamma-signal of astrophysical origin and to search for indirect dark-matter signatures.

After a synthetic review of the performances of the detector the main scientific results of the first two years of data taking will be presented in this paper.

2. On-board instruments

In Fig. 1 the four sub-detectors of the DAMPE experiment are shown:

- PSD - a Plastic Scintillator strip Detector used to measure the charge of cosmic rays and to select gamma-events (events in anti-coincidence with the PSD signal),
- STK - a Silicon-Tungsten tracKer-converter to reconstruct the direction of incident particles, to repeat the charge measurement and to convert photons in electron-positron pairs,

- BGO - a bismuth-germanate imaging calorimeter (14 layers) of about 32 radiation lengths to measure the energy with high resolution and to distinguish between hadron and electromagnetic showers,
- NUD - a NeUtron Detector to further increase the hadronic-shower rejection power.

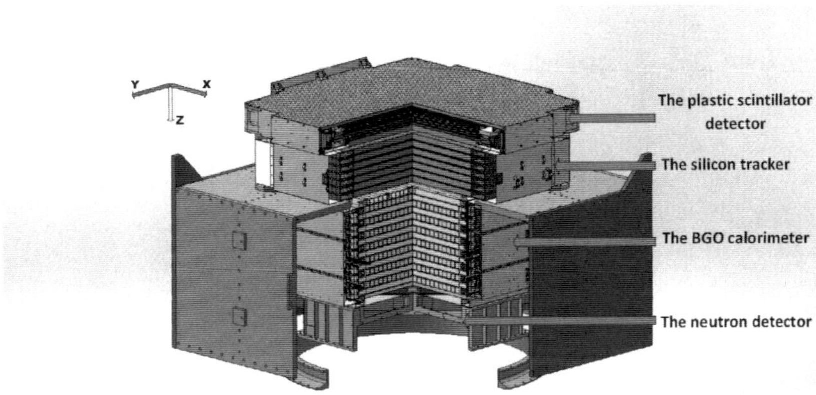

Figure 1. The DAMPE detector.

3. Ground tests

The detector design was performed looking at the proper compromise between research goals and limitations on geometry, power consumption, and weight. DAMPE performances were verified by a series of beam tests at CERN. The wide energy range of electron beam and the high-purity of proton beam were sufficient to check energy resolution (Fig.s 2 and 3), linearity and e/p separation. Also beams of argon and lead fragments were used to check the PSD capability to measure Z for heavy ions (Fig. 4 for the argon beam). Details of the beam-test results can be found in [1, 3, 4].

Furthermore sea-level muon test have been performed during different stages of the DAMPE assembly. Then the energy response to MIPs, the efficiency and the detector alignment have been calibrated with a relatively large sample of CR muon events.

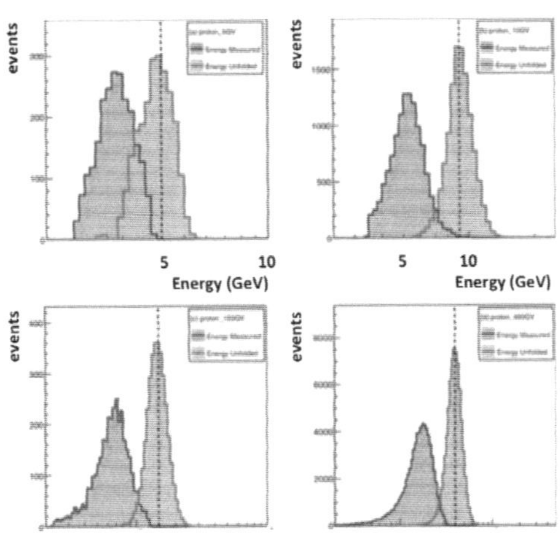

Figure 2. Beam test - Energy deposited in the BGO-calorimeter (blue histograms) and energy reconstructed (red histograms) by means of an unfolding procedure for different proton beams (5, 10, 150, 400 GeV/c).

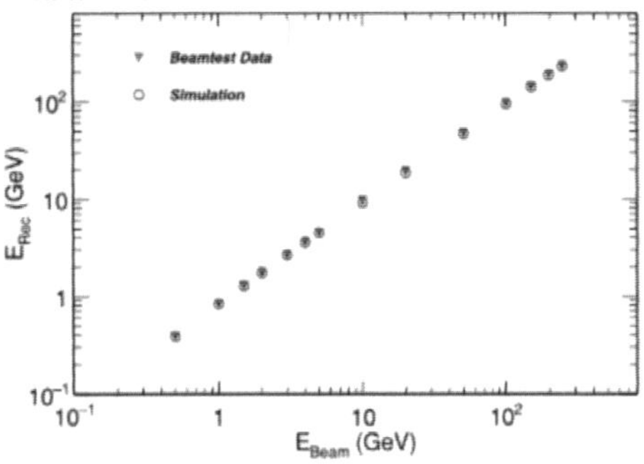

Figure 3. Beam test - Reconstructed energy versus electron beam energy.

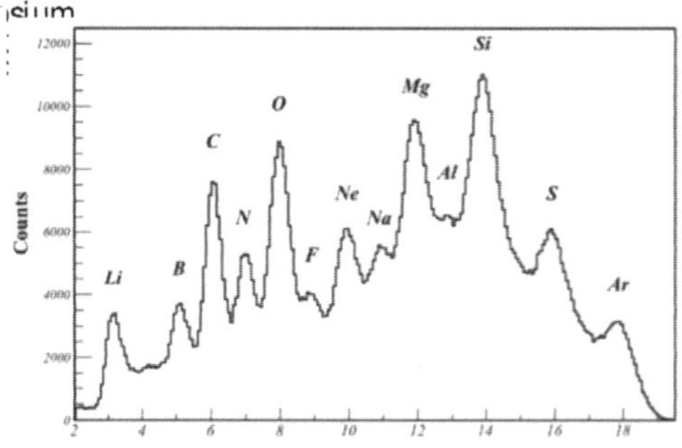

Figure 4. Beam test - PSD charge spectrum of fragments of an argon-beam (helium peak has been removed).

4. In-flight DAMPE performances

During on-orbit operation, calibration data are acquired in two different modes, including the dedicated mode for electronics linearity and pedestal calibration and the continuous observation mode for scientific data taking. The data acquisition system runs continuously in observation mode for most of the time and switches to dedicated mode for 40 seconds for pedestal calibration twice per orbit.

All the on-orbit calibration results (pedestal values, dynode ratios, energy gains, trigger thresholds, and so on) are consistent with the ground ones and are very stable after temperature correction. The STK alignment allows a significant improvement of the tracking resolution [5]. In Fig. 5 the PSD-signal due to Minimum Ionizing Particles (MIPs) is plotted as a function of time. The same is shown in Fig. 6 for the BGO-signal. In both cases the detector response is very stable.

As an example of the excellent DAMPE in-flight performances we can present the definition of the South Atlantic Anomaly (SAA) region. SAA boundaries are found looking at the hit rate given by the top layers of the BGO calorimeter with a low trigger threshold (Fig. 7). Then events collected in SAA region are rejected in the DAMPE analyses and the live-time is reduced.

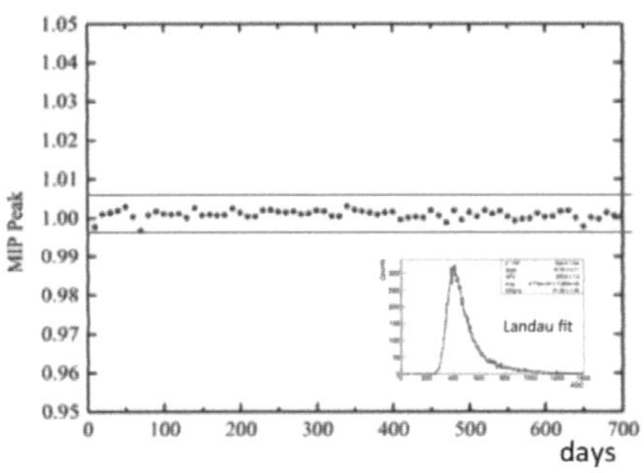

Figure 5. In-flight - The deposited energy in PSD for proton is fitted with the Landau function. The Most Probable Value (MPV) of the Landau fit is shown as a function of time.

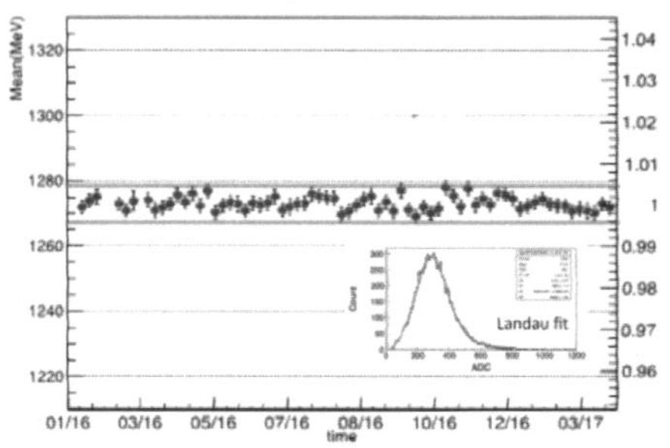

Figure 6. In-flight - The deposited energy in BGO for helium is fitted with the Landau function. MPV of the Landau fit is shown as a function of time.

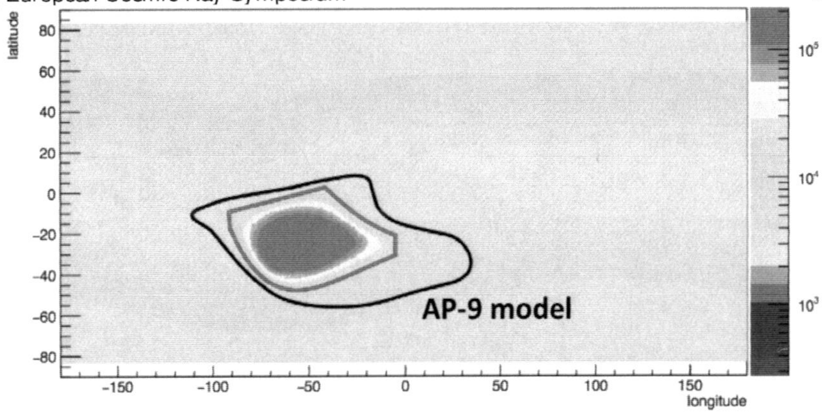

Figure 7. Trigger rate versus geographic longitude and latitude. The SAA is well recognized (red contour) according the requirement of the DAMPE data-taking [6]. The black contour is defined requiring more than 1 proton/cm^2/s above 10 MeV (AP-9 model).

5. Cosmic rays

The cosmic-ray (CR) studies are based on the charge measurement (Z^2 is proportional to the energy release in the PSD), the STK-tracking and the BGO-imaging calorimeter. Fig. 8 shows the CR elemental spectrum after alignment, light-attenuation, and equalization corrections, taking also into account the loss of linearity (Birks' law). The event shape in the calorimeter allows to distinguish electromagnetic showers from hadronic ones. Furthermore the unprecedented thickness of the BGO-calorimeter allows to push the direct measurement of CR-flux to highest energies with excellent energy resolution. Indeed DAMPE CR-measurements are going to reach energies of some hundreds of TeV.

The template fit for protons and helium in Fig. 9 shows an excellent data-MC agreement and allows to estimate the pollution in each peak. The preliminary measurement of proton and helium fluxes are presented in Fig. 10.

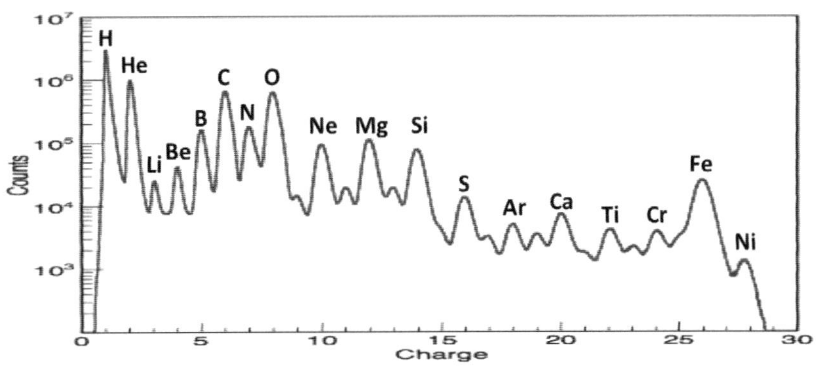

Figure 8. Charge spectrum reconstructed on the basis of the deposited energy in the PSD (two years of on-orbit data). Almost all Z-peaks are visible up to nickel.

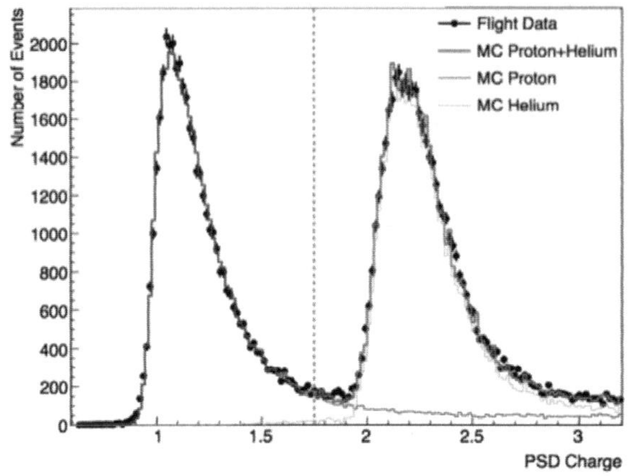

Figure 9. Combined PSD signal for protons and helium nuclei, for BGO energies in the range 447-562 GeV. The color codes to distinguish on-orbit and simulated data are shown in the plot. The cut to select proton candidates is represented by the vertical line.

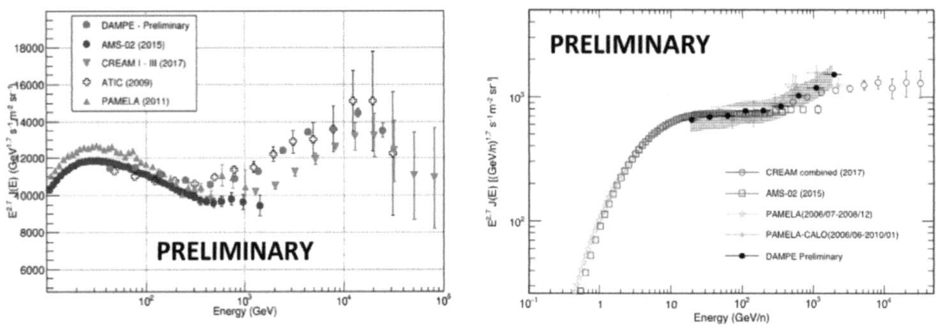

Figure 10. Preliminary - Proton (left) and helium spectrum (right).

6. Gamma astronomy

Another aim of the DAMPE mission is the gamma astronomy by collecting events without hits on the PSD. In less than 1 year DAMPE is able to scan the whole sky by looking for bright gamma-sources, detecting their fluxes and possible variability. The gamma-periodicity of two brightest pulsars (Geminga and Vela) has been measured in an energy range from 1 GeV to 100 GeV [7]. Also the variable high energy gamma emission from two Active Galactic Nuclei (CTA102, 3C454.3) has been detected [8].

7. Measurement of electron and positron flux

In the last years some measurements [9, 10, 11, 12, 13, 14, 15] reported an unexpected shape of the flux for CR electrons and positrons (CRE). Furthermore evidence for a spectral break in the TeV-range of the CRE flux has been provided by indirect measurements [16, 17], although the results were affected by significant uncertainties.

In 24 months of data-taking DAMPE was able to make a direct measurement of CREs in the range 25 GeV - 4.6 TeV [18] with unprecedented accuracy (high energy resolution and low background). Indeed the imaging properties of the BGO-calorimeter allows to distinguish a shower initiated by an electron w.r.t. a hadronic one. The identification algorithm is based on the spread of the BGO hits (*spread*) and on the fraction (F_{last}) of the energy on the lower layer w.r.t. the total BGO-energy.

The quantities *spread* and F_{last} are combined in the parameter

$$\zeta = F_{last}\frac{(spread/mm)^4}{8 \times 10^6} \tag{1}$$

used to select CRE sample by requiring $\zeta < 8.5$ (see Fig. 11). The CRE flux measured by DAMPE as a function of energy is shown in Fig. 12 with measurements by others experiments. A "smoothly broken power-law" is preferred to fit the largest part of the spectrum instead of a single power-law model. More details in [18].

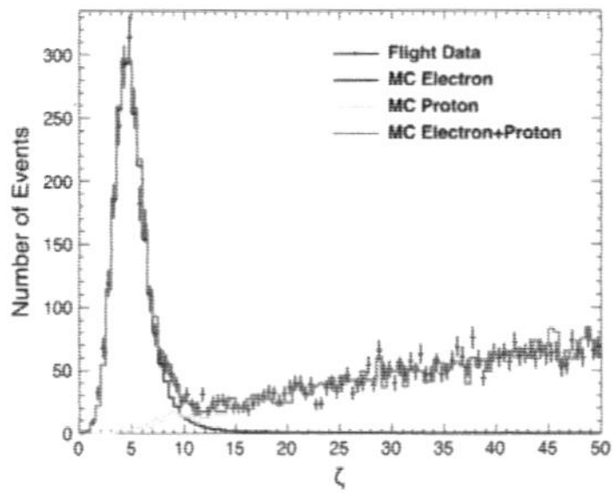

Figure 11. Distribution of the ζ parameter used to select the CRE sample requiring $\zeta < 8.5$.

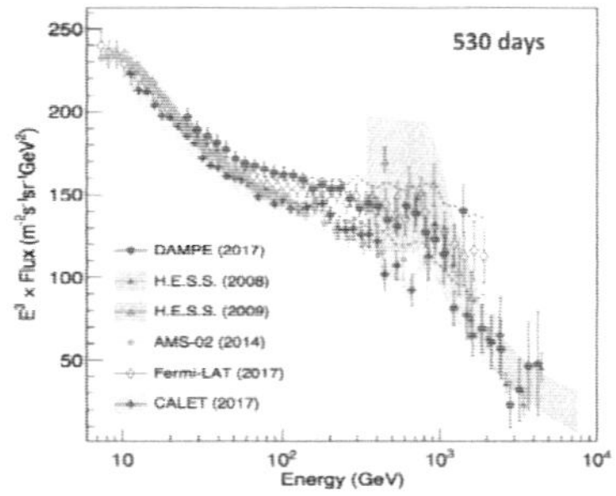

Figure 12. Electron and positron spectrum measured by different experiments (DAMPE: red dots, HESS systematic errors: grey band).

8. Conclusions

DAMPE detector works extremely well since its launch more than 2.5 years ago and it is expected to operate stably in the next few years. Many analyses based on DAMPE data are going on successfully on cosmic rays and gamma astronomy. The measurement of CRE spectrum is particularly remarkable. Indeed the direct detection of a spectral break at ~0.9 TeV confirms what found by previous indirect measurements, clarifies the shape of the CRE spectrum at energies above 1 TeV and sheds light on the physical origin of the sub-TeV CREs.

References

[1] Chang J 2014 *Chinese J. Space Science* **34** 550 (available at www.cjss.ac.cn/CN/10.11728/cjss2014.05.550)
[2] Chang J et al. 2017 *Astropart. Physics* **95** 6
[3] Azzarello P et al. 2016 *Nuclear Instruments and Methods* A **831** 378
[4] Zhang Z et al. 2016 *Nuclear Instruments and Methods* A **836** 98
[5] Tykhonov A et al. 2018 *Nuclear Instruments and Methods* A **893** 43
[6] Jiang W et al. 2017 *PoS(ICRC2017)* 228
[7] Salinas MM et al. 2017 *PoS(ICRC2017)* 709
[8] Yuan Q et al. 2017 *PoS(ICRC2017)* 617
[9] Chang J et al. (The ATIC Collaboration) 2008 *Nature* **456** 362
[10] Abdo AA et al. (The Fermi-LAT Collaboration) 2009 *Phys. Rev. Lett.* **102** 181101
[11] Abdollahi S et al. (The Fermi LAT Collaboration) 2017 *Phys. Rev.* D **95** 082007
[12] Aguilar M et al. (The AMS-02 Collaboration) 2014 *Phys. Rev. Lett.* **113** 221102
[13] Adriani O et al. (The PAMELA collaboration) 2011 *Phys. Rev. Lett.* **106** 201101
[14] Adriani O et al. (The CALET Collaboration) 2017 *Phys. Rev. Lett.* **119** 181101
[15] Adriani O et al. (The CALET Collaboration) 2018 *Phys. Rev. Lett.* **120** 261102
[16] Aharonian F et al. (The H.E.S.S. Collaboration) 2008 *Phys. Rev. Lett.* **101** 261104
[17] Aharonian F et al. (The H.E.S.S. Collaboration) 2009 *Astron. Astrophys.* **508** 561
[18] Ambrosi G et al. 2017 *Nature* **552** 63

e-ASTROGAM: a space mission for MeV-GeV gamma-ray astrophysics

R Rando[1,2], A De Angelis [2,3,4,5], M Mallamaci[2], on behalf of the e-ASTROGAM collaboration‡

[1] Dipartimento di Fisica e Astronomia G. Galilei, Università di Padova, I-35131 Padova, Italy

E-mail: `riccardo.rando@pd.infn.it`

[2] INFN, Sezione di Padova, I-35131 Padova, Italy

[3] INAF - Osservatorio Astronomico di Padova, I-35122, Padova, Italy

[4] Dipartimento di Matematica, Informatica e Fisica, Università di Udine, I-33100 Udine, Italy

[5] Laboratorio de Instrumentacao e Particulas and Instituto Superior Tecnico, Lisboa, Portugal

Abstract.

e-ASTROGAM is an observatory space mission dedicated to the study of the gamma radiation in the range from 0.3 MeV to 3 GeV. The detector is composed by a Silicon tracker, a calorimeter, and an anticoincidence system. The mission is based on an advanced space-proven detector technology, with unprecedented sensitivity, angular and energy resolution, combined with polarimetric capability. Thanks to its performance in the MeV-GeV domain, substantially improving its predecessors, e-ASTROGAM will open a new window on the non-thermal Universe. In particular it will determine the origin of key isotopes fundamental for the understanding of supernova explosions and the chemical evolution of our Galaxy. It will also shed light on the processes behind the acceleration of cosmic rays in our Galaxy.

1. The scientific objectives

e-ASTROGAM [1] is a concept for a gamma-ray space observatory operating in the energy range from 0.3 MeV to 3 GeV, aiming to provide gamma-ray data to a broad astronomical community in a decade of powerful observatories for multiwavelength astronomy and for the detection of gravitational waves, neutrinos and ultra-high-energy cosmic rays (UHECRs). The expected scientific return for such an observatory is substantial [2]. In particular, e-ASTROGAM will address the outstanding issue of the origin and propagation of cosmic rays. It will measure cosmic-ray diffusion in interstellar clouds and their impact on gas dynamics and state, thus providing crucial diagnostics about the impact of cosmic rays on star formation, structures of the interstellar medium, galactic winds and outflows. The improved sensitivity and angular resolution will be

‡ See `http://eastrogam.iaps.inaf.it`

e-ASTROGAM

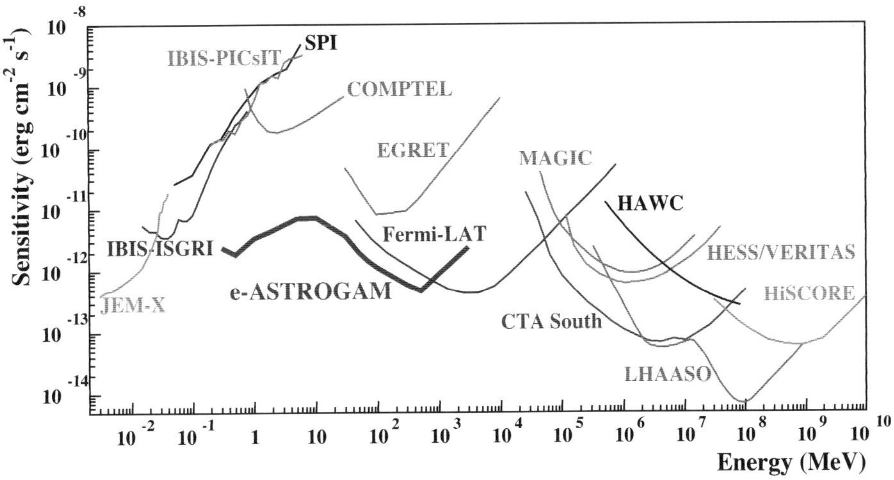

Figure 1. Point source continuum differential sensitivity of different X- and γ-ray instruments, from [2]

crucial to probe the interplay between cosmic rays and the turbulent medium of star forming regions (e.g. the Cygnus Cocoon) during the early steps of their Galactic voyage. Sensitive e-ASTROGAM observations of a set of cosmic-ray sources, such as young supernova remnants, will allow for the first time to distinguish the emission produced by the interactions of cosmic-ray nuclei with the ambient gas and the non-thermal emission from cosmic-ray electrons, helping in the discrimination among models of hadronic and leptonic production.

2. The instrument

The e-ASTROGAM instrument operates over more than four orders of magnitude in energy (from about 150 keV to 3 GeV) by detecting photons in both the Compton (0.15 – 30 MeV) and pair-production (> 10 MeV) regimes. The telescope is made up of three detection systems: a silicon tracker, a scintillator calorimeter and an anti-coincidence system, for a mass of 1.2 tons.

The Si tracker comprises 5600 double-sided strip detectors (DSSDs) arranged in 56 layers, divided in four units of 5×5 DSSDs. Each DSSD has a geometric area of 9.5×9.5 cm^2, a thickness of 500 μm, and a strip pitch of 240 μm. Stacking relatively thin detectors enables efficient tracking of the electrons and positrons produced by pair conversion, and of the recoil electrons produced by Compton scattering. The DSSD signals are read out by 860,160 independent, ultra low-noise and low-power electronics channels with self-triggering capability.

The calorimeter is a pixelated detector made of 33,856 Thallium-activated Cesium Iodide bars of 8 cm length and 5×5 mm^2 cross section. Each element is read out by silicon drift detectors (SDDs) at both ends, the depth of interaction along each

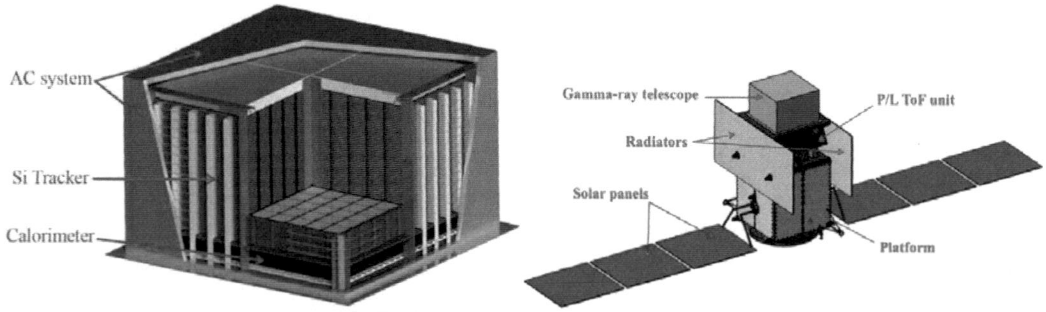

Figure 2. *Left panel* – overview of the scientific intrument. *Right panel* – the scientific intrument on top of the platform. From [1].

crystal is measured from the different light amount collected at both ends. The accurate measurement of the 3D position and deposited energy of each interaction is essential for a proper reconstruction of the Compton events. The total thickness – 4.3 radiation lengths – guarantees an 88% absorption probability for a 1-MeV photon on-axis.

The anti-coincidence system is composed of two parts: a standard anti-coincidence detector (Upper-AC), made of segmented panels of plastic scintillators covering the top and four lateral sides of the instrument, and a time-of-flight system (ToF) placed on the bottom of the instrument, to reject the particle background produced by the mass in the platform. The Upper-AC detector is segmented in 33 plastic tiles (6 tiles per lateral side and 9 tiles for the top) coupled to silicon photomultipliers (SiPM) by optical fibers, the ToF unit is composed of two plastic scintillator layers separated by 50 cm, read out by SiPMs with a timing resolution of 300 ps.

3. Performance

e-ASTROGAM is to be launched into a quasi-equatorial (inclination $i < 2.5°$) low Earth orbit at a typical altitude of 550 – 600 km. Extensive simulations based on the mass model of the satellite and on the model of the background environment [3, 4] have demonstrated the performance of the instrument [1]:

- broad energy coverage (∼0.15 MeV to 3 GeV), with nearly two orders of magnitude improvement of the continuum sensitivity in the range 0.15 – 100 MeV compared to previous missions (Figure 1);
- excellent sensitivity for the detection of key gamma-ray lines e.g. sensitivity for the 847 keV line from thermonuclear supernovae 70 times better than that of the *INTEGRAL* spectrometer (SPI);
- unprecedented angular resolution, improving not only on the angular resolution of CGRO-COMPTEL in the MeV regime, but also on that of *Fermi*-LAT in the GeV regime (68% containment radius at 1 GeV is 9').

e-ASTROGAM

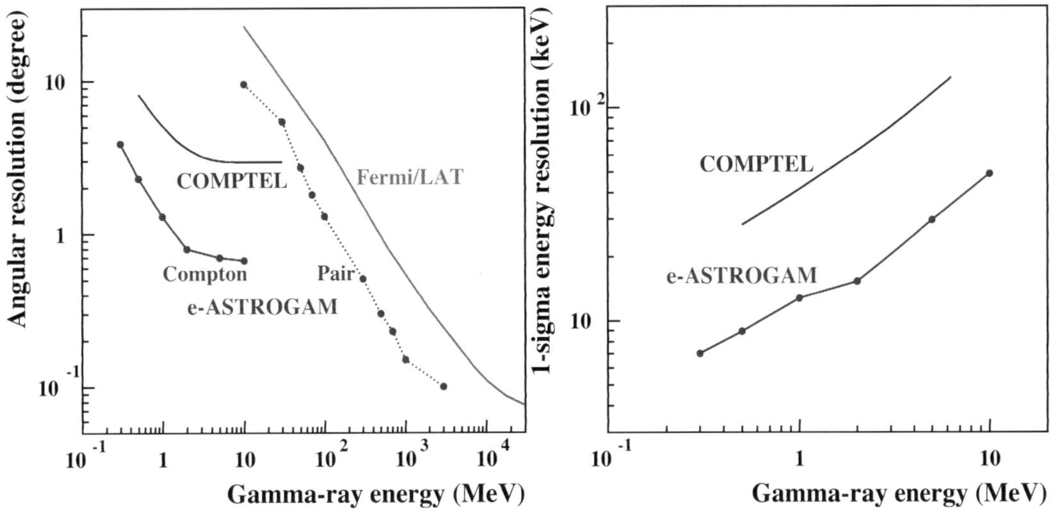

Figure 3. *Left panel* – e-ASTROGAM on-axis angular resolution compared to that of COMPTEL and *Fermi*/LAT. *Right panel* – 1σ energy resolution of COMPTEL and e-ASTROGAM in the Compton domain. From [1].

- large field of view (> 2.5 sr), ideal to detect transient Galactic and extragalactic sources, such as X-ray binaries and GRBs;
- timing accuracy of 1 μs (at 3σ), ideal to study the physics of magnetars and rotation-powered pulsars, as well as the properties of terrestrial gamma-ray flashes;
- pioneering polarimetric capability for both steady and transient sources.

References

[1] De Angelis A, Tatischeff V, Tavani M et al. 2017 *Exp. Astr.* **44** 25–82
[2] De Angelis A, Tatischeff V, Grenier I A et al. submitted to *J. High Energy Phys.* arXiv:1711.01265
[3] Zoglauer A, Andritschke R and Schopper F 2006 *New Astr. Rev.* **50**, 629–632
[4] Bulgarelli A, Fioretti V, Malaguti P, Trifoglio M and Gianotti F 2012 *Proc. SPIE* **8453** 845335

First results of the tracking system calibration of the TAIGA-IACT telescope

D. Zhurov[1,2], O. Gress[1], D. Sidorov[2], I. Astapov[3], P. Bezyazeekov[1], V. Boreyko[4], A. Borodin[4], N. Budnev[1], M. Brueckner[5], A. Chiavassa[6], A. Dyachok[1], O. Fedorov[1], A. Gafarov[1], A. Garmash[7], N. Gorbunov[4,8], V. Grebenyuk[4,8], T. Gress[1], O. Grishin[1], A. Grinyuk[4], D. Horns[9], A. Ivanova[1], N. Kalmykov[10], Y. Kazarina[1], V. Kindin[3], P. Kirilenko[7], S. Kiryuhin[1], R. Kokoulin[3], K. Kompaniets[3], E. Korosteleva[10], V. Kozhin[10], E. Kravchenko[7,11], M. Kunnas[9], L. Kuzmichev[10,1], Yu. Lemeshev[1], V. Lenok[1], B. Lubsandorzhiev[10,12], N. Lubsandorzhiev[10], R. Mirgazov[1], R. Mirzoyan[13,1], R. Monkhoev[1], R. Nachtigall[9], E. Osipova[1], A. Pakhorukov[1], M. Panasyuk[10], L. Pankov[1], A. Petrukhin[3], V. Poleschuk[1], M. Popescu[14], E. Popova[10], A. Porelli[5], E. Postnikov[10], V. Prosin[10], V. Ptuskin[15], E. Rjabov[1], G. Rubtsov[12], A. Pushnin[1], Y. Sagan[4], B. Sabirov[4], V. Samoliga[1], Yu. Semeney[1], A. Silaev[10], A. Silaev(junior)[10], A. Sidorenkov[12], A. Skurikhin[10], V. Slunecka[4], A. Sokolov[7,11], C. Spiering[5], L. Sveshnikova[10], V. Tabolenko[1], B. Tarashansky[1], A. Tkachenko[4], L. Tkachev[4,8], M. Tluczykont[9], R. Wischnewski[5], A. Zagorodnikov[1], V. Zurbanov[1] and I. Yashin[3]

[1]Institute of Applied Physics ISU, Irkutsk, Russia
[2]Irkutsk National Research Technical University, Irkutsk, Russia
[3]National Research Nuclear University MEPhI (Moscow Engineering Physics Institute), Moscow, Russia
[4]JINR, Dubna, Russia
[5]DESY, Zeuthen, Germany
[6]Dipartimento di Fisica Generale Universiteta di Torino and INFN, Torino, Italy
[7]Novosibirsk State University, NSU, Novosibirsk, Russia
[8]Dubna State University, Dubna, Russia
[9]Institute for Experimental Physics, University of Hamburg, Germany
[10]Skobeltsyn Institute of Nuclear Physics MSU, Moscow, Russia
[11]Budker Institute of Nuclear Physics SB RAS, Novosibirsk, Russia
[12]Institute for Nuclear Research of RAN, Moscow, Russia
[13]Max-Planck-Institute for Physics, Munich, Germany
[14]ISS, Bucharest, Romania
[15]IZMIRAN, Moscow, Russia

E-mail: `sidney28@yandex.ru`

Abstract. In TAIGA Observatory (Tunka Advanced Instrument for cosmic ray physics and Gamma-ray Astronomy) we are commissioning the first Imaging Atmospheric Cherenkov Telescope (IACT). The telescope has an alt-azimuth mount and 17-bit shaft encoder for each axis, stepper motors are used for axis control. For the pointing calibration of the telescope a CCD-camera is installed on the dish of the telescope and its position allows to capture simultaneously both the Cherenkov camera with LEDs and the sky with observed source. Since October 2017, the telescope has been operating in tracking mode. In this work the TAIGA-IACT telescope pointing calibration approach and first results of the tracking operations are described.

1. Introduction
The TAIGA observatory [1] (Tunka Advanced Instrument for cosmic ray physics and Gamma Astronomy) is located in the Tunka valley at a distance of 50 km from Lake Baikal and is developed to solve problems of high energy gamma-ray astronomy and cosmic-ray physics. This unique complex system of ground-based detectors is designed to detect extensive air showers (EAS) generated by high-energy gamma-quanta and cosmic-ray particles. The TAIGA observatory facility consists of various detectors, including wide-angle detectors of Cherenkov radiation, radio-antennas, detectors of electrons and muons of EAS [2]. In addition, the TAIGA-IACT (Imaging Atmospheric Cherenkov Telescope) is installed in Tunka valley and it has been operating in tracking mode since October 2017. It is planned to install more IACT telescopes by 2019 [3]. In contrast to other methods, IACT technique provides a robust separation of hadron and gamma showers by their image of Cherenkov radiation [4]. For the correct operation of the TAIGA-IACT telescope the high pointing and tracking accuracy for long exposure time is required.

2. The TAIGA-IACT telescope hardware
The first TAIGA-IACT telescope (see Figure 1) is installed near the TAIGA facility center. It has an alt-azimuth mount and a camera in the focus of the segmented Davis-Cotton design reflector with diameter of 4.3 m. The camera consists of 560 photomultiplier tubes (PMTs) each in a Winston cone and has a field of view of 9.5^o (0.36^o per pixel). On the camera of the telescope 8 positioning LEDs are installed along the perimeter to determine the position of the camera by the CCD-camera.

The CCD-camera Prosilica GC1380 installed at a distance of 1 m from the telescope optical axis on the dish near the mirrors. The position of the CCD-camera allows us capturing on the same image the sky with the observed source and the telescope camera with the positioning LEDs (see Figure 2). The CCD-camera is used to determine the telescope pointing direction by its images. It has 1360x1024 pixels resolution and 31.4^ox23.6^o field of view, i.e. 83 arc-second per pixel.

Each axis of the telescope is equipped with a Phytron hybrid stepper motor, a 17-bit shaft encoder and limit switches connected to the PhyMOTION control unit. Taking into account the gear ratio the telescope the axes rotate by $9 \cdot 10^{-4}$ degree per 1 full step of the motor. The control unit supports micro-step mode that is set up to 1/20 value to perform smooth tracking.

The telescope hardware is connected to the TAIGA data acquisition (DAQ) center over the Ethernet and can be controlled remotely.

3. The telescope software
As framework for the software development of the TAIGA-IACT telescope pointing and control system EPICS software tools are used [5]. The main control software runs on the computer in the DAQ center. For remote operation of the telescope, a graphical user interface based on the EPICS Qt Framework was developed.

Figure 1. The TAIGA-IACT telescope located in the Tunka valley. The telescope is on the parking position and directed to the north.

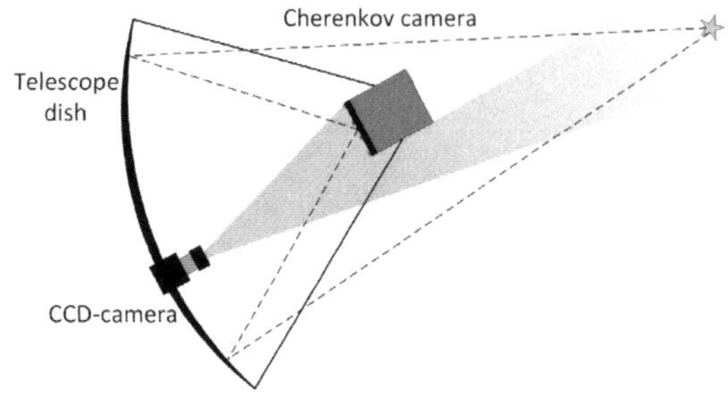

Figure 2. CCD-camera field of view that allow capturing both the telescope camera with the positioning LEDs and the observed source.

3.1. Telescope tracking software

The drive system control software uses the appropriate drivers for the motor controller. The observed source trajectory in horizontal coordinate system is calculated with 1 second step using the SOFA software[6]. After that, to transform trajectory to the telescope's coordinate system, a pointing model is applied and the result is loaded to the driver, where the trajectory is interpolated using cubic splines. To provide smooth and accurate tracking the driver controls velocity of the motors with feedback from the shaft encoders. Applying a corrections to the trajectory, while the telescope is moving, is allowed.

3.2. CCD-camera images processing

For accurate determination of the telescope position, CCD-camera images are used. Images from the CCD-camera (see Figure 3) are divided into two regions: the camera region and the sky region.

From the sky region the positions of all sources are extracted using SExtractor software [7]. Then WCS (World Coordinate System) parameters for this region are determined by Astrometry.net software [8], using astronomical catalogue Tycho-2 [9].

The LED positions are extracted from the camera region. The position of an LED is estimated as the center of gravity of its light distribution taking into account pixels that exceeded the established threshold above the background. The position of the camera is calculated as mean offset of LEDs from their reference position plus reference camera position.

For the telescope pointing direction determination the position of the central pixel of the camera is transformed to the sky region. The central pixel coordinates are transformed to the LED coordinate system and then to the sky region. To transform a point from LED coordinate system to the sky region, a linear regression model depending on source position, altitude and camera position is used.

The calculated telescope direction can be used for offline analysis or for on-the-fly corrections of the tracking trajectory. An on-the-fly correction algorithm was used for the TAIGA-IACT telescope operation in 2017-2018 observation season.

4. CCD-camera calibration

All transformation parameters and reference points must be determined for telescope pointing direction calculations. This includes the following measurements:

(i) reference positions of LEDs and camera;
(ii) parameters of transformation between focal plane and LED coordinate system;
(iii) parameters of transformation between LED coordinate system and the sky region.

In order to determine the reference positions and the transformation between focal plane and LED coordinate system, measurements are performed, that are started in twilight with open camera lid and switched-on positioning LEDs, so LEDs and Winston cones can be seen on the images. The positions of the Winston cones can be determined using the convectional template matching (as shown in Figure 4) and are used to determine the parameters of the transformation.

The white screen in the focus of the telescope optical system is installed (see Figure 3) to determine the parameters of the transformation between LED coordinate system and the sky region. To perform these measurements the telescope tracks a number of bright sources at different elevations. For each source several images are taken with different offsets from the camera center to scan the central region of the focal plane.

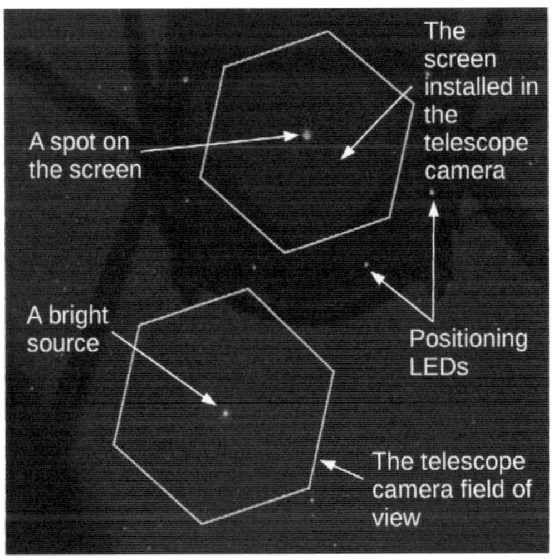

Figure 3. A CCD-camera image during calibration. On the one image we see the telescope camera with positioning LEDs and the tracked source. The white hexagons show the telescope camera PMTs region and field of view.

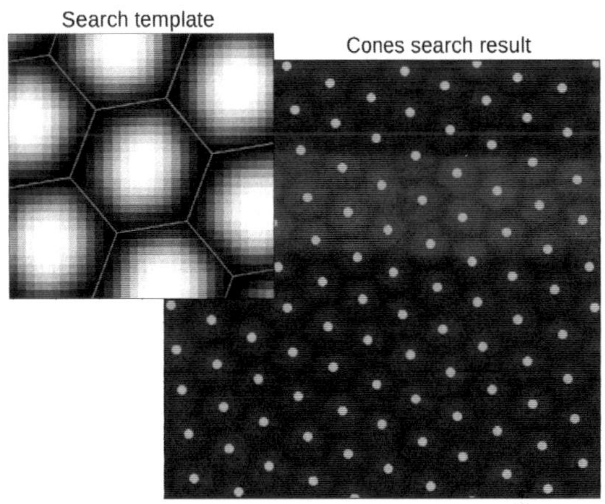

Figure 4. An example of result of the Winston cone detection algorithm. In the left upper corner the search template is shown. Detected positions of the Winston cones with PMTs are shown by dots on the right image. The performance of the algorithm should be better then one CCD-camera pixel.

On the obtained images, the positions of LEDs, of the bright source in the sky region and the spot on the screen are extracted. The parameters of linear regression model for transformation between LED coordinate system and the sky region are fitted using these data. The model also depends on telescope elevation (deformation of the construction caused by gravitation) and camera position on the image (small CCD-camera shifts).

5. The telescope tracking during the season

After the installation of the shaft encoder on the axis in September 2017, the commissioning of the TAIGA-IACT telescope tracking system was started and first source observations in tracking mode were performed in October 2017. This mode is tracking with feedback from the shaft encoders and applying a telescope pointing model. It is the standard approach for IACT telescope operation in other experiments [10, 11, 12]. Unfortunately, after a month of telescope operation it became clear that due to technical problems related to hardware this approach did not allow to achieve sufficient accuracy for this season.

The solution was to apply telescope trajectory corrections which are calculated on-the-fly using the CCD-camera images. Since January 2018 this approach with double feedback from the shaft encoders and the CCD-camera was in operate. An example of the online tracking errors using this approach is shown in Figure 5. This plot shows that tracking precision can be quite accurate (better then 0.01^o, except for some positions, where errors are caused by the backlash of the axis). Also telescope pointing precision is assuming be quite accurate in condition of a correct CCD-camera calibration.

The first model of transformation between the LED coordinate system and the sky region didn't take into account CCD-camera shifts, since the measured offsets of the positioning LEDs didn't exceed 0.02 degrees in the beginning [13]. Further study of the deviations showed the need to take into account the displacement of the CCD-camera. The maximum deviation was observed at extremely low temperatures $< -40^oC$ (see Figure 6).

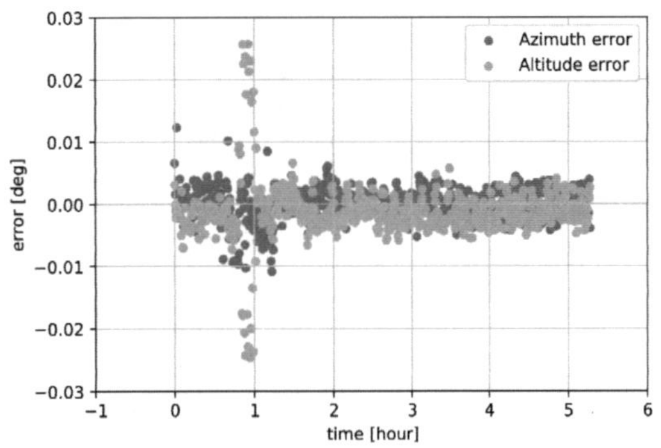

Figure 5. An example of double feedback telescope operation in 13 February 2018. Tracking errors are calculated on-the-fly by the CCD-camera image processing algorithm. The source was tracked for 5 hours. The maximum deviation from target trajectory was during the culmination of the observed source. It is caused by backlash of the drive system.

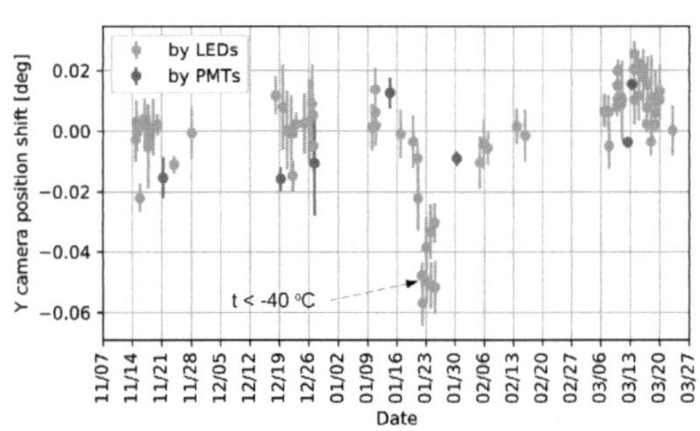

Figure 6. The offsets of telescope camera positions on CCD-camera images from reference position. The maximum deviation was observed at temperature $< -40^oC$. Blue points were measured by Winston cone detection algorithm. Orange points were measured by positioning LEDs.

At the end of February a test was conducted, which showed that the transformation parameters for the CCD-camera are not accurate enough. The test consisted in measuring the anode current on the central PMT that was measured for various offsets of the source. The idea of this experiment is similar to the approach that was used for the HERA telescope calibration [14], but the position of a bright source on the CCD image instead of the encoders was used to test the calibration. The difference of the position of the central pixel for a previous CCD-camera calibration and the current test was up to 0.09^o. Using this fact, a correction for the telescope CCD-camera calibration was applied for the March measurements. In April the CCD-camera was calibrated again taking into account these effects and the new calibration was applied also to previous data for offline analysis. The difference between the new CCD-camera calibration and the current test were less then 0.04^o. After obtaining the new CCD-calibration in April, all telescope tracking data was reprocessed. The maximal telescope tracking error for the observations in March, corrected by this procedure, was found to be 0.05^o.

6. Conclusion

During the 2017-2018 season the TAIGA-IACT telescope operated in tracking mode and a approach with double feedback from the shaft encoders and the CCD-camera was used. Instead of the standard technique this approach was used in cause of technical problems which are taken into account for future operations. The telescope pointing accuracy was better 0.05^o in March for the observations (mostly Mrk 421) with corrected calibrations and two times worse for the other observations in the season (Crab Nebula and Mrk 421). The analysis of the telescope camera data is presented in [15]. Pointing accuracy for the upcoming season is expected to be better than 0.05^o.

Acknowledgements

This work was supported by the Russian Federation Ministry of Education and Science (Tunka shared core facilities, unique identifier RFMEFI59317X0005, agreements: 3.9678.2017/8.9, 3.904.2017/4.6, 3.6787.2017/7.8, 3.6790.2017/7.8, 3.12739.2018/12.2), the Russian Foundation for Basic Research (grants 16-02-00738, 17-02-00905, 18-32-00460), the German Helmholtz Association (grant HRJRG-303), the DFG (grant TL 51-3) and the European Union's Horizon 2020 program under grant agreement No. 653477.

References

[1] Budnev N et al 2018 The TAIGA experiment from cosmic ray physics to gamma astronomy in the Tunka valley *Phys. Part. Nuclei* **49** 589
[2] Budnev N et al 2017 The TAIGA experiment - a hybrid detector for very high energy gamma-ray astronomy and cosmic ray physics in the Tunka valley *Proceeding of science* PoS(ICRC2017)768
[3] Kuzmichev L et al 2018 TAIGA Gamma Observatory: Status and Prospects *Physics of Atomic Nuclei* **81**.4 506
[4] Kunnas M et al 2015 *J. Phys.: Conf. Ser.* **632** 012040
[5] EPICS - Experimental Physics and Industrial Control System: http://www.aps.anl.gov/epics/
[6] SOFA - Standards Of Fundamental Astronomy: http://www.iausofa.org/
[7] SExtractor - Source Extractor: https://www.astromatic.net/software/sextractor
[8] Lang D, Hogg DW, Mierle K, Blanton M, Roweis S 2010 Astrometry.net: Blind astrometric calibration of arbitrary astronomical images *The Astronomical Journal* **139** 1782
[9] Høg E et al 2000 The Tycho-2 catalogue of the 2.5 million brightest stars *Astronomy and Astroph.* **355** 27
[10] Bretz T et al 2009 The drive system of the major atmospheric gamma-ray imaging Cherenkov telescope *Astroparticle Physics* **31** 2
[11] Braun I, Werner H 2007 Improving the pointing precision of the HESS experiment *Diss. Ruprecht-Karls Universitt Heidelberg* p.41
[12] Anderhub H et al 2013 Design and operation of FACT-the first G-APD Cherenkov telescope *Journal of Instrumentation* **8** 06
[13] Zhurov D et al 2017 Software design for the TAIGA-IACT telescope pointing and control system *Proceeding of science* PoS(ICRC2017)785
[14] Pühlhofer G et al 1997 Locating TeV γ-ray sources with sub-arcminute precision: the pointing calibration of the HEGRA system of Imaging Atmospheric Cherenkov Telescopes *Astroparticle Physics* **8**.1-2 104
[15] Sveshnikova L et al 2018 The first season operation of hybrid imaging/timing Cherenkov array TAIGA *26-th ECRS* to be pulished in *Bull. Russ. Acad. Sci. Phys.*

Constraints on the properties of the turbulent magnetic field around Geminga using HAWC measurements

Gwenael Giacinti[1] and Rubén López-Coto[2]

[1] Max-Planck-Institut für Kernphysik, Postfach 103980, D-69029 Heidelberg, Germany
[2] Università di Padova and INFN, I-35131, Padova, Italy

E-mail: Gwenael.Giacinti@mpi-hd.mpg.de

Abstract. We place constraints on the properties of the interstellar turbulence that surrounds Geminga pulsar, using the recent measurements from the HAWC Observatory in this region [1]. We propagate very-high-energy electrons in realizations of 3D isotropic Kolmogorov or Kraichnan turbulence, calculate their γ-ray emission, and compare with HAWC measurements. We show that the measurements can be well fitted for both models of the turbulence and for reasonable values of its strength, $B_{\rm rms}$, and coherence length, $L_{\rm c}$. Our best fits are obtained for $B_{\rm rms} \simeq 3\,\mu$G and $L_{\rm c} \simeq 1$ pc. Furthermore, the absence of strong asymmetries in the observed emission favours $L_{\rm c} \leq 5$ pc.

1. Introduction

Extended gamma-ray emissions around Galactic sources of cosmic-rays (CR) can provide insights into the properties of the turbulent magnetic fields that surround these sources, see e.g. [2, 3]. Such an emission has been recently detected by HAWC around Geminga, and is thought to be due to ~ 100 TeV electrons and positrons diffusing and cooling around the pulsar [1]. HAWC Collaboration measured the value of the diffusion coefficient of these electrons: $D_{100\,{\rm TeV}} = (4.5 \pm 1.2) \times 10^{27}\,{\rm cm^2\,s^{-1}}$ at 100 TeV, and noted that its extrapolation to lower energies is about two orders of magnitude smaller than the Galactic average for CRs as deduced from B/C measurements. In the present paper, we put constraints on the turbulence within ≈ 25 pc from Geminga, using the measurements published in Ref. [1]. We find that isotropic Kolmogorov or Kraichnan turbulence with a coherence length $L_{\rm c} \simeq 1$ pc provide a good fit to the measurements [4].

2. Method

We describe here how we produce the synthetic γ-ray maps that are compared with HAWC measurements in the next section. Instead of using the diffusion-loss equation, we propagate individual very-high-energy electrons (5000 for each map) in realizations of 3D isotropic Kolmogorov or Kraichan turbulence. This technique allows us to take into account effects than cannot be described properly within the standard isotropic diffusion approximation, such as highly anisotropic propagation of CRs along magnetic field lines. In particular, when most electrons escaping from a source are still located at distances smaller than $\approx L_c$ from it, their γ-ray emission should highlight field lines around the source and, therefore, appear filamentary.

Measurements of the γ-ray spectrum of Geminga by Ref. [1] shows that it follows a power law $dN/dE \propto E^{-2.34}$ between 8 TeV and 40 TeV. This is compatible with electrons being injected with a spectrum $dN_e/dE \propto E^{-2.24}$ at the pulsar. In the simulations, we inject the electrons at the location of the source, which we assume to be point-like. The initial energies of these electrons are chosen between 40 TeV and 500 TeV and follow a power-law spectrum $\propto E^{-2.24}$. We take into account their synchrotron and inverse Compton energy losses during propagation, which are the dominant losses at the energies we consider, see Ref. [5]. We neglect the effect of infrared and optical photons. Following Ref. [6], the energy loss per time unit of an electron with energy E_e and placed in a magnetic field B reads:

$$\left|\frac{dE_e}{dt}\right| \simeq 2.53 \times 10^{-15}\,\mathrm{TeV/s}\left[\left(\frac{B}{1\,\mu\mathrm{G}}\right)^2 + \frac{10.1}{(1+E_e/(99\,\mathrm{TeV}))^{1.5}}\right]\left(\frac{E_e}{1\,\mathrm{TeV}}\right)^2 . \tag{1}$$

We generate the 3D turbulent magnetic fields in which we propagate these particles with the method presented in Ref. [7]. We test both isotropic Kolmogorov, $P(k) \propto k^{-5/3}$, and Kraichnan turbulence, $P(k) \propto k^{-3/2}$, with root-mean-square strengths in the range $B_{\mathrm{rms}} = (2-5)\,\mu$G and coherence lengths in the range $L_c = (0.1 - 40)$ pc. Trajectories are stopped once the particles reach 39 TeV, because lower energy electrons do not contribute to the range of photon energies considered here. Our turbulence contains fluctuations down to scales smaller than the smallest electron gyroradius present in our simulations. Since HAWC measurements do not show any strong asymmetry in the emission, we do not add any large-scale magnetic field to our turbulence, see the discussion in Section 4. Since the typical cooling time of ~ 100 TeV electrons is only ~ 10 kyr, we assume that the electrons have been injected steadily on these time scales. Instead of injecting the electrons continuously in the simulation, we inject them at $t = 0$, and record their momenta and positions at equally-spaced intervals in time, every Δt, and consider each recording as a new particle for the total emission. We verified that $\Delta t = 20$ yr gives correct results. The gamma-ray emission is dominated here by the upscattering of CMB photons. We calculate it, using the `gamera` [8] and `edge` [5] libraries. Finally, since the number of electrons in our simulations is much smaller than that present around Geminga in reality, we normalize our total γ-ray emission to that measured by HAWC.

3. Results

In Fig. 1, we plot the simulated γ-ray surface brightness, as viewed from the Earth, in a 10°-radius region around Geminga. The pulsar is located in the centre of each plot. In all six panels, we use isotropic Kolmogorov turbulence with root-mean-square strength $B_{\mathrm{rms}} = 3\,\mu$G. In each panel, we set the coherence length to a different value: $L_c = 0.25$ pc (top left panel), $L_c = 1$ pc (top right), $L_c = 5$ pc (middle left), $L_c = 10$ pc (middle right), $L_c = 20$ pc (bottom left), and $L_c = 40$ pc (bottom right). Red and yellow regions are the brightest ones, see the colourbars on the side of the panels for the values of the surface brightness. The emission clearly becomes increasingly asymmetric (with respect to rotations around the position of the pulsar) for increasing values of L_c. This is due to the fact that neighbouring magnetic field lines close to the source remain close to one another typically up to distances $\approx L_c$ from the source. For small values of L_c (approximately ≤ 5 pc in these plots), magnetic field lines are tangled on scales that are significantly smaller than the size of the γ-ray emitting region. Therefore, even if CRs follow magnetic field lines individually, the resulting γ-ray emission around the source looks quite symmetric to the observer. In contrast, for larger values of L_c (approximately ≥ 10 pc in these plots), L_c is not small compared with the size of the γ-ray emitting region, and the bulk of escaping CRs is mostly confined in a magnetic flux tube of length $\leq L_c$ around the source.

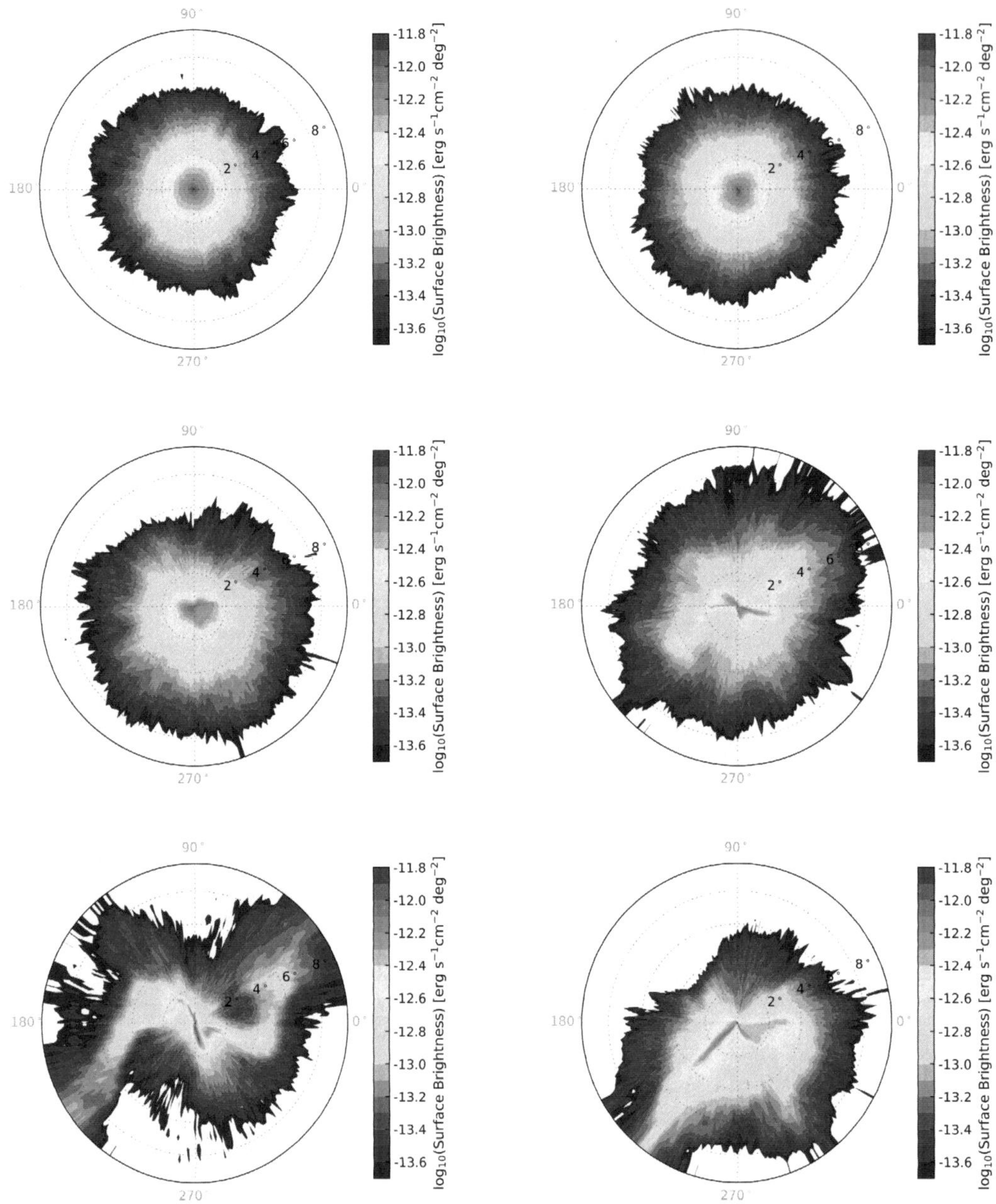

Figure 1. Plots of the simulated gamma-ray surface brightness around Geminga (located at the centre of each plot), as viewed from the Earth. Electrons are injected continuously at the position of Geminga in given realizations of 3D isotropic Kolmogorov turbulence, with $B_{\rm rms} = 3\,\mu$G and $L_{\rm c} = 0.25\,$pc (top left), 1 pc (top right), 5 pc (middle left), 10 pc (middle right), 20 pc (bottom left), or 40 pc (bottom right). The polar angle is written in grey, and the angular distance from the pulsar in black.

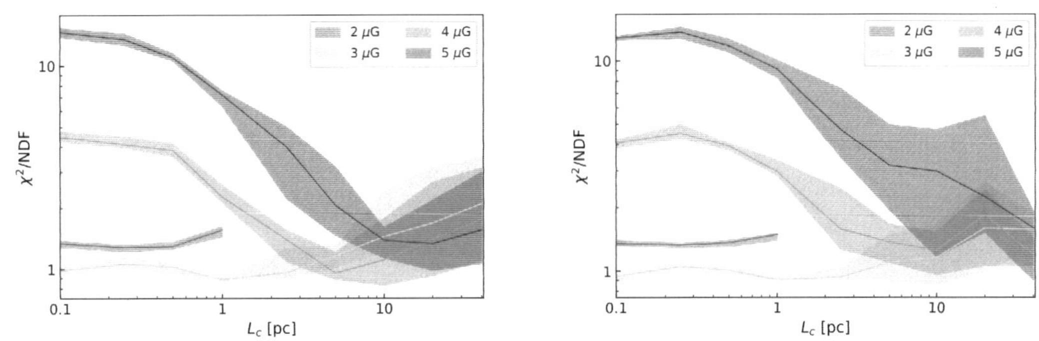

Figure 2. χ^2/ndf versus L_c for our fits to HAWC measurements from Ref. [1], for Kolmogorov (left panel) and Kraichnan (right panel) turbulence. Each colour band corresponds to a different value of B_rms, see the values in the keys. The thick solid lines represent the median χ^2/ndf, and the band width estimates the fluctuations from one realization of the turbulence to another, see the text for details.

The filamentary structures visible in γ-rays highlight the local magnetic field lines. Therefore, the presence or absence of asymmetries in the emission can be used to put constraints on the coherence length of the turbulence. Since no strong asymmetry has been detected by HAWC yet, the results in Fig. 1 allow us to set an upper limit on L_c, of approximately 5 pc. The simulated emissions in the plots with $L_\mathrm{c} \geq 10$ pc are too asymmetric to account for the measurements.

In each panel of Fig. 1, the turbulence has a given configuration. We redo these simulations for ten different realizations of the turbulence, while keeping the same $P(k)$, and the same values of B_rms and L_c. We find that the simulated emissions almost do not vary for small L_c. In contrast, the shapes of the filaments at large L_c strongly depend on the realization. This is unsurprizing, because the geometry of field lines around the source is completely different from one realization to another. For small L_c, magnetic field lines are too tangled on the scale of the emission for any substantial difference to be visible in the emission.

For each set of parameters and each realization, we then integrate the γ-ray surface brightness over all azimuthal angles, plot it versus the angular distance to the pulsar, and fit it to HAWC measurements [1]. The results for the χ^2/ndf of these fits versus L_c are presented in Fig. 2 for Kolmogorov (left panel) and Kraichnan (right) turbulence. Each line colour corresponds to a different value of B_rms: 2, 3, 4, and 5 μG — see the keys for the colour code. The width of the bands quantifies the fluctuations from one realization of the turbulence to another. The shaded region corresponds to the interval between the 18th and 82nd percentiles of all realizations. The increase of the widths of the bands with L_c is due to the reason discussed above. The thick lines inside each band correspond to the median value of χ^2/ndf. Both for Kolmogorov and Kraichnan turbulence, the best fits are obtained for $B_\mathrm{rms} = 3\,\mu$G and $L_\mathrm{c} = 1$ pc, where $\chi^2/\mathrm{ndf} < 1$. The value of L_c at which the best fit is obtained for each value of B_rms increases with B_rms. Too weak, $B_\mathrm{rms} \leq 2\,\mu$G, or too strong, $B_\mathrm{rms} \geq 5\,\mu$G, magnetic fields give bad fits. The fact that the measurements are integrated over all azimuthal angles does not allow us to take into account the constraint from the symmetry of the emission in this fit. In practice, the regions at $L_\mathrm{c} \geq 10$ pc in these plots are excluded for the aforementioned reason. Finally, we note that there is no substantial difference between the two panels of Fig. 2, which shows that one cannot firmly distinguish between the two power-spectra $P(k)$ with the current measurements. However, future analyses of the dependence of the emission on γ-ray energy should be able to provide stronger constraints on the power-spectrum of the turbulence.

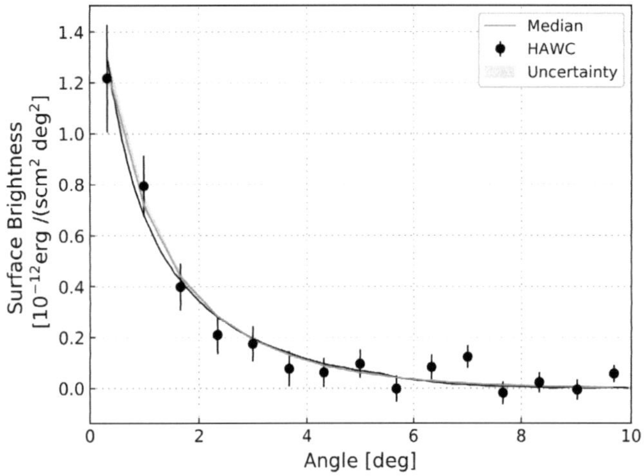

Figure 3. γ-ray surface brightness, integrated over all azimuthal angles, versus the angular distance to the pulsar for our best fit (orange line) to HAWC measurements from Ref. [1] (black dots): Kolmogorov turbulence with $B_{\rm rms} = 3\,\mu$G and $L_{\rm c} = 1\,$pc. The black line corresponds to the fit from the HAWC Collaboration [1].

In Fig. 3, we plot with an orange line the γ-ray surface brightness (integrated over all azimuthal angles) versus the angular distance to the pulsar for our best fit for Kolmogorov turbulence, i.e. for $B_{\rm rms} = 3\,\mu$G and $L_{\rm c} = 1\,$pc. HAWC measurements from Ref. [1] is plotted with black dots, and the black line corresponds to the fit from the HAWC Collaboration [1].

4. Discussion

We have not added any regular magnetic field in the above calculations, because the apparent symmetry of the emission detected by HAWC around Geminga hints at a strongly turbulent field in this region, i.e. the amplitude of the large-scale magnetic field in this region should be smaller than that of the turbulent field. In the presence of a strong regular field, the γ-ray emission would be elongated along its direction. Such a scenario would be compatible with observations only if this field points in our direction, so that the γ-ray emission does not appear asymmetric when viewed from the Earth. Although unlikely, this is nonetheless an interesting possibility because it could help reconcile the small extent of the γ-ray emission detected by HAWC with a larger CR diffusion coefficient along the line of sight, matching the Galactic average as deduced from e.g. the boron-to-carbon ratio. If the regular field, $\mathbf{B}_{\rm reg}$, points towards us, the size of the γ-ray emission would appear relatively small because it would be controlled by perpendicular diffusion. In contrast, electrons would diffuse much faster along $\mathbf{B}_{\rm reg}$ due to parallel diffusion. The resulting asymmetry of the electron distribution around Geminga would be undetectable by an observer at Earth as long as the angle between $\mathbf{B}_{\rm reg}$ and the line of sight remains smaller than $\theta_{\rm max} \approx \tan^{-1}(\sqrt{D_\perp/D_\parallel})$, where $D_\parallel$ and $D_\perp$ are the parallel and perpendicular diffusion coefficients. Turbulence levels $B_{\rm rms}/|\mathbf{B}_{\rm reg}| \leq 0.5$ are needed for $D_\parallel$ to reach "Galactic average" values with Kolmogorov turbulence, as can be seen in Fig. 3 of Ref. [9]. This corresponds to $D_\parallel/D_\perp \geq 200$, and thence $\theta_{\rm max} \leq 4°$. Such a near-perfect alignment of $\mathbf{B}_{\rm reg}$ with the line-of-sight, and therefore the presence of such a strong regular field in this region, are quite unlikely. Also, $B_{\rm rms}/|\mathbf{B}_{\rm reg}|$ is thought to be greater than 1 in the Galactic disc, and the regular field is thought to follow spiral arms, which is in tension with the fact that the direction to Geminga is not aligned with that of the Orion Spur.

Finally, we note that $D_{100\,{\rm TeV}}$ is not far from the Bohm value, and that the electrons may be probing turbulence generated by CRs. CR self-confinement around their sources has been studied by a number of authors, see e.g. [10–13], and Ref. [14] has recently suggested that it may be the reason for the low value of $D_{100\,\mathrm{TeV}}$ measured by HAWC around Geminga. Such a scenario is possible and very interesting. Our study nonetheless shows that the current HAWC measurements can still be fitted with Kolmogorov or Kraichnan turbulence and does not require such an explanation at the present time. Radio observations suggest that the coherence length of the turbulence in the spiral arms of our Galaxy is equal to only a few parsecs, which is very close to our best fit value. See e.g. Refs. [15, 16] where the outer scale ($= 5L_c$ for Kolmogorov turbulence) is found to be $\leq 20\,$pc. Future studies of the gamma-ray emission, especially at lower energies, should be able to clarify the situation. We will address the discrepancy between $D_{100\,\mathrm{TeV}}$ and the "Galactic average" value from B/C measurements in a future publication.

5. Conclusions and perspectives

We have shown that the extended γ-ray emission detected by HAWC around Geminga is compatible with that from electrons propagating and cooling in Kolmogorov or Kraichnan turbulence with reasonable strengths and coherence lengths, even if the diffusion coefficient of these electrons is substantially smaller than the values usually inferred from the boron-to-carbon ratio for the Galactic average. Our best fits are obtained for turbulent fields with a root-mean-square strength of $3\,\mu$G and a coherence length of about 1 pc. Magnetic field strengths smaller than $5\,\mu$G are favoured. Due to the lack of strong asymmetries in the observed emission, we can exclude coherence lengths greater than about 10 pc in this $\simeq 25\,$pc-radius region around Geminga. Even though the power-spectrum of the turbulence cannot be well constrained at the present time, we expect that one could place more stringent constraints in the future by studying the energy-dependence of the morphology of the emission.

Acknowledgments

We thank Jim Hinton and the HAWC Collaboration for useful discussions. The research of GG was supported by a Grant from the GIF, the German-Israeli Foundation for Scientific Research and Development.

References

[1] Abeysekara A U *et al* 2017 *Science* **358** 911–914
[2] Giacinti G, Kachelrieß M and Semikoz D V 2012 *Phys. Rev. Lett.* **108** 261101 (*Preprint* 1204.1271)
[3] Giacinti G, Kachelrieß M and Semikoz D V 2013 *Phys. Rev. D* **88** 023010 (*Preprint* 1306.3209)
[4] López-Coto R and Giacinti G 2018 *Mon. Not. Roy. Astron. Soc.* **479** 4526–4534
[5] López-Coto R, Hahn J, BenZvi S, Dingus B, Hinton J, Nisa M U, Parsons R D, Greus F S, Zhang H and Zhou H 2018 *Astropart. Phys.* **102** 1–11 (*Preprint* 1709.07653)
[6] Moderski R, Sikora M, Coppi P S and Aharonian F 2005 *Mon. Not. Roy. Astron. Soc.* **363** 954–966 (*Preprint* astro-ph/0504388)
[7] Giacinti G, Kachelrieß M, Semikoz D V and Sigl G 2012 *Journ. Cosmol. Astropart. Phys.* **7** 031 (*Preprint* 1112.5599)
[8] Hahn J 2015 *Proceedings of the 34th International Cosmic Ray Conference, id 917*
[9] Giacinti G, Kachelriess M and Semikoz D V 2018 *J. Cosmol. Astropart. Phys.* **7** 051 (*Preprint* 1710.08205)
[10] Ptuskin V S, Zirakashvili V N and Plesser A A 2008 *Advances in Space Research* **42** 486–490
[11] Malkov M A, Diamond P H, Sagdeev R Z, Aharonian F A and Moskalenko I V 2013 *Astrophys. J.* **768** 73 (*Preprint* 1207.4728)
[12] D'Angelo M, Blasi P and Amato E 2016 *Phys. Rev. D* **94** 083003 (*Preprint* 1512.05000)
[13] Nava L, Gabici S, Marcowith A, Morlino G and Ptuskin V S 2016 *Mon. Not. Roy. Astron. Soc.* **461** 3552–3562 (*Preprint* 1606.06902)
[14] Evoli C, Linden T and Morlino G 2018 *Phys. Rev. D* **98** 063017 (*Preprint* 1807.09263)
[15] Haverkorn M, Brown J C, Gaensler B M and McClure-Griffiths N M 2008 *Astrophys. J.* **680** 362-370 (*Preprint* 0802.2740)
[16] Iacobelli M *et al* 2013 *Astron. Astrophys.* **558** A72 (*Preprint* 1308.2804)

IOP Conf. Series: Journal of Physics: Conf. Series **1181** (2019) 012047 doi:10.1088/1742-6596/1181/1/012047

The relevance of fluorescence radiation in Cherenkov telescopes

F. Arqueros, J. Rosado, D. Morcuende, J. L. Contreras

High Energy Physics Group GAE and UPARCOS. Facultad de Ciencias Físicas. Universidad Complutense de Madrid. E-28040 Madrid, Spain.

E-mail: arqueros@ucm.es

Abstract. Cherenkov telescopes are also sensitive to the atmospheric fluorescence produced by the extensive air showers. However this contribution is neglected by the reconstruction algorithms of imaging air Cherenkov telescopes IACTs and wide-angle Cherenkov detectors WACDs. In this paper we evaluate the fluorescence contamination in the Cherenkov signals from MC simulations in both kinds of Cherenkov telescopes and for some typical observational situations. Results for an observation level of 2200 m a.s.l. are shown. In addition, the feasibility and capabilities of IACTs working as fluorescence telescopes are discussed with the assistance of some geometrical calculations.

1. Introduction

An extensive air shower (EAS) generates radiation in the optical range due to the Cherenkov effect as well as to the fluorescence emission after the de-excitation of air molecules excited by the charged particles of the shower. Cherenkov and fluorescence photons emitted from any point of an EAS are indistinguishable for a ground Cherenkov telescope because they arrive simultaneously and within the same wavelength range of around $300-450$ nm. On the other hand, fluorescence emission is less efficient than Cherenkov production (e.g., a 1 GeV electron near ground generates in one meter about 30 Cherenkov photons but only about 4 fluorescence photons). In addition, the fluorescence light is emitted isotropically, while Cherenkov radiation is peaked at a small angle of $\sim 1^\circ$ around the shower axis. Therefore, the signals recorded by a telescope pointing to the direction of the shower axis (on-axis observation) are dominated by the Cherenkov component as long as the telescope is not far away from the impact point (i.e., within the so-called Cherenkov light pool). This is the case of imaging air Cherenkov telescopes (IACTs) [1] used for VHE γ-ray astronomy. Arrays of wide-angle Cherenkov detectors (WACDs) can also be used as non-imaging telescopes for both γ- and cosmic-rays [2]. In this case, the reconstruction of the shower direction relies on the time structure of the front of Cherenkov photons reaching the stations of the array. The fluorescence light also entails a contamination to the recorded signals.

In this work, we have evaluated the fluorescence contamination in Cherenkov telescopes, both IACTs and WACDs, using CORSIKA[3]. This code includes the generation and transportation to ground of the Cherenkov photons generated by the showers. Fluorescence emission is not yet included in the official version of CORSIKA, but we have developed an implementation described in [4], which follows a procedure similar to that of the Cherenkov production. With the assistance of this tool, the photon density (photons/m^2) of both Cherenkov and fluorescence light at ground has been calculated as a function of the distance to the impact point (intersection of the shower axis with the ground plane) for a sample of EASs initiated by γ-rays in the energy range 100 GeV – 1 PeV. Results on fluorescence contamination will be shown in section 3.

Detection of atmospheric fluorescence is the basis of a well established technique for the study of UHE cosmic rays [5]. In this case, fluorescence telescopes register air-showers transversely to the shower axis. The working energy range of these telescopes is limited by several factors like the night sky background and the telescope field of view. Although IACTs are designed for the detection of Cherenkov light from γ-ray sources, in principle, they could also be used with minor modifications in "fluorescence mode" for the detection of γ- or cosmic-ray showers [6]. In section 4 we will discuss on geometrical aspects of interest for this topic.

2. MC simulations

The main options and parameters used in CORSIKA for these simulations have been the following. For hadronic interactions the QGSJET-01C and GHEISHA 2002d models have been chosen for high- and low-energy interactions, respectively. The energy thresholds were set to 0.3 GeV for hadrons, 0.1 GeV for muons and 0.02 GeV for the electromagnetic component. The observation level was set to an altitude of 2200 m a.s.l. and photons reaching a square 5×5 km^2 centered at the impact point were stored.

MC data were analyzed with dedicated Python scripts. The detection area was discretized into a dense grid and the number of photons in each grid element was counted. As an example, the average radial distribution of both Cherenkov and fluorescence photon densities for vertical 10 TeV γ-ray showers is shown in figure 1. It can be observed that Cherenkov light concentrates on the pool region, which extends up to a radial distance of $\sim$ 120 m in this case, while the photon density sharply drops outside. On the other hand, the fluorescence photon density decreases much less steeply with increasing core distance due to the isotropic nature of the fluorescence emission. In fact, the fluorescence photon density becomes larger than the Cherenkov one at very large core distance.

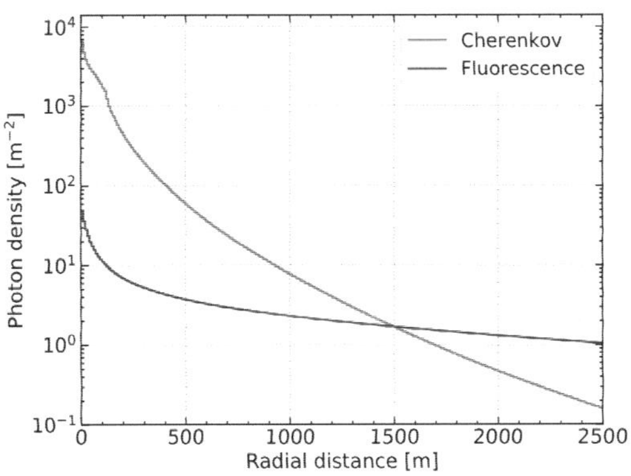

Figure 1: Average radial distributions of Cherenkov (red) and fluorescence (blue) light for vertical 10 TeV γ-ray showers. All photons (without any angular cut) have been used here.

3. Fluorescence contamination in Cherenkov telescopes

Two different observational techniques have been studied: IACTs working in on-axis mode with a FoV of 10° in diameter (see figure 2) and WACDs pointing to the zenith for which we have assumed a FoV of 60° in diameter (see figure 4). The pointing direction and FoV of these telescopes have been simulated by applying the appropriate cuts on the arrival angle of the photons. No cut in the arrival time was applied for this work.

For inclined showers we have calculated the photon densities only along the x axis, defined as the projection of the shower axis on the ground plane. The fluorescence contamination $R_{\rm FC}$ at a given position on ground is defined in this work as the ratio between the fluorescence and Cherenkov photon densities and it has been evaluated for both IACTs and WACDs as a function of the x distance, the shower energy E and zenith angle θ.

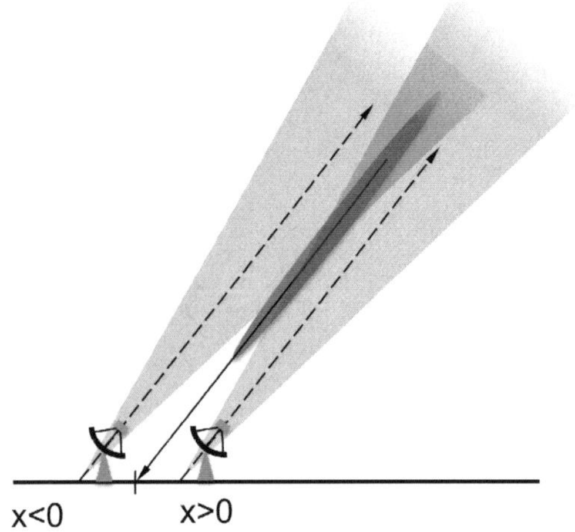

Figure 2: Schematic representation of the geometry for IACT observations.

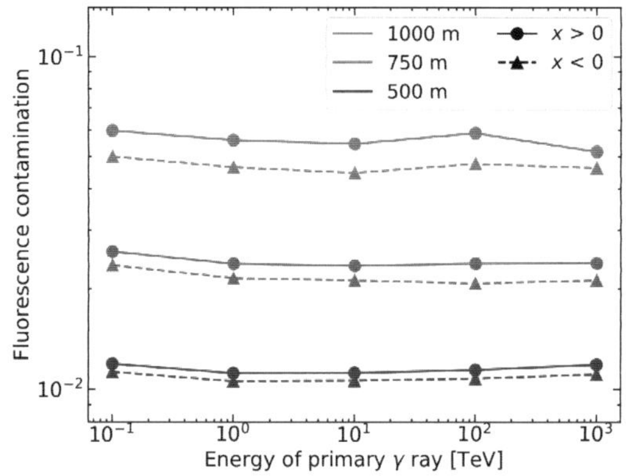

Figure 3: $R_{\rm FC}$ versus E for γ-ray showers of 20° zenith angle at three distances of the impact point.

IACT case

The fluorescence contribution is strongly suppressed in telescopes with a narrow FoV in such a way that the contamination is negligible within the light pool. Results for 20° γ-showers are displayed in figure 3 showing that $R_{\rm FC}$ has a very weak dependence with energy. It is also nearly independent of the x sign as expected from symmetry reasons (see figure 2).

The θ dependence of $R_{\rm FC}$ has been found to be negligible if evaluated in a plane perpendicular to the shower axis. However, for a horizontal plane (i.e., the ground), the contamination for a fixed x value decreases with the shower inclination, becoming negligible beyond about 40°.

It should be stressed that the fluorescence contamination becomes increasingly significant at large core distances (e.g., > 5% at 1000 m), as can be observed in figure 3. Thus ignoring this effect would have an impact on VHE γ-ray observations carried out by arrays of IACTs that collect measurable signals at large core distances (e.g., CTA [1]).

WACD case

Figure 5 shows the results for 20° γ-showers. Notice that $R_{\rm FC}$ depends significantly on the x sign, because of the evident asymmetry of this configuration (see figure 4). For a given $|x|$ value, the detection of the isotropic fluorescence light is favored in the positive side, while this is not the case for the directional Cherenkov radiation. The effect becomes more important as the shower energy increases. Besides, we have found that $R_{\rm FC}$ is weakly dependent of the shower inclination in the x positive side, while decreases with θ in the negative one.

Our results indicate that the signals of a detector located at large distance from the impact point can be contaminated with a non-negligible fraction of fluorescence light. In particular, the contamination may be as large as 45% (i.e., $R_{\rm FC} = 0.8$) at $x = 1000$ m in the PeV energy range.

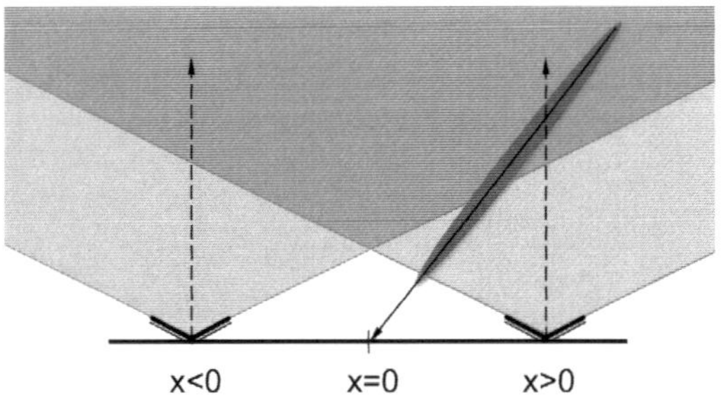

Figure 4: Schematic representation of the geometry for WACDs.

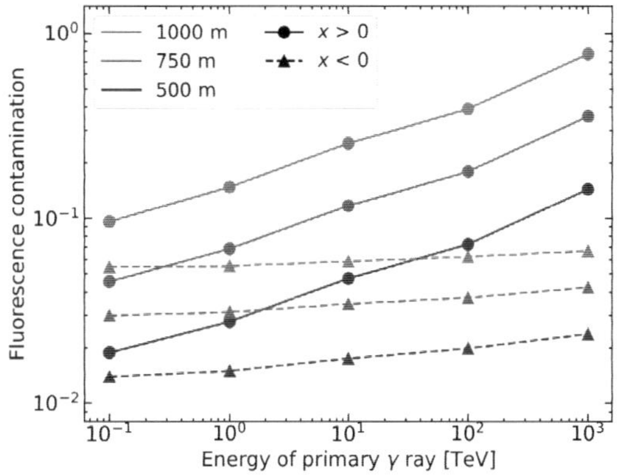

Figure 5: The same as figure 3 for WACDs.

4. IACTs in fluorescence mode

In a previous paper [6], we discussed for the first time about the feasibility of a large array of IACTs, like CTA [1], to work simultaneously as a fluorescence observatory. In this scenario the telescopes should fulfill several requirements related to sensitivity (e.g., the trigger configuration). Also there are some general geometrical constraints that we discuss next. Figure 6 is a sketch showing the definition of the relevant geometric parameters. In the first place,

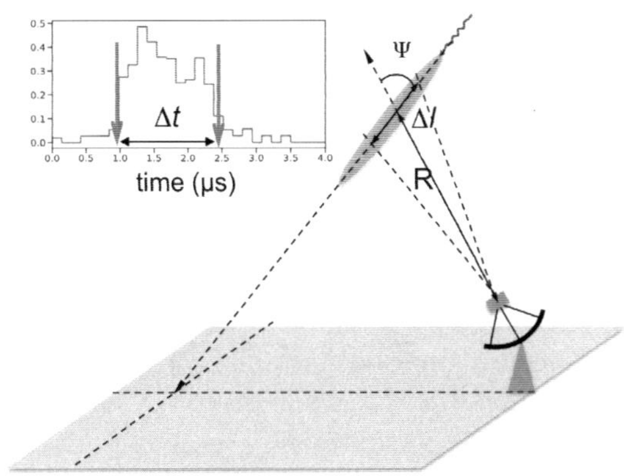

Figure 6: Geometrical configuration in the fluorescence technique. For simplification we will assume that the optical axis crosses the shower axis and thus the track coincides with a diameter of the camera. R is the shower-to-telescope distance measured along the optical axis, Ψ is the angle between both axes and Δl is the length of the track within the telescope FoV. Δt is the pulse duration in the camera.

the effective time window of the camera has to be wide enough to record the passage of the shower track through the telescope FoV (i.e., larger than Δt). In addition, the length Δl of the track observed by the telescopes should be sufficient for reconstruction purposes. The results of simple geometrical calculations for a telescope with a FoV of 9° in diameter are displayed in figures 7 and 8. From these results, we infer that it would be possible to record a large fraction ($\Delta l > 5$ km) of the track of showers developed at tens of km distances in case that a time window of tens of μs is available. A time window of about 1 μs could be sufficient to reach showers at distances of a few km (full track at $\Psi < 30°$). Even though IACTs are designed to record Cherenkov flashes that are much shorter (< 100 ns), some camera types could be adapted to register traces over time intervals of some μs [7]. A modest time resolution would be enough to store those traces, keeping the high time resolution for prompt Cherenkov signals.

Stereoscopic observation by several IACTs of the array would be necessary for an accurate shower reconstruction. In addition, this may allow the registration of the full shower track even though each telescope only images a small Δl portion of the shower because of its limited FoV.

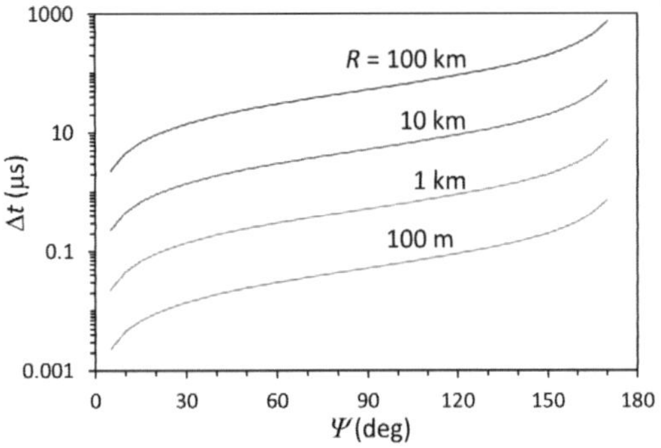

Figure 7: Duration of fluorescence pulse traversing a telescope FoV of 9° versus Ψ for several values of R.

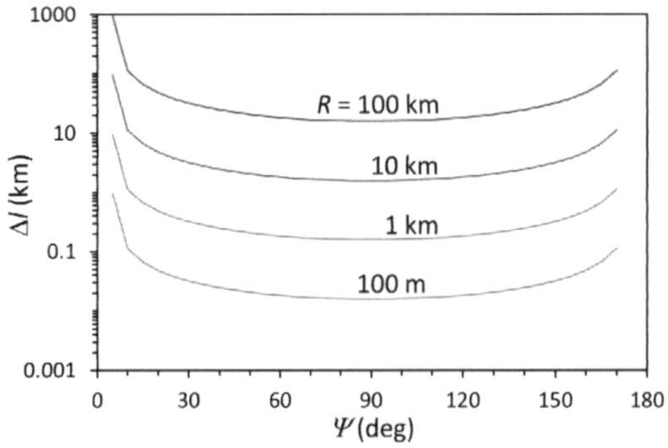

Figure 8: The same as Figure 7 for the length of the track observed by the telescope.

IACTs would have many advantages in comparison with classical fluorescence telescopes. The small angular size of the pixels of the camera (typically a factor of 10 smaller) leads to a much improved angular resolution. This feature together with the large size and good quality of their mirrors would lower the threshold for detecting light pulses over the night sky background.

The exploitation of fluorescence signals could be applied to several fields. In regard with VHE gamma-ray astronomy, it would be possible to enlarge the effective area of IACTs by the combined detection of Cherenkov and fluorescence light in showers with low Ψ angle. This would require modest extensions of the time windows. If a time window of microseconds is available, an array of IACTs covering a wide FoV could be used in pure fluorescence mode for the stereoscopic detection of EASs. It would allow for registering cosmic-ray showers with unprecedented resolution in the radial distribution along the whole longitudinal development, providing very valuable information for studies of hadronic interactions in atmospheric showers. Large time windows could also open the detection of extremely energetic showers at hundreds of kilometers, yielding huge effective areas.

We plan to make a quantitative evaluation of the feasibility of these potential applications, including the calculation of energy thresholds, effective areas, primary discrimination, etc. These tasks will require detailed simulations for which our implementation of the fluorescence emission in CORSIKA will be the ideal tool.

Acknowledgements

We gratefully acknowledge support from Spanish MINECO (contracts FPA2015-69210-C6-3-R and FPA2017-82729-C6-3-R) and the European Commission (E.U. Grant Agreement 653477). D. Morcuende acknowledges a predoctoral grant UCM-Harvard University (CT17/17–CT18/17) from *Universidad Complutense de Madrid*.

References

[1] Acharya B S *et al.* 2013 *Astropart. Phys.* **43** 3
[2] Tluczykont M *et al.* 2014 *Astropart. Phys.* **56** 42
[3] D. Heck *et al.* 1998 *Forschungszentrum Karlsruhe Report* FZKA 6019.
[4] Morcuende D *et al.* 2018 submitted to *Astropart. Phys.*. arXiv:1806.04790
[5] Abraham J *et al.* 2010 *Nucl. Instrum. Meth. Phys. Res. Sect. A* **620** 227
[6] Contreras J L *et al.* *PoS(ICRC2015)* 993, DOI: https://doi.org/10.22323/1.236.0993
[7] Pühlhofer G *et al.* *PoS(ICRC2015)* 1039, DOI: https://doi.org/10.22323/1.236.1039

IOP Conf. Series: Journal of Physics: Conf. Series **1181** (2019) 012048 doi:10.1088/1742-6596/1181/1/012048

Gamma/Hadron Separation in Imaging Air Cherenkov Telescopes Using Deep Learning Libraries TensorFlow and PyTorch

E B Postnikov[1,*], A P Kryukov[1], S P Polyakov[1], D A Shipilov[2], and D P Zhurov[2]

[1] Lomonosov Moscow State University Skobeltsyn Institute of Nuclear Physics (MSU SINP), Leninskie gory 1(2), GSP-1, Moscow, 119991, Russia

[2] Irkutsk State University (ISU), Karl Marx street 1, Irkutsk, 664003, Russia

E-mail: *evgeny.post@gmail.com

Abstract. In this work we compare two open source machine learning libraries, PyTorch and TensorFlow, as software platforms for rejecting hadron background events detected by imaging air Cherenkov telescopes (IACTs). Monte Carlo simulation for the TAIGA-IACT telescope is used to estimate background rejection quality. A wide variety of neural network algorithms provided by both libraries can easily be tested on various types of data, which is useful for various imaging air Cherenkov experiments. The work is a component of the Astroparticle.online project, which collaborates with the TAIGA and KASCADE experiments and welcomes any astroparticle experiment to join.

1. Introduction

Gamma-rays have extremely low photon fluxes at very-high energies, so that the appropriate detector size would be too large for current space-based instruments, and, therefore, only ground-based measurements are accessible. With the imaging air Cherenkov technique [1], gamma-rays are observed on the ground optically via the Cherenkov light emitted by extensive showers of secondary particles in the air when a very-high-energy gamma-ray strikes the atmosphere.

However, very energetic gamma-rays contribute only a minuscule fraction to the flux of cosmic rays (below one per million [2]), and that is why a very important problem of data analysis is discrimination between gamma-rays and cosmic rays, which are hadrons (mostly protons), given a telescopic image of the particle shower.

The conventional way to solve this problem is a parametrization of the image using empirical variables named after Hillas [3]: length, width, orientation, etc. Their combinations, also empirical, serve as a discriminator between gamma-like and hadron-like images taking into account that on average gamma-rays produce narrower images than cosmic rays, and gamma-ray images are orientated towards the point of the gamma-ray source projection on an image plane whereas cosmic-ray images are randomly oriented. For example, consecutive cuts on various Hillas parameters can be used to remove background events step-by-step, with optimal cut values determined by Monte Carlo simulation of the telescope [4]. Various attempts of combining these empirical parameters and derived quantities into machine learning algorithms have been carried out [5, 6, 7, 8, 9], but only three very recent efforts of a deep learning approach, which means fully automatic choice of the image features instead of empirical ones, have been made so far [10, 11, 12], of which the first one is an attempt not to distinguish primary gamma-rays from cosmic rays, but to select a special kind of images produced by secondary muons ('Muon Hunter' project [13]).

In this work we check the possibility of solving the above mentioned problem by Monte Carlo simulation of the TAIGA-IACT telescope [14]. The work is a component of the Karlsruhe-Russian astroparticle data life cycle initiative (also known as Astroparticle.online). This initiative aims to develop an open science system for collecting, storing, and analyzing astroparticle physics data. Currently it works with the TAIGA [15] and KASCADE [16] experiments and welcomes any other astroparticle experiments to join.

2. Monte Carlo simulations

Simulation was performed for primary gamma-rays and protons of power-law energy spectra with the experimental value of the slope. The IACT pointing direction spanned the zenith angle range of 30–40° corresponding to the Crab Nebula observations of the TAIGA-IACT telescope. Protons, which are background cosmic ray particles in the imaging air Cherenkov technique, were incident within up to $\pm 10°$ around the fixed pointing direction of the IACT. Gamma-rays were incident within up to $\pm 0.05°$ around the IACT pointing direction, which is the expected IACT pointing and tracking accuracy.

At the first step, the shower development in the atmosphere was simulated with the CORSIKA package [17]. The response of the IACT system was simulated at the second step using the OPTICA-TAIGA software developed for this task at JINR, Dubna [18]. The mirror of the TAIGA-IACT consists of 29 segments and has an area of about 8.5 m^2 and a focal length of 4.75 m. The camera is located at the focus, it consists of 560 photomultipliers with the total field of view (FOV) diameter $\sim 9.6°$, and the single pixel FOV 0.36°. Cherenkov photons of the shower were traced through the IACT optical system, the number of corresponding photoelectrons (p.e.) in each pixel of the camera was counted, and the arrival time of each of them was calculated.

Next, an output image was formed using a dedicated software developed at MSU SINP. For that purpose, photoelectrons in each photomultiplier were transformed into electric pulses using signal pulse-shape parametrization, and triggering and readout procedures were simulated for the data acquisition system. Upon readout, the Cherenkov signals are merged with night sky background (NSB) fluctuations, from a dedicated NSB simulation. A final image to be analyzed is obtained by subtracting the average NSB value (pedestal) from the pulse amplitudes (including NSB fluctuations) after readout in each photomultiplier.

3. Image cleaning

The procedure of image reconstruction above NSB is called image cleaning [19]. Conventional image cleaning procedure is two-parametric: it excludes from subsequent analysis all image pixels except the 'core pixels', i.e. those with the amplitude above a 'core threshold' and at least one neighbour pixel above a 'neighbour threshold', and the neighbour pixels themselves.

An example of the simulated image before and after cleaning with a low threshold (core threshold is 6 p.e. and neighbour threshold 3 p.e., whereas an average NSB amplitude is 2.6 p.e.) is given in figure 1. Cluster structure of the telescope camera is also shown (black lines demarcate clusters of pixels), and the triggering procedure feature is also clearly visible: only the part of the image within the triggered clusters is registered.

In this work we tried deep learning both with and without image cleaning.

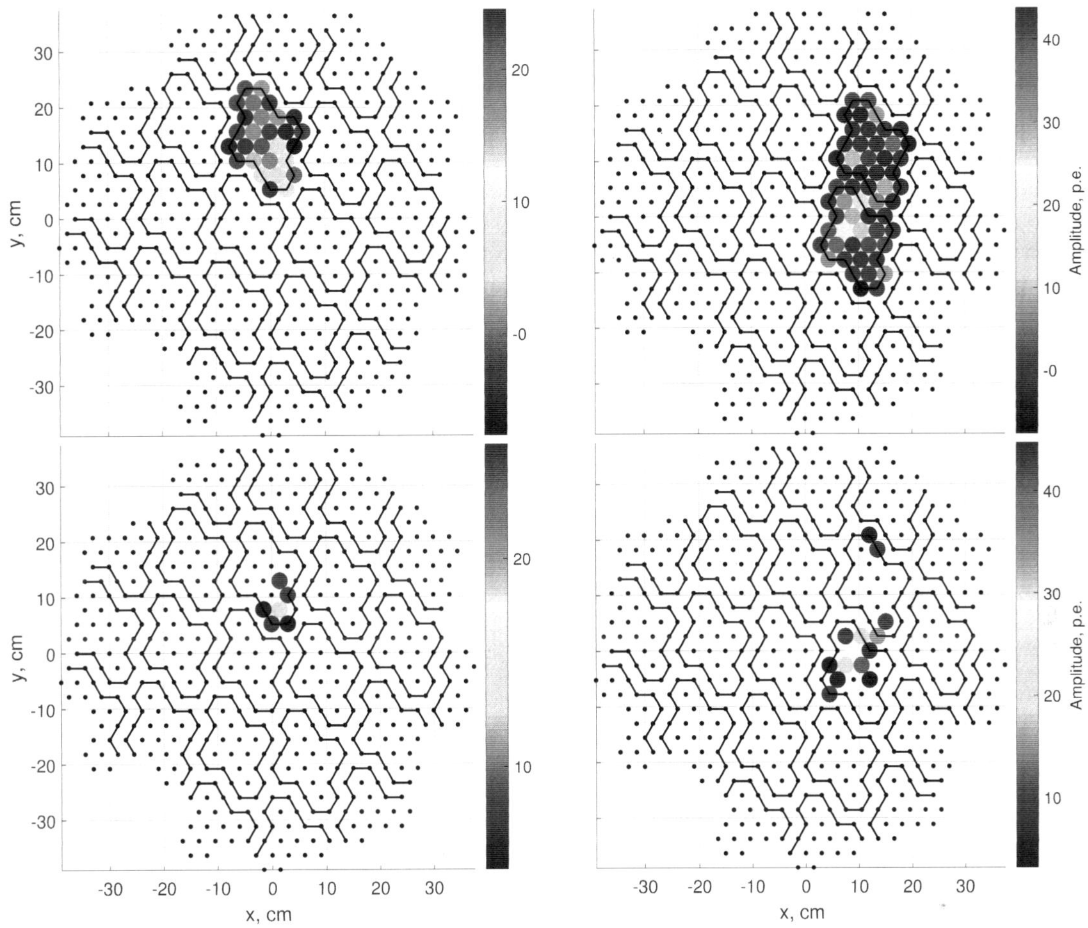

Figure 1. Gamma-ray (left panel) and proton (right panel) images before image cleaning (top panel) after image cleaning with a low threshold (bottom panel).

4. Hillas parameters for comparison with the deep learning technique

For the reference technique, two Hillas parameters [3] most informative for discrimination between gamma-rays and protons were calculated: the image width w and angle α between the image orientation and the direction to the camera centre. Since on average gamma-rays produce narrower images than cosmic rays, and gamma-ray images are orientated towards the gamma-ray source projection on camera (in our case it's the centre of the camera, (0; 0) point in figure 1), these two parameters allow discriminating gamma-ray images (as those with small w and α) against proton background.

The reference gamma/proton separation technique to compare with the deep learning approach was a set of two consecutive cuts (on w and α) with the optimal cut values found as maximizing Q value (section 5.2) on the learning samples of simulated gamma-rays and protons under the condition that the number of incorrectly identified gamma-rays is below 50%.

5. Methods

5.1. Convolutional neural network (CNN)

Under the deep learning approach, the so cold convolutional neural network (CNN) was chosen as it's a kind of artificial neural network that uses a special architecture, which is particularly well-adapted to classify images. Today, CNN or some close variant is used in most neural networks for image recognition, and it's also an independent choice for all other published deep learning applications to IACT technique [10, 11, 12]. Besides its discriminative power, an advantage of CNN is a fully automatic algorithm, including automatic choice of image features ('capable of classifying IACT images without any prior parametrization', [11]). The core principle behind CNN is that its convolutional layers apply a convolution operation (cross-correlation, or simply filtering) to the input, passing the result to the next layer, and so on. Special features of feedback (dropout [20] etc.) avoid overfitting that was the problem for conventional artificial neural network.

CNN is realized in free convenient software packages: PyTorch [21] and TensorFlow [22], and presently we are working with both of them. In contrast to work [11], pixels of the TAIGA-IACT camera are not square, but hexagonal. This special feature has not been fully taken into account yet, but only an approximation of the regular square grid using oblique coordinates with angle 60° was used.

Training datasets contained gamma-ray and proton images (Monte Carlo of TAIGA-IACT, energy distributions in the range 2–60 and 3–100 TeV respectively with the spectral slope -2.6); NSB, trigger procedure and detector response added. Test datasets (different from training ones) of gamma-ray and proton images in random proportion (blind analysis) were classified by each of the packages: TensorFlow and PyTorch. The output of each of both packages is a scalar parameter, 'probability' of being gamma or hadron; the term 'hadronness' [23] was introduced by the MAGIC collaboration: it 'spans a range between 0 and 1 and characterizes the event images being less or more hadronlike').

5.2. Quality criterion

As a quality criterion of particle separation the selection quality factor Q was estimated. This factor indicates an improvement of a significance S_0 of the statistical hypothesis that the events do not belong to the background. For Poisson distribution (that is for a large number of events), the selection quality factor is:

$$Q = \epsilon_\gamma / \sqrt{\epsilon_{bckgr}}, \tag{1}$$

where ϵ_γ and ϵ_{bckgr} are relative numbers of gamma-ray events and background events after selection. For our task we consider protons as background and aim to select gamma-rays above this background.

5.3. Gamma/proton discrimination without image cleaning

This basic advantage of CNN made us first trying gamma-ray separation from proton background using Monte Carlo images without image cleaning (section 3) at all. For that purpose training and test samples with neither cleaning nor preselection applied were given for analysis to both CNN packages (PyTorch, TensorFlow).

A conventional Hillas analysis in a simplified version using only two basic cuts (section 4): image width w and angle α, was used as a reference method. Conventional procedure of image cleaning (in our case, low-threshold cleaning with 6 p.e./3 p.e. threshold values) was also applied to the data before applying cuts; however, the images that were completely removed by image cleaning were also taken into account as 'protons' – for a fair comparison.

5.4. Gamma/proton discrimination after image cleaning

Training and test samples after image cleaning with low thresholds (6 p.e./3 p.e.), were also analyzed by CNN in both software packages. In case the image had less than 3 pixels after cleaning, it was removed from the analysis before calculating quality factor. The same reference method as in the previous section was used with a slight difference that now the images removed by image cleaning were not taken into account – as well for a fair comparison of both techniques.

6. Results

Quality factor (1) values obtained by the best CNN configuration among all the trained networks are assembled in table 1 together with the quality factor for the reference technique (sections 5.3, 5.4). Quality of the reference method of separation changed only slightly, because the reference technique always includes image cleaning, so the difference was only in the test sample size – the first table row contains the value for all events, whereas the last row contains the value for all events except those removed after image cleaning (section 3).

Table 1. Quality factor for gamma/proton separation.

	Reference(Hillas analysis)	CNN(PyTorch)	CNN(TensorFlow)
Without image cleaning	1.76	1.74	1.48
With image cleaning	1.70	2.55	2.99

7. Discussion

A fully automatic algorithm without image cleaning demonstrated the same quality as the simplest version of the conventional technique, Hillas analysis [3]. This means that one of more elaborated versions of this technique [5, 6, 7, 8, 9] would definitely overwhelm our CNN result. Therefore, some image cleaning procedure is necessary before the CNN implementation; however, the one we used is not optimal for CNN quality. Besides, some more elaborated CNN can also be applied to this task in future, so the conclusion of image cleaning necessity before CNN application should be considered preliminary.

Comparison of different CNN versions for both CNN packages (PyTorch and TensorFlow) is illustrated in figure 2. Overall PyTorch had more stable results in a wide range of CNN output parameter values (section 5.1). However, the TensorFlow CNN version trained by a 'modified' training sample, which contains additional images obtained by rotating 70% of images from the initial sample by 60° (symmetry angle of hexagonal structure), and thereby has $\sim$180000 events instead of $\sim$30000, shows significant improvement. Therefore, we think that these packages demonstrate approximately equal performance, and training sample size has the strongest influence on it. (PyTorch CNN has not been tested with the modified training sample yet.) Another issue is a problem of choosing the output parameter value; to get better quality, this value should be as high (close to 1) as possible (figure 2 (left)), but not as high as to lead to misidentifying more than 50% of gamma-rays (that happens when the curves in figure 2 (right) abruptly fall down below the ϵ_γ=0.5 value).

8. Conclusions

Both CNN packages were tested and showed approximately equal results. Two algorithms were found to significantly improve CNN separation quality: image cleaning, even with a low threshold (section 3), and additional image rotation in training sample, which allows increasing sample size. Problem of choosing the CNN output parameter value is still unsolved; however, there is still considerable potential to further improve the results by taking into account the hexagonal pixel shape and increasing training sample size by one order of magnitude.

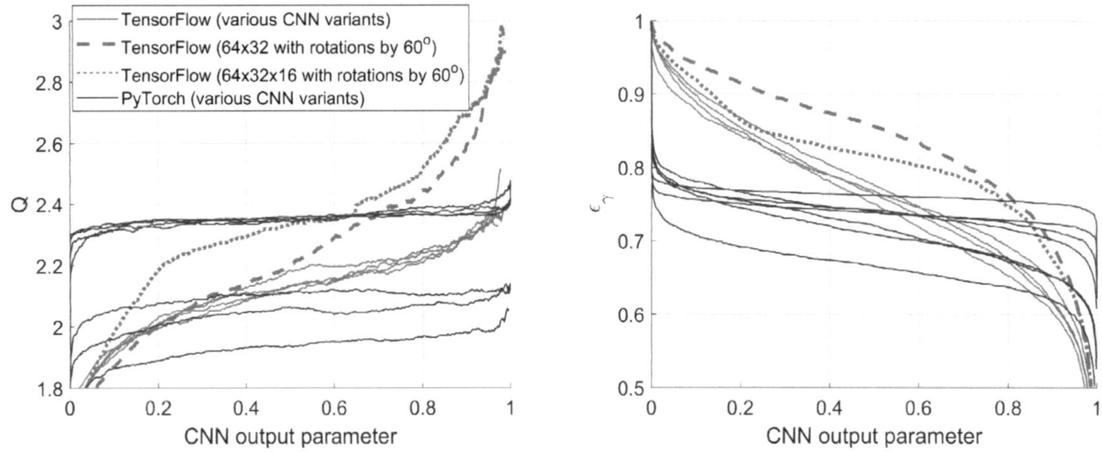

Figure 2. Quality factor (left) and percentage of correctly identified gamma-rays (right) vs CNN output parameter.

Acknowledgments
The work was supported by the Russian Science Foundation, grant #18-41-06003.

References

[1] Weekes T C *et al.* 1989 *Astrophys. J.* **342** 379–395
[2] Lorenz E, Wagner R 2012 *EPJ H* **37** 459–513
[3] Hillas A M 1985 *Proc. 19th Int. Cosmic Ray Conf. (La Jolla)* vol 3 (La Jolla: NASA Conf. Publ.) pp 445–448
[4] Reynolds P T *et al.* 1993 *Astrophys. J.* **404** 206–218
[5] Bock R K *et al.* 2004 *Nucl. Instrum. Meth. A* **516** 511–528
[6] Albert J *et al.* 2008 *Nucl. Instrum. Meth. A* **588** 424–432
[7] Boinee P, Barbarino F, De Angelis A, Saggion A, Zacchello M 2005 *Proc. 6th Int. Symp. Frontiers of Fundamental and Computational Physics (Udine, Italy)* (Dordrecht: Springer) pp 297–302
[8] Ohm S, van Eldik C, Egberts K 2009 *Astropart. Phys.* **31** 383–391
[9] Konopelko A, Chilingarian A, Reimers A 2006 *J. Phys. G: Nucl. Part. Phys* **32** 2279
[10] Feng Q, Lin T T Y for the VERITAS Collaboration 2017 *Proc. Int. Astronomical Union, IAU Symposium 325 (Sorrento)* (Cambridge: Cambridge University Press) **325** pp 173–179
[11] Nieto D, Brill A, Kim B for the CTA Consortium 2017 *Proc. 35th Int. Cosmic Ray Conf. (Busan)* (Busan: Proceedings of Science) PoS(ICRC2017)809
[12] Shilon I, Kraus M, Bchele M, Egberts K, Fischer T, Holch T L, Lohse T, Schwanke U, Steppa C, Funk S 2018 arXiv:1803.10698
[13] Feng Q for the VERITAS Collaboration, Jarvis J 2017 *Proc. 35th Int. Cosmic Ray Conf. (Busan)* (Busan: Proceedings of Science) PoS(ICRC2017)826
[14] Postnikov E *et al.* 2017 *Proc. 35th Int. Cosmic Ray Conf. (Busan)* (Busan: Proceedings of Science) PoS(ICRC2017)756
[15] Budnev N *et al.* 2017 *J. Instrum.* **12** C08018
[16] Antoni T *et al.* 2003 *Nucl. Instrum. Meth. A* **513** 490–510
[17] Heck D, Knapp J, Capdevielle J N, Schatz G, Thouw T 1998 *Report FZKA 6019* (Forschungszentrum Karlsruhe)
[18] Postnikov E B, Grinyuk A A, Kuzmichev L A, Sveshnikova L G 2017 *J. Phys.: Conf. Series* **798** 012030
[19] Bretz T and the MAGIC collaboration 2005 *Proc. 29th Int. Cosmic Ray Conf. (Pune)* vol 4 (Pune: Tata Institute of Fundamental Research) pp 315–318
[20] Srivastava N, Hinton G, Krizhevsky A, Sutskever I, Salakhutdinov R 2014 *Journal of Machine Learning Research* **15** 1929–1958
[21] Ketkar N 2017 *Deep Learning with Python* (Berkeley, CA: Apress) pp 195–208
[22] Martín A *et al.* 2016 *Proc. 12th USENIX Symposium on Operating Systems Design and Implementation (Savannah, GA)* (Savannah, GA: USENIX Association) pp 265–284
[23] Albert J *et al.* 2007 *Astrophys. J.* **663** 125–138

IOP Conf. Series: Journal of Physics: Conf. Series **1181** (2019) 012049 doi:10.1088/1742-6596/1181/1/012049

Intergalactic γ-ray propagation: basic ideas, processes, and constraints

Timur Dzhatdoev[1,2], **Emil Khalikov**[1], **Egor Podlesnyi**[3,1] **and Anastasia Telegina**[3]

[1] Skobeltsyn Institute of Nuclear Physics, Moscow State University, Moscow, 119991 Russia
[2] Institute for Cosmic Ray Research, University of Tokyo, 5-1-5 Kashiwanoha, Kashiwa, Japan
[3] Faculty of Physics, Moscow State University, Moscow, 119991 Russia

E-mail: timur1606@gmail.com, nanti93@mail.ru, podlesnyi.ei14@physics.msu.ru, nastias.skateme@gmail.com

Abstract. We review extragalactic γ-ray propagation models with emphasis on the electromagnetic (EM) cascade process in the magnetized expanding Universe. We consider cascades initiated by primary protons of ultra-high energy accelerated by blazars and show that the observable spectrum is similar to the universal spectrum of a purely EM cascade. We also present a detailed calculation of the observable angular distribution for the case of EM cascades developing from relatively nearby (<20 Mpc) sources. Finally, we calculate the point-like source differential sensitivity of a novel liquid Argon time projection chamber γ-ray telescope and show that its sensitivity is several times better than the Fermi LAT sensitivity in the 100 MeV – 100 GeV energy range.

1. Introduction

Observable γ-ray spectral, angular, and temporal distributions for extragalactic sources may be significantly transformed during intergalactic propagation. The most basic elementary processes involved are:
1) electron-positron pair production (PP) $\gamma\gamma \rightarrow e^+e^-$ on extragalactic background light (EBL) [1, 2] and cosmic microwave background (CMB) photons [3] for γ-rays;
2) inverse Compton (IC) scattering $e\gamma \rightarrow e'\gamma'$ for electrons and positrons[1] that occurs mainly on the CMB (see [4] and references therein);
3) Bethe-Heitler pair production (BHPP) and pion photoproduction (PIP) for protons and nuclei[2]. For ultrahigh energy (UHE, $E > 1/(1+z)$ EeV, z is redshift) BHPP is usually dominated by interactions on the CMB. For $1/(1+z)$ EeV $< E < 50/(1+z)$ EeV PIP is subdominant; at higher energies it is the main proton energy loss process, occuring mainly on the CMB (see [5] and references therein);
4) adiabatic losses (AL).
For especially high values of center-of-mass energy, additional processes also set in, such as muon pair production $\gamma\gamma \rightarrow \mu^+\mu^-$ [6] and double pair production $\gamma\gamma \rightarrow e^+e^-e^+e^-$ [7] for

[1] hereafter simply "electrons"
[2] below we do not consider primary nuclei, thus the photodissociation process is not relevant

γ-rays. Additionally, there is a quest for New Physics processes, such as γ-axion-like particle (ALP) mixing [8]–[11] or Lorentz invariance violation (LIV) [13]–[15].

Models involving production and propagation of γ-rays in the intergalactic volume may be classified as belonging to one of the following three types [16, 17]:
1) the absorption-only model, which accounts for only the PP and AL processes;
2) cascade models, which also include the IC process, and BHPP and PIP for primary protons;
3) exotic models.
Following the lines of argument presented in [17], in this paper we concentrate on the cascade models. To recall, the physics behind the absorption-only model is already very well-known. Many exotic models are already ruled out [18] [19], and there is growing evidence against some other options such as γ-ALP mixing [20, 21]. On the other hand, all existing possible deviations from the absorption-only model, such as [22, 23] (see also [24, 25]), could be qualitatively accomodated in the framework of the cascade models [26]. A caveat is that these anomalies are still not very well established [27, 28] (see also [29]).

The present paper is organised as follows. In section 2 we consider two specific cases where intergalactic electromagnetic (EM) cascades are important, namely: secondary γ-rays from 1) ultra-high energy cosmic rays emitted by blazars (subsection 2.1) and 2) relatively nearby sources such as galaxy clusters with the source-observer distance less than 20 Mpc (subsection 2.2). A number of other case studies could be found in [26, 16, 30]. In section 3 we discuss a new experimental technique called on to measure the extragalactic magnetic field (EGMF), namely, the liquid argon time projection chamber (TPC) approach. Finally, we conclude in section 4. All figures presented below were produced with the ROOT software [31].

2. Specific cases

2.1. Intergalactic hadronic cascade model

γ-rays from EM cascades initiated by primary protons accelerated in blazars may contribute to the observable emission (see e.g. [32, 33]). These protons may be strongly deflected on large scale structure (LSS) filament magnetic fields; below we show that this effect impacts the observable γ-ray spectrum (see also [26]). A simplified scheme of the corresponding geometry is shown in figure 1. A source (S) emits UHE protons that first propagate through a void (underdense region of space) with the diameter L_V and then are deflected on a filament (denoted by twin dashed lines). The proton deflection angle is denoted as δ, θ_{obs} is the angle between the direction from the source to the observer (O) and the direction of an incoming observable γ-ray. L is the distance between the source and the observer, $L_{int} = L_V + L_f$ is the characteristic proton interaction length.

Let's extrapolate the observable γ-ray path until we have a triangle with a right angle. The shortest side of that triangle can be represented by the following equation:

$$d = \sin(\theta_{obs}) \cdot (L - L_V) = \sin(\frac{\pi}{2} - \alpha) \cdot L_f \quad (1)$$

At the same time

$$\delta + \alpha = \frac{\pi}{2} - \theta_{obs} \rightarrow \frac{\pi}{2} - \alpha = \delta + \theta_{obs} \quad (2)$$

Assuming that $\theta_{obs} \ll \frac{\pi}{2}$ and $\delta \ll \frac{\pi}{2}$ we obtain the following:

$$\theta_{obs} \cdot (L - L_V) \approx (\delta + \theta_{obs}) \cdot L_f \quad (3)$$

$$\theta_{obs} \cdot (L - L_f - L_V) \approx \delta \cdot L_f \quad (4)$$

And finally

$$\theta_{obs} \approx \frac{\delta \cdot L_f}{L - L_f - L_V} = \delta \frac{L_{int} - L_V}{L - L_{int}}. \quad (5)$$

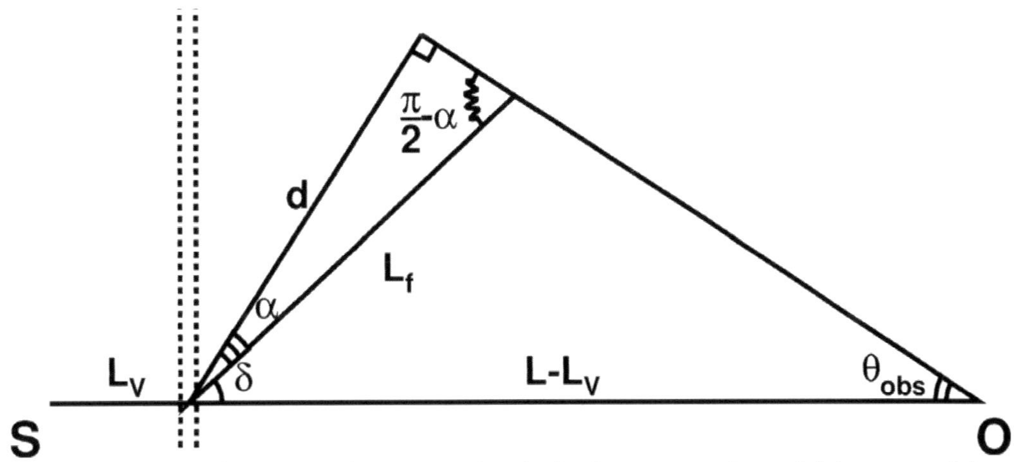

Figure 1. A scheme of geometry for the hadronic cascade model (not to scale).

Now let's estimate θ_{obs} assuming L = 750 Mpc and L_V = 50 Mpc. Let's assume L_{int} to be one of the following set of values: 100, 200 and 500 Mpc. The proton deflection angle δ may be estimated as follows [34]:

$$\delta \approx 1^\circ \frac{B}{nG} \frac{40EeV}{E/Z} \frac{\sqrt{L_B l_c}}{Mpc}, \tag{6}$$

where B is the magnetic field strength, E and Z are the energy and charge of the primary particles (protons in our case), L_B is the thickness of the filament and l_c is the coherence length of that magnetic field. For B = 1 nG, E = 40 EeV, L_B = 1 Mpc, l_c = 1 Mpc, we get δ= 1 ° and $\theta_{obs} = 0.077°, 0.27°, 1.8°$ for the three values of L_{int}. The typical extension of an imaging atmospheric Cherenkov telescope (IACT) point spread function (PSF) is about 0.1°, thus a part of observable γ-ray flux in our case will not fit into the point-like image of the source. In a more realistic case of lower proton energies, lesser values of L_V, and many filaments on the line of sight, this leads to the conclusion that most of observable γ-rays produced farther than 100 Mpc from the source would not contribute to the observable spectrum of the source. This, in turn, leads to an effective cutoff at the highest observable energies, as pointed out in [26]. In effect, the observable spectrum is almost the same as in the framework of the electromagnetic cascade model in the universal regime [35].

2.2. EM cascades from nearby sources

Another interesting, rarely discussed case is that of relatively nearby sources (L <20 Mpc). Here we consider an example of the Virgo cluster located at L=16.8 Mpc from the Earth. Very recently, [36] considered this source in context of very heavy dark matter searches. In particular, the authors of [36] estimated the secondary electron deflection angle (see their equation (3.7)). We note that this equation was obtained following e.g. [37]. The latter work assumed blazars as sources of primary γ-rays. Blazars typically have z >0.03, therefore the energy of last-generation cascade electrons is modest (~10 TeV or less), making the Thomson approximation used by [37] fully applicable in this case. For nearby sources, however, such electrons may have the energy well in excess of 10 TeV, where the Thomson approximation is not valid. Figure 2 shows the electron mean free path (black curve), as well as the characteristic electron energy loss length $L_{E-e} = c \cdot E/(dE/dt)$ (blue curve). These quantities were computed using the approximation of [38]. For comparison, red line denotes L_{E-e} in the Thomson approximation. We note that for electron energy in excess of 50 TeV the Thomson approximation fails and thus a new calculation of observable spectral and angular distributions is required.

Using the publicly-available code of [39] and assuming the EGMF strength 1 nG on coherence length 1 Mpc, we have performed a detailed calculation of the observable angular distribution. Such EGMF parameters were promtped by the fact that we do not expect a large void with low magnetic field strength in the local Universe. We also assumed the EBL model of [40] which is consistent with current constraints such as [41]. Figure 3 shows the observable angular distribution for three ranges of primary energy: 20–36 TeV (black histogram), 97–158 TeV (red histogram), and 1–3 PeV (blue histogram). In the first case the primary γ-ray mean free path $L_\gamma > L$; the second case corresponds to the one-generation regime, and the last case — to the universal regime (see [26] for detailed discussion of the latter two regimes). We note that the typical total electron deflection angle is well in excess of π; therefore, EM cascades develop an almost isotropic cloud around the source — the so-called pair halo (PH) [42].

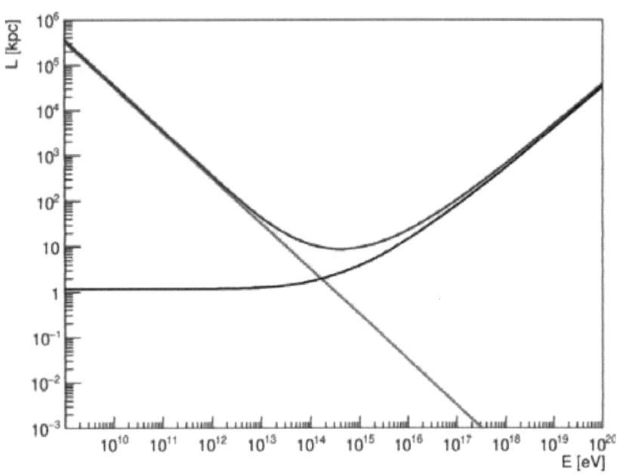

Figure 2. Characteristic lengths for electron in the intergalactic medium at z=0 vs. energy.

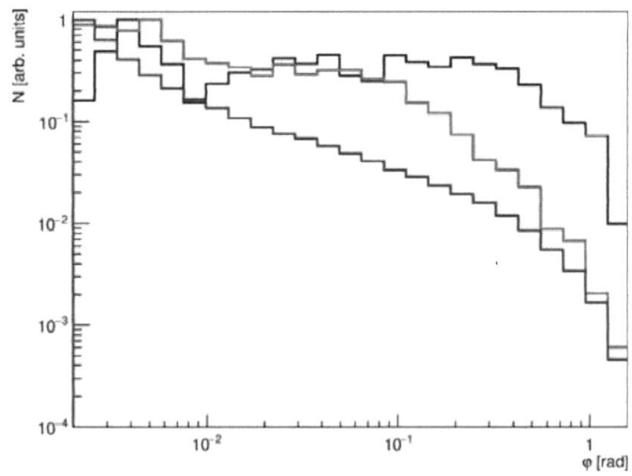

Figure 3. Observable angular distributions for various primary energy ranges.

3. Point-like source differential sensitivity of liquid Argon TPC γ-ray telescope

The main uncertainty of extragalactic γ-ray propagation models is the unknown strength B and structure of the EGMF (for a review, see [43, 44]). Possible values of B in voids range from 1 nG to 0.01 fG. Additionally, the EGMF may be highly inhomogeneous (see, e.g. [45]).

In [30] we argue that the most robust method to measure weak (B <10 fG) EGMF in voids is to detect magnetically broadened cascade (MBC) emission from blazars. If B <1 fG, the angular width of the MBC is typically comparable or smaller than the PSF width of currently operating γ-ray instruments. Therefore, a new experimental technique with better angular resolution is required in order to measure weak EGMF. In [30] we propose, for the first time, to utilize a novel technique of liquid argon time projection chamber (TPC) [46] to constrain the EGMF parameters. Here we provide an estimate of point-like source differential sensitivity vs. energy for such a detector following [46].

We assume a 2.0 m × 2.0 m × 0.8 m sensitive volume with 100 mkm longitudinal sampling step, filled with liquid Argon, segmented into 8 layers, each 10 cm thick. The total sensitive mass of the detector is 4480 kg, its total thickness is 110 g/cm^2 corresponding to 5.7 radiation lengths and optical depth τ =3.3 for 100 MeV γ-rays. Currently we are working to develop an energy reconstruction technique utilizing such a detector of medium thickness.

Figure 4 shows sensitivity for this γ-ray instrument assuming 1.7 years of continious observation of a source (thick red curve) or 10 years of continious observation (blue thick curve) for 5 σ detection of the source. Additionally, a minimum of 10 signal events are required; we also require that the number of signal events exceeds the number of background events. Other sensitivity curves were taken from [47] (see also [48]) and denote the differential sensitivity of Fermi LAT and atmospheric Cherenkov telescopes (ACT) [49, 50] (see Figure 5 (left) of [16] for the meaning of these curves). We conclude that the proposed new γ-ray instrument has the sensitivity several times greater than the Fermi LAT one in the wide energy range 100 MeV–100 GeV and could provide a natural low-energy extension for IACT such as the CTA array [49, 50] or LHAASO [51].

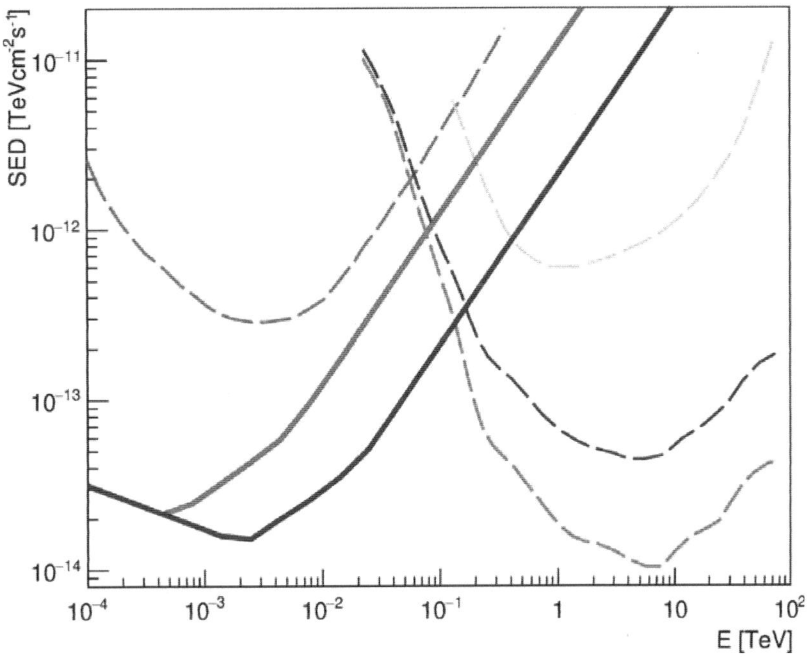

Figure 4. Differential sensitivity for point-like sources for liquid argon TPC and other γ-ray instruments.

4. Conclusions

In this work we continued our review of extragalactic γ-ray propagation phenomenology concentrating on intergalactic EM cascades. We have shown that the typical observable spectrum in the framework of the intergalactic hadronic cascade model practically coincides with the universal spectrum of the electromagnetic cascade model. This is due to deflection of primary protons in LSS filaments, causing the broadening of the observable angular distribution. We note that this effect is relevant even for isotropic sources. We also considered EM cascades from relatively nearby (L <20 Mpc) sources such as galaxy clusters and calculated the observable angular distribution in this case. Finally, we have estimated the point-like source differential sensitivity for a novel liquid Argon TPC γ-ray telescope, finding out that it has the sensitivity several times better than Fermi LAT in the 100 MeV – 100 GeV energy range.

Acknowledgements

This work is supported by the Russian Science Foundation (RSF) (project No 18-72-00083).

References

[1] Nikishov A I 1962 *Sov. Phys. JETP* **14** 393
[2] Gould R J and Shreder G 1967 *Phys. Rev.* **155** 1408
[3] Jelley J V 1966 *Phys. Rev. Lett.* **16** 479
[4] Blumenthal G 1970 *Phys. Rev. D* **1** 1596
[5] Berezinsky V, Gazizov A and Grigorieva S 2006 *Phys. Rev. D* **74** 043005
[6] Li Z and Waxman E 2007 EeV neutrinos associated with UHECR sources *Preprint* arXiv:0711.4969
[7] Brown R W et al. 1973 *Phys. Rev. D* **8** 3083
[8] Raffelt G and Stodolsky L 1988 *Phys. Rev. D* **37** 1237
[9] Kartavtsev A et al. 2017 *JCAP* **01** 024
[10] Montanino D et al. 2017 *Phys. Rev. Lett.* **119** 101101
[11] Anastassopoulos V et al. 2017 *Journal of Instrumentation* **12** P11019
[12] Coleman S and Glashow S 1999 *Phys. Rev. D* **59** 11
[13] Kifune T 1999 *ApJ* **518** L21
[14] Tavecchio F and Bonnoli G 2016 *A&A* **585** A25
[15] Abdalla H and Boettcher M 2018 *ApJ* **865** 159
[16] Dzhatdoev T 2017 *Proc. of the Moriond-2017 Very High Energy Phenomena in the Universe conf.* La Thuile
[17] Dzhatdoev T, Khalikov E and Kircheva A 2017 in *Proc. of Science, the 35th Int. Cosmic Ray Conf. 2017 (ICRC-2017)* 866 Busan
[18] Aharonian F A et al. 2000 *ApJ* **543** L39
[19] Rubtsov G, Satunin P and Sibiryakov S 2017 *JCAP* **1705** 049
[20] Ajello M et al. (Fermi-LAT Collaboration) 2016 *Phys. Rev. Lett.* **116** 161101
[21] Malyshev D et al. 2018 Improved limit on axion-like particles from gamma-ray data on Perseus cluster *Preprint* arXiv:1805.04388
[22] Horns D and Meyer M 2012 *JCAP* **1202** 033
[23] Korochkin A, Rubtsov G and Troitsky S 2018 Distance-dependent hardenings in gamma-ray blazar spectra corrected for the absorption on the extragalactic background light *Preprint* arXiv:1810.03443
[24] Furniss A et al. 2015 *MNRAS* **446** 2267
[25] Chen W et al. 2015 *Phys. Rev. Lett.* **115** 211103
[26] Dzhatdoev T, Khalikov E, Kircheva A and Lyukshin A 2017 *A&A* **603** A59
[27] Biteau J and Williams D 2015 *ApJ* **812** 60
[28] Dominguez A, Ajello M 2015 *ApJ* **813** L34
[29] Fermi-LAT Collaboration and Biteau J 2018 *ApJ Supplement Series* **237** 32
[30] Dzhatdoev T et al. 2018 Intergalactic electromagnetic cascades in the magnetized Universe as a tool of astroparticle physics *Preprint* arXiv: 1808.06758
[31] Brun R and Rademakers F 1997 *Nucl. Instrum. Methods Phys. Res. Sec. A* **389** 81
[32] Uryson A 1998 *JETP* **86** 213
[33] Essey W et al. 2011 *ApJ* **731** 51
[34] Harari D, Mollerach S and Roulet E 2016 *Phys. Rev. D* **D93** 063002
[35] Berezinsky V and Kalashev O 2016 *Phys. Rev. D* **94** 023007
[36] Blanco C, Harding J P and Hooper D 2018 *JCAP* **1804** 060
[37] Neronov A and Semikoz D 2009 *Phys. Rev. D* **80** 123012
[38] Khangulyan D, Aharonian F A and Kelner S R. 2014 *ApJ* **783** 100
[39] Fitoussi T et al. 2017 *MNRAS* **466** 3472
[40] Gilmore R, Somerville R, Primack J and Dominguez A 2012 *MNRAS* **422** 3189
[41] Korochkin A and Rubtsov G 2018 *MNRAS* **481** 557
[42] Aharonian F, Coppi P and Voelk H 1994 *ApJ* **423** L5
[43] Durrer R and Neronov A 2013 *A&A Rev.* **21**, 62
[44] Han J L 2017 *Annual Review of Astronomy and Astrophysics* **55** 111
[45] Hackstein S et al. 2018 *MNRAS* **475** 2519
[46] Bernard D 2013 *Nucl. Instrum. Methods Phys. Res. Sec. A* **701** 225 [Erratum: Bernard D 2013 *Nucl. Instrum. Methods Phys. Res. Sec. A* **713** 76]
[47] Funk S and Hinton J A 2013 *Astropart. Phys.* **43** 348
[48] Bernlohr K et al. 2013 *Astropart. Phys.* **43** 171
[49] Actis M et al. 2011 *Experimental Astronomy* **32** 193
[50] Acharya B et al. 2013 *Astropart. Phys.* **43** 3
[51] Cui S et al. (on behalf of the LHAASO Collaboration) 2014 *Astropart. Phys.* **54** 86

IOP Conf. Series: Journal of Physics: Conf. Series **1181** (2019) 012050 doi:10.1088/1742-6596/1181/1/012050

The sources of long GRBs: population inhomogeneity or possibility its using as standard candles

I V Arkhangelskaja

[1]National Research Nuclear University MEPhI (Moscow Engineering Physics Institute), Kashirskoe highway 31, Moscow, 115409, Russia

E-mail: irene.belousova@usa.net

Abstract. Large articles amount refers to GRBs (at least long ones) as standard candles and then concludes various cosmological consequences. But firstly long GRBs sources population homogeneity should be proved. Redshift distribution analysis should represent GRBs population homogeneity if its shape similar to one of objects with real uniform distribution in our Metagalaxy, such as SN1a (subsets of 42 and 53 ones used for its Ω and Λ definition). There are one maximum in each subset redshifts distributions. Typically considered short and long GRBs classes separated by t_{90}=2s. Moreover, registered event duration strongly depends on detector energy band and method used for temporal profile analysis and the same events could have different duration on various detectors data. GRBs mostly located at high redshifts and cosmological correction should be used in duration investigation. Redshift distribution for registered by Swift/BAT GRBs with t_{90}>1s (corrected to z) was analyzed with short GRBs exclusion. But its redshift distribution is sufficiently differs from ones for SN1a: bursts deficit occur at z~2, possibly caused by long GRBs population inhomogeneity. So, preliminary analyses results allow conclude long GRBs population inhomogeneity. Thus using GRBs as "standard candles" previously required events advanced classification completion and consideration only certain GRBs subsets.

1. Introduction: GRBs typical characteristics – short and long events.

Vela series satellites launched in middle of 1960th [1]. The first GRB with duration ~1s was observed on July 2, 1967 by detectors onboard Vela-4A in energy range 0.1 - 1 MeV [2]. GRBs characteristics vary in very large intervals. For example (see [3] and references therein), bursts duration lies in the interval 10^{-2} –10^{3} s, registered near the Earth fluencies varies in the range 10^{-8} –10^{-3} erg/(cm^2×s). Several thousand GRBs were detected up to now (see, for example, [3]) by more than 30 instruments onboard various satellites in both hear-Earth and interplanetary space.

GRBs duration distribution was firstly analyzed using data of BATSE [4] (Burst and Transient Source Experiment) onboard the Compton Gamma Ray Observatory (CGRO). CGRO [5] was launched on April 5, 1991 and finished its functioning on June 4, 2000. BATSE registered GRBs temporal profiles in four energy bands (25 - 50 keV, 50 - 100 keV, 100 – 300 keV and > 300 keV) due 8 large area detectors (LAD) based on NaI(Tl) scintillation crystal 50.8 cm in diameter and 1.27 cm thick each also with SD detectors for spectroscopic measurements - see, for example, [4].

GRBs duration characterized by t_{90} and t_{50} accordingly to BATSE data analysis [4, 6]. These two parameters are the times of accumulation for 90% and 50% of burst statistics (i.e. duration of intervals where the integrated counts from the GRB raise from 5% to 95% and for 25% to 75%

correspondingly). The GRBs duration distribution analysis had shown two bursts classes existence: long and short GRBs separated by $t_{90} \approx 2$ s. But results of similar distributions for bursts observed by other detectors have shown shifting of boundary between short and long events from value of 2 s. For example, durations of 1 s more likely correspond to this separator point for Swift/BAT GRBs subset distribution [7]. During 4B current BATSE catalogue [9] analysis group of intermrdiate duration events located at 99% confidence level in duration interval of $0.8\ \text{s} \leq t_{90} \leq 50$ s with $\langle t_{90} \rangle \approx 3$ s taking into account duration and duration-hardness distributions - see, for example, [8] and references there're. Additional criterion of burst spectrum description was hardness H_{32}, introduced using BATSE data analysis results. It was defined as the ratio of fluence in third and second energy channels. Taking into account duration-hardness distributions analysis the following criteria were obtained for short ($t_{90} < 3$ s, $\langle t_{90} \rangle \approx 0.5$ s, $H_{32} > 6.00$) and long ($t_{90} > 5$ s, $\langle t_{90} \rangle \approx 30$ s, $H_{32} < 1.85$) GRBs [10]. Also this type bursts presents in all BATSE catalogues (for example, 1B [9] and 3B [11]) in spite of it's subsets volumes. Intermediate GRBs are more intensive than short and long events [7, 10] (such bursts were absent in catalogue of faint events separated by ground analysis [12, 13]). Figure 1 presents this subgroup appearance in BATSE data [7]. Similar subgroup presents are widely discussed – see, for instance, [14, 15].

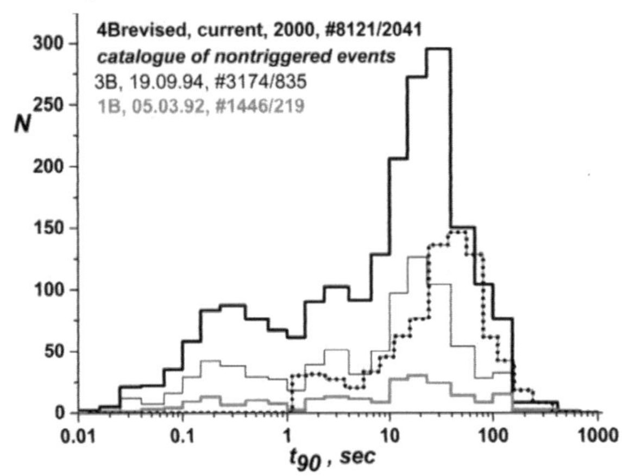

Figure 1. Intermediate GRBs subgroup appearance on BATSE data in duration distributions for 1B (thick grey), 3B (thin black) and 4B revised (thick black) catalogues of triggered bursts and its absence in non-triggered event catalogue (dotted histogram) [7]. Catalogues of triggered bursts contain 219, 835 and 2041 bursts correspondingly, last burst number marked as #. Non-triggered event one include 873 bursts and boundary between short and long events shifted at ~4.5 s.

However GRBs sources' origins nature is cosmological – see redshift corresponding columns in catalogues. Therefore correction to cosmological dilation of GRBs duration should be consider because of real cosmological sources time properties should be investigated only taking into account its redshift.

2. GRBs detection on Swift/BAT and *Fermi/GBM data.*

The Burst Alert Telescope (BAT) onboard Swift satellite [16] is a highly sensitive coded aperture imaging instrument designed to provide GRBs observations with large field of view (1.4 steradian half coded) and 4-arcmin positions definition in energy bands 15-150 keV for imaging mode and up to 500 keV with a non-coded response. It was launched November 20, 2004 [14]. Three GRBs catalogues are published due to BAT data analysis. Now Swift/BAT catalogue contains ~ 1200 bursts and 358 ones with known redshift [17].

GBM (Gamma-ray Burst Monitor) consist of 12 NaI based low-energy detectors for burst temporal profile registration and two high-energy ones made on BGO provide event spectra analysis – see, for example, [18]). It installed onboard the Fermi Gamma-Ray Space Telescope (formerly GLAST) [19] was launched on June 11, 2008 and operates up to now. Characteristics of NaI and BGO detectors of GBM instrument are similar to ones of BATSE LAD and SD combination but cover a wider energy range despite of its smaller collection area. Four GRB catalogues are published based on GBM data analysis – see, for example, [20]. Current GBM catalogue listed in [21].

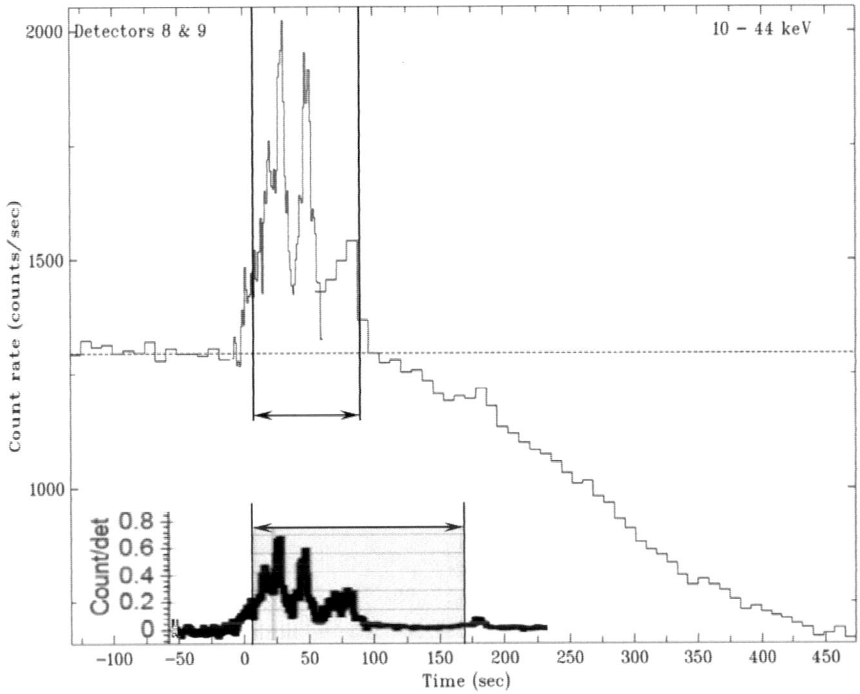

Figure 2. Difference in duration of GRB170405A (18:39:48 UT) in energy range 15-350 keV on Swift/BAT data (t_{90}=165 ± 32 s marked by arrow and filed light gray on the lower part of figure) and in energy band 50-300 keV on Fermi/GBM data (t_{90}= 78.593 ± 0.572 s marked by arrows on the upper part of figure).

Unfortunately, duration of registered event strongly depends on detector energy band and method used for temporal profile analysis. These reasons lead the same events could have different duration on data of various detectors. For example, GRB170405A (18:39:48 UT) had t_{90}=165 ± 32 s in energy range 15-350 keV on Swift/BAT data [17] but t_{90}= 78.593 ± 0.572 s in energy band 50-300 keV on Fermi/GBM data [21] – see figure 2. There are several peaks at this burst temporal profile and last peak was located at more than 50 s from main three-peaks episode. Moreover, it appears only at low energies. This peak is sufficiently slight than first episode: ten times smaller than first peak on Swift/BAT data in energy range 15-100 keV and is absent in energy band 100-350 keV. On Fermi/GBM data in energy band 10-44 keV, its intensity is only about 50 counts/s over background count rate 1200 counts/sec against ~2000 counts/s over background count rate ~1300 counts/s during first peak. And this peak separation cause difficulties just because of geomagnetic modulation of Fermi/GBM background. Thus Fermi/GBM burst table doesn't contain data about this episode of GRB170405A.

3. Duration cosmological correction

In addition, cosmological correction should be taken into account in event classes' separation due to duration analysis because of sufficient amount of GRBs located at high redshift – see, for example, [22]. GRBs distribution on t_{90} and corrected to cosmological extension t_{90} is presented at Figure 3.

Thus, GRB110731A (11:09:30 UT) had t_{90}= 7.485±0.923 s in energy band 50-300 keV on Fermi/GBM data [21] but t_{90}= 38.8 ±32 s in energy range 15-350 keV on Swift/BAT ones [17]. After cosmological correction (its z= 2.83) t_{90}~10.1 s on Swift/BAT data (long burst accordingly standard classification) and t_{90}~1.95 s on Fermi/GBM data (short event accordingly standard classification).

Moreover, separation point between short and long GRBs shifted to ~1s on preliminary results of analysis of Swift/BAT events with known z and duration (358 GRBs) and intermediate group of events occurs in range $1<t_{90}<10$ s. The last facts can indicate inhomogeneity of long GRBs population. Unfortunately analogous Fermi/GBM subset is insufficient for statistical analysis. It contains only ~50 events because of more than 55% of Fermi/GBM bursts has 1-sigma statistical uncertainty in the t_{90} duration bigger than 10%.

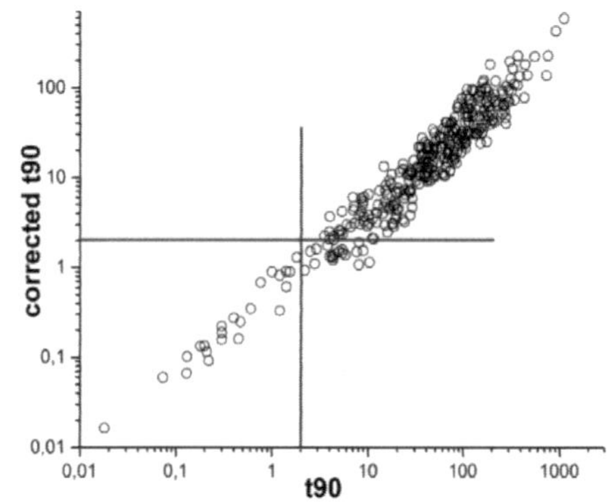

Figure 3. GRBs distribution on t_{90} and corrected to cosmological extension t_{90}. Vertical and horizontal lines illustrates location of point with z<0.1 and t90~2 sec, i.e. separation point between short and long events not shifted after taking into account cosmological extension only for low redshift bursts.

4. GRBs distribution on duration and t_{90}

As i. was mentioned above, the volume of Fermi/GBM bursts subset with known redshift and duration is insufficient for analysis because of big 1-sigma statistical uncertainty in the t_{90} duration. Thus redshift distribution for subset of registered by Swift/BAT GRBs (corrected to z) was analyzed. Fi ire 4 presents redshift (a) and redshift-duration (b) distributions for total Swift/BAT GRBs popul...ion taki..g into account cosmological extension. Short bursts with $t_{90}<1$ s mostly located at z<2, but additional 9 events with longer duration should be included to this group due preliminary results of population analysis. There is deficit of events with z~2 and shape of distribution differ than one of objects with real uniform allocation in our Metagalaxy, such as SN1a – see, for example, [10].

Than Swift/BAT GRBs subset was investigated but mentioned at figure 4 short bursts are excluded – see figure 5. But its redshift distribution is sufficiently differs from ones for SNIa too: deficit of bursts also occur at z~2. It can be caused by long GRBs population inhomogeneity and presence of intermediate duration group of bursts (marked triangles at figure 5b).

Also there are several differences of magnitude dependence on redshift for GRBs, usually used as standard candles SNIa and SNIc (especially connected with GRBs) types supernovae. The preliminary analysis results of Latest supernovae project data [23] (samples of 5918 SNIa and 374 SNIc) presented at Figure 6.

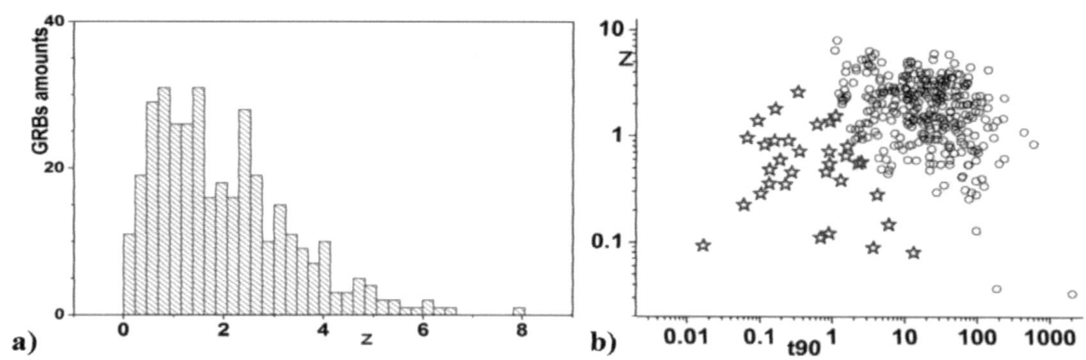

Figure 4. Redshift (a) and redshift-duration (b) distributions for total Swift/BAT GRBs population with correction to cosmological extension. Stars shows short bursts.

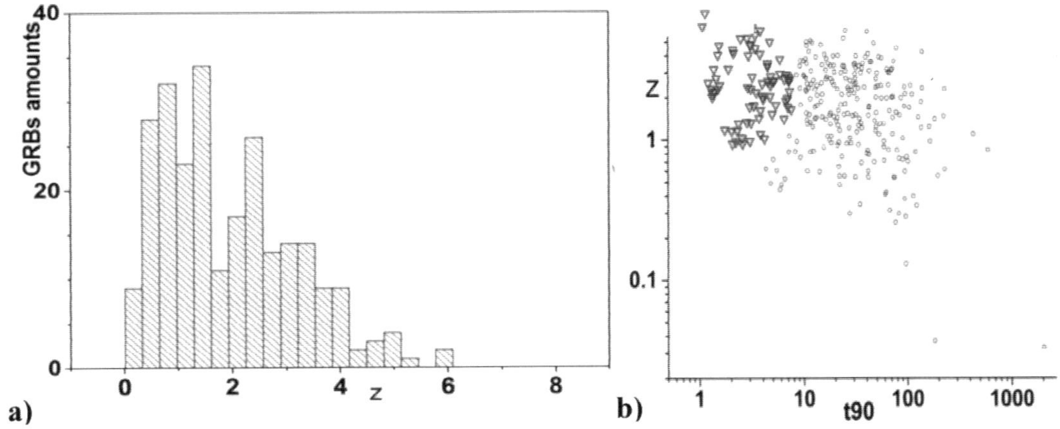

Figure 5. Redshift (a) and redshift-duration (b) distribution for Swift/BAT GRBs subset with removed marked by stars at figure 4b short events. GRBs marked triangles at panel (b) possible correspond intermediate bursts.

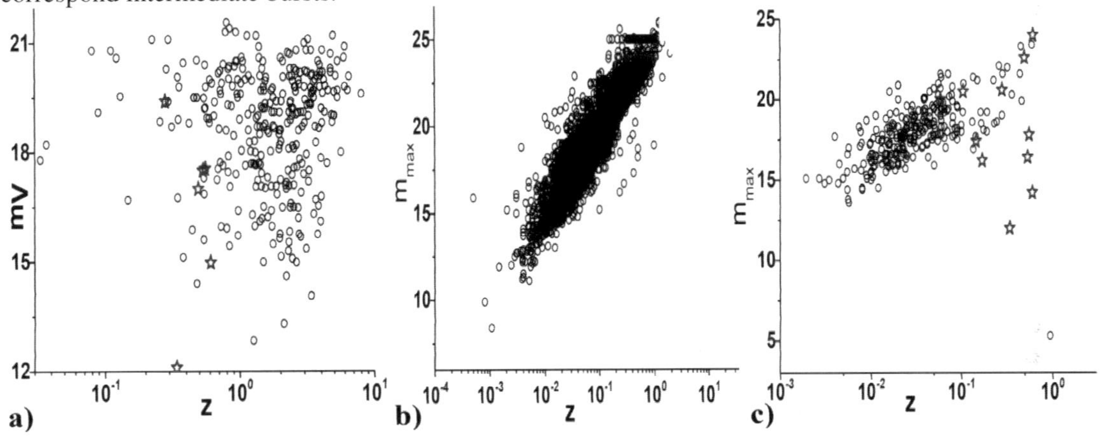

Figure 6. Magnitude dependence on redshift for Swift/BAT GRBs, SNIa and SNIc. Twelve connected SNIc with GRBs ones marked stars at panel (c), eight bursts associated with SNIc marked as stars at panel (a).

Also there are several differences of magnitude dependence on redshift for GRBs, usually used as standard candles SNIa and SNIc (especially connected with GRBs) types supernovae. The preliminary analysis results of Latest supernovae project data [23] (samples of 5918 SNIa and 374 SNIc) presented at Figure 6.

Conclusion

In discussion of possibility using of any sources as "standard candles" firstly redshift distribution analysis should be made. If redshift distribution shape of analysed subset is similar to one of objects with real uniform distribution in our Metagalaxy than we could use sources of this subset as "standard candles". Usually SN1a consider as standard candles (for example, subsets of 42 and 53 ones used by group of S. Perlmutter for definition of Ω and Λ for our Metagalaxy [24]). Investigation of these subsets redshifts distributions shows one maximum each.

Redshift distribution of subgroup of Swift/BAT GRBs with known z and duration have been analysed. But its redshift distribution is sufficiently differs from ones for SN1a both for total subset and one with excluded short GRBs: there is deficit of bursts at z~2. It can be caused by long GRBs population inhomogeneity because of all selection effects should be appear in results of analysis of subgroups of other sources too. Also, there are several variations of magnitude dependence on redshift for GRBs, SNIa and SNIc, moreover, sample of usual SNIc differ from one associated with GRBs.

So, results of preliminary analysis allow make conclusion about inhomogeneity of long gamma-ray bursts population. Thus using GRBs as "standard candles" previously required completion of events advanced classification and then taking into account only certain GRBs subsets. Inhomogeneities of long GRBs population also require new theoretical implications of its sources models.

Acknowledgments
Author thank for the support from National Research Nuclear University MEPhI in the framework of the Russian Academic Excellence Project (contract No. 02.a03.21.0005, 27.08.2013)

References
[1] Klebesadel P et al 1973 Astrophys. J. Lett. **182** 85
[2] Strong I B et al 1974 Astrophys. J. Lett. **188** 1
[3] http://nssdc.gsfc.nasa.gov/nmc/
[4] Fishman G J et al. 1994 Astrophys. J. Suppl. Series **92** 229
[5] Varendorff M 1995 SIF Conf. Proc. **47** 183
[6] Kouveliotou C 1995 Ann. N.Y. Acad. Sci. **759** 411
[7] Arkhangelskaja I V 2017 J. of Phys.: Conf. Ser. **798** 012007
[8] Belousova I V et al 1999 Astronomy Reports **43(11)** 734
[9] Paciesas W S et al. 1999 Astrophys. J. Suppl. Series **122** 465
[10] Arkhangelskaja I V 2015 Phys. Proc. **74** 246
[11] Meegan C A et al. 1996 Astrophys. J. Suppl. Series **106** 65
[12] Kommers J M et al. 2000 arXiv: 0012519v1 [astro-ph]
[13] Stern B E et al. 2000 Ap J.Lett. **538** L21
[14] Horváth I et al. 2006 Astron. Astrophys. **447** 23
[15] Tarnopolski M 2016 Astrophys. Space Sci. **361** 125
[16] Barthelmy S D et al. 2005 Sp Sci. Rev. **120** 143
[17] http://swift.gsfc.nasa.gov/results/batgrbcat/index.html
[18] Atwood W B et al. 2009 Astrophys. J. **697** 1071
[19] Paciesas W S 2011 AIP Conf. Proc. **1366** 165
[20] Ackermann, M. et al. 2014 Astrophys. J., **787** 15
[21] http://heasarc.gsfc.nasa.gov/W3Browse/fermi/ fermigbrst.html
[22] Arkhangelskaja I V 2015 Bulletin of the Russian Academy of Sciences: Physics **79** 413
[23] http://www.rochesterastronomy.org/snimages/snredshiftall.html
[24] Perlmutter S et al 1999 Astrophys. J. **517** 565

IOP Conf. Series: Journal of Physics: Conf. Series **1181** (2019) 012051 doi:10.1088/1742-6596/1181/1/012051

The puzzle of muons: novel particles or novel properties

Maciej Rybczyński and Zbigniew Włodarczyk
Institute of Physics, Jan Kochanowski University, PL-25406 Kielce, Poland
E-mail: Zbigniew.Wlodarczyk@ujk.edu.pl

Abstract. In order to examine a muon excess observed by the Pierre Auger Observatory, detailed Monte Carlo simulations were carried out, assuming the existence of strangelets (hypothetical stable lumps of strange quark matter) in the primary cosmic rays. We obtain a rough agreement between the simulations and the data for ordinary nuclei without any contribution of strangelets in primary flux of cosmic rays.

1. Introduction

Ultra-high energy ($E > 10^{18}$ eV) cosmic rays (UHECR) provide a formidable beam to study particle collisions at center-of-mass energies and kinematical regimes not accessible at terrestrial accelerators [1]. UHECR can only be observed indirectly, through air showers. The slant depth of the shower maximum X_{max} and the number of muons in an air shower N_{μ} are powerful tracers of the mass composition. Detailed simulations show dependences of both observables on hadronic interaction properties, such as multiplicity, inelasticity, charge ratio and baryon-antibaryon pair production [2, 3].

Over the past few decades, it has been suspected that the number of registered muons at the surface of the Earth is tens of percentage points higher than what it should be, according to existing hadronic interaction models [4, 5]. Recently, a study from the Pierre Auger Collaboration (Auger) has strengthened this suspicion, using a novel technique to mitigate some of the measurement uncertainties of earlier methods [5, 6]. The new analysis of Auger data suggests that the hadronic component of showers with primary energy $E > 10^{18}$ eV contains about 30% to 60% more muons than expected. Also the number of muons with energies above 0.75 GeV, determined by the Sydney University Giant Air-shower Recorder (SUGAR), exceeds the simulated one by the factors $\sim$ 1.67 and $\sim$ 1.28 for 10^{17} eV proton and iron primaries, respectively [7]. On the other hand, very recently, no muon excess in EAS has been reported [8]. To explain the muon excess, several new models have been proposed, exploring new physics [9, 10, 11] or new forms of matter, namely strange quark matter (SQM) [12]. Roughly, two possibilities arise: either we are dealing with novel particles or known particles have novel properties of multiparticle production processes.

In this paper we adopt a purely phenomenological approach to develop an SQM scheme. In sharp contrast to previous models our approach is based on the assumption that ultra-high energy cosmic rays are very heavy strange quark objects, i.e. strangelets. To examine this scenario in the shower development, as observed by the Pierre Auger Observatory, detailed Monte Carlo simulations were carried out.

2. Strangelets in cosmic rays

The analysis of the EAS data offers a unique possibility to observe possible imprints of strangelets arriving from the outer space. They are lumps of SQM, a new possible stable form of matter (cf. [13, 14, 15] for details). Following [16, 17] it is fully reasonable to search for SQM in cosmic ray experiments because the specific features of strangelets allow them to penetrate deep into the atmosphere [18]. The point is that there is a certain critical size of the strangelet, given by the critical value of its mass number $A = A_{crit} \simeq 300 - 400$, such that for $A > A_{crit}$ strangelets are absolutely stable against neutron emission. Below this limit strangelets decay rapidly evaporating neutrons. The spatial radii of strangelets turn out to be comparable to the radii of ordinary nuclei [16], i.e., their geometrical cross sections are similar to the normal nuclear ones. To account for their strong penetrability one has to accept that strangelets penetrating deeply into the atmosphere are formed in many successive interactions with the air nuclei by the initially very heavy lumps of SQM entering the atmosphere and decreasing due to the collisions with air nuclei (until their A reaches the critical value A_{crit} [16]). In this scenario the interaction of a strangelet of mass A with an air nucleus of mass A_{air} involves all quarks of the target located in the geometrical intersection of the colliding air nucleus and the strangelet. So, up to $3 \cdot A_{air}$ quarks from the strangelet could be used making its mass drop to a value of $A - A_{air}$.

There are several reports suggesting the existence of direct candidates for SQM (characterized mainly by their very small Z/A ratios). All of them have mass numbers A near or slightly exceeding A_{crit}. Analysis of these candidates for SQM show [16, 18, 17] that the abundance of strangelets in the primary cosmic ray flux is $F_{SQM}(A = A_{crit})/F_{tot} \simeq 2.4 \cdot 10^{-5}$ at the same energy per particle. For a normal flux of primary cosmic rays the expected flux of strangelets is then equal to $F_{SQM} \simeq 7 \cdot 10^{-6}$ m^{-2}h^{-1}sr^{-1} for the energy above 10 GeV per initial strangelet.

The experimental data mentioned above lead to a flux of strangelets which is consistent with the astrophysical limits and with the upper limits given by the experiment [19]. The data follows the $A^{-7.5}$ behaviour, which coincides with the behaviour of abundance of normal nuclei in the Universe [17]. Interpretation of indirect observations (anomalous events observed in emulsion chambers, and also results from the measurements of EAS) can provide signals of strangelets. In particular, the excess of muon bundles has been observed by the ALICE detector at LHC, in its dedicated cosmic ray run [20] (confirming similar findings from the LEP era at CERN [21]). Arguing that muonic bundles of highest multiplicity are produced by strangelets, which are hypothetical stable lumps of strange quark matter infiltrating our universe, we successfully described data from CERN experiments [22]. All these considerations motivate further experimental search for the SQM and for its cosmological and elementary particle physics aspects.

3. EAS simulations

For the simulation of the propagation of extensive air showers in the Earth's atmosphere we have used a suitably modified SHOWERSIM [23] modular software. In our simulations we used the F00 model which consists in a simple non-scaling extrapolation of the inclusive data at ISR and SPS energies [24]. In this version, the leading particle remembers its charge, and the x-distribution of secondaries does not depend on the elasticity.

In the F00 model the identity of secondary particle is generated with the probability

$$p\left(x, E\right) = \alpha\left(x\right) + \beta\left(x\right)\ln\left(E\right), \tag{1}$$

where $x = p, \pi, K$ denotes various secondaries at 1 TeV. Probability of the emission of charged particle $(\pi^{\pm}, K^{\pm})$ is:

$$p\left(N_{ch}, E\right) = 1 - 0.5\left[\alpha\left(K\right) + \beta\left(K\right)\ln\left(E\right)\right] - \alpha\left(\pi^0\right) - \beta\left(\pi^0\right)\ln\left(E\right). \tag{2}$$

In the standard version of F00 model $\alpha(\pi^0) = 0.3$, $\alpha(K) = 0.1$, $\beta(\pi^0) = 0.0128$, and $\beta(K) = 0.0.0028$. The fraction of produced neutral pions increase by 0.0295 per energy decade and fraction of produced kaons increases by 0.006 per decade. Changing the value of parameter β we can obtain a different number of charged particles, $\langle N_{ch}\rangle$ for a fixed number of secondaries, $\langle N\rangle = \text{const}$. In figure 1 we show the mean multiplicity of charged particles produced in proton-proton interactions as a function of the center-of-mass energy for our F00 model (full line). Again, EPOS LHC (dotted line) and QGSJETII-04 (dash-dotted line) predictions are taken from [25].

In the F00 model the inelasticity K, i.e. the relative energy used for production of secondaries, is sampled from a uniform distribution over an (0,1) interval for interacting nucleons and over an (0.333,1) interval for meson interactions. Therefore, the average inelasticity $\langle K\rangle = 0.5$ for nucleons and $\langle K\rangle = 0.667$ for mesons. In figure 2 we show the average inelasticity in proton-proton interactions as a function of the center-of-mass energy for our F00 model (full line). EPOS LHC (dotted line) and QGSJETII-04 (dash-dotted line) predictions are taken from [25].

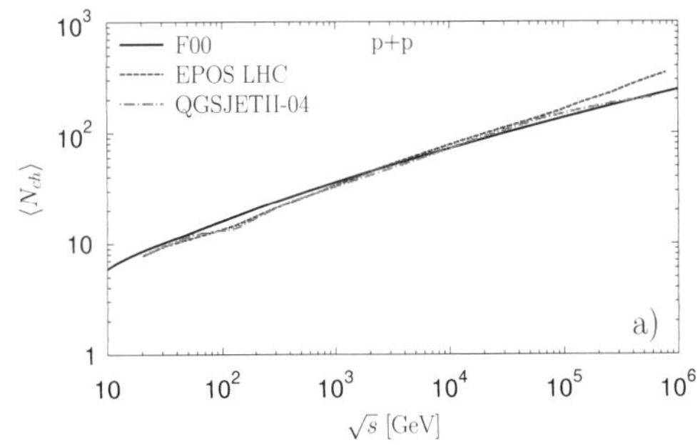

Figure 1. Mean multiplicity of charged particles produced in proton-proton interactions as a function of center-of-mass energy.

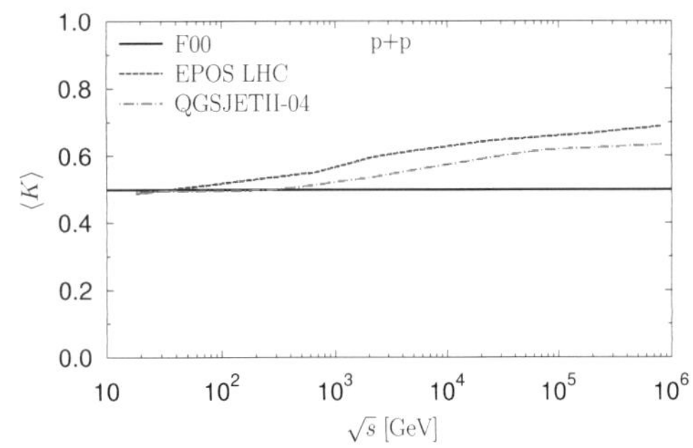

Figure 2. Average inelasticity in proton-proton interactions as a function of the center-of-mass energy.

For collisions with air nuclei, the inelastic cross-section dependence on energy is given by: $\sigma_{inel} = \sigma_0 + \delta \cdot [\ln(E/E_0)]^{\alpha}$ for $E \geq E_0$ and $\sigma_{inel} = \sigma_0$ for $E < E_0$, where α =1.8, $\sigma_0 = 280$, 196, 178 mb, $\delta = 2.5$, 1.7, 1.6 mb and $E_0 = 0.1$, 0.067, 0.067 TeV for primary nucleon, pion and kaon respectivly.

In the development of EAS, the inelasticity K and the cross section for interactions σ are strongly correlated. The attenuation of hadrons or the depth of the shower maximum X_{max} are actually the measure of combinations of K and σ, and the effect of these two parameters is extremely difficult to disentangle. From the above we can see that $\langle K\rangle$ and σ_{inel} differ significantly for different models. Nevertheless, the $\langle X_{max}\rangle$ for these models are quite similar. The average positions of showers maxima as a function of primary energy are shown in figure 3. Obviously, the simulated values for different primaries follow a logarithmic trend, thus were fitted using the formula $\langle X_{max}\rangle = c + d \cdot \ln(E/10^{19}\text{ eV})$, with $c = 801.4$, 724.3, 683.8 and $d = 24.7$, 28.0, 27.7 for protons, iron nuclei and SQM, respectively.

The deep tail of the depth of maximum distribution, which has an exponential behaviour: $dN/dX_{max} \sim \exp(-X_{max}/\Lambda_X)$ and depends on the proton interaction length $\lambda \sim 1/\sigma_{inel}$ via shower maxima attenuation length: $\Lambda_X \simeq 0.8\lambda/\langle K\rangle$ [g/cm^2]. Figure 4 shows the shower maxima attenuation length as a function of the center-of-mass energy for our F00 model (full line). EPOS LHC (dotted line) and QGSJETII-04 (dash-dotted line) predictions are taken from reference [25]. Despite the very different energy dependence of $\langle K\rangle$ and σ_{inel}, the shower maxima attenuation lengths Λ_X are very similar. We evaluate the attenuation of the showers

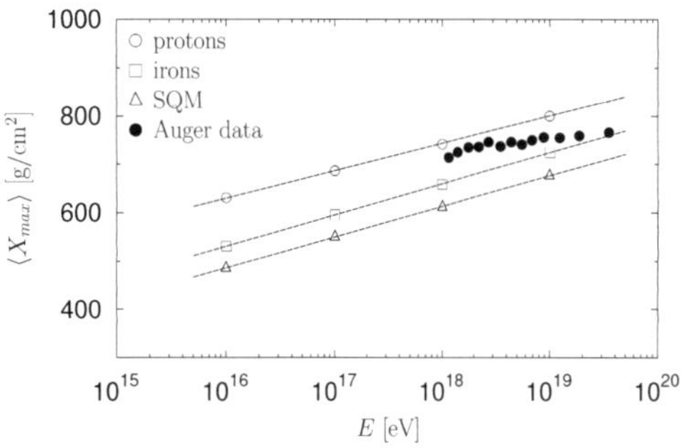

Figure 3. Average position of the shower maximum as a function of the primary energy for EAS generated by protons, iron nuclei, and SQM. Pierre Auger Collaboration measurements [26] are indicated by black circles.

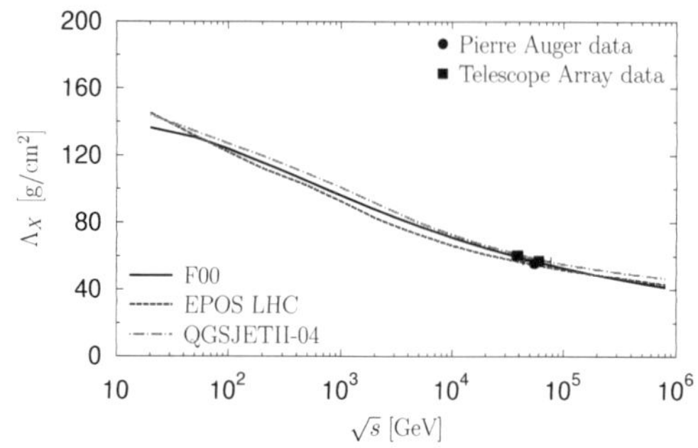

Figure 4. Attenuation length as a function of the center-of-mass energy. Pierre Auger Collaboration measurements [27] are indicated by black circles while Telescope Array data [28] by black squares.

maxima, Λ_X, for protons at 10^{18} eV obtaining $\Lambda_X = 58.5 \pm 1$ g/cm^2, which agrees nicely with the experimental data $\Lambda_X = 55.8 \pm 2.3$ g/cm^2 at $10^{18.2}$ eV, reported by the Pierre Auger Collaboration in [27], $\Lambda_X = 57.4 \pm 1.8$ g/cm^2 at $10^{18} - 10^{18.5}$ eV, and $\Lambda_X = 60.7 \pm 2.1$ g/cm^2 at $10^{17.8} - 10^{18}$ eV, given by the Telescope Array experiment [28].

We performed Monte Carlo simulations of the EAS generated by primary nuclei (protons and iron nuclei) and by primary strangelets with mass $A \geq A_{crit} = 320$ taken from the $A^{-7.5}$ distribution. In this study we have generated the inclined (primaries with zenith angle Θ from 60 deg to 80 deg were sampled) events with energies in the interval $10^{16} < E < 10^{19}$ eV. In the analysis we have focused on the muons in the nuclear cascade with energies larger than 0.3 GeV, which is the Cherenkov threshold for muons in water, reaching the Pierre Auger South Laboratory surface detector placed at the altitude 1425 m above sea level ($X_{atm} = 750$ g/cm^2).

4. Results

In this section we provide a comparison of our SHOWERSIM simulations with the Pierre Auger Collaboration results presented in reference [5]. We focus at the showers generated with average zenith angles $\langle \Theta \rangle = 67$ deg. Following [5] the muon content R_μ is defined as:

$$R_\mu = \frac{N_\mu}{N_{\mu,19}}, \tag{3}$$

where N_μ is the total number of muons at the ground in EAS generated by primary cosmic rays at different primary energies, and the scale factor $N_{\mu,19}$ is the total number of muons in EAS generated by primaries with energy $E = 10^{19}$ eV. We used the power-law parametrization:

$$\langle R_\mu \rangle = a \cdot \left(E/10^{19}\ \mathrm{eV}\right)^b \tag{4}$$

with parameters $a = 1.841$ and $b = 1.029$ fitting Auger experimental events above $4 \cdot 10^{18}$ eV. At zenith angle $\Theta = 67$ deg the muon content $R_\mu = 1$ corresponds to $N_\mu = 1.455 \cdot 10^7$ muons at the ground with energies above 0.3 GeV. For model comparisons, as described in [5], the simulated number of muons should be then divided by $N_\mu = 1.455 \cdot 10^7$ to obtain R_μ, which can be directly compared to Auger measurements. In figure 5 we show the simulated muon content $\langle R_\mu \rangle$ of individual showers generated by primary protons, iron nuclei, and SQM, as a function of the primary energy E together with the power-law fits (prepared using equation 4 with the parameters a= 1.5, 2.5, 3.2 and b= 0.9, 0.9, 0.92 for protons, iron nuclei, and SQM respectivly. It is noticeable that Auger results are located between the simulation of protons and iron nuclei. The dependence of the average muon content $\langle R_\mu \rangle$ on the position of the showers maximum is

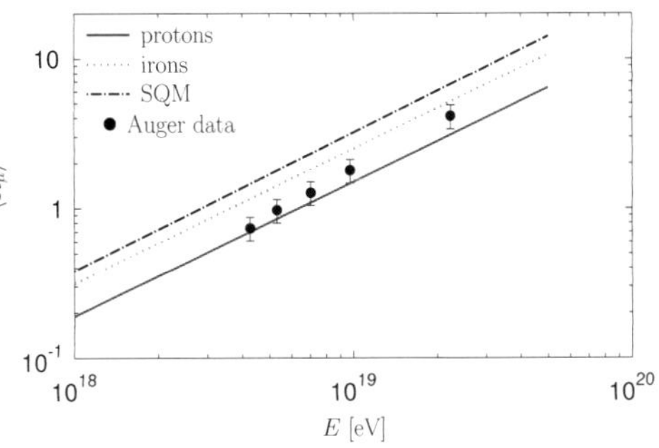

Figure 5. Average muon contents $\langle R_\mu \rangle$ of individual showers generated by primary protons, iron nuclei, and SQM, plotted as a function of the primary energy E, together with the Auger results taken from [5].

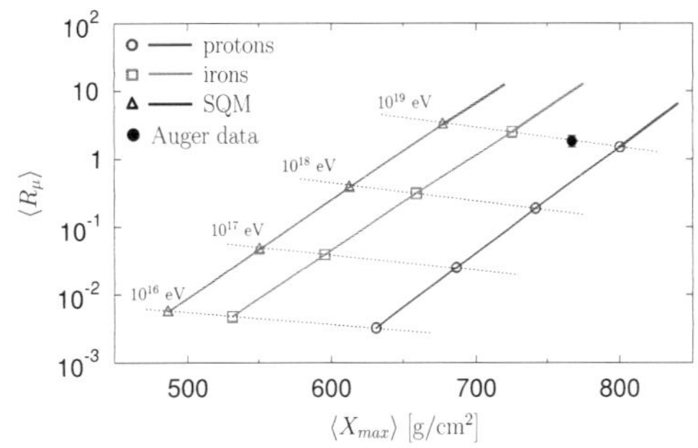

Figure 6. Average muon content $\langle R_\mu \rangle$ of individual showers generated by primary protons, iron nuclei, and SQM as a function of the average position of the shower maximum. The full circle shows the Auger result [5].

presented in figure 6. It is remarkable that no heavier component than iron is needed to describe the experimental values of $\langle R_\mu \rangle$ at 10^{19} eV. We can explain experimental data without strange quark matter.

The muon content in the air shower, R_μ, is a quantity related to the atomic mass A of the primary cosmic ray. A possible implication for the mass composition is demonstrated on the example, taking into account the obtained experimentally $\langle R_\mu \rangle = 1.82 \pm 0.38$ and the relative standard deviation $\omega = \sigma(R_\mu)/\langle R_\mu \rangle = 0.20 \pm 0.01$ [5]. For a single component, the values of ω can vary between $\omega = 0.04$ for pure iron nuclei and $\omega = 0.13$ for protons. More than two components (proton and iron) are needed to describe the first two moments of the R_μ distribution. The best description of the data is obtained with four components (40% protons, 20% helium, 35% nitrogen, and 5% iron nuclei) while the addition of more species does not improve the quality of the fit. The comparatively small abundance of iron nuclei has been evaluated from the analysis of the $\langle X_{max} \rangle - \sigma(X_{max})$ difference [29, 30]. Our results suggest roughly $\langle \ln A \rangle = 1.4$ with standard deviation $\sigma(\ln A) = 1.3$ at energies above $4 \cdot 10^{18}$ eV.

5. Concluding remarks

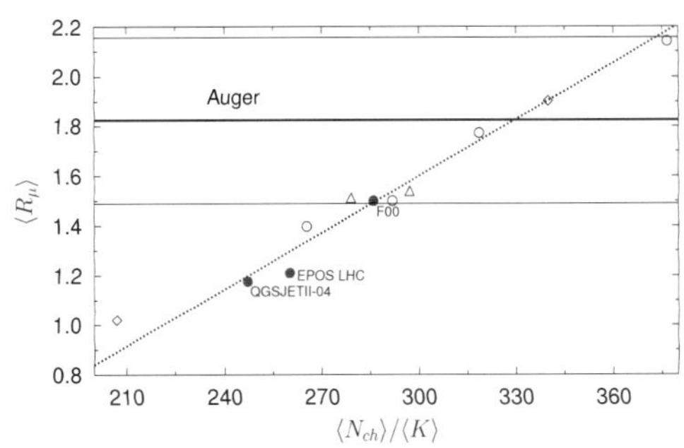

Figure 7. Average muon content $\langle R_\mu \rangle$ of showers generated by primary protons at 10^{19} eV plotted as a function of the ratio $\langle N_{ch} \rangle / \langle K \rangle$. The Auger result (with error band), indicated by a full line, is taken from [5]. The dotted line shows a linear fit given by equation 5. Open symbols show the results obtained for different modifications of $\langle N_{ch} \rangle$ in the F00 model.

Very surprisingly the F00 model included in the SHOWERSIM simulation package [23] describe nicely the muon content in EAS. Ordinary nuclei, without any contribution from strange quark matter in the primary flux of cosmic rays, can describe experimental data. Even if the strangelets contribute with a small amount in the primary flux and generate high multiplicity muon bundles, as we advocated recently [22], their influence on the average muon content $\langle R_\mu \rangle$ in EAS is negligible.

Apparently, our model differs from modern high-energy hadronic models such as EPOS LHC or QGSJETII-04, which cannot fit the experimental values of $\langle R_\mu \rangle$. Comparing different characteristics, we find that in the discussed models, $\langle R_\mu \rangle$ and the ratio of charged multiplicity to inelasticity $\langle N_{ch} \rangle / \langle K \rangle$ follow the same ordering. This is illustrated in figure 7. In addition to standard F00 model (shown by full circle) we present results for different modified $\langle N_{ch} \rangle$ according to equation 2 (shown by open symbols: circles correspond to different $\beta\,(K)$ and $\beta\,(\pi^0)$, triangles corresponds to different $\beta\,(K)$ and diamonds correspond to different $\beta\,(\pi^0)$). Roughly the number of muons depends linearly on the charged particle multiplicity:

$$\langle R_\mu \rangle = 0.0076 \langle N_{ch} \rangle / \langle K \rangle - 0.68. \tag{5}$$

We have shown in our simple model that we are able to reproduce the muon content in EAS. The charged particle multiplicity to inelasticity ratio seems to be crucial for understanding the muon excess. The muon component of EAS provides not only a powerful key for primary mass measurement but also provides information regarding hadronic interactions.

Acknowledgments

MR was supported by the Polish National Science Centre (NCN) grant 2016/23/B/ST2/00692.

References

[1] Anchordoqui L A, Dova M T, Epele L N and Sciutto S J 1999 *Phys. Rev.* D **59** 094003
[2] Pierog T and Werner K 2008 *Phys. Rev. Lett.* **101** 171101
[3] Ulrich R, Engel R and Unger M 2011 *Phys. Rev.* D **83** 054026
[4] Abu-Zayyad T *et al* [HiRes and MIA Collaborations] 2000 *Phys. Rev. Lett.* **84** 4276
[5] Aab A *et al* [Pierre Auger Collaboration] 2015 *Phys. Rev.* D **91** 032003
[6] Aab A *et al* [Pierre Auger Collaboration] 2016 *Phys. Rev. Lett.* **117** 192001
[7] Bellido J A *et al* 2018 *Phys. Rev.* D **98** 023014
[8] Fomin Y A *et al* 2017 *Astropart. Phys.* **92** 1
[9] Farrar G R and Allen J D 2013 *EPJ Web Conf.* **53** 07007
[10] Allen J and Farrar G 2013 *Proc. 33rd International Cosmic Ray Conference (Rio de Janeiro)* vol 2 p 827
[11] Alvarez-Muniz J *et al* 2012 arXiv:1209.6474 [hep-ph]
[12] Anchordoqui L A, Goldberg H and Weiler T J 2017 *Phys. Rev.* D **95** 063005
[13] Klingenberg R 1999 *J. Phys.* G **25** R273
[14] Witten E 1984 *Phys. Rev.* D **30** 272
[15] Alcock C and Olinto A 1988 *Ann. Rev. Nucl. Part. Sci.* **38** 161
[16] Wilk G and Wlodarczyk Z 1996 *J. Phys.* G **22** L105
[17] Rybczynski M, Wlodarczyk Z and Wilk G 2005 *Acta Phys. Pol.* **33** 277
[18] Wilk G and Wlodarczyk Z 1997 *Nucl. Phys. Proc. Suppl.* **52B** 215
[19] Sahnoun Z *et al* [SLIM Collaboration] 2009 *Radiation Measurements* **44**, 894
[20] Adam J *et al* [ALICE Collaboration] 2016 *JCAP* **1601** 032
[21] Avati V *et al* 2003 *Astropart. Phys.* **19** 513
[22] Kankiewicz P, Rybczynski M, Wlodarczyk Z and Wilk G 2017 *Astrophys. J.* **839** 31
[23] Wrotniak A 1984 *SHOWERSIM/84* (University of Maryland Preprint) p 85
[24] Wrotniak J A and Yodh G B 1985 *Proc. 19th International Cosmic Ray Conference (La Jolla)* vol 6 p 56
[25] Pierog T 2017 *EPJ Web Conf.* **145** 18002
[26] Abraham J *et al* [Pierre Auger Collaboration] 2010 *Phys. Rev. Lett.* **104** 091101
[27] Abreu P *et al* [Pierre Auger Collaboration] 2012 *Phys. Rev. Lett.* **109** 062002
[28] Abbasi R 2016 *EPJ Web Conf.* **120** 04005
[29] Wilk G and Wlodarczyk Z 2011 *Proc. 32th International Cosmic Ray Conference (Beijing)* vol 2 p 27
[30] Wilk G and Wlodarczyk Z 2011 *J. Phys.* G **38** 085201

IOP Conf. Series: Journal of Physics: Conf. Series **1181** (2019) 012052 doi:10.1088/1742-6596/1181/1/012052

Gamma-ray counterpart of the IceCube neutrinos

M. Kachelrieß[1], A. Neronov[2], D. V. Semikoz[3,4]

[1]Institutt for fysikk, NTNU, Trondheim, Norway
[2]Astronomy Department, University of Geneva, Versoix, Switzerland
[3]APC, Universite Paris Diderot, Observatoire de Paris, Sorbonne Paris Cite, Paris, France
[4]National Research Nuclear University MEPHI (Moscow Engineering Physics Institute)

Abstract. A γ-ray excess at high Galactic latitudes starting at energies 300 GeV was recently discovered in the data of the Fermi telescope. The multi-TeV γ-ray emission found has spectral characteristics at both low and high Galactic latitudes compatible with those of the IceCube neutrinos in the same sky regions. This suggests that these γ-rays are the counterpart of the IceCube neutrino signal, implying that a sizeable part of the IceCube neutrino flux originates from the Milky Way. The diffuse neutrino and γ-ray signal at high Galactic latitudes may originate either from a nearby cosmic ray "PeVatron" cosmic ray source, an extended Galactic cosmic ray halo or from decays of heavy dark matter particles.

1. Introduction

The discovery of an extraterrestrial neutrino signal in the TeV–PeV energy range by the IceCube collaboration has opened the era of multi-messenger astronomy [1]. The source(s) of this neutrino signal have remained unidentified so far. Since the production of high-energy neutrinos is accompanied by γ-rays, these neutrino sources could be identified using a "multi-messenger" approach by combining neutrino and γ-ray data. The TeV–PeV γ-ray flux from distant sources is suppressed by electron-positron pair production in interactions with low-energy photons of the extragalactic background light and the cosmic microwave background. Therefore, the presence or absence of a γ-ray counterpart can be used to clarify the origin of the neutrino signal: If the signal originates from extragalactic sources at cosmological distances, no γ-ray counterpart is expected in the multi-TeV to PeV band. In contrast, a Galactic origin implies the presence of a comparable multi-TeV γ-ray flux.

The search for the γ-ray counterpart of the neutrino signal is challenging with both ground and space-based γ-ray telescopes. Ground-based telescopes or air shower arrays suffer from a high background of events produced by charged CRs. The arrival directions of the CR background events are distributed over large angular scales, similar to the expected γ-ray counterpart of the neutrino signal. Space-based telescopes like the Fermi Large Area Telescope (LAT) achieve a much better suppression of the charged CR background, but they have small collection areas which severely limit the signal statistics.

In Ref. [2], we reported a study of the TeV diffuse gamma-ray sky based on the data of Fermi/LAT. We showed that the γ-ray flux and spectrum at low and high Galactic latitudes are compatible with the flux of the measured neutrino signal, in the energy range where the two signals overlap. This suggests that the γ-rays in the multi-TeV band are the counterpart of the soft part of the IceCube neutrinos, while the part with an $1/E^{2.2}$ slope at the highest energies has an extragalactic origin.

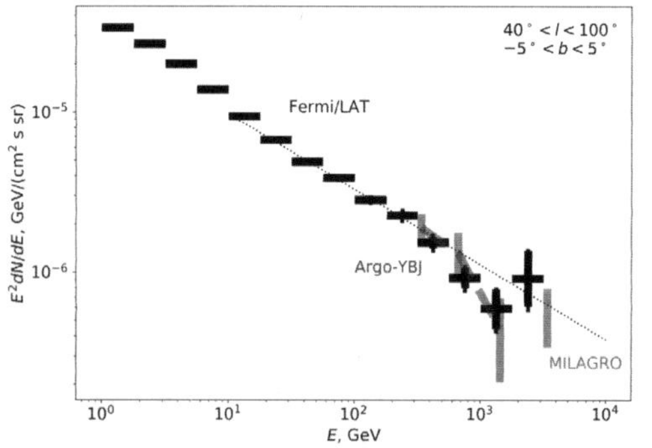

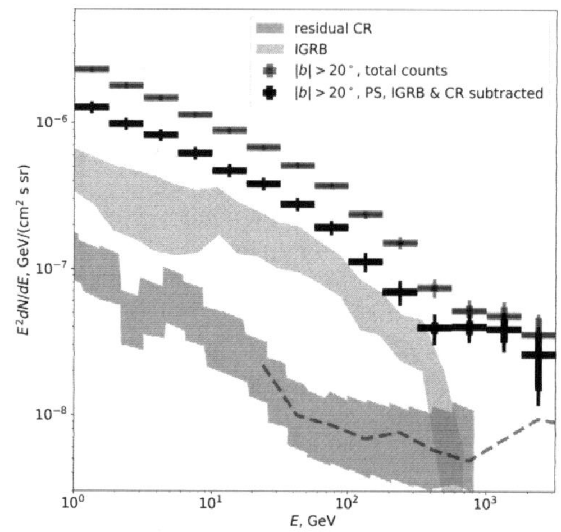

Figure 1. Left: Combined Fermi/LAT, ARGO-YBJ (blue) [5] and MILARGO (green) [6] spectrum of the $40° < l < 100°$ stretch of the Galactic plane. Right: Fermi/LAT spectrum of diffuse γ-ray emission from high Galactic latitudes compared to the residual CR background (dotted thin line). The green shaded band shows the uncertainty of the IGRB derived in Ref. [7]. Grey data points below 3 TeV show total high Galactic latitude flux without subtraction of catalogue sources, IGRB and residual CR background contributions.

2. Cross-calibration of the LAT data in the multi-TeV band

The energy resolution and the calibration of the effective area of Fermi/LAT degrade in the TeV band [3, 4]. Therefore we performed an additional cross-calibration of the Fermi/LAT flux measurements with those of ground-based γ-ray telescopes via a comparison of stacked spectra of selected calibration sources, see Ref. [2] for details. We found that a cross-calibration factor $\kappa = 1 - c\log(E/100 \text{ GeV})$ with $c = 0.25 \pm 0.12$ applied to the LAT flux measurements above 300 GeV leads to a better consistency with the ground-based telescope measurements. The uncertainty of the parameter c is taken into account as an additional systematic error. We verified that the cross-calibration factor also assures the consistency of the Fermi/LAT measurements of diffuse TeV emission from large regions of the sky with the measurements by the ground-based air shower arrays ARGO-YBJ [5] and MILAGRO [6], as shown in the left panel of Fig. 1.

3. Diffuse TeV γ-ray signal

Figure 2 compares Fermi/LAT γ-ray spectra of the full sky (upper panel), of the Galactic plane $|b| < 10°$ (middle panel) and at Galactic latitudes $|b| > 10°$ (lower panel) with the neutrino spectra of the same sky regions [8, 9, 10, 11]. In the spectra of the all-sky and the $|b| > 10°$ region we removed residual CR background, while the Galactic plane spectrum is calculated by subtracting high Galactic latitude background and residual cosmic ray contributions (see Ref. [2] for details). The γ-ray and neutrino all-sky flux and spectral slope measurements agree in the overlapping multi-TeV band. Figure 2 also shows the model of diffuse γ-ray emission from pion decays derived from an all-sky analysis of the LAT data [12]. It is this component which is expected to have the neutrino counterpart, since pion decays produce simultaneously γ-rays (π^0 decays) and neutrinos ($\pi^{\pm}$ decays).

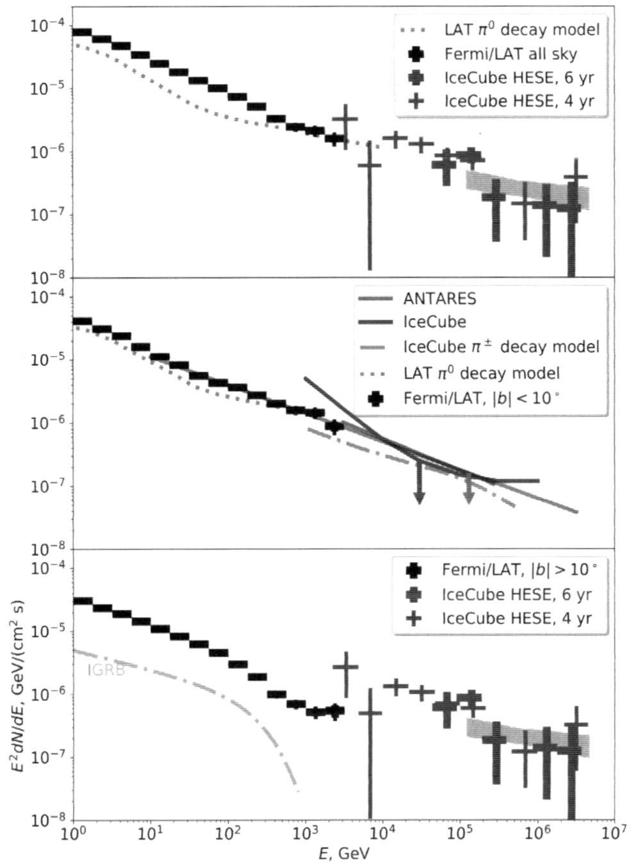

Figure 2. Top: the multi-messenger spectrum of the full sky: Fermi/LAT (black), IceCube (blue data and green bow-tie (from Ref. [8]) together with a model (dash-dotted) for the Galactic diffuse hadronic emission [12]. Middle: Fermi/LAT spectrum for $|b| < 10^\circ$ with model-dependent upper limit on neutrino flux [10]. Bottom: Fermi/LAT spectrum of $|b| > 10^\circ$ region, compared to the IceCube neutrino flux measurements. The dash-dotted curve shows the best-fit model of the IGRB [7].

The γ-ray flux below TeV is dominated by the emission from the Galactic plane, while only a moderate fraction of the neutrino flux in the 100 TeV range comes from the Galactic plane [9, 10, 8, 11]. Therefore, a multi-TeV γ-ray flux as counterpart of the neutrino signal should have a harder spectrum at high Galactic latitudes than at the Galactic plane so that its relative contribution to the all-sky flux can grow with increasing energy.

This hardening appears more pronounced in the analysis of the spectrum of the part of the sky at higher Galactic latitude, $|b| > 20^\circ$, shown in Fig. 3. In this figure we have removed contributions from resolved point sources, extragalactic isotropic diffuse γ-ray background (IGRB) and residual CR backgrounds thus leaving only the Galactic diffuse emission. The hardening of the spectrum of diffuse emission at high Galactic latitudes starts at 300 GeV and it can not be explained by instrumental effects (see the right panel of Fig. 1 and Ref. [2] for details). Below 300 GeV the spectrum is well fit by a smoothly broken power-law with the slope $\Gamma = 2.906 \pm 0.015$ in the 30–300 GeV range. The spectrum in the 0.3–3 TeV range has the slope $\Gamma = 2.09 \pm 0.09$.

The most significant excess above the extrapolation of the power-law valid below 300 GeV is in the energy bin 1–1.7 TeV. The model prediction of the number of photon counts in this bin is 16.4. The observed number of counts is 39. The chance probability of such an excess is 1.5×10^{-6}. In the energy bin 1.7–3.16 TeV the expected number of counts is 3.8, while the

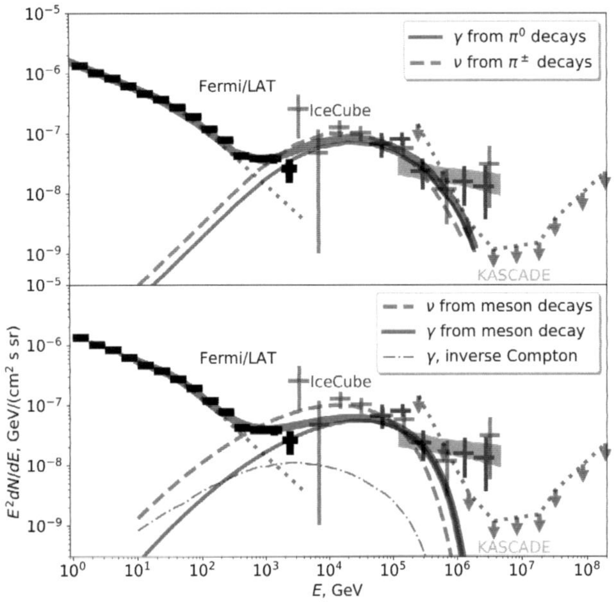

Figure 3. High Galactic latitude emission for the local PeVatron (top) and DM (bottom) models. Thick and thin errorbars of Fermi/LAT data points (black) show statistical and systematic uncertainties, including the uncertainties of subtraction of IGRB and residual CR backgrounds. Vertical arrows show KASCADE upper limits on the γ-ray flux from Northern sky [13]. Solid thin lines show the gamma-ray emission from the additional hard component. Dotted line shows a broken power-law fit to the sub-TeV γ-ray spectrum.

observed one is 10. The chance probability of such an excess is 5.8×10^{-3}. In the energy bin 0.3-1 TeV, the model predicts less than 66.5 counts while the observed signal is 100 counts. The chance probability of the excess in this bin is 8×10^{-5}. The energy-binning independent combined chance probability of the excess above 300 GeV is less than 8×10^{-10}.

4. Interpretation

The conventional high Galactic latitude diffuse emission components have soft spectra in the TeV range [7] and can not explain the observed spectral hardening above 300 GeV. The same is true for the IGRB, which is dominated by the cumulative flux of blazars [14], a special class of active galactic nuclei which do not provide the dominant contribution to the neutrino signal [15, 16]. Thus, the observed hardening of the γ-ray spectrum has to be interpreted as due to the presence of a new Galactic γ-ray flux component above 300 GeV. It is this component which is the counterpart of the neutrino signal with comparable flux in the multi-TeV range.

Only few source types could produce multi-TeV multi-messenger emission on large angular scales at high Galactic latitude with a hard spectrum. One possibility is interactions of CRs forming a previously unknown component of the Galactic CR population. If this new component would reside everywhere in the Galactic disk, an equivalent spectral hardening would be observed in the spectrum of the Galactic plane—which is not the case. Instead, the hard spectrum CRs could either reside in our local Galactic environment, or be a part of a very large halo.

The local source of CRs with a hard spectrum reaching PeV energies (a "PeVatron") should be a recent and nearby source, like e.g. the Vela supernova [17]. It should have injected CRs less than 10^5 year ago at a distance d not larger than several hundred parsecs. These two conditions are required for the presence of PeV CRs which produce 10–100 TeV neutrinos and the large angular extent Ω of the multi-messenger emission [18]. Cosmic rays with total energy $U_{CR} \sim 10^{50}$ erg injected by the PeVatron and loosing their energy on the time scale $t_{\rm pp} \simeq 1.5 \times 10^8 \left(n_{\rm ISM}/0.5 \text{ cm}^{-3}\right)$ yr in interactions with the interstellar medium of the density

$n_{\rm ISM} \sim 0.5$ cm^{-3} produce the γ-ray and neutrino flux $F = U_{CR}/(4\pi d^2\Omega t_{pp})$ with magnitude

$$F \sim 2\times 10^{-7}\left(\frac{\Omega}{2\pi \rm sr}\right)^{-1}\frac{n_{\rm ISM}}{0.5/\rm cm^3}\left(\frac{d}{0.3\rm kpc}\right)^{-2}\frac{\rm GeV}{\rm cm^2\,s\,sr}.$$

This flux estimate matches the observed level, cf. with Fig. 3. Otherwise, the high Galactic latitude emission could be from a very large (hundred kiloparsec) CR "storage" around the Milky Way disk [19].

The local PeVatron model predicts strong variability of the multi-messenger signal across the sky. This variability is determined by the peculiarities of the energy-dependent spread of the CRs and of the matter distribution in the local Galaxy. Low energy CRs which had no time to escape from the source region would not contribute to the large angular scale emission. This leads to a low-energy hardening of the spectrum, as shown in the top panel of Fig. 3 [18]. In contrast, the signal is not expected to experience neither strong fluctuations nor a low-energy hardening in the large scale halo model [19].

An alternative possibility shown in the bottom panel of Fig. 3 is that decays of metastable DM particles X with mass $m_X \simeq 5$ PeV generate photons and neutrinos [20, 21, 22]. The spectral shape of the decay mode $X \to \bar{q}q \to$ hadrons is determined by Quantum Chromodynamics (QCD). Since at the end of the QCD cascade quarks combine more easily to mesons than to baryons, mainly neutrinos and photons from pion decays are produced. The γ-ray and neutrino flux measurements constrain the X particle lifetime to be $\tau_X \sim 2\times 10^{27}\,(\Omega_X/\Omega_{DM})^{-1}$ s, where Ω_X/Ω_{DM} is the fraction of the DM in the form of X particles [22, 21]. Since the mass m_X is above the unitarity limit, the X particles were never in thermal equilibrium. They should have been produced by gravitational interactions or other non-thermal processes and may serve as a tool to study the earliest phases of the Universe.

The DM decay neutrino signal has a sizeable extragalactic contribution, while its γ-ray component in the TeV-PeV range has only the Galactic part. This leads to a systematically lower normalisation of the multi-TeV γ-ray component. The same is true for the large scale CR halo, which should be present around all galaxies, so that the neutrino flux is expected to have a significant extragalactic contribution. To the contrary, the neutrino and γ-ray components in the local PeVatron model both originate from the Milky Way. The absence of the extragalactic component leads to similar γ-ray and neutrino fluxes (see top panel of Fig. 2).

The DM halo of the Galaxy is denser in the direction of the inner Galaxy. This means that in the DM model, the flux from the inner Galaxy should be stronger than that from the outer Galaxy. However, the signal from the Galactic plane shown in Fig. 3 contains both the direction toward the Galactic center and the anticenter, from which the strongest and the weakest DM decay signal should be observed. We have verified that the expected excess of the DM decay signal from the Galactic Plane is consistent with the IceCube upper bounds on the Galactic plane flux. The fraction of the DM decay signal from the region $|b| < 10^\circ$ is 0.22. Combining the information form Fig. 3 and Fig. 3, one can see that the neutrino flux from the high Galactic latitude region which is supposed to account for the full neutrino signal at high Galactic latitude at 100 TeV is at the level 6×10^{-7} GeV/cm^{2}s at this energy (cf. with the bottom panel of Fig. 3). Re-scaling it by a factor $0.22/0.78 \simeq 0.3$, one could check that the expected DM decay flux from the direction of the Galactic plane is at the level of 2×10^{-7} GeV/cm^{2}s, i.e. marginally consistent with the IceCube upper limit on the neutrino flux from the Galactic plane (the IceCube upper limit is exactly at the level of the flux estimate, which means that the signal of DM origin should soon reveal an excess toward the inner Galaxy). There is, however, one important reservation which should be added. The IceCube upper limit on the Galactic emission is derived assuming certain spatial template for the signal distribution. This template does not correspond to the spatial template of the DM signal. Thus, the IceCube limit on the Galactic emission is not directly comparable to the DM model prediction.

For the local PeVatron model, there is no fixed spatial template because the source morphology is not known. No excess toward the Galactic Plane is generically expected. In this respect, the IceCube limit on the Galactic emission component does not provide constraints on the local PeVatron model.

The distinction between possible models of the multi-messenger signal based on spectral or spatial characteristics will be possible with next generation instruments neutrino telescopes and the space-based γ-ray telescope HERD which will accumulate higher signal statistics. The detection of the γ-ray part of the signal by ground-based telescopes requires a sufficiently high ($\sim 10^5$) rejection level of the CR background.

5. Conclusions

We have demonstrated that the properties of the large scale diffuse Galactic γ-ray flux in multi-TeV band are compatible with the flux and spectrum of the neutrino signal in 1-100 TeV range, so that the two signals may be considered as different components of one and the same "multi-messenger" signal in the multi-TeV sky. The γ-ray flux at high Galactic latitude exhibits a pronounced hardening above 300 GeV, while no hardening is observed in the low Galactic latitude flux. This effect explains the lower contribution from the Galactic plane to the neutrino signal at higher energies, as observed by IceCube. We have suggested three possible models which could explain the observed hard spectrum high Galactic latitude multi-messenger emission above 300 GeV: (i) interactions of CRs injected by a recent nearby cosmic PeVatron, (ii) CR interactions in a large halo around the Milky Way, or (iii) decays of DM particles.

References

[1] IceCube Collaboration 2013 *Science* **342** 1242856 (*Preprint* 1311.5238)
[2] Neronov A, Kachelrieß M and Semikoz D V 2018 *Phys. Rev.* **D98** 023004 (*Preprint* 1802.09983)
[3] Bruel P 2012 *J. Phys. Conf. Ser.* **404** 012033 (*Preprint* 1210.2558)
[4] Atwood W *et al.* 2013 *arXiv:* **1303.3514** (*Preprint* 1303.3514)
[5] Bartoli *et al.* 2015 *Ap.J.* **806** 20 (*Preprint* 1507.06758)
[6] Atkins R *et al.* 2005 *Physical Review Letters* **95** 251103 (*Preprint* astro-ph/0502303)
[7] Ackermann M *et al.* 2015 *Ap.J.* **799** 86 (*Preprint* 1410.3696)
[8] IceCube Collaboration (IceCube) 2017 (*Preprint* 1710.01179)
[9] Neronov A and Semikoz D 2016 *Astroparticle Physics* **75** 60–63 (*Preprint* 1509.03522)
[10] IceCube Collaboration 2017 *Ap.J.* **849** 67
[11] Albert A *et al.* 2017 *Phys. Rev. D* **96** 062001 (*Preprint* 1705.00497)
[12] Acero F *et al.* 2016 *Ap.J.Supp.* **223** 26 (*Preprint* 1602.07246)
[13] Apel *et al.* 2017 *Ap.J.* **848** 1 (*Preprint* 1710.02889)
[14] Ackermann *et al.* 2016 *Physical Review Letters* **116** 151105 (*Preprint* 1511.00693)
[15] Neronov A, Semikoz D V and Ptitsyna K 2017 *A&A* **603** A135 (*Preprint* 1611.06338)
[16] Aartsen M G *et al.* 2017 *Ap.J.* **835** 45 (*Preprint* 1611.03874)
[17] Cha A N, Sembach K R and Danks A C 1999 *Ap.J.Lett* **515** L25–L28 (*Preprint* astro-ph/9902230)
[18] Andersen K J, Kachelrieß M and Semikoz D V 2018 *Astrophys. J.* **861** L19 (*Preprint* 1712.03153)
[19] Taylor A M, Gabici S and Aharonian F 2014 *Phys. Rev. D* **89** 103003 (*Preprint* 1403.3206)
[20] Berezinsky V, Kachelrieß M and Vilenkin A 1997 *Phys. Rev. Lett.* **79** 4302–4305 (*Preprint* astro-ph/9708217)
[21] Feldstein B, Kusenko A, Matsumoto S and Yanagida T T 2013 *Phys. Rev.* **D88** 015004 (*Preprint* 1303.7320)
[22] Esmaili A and Serpico P D 2013 *JCAP* **1311** 054 (*Preprint* 1308.1105)

IOP Conf. Series: Journal of Physics: Conf. Series **1181** (2019) 012053 doi:10.1088/1742-6596/1181/1/012053

On energy estimation of high energy muon events in KM3 detectors based on a more exact range fluctuations of high energy muons

N Takahashi[1], Y Okumura[1], T Tanemori[2] and A Misaki[3]

[1] Faculty of Science and Technology, Hirosaki University, Hirosaki, 036-8561, Japan
[2] Department of Physics, Saitama University, Saitama, 338-8570, Japan
[3] Graduate School of Science and Engineering, Saitama University, Saitama, 338-8570, Japan

amisaki@sand.ocn.ne.jp

Abstract. The analyses of muon events above 1 TeV in high energy neutrino astrophysics are under progress in IceCube, ANTARES and Baikal-GVD. In these analyses, the verification of validity of the utilized procedures is particularly important. In present report, an analysis procedure for the KM3 detectors is proposed. The procedures thus utilized are composed of two parts, namely, (1) the exact treatment towards the parent muons and (2) the exact treatment of Cherenkov light due to both the parent muon and their dissipated energies. As for the former, we check the validity, based on the investigation of range fluctuations of high energy muon, and, for the latter, we check the verification on three dimensional cascade showers which are the origin of the main part of the Cherenkov light. Every process due to the muon and accompanying cascade shower are exactly investigated in stochastic manner. Thus, by using the established procedure, we analyse *Fully Contained Event* in the KM3 detector exclusively. The comparisons are made of the relation between the generated Cherenkov Yields and the observed ones.

1. Introduction
In the analyses of the neutrino events above 1 TeV in high energy neutrino astrophysics, such as IceCube, ANTARES, Baikal-GVD, frequencies for the muon neutrino events is far higher than those of electron neutrino events due to their big difference in effective lengths in high energy neutrino events.

Therefore, the reliable energy estimation of the muon neutrino events has become particularly important among those of high energy neutrino astrophysics. Originally, neutrino events in higher energies are rare events. From the energy estimation of such rare events, we are forced to construct the physical images due neutrinos. This is the reason why we should pursue the reliable energy estimation as exactly as possible.

2. Calculation procedures
This paper is a natural extension of the preceding paper [1]. The higher energy neutrino events are generated mainly due to the deep inelastic scattering of the neutrinos, such as

$$\nu_\mu + N \rightarrow \mu + X \quad (1)$$

For muon neutrino interactions, the energies of hadronic parts (hadronic showers) are smaller than those of the generated muons and the traversed lengths of muons are far longer than those of the hadronic parts. The muons thus produced traverse in water (ice) 1 kilometer or more, dissipating their energies due to either direct pair production, or bremsstrahlung or nuclear interaction. These processes are in nature of stochastic characters. The behaviors of these muons are investigated under the concept of the range fluctuations of the muons.

In previous paper [1], we treat the range fluctuations of muons as exactly as possible and examine the application limit of Stanev and Lipari theory [2] and similar theories [3] which are the basement of the muon analysis in the present high energy astrophysics, such as IceCube, Antares and Baikal-GVD.

2.1. The validities of the Monte Carlo methods concerned

Nowadays, the range fluctuations of high energy muons are investigated exclusively by Monte Carlo method. However, it is essentially difficult to verify the validity of the Monte Carlo method concerned due to inherent nature involved in the method.

As far as the validity of our Monte Carlo method on the range fluctuations of muons is concerned, its validity is confirmed by the comparison of our method with the analytical theory, methodologically independent one of the Monte Carlo procedure, and the agreement between them is confirmed [1, 4].

In the present analysis of the muon events, their dissipated energies due to stochastic characters, namely, due to direct pair production, bremsstrahlung and nuclear interaction are treated as ensemble of three-dimensional electromagnetic cascade showers - hereafter, simply, cascade shower - whose incident particles are different in different interactions. Also, these cascade showers are treated by an exact Monte Carlo method.

The dissipated energies due to respective interactions are converted into the cascade showers. The validity of our Monte Carlo method in the calculation of three-dimensional cascade showers is verified by the comparison of the mean square lateral and angular spread of shower particles obtained by our Monte Carlo method with corresponding quantities obtained by the analytical theory [5] based on Nishimura-Kamata theory [6], the methodologically independent of the Monte Carlo method. We confirm the agreement between them. The details of the certification of the validities will be described in detail in a full paper.

2.2. The Cherenkov lights due to the segments of electrons in the cascade showers

The shower particles in a cascade shower are pursued exactly in stochastic manner including the ionization loss three-dimensionally up to 1 MeV under Approximation B in the traditional cascade shower theory [6, 7]. The cascade showers due to the dissipated energies from the parent muons develop three-dimensionally, taking into account of the multiple scattering of shower particles along the direction of the muons whose multiple scatterings are neglected.

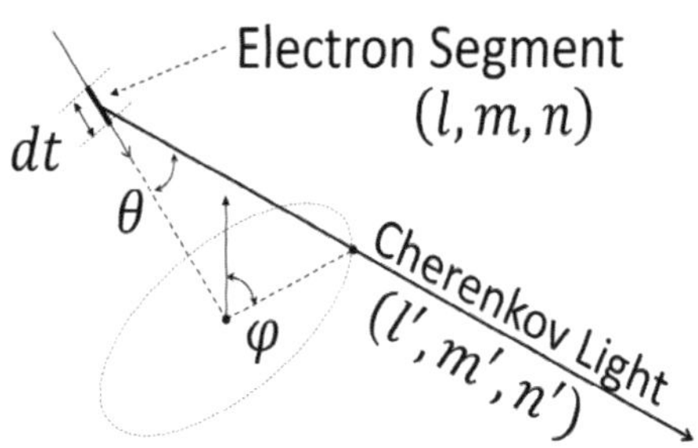

Figure 1. The geometrical relation between cosine direction of *dt*, an electron segment and those of the Cherenkov lights.

The Cherenkov lights due to *dt*, a segment in shower electrons are emitted with the direction cosine $(l' \quad m' \quad n')$ for the dt, whose direction cosine $(l \quad m \quad n)$ is shown in figure 1.

Here, dt denotes an electron segment in a cascade shower with the direct cosine $(l \quad m \quad n)$ at the depth t in a three-dimensional Monte Carlo electromagnetic cascade shower. The Cherenkov lights due to the dt radiate the direction with θ the emitted angle of the Cherenkov light, and φ, azimuthal angle, with the cosine direction $(l' \quad m' \quad n')$ as shown in figure 1.

Furthermore, the relation between two kinds of the direct cosine is given as

$$\begin{pmatrix} l' \\ m' \\ n' \end{pmatrix} = \begin{pmatrix} ln(l^2+m^2)^{-1/2} & -m(l^2+m^2)^{-1/2} & l \\ mn(l^2+m^2)^{-1/2} & l(l^2+m^2)^{-1/2} & m \\ -(l^2+m^2)^{1/2} & 0 & n \end{pmatrix} \times \begin{pmatrix} \sin\theta\cos\varphi \\ \sin\theta\sin\varphi \\ \cos\theta \end{pmatrix} \quad (2)$$

The Cherenkov lights radiated from dt, an electron segment, which are absorbed at the most adjacent detector. Let us the absorption coefficient of the Cherenkov light be λ, then the Cherenkov light absorbed at the detector is given as

$$N = N_0 \times \exp\left(-\frac{t}{\lambda}\right) \quad (3)$$

where N_0 denotes the generated Cherenkov lights due to the dt and N denotes the observed Cherenkov light at the detector and t denotes the distance from the dt to the place for detection for Cherenkov lights. Here, λ is taken as 40 meter.

In our calculation, we examine two cases of the total Cherenkov lights, namely the generated Cherenkov lights and the observed ones.

Related to the mutual relations between [generated points] and [terminated points] among the muon events, these events are classified in the four types, namely, [*Fully Contained Event*], [*Partially Contained Events*], [*Stopping Event*] and [*Passing through Events*].

In present paper, we examine *Fully Contained Events* only.

3. The analyses of *Fully Contained Events*

3.1. Do the Fully Contained Events guarantee the reliable energy estimations ?

It is needless to say that only *Fully Contained Events* may give the most reliable energy estimations of the muon events concerned compared with other three types of events, from the point of view of energy conservation.

In present calculations, we could examine *the Fully Contained Events* whose muon energies are below 1 TeV, while in higher energies above ~1 TeV, we could hardly observe *Fully Contained Events*.

Here, let us examine characteristics of the three *Fully Contained Events*.

Table 1. The generated Cherenkov lights for three *Fully Contained Events*. B, N, D, B+N+D denote the kinds of interactions, namely, bremsstrahlung, nuclear interaction, direct production, and the total number of interactions, respectively. M (muon), E (shower), M+E denote generated Cherenkov lights due muon, electron shower, muon + electron shower (generated total Cherenkov lights), respectively.

	Traversed depth	Number of Interactions				Generated Cherenkov		
	(m)	B	N	D	B+N+D	M: muon	E: shower	M+E
#FC-4	491.8	1	0	46	47	1.23×10^7	7.27×10^7	8.50×10^7
#FC-390	603.8	1	3	34	38	1.51×10^7	6.63×10^7	8.14×10^7
#FC-401	975.6	3	1	64	68	2.44×10^7	6.10×10^7	8.53×10^7

In table 1, we summarize the characteristics of three *Fully Contained Events*. In spite of the big difference of the traversed lengths (491.8m, 603.8m and 975.6m) among three events, it should be noticed that totally generated Cherenkov lights is nearly the same, as they must be, because total

Cherenkov lights exactly correspond to the primary energies. Total Cherenkov light generated exactly corresponds to the total track lengths for electrons segments in the cascade showers and the parent muon. However, in real measurements, it should be noticed that the generated Cherenkov lights are never observed, because the some part of generated Cherenkov lights are lost due to their absorption (see equation (3)).

In table 2, we give the ratios of the observed Cherenkov lights to the generated Cherenkov ones in the case of $\lambda = 40$ m for the different unit for the detection of Cherenkov lights for the event [FC-4].

Here, for examples, [60 c.u.] denotes that each measurement is made in the unit of 60 c.u. over the distance from the beginning to the end. Namely, we measure Cherenkov light at 60 c.u., 120 c.u., 240 c.u. and so on, until the muon events concerned terminate and the ratio of the observed Cherenkov lights to corresponding generated ones are obtained at each measurement point. The ratio may be a good measure for estimation of the reliability on energies of the muon events concerned. It is easily understood from the table that reliability of the energy estimation decreases as the distance for the measurements increase even for *Fully Contained Events*.

Table 2. The dependence of the ratios of observed Cherenkov lights to generated ones in the unit length for their measurement given for one example of Fully Contained Events (#FC-4).

Observed Yield	Total	$\Sigma E_{\_shower}$	$\Sigma\mu$
60 c.u.	6.92×10^{7}	6.05×10^{7}	8.69×10^{6}
Ratio	8.15×10^{-1}	8.33×10^{-1}	7.07×10^{-1}
120 c.u.	6.68×10^{7}	6.03×10^{7}	6.44×10^{6}
Ratio	7.86×10^{-1}	8.30×10^{-1}	5.24×10^{-1}
240 c.u.	1.74×10^{7}	1.35×10^{7}	3.85×10^{6}
Ratio	2.04×10^{-1}	1.48×10^{-1}	3.13×10^{-1}
480 c.u.	1.51×10^{7}	1.33×10^{7}	1.83×10^{6}
Ratio	1.78×10^{-1}	1.83×10^{-1}	1.49×10^{-1}

3.2. The behaviours of a Fully Contained Event: #FC-4 : Longitudinal development for Cherenkov Lights

Here, we show several different graphs for different physical quantities for one *Fully Contained Event* (#FC-4) to grasp the physical image of the event concerned as a whole. The characteristics of this event are as follows: The total traversed length is 491.8 meter, total number of interactions ever experienced is 47. The details are 46 direct pairs, 1 bremsstrahlung and 0 nuclear interaction. The biggest energy loss by the muon is due to the bremsstrahlung whose energy is 8.41×10^{11} eV and whose generation point is 2.94×10^{4} g/cm^2 from the starting point of the incident muon.

In figure 2, we give two cascade curves for Cherenkov light yields as the function of the depth. One is the generated Cherenkov light (dotted line) and the other is the observed one (continuous line) in the unit of 60 c.u. It is clear from both curves that the both peaks from both curves correspond to the catastrophic energy loss due to bremsstrahlung mentioned above.

3.3. The Lateral distribution for Cherenkov light yields from both the muon and cascade shower particles integrated by the depth

In figure 3, the lateral distribution for the generated Cherenkov light yields integrated by the depth is given. The contribution from both cascade shower electrons and the incident muon are shown in the figure. It is clear from the figure that the contribution from the incident muon is far small compared with those from electrons, which is also shown in table 1. It is estimated from figure 3 that the average spread of the 1 TeV muon event (#FC-4) is of 100 c.u. (36 meters). The spreads of the muon events come mainly from cascade shower particles due to their multiple scattering and they decrease as their primary energies increase.

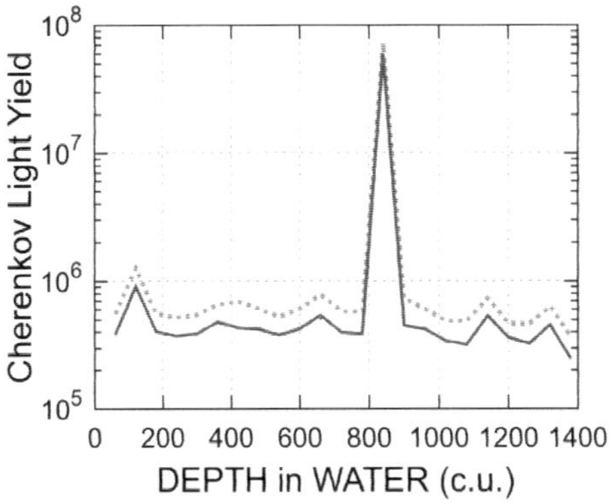

Figure 2. Transition curves for Cherenkov light in the cases of both generated Cherenkov lights and observed ones ($\lambda = 40$ m).

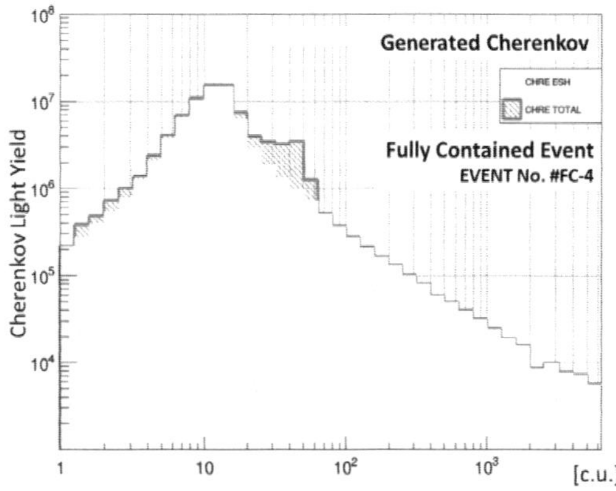

Figure 3. The lateral distribution for generated Cherenkov Lights integrated by depth.

3.4. *The* cosθ *distribution for both the muon and cascade shower particles integrated by depth*

In figure 4, we give the cosθ distribution for total Cherenkov lights with regard to the direction of the parent muon, for the parent muon and electrons from cascade showers in the case of generated Cherenkov light. In our calculation, we adopt the cosθ of the muon is of cosθ=0.752(=1/n, n=1.4, refraction index in water). If we assume the Cherenkov lights due to the dissipated energies from the incident muon are generated along the same direction as the Cherenkov lights by the muon, then we should expect the delta function type distribution for around cosθ=0.752. However, the real distribution for figure is quite different from the delta function type distribution. This is due to the multiple scattering of cascade electrons as shown in figure 1. It is also noticed that the contributions from the incident muon are far smaller than those from cascade electrons.

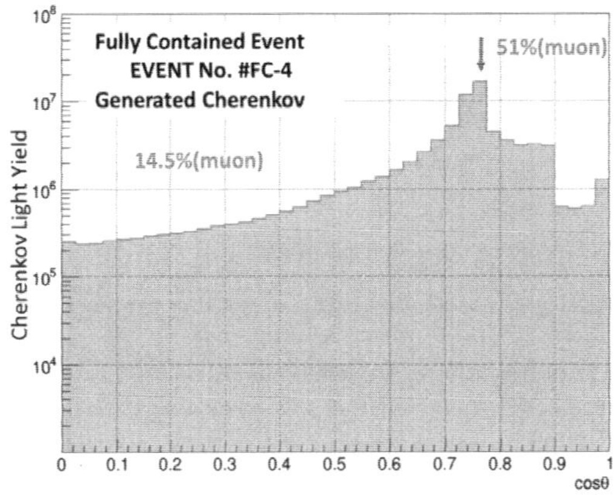

Figure 4. cosθ distribution of the generated Cherenkov lights for the parent muon and cascade shower electrons integrated by depth.

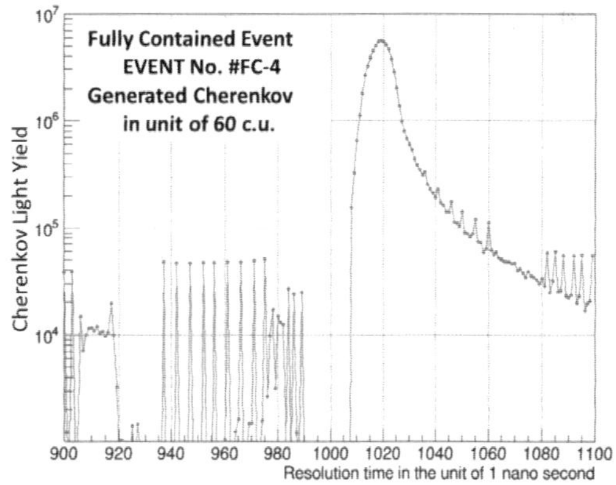

Figure 5. A part of the arrival time distribution of the generated Cherenkov light. The Cherenkov light yields due to bremsstrahlung are clearly registered. The time is measured from the generation points for the parent muon. The time resolution is one nanosecond.

3.5. The arrival time distribution for the Cherenkov light yields from the both the muon and cascade shower particles integrated by the depth

Finally, we give the arrival time distribution for the total Cherenkov light whose time resolution is 1 nanosecond. The peak at around 1020 nanosecond corresponds to the peak of the Cherenkov light in figure 2. It is clear from figure 2 and figure 5 that we could find easily the phenomena whose dissipated energies are so high. The higher dissipated energy comes mainly from the catastrophic energy loss due to bremsstrahlung.

4. The range distribution for muons with different primary energies

In figure 6, we show the range distributions for muons whose primary energies cover from 1 TeV to 1 PeV. Sampling number for each primary muon is 1000. It is clear from the figure that fluctuations with regard to the depth increase rapidly as primary energies increase. It is easily concluded from the figure the following. Namely, it is almost impossible for us to find *Fully Contained Events* whose primary energies exceed 10 TeV within the interior of the KM3 detector. Even if one is interested in the muon events with ~1 TeV, it should be noticed that the majority is occupied by *Partially Contained Events*. Only several percent of muon events with 1TeV are classified as *Fully Contained Events*.

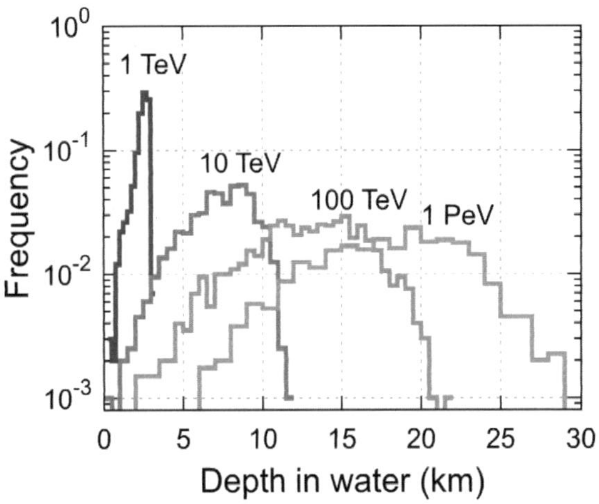

Figure 6. Range distributions for the incident muons with incident energies 10^{12} eV to 10^{15} eV. The minimum observation energies are taken as 10^9 eV. Each sampling number is 1000.

5. Conclusion

For the moment, we restrict our theme to the *Fully Contained Events* with 1 TeV only. In the present paper, we have clarified that it is not so easy to estimate energy of the muon events even for *Fully Contained Events* with 1 TeV. In the next paper, we analyse the *Partially Contained Events* whose primary energies exceed 10 TeV from the view point of reliable energy estimation.

References

[1] Okumura Y, Takahashi N and Misaki A 2016 *Astroparticle Physics* **72** 95-108
[2] Lipari R and Stanev T 1991 *Phys. Rev.* D **44** 3543-54
[3] Dutta S I, Reno M H, Sarcevic I and Seckel D 2001 *Phys. Rev.* D **63** 094020;
Klimushin S I, Bugaev E V and Sokalski I A 2001 *Phys. Rev.* D **64** 014016;
Sokalski I A, Bugaev E V and Klimushin S I 2001 *Phys. Rev.* D **64** 074015
[4] Misaki A and Nishimura J 1977 *ICR-Report* 45574-4 (Tokyo: University of Tokyo)
[5] Misaki A 1964 *Suppl. Prog. Theor. Phys.* **32** 82-103
[6] Kamata K and Nishimura J 1958 *Suppl. Prog. Theor. Phys.* **6** 93-155
[7] Rossi B and Greisen K 1941 *Rev. Mod. Phys.* **13** 240-309

High-energy atmospheric muon flux calculations in comparison with recent measurements

A A Kochanov[1,2], A D Morozova[3,4], T S Sinegovskaya[5] and S I Sinegovsky[2,4]

[1] Institute of Solar-Terrestrial Physics SB RAS, Lermontov str. 126a, Irkutsk, 664033, Russia
[2] Irkutsk State University, Gagarin blvd. 20, Irkutsk, 664003, Russia
[3] Lomonosov Moscow State University, Leninskie Gory 1, Moscow, 119991, Russia
[4] Joint Institute for Nuclear Research, Joliot-Curie 6, Dubna, Moscow region, 141980, Russia
[5] Irkutsk State Transport University, Chernyshevskiy str. 15, Irkutsk, 664074, Russia

E-mail: kochanov@iszf.irk.ru, refriz@yandex.ru, tanya@api.isu.ru, sinegovsky@jinr.ru

Abstract. Recently the atmospheric muon spectra at high energies were reconstructed for two ranges of zenith angles, basing on the events collected with the IceCube detector. These measurements reach high energies at which the contribution to atmospheric muon fluxes from decays of short-lived hadrons is expected. Latest IceCube measurements of the high-energy atmospheric muon spectrum indicate the presence of prompt muon component at energies above 500 TeV.

In this work, the atmospheric conventional muon flux in the energy range 10 GeV – 10 PeV is calculated using a set of hadronic models in combination with known parameterizations of the cosmic ray spectrum by Zatsepin & Sokolskaya and by Hillas & Gaisser. The calculation of the prompt muons with use of the quark-gluon string model (QGSM) reproduces the muon data of IceCube experiment. Nevertheless, an additional contribution to the prompt muon component is required to describe the IceCube muon spectra, if a charm production model predicts the appreciably lower prompt lepton flux as compared with QGSM. This addition apparently originating from rare decay modes of the short-lived unflavored mesons $\eta, \eta', \rho, \omega, \phi$, might ensure the competing contribution to the high-energy atmospheric muon flux.

1. Introduction

High-energy atmospheric muons and neutrinos arise in weak decays of mesons produced in cosmic ray collisions with the Earth's atmosphere. Atmospheric muon flux is a tool to study the cosmic ray spectrum and elemental composition if there is a reliable model for hadron-nucleus interactions at high energies. Otherwise, if the elemental spectra of primary cosmic ray are well studied, a comparison of calculated high-energy cosmic ray muon spectra with the experimental data allows one to get characteristics of the hadron-nucleus interactions.

The atmospheric muon spectra at high energies were reconstructed for two ranges of zenith angles basing on the events collected with IceCube detector [1]. These measurements are of strong interest because they reach the high energies at which atmospheric muons from decays of short-lived hadrons are expected. The IceCube + IceTop combined analysis [2] resulted in the reconstruction of the high energy muon spectrum separately, near vertical direction and close to horizontal, that allows: (i) the comparison with the muon data of the ealier experiments; (ii) the search of a reliable model for high-energy hadronic interactions; (iii) the detection of the prompt muons component.

We calculate the atmospheric muon spectrum in the energy range 10 GeV-10 PeV using the hadronic models Kimel & Mokhov [3, 4], QGSJET-II [5, 6], SIBYLL 2.1[7], EPOS-LHC [8, 9], in combination with known parametrizations of the cosmic ray spectrum by Zatsepin & Sokolskaya (ZS) [10] and by Hillas & Gaisser (H3a) [11]. The prompt muon contribution calculated with use of the quark-gluon string model (QGSM) [12] is compatible with recent IceCube measurements up to PeV energies. However in case of using the updated version of the QGSM [13, 14] or other models of the charm production, which lead to appreciably lower prompt lepton flux, we find that an additional contribution to the prompt muon component is required to describe the IceCube muon spectra. This component can originate from rare decay modes of the short-lived unflavored vector mesons ($\eta, \eta', \rho, \omega, \phi$) [15, 16] which contribute to the high-energy atmospheric muon flux.

2. Methods of the calculation

The muon flux computations are performed with $\mathcal{Z}(E,h)$-method for solution of atmospheric hadron cascade equations [17, 18]. The method enables one to calculate atmospheric fluxes of mesons, nucleons, muons [19, 20] and neutrinos [21] for non-power primary cosmic ray spectra, non-scaling inclusive cross sections and rising cross sections of inelastic hadron-nucleus collisions. Besides, we have performed the calculation with different approach, Matrix Cascade Equations ([22, 23], using the free access package MCEq [24] (details of the comparison with $\mathcal{Z}(E,h)$-method see in [25, 26, 27]). The calculations are performed for a set of hadron-nucleus interaction models using the parametrizations of the primary spectrum based on experiments. Besides parametrizations of cosmic ray spectra by Zatsepin & Sokolskaya and Hillas & Gaisser (H3a) we use also the Nikolsky, Stamenov and Ushev model (NSU) [28] and the Elrykin, Krutikova and Shabelsky one (EKS) [29], as well as the toy model by Thunman, Ingelman and Gondolo (TIG) [30], in order to compare results of updated calculations with those of past years.

3. Energy spectra of atmospheric muons

The spectrum of atmospheric muons near the vertical is shown in figure 1. Data of last decades measurements at $\theta = 0^\circ$ (points) and calculation results (set of curves) are presented. Experimental data are taken from Refs. [18, 19, 20] (see references therein). Muon data of the BUST experiment (Baksan, 2009) were taken from Refs. [31, 32]. Earlier predictions for the conventional muons [18] are represented by lines 2, 3 obtained with the comsic ray spectra NSU and EKS correspondingly. Line 1 illustrates the prompt muons contribution calculated with the early version of QGSM [12]. The rest lines show the calculations with hadronic models Kimel & Mokhov (KM), SIBYLL 2.1, QGSJET II-03 combined with CR spectra by Zatsepin & Sokolskaya (ZS) and by Hillas & Gaisser (H3a). Filled small squares represent the recent results of the combined IceCube + IceTop measurements [2].

In figure 2, the calculated muon spectrum is compared with the best fit [2] for IceCube + IceTop data, valid in the energy range 6 − 400 TeV for near vertical directions ($\cos\theta > 0.88$) (solid line):

$$\frac{d\Phi_\mu}{dE_\mu} = (9.0 \pm 0.3) \cdot 10^{-17} \left(\frac{E_\mu}{50 \text{ TeV}}\right)^{-3.74 \pm 0.03}, \text{ cm}^{-2}\text{s}^{-1}\text{sr}^{-1}\text{GeV}^{-1}.$$

Our calculations with EPOS-LHC and KM combined with H3a spectrum lead to the muon spectrum compatible with the fit, excepting the energy range beyond 200 TeV.

The IceCube muon spectra [1] for zenith angles $\theta < 60^\circ$ and $\theta > 60^\circ$ are shown in figures 3, 4 along with our calculations. These figures display the clear evidence in favor of the muon

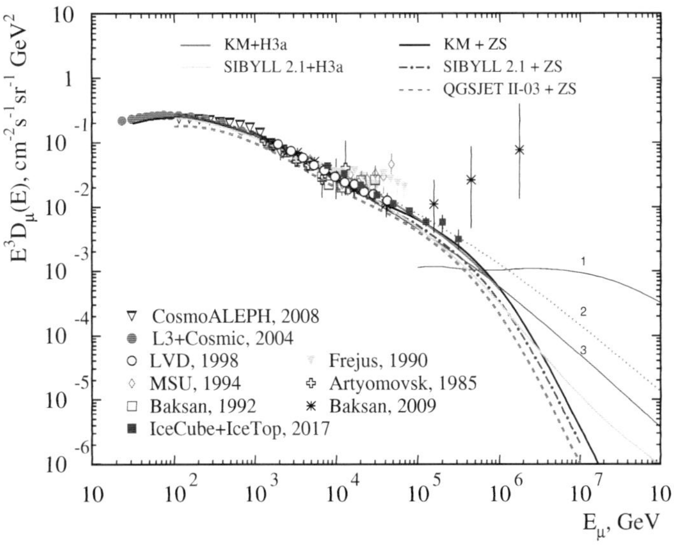

Figure 1. Calculated muon flux at $\theta = 0°$ (curves) vs. the experimental data of last decades measurements (see text) and latest results of the IceCube + IceTop recontruction [2].

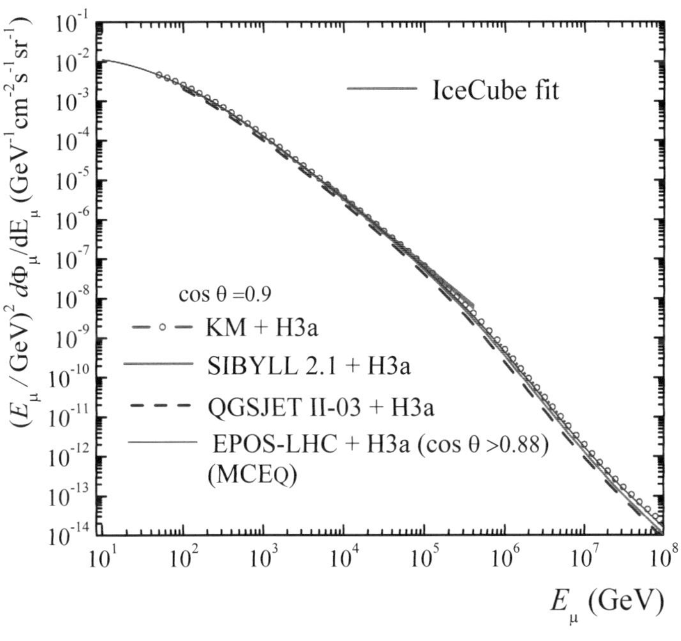

Figure 2. Calculated muon spectrum near the vertical and the best fit [2] for IceCube + IceTop data in the energy range 6 – 400 TeV (red line). Curves: EPOS-LHC, KM, and SIBYLL 2.1 combined with the cosmic-ray spectrum H3a.

component with harder spectrum than is expected for the conventional flux. Such behavior is consistent with sizable contribution of prompt muons, and thus the IceCube experiment reveals the hard component, inspite of large errors. As can be seen from the calculated curves in figure 3, at energies above 100 TeV the slope of the conventional muon spectrum increases by ~ 0.3 due to the knee of the cosmic ray spectrum. At angles $\theta > 60°$ (figure 4), the difference between expected conventional flux and measured one is less as compared with that in figure 3, because of zenitn-

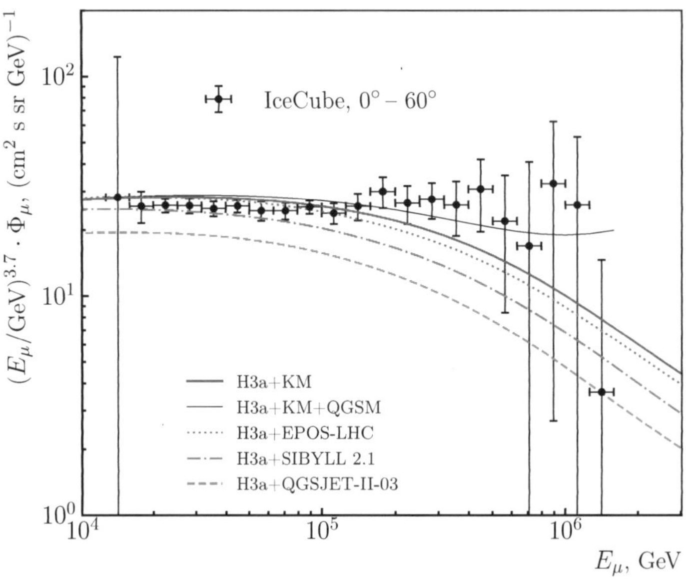

Figure 3. IceCube data measurements [1] for zenith angles $\theta < 60°$. Curves: calculations for hadronic models QGSJET II-03, SIBYLL 2.1, EPOS-LHC and KM combined with the cosmic-ray spectrum H3a.

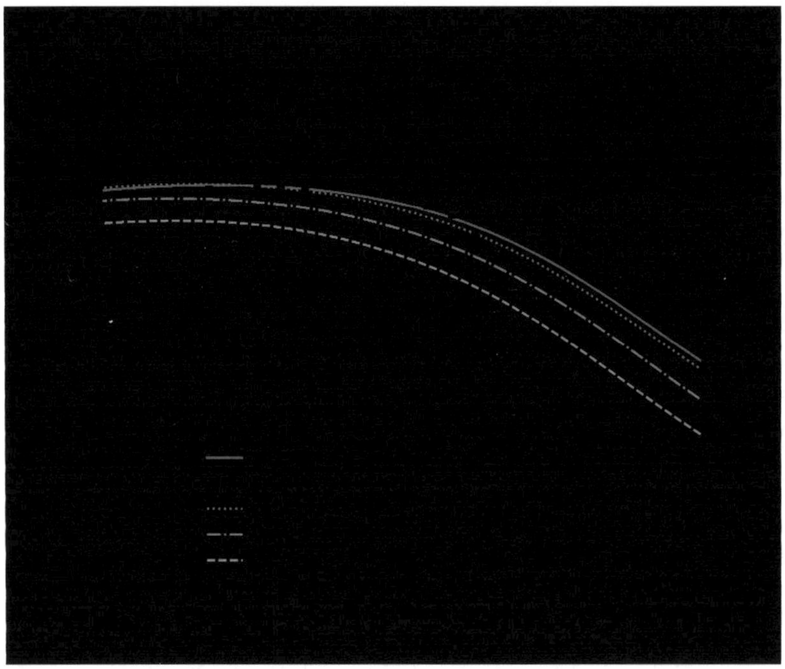

Figure 4. IceCube data measurements for $\theta > 60°$ and calculations. The same notation as in figure 3.

angle enhancement of the conventional muon flux. The prompt muon component calculated with QGSM, being added to the conventional flux, agrees well with the IceCube data in both angle intervals.

The all-sky reconstruction of the atmospheric muon spectrum in IceCube experiment [1] is

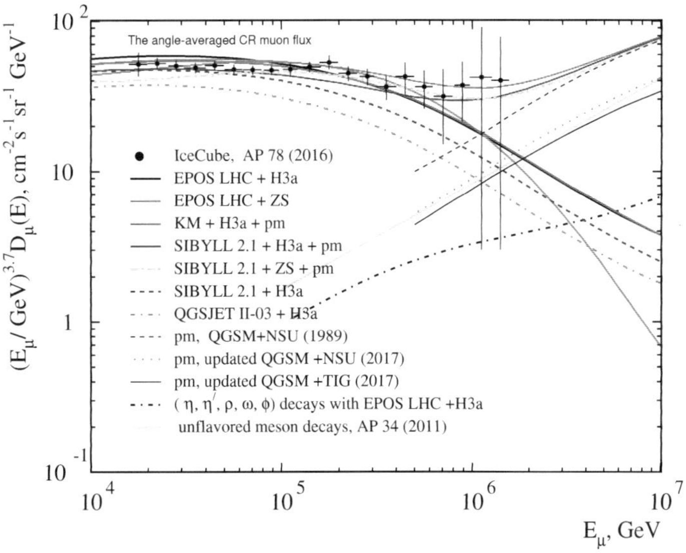

Figure 5. IceCube data (points) and calculations using EPOS-LHC, KM, SIBYLL 2.1, and QGSJET II-03 combined with cosmic-ray spectra models H3a and ZS. Contribution of vector mesons decays $(\eta, \eta', \rho, \omega, \phi)$: solid cyan line – [16]; black dash-dotted – this work (EPOS-LHC).

shown in figure 5. The set of curves shows the muon spectra calculated using models EPOS-LHC, KM, SIBYLL 2.1, and QGSJET II-03 with the CR spectra ZS and H3a. Also shown are the preceding QGSM calculations [12, 19] of the prompt muon flux (black dashed line) and that with updated version of QGSM [14], performed for the same choice of the CR spectrum (NSU) (red dotted), and beside, for the TIG spectrum (black thin line). The upper three curves present the sum of conventional and prompt muons (pm) [19]: the red solid (top) line corresponds the model KM + H3a + pm; blue solid – the same for SIBYLL 2.1; green dash-dotted – SIBYLL 2.1 + ZS + pm. The result derived for EPOS-LHC (black solid line represents the conventional flux) is close to that for KM model. Solid cyan line corresponds to the contribution from decays of the unflavored mesons, taken from Ref. [16] (SIBYLL 2.1 + TIG). Black dash-dotted line corresponds to this work calculation with EPOS-LHC.

We may summarize: the calculation of the all-sky muon spectrum with use of models KM or EPOS-LHC (conventional flux) in combination with QGSM (prompt muons), describes well the IceCube measurement data.

4. Conclusions

The comparison of the calculated high-energy atmospheric muon spectra with recent measurements of IceCube experiment, as well as with prior ones, justifies the proper consideration of the muon production in hadron showers, and reliability of the performed calculations. Atmosperic muon spectra calculated with hadronic models, EPOS-LHC, Kimel & Mokhov, SIBYLL 2.1 are consistent with recent experimental results. The computation for EPOS-LHC combined with the Hillas & Gaisser cosmic ray spectrum leads to closest agreement with the best fit of IceCube [2] in the energy range 6 – 400 TeV.

High-energy spectra of atmospheric muons for two zenith-angle ranges, reconstructed in IceCube experiment, evidence the prompt muons contribution. In spite of rather large errors, the experimental points against the background of the conventional atmospheric muons unambiguously indicate the presence of prompt muons at energies beyond 500 TeV. The discovery of the prompt muon component is the most important result of the latest measurement of the cosmic-ray muon spectrum in IceCube experiment.

The prompt muon flux calculated with use of the quark-gluon string model [12] is compatible with these IceCube measurements up to PeV region. However, usage of the updated version of the QGSM [13, 14], as well as different "low-flux" models of the charm production, leads to the appreciably lower prompt lepton flux as compared with Ref. [12]. In that case, an additional contribution to the prompt muon component is required to describe the IceCube measured muon spectra.

Acknowledgements
This work was supported partly by the Russian Federation Ministry of Science and Higher Education, agreement 3.9678.2017/8.9. A. Kochanov is supported with budgetary funding of Basic Research program II.16 at ISTP SB RAS.

References

[1] Aartsen M G *et al* (IceCube Collaboration) 2016 *Astropart. Phys.* **78** 1-27
[2] Tenholt F *et al* (IceCube Collaboration) 2017 High-energy atmospheric muons in IceCube and IceTop 35th Int. Cosmic Ray Conf. (ICRC2017 Bexco, Busan, Korea) PoS (ICRC2017) 317 (*Preprint* 1710.01194)
[3] Kimel L R and Mokhov N V 1974 *Izv. Vyssh. Uchebn. Zaved.* **10** 17
[4] Kalinovsky A N, Mokhov N V and Nikitin Yu P 1989 *Passage of high-energy particles through matter* (New York: AIP)
[5] Ostapchenko S 2006 *Phys. Rev.* D **74** 014026
[6] Ostapchenko S 2011 *Phys. Rev.* D **83** 014018
[7] Ahn E J, Engel R, Gaisser T K, Lipari P, Stanev T 2009 *Phys. Rev.* D **80** 094003
[8] Pierog T, Karpenko I, Katzy J M, Yatsenko E and Werner K 2015 *Phys. Rev.* C **92** 034906
[9] Pierog T 2017 *EPJ Web Conf.* **145** 18002
[10] Zatsepin V I and Sokolskaya N V 2006 *Astron. Astrophys.* **458** 1-5
[11] Gaisser T K 2012 *Astropart. Phys.* **35** 801-806
[12] Bugaev E V, Naumov V A, Sinegovsky S I and Zaslavskaya E S 1989 *Nuovo Cim.* C **12** 41-73
[13] Sinegovsky S I and Sorokovikov M N 2017 *Rus. Phys. J.* **60** 1189
[14] Sinegovsky S I and Sorokovikov M N 2018 Prompt atmospheric neutrinos in the quark-gluon string model *Preprint* P2-2018-4 JINR, Dubna
[15] Volkova L V 2011 *Phys. Atom. Nucl.* **74** 318-323
[16] Illana J I, Lipari P, Masip M and Meloni D 2011 *Astropart. Phys.* **34** 663-673
[17] Naumov V A and Sinegovskaya T S 2000 *Phys. Atom. Nucl.* **63** 1927-1935
[18] Kochanov A A, Sinegovskaya T S and Sinegovsky S I 2008 *Astropart. Phys.* **30** 219
[19] Sinegovsky S I, Kochanov A A, T.S. Sinegovskaya, Misaki A and Takahashi N 2010 *Int. J. Mod. Phys.* A **25** 3733
[20] Kochanov A A, Sinegovskaya T S and Sinegovsky S I 2013 *J. Exp. Theor. Phys.* **116** 395
[21] Sinegovskaya T S, Morozova A D and Sinegovsky S I 2015 *Phys. Rev.* D **91** 063011
[22] Fedynitch A, Engel R, Gaisser T K, Riehn F and Stanev T 2015 Calculation of conventional and prompt lepton fluxes at very high energy *EPJ Web Conf.* **99** 08001 (*Preprint* 1503.00544)
[23] Fedynitch A, Engel R, Gaisser T K, Riehn F and Stanev T 2015 *PoS* (ICRC 2015) 1129
[24] Fedynitch A 2017 https://github.com/afedynitch/MCEq
[25] Morozova A D, Kochanov A A, Sinegovskaya T S and Sinegovsky S I 2017 *J. Phys. Conf. Ser.* **798**, 012101
[26] Morozova A D, Kochanov A A, Sinegovskaya T S and Sinegovsky S I 2017 *J. Phys. Conf. Ser.* **934**, 012008
[27] Fedynitch A, Engel R, Gaisser T K, Riehn F and Stanev T 2018 The hadronic interaction model Sibyll 2.3c and inclusive lepton fluxes *Preprint* 1806.04140
[28] Nikolsky S I, Stamenov J N and Ushev S Z 1984 *Sov. Phys. JETP* **60** 10-21
[29] Erlykin A D, Krutikova N P and Shabelskii Yu M 1987 *Yad. Fiz.* **45** 1075
[30] Thunman M, Ingelman G and Gondolo P 1996 *Astropart. Phys.* **5** 309
[31] Bogdanov A G, Kokoulin R P, Novoseltsev Yu F, Novoseltseva R V, Petkov V B and Petrukhin A A 2009 Energy spectrum of cosmic ray muons in 100 TeV energy region reconstructed from the BUST data *Preprint 0911.1692*
[32] Bogdanov A G, Kokoulin R P, Novoseltsev Yu F, Novoseltseva R V, Petkov V B and Petrukhin A A 2012 *Astropart. Phys.* **36** 224

The atmospheric neutrino flux from decays of charmed particles

S I Sinegovsky[1,2] and M N Sorokovikov[1,2]

[1]Joint Institute for Nuclear Research, Joliot-Curie, 6, Dubna, Moscow region, 141980, Russia
[2]Irkutsk State University, Gagarin blv. 20, Irkutsk, 664003, Russia

E-mail: sinegovsky@jinr.ru, waleralincoln@mail.ru

Abstract. We calculate the high-energy component of the atmospheric neutrino flux which arises from decays of the charmed particles. Cross sections of the D mesons and Λ_c^+ baryons production in pA and πA collisions are calculated within the framework of the quark-gluon string model (QGSM). The recent data on the cross sections of charmed meson production at high energies, obtained in experiments at the LHC, allow an improvement of the QGSM free parameters and new estimate of the prompt atmospheric neutrino flux. The newly-calculated flux of the prompt neutrinos is three times smaller than the flux predicted with the previous version of QGSM. We compare our result with the constraint derived from IceCube experiment as well as with predictions obtained with different charm production models.

1. Introduction

High-energy neutrinos from decays of mesons and baryons produced in collisions of cosmic rays with the Earth's atmosphere form an unavoidable background in the detection of neutrinos from distant astrophysical sources. The detection of several tens of events in the IceCube experiment with energy deposition in the range 30 TeV – 2 PeV from neutrinos of cosmic origin [1, 2] and recent announcement of the blazar TXS 0506 + 056 as likely source of high energy neutrinos [3] increase significance of the calculation of atmospheric neutrino background.

Atmospheric neutrino fluxes consist of two components, which have different zenith-angle distribution and energy spectrum. The anisotropic component is produced in muon, pion and kaon decays and has the softer spectrum – these are the conventional (π, K)–neutrinos. The second one, quasi-isotropic and hard component is produced at high energies mainly in decays of short-lived heavy charmed mesons and baryons (D, Λ_c^+). Such neutrinos are called as prompt atmospheric neutrinos. The most uncertain contribution to the high-energy atmospheric neutrino flux arises from decays of charmed particles because of a wide spread in model predictions and lack of measurements of the charm production cross sections at high-energies.

We calculate the prompt neutrino flux using updated version of the quark-gluon string model (QGSM) [4]. The recent data on the cross sections of charmed particle production, obtained in experiments at the LHC [5, 6, 7, 8], allow an improvement of the QGSM free parameters. In the calculation we use the hadronic cascade model [9, 10, 11] and cross sections of D meson and Λ_c^+ baryon production in pA and πA collisions which were computed for updated parameters of QGSM. We compare our result with the constraint obtained in the IceCube experiment as well as with predictions of dipole model (ERS) [12] and NLO pQCD (BEJKRSS) calculation [13].

2. Production of charmed particles in the quark-gluon string model

QGSM is a nonperturbative model which describes soft hadronic processes: production of nucleons, pions, kaons and charmed hadrons. The model is based on Reggeon calculus, topological $1/N_f$-expansion of the amplitudes (planar diagrams correspond to Reggeon contributions and cylinder ones correspond to Pomerons) and color string dynamics [14, 16, 15]. QGSM is phenomenological model, it contains a number of free parameters determined by experimental measurements. Besides parameters, which are determined from global fit of experimental data, there are free parameters: intercept of $c\bar{c}$-trajectory Regge $\alpha_\psi(0)$ and parameter a_1 that provides both an enhancement and unified description of the kinematic regions $x \to 0$ and $x \to 1$ (x – Feynman variable) in case of the valence quark fragmentation. Now there are no clear arguments for choice of its value, and different authors apply various values in the range from $a_1 = 2$ [17] to $a_1 = 30$ [16]. As for the intercept of $c\bar{c}$-trajectory there are two possible values. The nonlinear trajectory with $\alpha_\psi(0) = 0$ is dictated by the perturbative quantum chromodynamics (pQDC). On the other hand by analogy with light quarks the Regge trajectory is linear and its intercept $\alpha_\psi(0) = -2.2$. Data on the production of charmed particles obtained in recent years indicate preferability of the value $\alpha_\psi(0) = -2.2$, which takes account of effects beyond the pQCD [18]. New measurements of the total cross sections of charmed meson production at high energies in the experiments ALICE [5, 6, 7], ATLAS [8] make it possible to check the predictions of the QGSM for these free parameters.

Figure 1 shows the total cross section of D mesons production in pp collisions calculated for four sets of free parameters in comparison to experimental data in a wide energy range [19, 20, 21, 22, 23, 24, 25, 26, 27] including LHC measurements [5, 6, 7, 8].

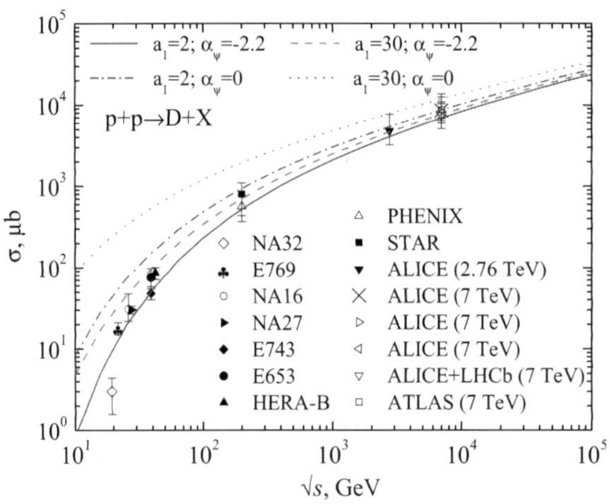

Figure 1. Total cross sections of $D/\bar{D}$ mesons production in pp collisions. QGSM calculations for $\alpha_\psi(0) = -2.2$ and $a_1 = 2$ (solid line) or $a_1 = 30$ (dashed); dash-dotted and dotted lines correspond to $\alpha_\psi(0) = 0$. Experimental data: ▼, × – [5]; ▷ – [6]; ◁, ▽ – [7]; □ – [8]; ◇ – [19]; ♣ – [20]; ● – [21]; ► – [22]; ♦ – [23]; ● – [24]; ▲ – [25]; △ – [26]; ■ – [27].

Calculation with $\alpha_\psi(0) = 0$ and $a_1 = 30$ does not agree with experimental data at $\sqrt{s} < 100$ GeV, while the calculations with $\alpha_\psi(0) = -2.2$ are in close agreement with the measurements in a wide energy range. At low energies the cross sections with $\alpha_\psi(0) = -2.2$ calculated for extreme values $a_1 = 2$ and $a_1 = 30$ differ by a factor $2 - 5$. With a rise of energy, the influence of the parameter a_1 decreases and becomes negligible at high energies ($\sqrt{s} > 1$ TeV). As can be seen in figure 2 the differential cross sections of D mesons production are described better for $\alpha_\psi(0) = -2.2$, while the computation with the intercept $\alpha_\psi(0) = 0$ is not compatible

IOP Conf. Series: Journal of Physics: Conf. Series **1181** (2019) 012055 doi:10.1088/1742-6596/1181/1/012055

with the experimental data [22].

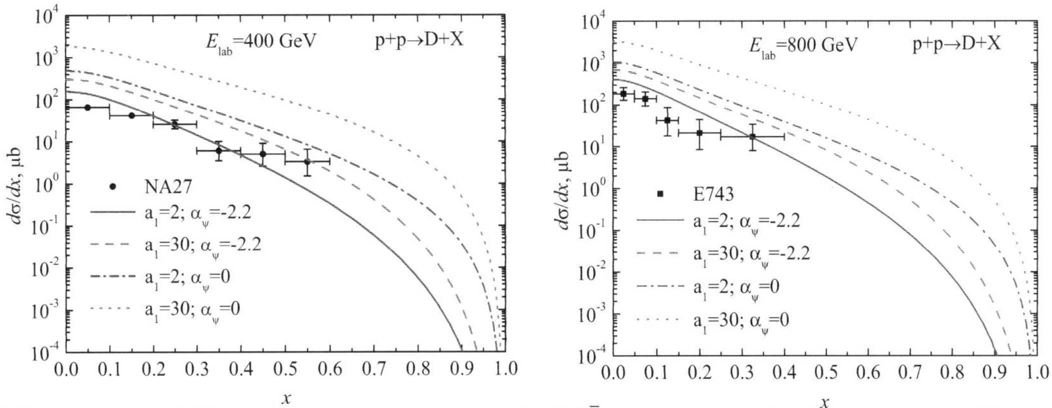

Figure 2. Differential cross sections of $D/\bar{D}$ mesons production in pp collisions at E_{lab} = 400 GeV (left) and E_{lab} = 800 GeV (right). Experimental data: ● – [22]; ■ – [23]. The same notation for lines as in figure 1.

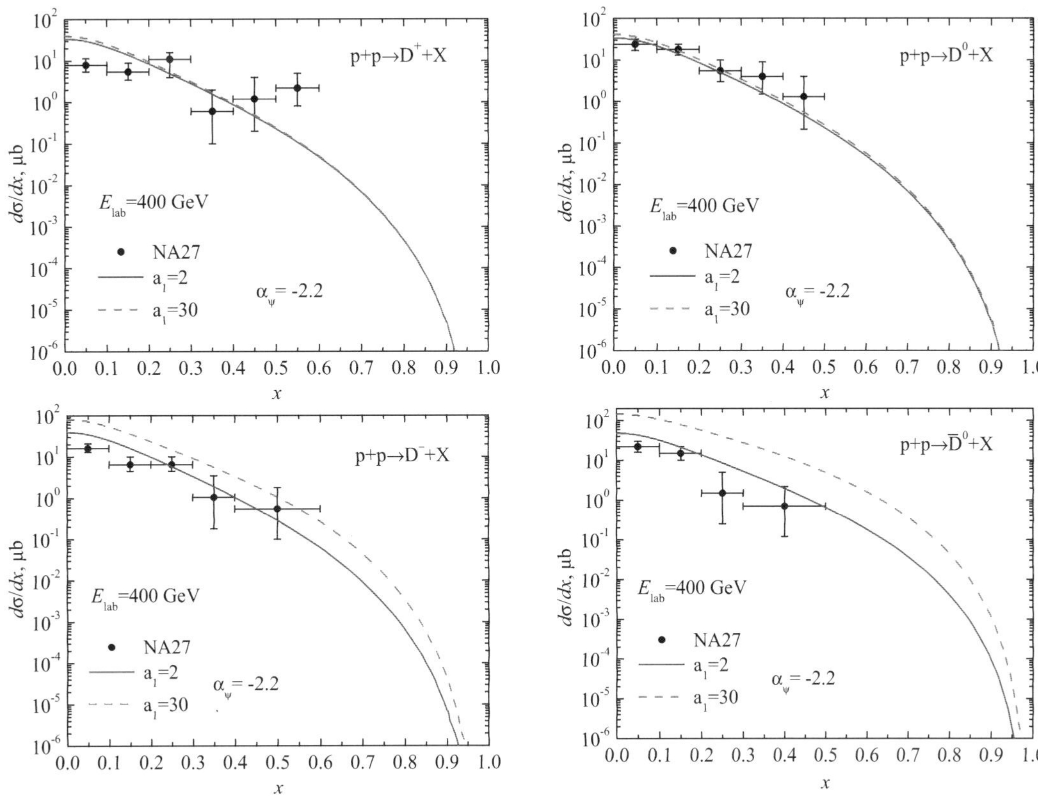

Figure 3. Differential cross sections of each type of D mesons ($D^+, D^-, D^-, \bar{D}^0$) production in pp collisions at E_{lab} = 400 GeV: calculations with $\alpha_\psi(0) = -2.2$ for $a_1 = 2$ (solid line) and $a_1 = 30$ (dashed). Experimental data: ● – [22].

Figure 3 presents comparison of the experimental measurements (pp collisions, 400 GeV) [22] with the inclusive cross sections for each type of D mesons production ($D^+, D^-, D^-, \bar{D}^0$)

performed for $\alpha_\psi(0) = -2.2$ and two values of the parameter a_1: $a_1 = 2$ and $a_1 = 30$. The cross sections of D^+ and D^0 mesons weakly depend on the parameter a_1, while the cross section of D^- and $\bar{D}^0$ mesons calculated for $a_1 = 2$ is smaller (by a factor $2 - 10$) than that for $a_1 = 30$ and leads to better agreement with the experimental data. Production of D^- and $\bar{D}^0$ in pp interactions has a higher probability because these mesons contain the valence quarks of colliding protons. The contribution of the leading fragmentation functions dominates in the x-distributions, and x-distribution of D^- and $\bar{D}^0$ is harder in comparison with D^+ and D^0. The influence of the parameter a_1 on the cross section of all D mesons production is also noticeable (figure 2).

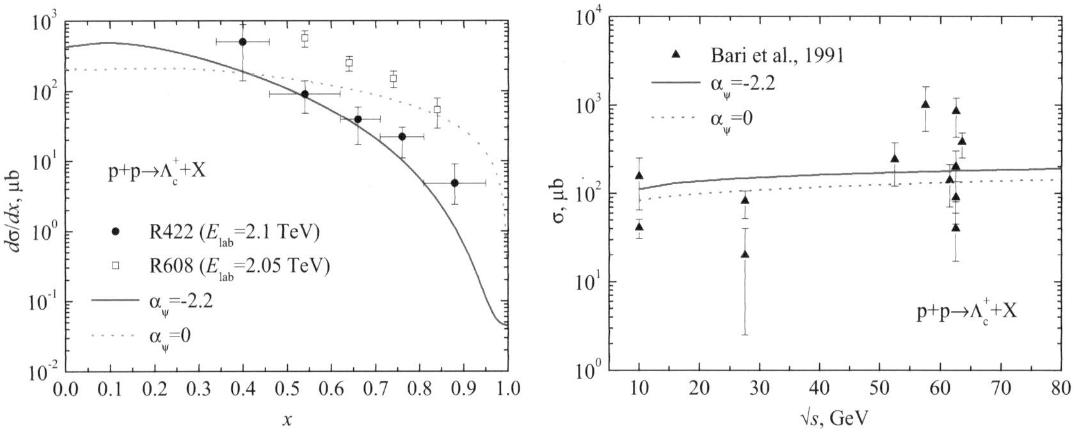

Figure 4. Differential (left) and total (right) cross sections of Λ_c^+ baryon production in pp collisions: calculations for $\alpha_\psi(0) = -2.2$ (solid curve) and $\alpha_\psi(0) = 0$ (dotted). Experimental data: ● – [28]; □ – [29]; ▲ – [28].

Comparison of the cross section of Λ_c^+ baryon production in pp collisions with experimental data is shown in figure 4. The differential cross section at $E_{\text{lab}} = 2.05$ TeV (left panel) is calculated for $\alpha_\psi(0) = -2.2$ and $\alpha_\psi(0) = 0$, the experimental data were obtained at energies $E_{\text{lab}} = 2.05$ TeV [28] and 2.1 TeV [29]. There is appreciable difference in the cross section data of these two experiments. The calculation with the parameter $\alpha_\psi(0) = -2.2$ agrees with the later experiment. The right panel of figure 4 shows the total cross section of Λ_c^+ production as a function of center-of-mass energy. The experimental points are taken from [28], the calculation is made for same parameter sets. The large spread of the total cross section measurement data does not allow making a definite choice of the intercept $\alpha_\psi(0)$.

3. Energy spectra of the atmospheric prompt neutrinos

In the present work the calculation of prompt neutrino fluxes is performed within method [9, 10] for QGSM and the parameterization of cosmic ray spectrum by Nikolsky, Stamenov, Ushev (NSU) [30] and the toy model by Thunman, Ingelman, Gondolo (TIG) [31]. The NSU spectrum which takes into account an elemental composition of primary cosmic rays was chosen in order to compare new results with the old one [10]. Simplified broken power law approximation TIG assuming only nucleon component of spectrum was chosen for comparison of our result with other calculations including dipole model prediction by ERS [12] which is used by IceCube as a benchmark model.

Figure 5 (left panel) shows the calculation of vertical flux of prompt atmospheric neutrinos: the band represents this work calculation with NSU-spectrum for the QGSM with the intercept $\alpha_\psi(0) = -2.2$. The band shows uncertainty due to change of the parameter a_1:

extreme values of a_1 lead to change of the neutrino flux by a factor 1.4 ($a_1 = 2$ corresponds to lower bound and $a_1 = 30$ to upper one). However, the intercept of Regge trajectory $\alpha_\psi(0)$ makes substantial influence: the replacement of $\alpha_\psi(0) = 0$ by $\alpha_\psi(0) = -2.2$ reduces the flux by a factor 3 as compared to previous result (solid curve) obtained for similar scheme (QGSM+NSU) with intercept $\alpha_\psi(0) = 0$.

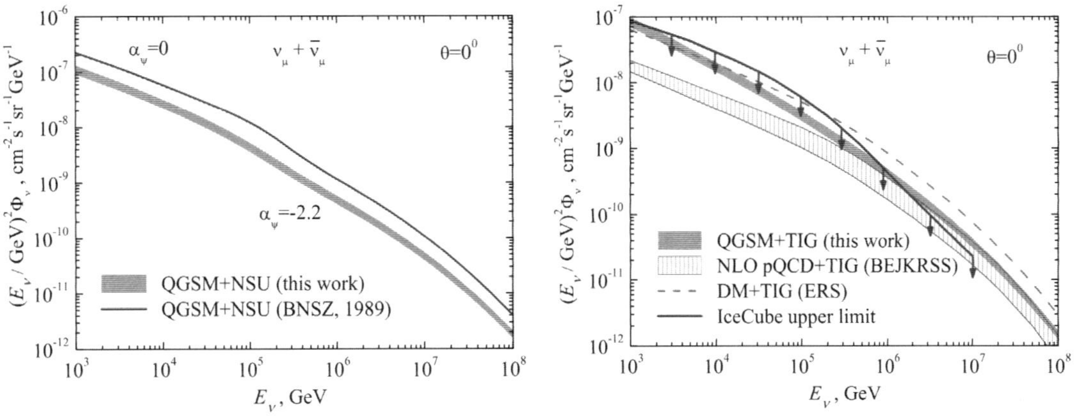

Figure 5. Vertical fluxes of prompt atmospheric neutrinos. Left: calculations of the QGSM with NSU cosmic ray spectrum: shaded band – this work ($\alpha_\psi(0) = -2.2$), solid curve – the result for $\alpha_\psi(0) = 0$ from Refs. [10, 11]. Right: calculations with TIG cosmic ray spectrum: shaded band – updated QGSM, hatched area – NLO pQCD (BEJKRSS) [13], dashed line – dipole model (ERS) [12], solid line – the IceCube upper limit [1].

An influence of charm production models on neutrino fluxes is shown in figure 5 (right panel). All results are obtained for the same cosmic ray spectrum TIG. QGSM flux (shaded band) is comparable with the NLO pQCD [13] calculation (hatched area) at energies above 200 TeV. In the important energy range above 1 PeV, where atmospheric neutrinos from the decay of charmed particles dominate, QGSM leads to appreciably lower flux as compared to the dipole model result [12] (dashed line). As already noted the dipole model was used as the benchmark model in data processing in the IceCube experiment, although the cross sections of this model were not verified by comparison with experimental data, and the prompt neutrino flux was calculated with the toy model (TIG) of the cosmic ray spectrum.

In recent analysis [1], the upper limit on the prompt atmospheric neutrino flux was obtained by IceCube using high statistics collected over six years. The prompt neutrino flux was constrained based on the dipole model [12], but with usage of more realistic cosmic ray flux parametrization H3p [32]. We have found that our result is compatible with the IceCube limit up to 2 PeV.

4. Conclusion

The recent data on cross sections of charmed particle production obtained in the LHC experiments allow an improvement of the QGSM free parameters $\alpha_\psi(0)$ and a_1. Analysis showed that the intercept of Regge trajectory $\alpha_\psi(0)$ appreciably affects the charm production cross section and so the prompt atmospheric neutrino flux. Updated version of QGSM leads to decrease of the prompt neutrino flux by a factor $\sim 2-3$ as compared to the former QGSM prediction using $\alpha_\psi(0) = 0$. In the energy region beyond 1 PeV, where atmospheric neutrinos from the decay of charmed particles dominate, the updated QGSM flux is noticeably lower in comparison with the dipole model which is used in the IceCube experiment as benchmark model. The QGSM flux obtained for intercept $\alpha_\psi(0) = -2.2$ is compatible with the NLO pQCD calculation at $E_\nu > 200$ TeV and does not contradict the IceCube limitation up to 2 PeV.

Acknowledgements

This work is supported in part by the Russian Federation Ministry of Science and Higher Education, agreement 3.9678.2017/8.9.

References

[1] Aartsen M G *et al* 2016 *Astrophys. J.* **833** 3
[2] Kopper C *et al* 2017 *Proc. of Science, The 35th Int. Cosmic Ray Conf. (Bexco, Busan, Korea)* p 981
[3] Aartsen M G *et al* 2018 *Science* **361** 147
[4] Sinegovsky S I and Sorokovikov M N 2018 Prompt atmospheric neutrinos in the quark-gluon string model *JINR preprint* P2-2018-4
[5] Abelev B *et al* 2012 Measurement of charm production at central rapidity in proton-proton collisions at $\sqrt{s} = 2.76$ TeV *J. High Energy Phys.* JHEP07 (2012) 191
[6] Adam J *et al* 2016 *Phys. Rev.* C **94** 054908
[7] Acharya S *et al* 2017 *Eur. Phys. J.* C **77** 550
[8] Aad G *et al* 2016 *Nucl. Phys.* B **907** 717
[9] Vall A N, Naumov V A and Sinegovsky S I 1986 *Yad. Fiz.* **44** 1240
[10] Bugaev E V, Naumov V A, Sinegovsky S I and Zaslavskaya E S 1989 *Nuovo Cim.* C **12** 41
[11] Naumov V A, Sinegovskaya T S and Sinegovsky S I 1998 *Nuovo Cim.* A **111** 129
[12] Enberg R, Reno M H and Sarcevic I 2008 *Phys. Rev.* D **78** 043005
[13] Bhattacharya A, Enberg R, Jeong Y S, Kim C S, Reno M H, Sarcevic I and Stasto A 2016 Prompt atmospheric neutrino fluxes: perturbative QCD models and nuclear effects *J. High Energy Phys.* JHEP11 (2016) 167
[14] Kaidalov A B and Ter-Martirosyan K A 1984 *Sov. J. Nucl. Phys.* **39** 979
[15] Kaidalov A B and Piskunova O I 1986 *Z. Phys.* C **30** 145
[16] Kaidalov A B and Piskunova O I 1986 *Sov. J. Nucl. Phys.* **43** 994
[17] Arakelyan G H 1998 *Phys. Atom. Nucl.* **61** 1570
[18] Kaidalov A B 2003 *Phys. Usp.* **46** 1121-1136
[19] Barlag S *et al* 1988 *Z. Phys.* C **39** 451
[20] Alves G A *et al* 1996 *Phys. Rev. Lett.* **77** 2388
[21] Aguilar-Benitez M *et al* 1984 *Phys. Lett.* B **135** 237
[22] Aguilar-Benitez M *et al* 1988 *Z. Phys.* C **40** 321
[23] Ammar R *et al* 1988 *Phys. Rev. Lett.* **61** 2185
[24] Kodama K *et al* 1991 *Phys. Lett.* B **263** 573
[25] Zoccoli A *et al* 2005 *Eur. Phys. J.* C **43** 179
[26] Adare A *et al* 2006 *Phys. Rev. Lett.* **97** 252002
[27] Adamczyk L *et al* 2012 *Phys. Rev.* D **86** 072013
[28] Bari G *et al* 1991 *Nuovo Cim.* A **104** 571
[29] Chauvat P *et al* 1987 *Phys. Lett.* B **199** 304
[30] Nikolsky S I, Stamenov J N and Ushev S Z 1984 *Sov. Phys. JETP* **60** 10
[31] Thunman M, Ingelman G and Gondolo P 1996 *Astropart. Phys.* **5** 309
[32] Gaisser T K 2012 *Astropart. Phys.* **35** 801

A search for neutrino bursts in the Galaxy at the Baksan Underground Scintillation Telescope

R V Novoseltseva, M M Boliev, I M Dzaparova, M M Kochkarov, A N Kurenya, Yu F Novoseltsev, V B Petkov, P S Striganov and A F Yanin

Institute for Nuclear Research of the Russian Academy of Sciences, 60th October Anniversary Prospect, 7a, 117312 Moscow, Russia

E-mail: kchkrv@rambler.ru, novoseltsev@inr.ru

Abstract. The experiment on recording neutrino bursts operates since the mid-1980. As the target, we use two parts of the facility with the total mass of 240 tons. Over the period of June 30, 1980 to December 31, 2017, the actual observational time is 32.1 years. No candidate for the stellar core collapse has been detected during the observation period. An upper bound of the mean frequency of core collapse supernovae in our Galaxy is 0.072 $year^{-1}$ (90% CL).

1. Introduction

The detection of neutrinos from the supernova SN1987A experimentally proved the critical role of neutrinos in the explosion of massive stars, as it was suggested more than 50 years ago [1, 2, 3].

Due to their high penetration power, neutrinos deliver information on physical conditions in the core of the star during the gravitational collapse. SN1987A has become the nearest supernova in the past several hundred years, which allowed the SN formation process to be observed in unprecedented detail beginning with the earliest time of radiation. It was the first time that a possibility arose for comparing the main parameters of the existing theory – total radiated energy, neutrino temperature, and neutrino burst duration – with the experimentally measured values [4, 5].

Since light (and electromagnetic radiation in general) can be partially or completely absorbed by dust in the galactic plane, the most appropriate tool for finding supernovae with core collapse is a large neutrino detector. In the past decades (since 1980), the search for neutrino bursts was carried out with such detectors as the Baksan Scintillation Telescope [6, 7], Super-Kamiokande [8], MACRO [9], LVD [10], AMANDA [11] and SNO [12]. Over the years, our understanding of how massive stars explode and how the neutrino interacts with hot and dense matter has considerably increased. At present the scale and sensitivity of the detectors capable of identifying neutrinos from a Galactic supernova have grown considerably so that current generation detectors [13, 14, 15] are capable of detecting of order ten thousand neutrinos for a supernova at the Galactic Center.

The Baksan Underground Scintillation Telescope (BUST) [16] is the multipurpose detector intended for wide range of investigations in cosmic rays and particle physics. One of the current tasks is the search for neutrino bursts. The facility has been uninterruptedly used for this purpose since the middle of 1980. The total galaxy observation time amounts to 90% of the calender time.

2. The method of neutrino burst detection

The BUST consists of 3184 standard autonomous counters arranged in four horizontal and four vertical planes. Each counter is $0.7 \times 0.7 \times 0.3$ m^3 in size, filled with an organic scintillator C_nH_{2n+2} ($n \simeq 9$), and viewed by one photomultiplier with a photocathode diameter of 15 cm [16]. A counter pulse channel has an operation threshold 8 MeV and 10 MeV for the horizontal and vertical planes, respectively (the most probable energy deposition of a muon in a counter is 50 MeV $\equiv$ 1 relativistic particle). The total scintillator mass is 330 t, and the mass enclosed in three lower horizontal layers (1200 standard counters) is 130 tons. The majority of the events recorded with the Baksan telescope from a supernova explosion will be produced in inverse beta decay (IBD) reactions

$$\overline{\nu}_e + p \to n + e^+ \tag{1}$$

If the mean antineutrino energy is $E_{\nu_e} = 12 - 15$ MeV [17, 18] the path of e^+ (produced in reaction (1)) will be confined, as a rule, in the volume of one counter. In such case the signal from a supernova explosion will appear as a series of events from singly triggered counters (one and only one counter from 3184 operates; below we call a such event "the single event") during the neutrino burst. The search for a neutrino burst consists in recording of single events cluster within time interval of $\tau = 20$ s (according to the modern collapse models the burst duration does not exceed 20 s).

The expected number of neutrino interactions detected during an interval of duration Δt from the beginning of the collapse can be expressed as:

$$N_{ev}^{H} = N_H \int_0^{\Delta t} dt \int_0^{\infty} dE\ F(E,t) \cdot \sigma(E)\eta(E), \tag{2}$$

here N_H is the number of free protons, $F(E,t)$ is the flux of electron antineutrinos, $\sigma(E)$ - the IBD cross section, and $\eta(E)$ is the detection efficiency. The symbol "H" in left side indicates that the hydrogen of scintillator is the target.

If one assumes the distance from the SN is 10 kpc, the total energy irradiated in neutrinos is

$$\varepsilon_{tot} = 3 \times 10^{53}\ erg \tag{3}$$

and the target mass is 130 t (three lower horizontal layers) the expected number of single events from reaction (1) (we assume that the antineutrino temperature is $k_B T = 3.5$ MeV) will be

$$N_{ev}^{H} \simeq 35 \tag{4}$$

Flavor oscillations are unavoidable of course. However, it was recognized in recent years that the expected neutrino signal depends strongly on the oscillation scenario (see e.g. [19, 20, 21, 22]). In the absence of a quantitatively reliable prediction of the flavor-dependent fluxes and spectra it is difficult to estimate the oscillation impact on ν_e- and $\overline{\nu}_e$ fluxes arriving to the Earth. Therefore we do not discuss the effects of flavor oscillations in this work.

Background events are i) radioactivity (mainly from cosmogeneous isotopes) and ii) cosmic ray muons if only one counter from 3184 hit. The total count rate from background events (averaged over the period of 2001 – 2017 years) is $f_1 = 0.0207$ s^{-1} in internal planes (three lower horizontal layers) and $\simeq 1.5$ s^{-1} in external ones. Therefore three lower horizontal layers are used as a target; below, we will refer to this counter array as the D1 detector (the estimation (4) has been made for the D1 detector).

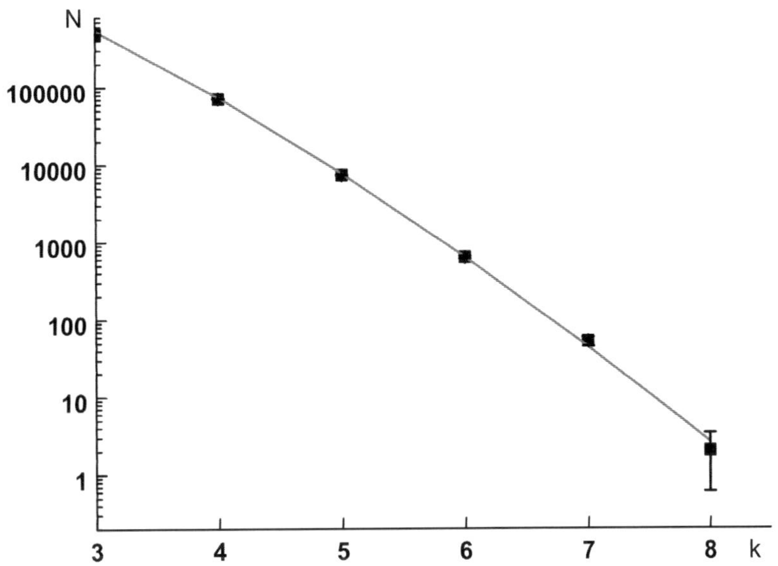

Figure 1. The number of clusters with k single events within time interval of $\tau = 20$ s. Squares are experimental data, the curve is the expected number according to the expression (5)

Background events can imitate the expected signal (k single events within sliding time interval τ) with a count rate

$$p(k) = f_1 \times exp(-f_1\tau)\frac{(f_1\tau)^{k-1}}{(k-1)!} \tag{5}$$

The treatment of experimental data (single events over a period 2001 - 2017 y; $T_{actual} = 14.5$ years) is shown by squares in figure 1 in comparison with the expected distribution according to the expression (5) calculated at $f_1 = 0.0207$ s^{-1}. Note that there is no normalization in figure 1.

According to the expression (5), background events create clusters with k = 8 with the rate 0.178 y^{-1}. The expected number of such clusters during the time interval T = 14.5 y is 2.58 that we observe (2 events). The formation rate of clusters with k = 9 background events is 9.2×10^{-3} y^{-1}, therefore the cluster with multiplicity $k \geq k_{th} = 9$ should be considered as a neutrino burst detection.

3. Two independent detectors

As it follows from the estimation (4) the "sensitivity radius" of the D1 detector is $R_s \simeq 20$ kpc. To increase the sensitivity radius (and to increase the number of detected neutrino events), we use those parts of external scintillator layers that have relatively low count rate of background events. The total number of counters in these parts is 1012, the scintillator mass is 110 tons. We call this array the D2 detector, it has the count rate of single events $f_2 = 0.12$ s^{-1}. The count rates of single events in D1 and D2 detectors and the operating stability have been shown in figure 2.

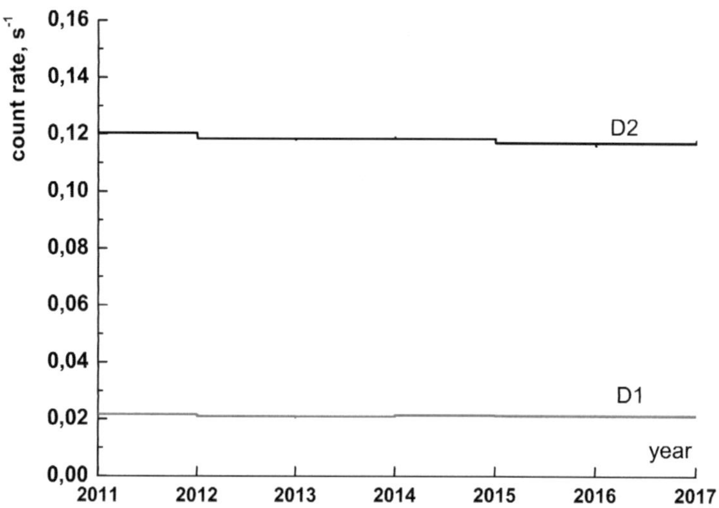

Figure 2. The count rates of single events in D1 and D2 detectors

The joint use of D1 and D2 detectors allows us to decrease the threshold multiplicity in D1 cluster ($k_{th} = 9$) and, consequently, to increase R_s.

We use the following algorithm: in case of cluster detection with $k1 \geq 6$ in the D1, we check the number of single events $k2$ in the 10-second time frame in the D2 detector. The start of the frame coincides with the start of the cluster in D1. Mass ratio of D2 and D1 detectors 1012/1200 = 0.843 implies that for the mean value of neutrino events $k1 = 6$ in D1, the mean number of neutrino events in D2 will be $\overline{k2} = 6 \times 0.843 * 0.8 = 4.05$ (factor 0.8 takes into account that the frame duration in D2 is 10 seconds instead of 20 seconds in D1). Since the background adds $f_2 \times 10$ s = 1.2 events, we obtain finally $\overline{k2}(\ \overline{k1} = 6) = 4.05 + 1.2 = 5.25$.

The D1 and D2 detectors are independent, therefore the imitation probability of clusters with multiplicities k1 in D1 and k2 in D2 by background events is the product of appropriate probabilities

$$P(k1, k2) = P1(k1) \times P2(k2) \tag{6}$$

and we obtain $P(6,5) = 0.23$ y^{-1}, $P(6,6) = 0.045$ y^{-1} (note $P1$ is determined according to the expression (5) and $P2$ is the Poisson distribution).

Therefore the events with $k1 \geq 6$, $k2 \geq 6$ should be considered as candidates for a neutrino burst detection (since mean values of k1 and k2 are significantly exceeded in two independent detectors simultaneously and the imitation probability of such events by background is very small). Thus we decrease the threshold value of $k1$ from 9 to 6 and increase the sensitivity radius up to $R_s \simeq 23$ kpc.

4. Conclusion

The Baksan Underground Scintillation Telescope operates under the program of search for neutrino bursts since June 30, 1980. As the target, we use two parts of the BUST (the D1 and D2 detectors) with the total mass of 240 tons. The "sensitivity radius" of the BUST (for a recording of neutrino bursts from supernovae) is $R_s \simeq 23$ kpc.

Background events are 1) decays of cosmogeneous isotopes (which are produced in inelastic interaction of muons with the scintillator carbon and nuclei of surrounding matter) and 2) cosmic ray muons if only one counter from 3184 hit.

Over the period of June 30, 1980 to December 31, 2017, the actual observation time was 32.1 years. This is the longest observation time of our Galaxy with neutrinos at the same facility. No candidate for the core collapse has been detected during the observation period. This leads to an upper bound of the mean frequency of gravitational collapses in the Galaxy

$$f_{col} < 0.072 \text{ y}^{-1}$$

at 90% CL. Recent estimations of the Galactic core-collapse SN rate give roughly the value $\simeq 2 - 5$ events per century (see e.g. [23]).

Acknowledgments
The work has been carried out at a unique scientific facility the Baksan Underground Scintillation Telescope (Common-Use Center Baksan Neutrino Observatory INR RAS) and was supported by the Program for Fundamental Scientific Research of RAS Presidium "Fundamental Interactions Physics and Nuclear Technologies".

References

[1] Gamow G and Shoenberg M 1940 *Phys.Rev.* **58** 1117
[2] Zeldovich Ya B and Guseinov O Kh 1965 *Dokl. Akad. Nauk SSSR* **162** 791
[3] Colgate S A and White R H 1966 *Astrophys. J.* **143** 626
[4] Loredo T J and Lamb D Q 2002 *Phys. Rev.* D **65** 063002
[5] Pagliaroli G, Vissani F, Costantini M L and Ianni A 2009 *Astropart. Phys.* **31** 163
[6] Alekseev E N, Alekseeva L N, Volchenko V I *et al* 1993 *J. Exp. Theor. Phys.* **77** 339
[7] Novoseltseva R V, Boliev M M, Dzaparova I M *et al* 2016 *Phys. Part. Nucl.* **47** 968
[8] Ikeda M *et al* (Super-Kamiokande Collab.) 2007 *Astrophys. J.* **669** 519
[9] Ambrosio M *et al* (MACRO Collaboration) 2004 *Eur. Phys. J.* C **37** 365
[10] Aglietta M *et al* (LVD Collaboration) 1992 *Nuovo Cimento* A **105** 1793
[11] Ahrens J *et al* (AMANDA Collaboration) 2002 *Astropart. Phys.* **16** 345
[12] Aharmim B, Ahmed S N, Anthony A E *et al* 2011 *Astrophys. J.* **728** 83
[13] Lund T, Marek A, Lunardini C *et al* 2010 *Phys. Rev.* D **82** 063007
[14] Bellini G *et al* (Borexino Collaboration) 2007 *Phys. Lett.* B **658** 101
[15] Eguchi K *et al* (KamLAND Collaboration) 2003 *Phys. Rev. Lett.* **90** 021802
[16] Alexeyev E N, Alexeyenko V V, Andreyev Yu M *et al* 1979 *Proc. 16th ICRC (Kyoto)* vol 10 (Tokyo: Institute for Cosmic Ray Research Press) p 276
[17] Imshennik V S, Nadezhin D K 1982 *Itogi Nauki i Tehniki, ser. Astronomy* **21** 63
[18] Hillebrandt W, Hoflish P 1989 *Rep. Prog. Phys.* **52** 1421
[19] Pantaleone J 1992 *Phys. Lett.* B **287** 128
[20] Sawyer R F 2005 *Phys. Rev.* D **72** 045003
[21] Duan H, Fuller G M, Carlson J *et al* 2006 *Phys. Rev.* D **74**, 105014
[22] Tamborra I, Raffelt G, Hanke F *et al* 2014 *Phys. Rev.* D **90** 045032
[23] Adams S M, Kochanek C S, Beacom J F *et al* 2013 *Astrophys. J.* **778** 164

The underground muon flux with 24 years of data of the LVD detector

C F Vigorito
on behalf of the LVD Collaboration
Dipartimento di Fisica, Universtità di Torino & INFN, via Pietro Giuria 1, 10125 Torino, Italia
E-mail: vigorito@to.infn.it

Abstract. The Large Volume Detector (LVD), in the INFN Gran Sasso National Laboratory (Italy), has been taking data since June 1992. The experiment, 1 kton of liquid scintillator at the equivalent depth of 3600 m w.e., has been mainly designed to observe low energy neutrinos from the core collapse of a supernova but allows also the measure of the atmospheric muon flux underground as well as the induced neutron production.

In this work we summarize the results of the analysis of the global LVD data set, 5.6 10^7 muons in a livetime of 8402 days collected during 24 years of continuos operations since 1994 up to 2017. The present measurement represents an unprecedented collection obtained by a unique experiment in a fixed location. The modulation in time of the flux and its correlation with the the effective temperature in the upper atmosphere are here discussed.

1. Introduction

The flux of muons detected in underground laboratories is directly related to the production of mesons in the stratosphere by hadronic interactions between cosmic rays and the nuclei of air molecules and to the probability that they decay before interacting. This flux shows time variations which are, at first approximation, seasonal and related to the air density fluctuations affecting the fraction of mesons decaying to high energy muons able to trigger the underground detector. This effect has been known and studied for many decades [1]. Various experiments at Laboratori Nazionali del Gran Sasso (LNGS), Italy [2, 3, 4, 5], and in other underground sites [6, 7, 8, 9, 10, 11, 12, 13, 14] were able to measure the muon flux underground and its variations on a time scale of few years at different depths. In this work we present the results of the monitoring of the muon flux by LVD, the largest dataset ever provided so far for undeground muons by a single detector.

2. Temperature effects on the muon rate

The variations of the temperature in the stratosphere causes variations in the air density, changing the probability that the mesons, produced by primary cosmic rays interactions, could interact or decay to high energy muons. In particular, an increase in the temperature of the stratosphere induces a decrease in air density, thus reducing the chance of meson interaction, which in turn results in a larger fraction decaying to produce high energy muons which can eventually reach underground sites. Deviations from the average muon flux measured underground, $\Delta I_\mu(t) = (I_\mu(t) - I_\mu^0)$, can be correlated to variations of the average atmospheric

IOP Conf. Series: Journal of Physics: Conf. Series **1181** (2019) 012057 doi:10.1088/1742-6596/1181/1/012057

temperature $\Delta T(X) = (T(X) - T^0(X))$ at a given depth (altitude) X. The net effect can be written as: $\Delta I_\mu(t) = \int dX\ W(X) \cdot \Delta T(X,t)$ where the weight $W(X)$ (see [15] for details on calculation) reflects the altitude dependence of the production of mesons in the atmosphere (and their decay into muons that can be observed deep underground) and the integral extends over the full atmospheric depth.
Assuming the atmosphere as a stratification of N layers with a continuos distribution of temperature and pressure is possible to define an effective temperature, T_{eff} defined as the weighted average over the atmospheric depth:

$$T_{eff} = \frac{\int dX\ T(X)W(X)}{\int dX\ W(X)} \simeq \frac{\Sigma_{n=0}^{N} \Delta X_n T(X_n) W(X_n)}{\Sigma_{n=0}^{N} \Delta X_n W(X_n)} \tag{1}$$

being the temperature profile measured at N discrete atmospheric levels, X_n. Figure 1 shows the average temperature profile for the period [1994-2017] at LNGS site and the corresponding weight $W(X)$ as function of the pressure levels X_n. Defining the effective temperature coefficient as $\alpha_T = \frac{T_{eff}}{I_\mu^0} \int dX\ W(X)$, the relation between effective temperature and muon intensity variations can be simplified as: $\frac{\Delta I_\mu}{I_\mu^0} = \alpha_T \frac{\Delta T_{eff}}{T_{eff}}$.

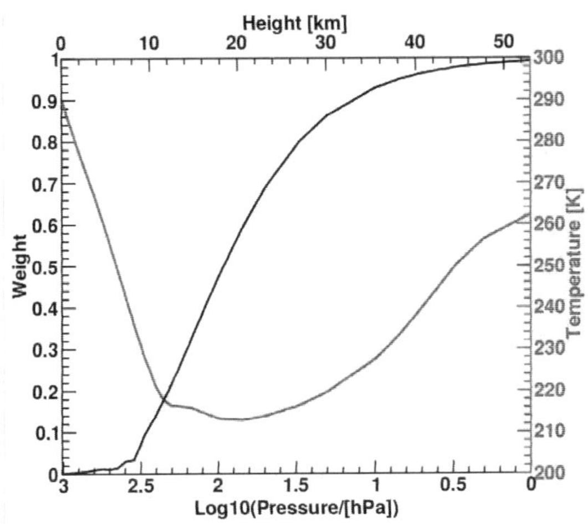

Figure 1. Distribution of the average [1994-2017] temperature profile at LNGS site (red line) and the calculated weight $W(X)$ (black line) as a function of the atmospheric depth X [hPa] or altitude [km]).

Figure 2. Front view of the LVD experiment in the hall A of the Laboratori Nazionali del Gran Sasso (Italy).

3. The LVD detector

LVD (see figure 2) is a 1000 t liquid scintillator experiment aimed at detecting neutrinos of astrophysical origin. It consists of an array of 840 scintillator self-triggering counters, 1.5 m^3 each, viewed from the top by three photomultipliers (PMTs). The counters are organized in a modular and compact geometry which allows to achieve a very high duty cycle. This feature is essential for the search of unpredictable sporadic events like neutrino bursts from Gravitational Stellar Collapses and it is valuable for the study of time variations of the muon flux underground. LVD has been in operation since 1992, June 9th after a short commissioning phase, its mass

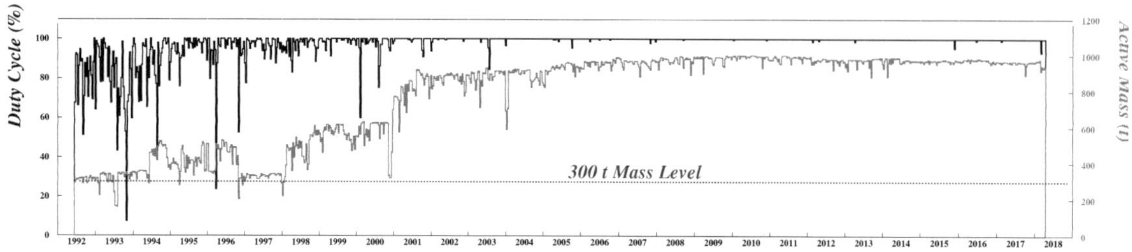

Figure 3. LVD duty cycle(black line) and active mass(red line) as a function of time from 1992, June 9th 1992 to 2017, May 28th.

increasing from 300 t to its final one, 1000 t, in January 2001. The LVD active mass and duty cycle evolution in time, updated to May 2018, are shown in figure 3. The LVD trigger logic (extensively described in [16]) is based on the 3-fold coincidence of the PMTs in a single counter corresponding to an energy threshold $E_{th} \sim 4\,\mathrm{MeV}$. The energy resolution of the counter is $\Delta E/E \sim 15\%$ at 10 MeV. The time stamp of triggered events has a relative time precision of 12.5 ns with an absolute accuracy of 100 ns.
For the purpose of this analysis, data collected in years 1992 and 1993 have been discarded because of scarce of contiguity and low duty cycle efficiency, both correlated to the deployment phases of the first part of the detector.

4. Temperature data

The temperature data has been obtained from the European Center for Medium-range Weather Forecasts (ECMWF ERA-Interim data) [17]. It exploits different type of observations (e.g. surface, satellite, and upper air sounding) at many locations, and then uses a global atmospheric model to interpolate them to a particular location. We considered in this analysis the precise coordinates of the LNGS underground halls (13.5333°E, 42.4275°N) to download data from the ECMWF grid at the highest possible resolution. The model provides atmospheric temperatures at 37 discrete pressure levels in the [1-1000] hPa range, four times a day at 00.00 h, 06.00 h, 12.00 h, and 18.00 h UTC. Based on this data set, the T_{eff} is calculated following equation 1 four times a day providing a mean daily value. The variance of the four daily values is assumed as estimate of the uncertainty of the mean. The daily value of T_{eff}, over the period considered in the present analysis, is shown in figure 4: the simple average gives $< T^{o}_{eff} >= (220.307 \pm 0.006)\,\mathrm{K}$.

5. LVD muon data

Muons are identified in LVD through the time coincidence of signals with an energy release $E \geq$ 10 MeV in two or more counters within a time window of 175 ns, which is large enough to include the time jitter of the PMT's transit time. Data selection is performed via the standard quality cuts that have been defined in the search for neutrino burst [18]. Additionaly, the muon-like events associated to the CNGS (CERN Neutrino to Gran Sasso [19]) neutrino beam in the period [2006-2012] have been excluded if in coincidence with the time of the beam spills inside a time window of $[-10, +20.5]\mu s$, being $10.5\mu s$ the spill duration. The resulting dataset consists of $5.6 \cdot 10^{7}$ muons for a total livetime of 8402 days. The measured daily muon rate is shown in figure 5 (pannel top figure) and it is mainly affected by the different detector configurations over the time, as shown in figure 3. The muon flux underground $I_{\mu}(t)$ is then obtained through the ratio between the measured daily muon rate and the simulated detector exposure (i.e the geometrical acceptance times the livetime) which changes according to the LVD active mass configurations

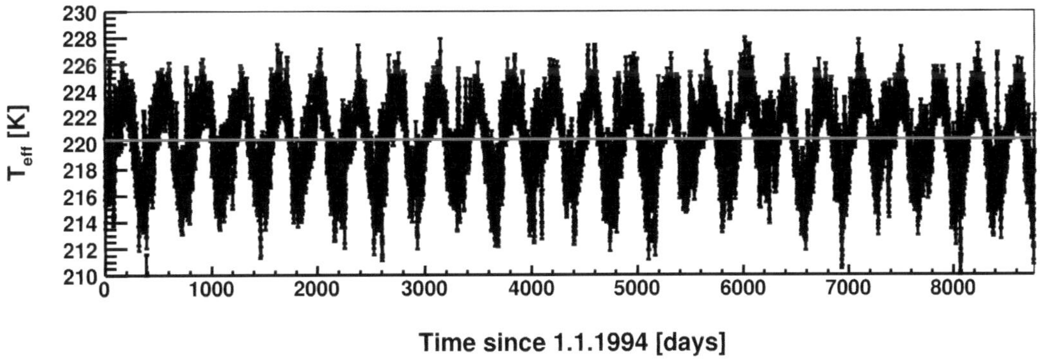

Figure 4. Mean daily values of T_{eff} from 1994 up to 2017 as obtained by using ECMWF Era-interim data: error bars account for the RMS dispersion in each single day. The red line is the overall average value.

and duty cycle as shown in figure 3. A detailed description of the simulation can be found in [20]. For the full detector configuration (i.e. $M = 1000$ t) the geometrical acceptance, averaged over the cosmic muon arrival directions in the hall A of LNGS, corresponds to $S = (298 \pm 3)\,\mathrm{m}^2$, where the uncertainty (1%) is mainly dominated by the systematic errors assumed in the muon direction. The obtained daily detector exposure and the muon flux are shown in figure 5 (pannel middle and bottom figures) . The bigger fluctuations which are observed till the beginning of 2001 (day 2557) are due to the lower active mass of the detector during construction phase. The average value of the muon flux is $I_\mu^0 = (3.35 \pm 0.0005^{stat} \pm 0.03^{sys}) \cdot 10^{-4}\,\mathrm{m}^{-2}\mathrm{s}^{-1}$, being the systematic error induced by the geometrical acceptance calculation.

6. Results

As a first approximation, the daily muon flux has been fitted with a sinusoidal function $I_\mu(t) = I_\mu^0 + \Delta I_\mu cos(\frac{2\pi}{T}(t - \phi))$ obtaining a period of $T_\mu = (365.1 \pm 0.2)\,\mathrm{d}$, which is compatible with what expected and also measured, in same way, in the effective temperature modulation $T_{T_{eff}} = (365.1 \pm 0.3)\,\mathrm{d}$. The amplitude ΔI_μ of the muon flux and the mean phase ϕ of the modulation are therefore better evaluated fixing $T = 365.25$ d, and including systematic errors of the measured daily values. We obtained $\Delta I_\mu = (4.7 \pm 0.3) \cdot 10^{-6}\,\mathrm{m}^{-2}\mathrm{s}^{-1}$ ($\Delta I_\mu / I_\mu^0 \sim 1.4\%$) and a phase $\phi = (183 \pm 1)\,\mathrm{d}$ (corresponding to the beginning of July).
From the daily muon flux (as in figure 5) and effective temperature (as in figure 4) the scatter plot of the relative daily variations values $\frac{\Delta I_\mu}{I_\mu^0}$ and $\frac{\Delta T_{eff}}{T_{eff}^0}$ are obtained, as shown in figure 6. The value of α_T is obtained performing a linear regression which accounts for error bars of both parameters. We got a value of $\alpha_T = 0.94 \pm 0.02$ (correlation coefficient $r = 0.56$), which is in agreement with the expected value of $\alpha_{T,LNGS} = 0.92 \pm 0.02$ at LNGS depth [6, 15]. This result confirms the values that were also measured by other detectors at LNGS.
Figure 7 shows a collection of the α_T values by various experiments at different depths, together with the expected theoretical prediction including muon production by both pions and kaons.

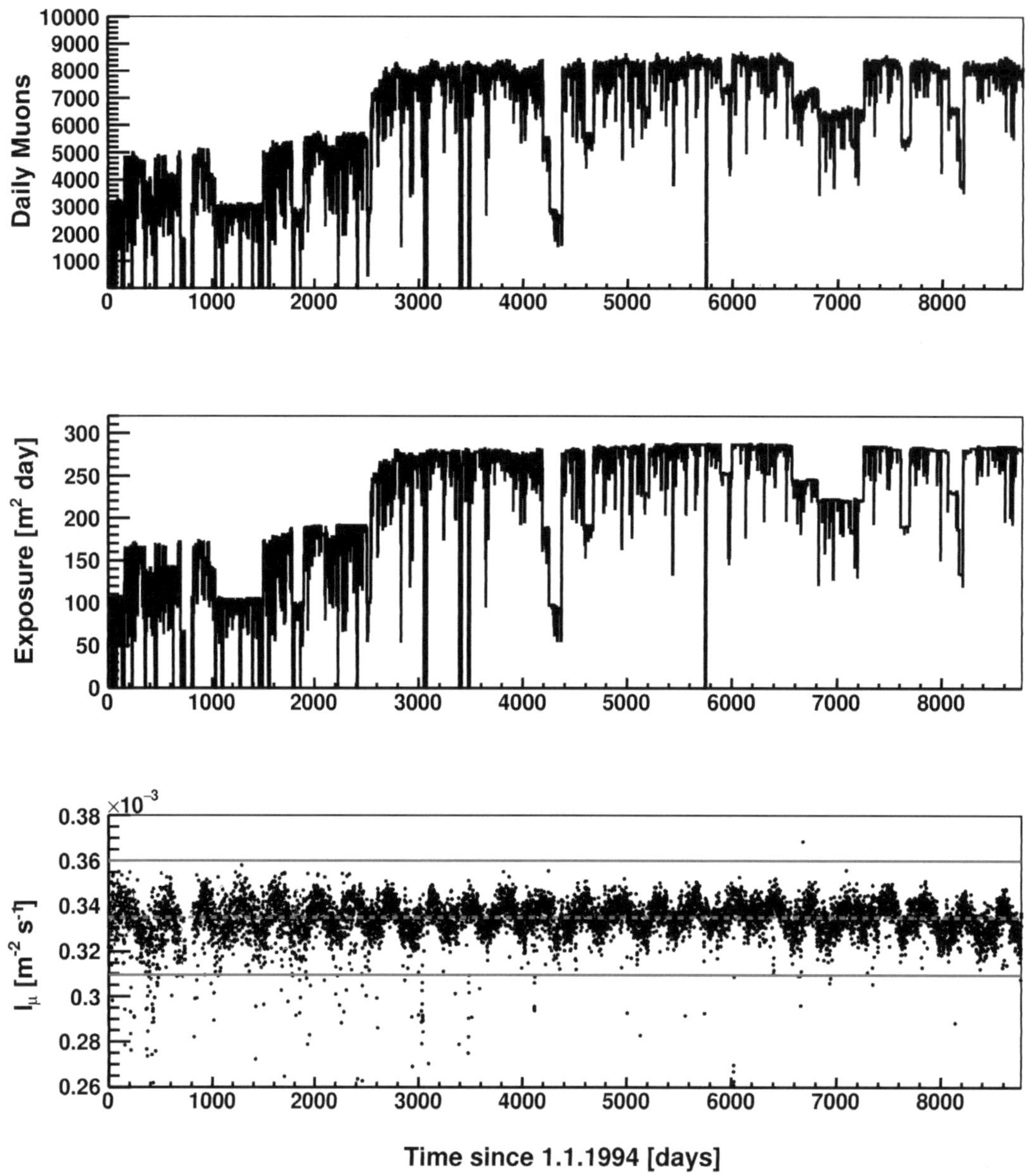

Figure 5. From top to bottom: daily muon rate, exposure and flux from 1994 up to 2017. The bigger fluctuations in the flux, observed till the beginning of 2001, are due to the lower active mass of the detector during construction.

7. Conclusions

We have analized 5.6×10^7 muon events detected by the LVD experiment over 8402 live days in the [1994-2017] period to search for variations of the muon flux underground in correlation with the atmospheric effective temperature variations at LNGS site.

The average value of the muon flux over 24 years of data is: $I_\mu^0 = (3.3335 \pm 0.002[stat.] \pm 0.03[sys])10^{-4}\mathrm{m}^{-2}\mathrm{s}^{-1}$, being the error dominated by the sistematic uncertainty on the geometrical acceptance correction (1%). A clear modulation with a time period $T = 1\,\mathrm{y}$ is observed with an amplitude of 1.4% and an average phase corresponding to beginning of July. Relative variations of the muon flux and of the effective temperature are correlated: the measured coef-

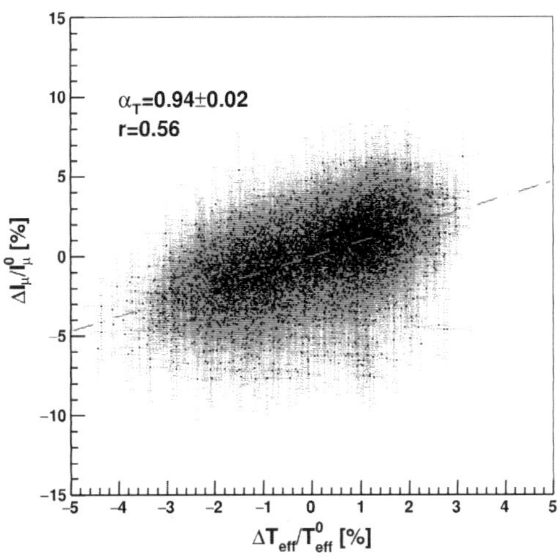

Figure 6. Scatter plot of daily values of $\Delta I_\mu/I^0_\mu$ and $\Delta T_{eff}/T^0_{eff}$: the dashed red line corresponds to the linear regression of the whole data set. The value of the Pearson coefficient R=0.56, being the DoF=8400, excludes the chance probability of a random correlation.

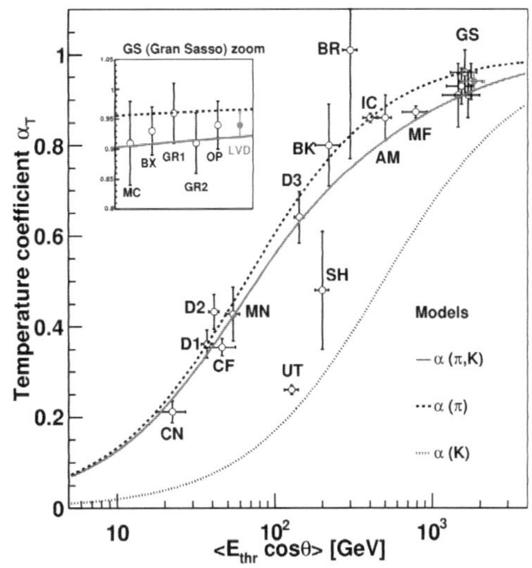

Figure 7. Measured values for the effective temperature coefficient α_T at different site depths. The results of this analysis (in red) compared with those of other detectors (see text for details & references). The red line is the predicted value including both muon production by pions and kaons, shown separately by the dashed lines (see [6] for details).

ficient $\alpha_T = (0.94 \pm 0.02)$ is well in agreement with previous measurements at LNGS site and also with the expected value for the LNGS depth.

References

[1] Barrett P H *et al* 1952 *Rev. Mod. Phys.* **24** 133
[2] Ambrosio M *et al* 1997 *Astropart. Phys.* **7** 109
[3] Bellini G *et al* 2012 *J. of Cosmol. and Astropart. Phys.* **5** 15
[4] Agostini M *et al* 2016 *Astropart. Phys.* **84** 29
[5] Tenti M 2017 *Proc. of Int. Workshop on Neutrino Telescopes* 13-17 March Venezia Italy (PoS) 046
[6] Adamson P *et al* 2010 *Phys. Rev. D* **81** 012001 & Adamson P *et al* 2014 *Phys. Rev.* **90** 012010
[7] Bouchta A 1999 *Proc. of 26th Int. Cosmic Rays Conference* 17-25 August Salt Lake City USA
[8] Desiati P 2011 *Proc. of 32nd Int. Cosmic Rays Conference* 11-18 August Beijing China
[9] Andreyev Y M 1991 *Proc. of 22nd Int. Cosmic Rays Conference* 11-23 August Dublin Ireland
[10] An F P *et al* 2018 *JCAP* **1** 1
[11] Abrahão T *et al* 2017 *JCAP* **2** 17
[12] Barrett P H *et al* 1954 *Mod. Phys* **95** 1573
[13] Sherman N 1954 *Phys. Rev.* **93** 208
[14] Cutler D J and Groom D E 1981*Proc. of 17th Int. Cosmic Rays Conference* 13-23 July Paris France
[15] Grashorn E W *et al* 2010 *Astropart. Phys.* **33** 140
[16] Agafonova N Y *et al* 2007 *Astropart. Phys.* **27** 254
[17] ECMWF, European Centre for Medium-Range Weather Forecast, http://www.ecmwf.int
[18] Agafonova N Y *et al* 2015 *Astrophys. J.* **802** 1 47
[19] Acquistapace G *et al* 1998 CERN-98-02, INFN/AE-98/05 & Acquistapace G *et al* 1999 CERN-SI/99-034(DI), INFN/AE-99/05
[20] Selvi M 2009 *Proc. of 31st Int. Cosmic Rays Conference* 7-15 Jul Lodz Poland

IOP Conf. Series: Journal of Physics: Conf. Series **1181** (2019) 012058 doi:10.1088/1742-6596/1181/1/012058

Search for electron neutrinos associated with gravitational-wave events at the Baksan Underground Scintillation Telescope

V B Petkov[1,2], M M Boliev[1], I M Dzaparova[1,2], M M Kochkarov[1], A N Kurenya[1], Yu F Novoseltsev[1], R V Novoseltseva[1], P S Striganov[1] and A F Yanin[1]

[1]Institute for Nuclear Research of the Russian Academy of Sciences, Moscow, Russia
[2]Institute of Astronomy of the Russian Academy of Sciences, Moscow, Russia

E-mail: vpetkov@inr.ru, makhti.kochkarov@yandex.ru

Abstract. Results of a search for electron neutrinos from the gravitational wave events GW150914, GW151226, GW170104, GW170608, GW170814, and GW170817 at the Baksan underground scintillation telescope (BUST) are presented. We searched for coincident neutrino candidates in the data recorded by the BUST detector. No neutrino candidates in temporal coincidence with the gravitational wave events were found. Upper bounds on neutrino and antineutrino fluence from astrophysical sources of gravitational bursts have been set.

1. Introduction

Astronomers with LIGO and its collaborating detector Virgo have registered six gravitational wave events from the merger of neutron stars and black holes: GW150914 [1], GW151226 [2], GW170104 [3], GW170608 [4], GW170814 [5], and GW170817 [6]. These events can result in a black hole plus accretion disk system that produces non-thermal, high-energy radiation and may have a neutrino component. The observation of gravitational wave sources with multiple cosmic messengers is a unique opportunity that enables the detailed study of the phenomena. The detection of gravitational wave events initiated the search for their astrophysical sources. No reliable signals have been detected from the first five gravitational wave bursts. On August 17, 2017, the Advanced LIGO and Advanced Virgo observatories recorded a GW signal, GW170817. Soon afterwards, Fermi-GBM and INTEGRAL detected a short GRB from a consistent location [7]. Subsequently, ultraviolet, optical, and infrared emission was observed from the merger. Optical observations allowed the precise localization of the merger. Neutrino observatories continuously monitor the whole sky making them well suited to study emission from GW sources.

Astrophysicists searched for high-energy neutrinos from the gravitational wave events in the energy band of [10^{11}eV, 10^{20} eV] using the ANTARES, IceCube, and Pierre Auger Observatories, as well as for MeV neutrinos with IceCube. Also, scientists searched for possible neutrino signals coincident with gravitational wave events in Borexino [8], Super-Kamiokande [9, 10] and KamLAND [11] experiments using a wide energy range from 3.5 MeV to 100 PeV [12, 13, 14, 15, 16, 8, 11, 9, 10]. In all experiments, neutrino events were sought in a time interval of ± 500 s, as the maximum time interval between a neutrino signal and corresponding

IOP Conf. Series: Journal of Physics: Conf. Series **1181** (2019) 012058 doi:10.1088/1742-6596/1181/1/012058

gravitational wave event [17]. Neutrino signals from gravitational wave events were not detected in any of the listed experiments. The number of neutrino candidate events observed in the search window can be converted to an upper limit on neutrino fluence for all of the gravitational wave events. This is done for all listed above neutrino experiments.

Electron antineutrinos were sought at the Borexino, Super-Kamiokande and KamLAND experiments through the inverse beta decay (IBD) reaction

$$\overline{\nu_e} + p \to n + e^+, \tag{1}$$

which has a low threshold $E_{e^+} = E_{\overline{\nu_e}} - 1.3$ MeV and the largest cross section in the range from 0.5 to 110 MeV. The search at the Borexino and Super-Kamiokande experiments was also performed for elastic scattering (ES)

$$\nu + e^- \to \nu + e^-, \tag{2}$$

which is sensitive to all types of neutrinos and has the largest cross-section for electron neutrinos.

Because of the background conditions at the BUST, to seek low-energy electron neutrinos and antineutrinos from gravitational wave events, the reactions of their interaction with carbon in a scintillator were used. In this work, we report upper limits obtained at the BUST on fluxes of electron neutrinos and antineutrinos with energies above 21 MeV from astrophysical sources of the GW150914, GW151226, GW170104, GW170608, GW170814, and GW170817 gravitational bursts.

2. Baksan underground scintillation telescope

The Baksan underground scintillation telescope (BUST) is a multipurpose detector designed for a wide range of investigations in the physics of cosmic rays, elementary particles, and neutrino astrophysics. The telescope is located in the Northern Caucasus (Russia) in an underground laboratory at an effective depth of 850 m.w.e.[18, 19]). BUST is one of the largest underground detectors operating at the Baksan Neutrino Observatory of the Institute for Nuclear Research. The detector is a 17 × 17 × 11 m parallelepiped with modular structure and two additional internal planes. The planes are covered by standard scintillator counters. The total number of counters in the BUST is 3184. The total mass of the scintillator is 330 t.

A standard scintillator counter is a $0.7 \times 0.7 \times 0.3$ m aluminium tank filled with liquid organic scintillator based on white spirit $C_nH_{2n} + 2(n \simeq 9)$. The scintillator volume is viewed by one FEU-49 photomultiplier tube with a photocathode diameter of 15 cm. The most probable energy deposition in a counter from muons is 50 MeV. Four signals are recorded from each counter. A signal from the anode of the photomultiplier tube is used to measure the plane trigger time and the energy deposition up to 2.5 GeV.

The current output (the signal from the anode of the photomultiplier tube through an integrating circuit) is used to adjust and control the gain of the photomultiplier tube. The signal from the 12th dynode is fed to the input of a discriminator (the so-called pulse channel) with a trigger threshold of 8 and 10 MeV for the horizontal and vertical planes, respectively. The signal from the fifth dynode of the photomultiplier tube is fed to the input of a logarithmic converter, where it is converted into a pulse whose length is proportional to the logarithm of the signal amplitude. The logarithmic channel allows the energy deposition in an individual counter to be measured in the range of 0.5 - 600 GeV.

The triggering of the pulse channel for any BUST counter is used as a trigger for the recording system. The trigger count rate is 17 s^{-1}. When a trigger appears, the entire information on this event is fed to an online computer, where the events are preprocessed in order to obtain information on the current status of the detection instrument. The GPS signal is used for synchronization with Universal Time with an accuracy of 0.2 ms.

IOP Conf. Series: Journal of Physics: Conf. Series **1181** (2019) 012058 doi:10.1088/1742-6596/1181/1/012058

3. Search methods

One of the important tasks performed by the BUST experiment, we should mention the detection of the neutrino burst from the collapse of the core of an exploded supernova. This was detected using the signals from three bottom horizontal planes comprising 1200 counters and containing 130 t of the scintillator.

The neutrinos are detected through the reaction of inverse beta-decay. Given that the mean energy of Supernova antineutrinos is $E_{\nu_e} \simeq 12 - 15$ MeV, the path of the positron produced in reaction (1) is comparable with the size of one detector module. The detector response to such an event referred to as the single firing, is the isolated pulse from a single counter. The supernova collapse is detected as a series of single-counter firings throughout the neutrino burst. The active shield is provided by the detectors external planes operating in anticoincidence mode. Three lower inner horizontal planes containing 1200 counters with 130 t of scintillator are used as targets. The count rate in these planes from background events is 0.02 s^{-1}. Because of such a count rate of background events, the inverse beta decay reaction cannot be used to seek neutrino events from gravitational bursts because a much wider time window of 1000 s is involved and the number of background events becomes unacceptably large.

For this reason, low-energy electron neutrinos (antineutrinos) from gravitational bursts are sought at the BUST through the reactions of their interaction with carbon in a scintillator:

$$\nu_e +^{12} C \to^{12} N + e^-, \tag{3}$$

$$\overline{\nu_e} +^{12} C \to^{12} B + e^+. \tag{4}$$

Events of production of unstable ^{12}N and ^{12}B isotopes (with the lifetimes $\tau(^{12}N) = 15.9$ ms and $\tau(^{12}B) = 29.1$ ms) in reactions (3) and (4) and their subsequent decay involve two successive counts of the same counter. The main (and quite well-studied) source of the background for the detection of low-energy electron neutrinos/antineutrinos is the interaction of neutrons with carbon nuclei $n +^{12} C \to^{12} B + e^+$ in the scintillator of the counters of the BUST with the production of the unstable ^{12}B isotope [20]. Thus, successive single counts of the same counter in the range of 150 ms ($> 5 \cdot \tau(^{12}B)$) are considered as candidates for low-energy neutrino events from gravitational bursts at the BUST. The selection of such events makes it possible to significantly reduce the background and, as a result, to increase the mass of the target scintillator to 240 t by adding 1020 counters of the outer planes of the BUST to 1200 counters of the inner planes.

No events that could be interpreted as signals from the interaction of electron neutrinos/antineutrinos with carbon nuclei of the target were detected at the BUST within the interval of ± 500 s from the GW150914, GW151226, GW170104, GW170608, GW170814, and GW170817 gravitational wave bursts. Furthermore, the minimum interval between a possible neutrino signal at the BUST and the GW151225 gravitational wave event is about 1 d. According to the absence of neutrino signals, we obtained the upper limits (90% C.L.) on the flux of electron neutrinos and antineutrinos with different energies under the assumption of a monoenergetic spectrum:

$$F(E_\nu) = \frac{n_{90}}{S(E_\nu)}, \tag{5}$$

where $n_{90} = 2.3$ and $S(E_\nu)$ is the effective area of detection of the electron neutrino/antineutrino, which can be represented in the form

$$S(E_\nu) = \epsilon \cdot N \cdot \sigma(E_\nu) \cdot [\delta_{in} \cdot P_{in}(E_\nu) + \delta_{out} \cdot P_{out}(E_\nu)] \tag{6}$$

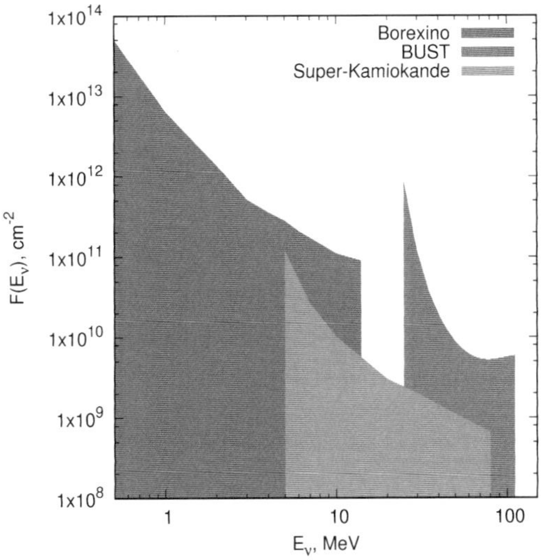

Figure 1. Upper limits (at 90% C. L.) on the electron neutrino flux from gravitational wave source during a ±500 s window centered on the gravitational wave trigger time.

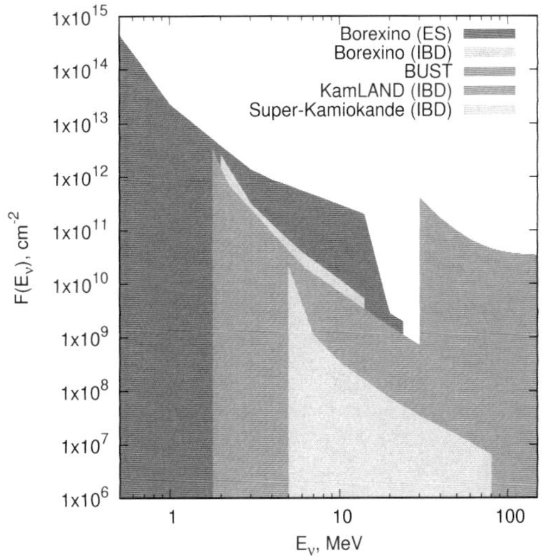

Figure 2. Upper limits (at 90% C. L.) on the electron antineutrino flux from gravitational wave source during a ±500 s window centered on the gravitational wave trigger time.

Here, $\sigma(E_\nu)$ is the cross section for the interaction of the electron neutrino/antineutrino with the carbon nucleus (reactions (3) and (4), [21, 22]), $N = 10^{31}$ is the total number of carbon atoms in the target, $\epsilon = 0.94$ is the fraction of the pure detection time during the search for neutrino signals from gravitational wave events, and δ_{in} and δ_{out} are the fractions of the target mass in the inner and outer planes of the BUST, respectively. The efficiencies of detection of events from the interaction of electron neutrinos/antineutrinos with carbon nuclei in scintillator are different for the inner ($P_{in}(E_\nu)$) and outer ($P_{out}(E_\nu)$) planes of the BUST because of different triggering thresholds of the counters of the inner (8 MeV) and outer (10 MeV) planes of the telescope. These efficiencies for reactions (3) and (4) were calculated taking into account the efficiency of detection of electrons/positrons in the counters of the BUST [23] and beta spectra of ^{12}B and ^{12}N[24].

4. Conclusion and results

We search for possible neutrino signals coincident with GW150914, GW151226, GW170104, GW170608, GW170814, and GW170817 in the BUST detector using an energy range from 21 MeV to 200 MeV. No neutrino candidate events are found in the search window of ±500 s around the LIGO detection time for all gravitational wave signals.

Figures 1 and 2 show the upper bounds (at 90% C.L.) obtained at the BUST on the fluxes of electron neutrinos and antineutrinos from gravitational wave events as functions of their energy (for a monoenergetic spectrum) in comparison with the Borexino, KamLAND, and Super-Kamiokande results. To detect electron neutrinos at the Borexino and Super - Kamiokande facilities, elastic scattering of neutrinos on electrons (2) was used. To detect electron antineutrinos at the Borexino facility, elastic scattering (at lower energies) and inverse beta decay (1) were used. Inverse beta decay was used to seek electron antineutrinos from gravitational wave events at the KamLAND and Super-Kamiokande facilities.

Upper limits obtained in our experiment are identical for all GW150914, GW151226, GW170104, GW170608, GW170814, and GW170817 gravitational wave events. The Borexino bounds shown in the figures are the average for three events GW150914, GW151226, and GW170104 [8]. Electron antineutrinos were sought at the KamLAND facility for the GW150914 and GW151226 events and the bound for the GW150914 event is shown in figure 2 as more stringent [11]. Bounds on the fluxes of electron neutrinos and antineutrinos were obtained at the Super-Kamiokande facility for the GW150914, GW151226, and GW170817 gravitational wave events and the most stringent bounds are also shown in figures 1 and 2 [9, 10].

Acknowledgments
The work has been carried out at a unique scientific facility the Baksan Underground Scintillation Telescope (Common-Use Center Baksan Neutrino Observatory INR RAS) and was supported by the Program for Fundamental Scientific Research of RAS Presidium "Fundamental Interactions Physics and Nuclear Technologies".

References

[1] Abbott B P *et al* 2016 *Phys. Rev. Lett.* **116**(6) 061102
[2] Abbott B P *et al* 2016 *Phys. Rev. Lett.* **116**(24) 241103
[3] Abbott B P *et al* 2017 *Phys. Rev. Lett.* **118**(22) 221101
[4] Abbott B P *et al* 2017 *The Astrophysical Journal Letters* **851** L35
[5] Abbott B P *et al* 2017 *Phys. Rev. Lett.* **119**(14) 141101
[6] Abbott B P *et al* 2017 *Phys. Rev. Lett.* **119**(16) 161101
[7] Abbott B P *et al* 2017 *The Astrophysical Journal Letters* **848** L12
[8] Agostini M *et al* 2017 *The Astrophysical Journal* **850** 21
[9] Abe K *et al* 2016 *The Astrophysical Journal Letters* **830** L11
[10] Hayato Y *et al* 2018 *Astrophys. J.* **857**
[11] Gando A *et al* 2016 *The Astrophysical Journal Letters* **829** L34
[12] Adrián-Martínez S *et al* 2016 *Phys. Rev.* D **93**(12) 122010
[13] Aab A *et al* 2016 *Phys. Rev.* D **94**(12) 122007
[14] Albert A, André M *et al* 2017 *Phys. Rev.* D **96**(2) 022005
[15] Albert A *et al* 2017 *Eur. Phys. J.* C **77** 911
[16] Albert A *et al* 2017 *The Astrophysical Journal Letters* **850** L35
[17] Baret B *et al* 2011 *Astroparticle Physics* **35** 1
[18] Novoseltseva R V *et al* 2016 *Physics of Particles and Nuclei* **47** 968
[19] Novoseltsev Y F *et al* 2017 *Journal of Experimental and Theoretical Physics* **125** 73
[20] Kochkarov M M *et al* 2016 *Physics of Particles and Nuclei* **47** 980
[21] Fukugita M, Kohyama Y and Kubodera K 1988 *Physics Letters* B **212** 139 – 144
[22] Kolbe E, Langanke K and Vogel P 1999 *Nuclear Physics* A **652** 91 – 100
[23] Alekseev E N *et al* 1993 *Journal of Experimental and Theoretical Physics* **77** 339
[24] Mayer-Kuckuk T and Michel F C 1962 *Phys. Rev.* **127**(2) 545–554

IOP Conf. Series: Journal of Physics: Conf. Series **1181** (2019) 012059 doi:10.1088/1742-6596/1181/1/012059

The search for rare events using Large Volume Detector

N Yu Agafonova, V V Ashikhmin, E A Dobrynina, R I Enikeev, O G Ryazhskaya, I R Shakyrianova and V F Yakushev for the LVD Collaboration

Institute for Nuclear Research of the Russian Academy of Sciences, Moscow, Russia
E-mail: vole4ka86@mail.ru

Abstract. The results of the searching for rare events in the Large Volume Detector (LVD) are presented. The rare events could be caused by neutrino interactions in the experimental setup. In the work, the experimental data for 2006-2017 are analyzed. A stability of LVD background counting rate during the specified period is shown. Candidates on neutrino bursts from collapsing stars in the sequence of the LVD events are not found.

1. Introduction

Correct background estimations are very important in the underground physics experiments. The main goals of the experiments are the search for neutrino bursts from collapsing stars, double beta-decay, proton decay, dark matter particles and other extremely rare phenomena. The search for neutrino bursts from gravitational collapses of stars is a long-term experiment. Within the work, research of background events in the LVD setup is done to estimate parameters of real neutrino events. At the moment, only the neutrino burst from SN1987A in the Large Magellanic Cloud has been registered.

The basic scheme of the experiment to search for a neutrino burst is very simple. A cluster of events in detector that could be caused by neutrino interactions is being looked for. The main problem of the neutrino interaction detectionis determination of background events. The event due to neutrino burst detection is selected using characteristics of a cluster (duration and multiplicity) depending on the counting rate of background events. The event caused by the detection of a neutrino burst by the detector is selected. The neutrino burst detection will allow to check theoretical gravitational star collapse models.

2. LVD detector

The LVD detector [1] is located at the LNGS underground laboratory (the Laboratori Nazionali del Gran Sasso, Italy) at a depth of 3620 m.w.e. The LVD detector is an underground iron-scintillator calorimeter with a total mass of 2 kt (1 kt of liquid scintillator and 1 kt of iron). The detector contains 840 independent scintillation counters. The modular structure of LVD is presented in figure 1. The structure contains 3 towers. 5 columns are arranged in each tower, each column consists of 7 portatank levels. Each portatank contains 8 counters (tanks). Modular structure of the LVD setup allows use external counters as an active shield from muons and other background events. Scintillator in tanks is based on white spirit. In addition to its extremely low cost, the parameters of this scintillator remain unchanged during long time under operating conditions [3].

Each counter is viewed by 3 photomultipliers tubes (PMT). The PMTs have different high voltage values, so the background from PMT after-pulses is minimized. Energy threshold of LVD detector is 5 MeV. Coincidence of signals from 3 PMT during event detection reduces effect of the PMT dark current. Trigger opens time gate for a period of 1 ms with an energy threshold of 1 MeV. It thus becomes possible to detect both positrons and gamma-quanta produced by the capture of neutrons generated in the reaction of the inverse beta decay.

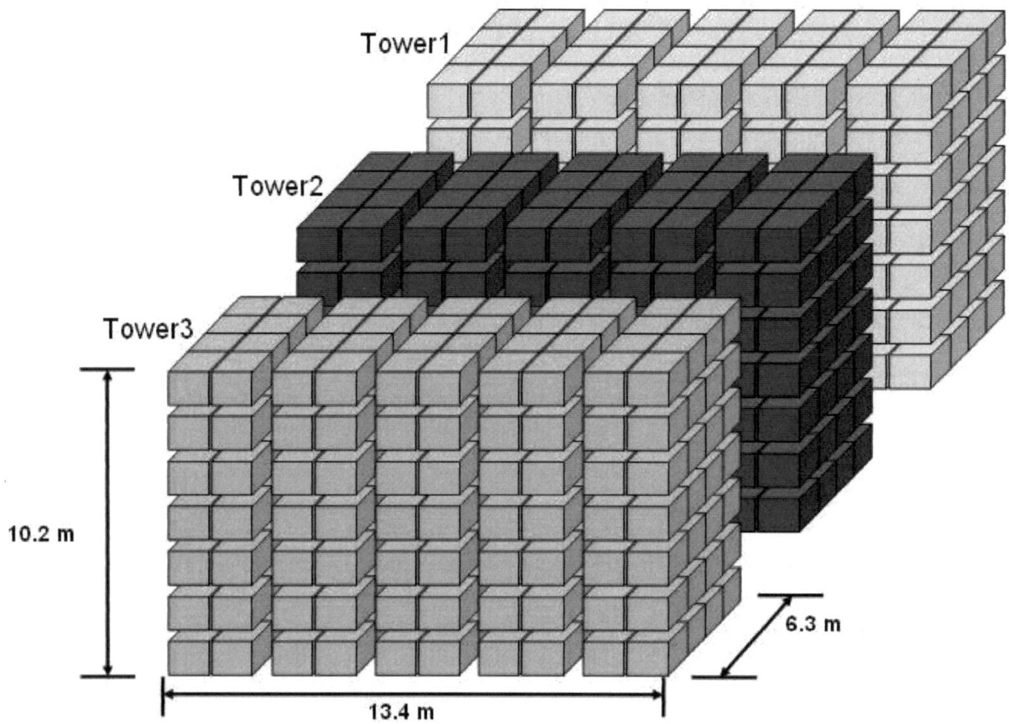

Figure 1. The modular structure of the LVD detector.

3. Events selection

The purpose of experimental data selection is the selection of events that could be caused by the neutrino interactions in the LVD detector. At the first stage, events caused by muon interactions in the detector are excluded. Such events are characterized by the response of two or more counters in a time window of less than 175 ns. Also external counters are excluded from the analysis, i.e. counters bordering on the surrounding volume of the LVD detector. In figure 1, these are counters that form an external rectangular parallelepiped. This limitation is caused by the different counting rate of background events in the counters. The counting rate of background events in external counters is several times higher than in internal ones. After excluding muon events and events in external counters, counters whose average counting rate exceeds the expected value are also excluded from the analysis.

Background events are random events. It means that time between them has an exponential distribution. In figure 2, the event time distributions inthe LVD detectorare presented. We can see that the time between events is distributed exponentially and experimental results do not contradict the calculations. Pseudorandom number generator described in the article of Marsaglia and Tsang [3] is used in the calculations. Average counting rate in the LVD detector is about 0.04 events per second.

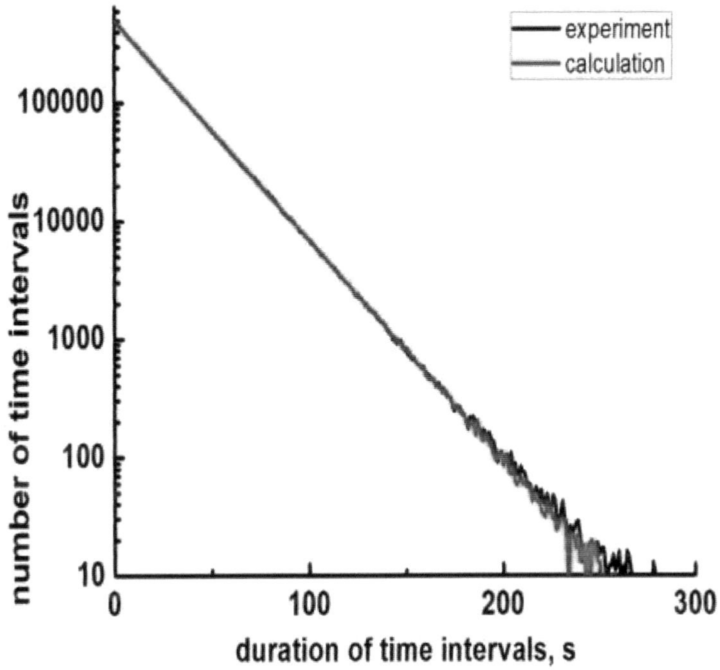

Figure 2. Time distributions of events in the LVD detector.

4. Searching for clusters in the LVD detector

Probability of detecting the cluster with multiplicity N and duration T has the form:

$$p_N = \frac{(\lambda T)^N}{N!} \cdot \exp(-\lambda T). \tag{1}$$

This is Poisson distribution, N is cluster multiplicity, T is cluster duration and λ is counting rate of background events. Counting rate of occurrence of events with cluster multiplicity N and duration T has the form:

$$\lambda_N = \lambda \frac{(\lambda T)^{N-1}}{(N-1)!} \cdot \exp(-\lambda T). \tag{2}$$

In that way, the average occurrence time of the cluster with multiplicity N and duration can be written as:

$$T_N = \frac{1}{\lambda_N}. \tag{3}$$

If we can estimate average occurrence time of the cluster, we can calculate its statistical significance. For example, we can select the cluster at its occurrence time. The search for clusters in the LVD detector was carried out as follows. The maximal cluster multiplicity is 30. So, it was done for all clusters with multiplicity no more than 30 and with duration less than 100 seconds. For each cluster, average time of its detecting and duration was calculated. Clusters of events from collapsing stars have not been detected in the LVD detector.

In future, it is planned to include in the analysis the experimental data from the external counters. Clusters will be searched in the same way as in the internal counters. The difference consists in the quantitative estimate of the average time of a cluster generation by background events. When calculating the average time, another value of the average rate of the background events will be used, it is lower than that in the internal counters. Therefore, the average time that the background event cluster generates in the internal counters is always greater than the average time that a similar cluster forms in the external counters. Furthermore, each counter in the LVD detector is an independent detection element, so the data from the external and internal counters can be considered as data from two independent detectors. The number of external and internal counters in the LVD detector is approximately the same. So when registering a neutrino burst, clusters of events with similar parameters (multiplicity and duration) in the external and internal counters should be simultaneously detected. After cluster obtaining in the internal and external counters, their coincidences will be searched for. The average time of formation of two background clusters at the same time will be calculated. This will enable clusters with lower multiplicity and longer duration to be included in the analysis.

5. Conclusion

In the article, a stable long-term (for about 11 years) functioning of the LVD detector is presented. The counting rate in the LVD detector is about 0.04 events per second. The practical invariability of the counting rate for background events enhances significantly the probability for identifying a neutrino burst. The experimental results do not contradict the calculations. In accordance with LVD data, the clusters of events from collapsing stars have not been detected. In future, the analysis will include data from the external counters of the LVD detector, which will reduce the requirements for the selection of clusters.

Acknowledgments

This work was partially supported by the Russian Foundation for Basic Research, projects ## 18-02-00064-a, 19-02-00262-a, and the joint program of investigations of the Presidium of the Russian Academy of Sciences "Physics of Fundamental Interactions and Nuclear Technologies".

References

[1] Aglietta M, Alpat B, Alyea E D, Antonioli P, Anzivino G, Badino G, Ban Y, Bari G, Basile M, Benelli A, Berezinsky V S, Bergamasco L, Bianco S, Bizzetti A, Bruni G, Cao Y *et al* 1992 *Nuovo Cimento* A **105** 1793
[2] Voevodskii A V, Dadykin V L and Ryazhskaya O G 1970 *Prib. Tekhn. Eksp.* No 1 85
[3] Marsaglia G and Wai Wan Tsang 2000 *J. Statistic. Software* **5** (8) 1

Searches for ultrahigh-energy neutrinos from gravitational wave events with the Pierre Auger Observatory

E Santos[1], for the Pierre Auger Collaboration[2]

[1] Institute of Physics (FZU) Czech Academy of Sciences, Na Slovance 2, Prague, Czech Republic

[2] Observatorio Pierre Auger, Av. San Martín Norte 304, 5613 Malargüe, Argentina

Full author list: http://www.auger.org/archive/authors_2018_07.html

E-mail: esantos@fzu.cz, auger_spokespersons@fnal.gov

Abstract. With the advent of the multi-messenger astronomy era, neutrinos open a unique window to the Universe allowing one to relate the most violent phenomena with the production and origin of the highest energy cosmic rays at distances beyond the GZK horizon. Neutrinos with $E > 0.1$ EeV of all flavors can be detected with the Surface Detector array of the Pierre Auger Observatory by means of their interaction in the atmosphere (downward-going ν) or in the Earth crust (Earth-skimming ν). The Pierre Auger Collaboration has searched for neutrino candidates in coincidence with gravitational wave events detected by LIGO/Virgo. Up to the present moment no neutrino candidates were found in any of these searches, which allows us to put competitive limits on the energy radiated in ultra-high energy neutrinos. The most stringent limits to the neutrino spectral fluence arise from the binary neutron star merger event GW170817 at a $D_s \sim 40$ Mpc. The non-detection of neutrinos from this event is compatible with the predictions of a short GRB observed at a large off-axis angle.

1. Introduction

The detection of the first gravitational wave transient GW150914, by the Advanced LIGO detectors, on September 14, 2015, at 09 : 50 : 45 UTC, marked the beginning of a new era. Its signal matched Einstein's predictions of General Relativity for a gravitational wave with $E_{GW} = 3.0^{+0.5}_{-0.5}\ M_{\odot}c^2$ produced by the merger of a binary black hole system at a luminosity distance of $D_s = 410^{+160}_{-180}$ Mpc [1]. Since then, several other events were detected by the Advanced LIGO/Virgo observatories [2, 3, 4] motivating astronomical multi-messenger observations to detect any other counterparts of these events. Presently, the most exciting results come from the event GW170817, detected both by the Advanced LIGO and Advanced Virgo Observatories on August 17, 2017, at 12 : 41 : 04 UTC. It was due to a binary neutron star inspiral in the host galaxy NGC 4993, just at $D_s = 40^{+8}_{-14}$ Mpc, which makes it the closest, most precisely localized and brightest event detected so far [5]. Only 1.7 s after the coalescence, the *Fermi*-GBM independently detected the short gamma-ray burst GRB 170817A associated to the merger. An unprecedented broadband follow-up carried out by tens of Collaborations worldwide resulted in the positive detection of several counterparts of the event GW170817 / GRB 170817A across the electromagnetic spectrum ranging from the radio to X-rays [4]. Additionally, several models predict that ultra-high energy cosmic rays and high-energy neutrinos could be produced in such mergers of compact objects [6, 7, 8]. If this is the case, neutrinos are the messengers by excellence of these interactions [9, 10]. Since they are neutral particles with extremely low cross-sections they can propagate through distances beyond the GZK horizon, while keeping their direction unaffected by the presence of matter and magnetic fields, allowing them to point back at their source.

Neutrinos of all flavors with $E > 0.1$ EeV can be detected by the Pierre Auger Observatory, located in the Argentinian province of Mendoza at ~ 1400 m a. s. l., ($X_{\rm ground} = 880$ g cm^{-2}). The Pierre Auger Observatory uses a hybrid detection technique combining a Surface Detector (SD) array, which samples the lateral density of particles at the ground, with a Fluorescence Detector (FD), which measures the fluorescence light produced by the excitation of the nitrogen molecules in the atmosphere. The SD array covers an area of 3000 km^2 with 1600 water-Cherenkov stations disposed in a triangular grid of 1.5 km spacing, while the FD is composed by four buildings placed at the array periphery, housing 6 telescopes each [11].

In this paper, the results of the neutrino searches done by the Pierre Auger Observatory in coincidence with gravitational wave events are presented.

2. Neutrino detection with the Pierre Auger Observatory

Each water-Cherenkov station of the SD array consists of a cylindrical polyethylene tank of 3.6 m diameter and 1.2 m height filled with 12 000 liters of purified water, and it is equipped with three 9″ photo-multiplier tubes (PMT) placed at the top. The signals of the PMTs are sampled by flash analogue to digital converters (FADC) with a frequency of 40 MHz [11]. These features make possible the detection of hadronic air showers with $E > 4$ EeV with zenith angles up to 80° [11, 12, 13].

Hadron primaries interact early in the atmosphere generating large numbers of secondary particles which, in order to be detected by the ground detectors, need to cross a mass overburden which grows with zenith angle as $X/\cos\theta$. This makes that the shower front of showers with $\theta > 60°$ is dominated by muons, as most of the electromagnetic component gets absorbed in the atmosphere. Thus, inclined air showers detected by the SD array, those with $60° < \theta < 80°$, typically present very flat shower fronts and narrow time distributions. Neutrinos, on the other hand, have much smaller cross-sections and can interact deeper, much closer to the ground. In this case, the FADC time traces of the SD stations would present a broad time structure, indicative of that of a "young" shower, with a high content of electromagnetic particles [14], as it is depicted in figure 1.

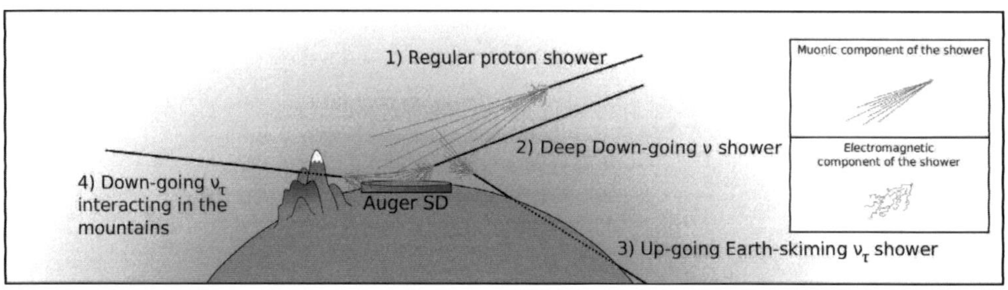

Figure 1. Different types of inclined showers which can be detected by the SD of the Pierre Auger Observatory [14]. 1) A common proton induced shower interacting high in the atmosphere. 2) A deep down-going neutrino shower interacting close to the ground. 3) An Earth-skimming ν_τ interacting in the Earth's crust generating a shower. 4) A down-going ν_τ crossing through the Andes and producing a τ which decays near the SD.

IOP Conf. Series: Journal of Physics: Conf. Series **1181** (2019) 012060 doi:10.1088/1742-6596/1181/1/012060

The discrimination of the neutrino signals from the hadronic background relies on the value of the Area over Peak (AoP) of the triggered stations, defined as the ratio of the integral of the FADC trace to its peak value, normalized to 1 for the average signal produced by a single muon [14].

Neutrino searches can be performed either in the Downward Going (DG), or in the Earth-skimming (ES) channels. The DG channel is subdivided into two regimes, the Downward Going Low (DGL), which is used for events with zenith angles between 60° and 75°, and the Downward Going High (DGH) for $75° < \theta < 90°$. In the DG channels the neutrino searches use a Fisher discriminant which combines up to 10 variables using the AoP of 4 (4 or 5) early (central) stations in the DGH (DGL) selections as described in detail in [15]. Finally, the ES channel applies for events in the zenith angle range $90° < \theta < 95°$, and it dominates the exposure of the Pierre Auger Observatory to neutrinos. The ES selection requires a minimum of three stations, a high eccentricity of the elliptic shape of the triggered area on the ground and an apparent speed of the trigger times between station pairs with an average value very close to c with a small spread [15]. This channel is only sensitive to tau neutrinos, and it is the one which presents the highest selection efficiency, as it is demonstrated in figure 2.

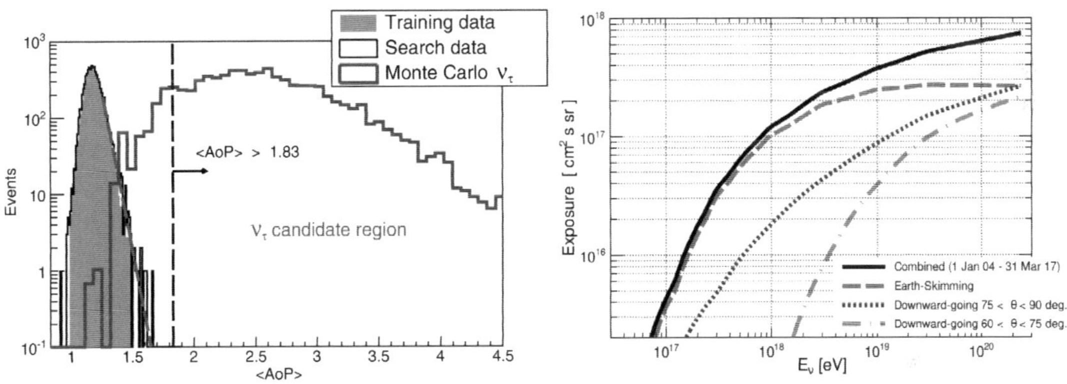

Figure 2. Left: Neutrino discrimination for the Earth-skimming channel. The cut in ⟨AoP⟩ (dashed line) ensures that there is less than one background event in 50 years of observation time. The selection efficiency for the Earth-skimming channel is 95% [15]. Right: Total exposure (black solid line) as a function of the neutrino energy and contributions of the ES (dashed red), DGH (blue dotted line) and DGL (chained magenta line) channels assuming equal fluxes for all neutrino flavors [16].

2.1. *Point-like sources*

The arrival directions of the events detected by the SD are calculated by the relative arrival time of the shower front to the triggered stations. The angular resolution of inclined events with $\theta < 80°$ is better than 2.5°, and improves with the number of triggered stations and the energy of the shower [12].

At the latitude of Auger, $\lambda = -35.2°$, a source with equatorial coordinates (α, δ) observed at a given sidereal time t is described, for a given zenith angle $\theta(t)$, by

$$\cos\theta(t) = \sin\lambda\sin\delta + \cos\lambda\cos\delta\sin(2\,\pi\,t/T - \alpha)\,, \tag{1}$$

where T is the duration of the sidereal day. Hence, the study of point-like neutrino sources is only possible within certain declination ranges, as it is illustrated in figure 3.

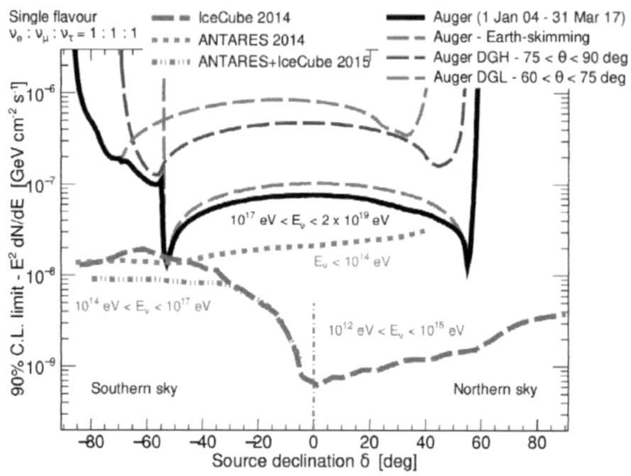

Figure 3. Upper limits at 90% C.L. for the sensitivity of the several neutrino channels as a function of the declination, namely: ES (red dashed line), DGH (blue dashed line), DGL (green dashed line), and the total exposure (black solid line) [16].

For the ES channel, the sky coverage of the Auger array ranges from $-54.5°$ to $59.5°$ in declination, while for the DG channels, it is possible to increase the declination range in the southern sky to $-84.5°$.

3. Neutrino searches in coincidence with gravitational wave events

The Pierre Auger Observatory participates in several multi-messenger studies, among which is the neutrino follow-ups of gravitational wave events. Currently Auger has participated in the neutrino searches in temporal and spatial coincidence of the gravitational wave events GW150914, the candidate event LVT151012, GW151226, and the GW170817 / GRB 170817A [17, 18]. A brief description of these events is summarized in table 1.

Table 1. List of gravitational wave events followed-up by the Pierre Auger Observatory. SNR represents the signal-to-noise ratio of the gravitational wave events [3, 4]. BBH stands for Binary Black Hole merger and BNS stands for Binary Neutron Star inspiral.

Event	Date and time of detection	D_s/Mpc	$E_{GW}/M_{\odot}c^2$	SNR	Category
GW150914	Sep. 14, 2015, 09 : 50 : 45 UTC	410^{+160}_{-180}	$3.0^{+0.5}_{-0.5}$	23.7	BBH
LVT151012	Oct. 12, 2015, 09 : 54 : 43 UTC	1000^{+500}_{-500}	$1.5^{+0.3}_{-0.4}$	9.7	BBH
GW151226	Dec. 26, 2015, 03 : 38 : 53 UTC	440^{+180}_{-190}	$1.0^{+0.1}_{-0.2}$	13.0	BBH
GW170817	Aug. 17, 2017, 12 : 41 : 04 UTC	40^{+8}_{-14}	> 0.025	32.4	BNS

The neutrino searches were performed in the energy range of $[0.1\,\text{EeV}, 25\,\text{EeV}]$, a complementary energy region to the searches done by IceCube/ANTARES [19]. Two time windows were used: one of ± 500 s around the time of the merger, to look for neutrinos coming from the prompt phase of gamma-ray bursts [20, 21], and an extended one of 1 day (14 days) after the merger of binary black hole (binary neutron star) systems, to search for longer lived processes [21].

Since the ES channel dominates the exposure of the Auger SD array to neutrinos while presenting the lowest background, it is desirable that the source position at the time of the merger lies in this region. From all the events followed by Auger, the event GW170817/GRB 170817A was the best located one since the source could be observed during the whole ±500 s time window in this channel, one reason more to highlight this event in a dedicated sub-section. Also, the events GW151226 and LVT151012 had some overlap with the ES channel during a part of the ±500 s time window. Finally, all the three binary black hole merger events GW151226, LVT151012 and GW150914 could be observed in the DGH channel during the ±500 s, and their follow-up could be carried on during a fraction of the 1 day time window. No neutrino candidates were found during the ±500 s and 1 day time windows [17].

3.1. The binary neutron star merger event GW170817/GRB 170817A

Thanks to the several electromagnetic counterparts, this was the most precisely located event at the equatorial coordinates $\alpha\,(2000) = 13^h09^m48^s.085$, $\delta\,(2000) = -23°22'53''.343$ [4]. Neutrino searches in the energy range of GeV to EeV in coincidence with GW170817/GRB 170817A were performed by the most sensitive neutrino observatories ANTARES, IceCube and Auger [18]. The sky map of these neutrino searches, as well as their results are shown in figure 4.

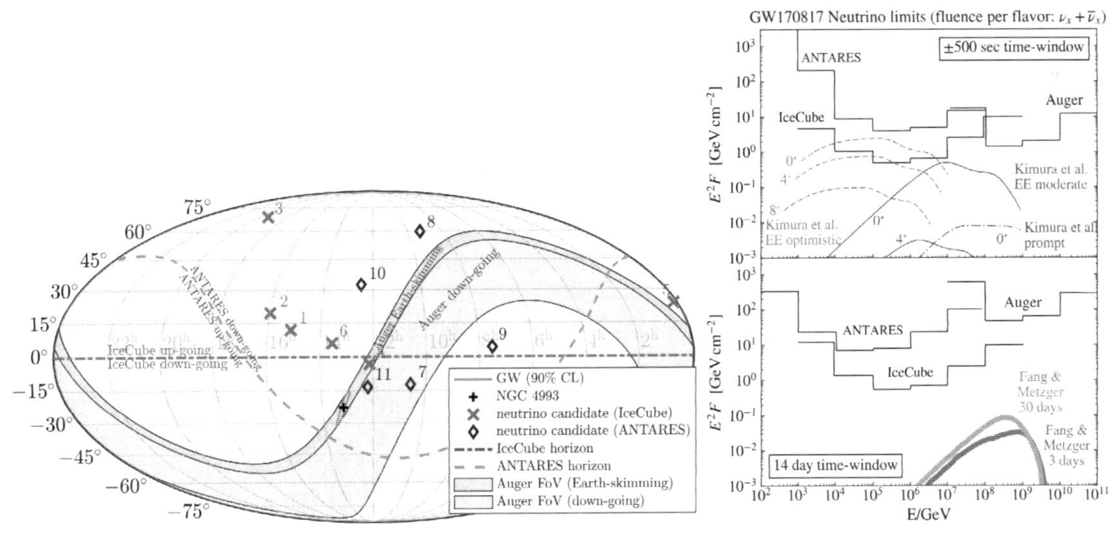

Figure 4. Left: Sensitive sky areas of ANTARES, IceCube and Auger at the time of the event GW170817 in Equatorial Coordinates. The red contour marks the 90% C.L. location of the event GW170817 [1, 18]. Right: Upper limits at 90% C.L. of the neutrino spectral fluence from the event GW170817 for a ±500 s time window (top panel) and in the following 14 days after the trigger (bottom panel). Some model predictions for the neutrino emission scaled to a distance of 40 Mpc are also shown in the plot.

In Auger, the whole ±500 s time window was observed in the ES channel field of view, the most sensitive channel to ultra-high energy neutrinos. In this period, the source of GW170817 transited from $\theta \sim 93°.3$ to $\theta \sim 90°.4$ as seen from the center of the array. The performance of the SD array (which is monitored each minute) was very stable, with an average number of active stations amounting to $\sim 95.8\% \pm 0.1\%$. No inclined showers passing the ES channel selection were detected. The estimated number of background events in this 1000 s window is $\sim 6.3 \times 10^{-7}$, for the cuts applied in the ES channel [15]. Assuming neutrinos are emitted steadily during this period, with an energy spectrum of E^{-2} [17], the non-detection of candidates allows us to put limits to its fluence (see figure 4 right). In the following 14 days searches were carried out both in the ES and DG channels. From the Auger coordinates, the zenith angle of the optical counterpart of the event oscillates daily between $\theta \sim 11^\circ$ to $\theta \sim 121^\circ$. The source is visible in the ES channel for $\sim 4\%$ ($90^\circ < \theta < 95^\circ$) of the day, in the DGL channel for $\sim 10.5\%$ ($60^\circ < \theta < 75^\circ$), and in the DGH for 11.1% ($75^\circ < \theta < 90^\circ$). No significative counterpart was found in any of the searches with any of the observatories, a result which is compatible with the expectations of a GRB observed off-axis or with a low luminosity GRB [18].

4. Conclusions

With the functioning of the LIGO/Virgo systems, and with the upcoming gravitational wave detectors, it is foreseeable an increase of the detection of gravitational wave events from nearby sources. The detection of neutrino candidates from gravitational wave events would allow to better understand the processes by which these ultra-high energy neutrinos, and possibly ultra-high energy cosmic rays are produced. Also, in the case of the non-detection of any other counterparts, it could help to better constrain the position of these sources in the sky with a precision, ranging from less than ~ 1 deg^2 to the order of 10 deg^2 [17]. The Pierre Auger Observatory, along with the neutrino detectors IceCube and ANTARES is committed in continuing the rapid searches of neutrino candidates from the follow-up of future gravitational wave events.

Acknowledgments

We wish to acknowledge the financial support from a number of funding agencies in the participating countries, listed at https://www.auger.org/index.php/about-us/funding-agencies. This work is funded by the MEYS under the project LTT 18004 and LM 2015038 and by the Czech Science Foundation under the project 14-17501S.

References

[1] The LIGO Scientific and Virgo Collaborations 2016 *Phys. Rev. Lett.* **116** 061102
[2] The LIGO Scientific and Virgo Collaborations 2016 *Phys. Rev. Lett.* **116** 241103
[3] The LIGO Scientific and Virgo Collaborations 2016 Phys. Rev. X **6** 041015
[4] Abbott B P *et al* 2017 *Astrophys. J.* **848** (2) L12
[5] The LIGO Scientific and Virgo Collaborations 2017 *Phys. Rev. Lett.* **119** 161101
[6] Murase K, Kashiyama K, Mészáros P, Shoemaker I and Senno N 2016 *Astrophys. J.* **822** (1) L9
[7] Kotera K and Silk J 2016 *Astrophys. J.* **823** (2) L29
[8] Gao H, Zhang B, Wu X F and Dai Z G 2013 *Phys. Rev.* D **88** 043010
[9] Berezinsky V S and Zatsepin G T 1969 *Phys. Lett.* B **28** 423
[10] Halzen F and Hooper D 2002 *Rept. Prog. Phys.* **65** 1025
[11] The Pierre Auger Collaboration 2015 *Nucl. Instrum. Meth.* **A798** 172
[12] The Pierre Auger Collaboration 2014 *JCAP* **1408** 019
[13] The Pierre Auger Collaboration 2015 *JCAP* **1508** 049
[14] The Pierre Auger Collaboration 2011 *Phys. Rev.* D **84** 122005; Erratum: [2011 *Phys. Rev.* D **84** 029902]
[15] The Pierre Auger Collaboration 2015 *Phys. Rev.* D **91** 092008
[16] Zas E for the Pierre Auger Collaboration 2017 *Proc. 35th Int. Cosmic Ray Conf. (Busan)* PoS (ICRC 2017) 972
[17] The Pierre Auger Collaboration 2016 *Phys. Rev.* D **94** 122007
[18] The ANTARES, IceCube, Pierre Auger, LIGO Scientific and Virgo Collaborations 2017 *Astrophys. J.* **850** (2) L35
[19] The ANTARES, IceCube, LIGO Scientific and Virgo Collaborations 2016 *Phys. Rev.* D **93** 122010
[20] Baret B *et al* 2011 *Astropart. Phys.* **35** 1
[21] Kimura S S , Murase K, Mészáros P and Kiuchi K 2017 *Astrophys. J.* **848** (1) L4

IOP Conf. Series: Journal of Physics: Conf. Series **1181** (2019) 012061 doi:10.1088/1742-6596/1181/1/012061

The upgraded GLE database includes assessment of radiation exposure at flight altitudes

A. Mishev[1,2], I. Usoskin[1,2], S. Tuohino[3] and A. Ibragimov[4]

[1]Space Climate Research Unit, University of Oulu, Finland
[2]Sodankylä Geophysical Observatory, University of Oulu, Finland
[3]Faculty of Science, University of Oulu, Finland
[4]Independent researcher, Helsinki, Finland

E-mail: alexander.mishev@oulu.fi

Abstract. The exposure to radiation due to cosmic rays at cruising aviation altitudes is an important topic in the field of space weather. While the effect of galactic cosmic rays can be easily assessed on the basis of recent models, assessments of the absorbed dose during strong solar particle events is rather complicated. A specific interest are events with energy about GeV/nucleon, which produce an atmospheric cascade registered by ground based detectors e.g. neutron monitors. Those events are known as ground level enhancements (GLEs) and can significantly enhance the radiation exposure at flight altitudes over the polar regions. A recent upgrade of the existing GLE database provides information on the estimated solar protons energy/rigidity spectra, the corresponding computed effective doses and the used bibliography. Using information retrieved from this upgrade we performed statistical analysis of maximum effective doses at commercial flight altitude of 35 kft during several GLEs, where the necessary information as energy/rigidity spectra of the solar protons is available. For the computations a recent model for assessment of effective dose due to cosmic ray particles was employed. A highly significant correlation between the maximum effective dose rate and neutron monitor peak count rate increase during GLEs is observed. Here, we propose to use the maximal count rate increase as a proxy for assessment of the effective dose at flight altitude during strong solar particle events.

1. Introduction

The Earth is continuously hit by high-energy particles with extra-terrestrial origin vs. cosmic rays (CRs). They collide with particles in the atmosphere and produce new, energetic particles, which also collide with other nuclei and in such a way produce cascade of secondary particles, i.e., an extensive air shower. The increased amount of particles compared to ground level determine the complex radiation environment at flight altitudes [1, 2]. Nowadays, the assessment of the exposure to the radiation in this complex field, henceforth exposure, is an important topic in the field of space weather [3, 4]. It is recommended to consider the exposure of aircrew as occupational [5]. Occasional, but important for the determination of the radiation environment source of particles, follows eruptive solar processes e.g. solar flares and coronal mass ejections (CMEs), leading to production of solar energetic particles (SEPs) [6, 7]. A special class of extreme SEPs [8] registered by ground based detectors such as neutron monitors (NMs) is called ground level enhancements (GLEs) [7, 9]. This type of events can dramatically enhance the radiation field in the Earth's atmosphere [4].

A database developed over the years by the research community provides information of NM count rates during GLEs, presently hosted by the University of Oulu [10]. This database was developed over several decades by the research community with main contribution of L. Gentile, M. Shea, D. Smart, M. Duldig, J. Humble, H. Moraal, A. Belov, E. Eroshenko, V. Yanke.

The estimation of the exposure at flight altitudes due to CRs of both galactic and solar origin is not trivial. It depends on a geographic position and altitude, solar activity etc...[1]. In order to fulfil this task, it is necessary to possess information about GLE particles energy spectra and angular distribution. In the last years several models for computation of the exposure at aviation altitudes are developed e.g. [11, 12, 13, 14, 15]. We would like to stress that different reconstructed GLE spectra result on considerably different exposures even for the same event as reported by [16, 17]. In this study we compute the exposure during GLEs, where spectral information is available. The computations are carried out at flight altitude of 35 kft (10668 m) in a region with cut-off rigidity $R_c < 1$ GV using a model, which full description is given in [18]. The computed exposure during GLEs allowed us to upgrade the existing GLE database, which now provides information on the SEP energy/rigidity spectra, the corresponding effective doses and the used bibliography sources [19].

2. Model for computation of effective dose rate at aviation altitudes

The employed model here for computation of the exposure at aviation altitudes is based on pre-computed effective dose yield functions. The effective dose rate at a given altitude h induced by primary CR particles is computed using the expression:

$$E(h, R_c, \theta, \varphi) = \sum_i \int_{E_{cut,i}(R_c)}^{\infty} \int_{\Omega} J_i(T) Y_i(T, h) d\Omega dT, \tag{1}$$

where h is the altitude above sea level, R_c is the local cut-off rigidity, θ and φ are the angles of incidence of the arriving particle, $J_i(T)$ is the differential energy spectrum of the primary CR arriving at the top of the atmosphere for i component (proton and/or $\alpha-$particle) and Y_i is the effective dose yield function for the corresponding primary CR component. The integration is over the kinetic energy above $T_{cut,i}(R_c)$, which is defined by R_c for a nuclei of type i by the expression:

$$T_{cut,i} = \sqrt{\left(\frac{Z_i}{A_i}\right)^2 P_c^2 + E_0^2} - E_0, \tag{2}$$

where $E_0 = 0.938$ GeV/c^2 is the proton's rest mass, $T_{cut,i}$ is given in [GeV], respectively P_c in [GV].

Accordingly the effective dose yield function Y_i is defined as:

$$Y_i(T, T^*, h) = \sum_j \int_{T^*} F_{i,j}(h, T, T^*, \theta, \varphi) C_j(T^*) dT^*, \tag{3}$$

where $C_j(T^*)$ is the fluence-to-effective dose conversion coefficient for a secondary particle of type j (neutron, proton, γ, e^-, e^+, μ^-, μ^+, π^-, π^+) with energy T^*, $F_{i,j}(h, T, T^*, \theta, \varphi)$ is the secondary particle fluence of type j, produced by a primary particle of type i (proton and/or $\alpha-$particle) with a given primary energy T. The conversion coefficients $C_j(T^*)$ are considered according to [20].

For GCR equations 1 and 3 lead to

$$E = 4\pi^2 \left[\int_{E_{cut}}^{\infty} J_p(T') Y_p(T') dT' + \int_{E_{cut}}^{\infty} J_\alpha(T') Y_\alpha(T') dT' \right]. \tag{4}$$

Equation 4 has two integral terms, the first reveals the contribution of CR protons. The second term describes accordingly $\alpha-$particles and includes effectively also heavier nuclei [21, 22]. During GLEs, the exposure is a superposition of GCRs and SEPs contributions. For GCRs we employ the force field model [23, 24, 25] and for SEPs we employ various reconstructions, the details are given in [19]. Note, that for GLE particles contribution to the exposure we are using only the first term in equation 3. In this study we used GLE spectra derived mostly on the basis of data from the global NM network, which is an integral detector and does not allow to obtain information about the mass composition of SEPs. In addition, we assume an isotropic SEP flux, which lead to a conservative estimation of the exposure, similarly to [26].

3. Upgrade of GLE database

The occurrence rate of GLEs is roughly one per year [27, 28]. Using the model described in Section 2 (equations 1–4) and multiple sets of derived energy/rigidity SEP spectra we compute the effective dose rate during selected GLEs. The computations are performed at flight altitude of 35 kft (10668 m) in a region with low cut-off rigidity $R_c < 1$ GV, where the exposure is maximal. The results are shown for various reconstructions of SEPs spectra, available in literature. This naturally lead to different results for one event. The computed exposures during GLEs were incorporated in the existing GLE database. The upgraded database provides, for each GLE event, where possible, information on reconstructed energy/rigidity spectra and the corresponding exposure `http://gle.oulu.fi/#/dose`. An example is shown in figure 1, where is presented the NM count rate increase during GLE 43 observed by several stations.

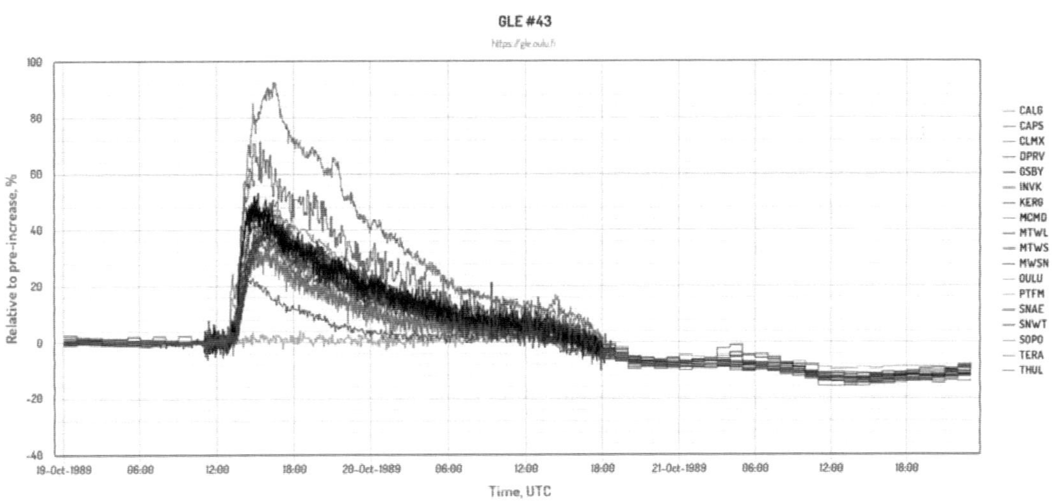

Figure 1. Global NM count rate increase during GLE 43 on 19 October 1989.

An illustration, namely computed exposures during GLE 45 is shown in figure 2 with the corresponding references.

Details about the upgrade of the GLE database, with full list of the events and used bibliography sources are given in [19]. Note, that the creation of such database is an open process. Therefore the database will be updated when new information and/or events occur and/or new information is retrieved including computation(s) with improved models when available.

IOP Conf. Series: Journal of Physics: Conf. Series **1181** (2019) 012061 doi:10.1088/1742-6596/1181/1/012061

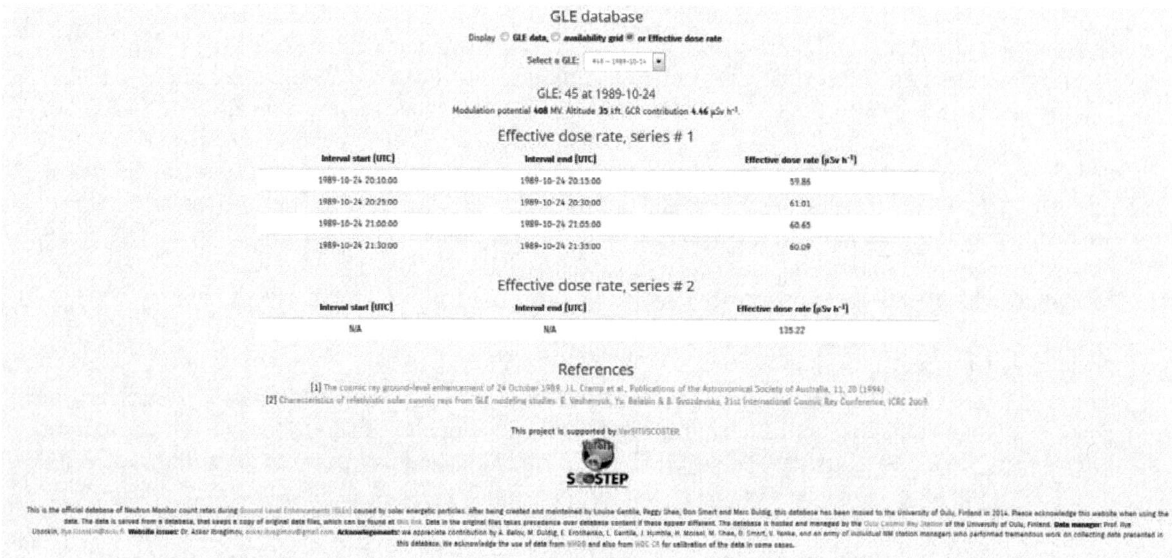

Figure 2. Computed effective dose rate at altitude of 35 kft a.s.l. during GLE 43 on 19 October 1989 with the corresponding SEP spectra and bibliography.

4. Application of upgraded GLE database

The assessed exposure during various stages of GLEs allow one to compute the distribution of the effective dose rate and to estimate the radiation hazard and radiation environment in the troposphere and stratosphere [29, 30, 31]. As example we present the distribution of the exposure over the globe during the main phase of GLE 70 on 13 December 20006 (figure 3). The SEP spectra are considered according to [32]. Here, the cut-off rigidities over the globe are obtained using the MAGNETOCOSMICS code [33], employing combination of Tsyganenko 1989 (external) [34] and IGRF (internal) [35] geomagnetic models, which allows precise and relatively straightforward computation [36, 37, 38]. The distribution of exposure depicts a maximum at polar and sub-polar regions and smoothly decreases at regions with higher cut-off rigidity.

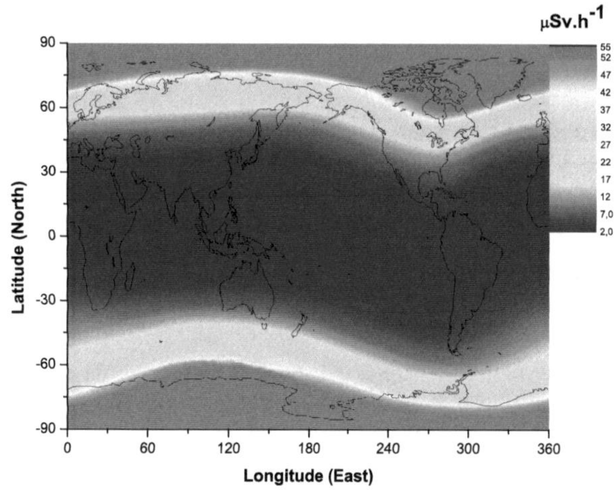

Figure 3. Map of the exposure during the main phase (03:30 UT) of GLE 70 on 13 December 2006.

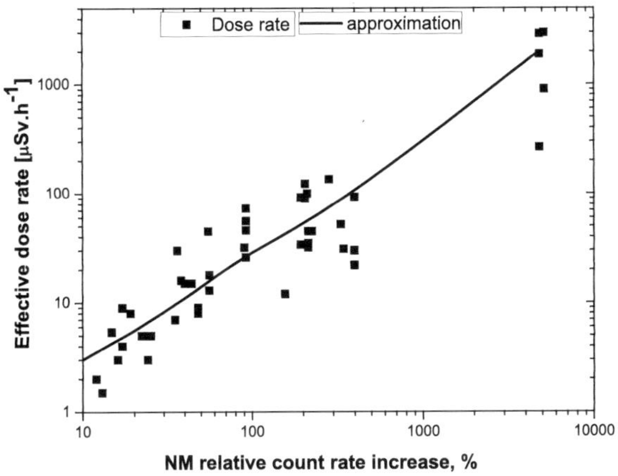

Figure 4. Peak effective dose rate during GLEs versus maximum NM count rate increase.

In addition, we performed a statistical study of a number of GLEs, where the information about their spectra was available. Using the GLE database and the computed exposures during GLEs, we computed the peak exposure at the aviation altitude of 35 kft during these events. In a such a way we study the peak exposure for 34 GLEs, which yields 47 different sets of assessments. The peak exposure due to GLE particles varies from several μSv.h^{-1} to tens or hundreds μSv.h^{-1}. For the two strongest GLE events, the estimated exposure was in the range 1–2.5 mSv.h^{-1}. Subsequently, we build a distribution: peak exposure – peak NM count rate increase during the event. A highly significant correlation (the Pearson's linear correlation coefficient of about 0.84, with high significance level $p-$ value $\ll 0.01$), between the peak exposure rate and peak NM count rate increase was found (figure 4). Therefore, the NM peak count rate increase can be used as a proxy to assess the peak exposure at flight altitudes [39].

5. Conclusions

We upgraded the existing GLE database by including information about the exposure during selected GLE events. The new database is linked to the existing GLE database `http://gle.oulu.fi/#/dose`. The database will be updated when new events occur and/or new information, namely GLE spectra for historical events is retrieved. The upgrade provides, for each observed GLE where data are available, information on the GLE spectra, the corresponding effective dose rates at cruise flight altitude of 35 kft and bibliographic source(s). The upgraded GLE database allows one to estimate the exposure over several solar cycles and gives the possibility to compare different models for estimation of the radiation hazard at flight altitude(s) and measurements.

The derived highly significant correlation between peak exposure – maximum NM count rate increases during GLEs, gives good basis to obtain a quick estimate of the exposure due to SEPs during GLE events on the basis of records from the global NM network, using mainly low cut-off rigidity stations. The peak exposure is due in most cases to the hard – prompt component of SEPs. Using the fact, that the most energetic SEPs from the prompt component arrive in the vicinity of Earth before the bulk of solar protons, a quick assessment of the exposure can be achieved instantly as in [40]. Therefore, the global NM network is a useful tool for assessment of space weather effect such as the radiation exposure of aircrew during strong SEP events.

Acknowledgments
This work was supported by the Academy of Finland (project No. 272157, Center of Excellence ReSoLVE and project No. 267186). This study was supported by VarSITI Program of ICSU Scientific Committee on Solar-Terrestrial Physics (SCOSTEP). We warmly acknowledge all the researchers and NM station managers who collected GLE data and performed the corresponding processing and analysis.

References

[1] Spurny F, Votockova I and Bottollier-Depois J 1996 *Radioprotection* **31** 275–280
[2] Shea M and Smart D 2000 *Space Science Reviews* **93** 187–205
[3] Lilensten L and Bornarel J 2009 *Space Weather, Environment and Societies* (Dordrecht: Springer) ISBN 978-1-4020-4332-1
[4] Vainio R, Desorgher L, Heynderickx D, Storini M, Flückiger E, Horne R, Kovaltsov G, Kudela K, Laurenza M, McKenna-Lawlor S, Rothkaehl H and Usoskin I 2009 *Space Science Reviews* **147** 187–231
[5] ICRP 2007 *Annals of the ICRP* **37**
[6] Reames D 2013 *Space Science Reviews* **175** 53–92
[7] Desai M and Giacalone J 2016 *Living Reviews in Solar Physics* **13** 3
[8] Gopalswamy N, Xie H, Yashiro S, Akiyama S, Mäkelä P and Usoskin I 2012 *Space Science Reviews* **171** 23–60
[9] Aschwanden M 2012 *Space Science Reviews* **171** 3–21
[10] Usoskin I, Ibragimov A, Shea M and Smart D 2015 *Proceedings of Science, Proc. of 34th ICRC Hague, Netherlands, 30 July - 6 August 2015*, 054
[11] Ferrari A, Pelliccioni M and Rancati T 2001 *Radiation Protection Dosimetry* **93** 101–114
[12] Sato T, Yasuda H, Niita K, Endo A and Sihver L 2008 *Radiation Research* **170** 244–259
[13] Matthiä D, Sihver L and Meier M 2008 *Radiation Protection Dosimetry* **131** 222–228
[14] Mishev A, Adibpour F, Usoskin I and Felsberger E 2014 *Advances in Space Research* **55** 354–362
[15] Copeland K 2017 *Radiation Protection Dosimetry* **175** 419–431
[16] Bütikofer R and Flückiger E 2013 *Journal of Physics: Conference Series* **409** 012166
[17] Bütikofer R and Flückiger E 2015 *Journal of Physics: Conference Series* **632** 012053
[18] Mishev A and Usoskin I 2015 *Journal of Space Weather and Space Climate* **5** A10
[19] Tuohino S, Ibragimov A, Usoskin I and Mishev A 2018 *Advances in Space Research* **62** 398–407
[20] Petoussi-Henss N, Bolch W, Eckerman K, Endo A, Hertel N, Hunt J, Pelliccioni M, Schlattl H and Zankl M 2010 *Annals of the ICRP* **40** 1–257
[21] Usoskin I and Kovaltsov G 2006 *Journal of Geophysical Research* **111** D21206
[22] Mishev A and Velinov P 2011 *Advances of Space Research* **48** 19–24
[23] Gleeson L and Axford W 1968 *Astrophysical Journal* **154** 1011–1026
[24] Burger R, Potgieter M and Heber B 2000 *Journal of Geophysical Research* **105** 27447–27445
[25] Usoskin I, Alanko-Huotari K, Kovaltsov G and Mursula K 2005 *Journal of Geophysical Research* **110** A12108
[26] Copeland K, Sauer H, Duke F and Friedberg W 2008 *Advances in Space Research* **42** 1008–1029
[27] Shea M and Smart D 1990 *Solar Physics* **127** 297–320
[28] Stoker P 1995 *Space Science Reviews* **73** 327–385
[29] Mishev A and Velinov P 2015 *Journal of Atmospheric and Solar-Terrestrial Physics* **129** 78–86
[30] Mishev A, Poluianov S and Usoskin S 2017 *Journal of Space Weather and Space Climate* **7** A28
[31] Mishev A and Velinov P 2018 *Advances in Space Research* **61** 316–325
[32] Mishev A and Usoskin I 2016 *Solar Physics* **291** 1225–1239
[33] Desorgher L, Flückiger E, Gurtner M, Moser M and Bütikofer R 2005 *Internationl Journal of Modern Physics A* **20** 6802–6804
[34] Tsyganenko N 1989 *Planetary and Space Science* **37** 5–20
[35] Langel R 1987 *Geomagnetism* (London: J.A. Jacobs Academic Press) chap 1, pp 249–512
[36] Kudela K and Usoskin I 2004 *Czechoslovak Journal of Physics* **54** 239–254
[37] Mishev A and Usoskin I 2013 *Journal of Physics: Conference Series* **409** 012152
[38] Nevalainen J, Usoskin I and Mishev A 2013 *Advances in Space Research* **52** 22–29
[39] Mishev A, Tuohino S and Usoskin I 2018 *Journal of Space Weather and Space Climate* **8** A468
[40] Latocha M, Beck P and Rollet S 2009 *Radiation Protection Dosimetry* **136** 286–290

Database capabilities for studying Forbush-effects and interplanetary disturbances

A A Abunin[1,2], M A Abunina[1], A V Belov[1], S P Gaidash[1], E A Eroshenko[1], I I Pryamushkina[1,3], L A Trefilova[1] and E I Gamza[1]

[1]Pushkov Institute of Terrestrial Magnetism, Ionosphere and Radio Wave Propagation (IZMIRAN), Troitsk, Moscow, Russia
[2]Kalmyk State University named after B.B. Gorodovikov, Elista, Russia
[3]Buguruslan Oil College, Buguruslan, Orenburg region, Russia

E-mail: abunin@izmiran.ru

Abstract. For a comprehensive study of Forbush-effects and their relationship to solar, interplanetary and geomagnetic disturbances, IZMIRAN researchers created (and continuously replenishes) a unique database of transient phenomena in cosmic rays and the interplanetary environment. In it, variations in the density and anisotropy of cosmic rays are combined with solar, interplanetary and geomagnetic parameters. The database includes a large number of different characteristics for about 7500 Forbush-effects, covering more than half a century of observations (1957-2017). In the work presented, some of the features of this tool are demonstrated.

1. Introduction

One of the actual fundamental tasks of solar-terrestrial physics is the advance prediction of space weather parameters and assessment of its influence on various natural, technological and biological systems located both on the Earth or in the near-earth space and at any point of the solar system [1].The space weathercondition is primarily determined by solar activity. The Sun is responsible for creating recurrent (high-speed streams from coronal holes) and sporadic (coronal mass ejections) disturbances of the interplanetary medium. Objects located in near-Earth space are most affected by influence of such disturbances, since the atmosphere is practically absent (or absent at all) and the geomagnetic field is weakened. During such disturbances on space objects, a number of problems can be observed, for example: failure of the spacecraft radio-electronic equipment due to the accumulation of surface and volume charge; single failures in the spacecraft radio electronic equipment due to damage by high-energy particles of solar and galactic origin [2]; violations in the operation of optical, magnetic and other sensors that cause failure in the performance of specified functions; deterioration and loss of communication (at any frequencies); a significant increase in the error of GPS and GLONASS systems; change of ballistic characteristics of orbits due to heating of the upper atmosphere and increase of its density and, as a consequence, loss of altitude and disorientation, unpredictable approach and possibility of collision with other space objects and elements of space debris, possibility of unauthorized uncontrolled descent from orbit; radiation impact on astronauts, etc.

Moreover, in addition to the space segment, space weather also affects earth-based objects, for example: deterioration and loss of communication between ground stations and controlled spacecraft of the space complex [3]; defeat of ground infrastructure objects when space debris elements fall to the Earth; appearance of induced currents in long conductors capable of causing accidents in power grids, underwater cables, pipelines and railway automation systems [4]; radiation impact on crews and passengers of flights (especially transpolar flights); failures in communication systems and avionics equipment; deterioration health of the people with cardiovascular diseases, etc.

Thus, it is clear that the impact of space weather on our daily lives cannot be underestimated. It is necessary to be able to predict this impact qualitatively, and then to consider, using earlier gained experience.

In general, forecasting of space weather condition reduces to forecasting solar and geomagnetic activity, as well as the fluxes of various particles (galactic and solar cosmic rays, relativistic electrons of magnetospheric origin, etc.). A prediction model for a given parameter can be constructed using a different approach. For example, it is possible to obtain the result theoretically by solving complex systems of equations describing the solar activity, transfer and transformation of the corresponding disturbances in interplanetary space and the interaction of magnetic inhomogeneities with the Earth's magnetosphere. But this way is very difficult and, often, because of the complexity of the overall picture of interactions, impossible even in the simplest cases.

Another, more simple way is to obtain relations between different parameters describing the space weather based on statistical, comparative or regression analysis. Moreover, the more events will be considered, the more accurately you can determine this connection. Examples of such studies, based on the analysis of a large number of events, will be considered further.

2. Data and methods

An unique database FEID (*database of Forbush-effects* (FEs) *and Interplanetary Disturbances*) was created and continuously updated by the IZMIRAN staff for comprehensive study of solar, interplanetary and geomagnetic disturbances [5-7]. It combines variations in the density and anisotropy of galactic cosmic rays with solar, interplanetary, and geomagnetic parameters. Galactic cosmic rays are taken as the basis of this tool. The fluxes of these ultra-energetic particles give us information about the magnetic inhomogeneities that they penetrate. In the database, cosmic rays are represented by the results of a global survey method according to the data of the world neutron monitors network for rigidity of 10 GV.Information on the solar wind is taken from the OMNI database (http://omniweb.gsfc.nasa.gov), geomagnetic activity data is taken from WDC Kyoto (http://wdc.kugi.kyoto-u.ac.jp) and GFZ Potsdam (ftp://ftp.gfz-potsdam.de).

Currently, FEID includes about 7500 events covering more than half a century of observations (1957-2017). Each event is described by more than one hundred parameters. It is safe to say that this is the largest and most complete database of interplanetary disturbances in the world.

Based on the database described above, a directory with similar information was developed and posted to the Internet (http://spaceweather.izmiran.ru/rus/dbs.html). In the scientific community you can find a lot of publications and dissertations, which are based on information about interplanetary disturbances, taken from this database.

3. Examples of use of information from FEID

The described database of transient phenomena in cosmic rays and interplanetary medium is not only a storage bank for information on interplanetary disturbances, but also a convenient tool for its processing, allowing to perform various types of analysis of these data (statistical, comparative, regression, etc.), giving the requested information in numerical and graphical forms. Several examples of using the database are given below.

3.1. Study of Forbush-effects with a sudden and gradual onset.

Isolated events (3455 events participated in the analysis) were divided into two groups in paper [8]: *S* – events that were accompanied by the arrival of the shock wave to the Earth (with SSC), and *NS* – events that were not accompanied by SSC (noSSC). The analysis revealed that these groups differ from each other not only quantitatively, but also qualitatively (figure 1 *a-c*), for example, in the structure of the disturbances. And more powerful events, on average, were in the *S*-group.

The obtained results show that different mechanisms of cosmic ray modulation predominate in the selected groups. The events of the *S* – group are largely due to coronal mass ejections, while a significant part of the events of the *NS*-group are due to high-speed streams from coronal holes.

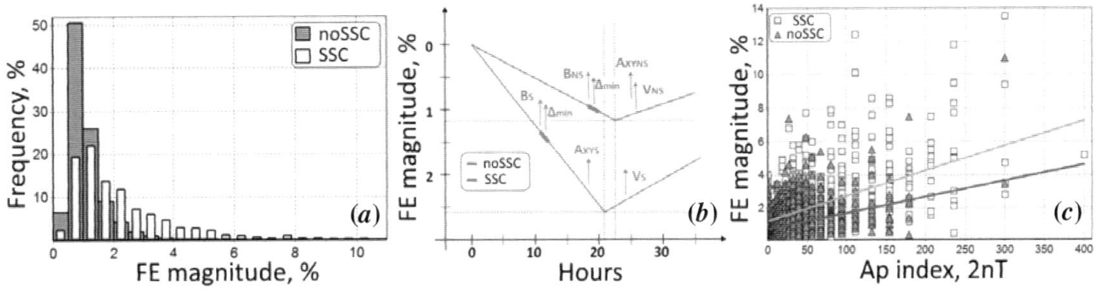

Figure 1. Distributions of FEs magnitude in groups with SSC and without SSC (*a*); a schematic representation of the average FEs in the groups *S* and *NS* (*b*); the dependence of the FEs magnitude on the Ap (*c*).

3.2. Relationship between Forbush-effect parameters and the heliolongitude of solar sources

Forbush-effects from the FEID database confidently identified with the corresponding solar source were analyzed (334 events) in [9]. All events were divided into five sectors.

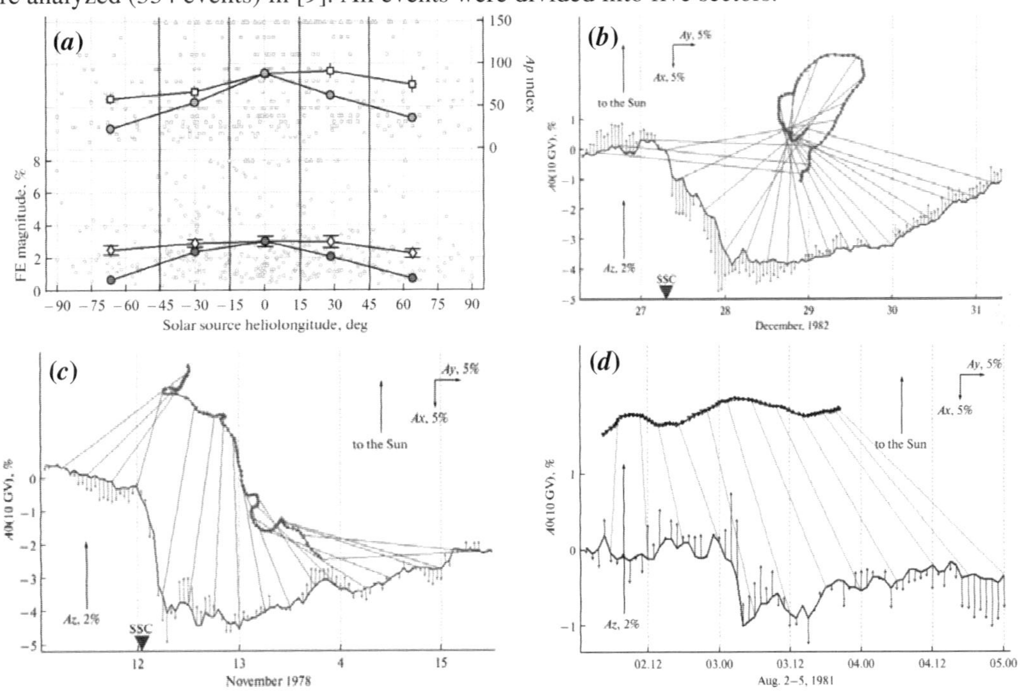

Figure 2. Dependence of the FE magnitude and geomagnetic activity (Ap) on the solar source longitude (□ and ◊ – observable results; o – calculated with accounting of the full number of events) (*a*); behaviour of the cosmic ray density and vector (daily) anisotropy during FEs caused by: a eastern source (*b*), the source from the region of central heliolongitude (*c*), the source from the western group (*d*).

It was shown that solar sources with different heliolongitudes have different effectiveness of creating Forbush-effects and geomagnetic storms (figure 2*a*). In addition, it was found that, the behavior of the anisotropy of galactic cosmic rays is significantly different, depending on the longitude of the solar source (figure 2*b-d*). Thus, having information only about galactic cosmic rays and being able to decipher it correctly, it is possible to determine a set of parameters of interplanetary disturbances.

4. Conclusion

Analysis of a large number of events allows us to obtain statistical relationships between different parameters characterizing the state of space weather. This makes it possible to estimate with a certain degree of probability the effect of space weather on space and ground-based objects. The database of transient phenomena in cosmic rays and the interplanetary medium created by IZMIRAN stuff is not the only development of this group of scientists. There are a number of databases that cover a large time interval and contain many parameters: geomagnetic disturbances (1868-2018, 103 parameter), solar activity (1975-2018, 64 parameters), cosmic ray variations (1957-2018, 45 parameters), coronal mass ejections (1996-2018, 25 parameters), coronal holes, etc.

Using these databases, statistical regularities have been obtained, that form the basis for the developed models for the realization of short, medium and long-term forecasts for each of the space weather directions. Continuous updating of information and timely recalculation of dependencies between parameters makes it possible to always keep databases and models up-to-date. And the the result of such analysis is used in the daily work of the IZMIRAN Space Weather Prediction Center to provide its customers with the necessary information for 20 years (since 1998).

Acknowledgments

The work was performed using the instruments of UNU "SCL Network". The work was partially supported by RFBR grant № 17-02-00508 and RSF №15-12-20001. We are also grateful to all the staff of the cosmic ray station network http://cr0.izmiran.ru/ThankYou.

References

[1] Gaidash S P, Belov A V, Abunina M A and Abunin A A 2017 *Geomagnetism and Aeronomy* **57** 869
[2] Belov A V, Dorman L I, Eroshenko E A, Levitin A E, Yanke V G, Villoresi J, Parisi M, Jucci N, Ptitsyna N G, Tyasto M I and Chizhenko V A 2004 *Geomagnetism and Aeronomy* **44** 461
[3] Zeleniy L M and Petrukovich A A 2015 *Priroda* **9** 31
[4] Eroshenko E, Belov A, Boteler D, Gaidash S, Lobkov S, Pirjola R and Trichtchenko L 2010 *Adv. Space Res.* **46** 1102
[5] Belov A V, Eroshenko E A and Yanke V G 1997 *Proc. 25th Int. Cosmic Ray Conf.* **1** 437
[6] Belov A V, Eroshenko E A and Yanke V G 1999 *Proc. 26th Int. Cosmic Ray Conf.* **6** 472
[7] Belov A, Abunin A, Eroshenko E, Abunina M, Oleneva V and Yanke V 2017 *VarSITI Newsletter* **14** 8
[8] Abunin A A, Abunina M A, Belov A V, Eroshenko E A, Oleneva V A and Yanke V G 2012 *Geomagnetism and Aeronomy* **52** 292
[9] Abunina M A, Abunin A A, Belov A V, Eroshenko E A, Asipenka A S, Oleneva V A and Yanke V G 2013 *Geomagnetism and Aeronomy* **53** 10

Long-term stability of the neutron monitors global network for overall monitoring period

A V Belov, R T Gushchina, E A Eroshenko, A A Abunin, P G Kobelev, V G Yanke

Pushkov Institute of Terrestrial Magnetism, Ionosphere and Radio Wave Propagation, RAS, ИЗМИРАН, Moscow, RF, 108840.

E mail: yanke@izmiran.ru

Abstract. Three independent methods for estimating the long-term stability of detectors of the world network of neutron monitors and muon telescopes are proposed. Conditions that determine the long-term stability of the detectors.

1. Introduction

To research physical processes in the interplanetary medium and to determine the structure of the heliosphere as a whole, such a powerful ground instrument as the world network of neutron monitors and muon telescopes has been used for six decades. To obtain the long-period characteristics of the interplanetary medium, the temporal stability of the functioning of the network as a whole and of its individual elements is extremely important, and estimating the quality of the data of each station in the network is the task of this article. Such a task was posed and earlier [1-4].

Several new aspects were taken into account in this paper: a) the data series under consideration was supplemented with data from the last decade, b) the analysis of the spectrum of variations was carried out for the 2009 database, and this allowed to increase the accuracy for the determined spectrum parameters, c) to solve the problem, new **ltvdb** database was developed d) the changes of the rigidity of geomagnetic cutoff were taken into account [5-6] for a network of cosmic ray stations for the period under review.

In order to quantify the quality of the device data, it is necessary to have a standard. Often an unreasonable approach is used, when a "reliably" working station is adopted as this standard. A number of stations really have "reliably working" status, but they are difficult to be used as a standard, since each station registers cosmic rays in its own range of cosmic ray spectrum.

In this paper, the model of cosmic rays variations developed in [7] is used as a standard. Naturally, the discrepancy with the model for each station is associated with the data quality of this station. Of course, this approach has its drawbacks, since the construction of an adequate model itself is a difficult task. But this problem can be solved by the method of successive approximations if, with appropriate modifications of the model, it is possible to describe the variations of cosmic rays in the heliosphere quite well over a sixty-year observation period.

Along with the model method, another independent method for estimating station stability is developed, which can be called as “the group” method. In this method, all stations are divided into several groups. When forming groups, it is necessary to satisfy rather contradictory requirements: in each group, the variations should be minimally different and the number of detectors should be as large as possible (at least three). Several approaches are possible here.

In the case of the method of groups/averages the average (or median) values of the variations in each group are considered as standard values. The method of groups/means (in contrast to the method of groups/medians) allows us to estimate the errors of the found efficiencies. Possible variations in this method are not excluded, and this is an its drawback. Another option, the group/relationship method is implemented by a much more complex algorithm, but the primary variations are automatically eliminated, which is its advantage. With the help of the developed mathematical algorithm, a given set of stations is divided into "reliable" and "unreliable" operating stations. The criterion of "reliability" is the confidence interval within which observed counting rates (which are freed from the primary variations) are changing. In this case, a group of stations that has the same variations and is defined as "reliably" operating is used as a standard, and they should be in the majority. This method for each station makes it possible to determine the efficiency of the detector, and, what is even more important, the accuracy of its evaluation.

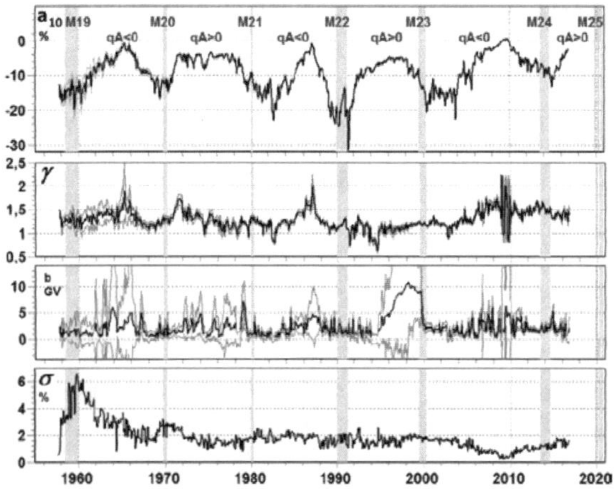

Figure 1. Parameters and their errors (red) of the cosmic rays variation spectrum: a_{10}, γ, b and the model discrepancy is shown at the bottom.

Each method has its advantages and limitations, but on their joint basis it is possible to obtain reliable quantitative estimates of the long-term stability. Current analysis was based on the data from the world network of neutron monitors, muon telescopes and stratospheric sounding stations, which can be found in the database of the monthly-averaged resolution data [8].

2. The principle of continuous evaluation of the stability of detectors

Let us figure out the dependence of the relative efficiency of the detector with the observed variations of the counting rate and with the variations of the standard detector. The efficiency of the counter ε is the probability of registering the particle, which penetrates the working volume of detector. In order to move from the observed counting rate N to the true counting rate, it is necessary to calculate N/ε. Therefore, the efficiency of the detector can also be defined as the number by which the observed counting rate needs to be divided in order to get rid of the variations associated with the internal changes of the detector itself or with the electronic path.

Therefore, proceeding from the definition of variations as $v = N/N_B - 1$, the variation of the measured counting rate of the detector N, taking into account its effectiveness ε, can be written as: $v = \dfrac{N/\varepsilon}{N_B} - 1$, where N_B is the detector's counting rate in the base period. Comparing the variations v in the counting rate of the detector with somehow chosen "standard" variations v_S, v, and taking into account detector's efficiency, we get

$$\frac{N/\varepsilon}{N_B} - 1 = v_S \quad \text{and} \qquad \varepsilon = \frac{v+1}{v_S+1}, \tag{1}$$

i.e effectiveness is determined only by observable v and "standard" v_S variations and this underlies of continuous evaluation of effectiveness. Starting from (3) and from the law of addition of random errors, the error of the found efficiency is

$$\sigma_\varepsilon(stat) = \frac{1}{(v_S+1)^2}\sqrt{(v_S+1)^2\sigma_v^2 + (v+1)^2\sigma_{v_S}^2} \tag{2}$$

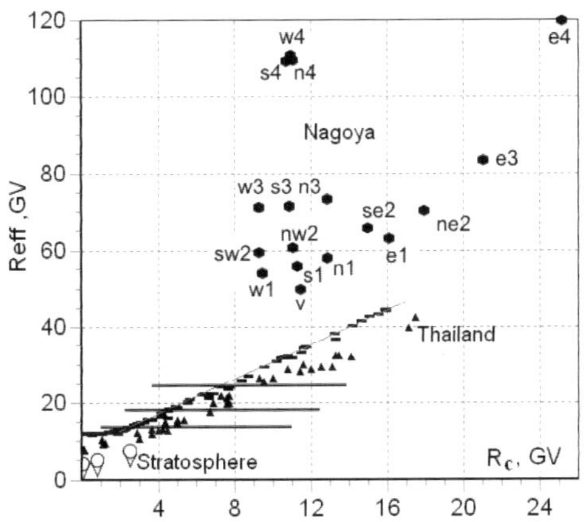

Figure 2. Effective rigidity for various detectors - neutron monitors (triangles), a muon telescope (Nagoya) and a telescope of charged particles of stratospheric sounding. Horizontal risks are the boundaries of groups.

where σ_v and σ_{v_S} are the statistical errors of the experimentally and of the "standard" detector determined variations. The statistical errors σ_v are

$$\sigma_v = N_B^{-2}\sqrt{N_B^2\sigma_N^2 + N^2\sigma_{N_B}^2}\,. \tag{3}$$

Taking into account that $\sigma_N^2 = N$ and $\sigma_{N_B}^2 = N_B/n$, where n is the number of intervals for the base value averaging (in our case n = 12), for the error of the experimentally found variations, we obtain

$$\sigma_v = \frac{(v+1)}{\sqrt{N}}\sqrt{1+\frac{v+1}{n}}\,. \tag{4}$$

The statistical errors of the calculated variations for the monthly-averaged intervals of averaging are negligibly small $\sigma_v \cong 10^{-6}$, statistical errors σ_{v_S} of the variations of the "standard" detector are determined by the method used and are given below.

In addition to statistical performance errors $\sigma_\varepsilon(stat)$, it is necessary to take into account difficult-to-evaluate systematic errors $\sigma_\varepsilon(sys)$, which may be significant.

The greatest contribution to systematic errors for the model method is, of course, due to the "incorrect" shape of the model spectrum used. This error can be only found by carrying out calculations for various forms of spectra, but it can be roughly estimated, assuming in the three-parameter spectrum the parameter b≡0, i.e. assuming the spectrum is purely power-law. The upper bound for the systematic error yields σ_{sys} = 0.005. Systematic errors are large enough for the group method. To ensure an acceptable number of stations in the group, it is necessary to form them from stations in a wide range of rigidities; which give a decisive contribution to systematic errors . For the groups under consideration, the maximum value is σ_{sys} =0.017.

Taking into account all sources of errors, efficiency is defined as affectivity=$\varepsilon \pm \sigma_{stat} \pm \sigma_{sys}$. As will be shown below, the total error is $\cong \pm 0.02$.

3. The model method

In [7], a spectrum of long-term variations of cosmic rays was determined using a specially improved for the researching of long-term variations variant of the global survey method [9]. The analysis was carried out on the database of monthly-averaged data of neutron monitors (45 stations), stratospheric detectors (3 stations) and the multidirectional muon telescope Nagoya. The observed variation $\delta N^i/N^i$ is $v^i_{\rm model} = \int_{R_c^i}^{\infty} W^i(R, R_c^i, h_0^i)\cdot\frac{\delta J}{J}(R)dR = a_{10}C_0^i$, where $\delta J/J(R)$ is the spectrum of isotropic variations, C_0^i is the reception coefficient of the zero harmonic for the station under consideration. The coupling functions $W^i(R, R_c^i, h_0^i)$ for the neutron component are taken from [10], for the muon component from [11] and for the stratospheric data from [12], here R_c^i is the geomagnetic threshold, h_0^i is the level of observation. In the model, the spectrum of variations is given in a three-parameter form and presented as: $\delta J/J(R) = a_1/(b+R)^\gamma$. Parameters changing area is $\gamma = 1.2 \div 2$

and $b = 0 \div 4$ GV. In [7] is shown that the three-parameter approximation of the spectrum of variations of galactic cosmic rays is suitable for describing the spectrum of long-term variations in the range of 5-50 GV. All variations are determined relative to the base period of 2009. Figure 1 shows the parameters found for the spectrum of variations of galactic cosmic rays a_{10}, b and γ. The lower panel shows the standard errors of the experimental data and model, which make it possible to estimate the adequacy of the used model of cosmic ray variations. In this case, the expected error of the variations of the "standard" detector σ_{v_S} in expression (4) defined as:

$$\sigma_{v_S} = \sqrt{\left(\frac{\partial v}{\partial a_{10}}\right)^2 \sigma_{a_{10}}^2 + \left(\frac{\partial v}{\partial \gamma}\right)^2 \sigma_{\gamma}^2 + \left(\frac{\partial v}{\partial b}\right)^2 \sigma_b^2} \quad (5)$$

Errors of the variations (5), determined by the model, are summed from three terms, with the terms contributing 0.005%, 0.05%, 5% respectively, i.e. the error is mainly determined by the error of parameter b.

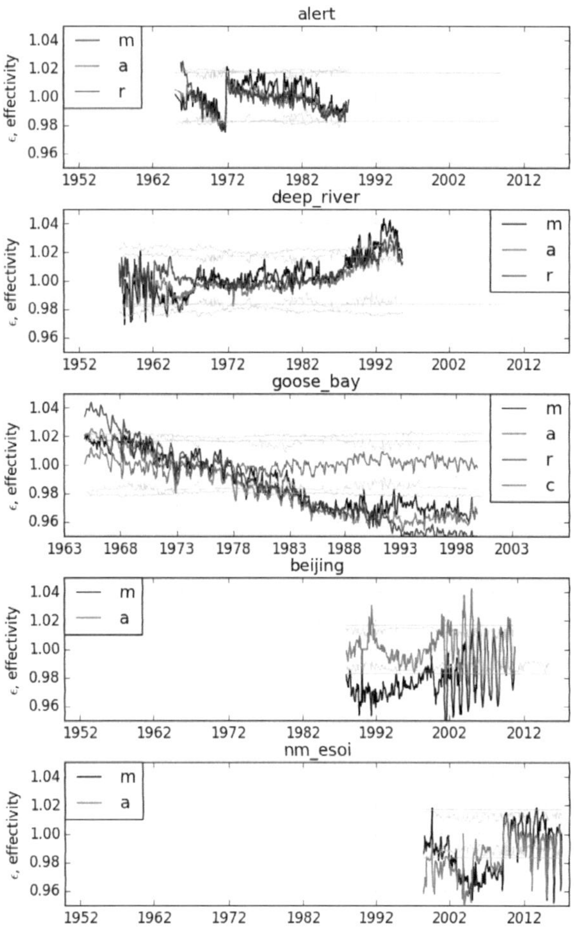

Figure 3. The efficiencies obtained by the model method (m), the group/average method (a) and the group/relation method (r), relative to the 2009 level.

The main drawback of the method is its model dependence, but if the model well describes the variations in the range of rigidity in question, this method makes it possible to estimate the efficiency of all detectors.

4. Method group/medium (median)

All detectors are divided into several groups, within which the variations of different detectors should differ only statistically. The average (or median) values of the variations in each group are considered as a "standard" values. Four groups of stations with effective rigidities are formed within the intervals: [0 ÷ 13), [13 ÷ 18), [18 ÷ 25) and [25 ÷ 50) GV (figure 2). The effective rigidities R_{eff} of all the detectors under consideration are compared in figure 2 (neutron monitors are divided into two groups: near the sea level-rectangles and mountain-triangles).

The fact that the group/medium method does not rely on any model is its merit. The disadvantage of this method is the need to consider groups of stations with close effective rigidities of the detected particles, which is not always possible to accomplish in practice.

5. Group/relationship method

This method was originally developed for the internal quality control of data from cosmic ray detectors. The basis of modern methods of internal control is the division of the detector into the largest possible number of identical and independent elementary detectors. This approach allows us to determine the relative efficiency of each elementary detector, i.e. it allows a continuous quality control of data [13]. The method of relations was adapted to analyze the long-term stability of network station detectors [1].

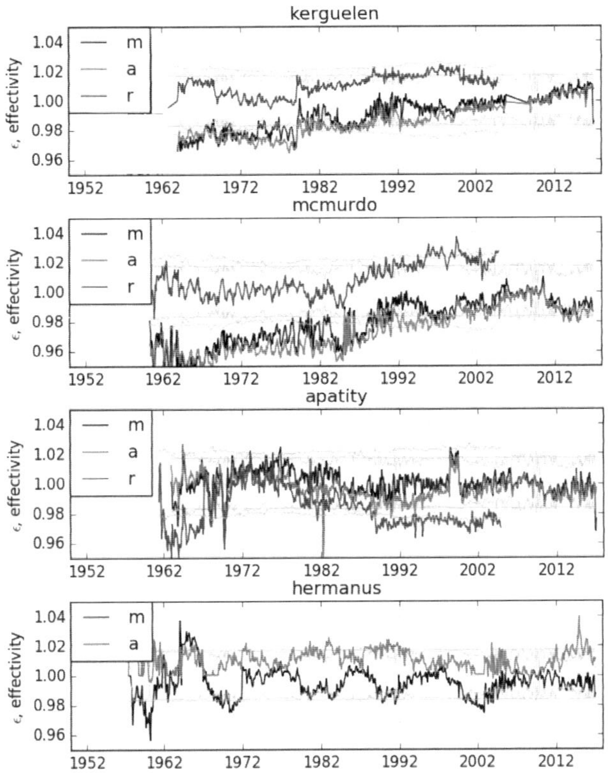

Figure 4. The efficiencies obtained by the model method (m), the group/average method (a) and the group/relation method (r), relative to the 2009 level.

The conditions for the applicability of the group/relationship method for the problem of analyzing the long-term stability of detectors are the same as for the group/average method, i.e. integration of stations into groups of stations with very close energy characteristics.

6. Discussion of the results

In figure 3-4, the efficiency changes for some stations obtained by using all three given methods: model, group/average and group/relations methods. The efficiencies obtained by the group/relations method were determined by us earlier [1]. In addition, estimates of the corridor of their errors were made. For all stations, the results are shown on the website [8].

The coincidence in the details of the efficiencies found by three independent methods is evidence in favor of all methods and their correct application.

An efficiency analysis shows that a constant drift of about 0.1 %/year is observed at many stations for a sufficiently long or even entire period of observation (figure 3). The largest and constant drift, around -0.4 %/year, is observed for the data of the station Goose Bay. Here is also shown the data corrected for the drift (curve c).

For some stations (figure 3, Beijing (and also Tibet, which is not listed here), an anomalously large annual temperature effect (2-4%) is observed. An insignificant part of this effect, <0.1%, is due to the actual temperature effect of the neutron component, while most of the temperature changes are due to local temperature changes affecting the elements of the electron path, although the Beijing station has quite good long-term stability. In the initial period of observations, the Beijing data were corrected for a local temperature effect.

ESOI station also has a large annual wave, but it is already connected with wet snow during the winter. In the initial observation period, the ESOI data were also corrected for the snow effect.

In figure 3 and figure 4 the moment of changing the IGY to NM64 (Deep River, Apatity) is clearly visible in the initial period of their work.

In general, changes in efficiency are sporadic in nature, apparently related to the human factor. Instrumental variations (or drifts) can be classified into: 1) daily and seasonal, associated with temperature changes; 2) long-term, associated with changing of the properties of sensors; 3) and sporadic instrumental variations.

If periodic variations are easily identified and corrected, it is extremely difficult to do this for long-term instrumental drift. The biggest error in the data appears from the pressure drift (up to 0.1 %/year). The use of a constant value of barometric coefficient leads to a false 11-year wave with an amplitude more than 0.1%.

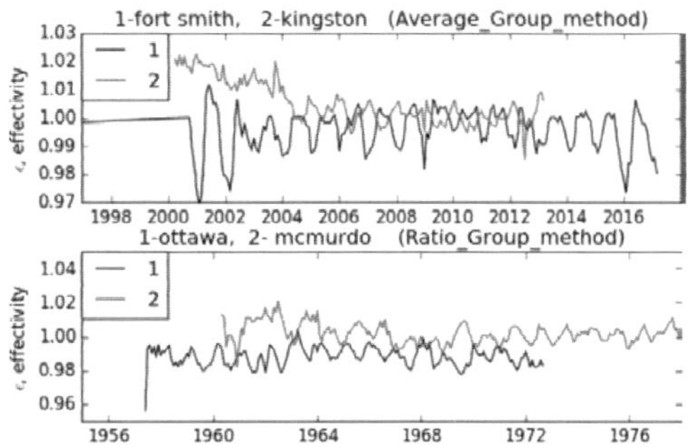

Figure 5. Illustration of the temperature effect of the neutron component by the example of two pairs of detectors.

Large sporadic changes in efficacy may be caused by at least two reasons. First reason is a leakage of charge (micro breakdowns) through the high voltage circuit. Another reason is the inadequate stability of high-voltage power supplies. In addition, for some mountain stations, the effect of snow is very important. This effect can lead to a complete distortion of the variations.

For high-latitude stations (figure 5), an annual wave with an amplitude of about 1%, opposite in phase in the northern and southern hemispheres (Fort Smith and Kingston, Ottawa and McMurdo) is clearly observed [14]. The annual wave is due to the temperature effect of the neutron component and is negligible for equatorial stations. This indicates a good sensitivity of the technique used. All the results obtained in this paper in digital and graphical form could be found in the resource [8].

7. Conclusion

1. Coincidence of the found efficiencies in details indicates the applicability of all methods (model, group/average method and group/relations method). In addition, our model of variations describes well the modulation of cosmic rays over the entire sixty-year observation period.
2. For "good" stations (about 10), the drift can reach 0.04 %/year, which is an order of magnitude less than the possible long-period variations ~ 1%/year.
3. The number of stations that operating for several decades and having stability more than 2% for the entire observation period is about 30. However, many of them are still characterized by sporadic changes in efficiency. The characteristic drift is about 0.1%/year.
4. For many stations (about 40), the drift of data is generally in the background; the quality of the data is determined by numerous sporadic changes.
5. An annual temperature wave with an amplitude of about 1% is viewed which is in antiphase in N/S.
6. For individual stations (ESOI), the snow effect is clearly visible; for other detectors (Beijing and Tibet.) a large local temperature effect is observed.

Acknowledgments

This work was partially supported by the program of the Presidium of RAS №3, RFBR 17-02-00508 and РНФ 19-72-20085, support the project USU "Russian network of ground stations of cosmic rays". We are grateful to all the staff of the World Network of CR stations http://cr0.izmiran.ru/ThankYou.

References

[1] Belov A, Gushchina R, Eroshenko E and Yanke V 2005 *Proc. 29th ICRC (Pune)* sh034 **2** 239
[2] Belov A V, Gushchina R T, Eroshenko E A et al 2007 *Geomagnetism and Aeronomy* **47** 251
[3] Asipenka A, Belov A, Gushchina R, Eroshenko E and Yanke V, 2009 *Proc. of the 31st ICRC* (Lodz)
[4] Belov A, Gushchina R and Eroshenko E et al 2017 *Proc. 35th ICRC* (Korea) PoS046
[5] Gvozdevskii B B, Belov A V and Yanke V G et al 2016 *Geomagnetism and Aeronomy* **56** 381
[6] Gvozdevsky B B, Belov A V, Gushchina R T et al 2017 *Nuclear Physics and Engineering* **8** 19
[7] Gushchina R T, Belov A V and Yanke V G 2013 *Bull. Russ. Acad. Sci.: Phys.* **77** 513
[8] LongTimeCRVariation ePub 2018 ftp://crsb.izmiran.ru/LongTimeCRVariation
[9] Belov A, Eroshenko E and Yanke V et al *2018 Solar Physics* **293** 23
[10] Aleksanyan T M, Belov A V, Yanke V G et al 1982 *Izv. RAS Ser. phys.* **46** 1689
[11] Fujimoto K, Murakami K, Kondo I and Nagashima K 1977 *Proc 15-th ICRC (Plovdiv)* **4** 321
[12] Stozhkov Yu I 1980 Doctoral dissertation FIAN
[13] Belov A V, Blokh Ya L, Klepach and E G, Yanke V G 1988 *Cosmic Rays (Moscow)* **25** 113
[14] Belov A, Dorman L, Gushchina R and Yanke V 1995 *Proc. 24th ICRC (Rome)* **4** 1141

IOP Conf. Series: Journal of Physics: Conf. Series **1181** (2019) 012064 doi:10.1088/1742-6596/1181/1/012064

The hardware function of the URAGAN muon hodoscope

Dmitry Peregoudov[1], Ivan Astapov[2], Natalia Barbashina[2], Alexei Gvishiani[1], Victor Getmanov[1], Anna Dmitrieva[2], Michael Dobrovolsky[1], Roman Sidorov[1], Anatoly Soloviev[1], Victor Shutenko[2] and Igor Yashin[2]

[1] Geophysical center of the Russian Academy of Sciences, Molodezhnaya St. 3, 119296 Moscow, Russia

[2] National Research Nuclear University MEPhI (Moscow Engineering Physics Institute), Kashirskoe shosse 31, 115409 Moscow, Russia

E-mail: d.peregoudov@gcras.ru

Abstract. The problem of the hardware response function calculation for the URAGAN hodoscope (MEPhI) is analyzed. The simplified hodoscope model is developed for which the response function is found out analytically. It is shown that the presence of non-detecting intervals (gaps between detectors) leads to the response function that includes azimuthal dependence. A numerical procedure of the response function calculation is developed for the URAGAN hodoscope. The result is in correlation with the averaged matrix data from the hodoscope. The application of response function analyses to detecting of reduced muon flux areas is demonstrated. Further refinement of response function calculation is discussed.

1. Introduction

Muon hodoscope is a device which is able to measure both individual muon tracks and, under continuous operation, angular distribution of muon flux in wide solid angle range. The design of hodoscope involves a number of particle detectors located at two or more parallel planes.

Muons are generated in upper atmosphere by primary cosmic rays as a product of secondary pion decay. Thus the observation of muon flux intensity and its angular distribution [1] provides important information about the processes in heliospere, particularly, may serve the basis for magnetic storms forecast. Muon flux is also imprinted by atmospheric processes and muon flux observations may be used for weather forecasts [2]. Recently muon hodoscope was successively used for detection of previously unknown cavity (probably chamber or gallery) in Khufu's Pyramid [3].

An important aspect of hodoscope data interpretation is a correct account for the distortions introduced by the device itself, so called hardware response function. It may be defined as angular distribution which would be measured in the case of purely isotropic muon flux. The direct calibration of hodoscope seems impossible since we don't possess any standard muon source which generates isotropic muon flux and is strong enough to accumulate sufficient statistical data. The natural muon flux is approximately isotropic over azimuth but has strong dependence on zenith angle, moreover it is subject to diurnal variations and the influence of atmosphere. Thus theoretical estimations of responce function are of great importance. In the present paper we consider theoretical estimation of response function for URAGAN muon hodoscope.

2. URAGAN hodoscope design

The detecting unit of URAGAN muon hodoscope [4, 5] is gas-discharge chamber of square cross-section (8.95x8.95 mm inner and 9.95x9.95 outer) and 3.5 m length. The inner surface of the chamber is covered with graphite, anod wire is spanned along the axis. The chambers are located in 8 parallel planes, 320 chambers in each plane, the coordinates of chamber's axes (in mm) equal to

$$x_{ij} = 9.95 \cdot i + 3.15 \cdot [i/8] + 4.47 \cdot [i/16], \quad i = 0, \dots, 319,$$
$$z_{ij} = 65.5 \cdot j, \quad j = 0, \dots, 7.$$

([...] denotes integer part, z axis is vertical, y axis points along the chamber length.) Thus there are non-detecting gaps after each 8-chambers set, the widths of odd gaps are 3.15 mm, the widths of even ones are 7.62 mm.

To prevent anod wire from sagging the so-called holders are used: 6 plastic plugs of 5 mm wide distributed homogeneously along the chamber length, coordinates of holder's centers (in mm) equal to

$$y_k = 525 + 490 \cdot k, \quad k = 0, \dots, 5.$$

The set of all holders in a given plane form 6 additional non-detecting gaps perpendicular to the gaps between chambers.

The chambers operate in streamer mode. The signal is inducted on so-called strips: aluminium bands located over and below each chambers plane. The strips parallel to chambers are called X-strips, their number is equal to chamber's number (320), their width is equal to chamber's width and they are located exactly over the chambers. The strips perpendicular to chambers are called Y-strips, their number equals 288, width equals 12 mm and their center's coordinates (in mm) equal to

$$y_m = 12 \cdot m + 3.05 \cdot [m/16], \quad m = 0, \dots, 287.$$

The location of Y-strips is not correlated with location of holders. Several X- and Y-strips in the same plane may be activated under oblique muon incidence.

The design of URAGAN hodoscope has two main features: discrete detectors of rectangular shape and irregular detector location with the presence of non-detecting gaps.

The raw hodoscope data is the set of activated strips. However the main purpose is to measure muon flux angular distribution so the raw data is converted first to muon track (the straight line that best fits the activated strips) and then to muon matrix: the 76 rows by 90 columns table with the number of muon's tracks in given 1° zenith and 4° azimuth angle intervals in each cell.

3. Averaged muon matrix features

Muon matrices averaged over extended time interval (hour and longer) exhibit, apart from zenith dependence, also weak azimuth variations. On figure 1 90° and 180° periodical azimuthal variations as well as rib-like ornament are clearly visible. In what follows we investigate the influence of two main hodoscope design features, the discreteness and rectabgular shape of detectors and the presence of non-detecting gaps on measured hodoscope data. For the purposes of such investigation the exact number of planes and vertical thickness of detectors as well as total horizontal dimensions of the device are in our opinion of secondary importance. Thus we consider two simple models in both detectors being located in two infinite parallel planes and have zero vertical thickness. In these simplified models, contrary to original design, both modelling of activated strips and muon track reconstruction are very simple since muon is able

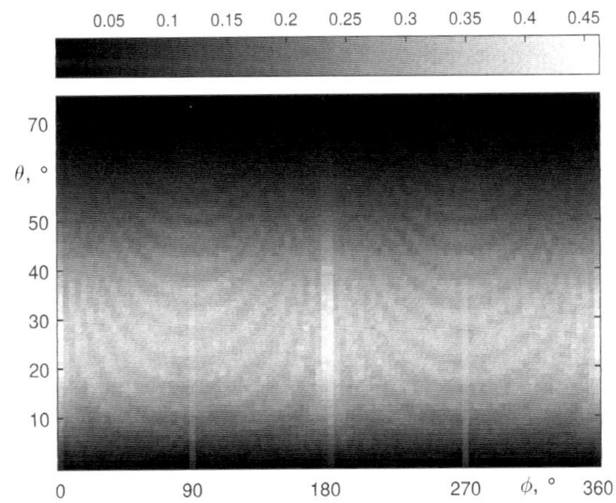

Figure 1. URAGAN hodoscope 1-hour muon matrix 22.06.2015 06:00–07:00.

to activate only one X- and one Y-strip in each plane. We assume detectors are rectangular with p and q dimensions, let the distance between the planes equals l. The first model assumes that detectors densely fill the plane, the second one assumes that periodical non-detecting gaps both in x- and y-directions occur. We find that neither rectangular detector's shape (in the case of high angular resolution, $l \gg p, q$) no rectangular shape of the device itself lead to azimuthal dependence. It is found that azimuthal dependence of hodoscope response function is due to presence of non-detecting gaps. We also consider numerical simulation of URAGAN response function completely taking into account all its design features. The result of this simulation is in accordance with measured hodoscope data. We discuss the possibilities of further refinement of response function evaluation.

4. The first simple model: two planes densely filled with rectangular detectors

Let two infinite parallel horizontal planes, one at distance l from another, be densely filled with detectors of $p \times q$ size. Let muon flux be isotropic in upper half-space, in other words, let $\cos\theta \sin\theta \, d\phi \, d\theta$ muons with tracks in zenith angle interval from θ to $\theta + d\theta$ and azimuth angle interval from ϕ to $\phi + d\phi$ pass the square $dx \, dy$ in one second (factor $\cos\theta$ accounts for oblique incidence, $\sin\theta \, d\phi \, d\theta$ is infinitesimal solid angle). Detectors in each plane may be enumerated with pairs of integer numbers (m, n). The event of muon registration is then described by quartet (i, j, m, n), in which first two numbers point to detector in one plane and next two in the other. Because of homogeneity we may restrict ourselves to events $(0, 0, m, n)$. The calculation of response function is divided into two steps. On the first step we calculate probabilities of events $(0, 0, m, n)$ with different m, n, on the second we reconstruct muon track. The second step is trivial: it is naturally to assume that the track passes through the centers of activated detectors, spherical angles (θ, ϕ) of the track satisfying $pm = l \tan\theta \cos\phi$, $qn = l \tan\theta \sin\phi$. The calculation of probability requires integration of muon flux distribution over the squares of detectors in one and another planes, the integral takes the form

$$P(0,0,m,n) = \frac{1}{pq} \int_0^p dx \int_0^q dy \int_0^p dx' \int_0^q dy' \frac{l^2}{[l^2 + (pm + x - x')^2 + (qn + y - y')^2]^2}.$$

Though the integration can be carried out in elementary functions it is much more illustrative to expand the result in powers of $(p, q)/l$. The reconstructed muon tracks distribution, $N(\theta, \phi) = P(0, 0, (l/p) \tan\theta \cos\phi, (l/q) \tan\theta \sin\phi)$, equals

$$N(\theta, \phi) = \frac{pq \cos^4\theta}{l^2} + \frac{pq \cos^6\theta}{3l^4} [6 \sin^2\theta (p^2 \cos^2\phi + q^2 \sin^2\phi) - (p^2 + q^2)] + \ldots \qquad (1)$$

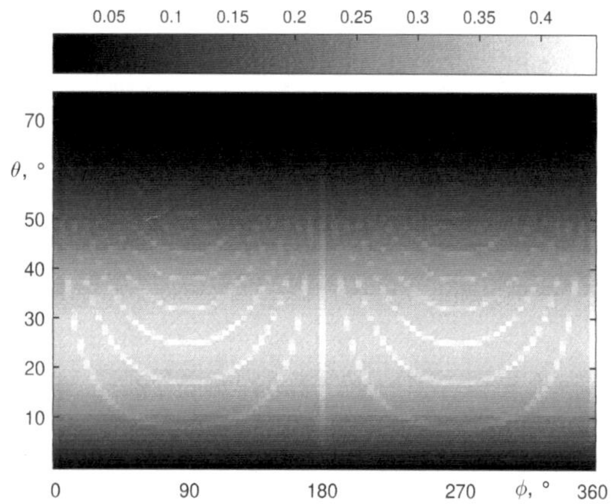

Figure 2. Response function for simplified two-plane model with non-detecting intervals. $A = 8$, $a = 0.8$, $B = 50$, $b = 0.5$, $l = 50$ cm.

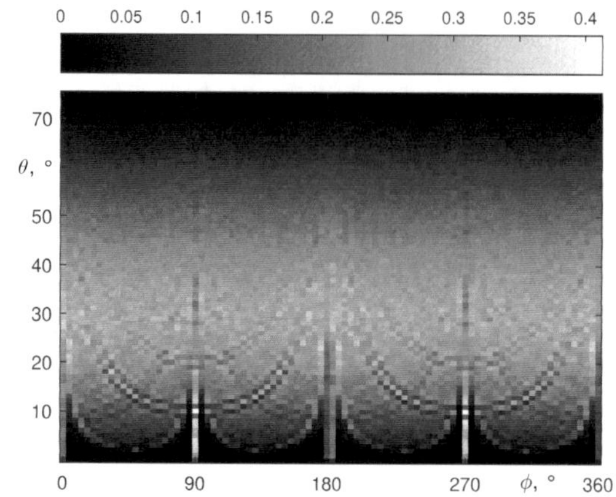

Figure 3. Numerically simulated response function for URAGAN hodoscope. About 10 millions tracks were used in the simulation.

The leading term does not depend on azimuth. As for altered zenith dependence it just comes from the change of variables $(p, q) \to (\theta, \phi)$ with Jacobian equal to $D(p, q)/D(\theta, \phi) = (l^2/pq)\sin\theta/\cos^3\theta$. The next-to-leading term, dependent on azimuth, is of order of magnitude 10^{-3} in comparison with leading term (for high angular resolution, $p, q \sim 1$ cm, $l \sim 50$ cm) and is far beyond hodoscope measuring accuracy.

The dependence on azimuth also comes from account for finite dimensions of device. Muon registered in the upper plane is not necessarily registered in the lower one in the case of oblique incidence. If hodoscope has rectangular shape with dimensions P and Q additional factor $\frac{1}{PQ}(P - p|m|)(Q - q|n|) = \frac{1}{PQ}(P - l\tan\theta|\cos\phi|)(Q - l\tan\theta|\sin\phi|)$ appears in formula for $N(\theta, \phi)$. However this factor is also unable to explain measured azimuthal dependence. We have to look for some other source.

5. The second simple model: periodical non-detecting gaps

Bearing in mind the experience of the first model consideration we do not take into account the discreteness of detectors and finite dimensions of the device while considering the model with non-detecting gaps. We just take the leading term from (1). We assume that non-detecting gaps are located periodically with period A along x axis and period B along y axis, the widths of gaps being a and b correspondingly. Then the presence of the gaps leads to additional factor equal to the ratio of the non-shadowed by gaps square to the full square of $A \times B$ cell. Calculating shadowed square one should keep in mind that the bars of one plane are shifted relatively to the bars of the other by the amounts of $l\tan\theta\cos\phi$ and $l\tan\theta\sin\phi$ along x and y axes correspondingly.

The result of calculation depends on if the bars overlap or not. Let's introduce periodical function (assuming $c > 2$)

$$g(x, c) = \begin{cases} x, & 0 < x \le 1, \\ 1, & 1 < x \le c - 1, \\ c - x, & c - 1 < x \le c. \end{cases}$$

Then response function can be written in the form

$$N(\theta,\phi)=\frac{pq\cos^4\theta}{l^2}\left(1-\frac{a}{A}-\frac{a}{A}g\Big(\frac{l}{a}\tan\theta\cos\phi,\frac{A}{a}\Big)\right)\left(1-\frac{b}{B}-\frac{b}{B}g\Big(\frac{l}{b}\tan\theta\sin\phi,\frac{B}{b}\Big)\right).$$

Figure 2 shows that this function reproduces the features of measured muon matrix. Thus we come to the conclusion that exactly the non-detecting gaps are responsible for azimuthal dependence of muon matrices.

6. Numerical simulation of URAGAN response function

Due to complexity of URAGAN design it is impossible to calculate its response function analytically, however it may be successively simulated numerically [6]. The result of simulation is shown on figure 3. The simulation was carried out as follows.

At the first step random muon track was generated. The track is a line segment with the ends in two planes above and below all detectors. Standard uniform random numbers on $[0,1]$ generator was used to generate four random numbers. The first number was multiplied by 2π and represented ϕ azimuth angle. The second random number F was used to calculate zenith angle $\theta=\arccos F^{1/(1+\alpha)}$. This angle is distributed with density $\cos^\alpha\theta$, $\alpha=0$ corresponds to uniform distribution and is used to simulate response function, $\alpha\approx 2.4$ is used to account for zenith dependence of incident muon flux in order to compare directly with hodoscope data. Two more numbers were multiplied by dimensions of the device and represented x and y coordinates of muon incicdent point in upper plane.

At the second step the segments of track inside detectors were determined. Under the detector we assumed the part of chamber between two neighbouring holders. The set of planes bounding the detectors were introduced: 8 pairs perpendicular to z axis, 320 pairs perpendicular to x axis and 7 pairs perpendicular to y axis. The determination was carried out sequentially: the segments of track between planes perpendicular to z axis were determined, then their subsegments between planes perpendicular to x and finaly subsubsegments between planes perpendicular to y.

At the third step we determine activated strips. We assume that the strip is activated if it overlaps one of the track segments inside detectors in the corresponding plane. The set of activated strips is a simulation of the hodoscope raw data.

The subsequent track reconstruction is performed in accordance with real data processing. The raw sets containing strips in less then 5 planes are omitted. If several X- (and/or Y-) strips in the same plane are activated we assume one X- (and/or Y-) strip is activated with coordinate equal to mean activated strips coordinates value. Least squares method is applied to reconstruct track direction separately in xz and yz planes. Namely, the track projection angle θ_x in xz plane is found by minimization of $\sum_n(x_n\cos\theta_x - z_n\sin\theta_x - a_x)^2$ with respect to fitting parameters θ_x and a_x, x_n and z_n being coordinates of activated X-strips, the same for yz plane. Track projection angles θ_x and θ_y are transformed into standard spherical angles θ and ϕ according to $\tan\theta_x=\tan\theta\cos\phi$, $\tan\theta_y=\tan\theta\sin\phi$. The tracks that don't cross all 8 detectors planes are omitted.

The result of simulation is in qualitative agreement with observed data though quantitative agreement is not that good. Our simulation was purely geometrical and we attribute the deviation to simplified assumption which strips are activated. This step of modelling requires more accurate account for physical processes in the gas-discharge chamber itself as well as the processes of strips signals induction. We hope that further development of these issues will improve quantitative agreement between simulations and observed data.

Figures 4 and 5 demonstrate how the response function can be used to detect the areas with reduced/enlarged muon flux. Typically the variation is very small, 5% and lower, and is shadowed by poor statistics. One can see that data to response function ratio has much better

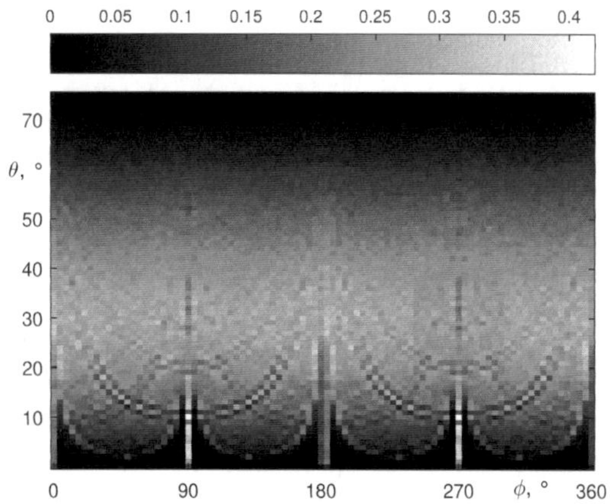

Figure 4. Artifically simulated data with reduced by 5% muon flux in the range $20 < \theta < 40$, $240 < \phi < 300$. The reduced flux area is shadowed by poor statistics.

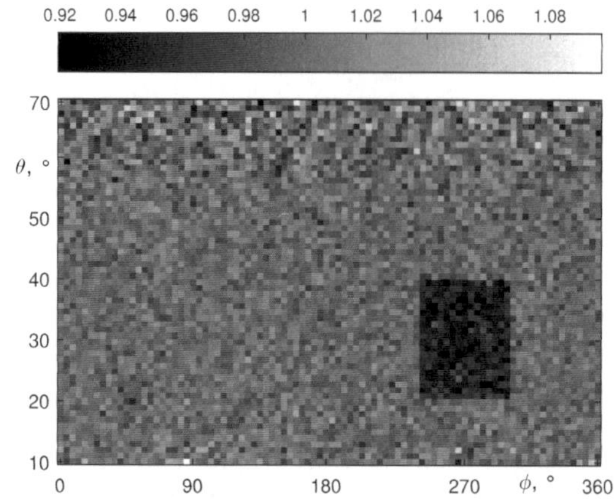

Figure 5. Data to response function ratio. The reduced flux area is clearly seen.

signal to noise ratio then the data itself. Further noise reduction may be achieved by filtering based on local rolling approximation models [7, 8].

7. Conclusion

URAGAN muon hodoscope response function theoretical estimation is considered. Two simplified hodoscope models are proposed for which the response function can be found analytically. These analytical solution demonstrate, on one hand, that azimuthal dependences of hodoscope data can not be attributed to the discreteness and rectangular shape of detectors as well as rectangular shape of the device itself, and, on the other hand, that the presense of non-detecting gaps leads to azimuthal dependence similar to that of observed data. The response function of URAGAN hodoscope is simulated numerically. The result is in agreement with muon matrices averaged over extended time interval. Further refinement of response function simulation requires more accurate account for physical processes in gas-discharge chambers and strips.

Acknowledgement

This study was supported by the Russian Science Foundation (grant no. 17-17-01215).

References

[1] Borog V V 2008 *Osnovy myuonnoy diagnostiki* (Moscow: MEPhI) [in Russian]
[2] Barbashina N S, Dmitrieva A N, Timashenko D V, Shutenko V V and Yashin I I 2008 *Muonnaya diagnostika magnitosfery i atmosfery* (Moscow: MEPhI) [in Russian]
[3] Morishima K *et al* 2017 *Nature* **552** 386
[4] Barbashina N S *et al* 2008 *Instruments and Experimental Techniques* **51** 180
[5] Yashin I I *et al* 2015 *Advances in Space Research* **56** 2693
[6] Kroese D P, Taimre T and Botev Z I 2011 *Handbook of Monte Carlo Methods* (Hoboken, NJ: John Wiley & Sons)
[7] Getmanov V G, Sidorov R V and Dabagyan R A 2015 *Meas. Tech.* **58** 1029
[8] Getmanov V G, Kryanev A V, Borog V V, Sidorov R V, Ulizko M S and Butyrsky Ye Yu 2018 *Physics of Particles and Nuclei Letters* **15** 189

IOP Conf. Series: Journal of Physics: Conf. Series **1181** (2019) 012065 doi:10.1088/1742-6596/1181/1/012065

Electronic configurable neutron monitor for studying of atmospheric shower

Yu.V. Balabin

Polar Geophysical Institute, Apatity, Russia

balabin@pgia.ru

Abstract. A high speed recording system operated on a neutron monitor (NM) is developed at the Laboratory of Cosmic Rays, Polar Geophysical Institute (PGI). The system records the time a neutron arrives to the NM, in a file, the precision of recording being of 1 μs. The number of the tube is also recorded. The data thus contain the exact time and the place of each of billions of pulses arrived. The recording system is placed on a number of stations, such as Apatity, Barentsburg, Moscow, Neitrino (the Northern Caucasus). The system is used to study the phenomenon of multiplicity on NM, as well as other effects of short-time occurrence. In data handling by pulse selection from a given set of tubes, there is a possibility to modernize a conventional NM to get a device of a different design, due to the actual task. That is, it is possible to get a modernized neutron monitor of a specified design without its stop, disassembling and transformation. A number of electronic neutron monitors have been developed to study the multiplicity events. These monitors differed in different distance between the tubes (with 1-15 tubes omissible). The spacing function for multiplicity events has been obtained. The spacing function has been analyzed to show that great values multiplicities are the result of atmospheric hadron showers, while moderate values multiplicities are of a conventional origin. All the pulses entering the events, are checked if there are any disturbances in order to filter the false multiplicity events off the periodic interference.

1. Introduction

A unique high speed recording system installed on a number of neutron monitors (NM) allows carrying out experimental measurements to determine the size of hadron showers arriving to NM. The time of each pulse arrival from NM is recorded by the above system with precision of 1 μs. Remind that one pulse means one neutron, and being registered with the tube, the neutron disappears. Also recorded is the channel (tube) number of NM. The exact and detailed data obtained allow carrying out experiments with the data recorded on NM, the actual construction being preserved unchanged. A conventional 18-NM-64 may be modified into a "virtual" or an electronic device of any design. To do this, it is necessary only to have a special program to select, from the total number of data, the pulses of the given channels and the time of their arrival. For example, to construct an NM based on one tube, of K in number, or based on a pair of neighboring tubes K and (K+1). Any of the given configurations is capable of selecting multiplicity events by the algorithm accepted.

2. Multiplicity study on a conventional NM

A scheme of multiplicity formation on NM is shown in Fig. 1 (left) [1, 2]. An energetic nucleon (of hundreds MeV and GeV) interact with the lead nucleus in the NM shield. As a result of inelastic interaction, either the excited lead nucleus emits some neutrons, or nucleon breaks the nucleus into fragments which also emit neutrons. These are of a much smaller energy if compared with that of a primary particle. They move in lead and polyethylene more slowly, up to thermal energies, and are absorbed by the active volume of the detector (tube). As a result, in the electric section of NM there is a pulse series, with small intervals Δt between them. Each pulse corresponds to one neutron recorded. The notion of infinitesimal here means that $\Delta t << \tau$, i.e., the average interval between NM pulses. Presented on the right is a GEANT-4 modeled interaction between a 10 GeV neutron and NM [3]. The model shows the same result and proves our suggestion. A single particle can't produce multiplicity a event in different (except neighboring) tubes.

It is known from earlier experiments and from the time intervals distribution deriving that the average time interval between pulses on NM τ = 17 μs. In addition, clearly seen within this distribution are the Poisson processes with the characteristic time of $\tau 1$= 110 μs $\tau 2$= 430 μs [4]. We suppose that these are the characteristic times of fast processes similar to multiplicity [5]. Based on the characteristic times determined in the previous experiments, the following criteria have been worked out to select the multiplicity events in a common data body. The algorithm of multiplicity events search (i.e., of an isolated cluster of M pulses) is as follows: 1) a rather great interval Tp before the beginning of multiplicity is present; 2) each subsequent pulse in a multiplicity event occurs not later time T0; 3) the multiplicity event terminates when the next pulse arrives in a period of time greater than T0. Tp = 3000 μs, T0 = 500 μs. According to the algorithm, millions of multiplicity events with M = 5-100 have been detected and selected [6].

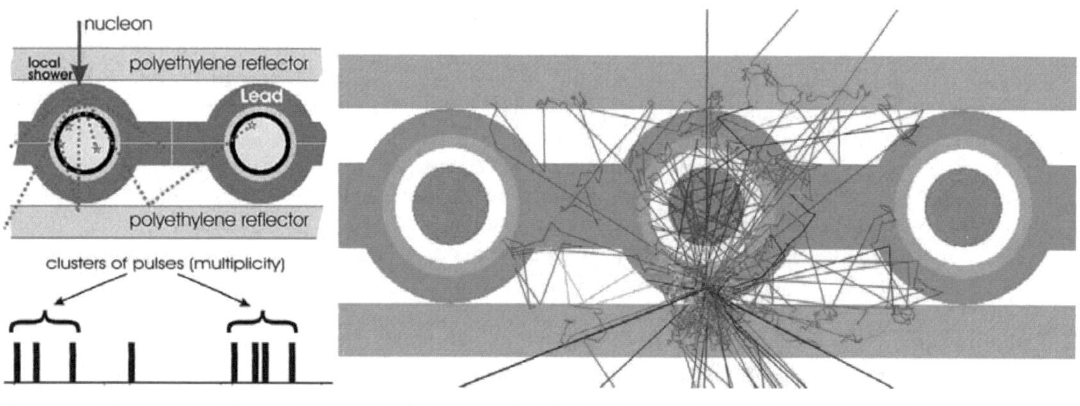

Figure 1. On the left. Here is the principle of multiplicity formation in NM. In interaction between the primary energetic particle and the lead nucleus there appear some neutrons. On the right, one can see a GEANT-4 modeled interaction between the energetic primary particle (10 GeV) and the lead nucleus. In dark blue, green and red lines are the trajectories of fast, moderate and thermal neutrons, correspondingly. In spite of high energy of the primary particle, the secondary neutrons are not able to pass through more than one tube.

We have obtained complete and detailed information on all the multiplicity events from M = 5 to M = 100. The next step in our study is to determine the average time profile of multiplicity events. Multiplicity events of the given M were chosen. Selected among these were the first intervals (the time interval between the first and the second pulses in a multiplicity), followed by the calculation of its average value. This procedure was carried out for the second, the third, etc. intervals. As a result, we obtain the average time profile of multiplicity M. This was the way the time profiles for multiplicities from M = 5 to M = 100 were determined. Figure 2 shows the time profiles of multiplicities M = 15, 25, 40, 60 and 80. The profiles have a common singularity. There is a flat section in the multiplicity.

For example, the flat section contains 15 first pulses for M = 25. Then gradual increase to the end of the event is observed. This is characteristic for all M. It would be noted that any multiplicity M has flat initial section on (M-10) pulses. It is quite important because similar gradual increase of intervals to the end of the event indicates that the process terminating multiplicity is similar in all the events. NM relaxation is supposed to be the result of the effect of a dense of neutron flux [7].

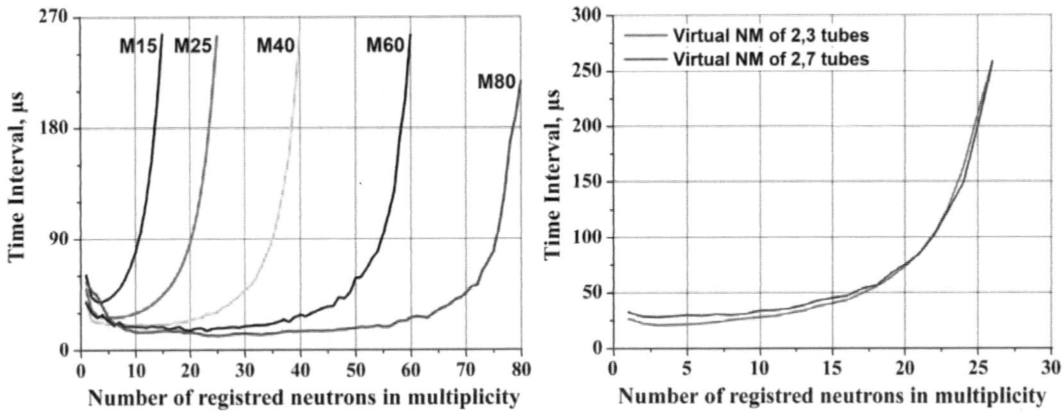

Figure 2. On the left. Multiplicity profiles on Barentsburg station. On the right. Multiplicity M=25 profile on two NMs of different design. NM of 2,3 tubes means that it was used only 2nd and 3rd tubes of conventional NM, NM of 2,7 tubes means that it was used 2nd and 7th tubes of conventional NM. Between 2nd and 7th tubes there are distance of 4 omitted tubes.

3. Neutron monitors of various design

With all the detailed information concerning each pulse (the time of its arrival, with an accuracy of 1 µs and the tube number)) being presented in the data, there is a possibility, in data processing, to additionally stipulate that pulses should be selected only from certain tubes, for example, only from tubes numbered as 2 and 4, as shown in Figure 3 (top, right). In other words, having the NM construction physically unchanged, in data processing it is possible to construct an "electronic" neutron monitor of any design, for example, that consisting of tubes 2 and 17, as shown in Fig.3, (bottom panel). In doing so, the number of tubes may be as great as possible, so it is possible to set any NM-configuration.

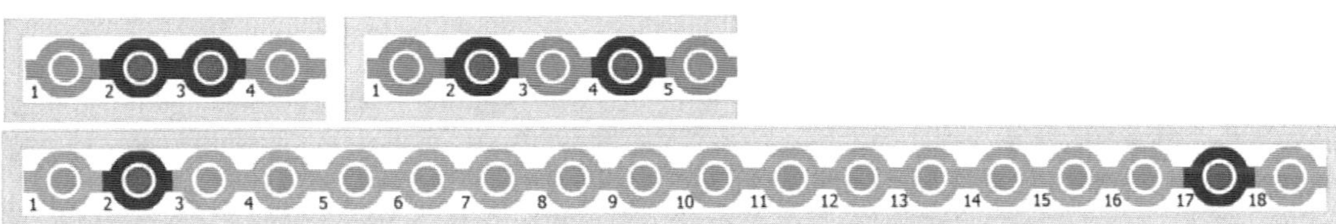

Figure 3. Top, left. An electronic neutron monitor consisting of two neighboring tubes (2 and 3). Operating tubes are in black, omitted (non-operating) tubes are in grey. The tubes are in dark blue, lead rings are in dark, polyethylene is in green. The picture is not in scale actually, lead rings are of 5 cm thick. Top, right: An electronic neutron monitor consisting of two tubes (2 and 4). The omitted tube 3 together with a lead ring is a neutron absorbent and a moderator. The operating tubes here are quite independent. Bottom: An electronic neutron monitor consisting of two tubes (2 and 17). The operating tubes are completely independent. There are no particles capable of passing from tube 2 to tube 17 through lead and the absorbent presented by non-operating tubes.

While studying typical multiplicity events with the help of a number of electronic devices consisting of tube K and tube (K+n) where n changes from 1 to the greatest possible one, it has been shown that there are two types of multiplicity events. The first type is multiplicity consisting of pulses that have been generated in two neighboring tubes. The second type is multiplicity into which many tubes are involved, including K and (K+n), where n>> 1. Multiplicities of the first type (not above of the neighboring tube) are generated by a nuclear cascade in lead, which is induced by a single energetic particle. Multiplicities of the second type seem to originate from hadron showers in the atmosphere above NM because tens of pulses (neutrons) are generated only during hundreds of μs in tubes designed on the edges of NM. Remember that the average time between pulses is 8-10 μs. Also carried out here were the experiments when tubes were designed in a more complicated way, which resulted in the spacing function. All the combination of the experiments allows making a conclusion concerning the fact that events formed by only one nuclear cascade in lead are recorded only up to the number of multiplicity ~7. If the number of multiplicities is greater, there occurs the effect of a hadron shower in the atmosphere. The size of these showers is determined by the spacing function. GEANT-4 modelled has carried out the same result.

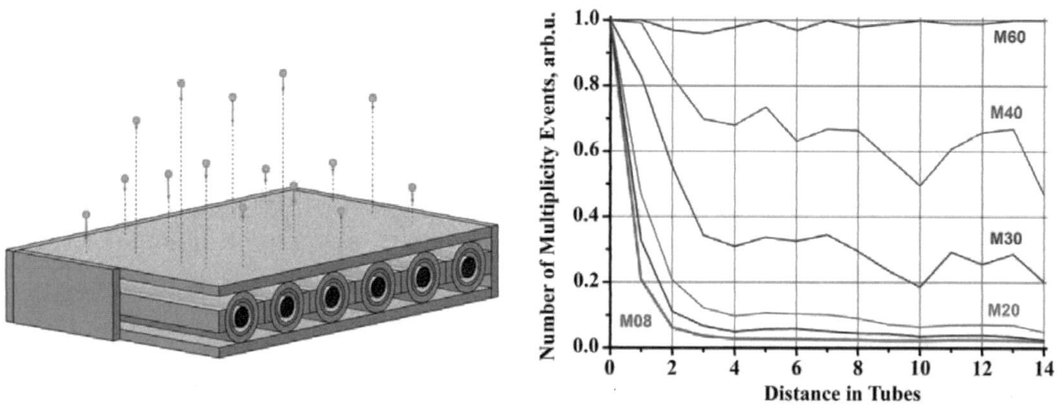

Figure 4. On the left. The sketch of local hadron shower falling on NM. Energetic nucleons (neutrons, mainly) are in green circles. All tubes in NM record neutrons during a short period of time (about hundreds of μs). It would be useful to obtain the spacing function for multiplicity, i.e. to study a number of NM with non-operating tubes. A non-operating tube means that pulses from it are not recorded, microseconds counting is going on until a pulse arrives from the tube marked as an operating one. The spacing function can really show the presence of atmospheric hadron showers. On the right. Multiplicities on electronic neutron monitors of various designs. Shown here is the relative amount of events of the specified multiplicity depending on the distance between tubes (the distance is given in tubes omitted). Multiplicities of M = 8, 15, 20 are actually available only for the electronic NM composed of the neighbouring tubes (tubes 2 and 3). Even one non-operating tube reduces the number of events by a factor of 2-3. The multiplicities of the specified values are practically absent on the electronic NMs with a great number of non-operating tubes. It is when M > 30 that the amount of events becomes considerable, growing and becoming practically uniform with M = 60

Fig. 4 (right) shows that most multiplicity events of M <20 occur only in the neighboring tubes. The distance equal only to one tube (one omitted tube; the NM-construction corresponds to that shown in Fig. 3, top right) reduces the number of events by a factor of 2-3, and the distance of more than two tubes reduces the multiplicity number to the background one. When M = 30, one can observe that a part of multiplicity events is constant and significant, showing no dependence on the distance between active tubes. When M = 60, all the events are uniformly distributed. In other words, the spacing function for M <20 has obviously sharp peak at 0 (two neighboring tubes are used), and at the distance equal to two tubes the spacing function is close to 0. On the contrary, the spacing function for M = 60 is uniformly distributed from 0 to 14 omitted tubes.

Narrow form at the moderate multiplicity means that such multiplicities are produces by a single particle which comes to given tubes. It generates some neutrons only in neighboring tubes. Large multiplicity events are produces by a lot of particles i.e. a particle shower (like it is on Fig.4 left). We can estimate a distance of absolute neutron cumber along NM: it is 2-3 tube distance. There were studied electronic NM of "pair-pair" design. For example, there were used tubes 2,3 (first pair) and 16,17 (second pair) and so on. There is no significant difference from it.

4. Conclusions

The high recording system for a neutron monitor is developed at Polar Geophysical Institute. Due to it short-time phenomena are possible to be studied as multiplicity. It is possible to develop a number of electronic neutron monitors based on the conventional one. The spacing function of electronic neutron monitors is obtained. This function confirms the presence of two types of the multiplicity origin. Moderate multiplicity is produced by a single energetic primary particle. Great multiplicity is produced by the local atmospheric hadron shower. It is worth noting here that the high speedy registration system developed in PGI allows different studies of fast processes occurring on NM and lasting units and tens of μs.

References

[1] Dorman L I 1972 *Meteorological effects of cosmic rays* (Moscow: Izdatel'stvo Nauka) p 212 (in Russian)

[2] Dorman L I 1975 *Experimental and theoretical foundations of cosmic ray astrophysics* (Moscow: Izdatel'stvo Nauka) p 402 (in Russian)

[3] Maurchev E.A., Balabin Yu.V. 2016 *Solar-Terrestrial Physics* **2** 3

[4] Goldansky V I, Kutsenko A V and Podgoretsky M I 1959 *Counting statistics for the registration of nuclear particles* (Moscow: Fizmatgiz) p 412 (in Russian)

[5] Balabin Yu.V., Gvozdevsk B.B., Maurchev E.A., Vashenyuk E.V., Dzhappuev D.D. 2011 *Astrophysics and Space Sciences Transactions* **7** 283

[6] Germanenko A.V., Balabin Yu.V., Vashenyuk E.V., Gvozdevsky 2011 *Astrophysics and Space Science* **7** 471

[7] Yu. V. Balabin, B. B. Gvozdevsky, and A. V. Germanenko 2015 *Bulletin of the Russian Academy of Sciences. Physics* **79** 654

IOP Conf. Series: Journal of Physics: Conf. Series **1181** (2019) 012066 doi:10.1088/1742-6596/1181/1/012066

The temperature effect of the muon component observed on A.I. Kuzmin cosmic ray spectrograph in Yakutsk

V.G. Grigorev[1], P.Yu. Gololobov[1], S.A. Starodubtsev[1], A.S. Zverev[1], M.A. Abunina[2], M.S. Preobrazhensky[2] and V.G. Yanke[2]

[1] Yu. G. Shafer Institute of Cosmophysical Research and Aeronomy of the Siberian Branch of the RAS (ShICRA), 31 Lenin ave., Yakutsk, Russia, 677980

[2] N.V. Pushkov Institute of Terrestrial Magnetism, Ionosphere and Radiowave Propagation of RAS (IZMIRAN), 4 Kaluzhskoye Highway, Troitsk, Moscow, Russia, 108840

E-mail: grig@ikfia.ysn.ru

Abstract. An estimation and accounting of a temperature effect in data of the complex of muon detectors of the Yakutsk cosmic ray spectrograph after A.I. Kuzmin are carried out. The results of determination of altitudinal distribution of the temperature from the database of world network of muon telescopes (MDDB), which are obtained using a model of the atmosphere developed in the National Centers for Environmental Prediction NCEP, are used. It is shown, that the vertical profiles of temperature distribution in the atmosphere that are presented in MDDB and the estimated temperature coefficients are allow qualitatively exclude the temperature effect from the data of registration of the muon component of secondary cosmic rays. The results of observation of the Yakutsk spectrograph and accounting of the temperature effect are presented in the Internet through the link *http://www.ysn.ru/ipm*.

1. The data of the Yakutsk spectrograph of CR and altitude distribution of the temperature

The Yakutsk cosmic ray (CR) spectrograph (61°59′N, 129°41′E) operates in ShICRA SB RAS and is designed to register the CR particles in the huge energy range from 2 to 300 GeV. The spectrograph consists of a neutron monitor 24-NM-64, ionization chamber ASK-1 and complex of 4 underground muon telescopes based on gas-discharge counters GMC-14. The data of detectors of the spectrograph are available through the link *http://www.ysn.ru/ipm/index.html* or the database of muon detectors [*http://crsa.izmiran.ru/phpmyadmin*, *http://cr0.izmiran.ru/gmdnet*]. The ionization chamber (IC) operated since 1953 to 2003. The counter telescopes at levels 0, 7, 20 meters of water equivalent (m w.e.) operated continuously since 1971 to 2003 and, after a pause, since 2009 till nowadays. The telescope at 60 m w.e., which worked since 1993 to 1999, was relocated at the level 40 m w.e. Since 2016 multidirectional scintillation detectors with efficient area of 8 m^2 had been being placed at all levels.

An analysis of the temperature effect requires engaging of data of altitudinal distribution of temperature in the atmosphere. In the work, the results of a model of the atmosphere, that was developed in the USA's National Centers for Environmental Prediction NCEP *http://www.nco.ncep.noaa.gov/pmb/products/gfs*, was used. The model based on all available data of global altitudinal probing (ground based, high-altitude, satellite) for their interpolation. The model allows to get both retrospective and prognostic data of the three-dimensional temperature field. The output of the model is the data on temperature at 17 standard isobaric levels in range 10-1000 mb for 4 times a day (0, 6, 12, 18 UT). In order to get the data with a hourly time resolution the interpolation using cubic splines are done. A detailed description of the method is presented in [1, 2]. By the data of altitudinal distribution of the temperature there are calculated an effective temperature T_{eff}, taking into account a contribution of different levels, an average-mass temperature T_m and an altitude of effective 100 mb level of muon generation H100 which are necessary for estimation of the temperature effect of the muon component.

2. Methods of determining the temperature effect

There are several methods for analysis the temperature effect of the muon component [3]. A universal integral method was developed in the beginning of 1950's [4, 5, 6]. Introduction of a function $W_T^{\mu}(h,\theta)$, which is temperature coefficient density, allows to determine the temperature variations as:

$$\left.\frac{dN}{N}\right|_T = \delta_T = \int_0^{h_0} W_t^{\mu}(h,\theta)\delta T(h)dh,$$

where $\delta T(h)$ is a change of the temperature profile in the atmosphere relative to the base temperature profile. Simultaneously, the method of effective temperature was developed [7], which could be considered as another representation of the integral method:

$$\left.\frac{dN_\mu}{N_\mu}\right|_{Temp} = \int_0^{h_0} W_t^{\mu}\delta T(h)dh = \alpha_T \delta T_{eff}.$$

The method is used for underground detectors.

The average-mass temperature method [8], which is based on definition of average-mass temperature in the atmosphere, is a particular case of the integral method. Since the density of the temperature coefficient for above ground detectors does not significantly vary with the depth of the atmosphere h, then an average value $\overline{\alpha_T}$ can be taken out from the integral sign, i.e.:

$$\left.\frac{dN_\mu}{N_\mu}\right|_{Temp} = \overline{\alpha_T}\int_0^{h_0} \delta T(h)dh = \overline{\alpha_T}\delta T_m,$$

where T_m is the average-mass temperature, which is calculated by the data of the probing or from the experiment.

3. Estimation of the temperature coefficients and their time dependence

The estimation of the temperature coefficients for every level of the muon telescope complex for 2009-2016 and the ionization chamber for 1953-2003 has been done. Except the level 60 m w.e., the calculations are well provided with statistics. In figure 1 (left panel) time dependencies of the temperature coefficients of the ionization chamber and the muon telescopes are presented. In values of the temperature coefficient for the ionization chamber clearly seen an 11-year wave and constant growth during 5 cycles of solar acitivity.

For every considered telescope (right panel) there is also shown a dependence of average values of the temperature coefficient on a zenith angle of registration. As expected, for the above ground telescope, the negative temperature effect rises with the increase of zenith angle.

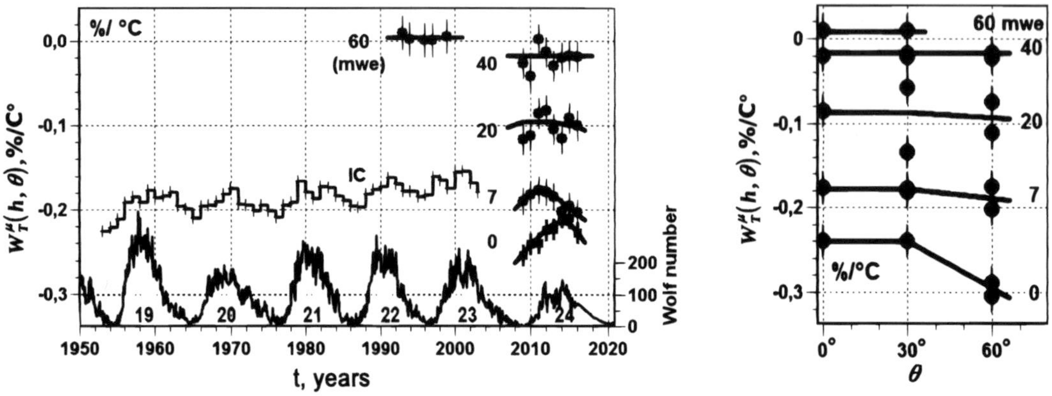

Figure 1. A time dependence of the temperature coefficients of ionization chamber and vertical directions of the muon telescopes (left panel) and their zenith-angular dependence during 2014 (right panel).

4. The results of data correction

As an example, in figure 2 there are shown uncorrected (Ip) and corrected (Ipt) on the temperature effect variations of CR intensity obtained by the data of vertical direction of the ground surface level muon telescope (MT0-ver) in 2010. Preliminarly, the data on Ip and Ipt are gotten rid of an influence of the atmospheric pressure. As seen from the figure, application of the temperature coefficients, that are obtained by the mentioned above methods, allows qualitatively take into account a seasonal change in the observed intensity of the muon component.

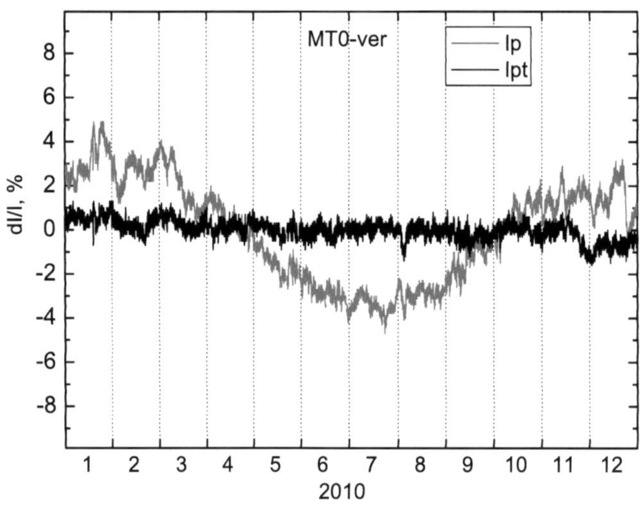

Figure 2. The uncorrected Ip and corrected Ipt on the temperature effect data of the vertical direction the muon telescope at 0 m w.e. (MT0-ver) in 2010.

In order to carry further estimations of the correctness of the accounting of temperature effect, the data of muon telescope MT0-ver are compared to the data of neutron monitor 24-NM-64. In figure 3 there are shown the data of the monitor 24-NM-64, corrected for pressure, and the data of the telescope MT0-ver, corrected for both pressure and temperature, in 2017. With the known relation between the amplitudes of CR variations, that are observed by the devices, and almost complete absence of the temperature effect in neutron component, the both data show good agreeement. It is necessary to note that the application of the data of the world network of muon telescopes that are corrected on the temperature effect will increase the potential of the ground-based measurements [9, 10] for monitoring of the space weather.

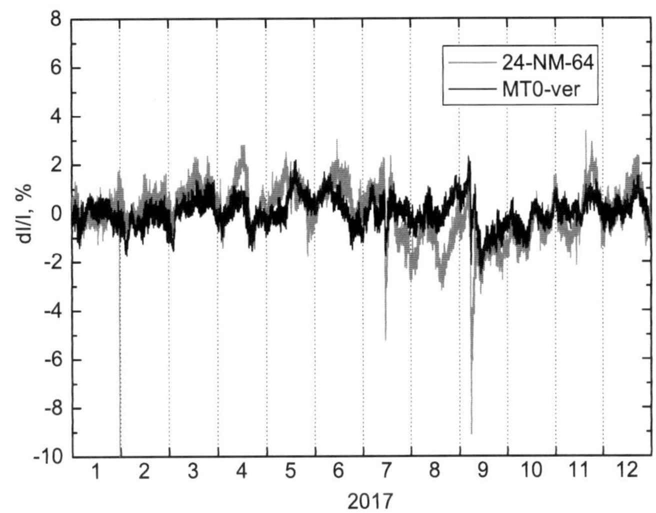

Figure 3. The corrected on pressure data of the neutron monitor (24-NM-64) and corrected on both pressure and temperature effect data of the muon telescope (MT0-ver) in 2017.

5. Results

1. Using the methods of average mass and effective temperatures, there are calculated the temperature coefficients for every level and direction of registration of the muon telescopes and ionization chamber of the Yakutsk spectrograph of CR. Results of the temperature effect corrections are presented in the Internet through the link $http://www.ysn.ru/ipm-T$.

2. The negative temperature effect decreases with increasing of the depth of telescope placement. For the surface level telescope, the rise of the zenith angle of registration is accompanied by growth of the negative temperature effect.

3. The vertical profiles of the distribution of atmospheric temperature and densities of temperature coefficients are allows qualitatively exclude the temperature effect from the data of observation of the Yakutsk CR spectrograph.

4. The value of the temperature coefficient of the ionization chamber reveals stable 11-year wave for 5 solar activity cycles.

6. Acknowledgements

The work was carried under the support of the RFBR grants Nos. $18-42-140002-r\_a$, $18-02-00451-a$ and using the data of Unique scientific installation No. 85 of the Russian national ground level network of cosmic ray stations.

References

[1] Berkova M D, Belov A V, Eroshenko E A *et al.* 2011 *Bull. Russ. Acad. Sci.: Physics* **75** 820–824
[2] Berkova M, Belov A, Eroshenko E *et al.* 2012 *Astrophys. Space Sci. Trans.* **8** 41–44
[3] Osipenko A S, Abunin A A, Berkova M D *et al.* 2015 *Bull. Russ. Acad. Sci.: Physics* **79** 662–666

[4] Dorman L I and Feinberg E L 1956 *Uspekhi Phys. Nauk.* **49** 189–228
[5] Maeda K 1960 *Journal Atm. Ter. Phys.* **19** 184–245
[6] Dorman L I 1972 *Metheorological effects of cosmic rays* (Moscow: Nauka)
[7] Barrett P H, Bollinger L M, Cocconi G *et al.* 1952 *Rev. Mod. Phys.* **24** 133–178
[8] Dvornikov V M, Krestyannikov Y Y and Sergeev A V 1976 *Geom. Aeron.* **16 (5)** 923–925
[9] Grigoryev V, Starodubtsev S and Gololobov P 2016 *Proceedings of Science* **PoS(ICRC2015)** 076
[10] Grigoryev V G, Starodubtsev S A and Gololobov P Y 2017 *Bull. of the RAS: Physics* **81** 200–202

IOP Conf. Series: Journal of Physics: Conf. Series **1181** (2019) 012067 doi:10.1088/1742-6596/1181/1/012067

Telescope Array observatory for the high energy radiation induced by lightning

Takeshi Okuda
Ritsumeikan University, Japan
E-mail: okuda@icrr.u-tokyo.ac.jp

Abstract. The Telescope Array (TA) experiment detects air-showers induced by ultra high energy cosmic rays. The TA ground Surface particle Detector array (TASD) observed some short-duration bursts of air-shower like events. These events are evidently correlated with lightning. Therefore, we have deployed lightning detectors in the vicinity of the TASD. We report the feature of these air-shower like events and the current study combined with these lightning detectors.

1. Introduction

The Telescope Array (TA) experiment detects air-showers induced by ultra high energy cosmic rays, located in midwest in Utah, USA (Figure 1). Some high energy radiation bursts of air-shower like events were observed by the Telescope Array Surface Detector (TASD), and correlated with lightning discharge detected by National Lightning Detection Network (NLDN) [1][2].

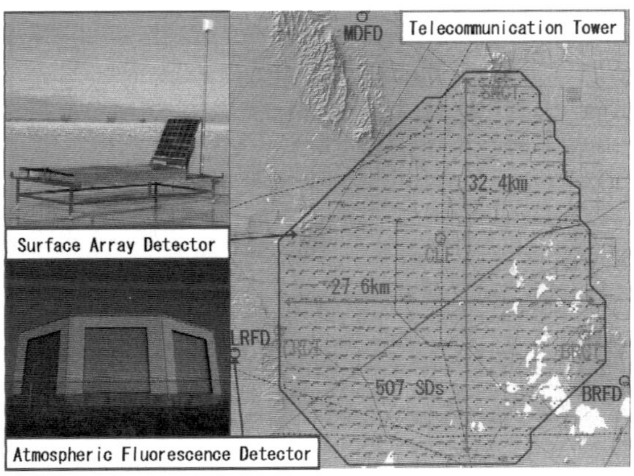

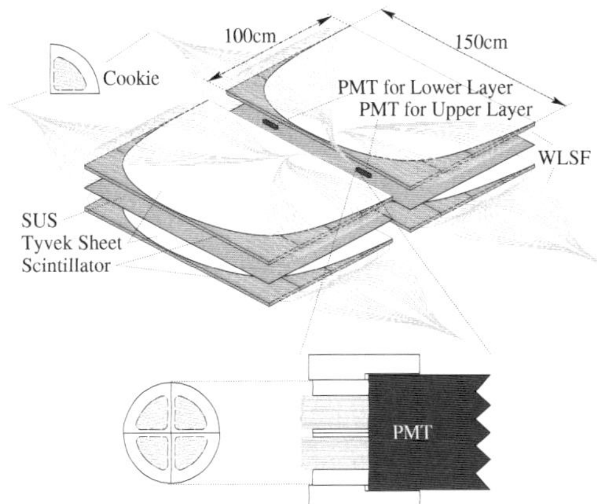

Figure 1. Telescope Array Experiment. Left: 507 ground surface particle detectors (tiny dots above) are surrounded by 3 atmospheric fluorescence telescope stations. Right: Particle detecting part of TASD consists of 2 layers of 3 m^2 plastic scintillators. Each scintillation layer has 1.2 cm thickness and provides information respectively via wavelength-shifting fibers [3].

There are some precedent observations for the relation between high energy radiation and identified lightning discharge. Their coverage area is several km^2 at most, whereas TASD is deployed 680 km^2. The precedent observations measure the energy of radiation and TASD measures spatial distribution of radiation. Therefore, TASD is unique and complementary to observe high energy radiation from lightning. In addition, TASD is the largest radiation detector with high sensitivity for electromagnetic component in the northern hemisphere. Near future, there is no promising plan to deploy radiation detectors in such a large area for radiation from lightning. Therefore, we deployed lightning detectors in the vicinity of the TASD to get coincidences with lightning discharge as many as possible.

2. Observed Burst Events

We searched for the burst of air shower events. The criteria of this search is more than 3 events within 1 ms, and we found 10 bursts of events in 5 years data [2]. In this search, we did not use position information. However, all these events for each bursts are very localized within ~1 km radius. Considering the events which could not generate shower trigger but are found in waveforms, the time gaps of events in a burst are distributed from several to a hundred microseconds. Some events of 5 bursts in 10 bursts were reconstructed.

An example comparison views of burst event with normal event are shown in Figure 2. All reconstructed burst shower fronts are much more curved than usual cosmic ray air showers (Figure 2 Middle). This means that these burst showers seems to start developing at lower altitudes in the sky than normal cosmic ray air showers. Waveforms of many burst events do not have sharp rising edges (less muonic) for which SDs are located at the nearest shower cores (Figure 2 Bottom). However, the time integrated energy deposit to SD is similar to the cosmic ray air shower (Figure 2 Top).

3. Correlation with Lightning

Our check of correlation between shower bursts and lightning shows that observed shower bursts are clearly synchronized (±1ms) with negative intracloud lightning. And the absolute peak current of synchronized lightning is much higher than other lightning. Therefore, these bursts are very rare phenomena. A geometrical example of reconstructed burst events and correlated lightning information is shown in Figure 3 Left. Two reconstructed shower axis tend to point small region at low sky, where is the intracloud lightning discharge. We also checked the correlation between lightning and all SD events, regardless of burst. The time difference between all SD events and lightning is shown in Figure 3 Right. Therefore, SD events correlated with lightning are associated with the initial processes of the flash. And the time gaps of events in a burst correspond to the stepped leader process of lightning.

4. Deployed lightning detectors and Future

We deployed different types of lightning detectors in the vicinity of the TASD. 4 lightning discharge antennas of LF band were deployed surrounding TASD, measuring timing, direction and amplitude of discharge. Time Of Arrival (TOA) and Magnetic Direction Finder (MDF) method can determine the position of the lightning discharge [4] (Figure 4). Another VHF band antennas were also deployed, which can measure the height of sferics. In addition to measure the pulse from lightning discharge, we deployed continuous atmospheric electric field antenna too. This result also shows the high energy radiation from lightning associates with the initial part of the lightning process [5] (Figure 5).

We will observe more high energy radiation bursts induced by lightning, with various lightning detectors.

IOP Conf. Series: Journal of Physics: Conf. Series **1181** (2019) 012067 doi:10.1088/1742-6596/1181/1/012067

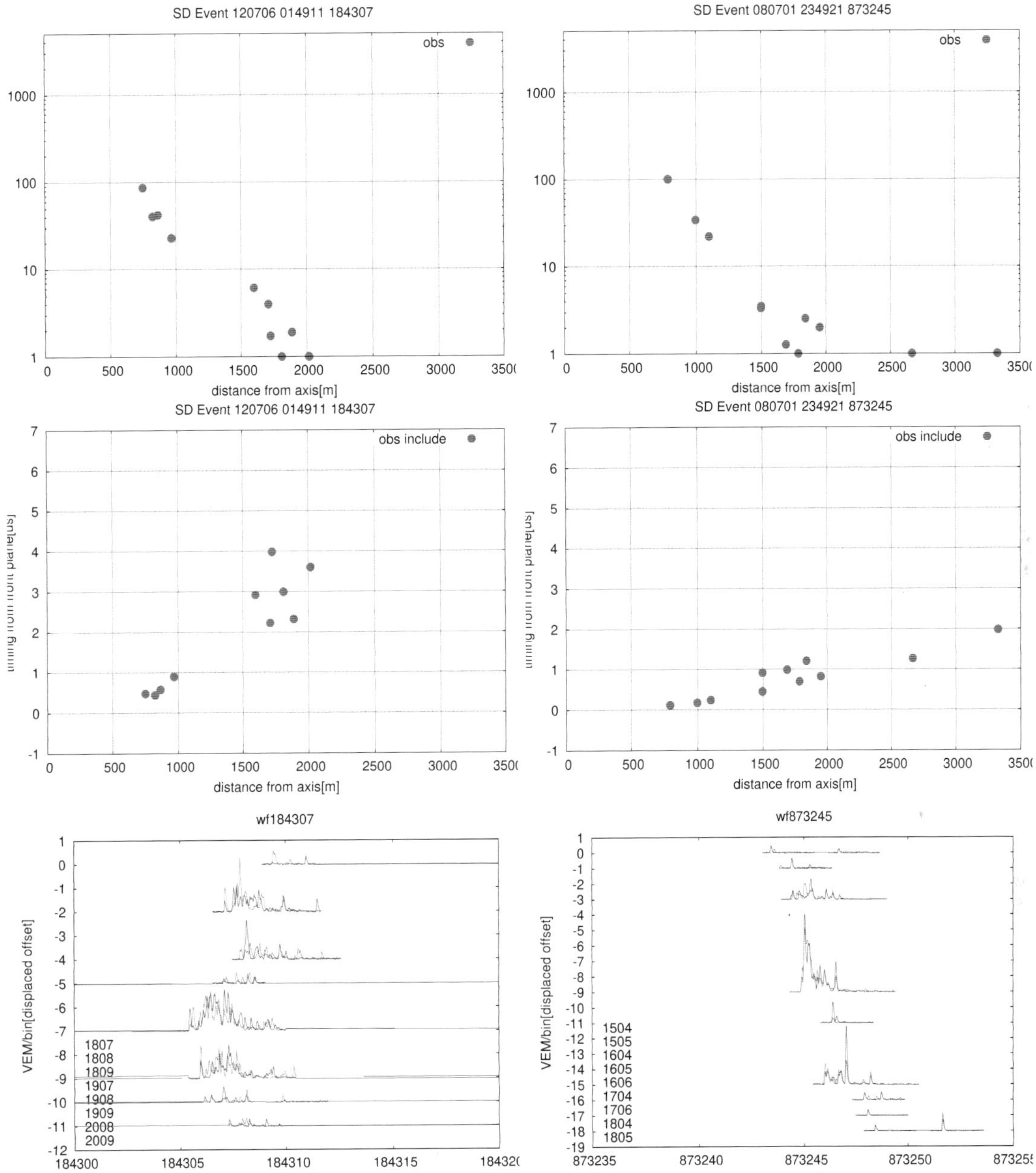

Figure 2.
Left: An event in a burst. Right: A normal air-shower event. This normal air-shower event has almost same zenith angle and shower size of burst event shown in left. Top: Lateral number of detected particle distribution. Middle: Lateral arrival timing distribution. Bottom: Signal waveforms from SDs in the vicinity of shower core. Red and Blue correspond two PMT channels for two layers of scintillator. Each horizontal level corresponds to individual SD. Vertical value has shifted to avoid superposition of waveforms from different SDs.

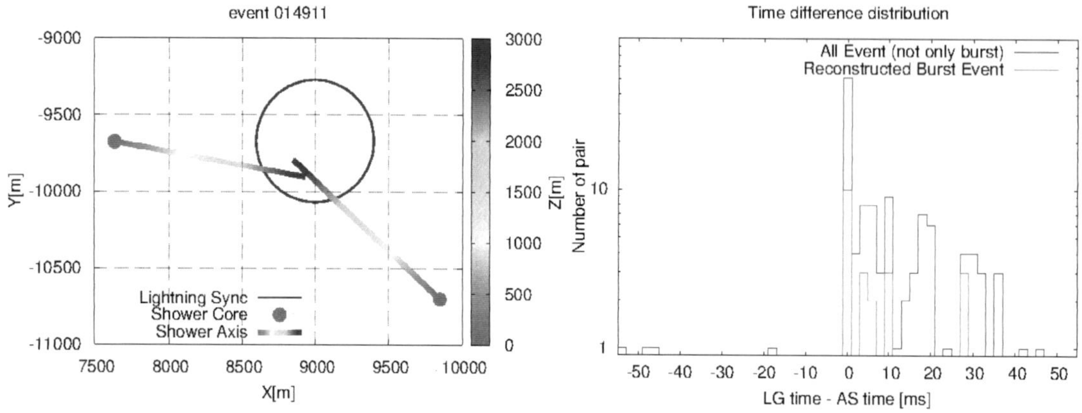

Figure 3. Left: Color shows the height from ground up to 3000 m. Red point shows shower core hit position. Circle shows lightning position with 300 m radius (NLDN position error). Synchronized (intracloud) lightning is set at z=3000 m. Right: Time difference between lighting and all SD events, not only burst. The synchronized lightning is at center bin.

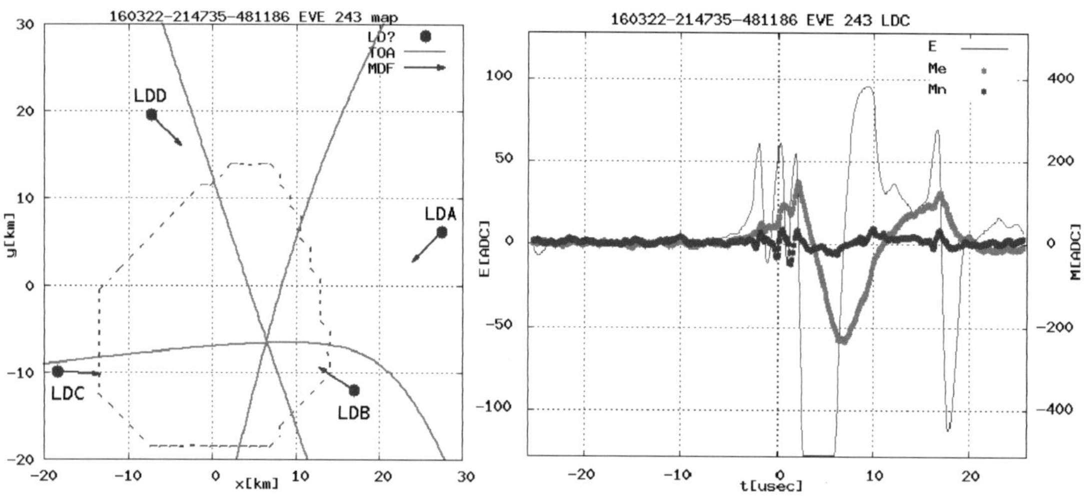

Figure 4. Left: Lightning location determination for a 4-fold coincidence event. Black dots show the locations of lightning detector. Red lines show hyperbolae by TOA on the basis of LDA. Blue arrows show the directions to the lightning discharge from detector by MDF. Right: Waveform of LDC for the left event. Black shows electric field. Red shows east-west magnetic field. Blue shows north-south magnetic field.

Acknowledgments

The Telescope Array experiment is supported by the Japan Society for the Promotion of Science through Grants-in-Aid for Scientific Research on Specially Promoted Research (21000002) "Extreme Phenomena in the Universe Explored by Highest Energy Cosmic Rays" and for Scientific Research (19104006); by the U.S. National Science Foundation awards PHY-0307098, PHY-0601915, PHY-0649681, PHY-0703893, PHY-0758342, PHY-0848320, PHY-1069280, PHY-1069286, PHY-1404495 and PHY-1404502; by the National Research Foundation of Korea (2007-0093860, R32-10130, 2012R1A1A2008381, 2013004883); by the Russian Academy

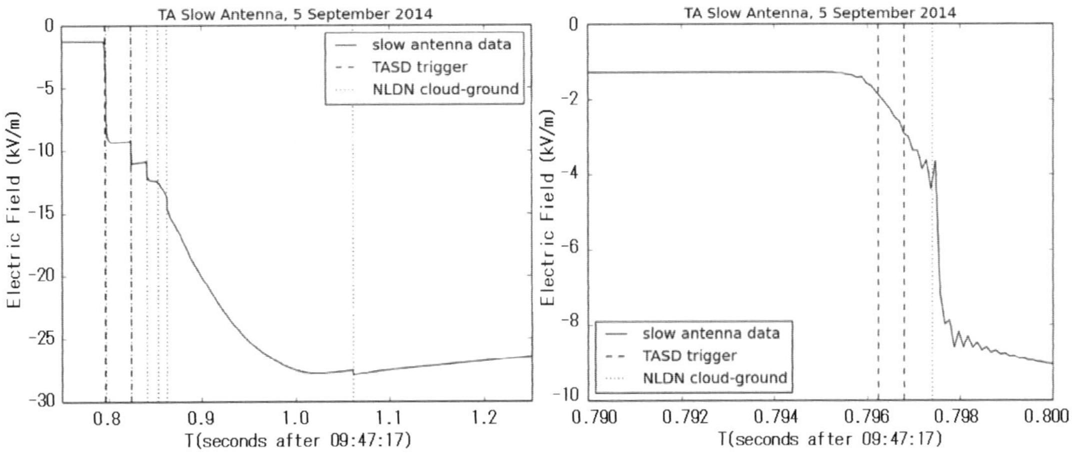

Figure 5. Left: Electric field slow antenna profile with TASD and NLDN event timing. Right: Enlarged initial part of left.

of Sciences, RFBR grants 11-02-01528a and 13-02-01311a (INR), IISN project No. 4.4502.13, and Belgian Science Policy under IUAP VII/37 (ULB). The foundations of Dr. Ezekiel R. and Edna Wattis Dumke, Willard L. Eccles, and George S. and Dolores Doré Eccles all helped with generous donations. The State of Utah supported the project through its Economic Development Board, and the University of Utah through the Office of the Vice President for Research. The experimental site became available through the cooperation of the Utah School and Institutional Trust Lands Administration (SITLA), U.S. Bureau of Land Management, and the U.S. Air Force. We also wish to thank the people and the officials of Millard County, Utah for their steadfast and warm support.

The lightning data and observation is supported by Vaisala Inc., the joint research program of the Institute for Cosmic Ray Research, University of Tokyo, and the Mitsubishi Foundation.

References

[1] Okuda T et al 2015 *PoS* **ICRC2015** 298
[2] Abbasi R U et al 2017 *Phys. Lett. A* **381** 2565-72
[3] AbuZayyad T et al 2012 *NIM Phys. Res. A* **689** 87-97
[4] Okuda T et al 2017 *PoS* **ICRC2017** 438
[5] Abbasi R U et al 2018 *J. GeoPhys. Res. Atm.* **123** 6864-79

IOP Conf. Series: Journal of Physics: Conf. Series **1181** (2019) 012068 doi:10.1088/1742-6596/1181/1/012068

Satellite monitoring of atmospheric temperature profiles and cloud cover over the Yakutsk EAS array and the TAIGA observatory

A A Lagutin, N V Volkov, A P Zhukov, K M Makushev, A A Maslov, E Yu Mordvin, R I Raikin, T L Serebryakova and V V Sinitsin

Altai State University, 61 Lenin St., Barnaul, Russian Federation, 656049

E-mail: lagutin@theory.asu.ru

Abstract. We consider the problem of night-time atmosphere monitoring at the locations of the Yakutsk EAS array and the TAIGA observatory for the annual October–April period of observations at these facilities. It is shown that the use of data from instruments aboard Terra, Aqua, Suomi-NPP and NOAA-20 satellites, received by the ground stations of the Altai State University, Russia, and processed to Level 2 (retrieval of geophysical parameters of the atmosphere), gives the possibility of the near-real-time monitoring observations of the cloud cover structure and the temperature profiles at night with a spatial resolution and a frequency sufficient for meteorological correction of the detector readings.

1. Introduction

Today, extensive air showers (EAS) are the only source of information about physical characteristics of the very high energy primary cosmic radiation ($E > 10^{15}$ eV). Composition, energy spectrum and arrival direction of a primary particle can be retrieved on the basis of fluorescence/Cherenkov light and lateral distribution of secondary particles which are measured by EAS array detectors spread on large areas.

However, such a retrieving is interfered by variations of properties of the atmosphere where cascade processes develop. Experimental data for different observed showers correspond to different states of the atmosphere. For example, the lateral distribution of electrons generated by air shower is affected by the air density distribution $\rho(t)$ throughout the atmospheric depth t, or by temperature distribution due to relationship between density and temperature $\rho(t) \sim 1/T(t)$. Thus variations of the temperature profile change the lateral distribution of particle density. Regarding fluorescence/Cherenkov detectors readings, another essential factor is the presence of cloud structures in the atmosphere over the array area.

The main goal of this work is to study a capability to measure variations of temperature profiles of the atmosphere and presence of night-time cloud cover over the locations of the Yakutsk EAS array [1] and the TAIGA observatory [2] in the near-real-time using data from satellite instruments.

2. Satellite observations

Temperature profiles of the atmosphere required for the meteorological correction of experimental data at the locations of the Yakutsk EAS array (61.7°N, 129.4°E) and the TAIGA observatory (51.8°N, 103.1°E) have been obtained using data from the hyperspectral infrared radiometer AIRS (Atmospheric Infrared Sounder) [3] aboard the Aqua satellite [4]. AIRS was specially designed to solve the key problem of satellite meteorology — measurement of the Earth's atmospheric temperature profiles with retrieval accuracy of better than 1 K in the lower troposphere under clear and partly cloudy conditions on a global scale.

AIRS software packages v.5.0.22.100 (Level 1B) and v.6.0.12.101 (Level 2) have been used to retrieve the characteristics of the atmosphere and the underlying surface of the Earth. The input information for these packages is data in the PDS (Production Data Set) format. These PDS files are created in the Remote Sensing Center of the Altai State University after unpacking the full raw data of Aqua satellite, received by EOScan and UniScan-24 ground stations in broadcast mode followed by using the RT-STPS (Real-time Software Telemetry Processing System) package.

Night-time cloud cover monitoring over the EAS arrays mentioned above was carried out using MOD35/MYD35 and VIIRS Cloud Mask products obtained in the Remote Sensing Center from the data of the MODIS/Terra-Aqua [5] and VIIRS/SNPP-NOAA-20 [6, 7] instruments, respectively. Note that MODIS channels used for cloud mask retrieval have nadir spatial resolution of 1000 m, while VIIRS channels have resolution of 750 m.

3. Atmospheric temperature profiles from AIRS/Aqua data

As a result of daily satellite measurements, we have created a database of temperature profiles over the above mentioned EAS experiments for the period of October, 2016–April, 2017.

Figure 1a shows the results of measurements of temperature profiles for night swaths of AIRS/Aqua at the location of Yakutsk EAS array in January, 2017. The range of temperature variations in the atmosphere in this zone is shown in figure 1b. Similar data were obtained for the TAIGA observatory.

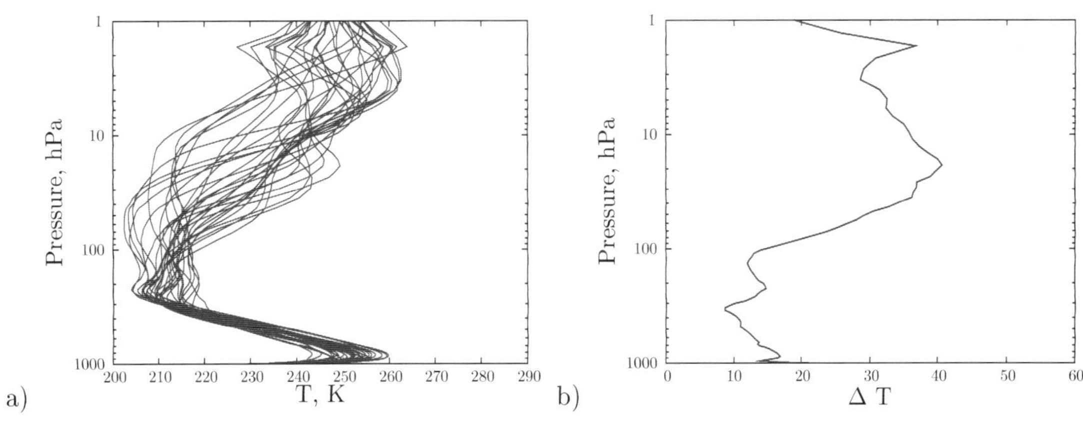

Figure 1. a) daily temperature profiles and b) the range of temperature variations at night-time over the Yakutsk EAS array for January, 2017

Variations of the average temperature values in a layer of 300 hPa above the observation level from November, 2016 to March, 2017 are shown in figure 2. AIRS/Aqua data show that such variations in the lower troposphere can reach up to 30 degrees.

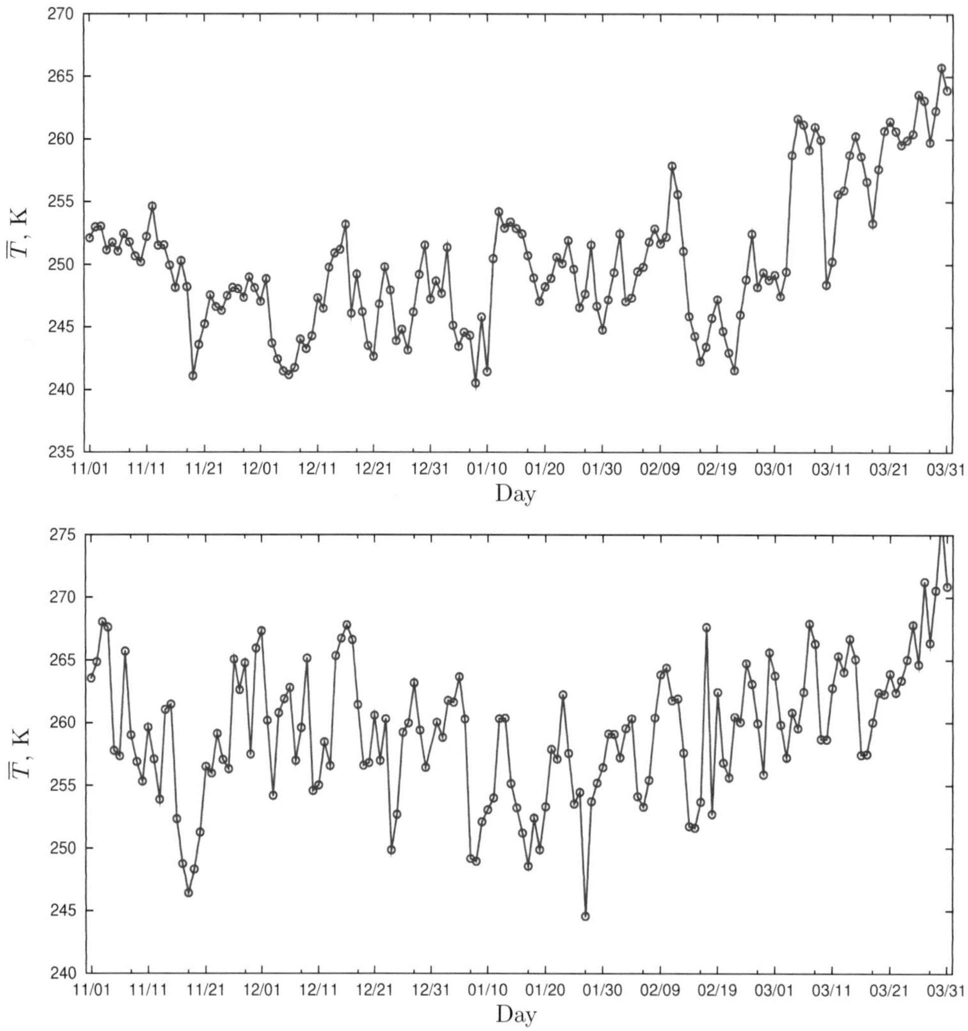

a)

b)

Figure 2. Variations of the average temperatures in a layer of 300 hPa above the observation level from November, 2016 to March, 2017: a) at the location of the Yakutsk EAS array, b) at the location of the TAIGA observatory

We have made additional comparisons of the satellite data with results of the ERA-Interim reanalysis [8]. The data presented in figure 3 show that satellite observations are in good agreement with the results of the reanalysis.

4. Cloudiness according to MODIS/Terra-Aqua and VIIRS/SNPP-NOAA-20 data

Figure 4 shows the time when the area of Yakutsk EAS array is observed by radiometers MODIS/Terra-Aqua and VIIRS/SNPP-NOAA-20 in the 16-day cycle of MODIS for the period 01.01.2018–16.01.2018. The analysis of that and others 16-day MODIS cycles showed that presently satellite instruments provide a sufficient frequency of cloud cover observations over the areas of the Yakutsk EAS array and the TAIGA observatory during the periods 13:00-14:00 UTC and 16:00-20:00 UTC. The period of 14:00-16:00 UTC should be controlled by surface instruments.

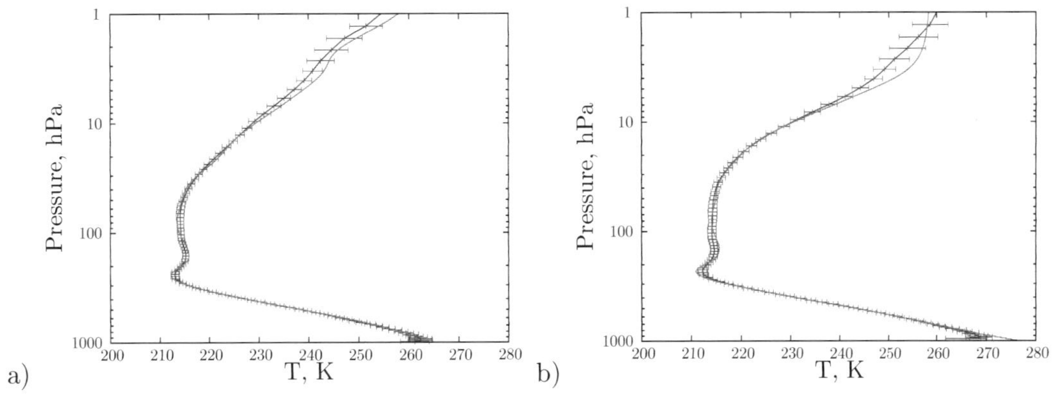

Figure 3. Average temperature profile of the atmosphere in March, 2017: a) at the location of the Yakutsk EAS array, b) at the location of the TAIGA observatory. The AIRS/Aqua data are plotted by the blue line; the results of the ERA-Interim reanalysis are plotted by red line. Intervals show retrieval errors of the monthly average temperature values according to the AIRS/Aqua

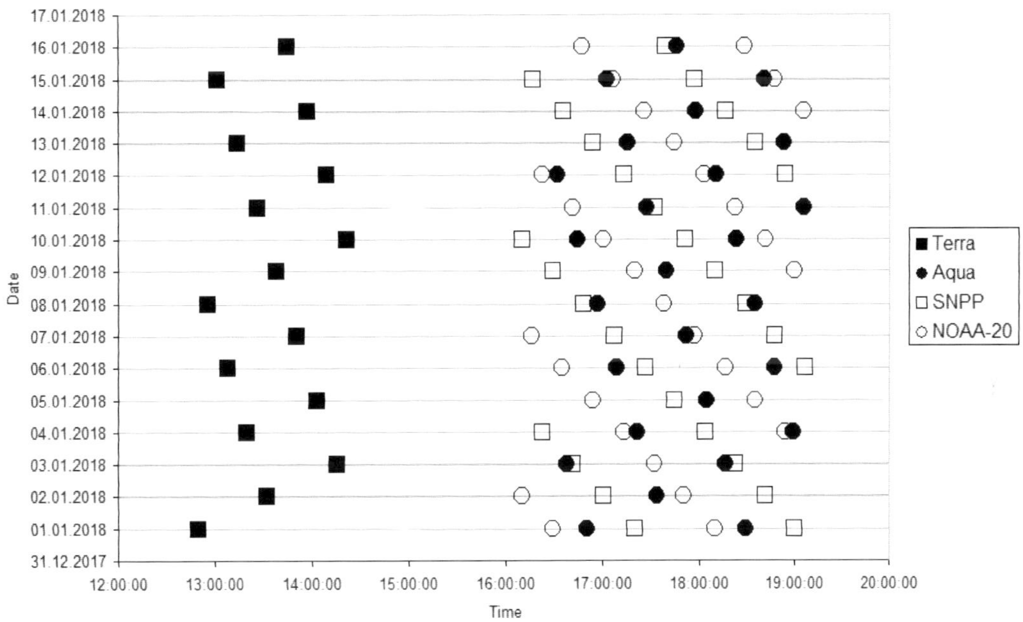

Figure 4. The time when the Yakutsk EAS array is covered by radiometers MODIS/Terra-Aqua and VIIRS/SNPP-NOAA-20 in the 16-day cycle of MODIS in the period of 13:00-19:00 UTC (22:00-04:00 local time)

Figure 5 shows the behavior of average cloud mask over 2 km by 2 km area covering TAIGA observatory in February 2018. It is easy to see that according to data of MODIS/Terra-Aqua about a half of measurements indicates cloudiness, while according to data of VIIRS/SNPP-NOAA-20 the most part of the area is free of clouds.

Note that although the definition of a cloud mask is one of four possible levels of confidence — confidently clear, possibly clear, uncertain (probably cloudy) and confidently cloudy, for the convenience of time-series analysis, in this paper the first two of them were combined into one — clear.

The behavior of average cloud mask over 3 km by 4 km area covering Yakutsk EAS array in February 2018 is shown on figure 6. For this area MODIS/Terra-Aqua and VIIRS/SNPP-NOAA-20 results comply with each other much better. According to satellite observations, this area was free of clouds during the most part of the period.

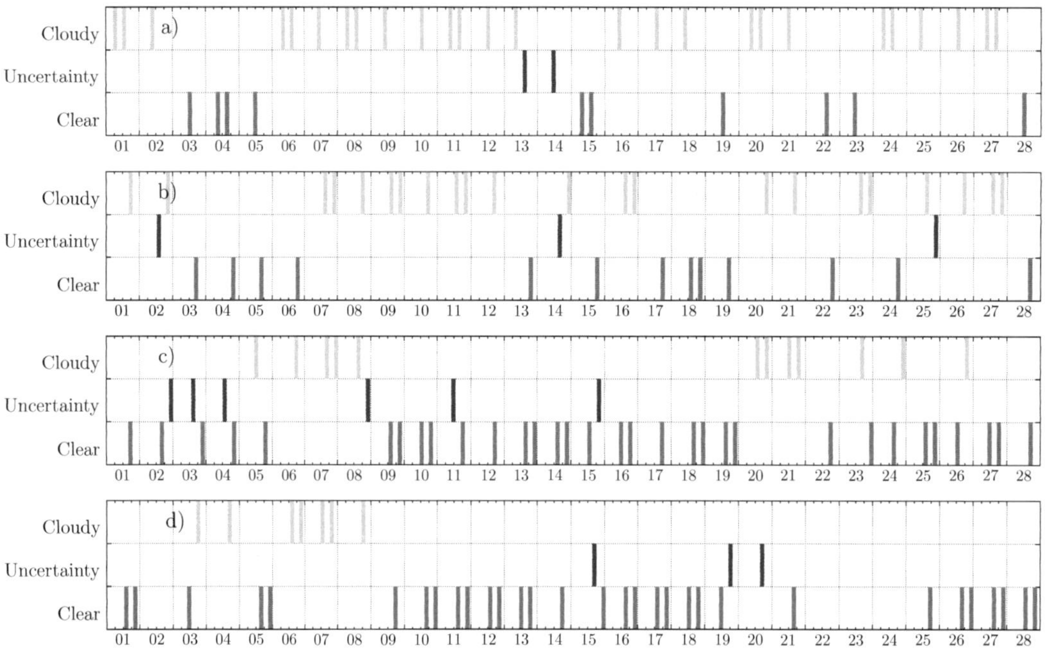

Figure 5. Cloud cover over 2 km by 2 km area over TAIGA observatory in February 2018 according to data of a) MODIS/Terra; b) MODIS/Aqua; c) VIIRS/SNPP; d) VIIRS/NOAA-20

5. Conclusions

Using data of hyperspectrometer AIRS/Aqua received by the Remote Sensing Center of Altai State University several times a day, the database of daily temperature profiles at the locations of the Yakutsk EAS array and the TAIGA observatory have been created for the period October, 2016–April, 2017. It was found that in the layer of 300 hPa above the observation level for the period from November 1st, 2016 to March 31st, 2017 the variations of the average atmospheric temperature were equal to 25 K and 30 K, respectively.

A problem of monitoring of cloud cover, which affects an appropriate detection of Cherenkov radiation, in the areas of Yakutsk EAS array and TAIGA observatory at night-time during annual observations from October to April has been discussed. The main sources of quantitative information on the state of the atmosphere were data from MODIS/Terra-Aqua and VIIRS/SNPP-NOAA-20 satellite instruments.

It is shown, that utilization of data from instruments aboard Terra, Aqua, SNPP and NOAA-20 satellites, received by ground stations of Altai State University and processed up to level 2 (retrieval of geophysical parameters of atmosphere), gives opportunity to carry out monitoring observations of cloud cover structure at night-time with high spatial resolution and frequency which is sufficient to make cloud correction of detector arrays readings.

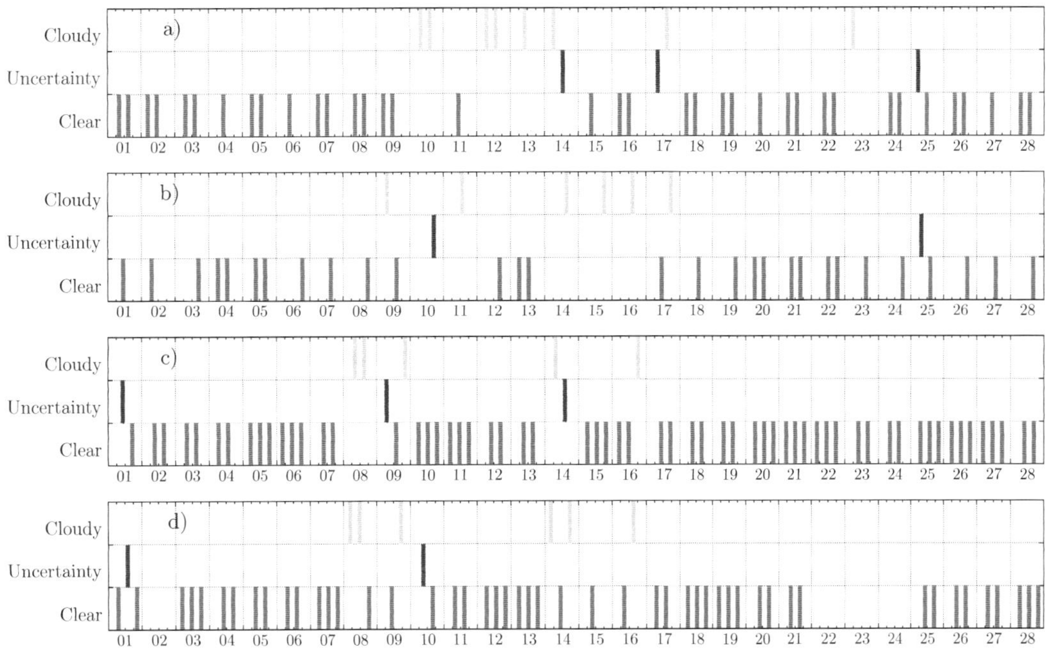

Figure 6. Cloud cover over Yakutsk EAS array area 3 km by 4 km in February 2018 according to data of a) MODIS/Terra; b) MODIS/Aqua c) VIIRS/SNPP; d) VIIRS/NOAA-20

Analysis of temporal sampling of satellite monitoring shows that the area of TAIGA observatory is covered by satellite observations 5-7 times during the night, while area of Yakutsk EAS array is observed by satellites 8-10 times per night.

Intercomparison of cloud mask products from MODIS/Terra-Aqua and VIIRS/SNPP-NOAA-20 algorithms leads to their significant difference for TAIGA observatory. However, for the area over Yakutsk EAS array the results correspond much better. In our view, an additional verification of these products in the areas of TAIGA observatory could be made by their comparison with VIIRS Day/Night Band data in full moon period.

Acknowledgments

This work was supported in part by the Ministry of Science and Higher Education of the Russian Federation (state assignment for the fundamental and applied research performed at Altai State University, project No 3.5939.2017/8.9)

References

[1] Ivanov A A *et al* (Yakutsk array group) 2013 *EPJ Web of Conferences* **53** 04003
[2] Kuzmichev L A, Astapov I I, Bezyazeekov P A *et al* 2018 *Phys. Atom. Nuclei* **81** 497–507
[3] Aumann H H, Chahine M T, Gautier C *et al* 2003 *IEEE Trans. Geosci. Remote Sens.* **41** 253–264
[4] Parkinson C L 2003 *Trans. Geosci. Remote Sens.* **41(2)** 173–183
[5] Salomonson V V, Barnes W L, Maymon P W *et al* 1989 *IEEE Trans. Geosci. Remote Sens.* **27** 145–153
[6] Hillger D, Kopp T, Lee T *et al* 2013 *BAMS* **94** 1019–1029
[7] Goldberg M D, Kilcoyne H, Cikanek H *et al* 2013 *J. Geophys. Res. Atmos.* **118** 13 463–13 475
[8] Dee D P, Uppala S M, Simmons A J *et al* 2011 *Q. J. R. Meteorol. Soc.* **137** 553–597

IOP Conf. Series: Journal of Physics: Conf. Series **1181** (2019) 012069 doi:10.1088/1742-6596/1181/1/012069

Characteristic energy-dissipating functions of muon penetrating through matters

Atsushi Iyono[1], **S Yamamoto**[1], **S Tsuji**[2], **K Okei**[2], **H Matsumoto**[2] **and T Nakatsuka**[3]

[1] Department of Fundamental Science, Okayama University of Science, Okayama 700-0005, Japan
[2] Department of Natural Sciences, Kawasaki Medical School, Kurashiki 701-0192, Japan
[3] Laboratory of Information Science, Okayama Shoka University, Okayama 700-8601, Japan

E-mail: iyono@das.ous.ac.jp

Abstract. The characteristic energy-dissipating functions of muon corresponding to positron-electron pair production, bremsstrahlung, and photonuclear interactions are expressed in series-expansions, to investigate transmission properties of muon effectively by analytical methods developed in the cascade shower theory. Accuracies of the new expression are examined by comparing one of the characteristic functions with that derived by the traditional analytical method. We apply the new-obtained characteristic functions to derive the survival probability of muon and compare the result with that of electron, discussing the discrepancies between the both.

1. Introduction

Dense structures of matter are now inspected by measuring atmospheric muons penetrating through the matter [1, 2, 3, 4], where the energy dissipations of muon during the passage should be accurately evaluated. The energy dissipation of muon among its passage is well predicted by the cascade theory improved by Nishimura [5], where the dissipation is described through M-function $\mathcal{M}(s, q, t)$ with the characteristic energy-dissipating functions $A(s)$ [5, 6].

Unfortunately, the characteristic functions $A(s)$ cannot be expressed by explicit functions applicable conveniently, so that we describe them for muon in series-expansions with the several coefficients specific to the matter. Contributions of the cross-section elements of positron-electron pair production (P), bremsstrahlung (B), and photonuclear interaction (N) to the characteristic functions are compared and discussed quantitatively and qualitatively. The difference of the characteristic functions between muon's and electron's is investigated by applying the functions to evaluate the survival probabilities.

2. Derivation of the characteristic energy-dissipating functions of muon

2.1. The characteristic energy-dissipating functions expressed in series-expansions

The characteristic energy-dissipating function is defined as

$$A(s) \equiv b^{-1} \int_0^1 \{1 - (1-v)^s\} \phi(v) dv, \tag{1}$$

where $\phi(v)dv$ denotes the probability of fractional energy loss per g/cm^2 and $b \simeq 4.0 \times 10^{-6}$ g^{-1}cm^2 denotes the reciprocal of the radiation unit of muon for Standard Rock ($Z = 11$, $A = 22$) [7]. As the probability density $\phi(v)$ for muon is additive as

$$\phi(v) \equiv \phi_{\mathrm{P}}(v) + \phi_{\mathrm{B}}(v) + \phi_{\mathrm{N}}(v) \tag{2}$$

among the cross-section elements of P, B, and N indicated in figures 1-3, the characteristic function $A(s)$ is also additive as

$$A(s) \equiv A_{\mathrm{P}}(s) + A_{\mathrm{B}}(s) + A_{\mathrm{N}}(s), \tag{3}$$

where each term $A_*(s)$ of the right-hand side is defined as

$$A_*(s) \equiv b^{-1} \int_0^1 \{1 - (1 - v)^s\}\phi_*(v)dv. \tag{4}$$

We express $A_*(s)$'s in series-expansions. Expanding $\{1 - (1 - v)^s\}$ with v at $v = 0$, we have

$$\{1 - (1 - v)^s\} = sv - \frac{s(s-1)}{2}v^2 + \frac{s(s-1)(s-2)}{6}v^3 - \cdots = -\sum_{k=1}^{\infty} \frac{(-)^k}{k!} s^{[k]} v^k, \tag{5}$$

$$\text{where} \qquad s^{[0]} \equiv 1 \quad \text{and} \quad s^{[k]} \equiv s(s-1)\cdots(s-k+1), \tag{6}$$

thus we have

$$A_*(s) \equiv \sum_{k=1}^{\infty} \frac{(-)^k}{k!} \alpha_{*,k} s^{[k]} \qquad \text{with} \qquad \alpha_{*,k} = -b^{-1} \int_0^1 v^k \phi_*(v)dv. \tag{7}$$

Likewise we can express the derivatives of $A_*(s)$ as

$$A'_*(s) \equiv \sum_{k=0}^{\infty} \frac{(-)^k}{k!} \alpha'_{*,k} s^{[k]} \qquad \text{with} \qquad \alpha'_{*,k} = -b^{-1} \int_0^1 v^k \phi_*(v) \ln(1-v)dv, \tag{8}$$

$$A''_*(s) \equiv \sum_{k=0}^{\infty} \frac{(-)^k}{k!} \alpha''_{*,k} s^{[k]} \qquad \text{with} \qquad \alpha''_{*,k} = -b^{-1} \int_0^1 v^k \phi_*(v) \ln^2(1-v)dv. \tag{9}$$

2.2. The characteristic functions for positron-electron pair production

The probability density $\phi_{\mathrm{P}}(E, v)$ of muon with energy E to dissipate fractional energy of v by positron-electron pair production is described by Bugaev *et al.* [8] as

$$\phi_{\mathrm{P}}(E, v) = \frac{16}{\pi} Z(Z+1)(\alpha r_e)^2 \frac{N}{A} \frac{1.7 \times 10^{-4}(v + 1.05 \times 10^{-4})}{v^2(v + 0.006)^2} \frac{1 + 2v}{1 + k/(vE)}$$
$$\times \left\{1 - \frac{\exp[-0.025 \ln^2(E/\mu)]}{1 + 0.323 \ln(10^3 v)}\right\}, \quad k = 0.02\mathrm{GeV}, \tag{10}$$

for $10^{-3} < v < 0.2$ as indicated in figure 1. This density is approximated at $E \to \infty$ by

$$\phi_{\mathrm{P}}(v) \simeq 2.20 \times 10^{-9} \frac{Z(Z+1)}{A} \frac{1}{v(v + 0.006)^2} \qquad \mathrm{g}^{-1}\mathrm{cm}^2, \tag{11}$$

regarding $1.05 \times 10^{-4} \ll v$, where we take 1 instead of $(1 + 2v)/(1 + k/(vE))$ at $0.2 < v$. Note that Kobayakawa took the same v-dependence in the probability density [9]. $\alpha_{\mathrm{P},k}$, $\alpha'_{\mathrm{P},k}$, and $\alpha''_{\mathrm{P},k}$ derived by Eqs. (7)-(9) for Standard Rock from $\phi_{\mathrm{P}}(v)$ are indicated in Table 1-3.

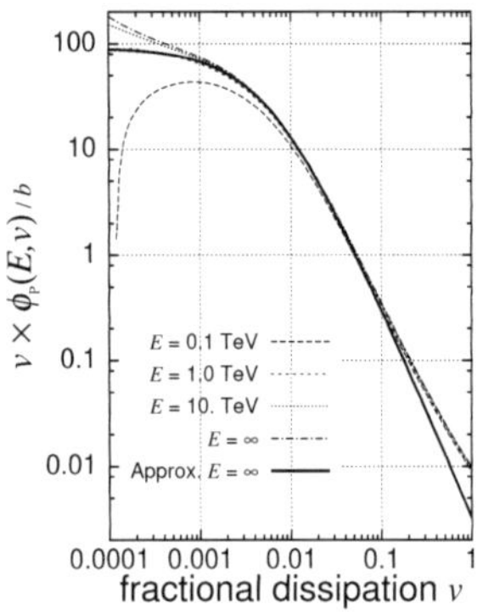

Figure 1. Energy-dissipating cross-section for P-process.

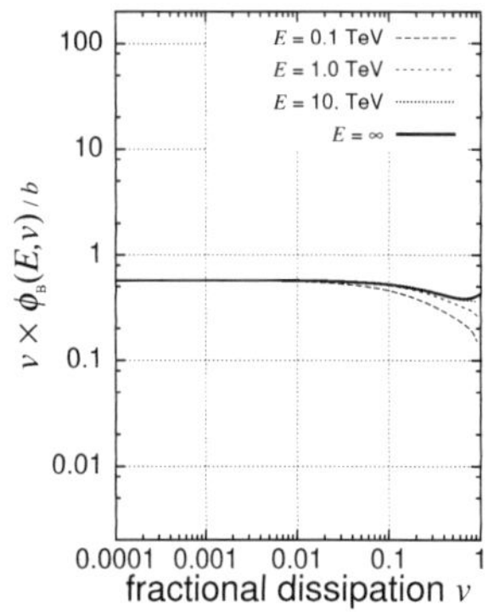

Figure 2. Energy-dissipating cross-section for B-process.

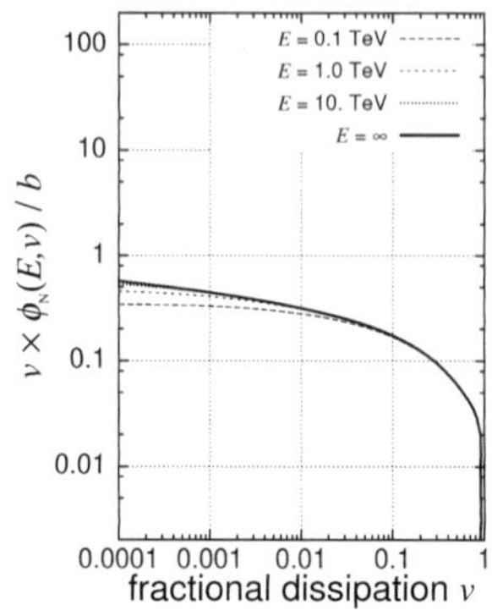

Figure 3. Energy-dissipating cross-section for N-process.

2.3. The characteristic functions for bremsstrahlung

The probability density $\phi_{\rm B}(E, v)$ of muon with energy E to dissipate fractional energy of v by bremsstrahlung is described in GEANT [10] based on Kelber *et al.* [11] as

$$\begin{aligned}\phi_{\rm B}(E,v) &= \frac{16}{3}\alpha N(\frac{m}{\mu}r_{\rm e})^2\frac{1}{vA}Z(Z\Phi_n+\Phi_e)\left\{1-v+\frac{3}{4}v^2\right\} \\ &= 3.27\times10^{-8}\ \frac{Z(Z\Phi_n+\Phi_e)}{A}\ \frac{(4/3)(1-v)+v^2}{v}\qquad {\rm g}^{-1}{\rm cm}^2, \end{aligned} \tag{12}$$

with

$$\Phi_n = \ln\frac{BZ^{-1/3}(\mu+\delta(D_n'\sqrt{e}-2))}{D_n'(m+\delta\sqrt{e}BZ^{-1/3})}, \tag{13}$$

$$\Phi_e = \ln\frac{B'Z^{-2/3}\mu}{(1+\delta\mu/(m^2\sqrt{e}))(m+\delta\sqrt{e}B'Z^{-2/3})}, \tag{14}$$

$$\begin{aligned}\delta &= \mu^2 v/2E(1-v),\quad D_n' = D_n^{(1-1/Z)},\quad D_n = 1.54A^{0.27}, \\ B &= 183,\quad B' = 1429,\quad \sqrt{e} = 1.648\cdots, \end{aligned} \tag{15}$$

Table 1. Coefficients $\alpha_{*,k}$ and α_k applied to the series expansion of the characteristic function $A(s)$.

k	$\alpha_{{\rm P},k}$	$\alpha_{{\rm B},k}$	$\alpha_{{\rm N},k}$	α_k
1	-0.54672	-0.43	-0.0817365	-1.05846
2	-0.0136222	-0.203056	-0.0237638	-0.240442
3	-0.00311685	-0.133778	-0.0119711	-0.148866
4	-0.00161211	-0.100333	-0.00749212	-0.109437
5	-0.00108054	-0.0805397	-0.0052516	-0.0868718
6	-0.000811975	-0.0674008	-0.00394304	-0.0721558
7	-0.000650217	-0.0580159	-0.00309962	-0.0617657
8	-0.000542168	-0.050963	-0.00251783	-0.054023
9	-0.000464899	-0.0454613	-0.00209615	-0.0480223
10	-0.000406902	-0.0410455	-0.00177882	-0.0432312

Table 2. Coefficients $\alpha'_{*,k}$ and α'_k applied to the series expansion of the characteristic function $A'(s)$.

k	$\alpha'_{\mathrm{P},k}$	$\alpha'_{\mathrm{B},k}$	$\alpha'_{\mathrm{N},k}$	α'_k
0	0.555921	0.692262	0.103377	1.35156
1	0.0168017	0.406111	0.0367687	0.459682
2	0.00520665	0.303588	0.0211748	0.329969
3	0.00323692	0.248126	0.0145632	0.265926
4	0.00243597	0.212372	0.0109579	0.225766
5	0.00198732	0.186987	0.00870106	0.197675
6	0.00169481	0.167829	0.00716098	0.176684
7	0.00148659	0.152749	0.00604624	0.160282
8	0.0013296	0.14051	0.0052042	0.147043
9	0.00120634	0.130338	0.00454725	0.136092
10	0.00110659	0.121726	0.00402153	0.126854

as indicated in figure 2. As they approach

$$\Phi_n \rightarrow \ln \frac{BZ^{-1/3}\mu}{D'_n m} \rightarrow 8.59, \tag{16}$$

$$\Phi_e \rightarrow \ln(B'Z^{-2/3}\mu/m) \rightarrow 11.0 \tag{17}$$

at $E \rightarrow \infty$, the density is approximated by

$$\phi_{\mathrm{B}}(v) = 1.72 \times 10^{-6} \frac{(4/3)(1-v)+v^2}{v} \quad \mathrm{g}^{-1}\mathrm{cm}^2. \tag{18}$$

$\alpha_{\mathrm{B},k}$, $\alpha'_{\mathrm{B},k}$, and $\alpha''_{\mathrm{B},k}$ derived by Eqs. (7)-(9) for Standard Rock from $\phi_{\mathrm{B}}(v)$ are indicated in Table 1-3.

2.4. The characteristic functions for photonuclear interaction

The probability density $\phi_{\mathrm{N}}(E, v)$ of muon with energy E to dissipate fractional energy of v by photonuclear interaction is described by Borog and Petrukhin [12] and applied to GEANT [10],

Table 3. Coefficients $\alpha''_{*,k}$ and α''_k applied to the series expansion of the characteristic function $A''(s)$.

k	$\alpha''_{\mathrm{P},k}$	$\alpha''_{\mathrm{B},k}$	$\alpha''_{\mathrm{N},k}$	α''_k
0	-0.0237038	-0.984192	-0.0631867	-1.07108
1	-0.0105516	-0.820185	-0.0415348	-0.872272
2	-0.0078061	-0.720482	-0.0312109	-0.759499
3	-0.00650595	-0.650553	-0.0250611	-0.68212
4	-0.00569665	-0.597565	-0.0209362	-0.624198
5	-0.00512585	-0.555418	-0.0179593	-0.578503
6	-0.00469349	-0.520758	-0.0157016	-0.541153
7	-0.00435046	-0.491548	-0.0139269	-0.509825
8	-0.00406926	-0.466466	-0.0124937	-0.483029
9	-0.00383309	-0.444607	-0.0113112	-0.459751
10	-0.00363095	-0.425324	-0.0103188	-0.439274

as

$$\phi_{\rm N}(E,v) = \frac{\alpha}{\pi}\frac{A_{\rm eff}N}{A}\sigma_{\gamma N}\frac{1}{v}\left\{v-1+\left[1-v+\frac{v^2}{2}\left(1+\frac{2\mu^2}{\Lambda^2}\right)\right]\ln\frac{\frac{E^2(1-v)}{\mu^2}\left(1+\frac{\mu^2v^2}{\Lambda^2(1-v)}\right)}{1+\frac{Ev}{\Lambda}\left(1+\frac{\Lambda}{2m}+\frac{Ev}{\Lambda}\right)}\right\}, \quad (19)$$

$$\text{where} \quad \Lambda^2 = 0.4 \text{ GeV}^2, \quad \sigma_{\gamma N} = 100\ \mu\text{b/nucl}, \quad \text{and} \quad A_{\rm eff} = 0.22A + 0.78A^{0.89}, \quad (20)$$

as indicated in figure 3. This density is approximated at $E \to \infty$ by

$$\phi_{\rm N}(v) \simeq 1.40\times10^{-7}\,\frac{A_{\rm eff}}{A}\,\frac{1}{v}\left\{v-1+[1-v+0.528\,v^2]\,\ln\left[1+\frac{1-v}{0.0279\,v^2}\right]\right\}\ \text{g}^{-1}\text{cm}^2. \quad (21)$$

$\alpha_{{\rm N},k}$, $\alpha'_{{\rm N},k}$, and $\alpha''_{{\rm N},k}$ derived by Eqs. (7)-(9) for Standard Rock from $\phi_{\rm N}(v)$ are indicated in Table 1-3.

3. Accuracies of the characteristic energy-dissipating function expressed by the series-expansion

We confirm accuracies of the reconstructed $A_*(s)$ of (7) by comparing that for bremsstrahlung with the exact $A_{\rm B}(s)$ derived analytically by applying Eq. (4) on the probability density (18);

$$A_{\rm B}(s) = \frac{1.72\times10^{-6}\ \text{g}^{-1}\text{cm}^2}{b}\left\{\frac{4}{3}\Big(\psi(s+2)-\psi(2)\Big)-\frac{1}{(s+1)(s+2)}+\frac{1}{2}\right\}, \quad (22)$$

indicated in figure 4 (thick line).

The reconstructed $A_{\rm B}(s)$ by Eq. (7) with k up to 8, 9, 10 are also indicated in the figure (broken, dot, and thin lines), which are approaching the exact one at $s > 0$ with the increase of k.

4. Comparison of the characteristic energy-dissipating function and the survival probability between muon and electron

P, B and N elements of $A_*(s)$ for muon are indicated in figure 5 (thin solid, broken and dot lines). P and B elements are comparable at $0 < s <\sim 1$; though P element exceeds B much at $\sim 1 < s$, increasing almost linearly. Almost linear increase of $A_{\rm P}(s)$ corresponds to almost continuous (less fluctuated) energy-dissipation of P process, as Kobayakawa obtained $A(s) \simeq s\, b_{\rm eff}$ in case without fluctuations [13]. This property is confirmed in narrower distribution of $\phi_{\rm P}(v)$ than $\phi_{\rm B}(v)$ and $\phi_{\rm N}(v)$ indicated in figures 1-3.

The characteristic energy-dissipating function $A(s)$ of muon and that of electron are indicated in the figure (thick solid and broken lines). $A(s)$ of muon is comparable with that of electron at $0 < s <\sim 1.5$, though the former exceeds the latter much reflecting the dominating linear increase of $A_{\rm P}(s)$ of muon.

We can evaluate the survival probability of muon at t of radiation length with its energy greater than E [6] by applying the characteristic energy-dissipating function $A(s)$ derived above,

$$\mu(E_0, > E, t) = \frac{1}{2\pi i}\int\frac{ds}{s}\left(\frac{E_0}{E}\right)^s e^{-A(s)t} \simeq \left(\frac{E_0}{E}\right)^{\bar{s}} e^{-A(\bar{s})t}\Big/\sqrt{2\pi\{1-\bar{s}^2A''(\bar{s})t\}} \quad (23)$$

$$\text{with} \quad t = \{\ln(E_0/E) - 1/\bar{s}\}/A'(\bar{s}) \quad \text{and} \quad 0 < \bar{s}, \quad (24)$$

where $\bar{s}$ denotes the saddle point. We compare in figure 6 the so obtained survival probability of muon ($\mu(E_0, > E, t)$, thick line) with that of electron ($\pi(E_0, > E, t)$, thin line) indicated in Eqs. (3.4) and (3.5) of [6]. We have to take $0 < \bar{s} <\sim 10$ for muon instead of $0 < \bar{s} <\sim 2$ for electron. The survival probability of muon decreases slower than that of electron at the early stage of penetration though the former drops abruptly at the late stage of penetration, which reflects a more continuous property of energy-dissipation for muon than for electron.

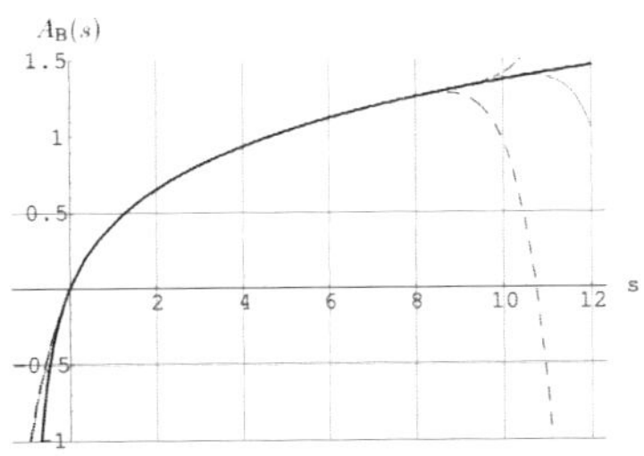

Figure 4. The characteristic function $A_B(s)$, expressed by series expansion with k up to 8, 9, and 10 (broken, dot, and thin lines), compared with the exact one (thick line).

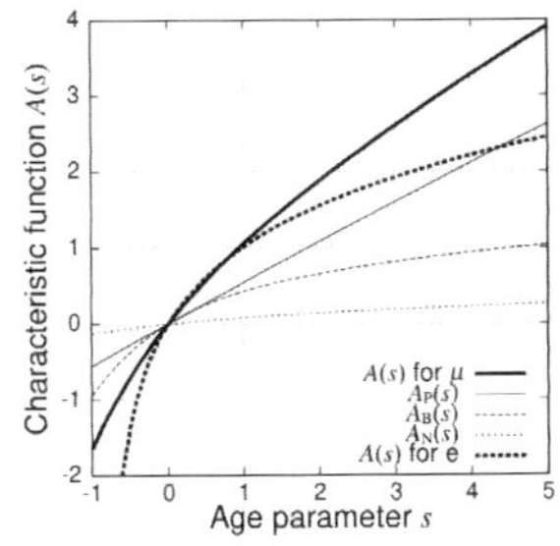

Figure 5. The characteristic functions of $A(s)$ for muon and electron (thick solid and broken lines), as well as P, B, and N elements $A_*(s)$ for muon (thin solid, broken, and dot lines).

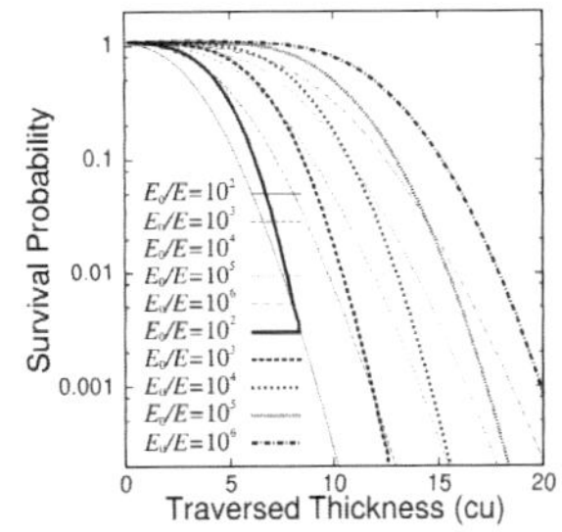

Figure 6. Dissipating curves of survival probability, $\mu(E_0, > E, t)$ and $\pi(E_0, > E, t)$, for muon and electron (thick and thin lines).

5. Conclusion

The characteristic energy-dissipating functions of muon are described with continuous functions with s expressed in series-expansions with the several coefficients indicated in Tables for Standard Rock.

Accuracies of the new expressed function are confirmed in case of bremsstrahlung element by comparing the result with the exact one derived analytically from the definition integral.

Properties of P, B and N elements of energy dissipation are investigated by comparing the characteristic functions among them. Differences of energy dissipation between muon and electron are investigated and discussed by comparing the survival probabilities of them evaluated with the characteristic functions expressed in the series.

Acknowledgments

Authors are very indebted to Prof. Jun Nishimura for valuable advices and encouragements.

References

[1] Alvarez L W *et al* 1970 *Science* **167** 832.
[2] Morishima K *et al* 2017 *Nature* **552** 386.
[3] Tanaka H K M, Nakano T, Takahashi S, Yoshida J and Niwa K 2007 *Nucl. Instr. and Meth.* A **575** 489.
[4] Borozdin K *et al* 2012 *Phys. Rev.* **109** L152501.
[5] Nishimura J 1967 *Handbuch der Physik, Band 46*, edited by Flügge S (Springer, Berlin) Teil **2** p 1.
[6] Iyono A, Yamamoto S, Tsuji S, Okei K, Matsumoto H and Nakatsuka T 2017 *Conference Papers of 35th International Cosmic Ray Conference, Busan* PoS (ICRC, 2017)565, https://pos.sissa.it/301/565/pdf.
[7] T. Nakatsuka, K. Kobayakawa, and T. Kitamura, Proc. of 20th ICRC (Moscow), 6(1987)261.
[8] Bugaev E V, Misaki A, Naumov V A, Sinegovskaya T S, Sinegovsky S I and Takahashi N 1998 *Phys. Rev.* D **58** 05401.
[9] Kobayakawa K 1973 *Proc. of 13th ICRC (Denver)* **5** 3156.
[10] http://geant4.web.cern.ch/geant4/UserDocumentation/UsersGuides/PhysicsReferenceManual/fo/PhysicsReferenceManual.pdf
[11] Kelner S R, Kokoulin R P, Petrukhin A A 1995 *Preprint MEPhI 024-95, Moscow* CERN SCAN-9510048.
[12] Borog V V and Petrukhin A A 1975 *Proc. 14th Int. Conf. on Cosmic Rays, Munich* vol **6** p 1949.
[13] Kobayakawa K 1967 *Nuovo Cim.* B **47** 156.

Investigation of exceptional solar activity in September 2017: GLE 72 and unusual Forbush decrease in GCR

L. Dorman[1,2], Y. Tassev[3], P. I. Y. Velinov[3], A. Mishev[4,5], D. Tomova[6] and L. Mateev[3]

[1] Cosmic Ray Department of IZMIRAN, Moscow, Russia
[2] Cosmic Ray and Space Weather Center of Tel Aviv University, Israel
[3] Institute for Space Research and Technology, Bulgarian Academy of Sciences, Bulgaria
[4] Space Climate Research Unit, University of Oulu, Finland
[5] Sodankylä Geophysical Observatory, University of Oulu, Finland
[6] Sofia University "St. Kliment Ohridski", Faculty of Mathematics & Informatics, Bulgaria

E-mails: alexander.mishev@oulu.fi, yktassev@outlook.com, tomova@fmi.uni-sofia.bg

Abstract. The exceptional solar activity in early September 2017 at minimum of solar cycle 24 is analyzed. Intensive solar-terrestrial disturbances was caused by Active Region AR2673, which produced four powerful eruptions class X, including the strongest flare X9.3 of Solar Cycle 24 on September 6, 2017, after which began G4 – Severe geomagnetic storm on 07-08.09.2017 with Ap = 96, and also the second strongest flare X8.2 of Solar Cycle 24 on September 10, 2017, which generated Ground Level Enhancement (GLE) of cosmic rays. This was GLE72 with increase of solar cosmic ray flux 6% in Oulu Station (Finland) (effective vertical geomagnetic cutoff rigidity: 0.8 GV), and increase 9% in DOMC Antartica and 14% in DOMB Antartica (in the latter case - lead free neutron monitors with effective vertical cutoff rigidity <0.01 GV). The GLE72 develops under the conditions of a deep Forbush decrease (around 15%) in South Pole cusp caused by September 7th Coronal Mass Ejection. The Forbush effect ends on September 11th (http://cosmicrays.oulu.fi). But cosmic ray measurements by flying balloons to the stratosphere over California show that after solar eruptions in September 2017 the radiation levels in stratosphere took more than two months to fully rebound to the conditions of minimal solar activity. This is interesting fact which deserves to be explored in detail. It is precisely the study and interpretation of this process that is concerned with this work.

1. Coronal mass ejections on September 4 – 6 and 10, 2017 in solar minimum

The cause of the intensive solar-terrestrial disturbances in early September 2017 at minimum of solar cycle 24 was the Active Region AR2673, which produced four powerful eruptions class X, including the strongest flare X9.3 of Solar Cycle 24 on September 6, 2017, after which began G4 - Severe geomagnetic storm on 07-08.09.2017 with A_p = 106, and also - the second strongest flare X8.2 of Solar Cycle 24 on September 10, 2017, which generated instantly the ground level enhancement of cosmic rays or Ground Level Event № 72 (GLE72) [1,2]. This GLE72 is the second GLE in the current solar cycle (Solar Cycle 24) after GLE71 on 17 May 2012 at solar maximum conditions [3, 4]. Therefore, the new GLE72 in solar minimum represents special interest.

2. The rapid development of the AR2673 active region for 24 hours

In the investigated period (02-08.09.2017), two CMEs were observed in sequence over two days. The first event was observed on September 2, 2017. There are four active regions in the solar photosphere.

Two of the active regions are in the eastern part of the Sun. The Active Regions AR2673 and AR2674 pass through the central meridian and occupy a frontal position with respect to the Earth. The double solar spot AR 2674 grows relatively rapidly, expanding both in terms of area and number of sunspots. As the number of sunspots increases, the magnetic field becomes more unstable. There is evidence of the development of a combined "beta-gamma" magnetic field, which is a prerequisite for outbreaks in the X-ray bursts of the solar spectrum, at least of class M [1, 2]. On September 3, it was found that while the AR2674 sunspot group grew for days, the AR2673 group increased its area fourfold, and the sunspots in just 24 hours. The fast-growing solar AR2673 has a "beta-gamma-delta" magnetic field that charges energy for M-class solar flares. On September 4, two eruptions have already been detected in the M1 class of X-ray area. On 4.09 at 19:00 UT there is a coronal mass ejection directed to the Earth.

CMEs on 4 and 6 September 2017 cause a geomagnetic storm, with the Dst geomagnetic index reaching –150 nT. As a result of the two events, there is a deep Forbush decrease in the galactic cosmic ray flux, which in the maximum phase reaches around 15% in South Pole cusp [1]. The decrease measured with the neutron monitor in OULU is -8% (figure 1). On September 6 was observed new CME. Going behind the apparent horizon of the solar disk, almost on the edge, the AR2673 solar plexus caused another CME accompanied by a powerful X9 X-ray burst at 16:06 UT, at 10.09, (figure 2). At the same time, this event reveals high energy proton emission (figure 3).

In the period September 4-9, 2017, the second largest geomagnetic storm was observed. It begins on Sept. 7 in the evening with a sudden increase of Kp = 8. On September 8 it continues with Kp = 8, passes through a brief decrease and again Kp = 8, after which the geomagnetic storm ceases until the end of the day (figure 4).

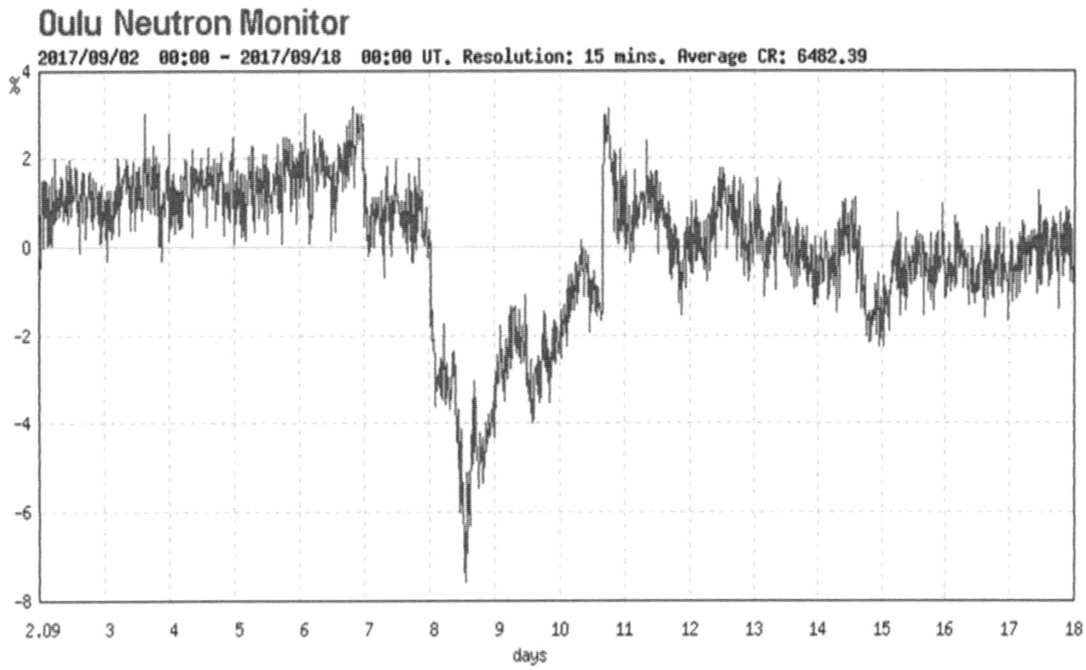

Figure 1. Oulu Neutron Monitor (NM) count rate variations in September 2017. It is clearly visible the Forbush decrease (–8%) of galactic cosmic rays on September 08, 2017 caused by CME on September 7th, and also the Solar Cosmic Ray increase – i.e. the Ground Level Event No 72 (GLE72) on September 10, 2017 caused by the flare X8.2 slightly earlier that day.

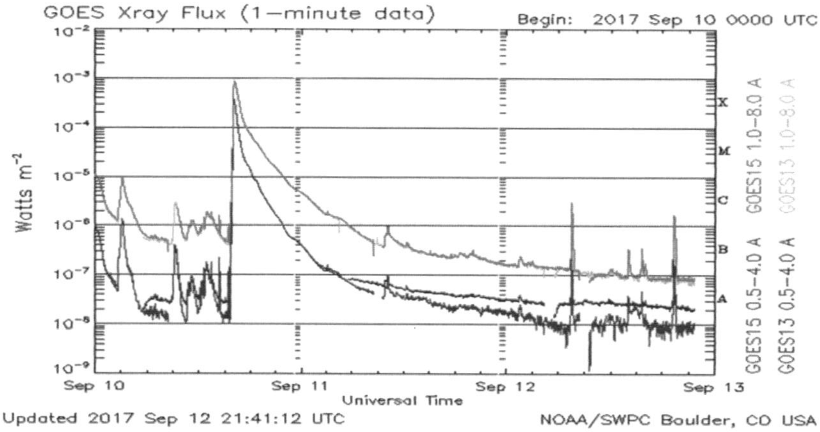

Figure 2. GOES observations of X-ray flux during 10 – 12 Sept., 2017. The X9 flare from 10 Sept. triggered the radiation storm in the Earth environ-ment.

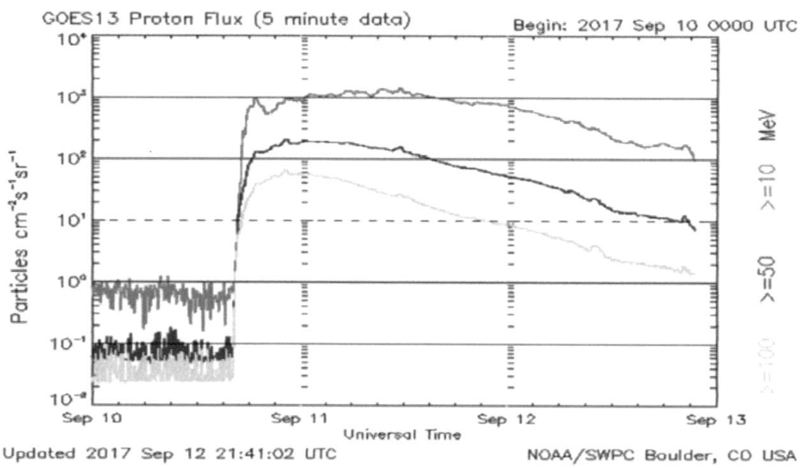

Figure 3. GOES observation: proton flux during GLE 72 from 10 September 2017.

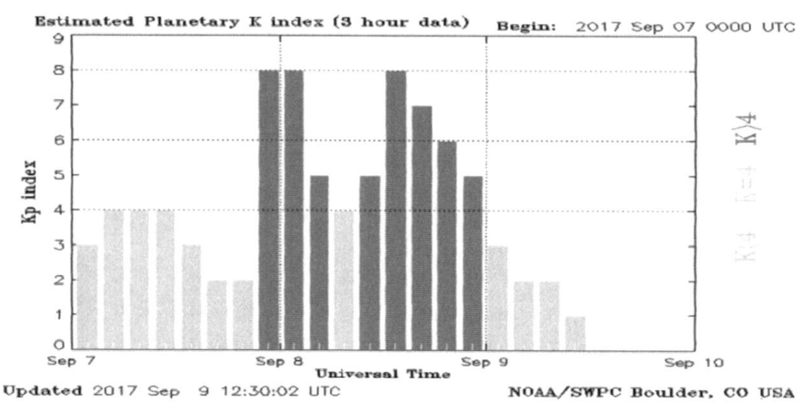

Figure 4. The 3-hour planetary Kp - index of the geomagnetic activity during storm on 07 and 08 Sept. 2017.

3. Energetic parameters of solar wind for the period September 2 – 15, 2017

Measurements of the basic parameters of the solar wind are taken for the period March 10 – 24, 2015 and September 2 – 15, 2017. These parameters are: the velocity ***V***, density or concentration ***N*** and temperature $\mathbf{T_p}$ of solar wind and the intensity of Interplanetary Magnetic Field (IMF) ***B***. In this study, measured solar wind parameters from the SOHO, ACE and WIND probes were considered. SOHO data for 2017 reveals growth of radial velocity, a more significant on 08.09.2017. At that time, the vector of IMF is in the south direction. There are all preconditions for the presence of geo-efficiency of those parameters [5]. For quantification of the different energies from the investigated CME events were calculated the following energetic parameters of solar wind [5]:

thermal energy $E_t = 3/2\ N\ k\ T_p$

magnetic energy $E_m = B^2 / 2\mu_0$

dynamic energy $E_k = 1/2\ N\ V^2$

where k is the constant of Boltzmann, T_p is the temperature of the protons, μ_0 is the magnetic permeability of vacuum (figure 5).

There are some unexplained experimental results related to stratosphere ionization. Since spring of 2015, Dr. Tony Phillips from Spaceweather.com and the students of Earth to Sky Calculus performed measurements of cosmic rays have by stratospheric balloons over California [6]. Soon after their monitoring program began, they quickly realized that radiation levels are increasing. The main reason for this behavior of stratosphere ionization is descending phase (2015-2018) of solar cycle 24 and the corresponding inverse course of cosmic rays, i.e. the increase of their intensity. In recent years, sunspot counts have plummeted as the Sun's magnetic field weakens, which allowed more galactic cosmic rays to penetrate in the Solar system. The latest measurements in 2017 show the radiation increase continuing apace–with an interesting exception in September – October after the G4 – Severe geomagnetic storm on September 08, 2017, and GLE72 on September 10, 2017. During the storm, the pace of the launch of the balloons was accelerated, and it was found that the radiation fell to levels that were not seen by 2015. The series of solar flares and CMEs actually pushed some cosmic rays away from Earth causing their strong modulation [7].

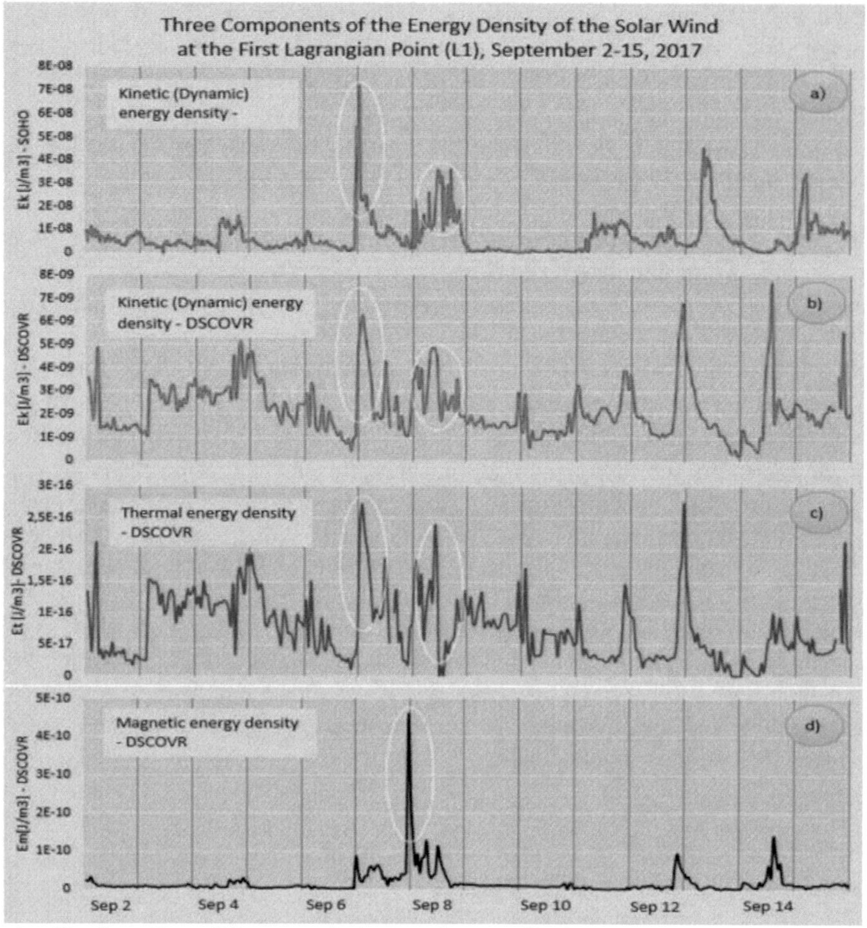

Figure 5. Calculations of the solar wind parameters from measurement by SOHO and DSCOVR space probes in the point of Lagrange: a, b) the kinetic energy density E_k - SOHO, DSCOVR. c) thermal energy density E_t and, d)magnetic energy density E_m during the investigated period September 02-15, 2017.

4. Explanation of decreased stratosphere radiation

The most reasonable and predictable explanation for the observed reduction in stratospheric radiation is the registered decrease in the intensity of the primary cosmic rays, the so-called Forbush effect [8-12] (figure 6).

The Forbush effect is usually observable by particle detectors on Earth within a few days after the a CME and the corresponding geomagnetic storm; the cosmic ray decrease takes place over the course of a few hours. A recovery is observed in the following several days. In this study the observed Forbush decrease of about 7% within the recovery can't explain the observed experimental results in stratosphere ionization [6]. That is why we are attracting a new additional physical mechanism that we call *cumulative Forbush effect.* However, the calculations and numerical evaluations performed by the methods presented in the works [8-12] show that the cumulative Forbush effect can't explain the observed long term decrease of stratospheric ionization in September-October 2017 [6].

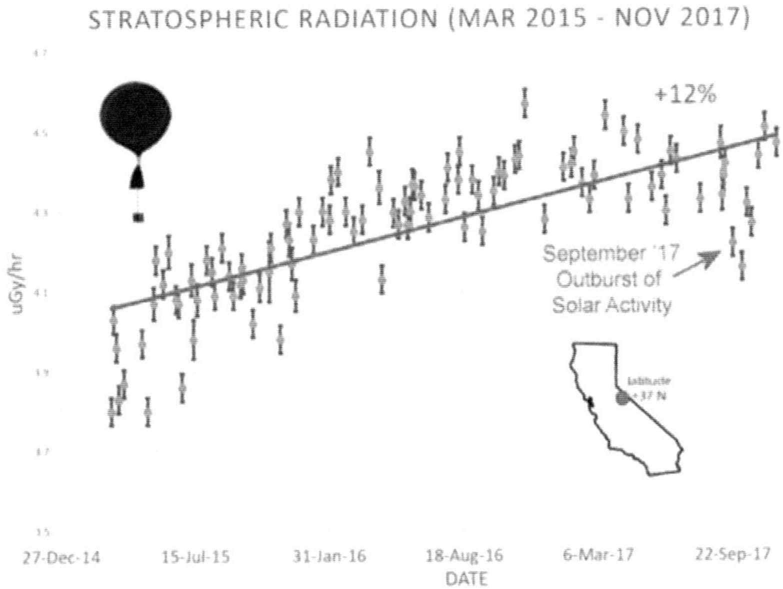

Figure 6. Stratosphere radiation measurement form March 2015 to November 2017. Toni Philips and his team show that radiation levels were increasing [6]. The main reason is the decreasing solar activity during the 24th cycle.

5. Variations of cosmic rays and Radiation Belts

First discovered before 60 years - at the dawn of the space age, the RBs are two collections of charged particles that circle Earth, held in place by the planet's magnetic field.

The pair of radiation belts surrounding Earth stop high-energy particles in their tracks. Data gathered by NASA's twin Van Allen Probes, which launched in August 2012, show that the donut-shaped RBs present a nearly impenetrable barrier to high-energy electrons, keeping them from hitting Earth. The barrier for the ultrafast electrons is a remarkable feature of the belts [13].

The variations of cosmic rays of galactic origin near the Earth is mainly influenced by heliosphere and geomagnetic factors. The nature of Galactic Cosmic Ray (GCR) variability has been explored in short and long-term scales and they reveal significant information on important processes related to the transport of GCRs from the source to the neutron monitor stations [7]. A significant factor is solar activity - the occurrence of CMEs on September 8-10, 2017, leads to significant variations of GCRs. On the one hand, this is reflected in a significant Forbush effect in the high energy region of GCRs, on the other it is associated with a general reduction in the intensity of the low energy region of GCRs. The latter is a fact, as mentioned above in stratospheric balloon measurements [6]. The first effect associated with the Forbush decreases of GCRs in the period of September 2017 can be relatively accurately interpreted as a process of shielding the GCR particles from the shock wave of the solar plasma stream of CME with a frozen magnet fields. This process is global for the Earth and has a classic first phase of GCR intensity decrease and a second subsequent recovery phase. The whole process lasts for three to five days [9-12]. More interesting is the fact of a reduction of the intensity of the low energy part of GCRs. This process is considerably longer in time, about two months, according to the measurements [6]. As we are talking about a low energy spectrum of GCRs, their effect will be observed with a decreased ionization in the stratosphere. Interpretation of such processes may be related to the passage of the GCRs through the heliosphere and their modulation. There are a few modulation mechanisms: 1) dispersion of the particles from the heterogeneity of interplanetary magnetic fields; 2) diffusion processes in interplanetary plasma; 3) adiabatic delay, etc. [13]. In addition, the geomagnetic factors generated by the radiation belts must also be taken into account. Radiation belts discovered in the dawn of the cosmic era still hide many of their secrets. In addition to the fact that they are special reservoirs of charged particles in the magnetosphere, they are also a shield against high energy electrons for the Earth. The data collected by NASA's Van Allen twins, launched in August 2012, show that Van Allen's radiation belts represent an almost impenetrable barrier for high energy electrons, preventing them from striking the Earth [14].

The cloud of the colder plasma, known as the Plasmosphere, sits in the middle of the conical belts and acts as a barrier to keeping fast-moving electrons away from Earth. But moreover, the information collected by the Van Allen probe shows that the inner edge of the outer belt is very abruptly defined. In solar events such as CME, this inner edge significantly shrinks to the Earth as a result of the shock wave. This leads to a particular reduction in the content of charged particles in the external radiation belt. This can also be interpreted as a fact of a decrease in ionization in the stratosphere after September 22, 2017, a few days after the coronal mass ejection of 8 and 10 September. Thus, the decrease in ionization in the stratosphere is not related to the Forbush reduction of the GCR and most likely to the modulation of the GCR stream from Van Allen's external radiation belts.

Acknowledgments

Acknowledgements are due to the NMDB: Real-Time Database for high-resolution Neutron Monitor measurements (www.nmdb.eu), founded under the 7th Framework Programme (FP7) of the European Union, as well as the IZMIRAN – Troitsk, Moscow region (Russian Federation) neutron monitor data base (http://cr0.izmiran.ru/common/ links.htm), and NOAA Space Weather Prediction Center – Boulder, Colorado (https://www.swpc.noaa.gov/) – for providing data.

References

[1] Tassev Y, Velinov P I Y, Mateev L and Tomova D 2017 *C. R. Acad. Bulg. Sci.* **70** 1437-1444.
[2] Tomova D, Velinov P I Y and Tassev Y 2017 *C. R. Acad. Bulg. Sci.* **70** 1567-1578.
[3] Velinov P I Y 1969 *C. R. Acad. Bulg. Sci.* **22** 33-36.
[4] Velinov P I Y 2017 *C.R. Acad. Bulg. Sci.* **70**, 995-1002.
[5] Schrijver C and Sicoe G 2009 *Heliophysics: Plasma Physics of the Local Cosmos* (New York: Cambridge University Press).
[6] Phillips T 2017 *Atmospheric radiation is increasing*; https://spaceweatherarchive.com/2017/12/11/atmospheric-radiation-is-increasing/
[7] Thomas S R, Owens M J, Lockwood M and Scott C J 2014 *Solar Physics* **289** 2653-2668.
[8] Velinov P I Y, Dorman L and Nestorov G 1969 *Geomagnetism and Aeronomy* **9** 813-817.
[9] Velinov P I Y, Dorman L and Nestorov G 1970. *Doklady Ross. Akad. Nauk* (*Proc. Russ. Akad. Sci.*) **190**, 1063-1065.
[10] Velinov P I Y, Nestorov G and Dorman L 1974 *Cosmic Ray Influence on the Ionosphere and on Radiowave Propagation* (Sofia: Publ. House of Bulgarian Academy of Sciences).
[11] Velinov P I Y, Mishev A, Mateev L and Dorman L 2008 *Proc. Int. Conf. on Fundamental Space Research (Sunny Beach, Bulgaria)* (Sofia: Publ. House of Bulgarian Academy of Sciences) p 431-434.
[12] Velinov P I Y, Dorman L and Mateev L 2009 Report S5.26 on ESWW6 16-20 November 2009, Brugge, Belgium, A. book, 66 pp 1-12; http://sidc.oma.be/esww6/
[13] Parker E N 1965 *Planet. Space Sci.* **13** 9–49.
[14] Redd N T 2014 *Radiation belts shield Earth from high-energy particles*; https://www.space.com/27979-van-allen-belts-barrier-particles.html

IOP Conf. Series: Journal of Physics: Conf. Series **1181** (2019) 012071 doi:10.1088/1742-6596/1181/1/012071

The BSUIN project – overview and same results.

Z Dębicki[1], K Jędrzejczak[1], M Kasztelan[1], W Marszał[1], J Orzechowski[1], J Szabelski[1] and P Tokarski[1]

[(1)]National Centre for Nuclear Research, Astrophysics Division, Cosmic Ray Laboratory, 28 Pułku Strzelców Kaniowskich 69, 90-558 Łódź, Poland,

E-mail: kj@zpk.u.lodz.pl

Abstract. The Baltic Sea Underground Innovation Network (BSUIN) [**1**] project aims to make the underground laboratories in the Baltic Sea region more accessible for innovation, business development and science by improving the information about the underground laboratories, the operation, user experiences and safety. BSUIN is EU funded by Interreg Baltic Sea [**2**] funding cooperation. The BSUIN consortium has 14 members from eight Baltic Sea countries. Six un-derground labs are looking for new collaboration in the project. One of the goals of the BSUIN project is to propose standard methods for the characterization of underground laboratories, including natural radiation background characterisation. We have proposed scheme for thermal neutron flux measurement: simple and low-cost but still very reliable. A pilot measurements were made in mines in Freiberg (Germany) Data analysis is in progress

1. Introduction

The purpose of the BSUIN project is to develop a service offering of Baltic Sea Region's (BSR) underground laboratories (UL) in order to develop their capability to offer technology transfer utilizing the facilities and research infrastructures of the ULs for business development. Currently the Baltic Sea ULs operate separately and their capacity to offer services to businesses is limited mostly to regional or national level. The project aims to develop the capabilities of ULs in order to improve their service offering as a capacity for innovation and to create a network of the Baltic Sea Regions Uls As a main outcome the BSUIN consortium will create a sustainable network organization, which will disseminate the technical, marketing, operational quality, training and other information about the BSR ULs created during the project.

2. BSUIN partners

Fourteen BSIUN partners from nine countries around the Baltic Sea are members of the BSUIN consortium. There are two partners from Finland, Sweden, Germany and Russia, three partners from Poland and a partner from Estonia, Latvia and Lithuania in the BSUIN consortium. There are also seventeen associated partners, who support BSUIN.

List of BSUIN partners:

(i) University of Oulu, Kerttu Saalasti Institute (Finland)
(ii) Oulu University of Applied Sciences (Finland)
(iii) University of Silesia in Katowice (Poland)

(iv) Swedish Nuclear Fuel and Waste Management Co (Sweden)
(v) KGHM Cuprum Research & Development Centre Ltd. (Poland)
(vi) TU Bergakademia Freiberg Technical University (Germany)
(vii) German Research Centre for Geosciences (Germany)
(viii) Vilnius University (Lithuania)
(ix) National Center for Nuclear Research (Poland)
(x) Baltic ScientificInstruments (Latvia)
(xi) Karelian Research Center of Russian Academy of Sciences (Russia)
(xii) Joint stock company "Khlopin Radium Institute" (Russia)
(xiii) Sotkamo Silver AB (Sweden)
(xiv) Tallinn University of Technology (Estonia)

There are six, existing or potential, underground laboratories involved in the BSUIN project. They are located in five countries, the exact location is shown in figure 1.

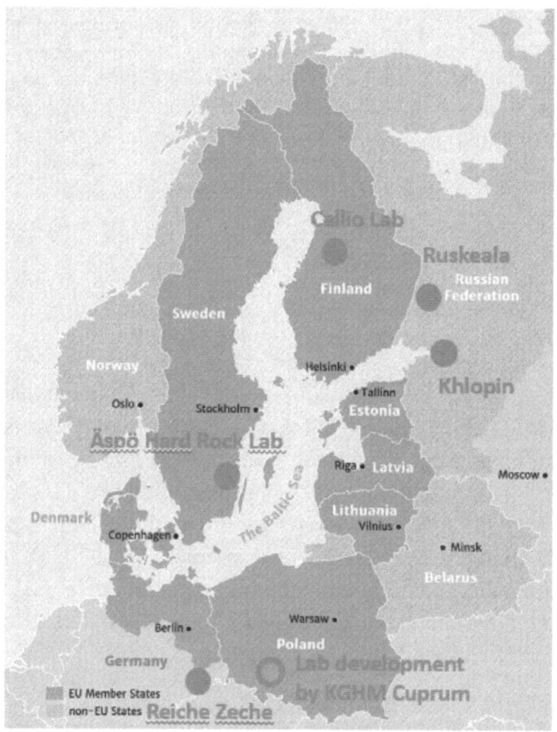

Figure 1. The map of underground laboratories involved in the BSUIN project. Underground laboratories are marked with circles. The INTERREG Baltic Sea [**2**] support area is also marked.

3. Work packages

Work at BSUIN is organized in five Work Packages: WP1 – "Project management and administration", WP2 – "Characterization of underground facilities", WP3 – "Service design, market design and branding of ULs as a capacity for innovation", WP4 – "Underground environment improvement", and WP5 – "Networking of BSR ULs and their users".

Tasks implemented in Work Package 2 are related to measurements of physical values. WP2 the package is divided into 4 activities:

- WP2.1 Geophysical characterization
- WP2.2 Natural radioactive background characterization
- WP2.3 Structural characterization
- WP2.4 Organizational characterization

The National Centre for Nuclear Research is the leader of activity WP2.2. The goal of this activity is the description of natural radioactivity background (NRB) in BSUIN underground locations. The extremely accurate characterization of NRB is essential in order to enable the utilization of the ULs as low NRB environments for product R&D, such as development of nuclear measurement or production of materials for nuclear industry or applied scientific research.

4. Minimal setup for thermal neutron flux measurements

4.1. The idea

As the first subject, we decided to consider what the simplest setup for thermal flux measurements should be. It is obvious that a setup containing many counters and sophisticated measuring electronics will be gave more reliable results than a smaller and simpler one, but in many cases it will be an overkill: unnecessarily raising the complexity and cost of the measurement. We would like to propose a setup as simple as possible, but still able to measure thermal neutron flux in underground laboratories with sufficient reliability.

In our opinion, the setup should have the following features:

- consist of at least two detectors, so that the measurement uncertainty can be determined by comparing the results
- the method of distinguishing real neutron recording from noise should exist
- the presence of the operator during the measurements should not be necessary, which will make very long-term measurements easier and cheaper

4.2. Our realisation

In our implementation, we used two proportional helium counters type ZDAJ NEM425A50. This type of counter is filled with helium 3He and efficiently registers thermal neutrons thanks to the significantly high cross section for reaction ^{3}He(n,p)^{3}H. Charged reaction products carry the released energy equal to 764 keV. A amplitude spectrum recorded by the counter has a characteristic shape with a peak corresponding to 764 keV energy and a tail of smaller amplitudes for cases when one of the reaction products escaped from the active volume. An example of amplitude spectrum, obtained with a relatively large flux of thermal neutrons is shown in the figure 2 .

ZDAJ NEM425A50 helium counter is made of a 50 cm long steel tube with a diameter of 2.5 cm. It is filled with helium under pressure ^{3} He at 4 atm. and natural krypton under pressure 0.5 atm Signals from individual counters were measured by independent measurement systems. These were simple devices consisting of an amplifier, ADC and a microcontroller, controlled and powered from a PC via a USB connector. Nevertheless, for every event a 50 samples long wave-form was collected (sampling rate 1.4 MHz = one sample per 700 ns). An example wave-form for a neutron is shown in the figure 2. This way of events registering allow us to distinguish the neutron from the electronics noises by analysing the shape of the wave-form

Our setup was remotely controlled via the Internet.

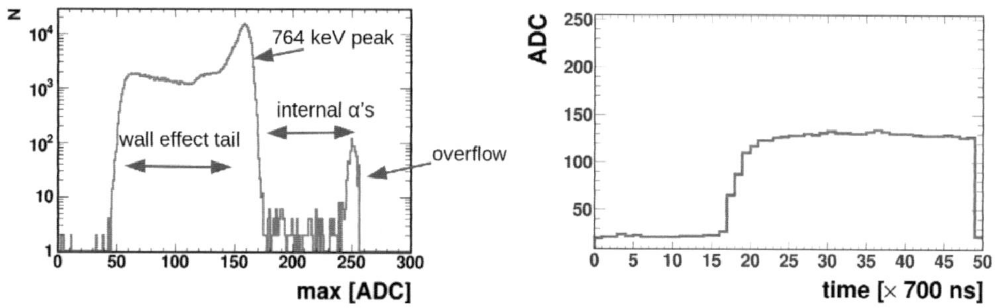

Figure 2. Left panel: The ZDAJ NEM425A50 type helium counter amplitude spectrum, obtained with a relatively large flux of thermal neutrons Right panel: An example wave-form for a neutron: 50 samples, one sample per 700 ns.

5. Data analysis method

The method of data analysis is based on the fact that the signals for real neutron have a longer rise time than the signals from electronic noise. In our case, instead of the rise time, it is more convenient to use the maximum time of the signal (step number of the time in which the maximum occurred). This is the same with the rise time because the position of the trigger is fix. On the two-dimensional histogram of the maximum signal position versus the maximum amplitude, real cases and noise are clearly separated into two "bands". Neutrons can be chosen thanks to the condition that the maximum position is greater than the 29th time step. After this cut, the spectrum of amplitudes will have the shape characteristic of the helium counter spectrum. The two-dimensional histogram the position of the maximum signal versus the maximum amplitude, and the spectrum of amplitudes before and after the cut is shown in the figure 3.

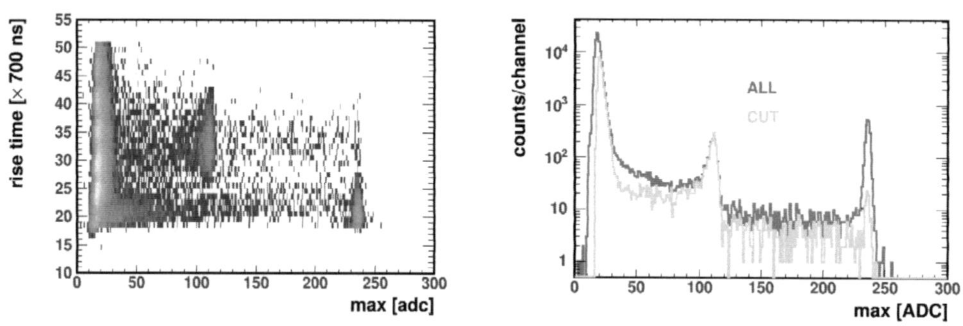

Figure 3. Data analysis method. On the two-dimensional histogram of the maximum signal position versus the maximum amplitude (left panel) real cases and the noise separates clearly into two "bands". Histogram of amplitudes after cutting the position maximum > 29 step of time (right panel) shows the shape characteristic for the spectrum from the helium counter. Data obtained in Freiberg, phase I of measurements.

6. Freiberg pilot program

We carried out pilot measurements of our method at the Freiberg mine (Germany). The measurements were carried out on the main level of the mine, at a depth of 250 m, near the Reiche-Zeche shaft. Measurements took place from March to June 2018 and consisted of two phases:

I: from Marth 6 to May 15 2018., life time 45.5 days, bare counters

II: from May 15 to June 27 2018r., life time 30.1 days, counters in the borax well

Setup in both phases is shown in the figure 4

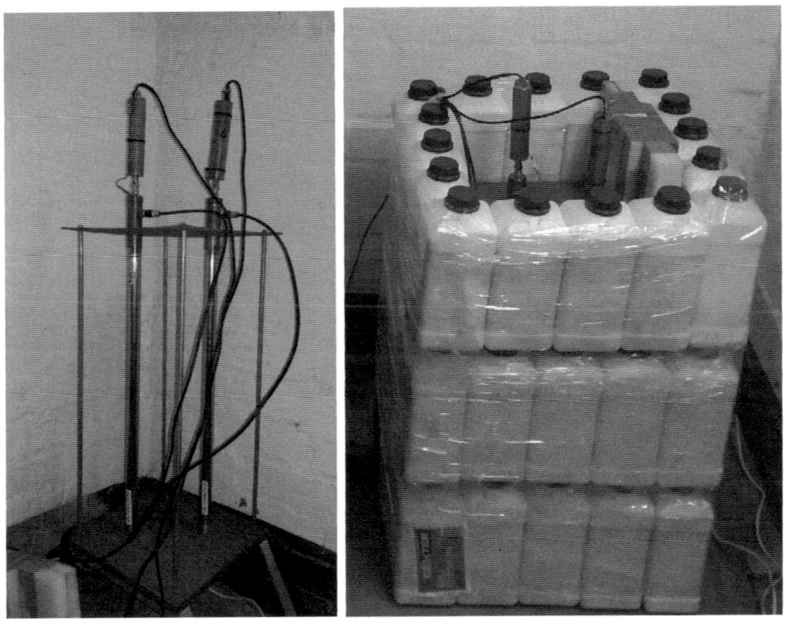

Figure 4. Configuration used for measurement in Freiberg. Left panel: Phase I, bare counters, panel Right: Phase II, counters inside the borax well.

The goal of Phase I was to measure of thermal neutron flux, while phase II was a test of our method. By surrounding the setup of phase I through about 100 kg of borax placed in plastic bottles, the flux of thermal neutrons reaching the counters was reduced. We expected that the rate of neutron counts will decrease, but counts from the internal alpha background will remain at the same level. Exactly this phenomenon was observed, as shown in figure 5 .

By comparing the rate of counts obtained in phase I with the predictions of Monte Carlo simulation thermal neutron flux can be determined. It is equal to:

- for counter No. 0 $(2.73 \pm 0.20) \times 10^{-6}$ $[cm^{-2}s^{-1}]$
- for counter No.1 $(3.07 \pm 0.27) \times 10^{-6}$ $[cm^{-2}s^{-1}]$

The distribution of results is consistent with the statistical uncertainty.

7. Summary

The BSUIN project, funded by the EU, links 14 institutions from 9 countries around the Baltic Sea. One of the tasks is the characterization of underground laboratories in terms of the natural radioactive background. In this context we have proposed a method for

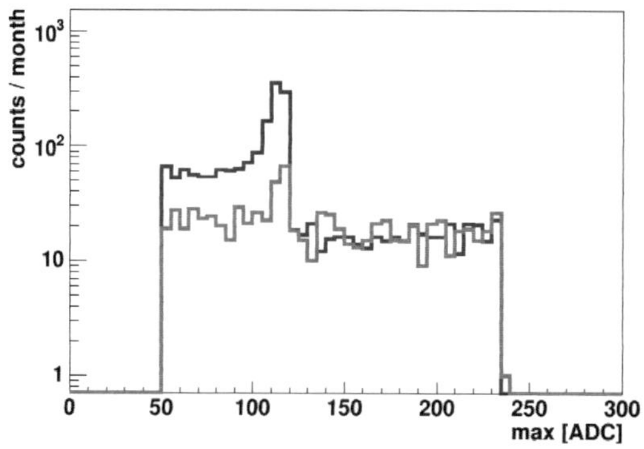

Figure 5. Comparison of the amplitude spectra registered in phase I and II. Spectra normalized on time.

measuring the thermal neutron flux. This method can become the standard used to characterize underground laboratories. We carried out pilot measurements at the Reiche-Zeche mine in Freiberg (Germany). Work will continue.

Acknowledgments
This work is supported by European Regional Development Fund of European Union and is a part of „Baltic Sea Underground Innovation Network" (BSUIN) Project. The BSUIN Project is funded by Interreg Baltic Sea funding cooperation.

References
[1] Baltic sea underground innovation network URL `http://bsuin.eu/`
[2] Interreg baltic sea URL `http://www.interreg-baltic.eu`

IOP Conf. Series: Journal of Physics: Conf. Series **1181** (2019) 012072 doi:10.1088/1742-6596/1181/1/012072

Fast multi channel acquisition system without dead time capable to measure oscilloscope-like signals.

Z Dębicki[1], K Jędrzejczak[1], M Kasztelan[1], W Marszał[1], R Nowak[2], J Orzechowski[1], J Szabelski[1] and P Tokarski[1]

(1) National Centre for Nuclear Research, Astrophysics Division, Cosmic Ray Laboratory, 28 Pułku Strzelców Kaniowskich 69, 90-558 Łódź, Poland,
(2) Lodz University of Technology, Institute of Automatic Control, Control Technology Division, B. Stefanowskiego Street 18/22, 90-924 Łódź, Poland.

E-mail: mk@zpk.u.lodz.pl

Abstract. Within the Baltic Sea Underground Innovation Network (BSUIN) project, we design and build a multichannel signal analysis and acquisition system. Our device can acquire signals from up to 32 detectors at the same time. Each channel has symmetric input, independent adjustable analog trigger, 10 MHz sampling rate, and 10 bits ADC resolution. Micro-controller (μC), FPGA, and micro-processor with Linux allow for different configuration of the system and for preliminary data analysis. In principle the signal's wave-forms from each channel are stored in memory, and a pulse shape analysis can be performed. Independent power suppliers are provided for each channel. We are going to use this measuring system for neutron registration with large number of neutron detectors (helium proportional counters, scintillation neutron detectors, and charged particle scintillation detectors). Especially, measurements of simultaneous measurement of coherent signals from a number of neutron detectors is the purpose for this data acquisition system.

1. Introduction

We want to present electronics that is able to collect the data with a very high degree of scalability in the number of measuring channels. High flexibility in the selection of acquisition parameters allow for optimal operation of connected detectors. Our multi channel acquisition system (photos of the prototype printed circuit boards on the figures 5 and 6) aims to be a universal and cost effective for multi purpose measurements. In particular, we are going to perform measurements at the same time with helium proportional counters, neutron scintillator detectors and with a classical plastic scintillators as well. One common acquisition system gives us consistent measurements with very good time resolution.

2. The BSUIN Project

Astrophysics Division of National Centre for Nuclear Research takes part in a multinational project called „Baltic Sea Underground Innovation Network" (BSUIN) [1]. This project is founded by UE programme „INTERREG Baltic Sea Region".

The main purpose of BSUIN is to combine underground laboratories (also those less known, and potential locations for underground laboratories, as well) in the Baltic Sea region. The main goal is the development of common standards for the description of the laboratories,

IOP Conf. Series: Journal of Physics: Conf. Series **1181** (2019) 012072 doi:10.1088/1742-6596/1181/1/012072

Interreg Baltic Sea Region — EUROPEAN UNION — EUROPEAN REGIONAL DEVELOPMENT FUND

Figure 1. „Baltic Sea Underground Innovation Network" logo (BSUIN)

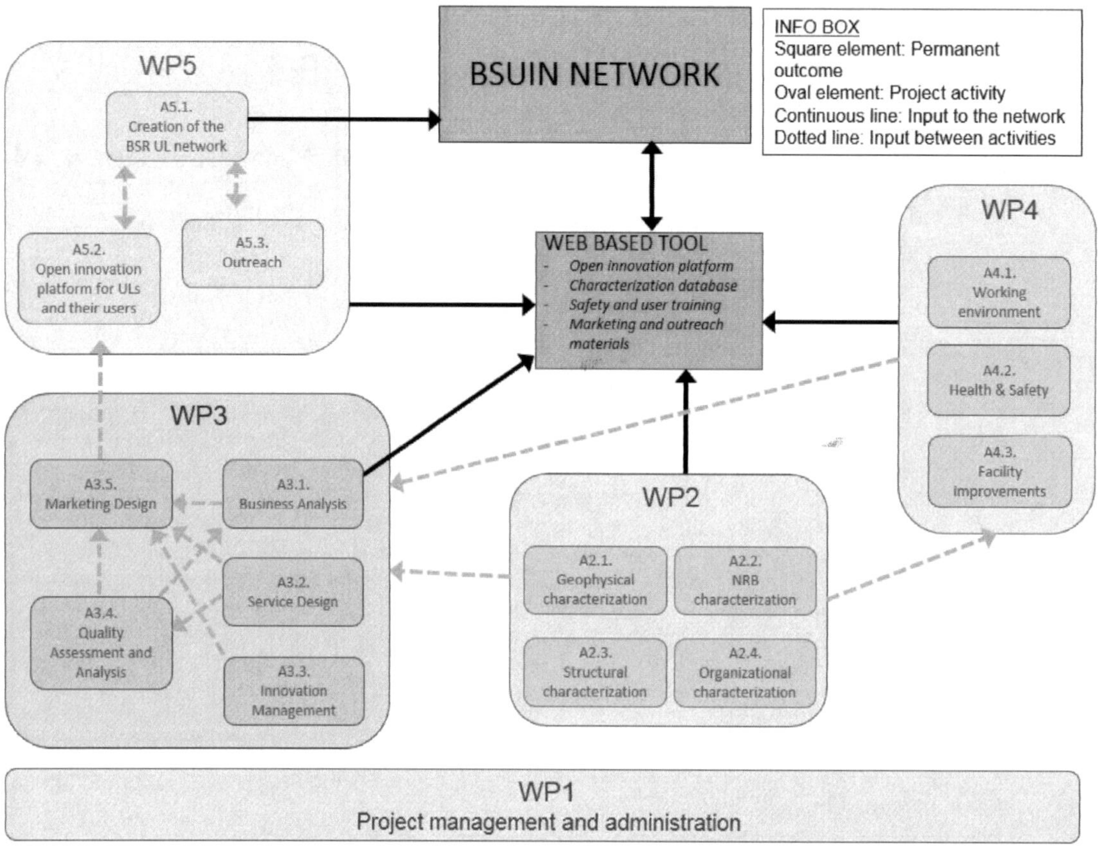

Figure 2. The BSUIN project work packages (WP) scheme. Schematically shown relations between different work packages and the activities. In each of the work packages, institutions from many countries of the Baltic Sea region participate.

security procedures, administration, advertising, etc. This is supposed to facilitate access to laboratories, and thus provide better ones to use. The Łódź group [2] is responsible in the project for the characteristics of the laboratories due to the background from ionizing radiation (indicated within BSUIN as WP2, A2.2, see figure 2), in connection with which we are planning several measurement campaigns in Pyhäsalmi mine [3] in Finland. The BSUIN consortium has 14 members from eight Baltic Sea countries. Six underground laboratories are involved in the BSUIN project.

3. Technical parameters

The biggest advantage of presented acquisition system (see figure 3) is an almost completely dead time free detection of oscilloscope like signals. It is possible due to a signal buffering system (see figure 4) with simultaneous memory Read&Write. Each ADC channel has the differential signal inputs, fully adjustable trigger parameters, the micro controller (μC) and the two regulated power supply units (PSU) for external detector equipment (e.g. ext. High Voltage

IOP Conf. Series: Journal of Physics: Conf. Series **1181** (2019) 012072 doi:10.1088/1742-6596/1181/1/012072

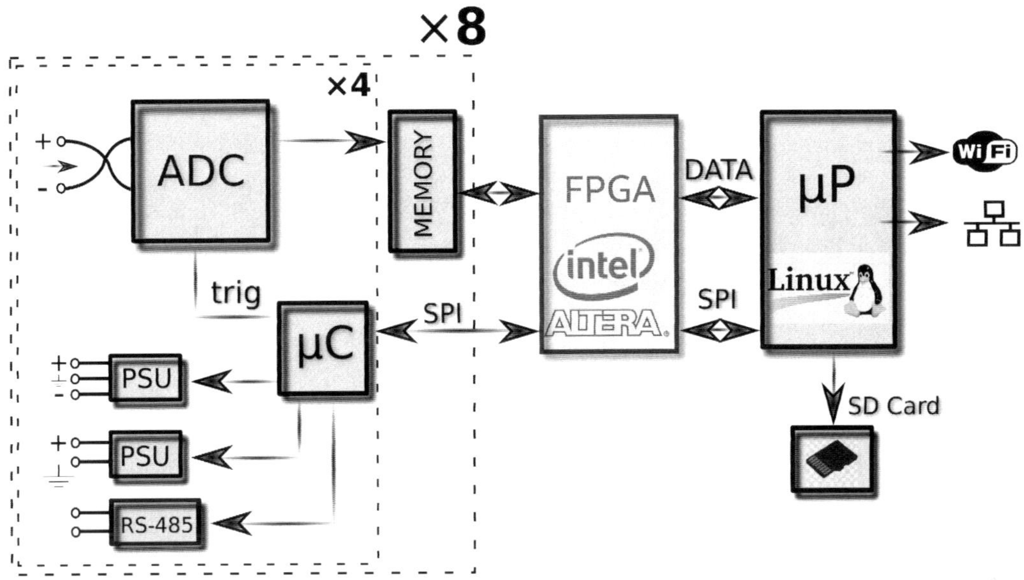

Figure 3. Schematic view of the multi channel acquisition system. On the left there are 8 modules (see figure 5), each with 4 ADC channels. All ADC modules are connected to the mainboard (see figure 6) with Altera chipset and embedded micro processor linux system.

PSU). Each channel can trigger (under independent adjustable conditions) acquisition data from all 32 ADC channels.
The micro processor (μP) system control module (right side of the figure 3) with Linux operating system manages acquisition from all channels and stores data onto built-in flash memory or transfers it to the external PC. The main work management system is based on the Field-Programmable Gate Arrays (FPGA) technology. In this design, the Altera programmable gate system was used [4]. Thanks to this, the modernization of the measurement system is much easier and does not require any hardware modifications. Communication between the μC to Linux acquisition system its done through the Serial Peripheral Interface (SPI) [5].

Table 1. Main technical parameters of the multi-channel analyzer.

ADC Channels	up to 32
Power Supply	2 per channel
ADC Resolution	10 bit
Sample rate	10 MHz (100ns)
FPGA	Altera (Intel)

4. Memory buffer

To store the oscilloscope like signals, we need fast ADC and the memory with low access time. When we aim to store all the data, we need a circular buffer (see figure 4) for each of 32 channels. Each buffer is divided into the several smaller circular buffers, which can be simultaneously written into and read from at the same time with high speed.
The signals are stored online in a series of the circular buffers for each channel. One circular buffer stores 32 samples ($3.2\mu s$ of the waveform). The signals from one channel can be stored in up to the 4096 circular buffers. When the trigger conditions are met (based on information from all channels) all the signals are stored in the memory. The length of stored wave-forms depends on signals amplitudes, newly emerged trigger signals and the specified time.

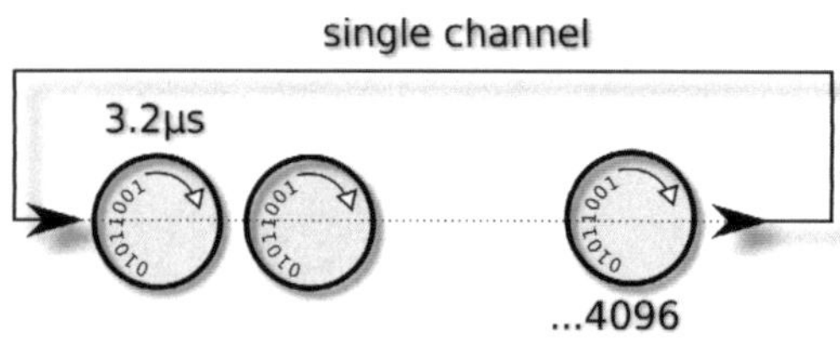

Figure 4. Schematic view of the memory circular buffer. The picture represents buffer for the single acquisition channel.

Writing and reading from each circular buffer is done almost at the same time (odd and even clock ticks), which allows to read data as soon as they are sampled by the ADC.

Figure 5. Electronic module with four ADC measuring channels, the micro-controller to control acquisition parameters and the power supplies.

Figure 6. The main-board with Field Programmable Array (Altera) and linux microprocessor system.

5. Summary

The new measurement data acquisition system (see figures 5 and 6) will allow to record the whole wave-forms without (almost) dead time. A large number of the measuring channels allows flexible selection of configurations and types of the detectors used in the measurements.

Differential inputs of the electrical signal from the detectors allow to avoid interference in the signal transmission from the detector to the ADC. Each measuring line has the ability to control two independent external power supplies that can be used to power up the detectors. The triggering parameters of individual data acquisition channels can be set independently in a very flexible way. Thanks to the programmable gate system (FPGA), an easy upgrade of this system is possible. An undoubted advantage is the use of the microprocessor system with its own operating system, thanks to which it is possible to carry out the measurements without an additional PC. It is especially important in locations where access to electricity could be very limited (e.g. in a mine). The low demand for electricity allows the use of a battery-powered power source, which is very important in the mines where the power supply can be very unstable and interfere with measurements.

Acknowledgments

This work is supported by European Regional Development Fund of European Union and is a part of ,,Baltic Sea Underground Innovation Network" (BSUIN) Project. The BSUIN Project is funded by Interreg Baltic Sea funding cooperation.

References

[1] Baltic sea underground innovation network URL `http://bsuin.eu/`
[2] National Centre for Nuclear Research, Astroparticle division, Łódź URL `http://zpk.u.lodz.pl/`
[3] Lubsandorzhiev S *et al.* 2018 *PoS* **ICRC2017** 1044
[4] Kuon, Ian and Tessier, Russell and Rose, Jonathan 2008 *Found. Trends Electron. Des. Autom.* **2** 135–253 ISSN 1551-3939 URL `http://dx.doi.org/10.1561/1000000005`
[5] Anand, Narinder and Joseph, George and Sam Oommen, Suwin and R, Dhanabal 2014 *Design and implementation of a high speed Serial Peripheral Interface* ISBN 978-1-4799-3543-7

GAMMA-400 experiment: perspectives of observation of the discrete astrophysical gamma-ray sources in the Milky Way disk

A V Bakaldin[1,2]**, A M Galper**[1,3]**, A A Leonov**[1,3]**, S I Suchkov**[1] **and N P Topchiev**[1]

[1] Lebedev Physical Institute of the Russian Academy of Sciences, Leninskiy prospect 53, Moscow, 119991, Russia
[2] Scientific Research Institute of System Analysis of the Russian Academy of Sciences (SRISA), Nakhimovskiy prospect 36/1, Moscow, 117218, Russia
[3] National Research Nuclear University MEPhI (Moscow Engineering Physics Institute), Kashirskoe highway 31, Moscow, 115409, Russia

E-mail: bakaldin71@mail.ru

Abstract. The GAMMA-400 space experiment is aimed for the study of gamma rays in the energy range from ~20 MeV up to several TeV. The observations will carry out with GAMMA-400 gamma-ray telescope installed onboard the Russian space observatory. GAMMA-400 has unique angular and energy resolutions for gamma rays with energy more than few GeV, which are significantly better than the Fermi-LAT ones. The orbit of the GAMMA-400 space observatory (near-Earth, circular with altitude about 200000 km) will be without the occultation by the Earth and outside the radiation belt. Thus it will be possible to carry out the continuous long-term observations of gamma-ray sources on the sky. In this work we present the expected statistics of gamma quanta that can be collected with GAMMA-400 from known discrete astrophysical gamma-ray sources in the disk of our galaxy during observations in orbit. 3FGL and 3FHL catalogs of discrete gamma-ray sources obtained in the Fermi experiment were used to calculate the number of gammas. It is shown that the accumulated statistics will allow us to carry out detailed investigation of characteristics (such as spatial distribution, energy spectrum, variability) of discrete gamma-ray sources in the Milky Way disk.

1. Introduction

The GAMMA-400 space mission is aimed for the study of gamma rays in the energy range from ~20 MeV up to several TeV. The experiment main scientific goals are: dark matter searching by means of gamma-ray astronomy; precise detailed observations of discrete gamma-ray sources and diffuse gamma-ray emission in the Milky Way disk (primarily, in the bulge) and beyond.

The observations will carry out with the GAMMA-400 gamma-ray telescope. The physical scheme of the instrument is presented in figure 1. GAMMA-400 consists of plastic scintillation anticoincidence top and lateral detectors (AC top and AC lat), converter-tracker (C), plastic scintillation detectors (S1 and S2) for the time-of-flight system (ToF), calorimeter (CC1, CC2), plastic scintillation detectors (S3, S4). The signals from anticoincidence detectors (AC top, AC lat) and S1, S2, S3 detectors are used for generation of the trigger. The instrument detailed description is presented in [1, 2]. Figure 2 shows the dependences of the instrument effective area A_{eff} for on-axis gammas vs. energy E and it efficiency of gamma-ray detection $\eta = A_{eff}(\theta)/A_{eff}(\theta=0°)$ vs. incident angle θ.

GAMMA-400 has unique angular and energy resolutions for gamma rays with energy more than few GeV, which are significantly better than the Fermi-LAT ones. The GAMMA-400 telescope will be installed onboard of the Navigator space platform, which is designed and manufactured by the Lavochkin Association. The GAMMA-400 observatory will be initially launched into a highly elliptical orbit (with an apogee of 300,000 km and a perigee of 500 km, with an inclination of 51.4°), with 7 days orbital period. After ~6 months the orbit will transform to about an approximately circular one with a radius of ~200000 km and will not suffer from the Earth's occultation and shielding by the radiation belts.

In this work we present the expected statistics of gamma quanta that can be collected with GAMMA-400 from known discrete astrophysical gamma-ray sources in the disk of our Galaxy during observations in orbit.

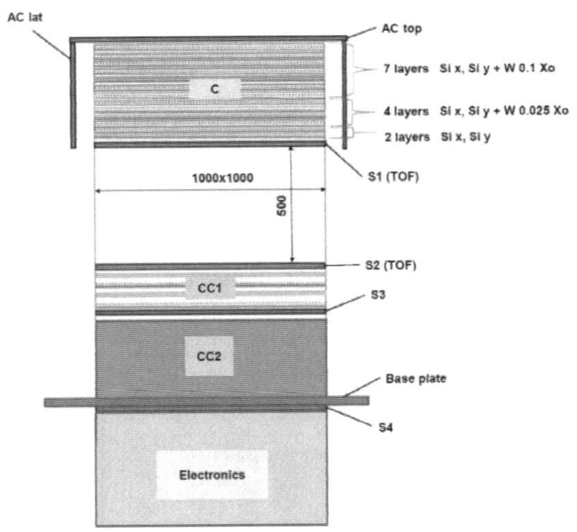

Figure 1. The GAMMA-400 telescope physical scheme. AC top, AC lat - anticoincidence top and lateral detectors; C - converter-tracker; S1, S2 – TOF detectors; CC1, CC2 – calorimeter; S3, S4 – scintillator detectors.

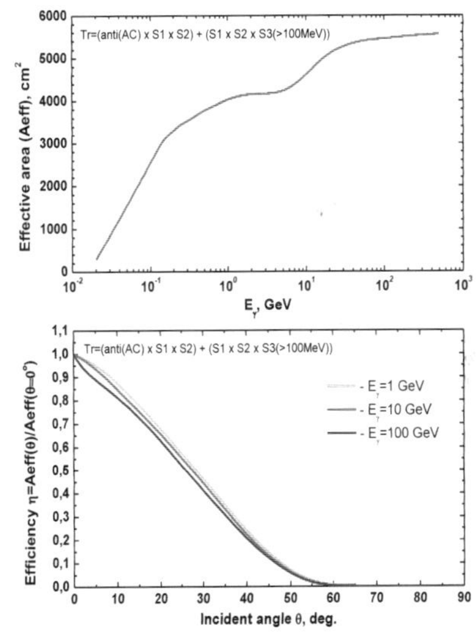

Figure 2. The dependences of the GAMMA-400 telescope effective area A_{eff} vs. energy E (top) and it efficiency $\eta=A_{eff}(\theta)/A_{eff}(\theta=0°)$ vs. incident angle θ (bottom).

2. The discrete gamma-ray sources

Since 2008 Fermi-LAT is operating in a near-Earth orbit in the scanning mode and surveying full sky every three hours. Up to time, the following catalogs of discrete gamma-ray sources have been published based on the Fermi-LAT observational results: 1FGL, 2FGL, 3FGL for the energy range from 100 MeV to 100 GeV; 1FHL for the energy above 10 GeV, 2FHL for the energy range of 50 GeV –2 TeV, and 3FHL for the energy range of 10 GeV –2 TeV.

In this work the 3FGL [3] and 3FHL [4] catalogs were used to estimate the expected statistics of gamma quanta that can be collected with GAMMA-400 from known discrete astrophysical gamma-ray sources. The main features of these catalogs are presented in table 1.

Table 1. The main features of 3FGL and 3FHL catalogs.

	3FGL	3FHL
Energy range	100 MeV – 100 GeV	10 GeV – 2 TeV
Number of sources	3034	1556
Period of observation	4 years	7 years

Data processing software version	Pass 7	Pass 8
Experimental energy spectrum bins	100 MeV – 300 MeV 300 MeV – 1 GeV 1 GeV – 3 GeV 3 GeV – 10 GeV 10 GeV – 100 GeV	10 GeV – 20 GeV 20 GeV – 50 GeV 50 GeV – 150 GeV 150 GeV – 500 GeV 500 GeV – 2 TeV
Interpolation energy spectrum types [a]	PowerLaw (2523 sources) LogParabola (395 sources) PLExpCutoff (110 sources) PLSuperExpCutoff (6 sources)	PowerLaw (1524 sources) LogParabola (32 sources)

[a]The mathematical formulas of interpolation energy spectra are given in [3, 4].

3. The considered GAMMA-400 observational programs

Taking into account the scientific goals of the GAMMA-400 experiment, we considered two possible programs of observations. The first one is the continuous rotation of the instrument axis in the Galactic plane with permanent angular velocity. The second one is the continuous orientation of telescope axis to the center of the Galaxy. The observational period was chosen the same and equal 1440 days.

Figure 3 illustrates the evaluated GAMMA-400 exposure at $E_\gamma = 10$ GeV in Galactic coordinates for two programs of observations and the 1440-day period. The units are equivalent on-axis observing time in seconds. The exposures for the other energies are close to the exposure at 10 GeV because the dependence of the instrument efficiency vs. gamma-ray incident angle demonstrates the weak dependence from the energy of gamma quanta (see figure 2).

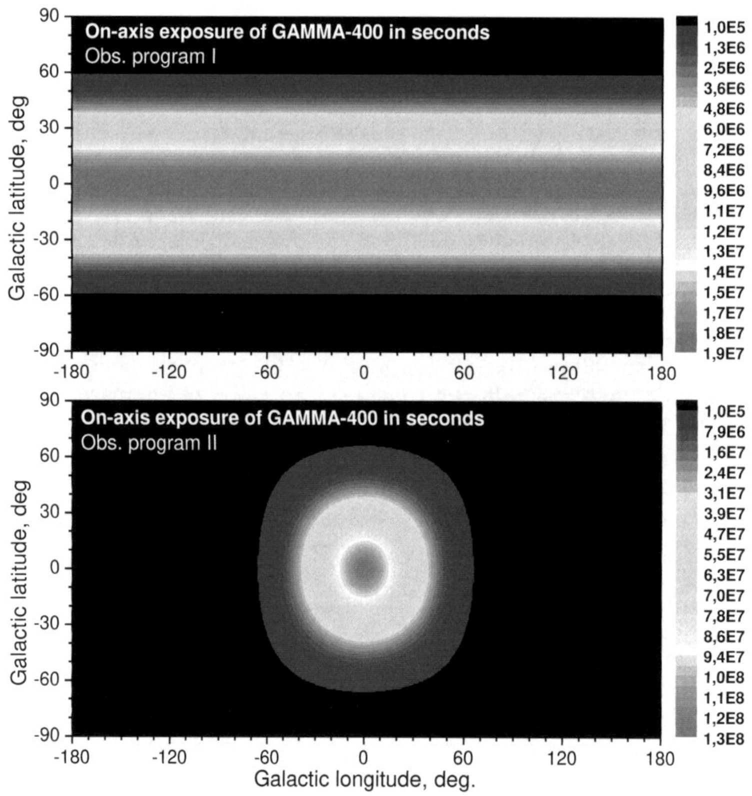

Figure 3. The evaluated GAMMA-400 on-axis exposure in seconds at $E_\gamma = 10$ GeV in Galactic coordinates for the first (*top*) and second (*bottom*) observational programs.

IOP Conf. Series: Journal of Physics: Conf. Series **1181** (2019) 012073 doi:10.1088/1742-6596/1181/1/012073

As it can be seen from figure 3, for the first program of observations GAMMA-400 will observe the Galactic disk with exposure about 20 Ms. For the second observational program GAMMA-400 will observe the center of the Galaxy only but the exposure will be in six times more – about 120 Ms.

4. The expected statistics of gamma-quanta from discrete sources

The estimated statistics of gamma quanta in 100 MeV - 100 GeV and 10 GeV - 2 TeV energy ranges from 3FGL and 3FHL discrete sources collected with GAMMA-400 telescope during the first and second observational programs are presented in figures 4 and 5, respectively.

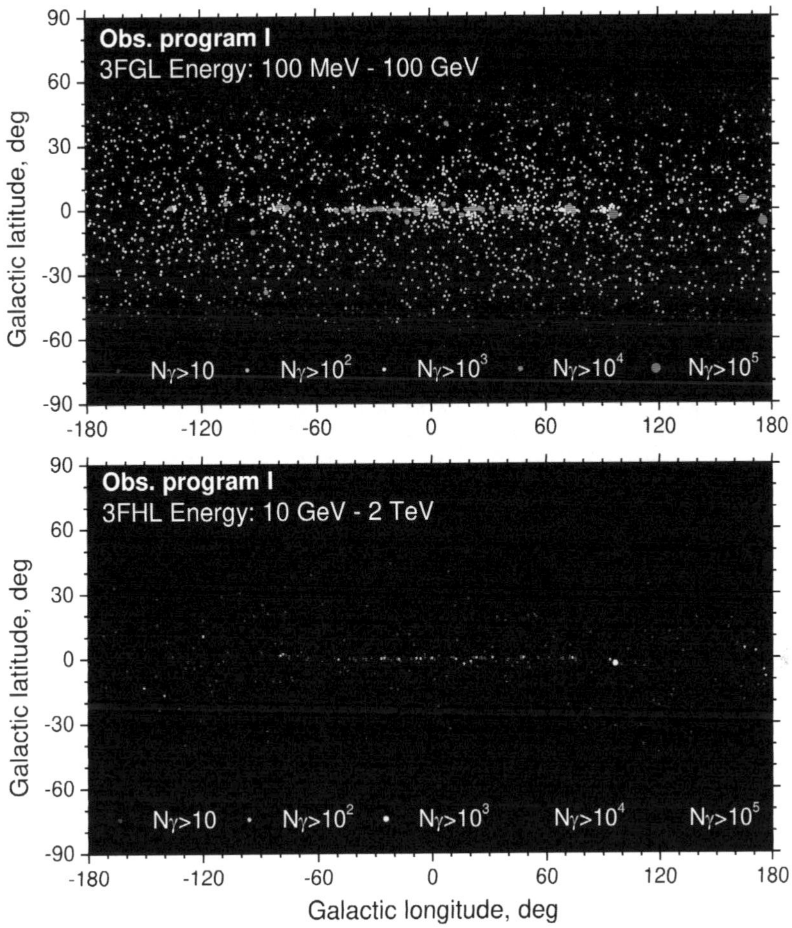

Figure 4. The estimated statistics of gamma quanta in 100 MeV - 100 GeV (top) and 10 GeV - 2 TeV (bottom) energy intervals from 3FGL and 3FHL discrete sources collected with GAMMA-400 during the first observational program.

For the first observational program the number of gamma quanta in the energy range 100 MeV - 100 GeV from single sources such as Vela, Crab, Geminga will be more than 10^5, and for the most sources in Galactic disk the number of gammas will be in the range 10^3 - 10^5. For the second program the number of gammas in 100 MeV - 100 GeV energy range from the several sources in the Galactic bulge will be > 10^5 and for the most sources the number of gammas will be more than 10^4. The statistics in the higher energy range 10 GeV – 2 TeV will be in about one hundred times less.

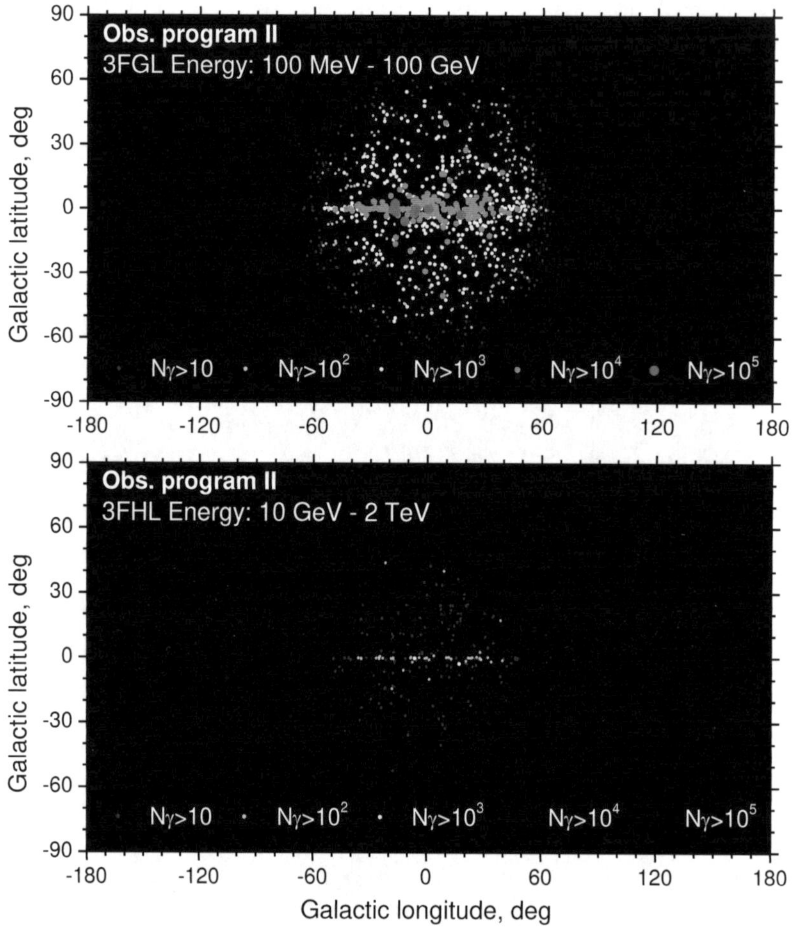

Figure 5. The estimated statistics of gamma quanta in 100 MeV - 100 GeV (top) and 10 GeV - 2 TeV (bottom) energy intervals from 3FGL and 3FHL discrete sources collected with GAMMA-400 during the second observational program.

Table 2 presents the number of sources with collected number of gammas more than 30 in the energy range from E_{min} to E_{max} for both observational programs.

The number of gamma quanta in nine bines in the 100 MeV – 2 TeV energy interval accumulated during the first and second GAMMA-400 observational programs is illustrated in figure 6.

Table 2. The number of sources with collected number of gammas more than 30 in the energy range from E_{min} to E_{max} for the first and second observational programs.

Catalog	Energy interval $E_{min} \div E_{max}$	Number of sources with $N_\gamma > 30$	
		Observation program I	Observation program II
3FGL	100 MeV – 100 GeV	2331	848
3FGL	300 MeV – 100 GeV	2039	775
3FGL	1 GeV – 100 GeV	1293	642
3FGL	3 GeV – 100 GeV	432	425
3FHL	10 GeV – 2 TeV	83	106
3FHL	20 GeV – 2 TeV	34	46
3FHL	50 GeV – 2 TeV	8	18

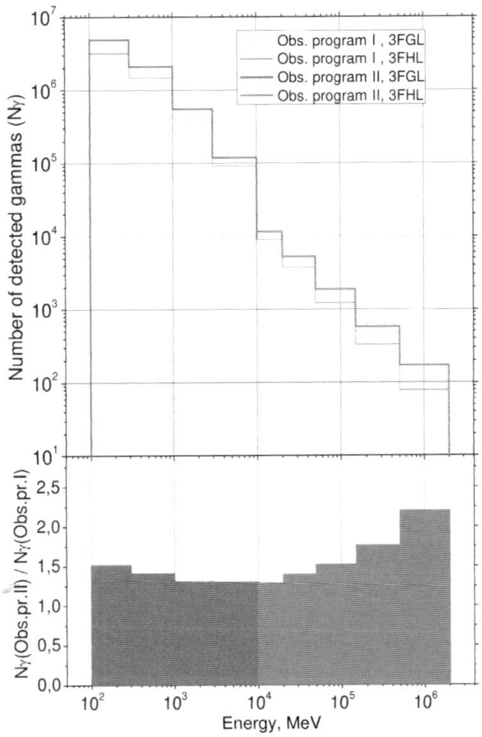

Figure 6. The total number of detected gammas in nine bins in the 100 MeV – 2 TeV energy range for the first and second observational programs.

As can be seen from table 2, for E_{min} < 1 GeV the number of observed sources for the first program is 2-3 times higher than the number of sources observed during the second program. At E_{min} > 3 GeV, more sources will be observed for the second program. In this case, the total number of observed gammas from sources for the second program will be ~1.5 times larger (see figure 6). Thus, the average number of gammas collected from separate sources during the second observational program will be 3-5 times greater than during the first program.

5. Conclusion

In this work we estimated the expected statistics of gamma quanta that can be accumulated with GAMMA-400 telescope from known discrete astrophysical gamma-ray sources in the disk of our Galaxy during observations in orbit. 3FGL and 3FHL catalogs of discrete gamma-ray sources obtained in the Fermi experiment were used to calculate the number of gammas. We considered two possible observational programs. The first program is the scanning of the Galactic plane, the second is the continuous observation of the Galactic center.

As it is followed from obtained results, from statistical point of view the second observational program is more effective. This program will allow us to observe about 800 discrete gamma-ray sources with unprecedented continuous exposure up to 120 Ms. The unique angular and energy resolutions of the GAMMA-400 instrument will permit to investigate in details these sources at very hard background conditions in Galactic bulge.

References

[1] Topchiev N P *et al.* 2016 *Journal of Phys.: Conference Series* **675** 032010
[2] Topchiev N P *et al.* 2016 *Journal of Phys.: Conference Series* **675** 032009
[3] Acero F *et al.* 2015 *The Astrophysical Journal Supplement Series* **218(2)** 23
[4] Ajello M *et al.* 2017 *The Astrophysical Journal Supplement Series* **232(2)** 18

Results of the missions within JEM-EUSO program

Mario E. Bertaina for the JEM-EUSO Collaboration

Department of Physics, University of Torino & INFN, Via P. Giuria, 1 - 10125 Torino, Italy

E-mail: bertaina@to.infn.it

Abstract. The JEM-EUSO program includes several missions employing fluorescence detectors to make a proof-of-principle of the UHECR observation from space and to raise the technological level of the instrumentation to be employed in a space mission like K-EUSO and POEMMA. EUSO-TA, installed at the Telescope Array (TA) site in Utah in 2013, has already detected 9 UHECRs in coincidence with TA Fluorescence Detector. EUSO-Balloon flew on board a stratospheric balloon in 2014. It measured the UV intensity on forests, lakes and the city of Timmins as well as proved the observation of UHECR-like events by shooting laser tracks. EUSO-SPB was launched on board a Super Pressure Balloon (SPB) in 2017. It proved the functionality of all the subsystems of the telescope on a long term; observed the UV emission on oceans and had a self-trigger system to detect UHECRs. A more ambitious mission (EUSO-SPB2) is planned. TUS, on board the Lomonosov satellite in orbit since 2016, has detected a few interesting signals in the UHECR trigger-mode. Mini-EUSO is in a final phase of integration and will be installed inside the International Space Station (ISS) in 2019. The main results obtained so far by these experiments are summarized.

1. Introduction

The origin and nature of Ultra-High Energy Cosmic Rays (UHECRs) remain unsolved in contemporary astroparticle physics [1]. To give an answer is rather challenging because of the extremely low flux of a few per km^2 per century at extreme energies such as $E > 5 \times 10^{19}$ eV (EECRs). The objective of the JEM-EUSO program (Joint Experiment Missions for Extreme Universe Space Observatory) is the realization of a space mission devoted to scientific research of EECRs [2]. Its super-wide-field telescope will look down from space onto the night sky to detect UV photons emitted from Extensive Air Showers (EAS) generated by EECRs in the atmosphere.

The JEM-EUSO program includes several missions from ground (EUSO-TA [3]), from stratospheric balloons (EUSO-Balloon [4], EUSO-SPB [5, 6]), and from space (TUS [7], Mini-EUSO [8]) employing fluorescence detectors to make a proof-of-principle of the EECR observation from space and to raise the technological level of the instrumentation to be employed in space missions such as K-EUSO [9] and POEMMA [10] (see Fig. 1). A space project devoted to the study of EECRs such as JEM-EUSO should have enough quality in terms of exposure determination and EAS parameter reconstruction to satisfy the scientific requirements of such an ambitious mission. A review of the key results obtained by each experiment of the JEM-EUSO program in this respect is here described (details can be found in [3] - [17]).

2. EUSO-TA

EUSO-TA is a ground-based telescope, installed at the Telescope Array (TA) site in Black Rock Mesa, Utah, USA. This is the first detector to successfully use a Fresnel lens based optical system

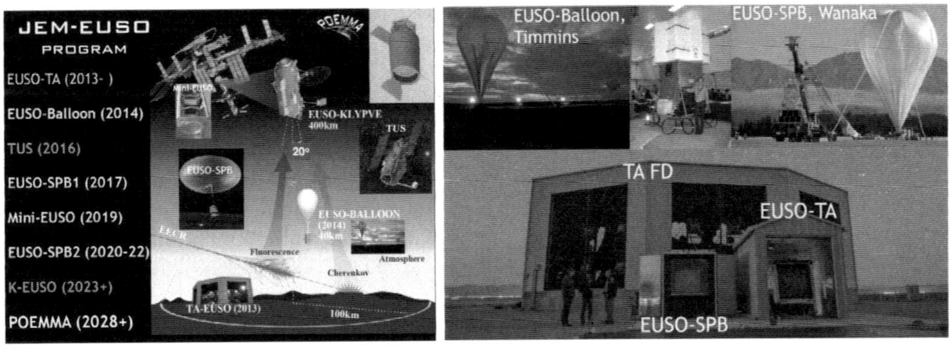

Figure 1. Roadmap and detectors of the JEM-EUSO program. See text for details.

and Multi-Anode Photomultipliers (MAPMT, 64 channels per tube, 2304 channels encompassing a 10.6° × 10.6° field of view - FoV) for detection of UHECRs. The telescope is located in front of one of the fluorescence detectors of the TA experiment (TAFD, see Fig. 1). Since its operation in 2013, the detector has observed several UHECRs and, in addition, meteors. The limiting magnitude of 5.5 on summed frames (~3 ms) has been established. Measurements of the UV night sky emission in different conditions and moon phases have been performed. These observations serve as a proof of concept for future application of this detector technology for space-based missions.

Fig. 2 shows an example of a UHECR which has been detected by EUSO-TA, using TA external trigger. It can be seen that re-binning of the images significantly increases the visibility of the tracks making EUSO-TA data more similar to those of ground-based UHECR telescopes, which have much larger pixel sizes. However, such a pixel size is not suited for space-based observations, to which EUSO-TA has been tuned. Simulations of the events made with the OffLine package [11] are also presented in Fig. 2. The shower image can be reproduced to very fine detail, taking into account the uncertainties in EAS reconstruction by TAFD and intrinsic modeling of the detector response. With the implementation of the external trigger, data is collected for each TAFD event. Therefore, an event is considered as detected if a linear trace is found in the EUSO-TA data and a corresponding event in TAFD results. To date, 9 UHECR events (see Fig. 2) have been identified in 130 hours of UHECR-dedicated observations. The distances of these events from the detector vary between approximately 1 and 9 km, while the energy is between $10^{17.7}$ - $10^{18.8}$ eV, according to the TAFD reconstruction. In 2016, simulations were performed, using the updated detector parameters and ESAF (EUSO Simulation and Analysis Framework) simulation code [11], resulting in 8 predicted events, consistent with the 9 UHECR observed to date. This result confirms the general understanding of the detector response through simulations. The proximity of the events and the dead time between frames makes 8 events visible in the detector for a duration of a single frame, and one event for two frames. EUSO-TA does not usually observe the EAS maximum, but a late stage of the shower development, and as such the number of registered photons corresponds to an EAS of lower energy than if the instrument was optimally pointed towards the shower maximum. Therefore, to estimate the instruments capabilities it was necessary to calculate the equivalent energies of the events (E_{eq}), corresponding to the reconstructed energy assuming that EUSO-TA observed the events shower maximum. This calculation is based on the parameters measured by TA for each individual shower. The corresponding points can be used to form a conservative estimate of the detectors energy threshold. In a very simplified approach, one can assume that the minimal number of counts on the focal surface for the UHECR to be detected is constant, proportional

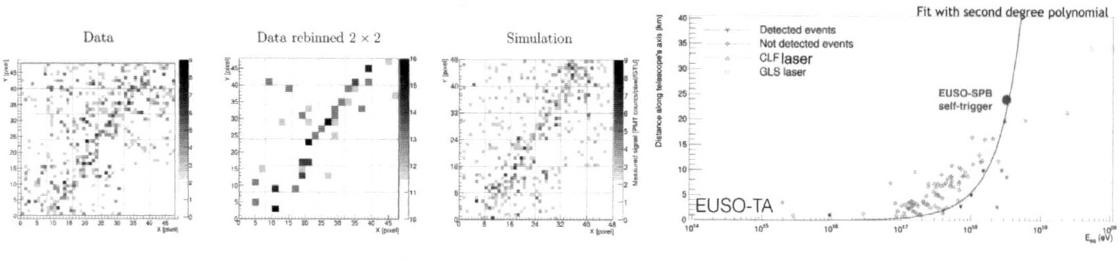

Figure 2. Left: Example of a UHECR observed by EUSO-TA (E $\sim 10^{18}$ eV, impact parameter 2.5 km with respect to the telescope, zenith angle of the axis of 35° and azimuth angle of 7°) with different pixel binning and comparison with simulations [11]. Right: All UHECRs detected by TAFD in the EUSO-TA FoV during its operation with non-detected events and laser shots superimposed. The plot shows the equivalent energy of the events as a function of their distance from the detector. The line suggests an estimate of EUSO-TA detection energy threshold while the blue point shows a similar estimate for EUSO-SPB (see text for details).

to its energy and reversely proportional to the square of the distance from the shower axis (R_p). Based on this assumption, $R_p = A \cdot \sqrt{(E_{eq})}$ is fit to detected events, where A is a free parameter. It can be seen that the strong signals of the TA Central Laser Facility (CLF, EAS equivalent energy of $\sim 10^{19.4}$ eV at a distance of 21 km) and Ground Laser System (GLS, EAS equivalent energy of $\sim 10^{19.7}$ eV and $\sim 10^{20}$ eV at a distance of 33 km) are on the right side of the curve, i.e. in the detectable region, as expected. Moreover, the ground-field tests performed with EUSO-SPB employing an internal trigger logic, indicate that its energy threshold lies on top of the fitting curve supporting the conclusions derived with such method, though EUSO-SPB overall detector efficiency is $\sim$2 better than EUSO-TA (see details in section 3).

3. EUSO-Balloon & EUSO-SPB

EUSO-Balloon [4] was launched by CNES from the Timmins base in Ontario (Canada) on the moonless night of August 25, 2014. After reaching the floating altitude of $\sim$38 km, EUSO-Balloon imaged the UV intensity in the wavelength range 290 - 430 nm for more than 5 hours before descending to ground. The refractor telescope consisted of a similar apparatus as EUSO-TA (two Fresnel lenses of ~ 1 m^2 size and a Focal Surface -FS- filled with MAPMTs). The spatial and temporal (Gate Time Unit - GTU) resolutions of the detector were 130 m and 2.5 μs, respectively. The full FoV in nadir mode was $\sim 11°$. During 2.5 hours of EUSO-Balloon flight, a helicopter circled under the balloon operating UV flashers and a UV laser to simulate the optical signals from UHECRs, to calibrate the apparatus and to characterise the optical atmospheric conditions. During flight EUSO-Balloon took more than 2.5 million images that have been analysed to infer different information: study of the performance of the detector; response of the detector to the UV flasher and laser events; UV radiance from the Earth atmosphere and ground in different conditions: clear and cloudy atmosphere, forests, lakes, as well as city lights.

The measurement of UV light intensity is relevant for a JEM-EUSO-like mission as it is one of the key parameters to estimate the exposure curve as a function of energy [12]. However, EUSO-Balloon uses a very different approach compared to previous measurements as it is based on an wide-FoV and wide-bandwidth optical refractive system with very fine spatial and temporal resolutions, which requires a careful computation of the optics and detector response to translate the detected counts into an absolute measurement. From the point of view of the capability of a space-based observatory for EECRs, the essential point is the number of counts per GTU at pixel level. Such pedestal should be dark enough to detect a EECR track on top. Fig. 3

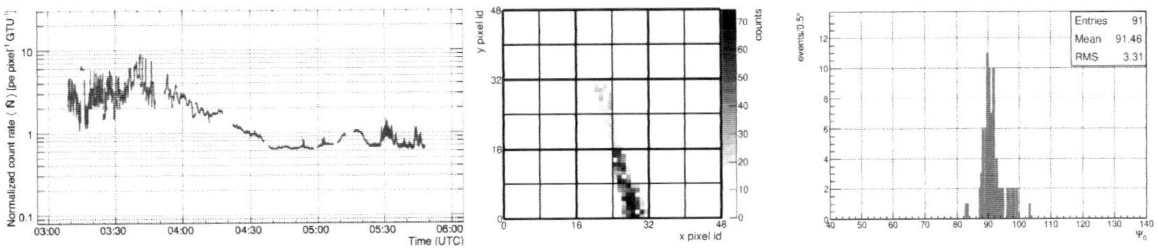

Figure 3. Left: Average normalized count rates $\langle \hat{N} \rangle$ as a function of the packet time [13]. Center: Example of a laser track detected by EUSO-Balloon. Right: Zenith angle reconstruction of the helicopter laser shots of energy 15 mJ with a 2-parameter fit method [14].

shows the average normalized count rates $\langle \hat{N} \rangle$ as a function of the packet time (breaks are due to technological tests foreseen for the flight). The count rate, $\hat{N}_0$, for clear atmosphere conditions is an input parameter to the EAS simulations used to estimate the reference aperture for EECR observations. The corresponding absolute intensity (I_0) is useful for the design of any space-based EECR observatory, independent of the configuration of the instrument. In EUSO-Balloon, only the back-scattered light from the airglow and extraterrestrial light contributes to the measured diffuse light. The reflectivity of the clouds is expected to be higher than clear atmospheric conditions. According to the pilot of the helicopter clear conditions were present between 04:38 and 04:52. Based on the average of the distribution in that time window, the reference $\hat{N}_0$ value is $\sim$0.65, the FWHM of the pixel distribution being $\sim$0.03. Between 04:20 and the end of measurement, when the artificial lights of Timmins and surroundings were out of the FoV, the count rate varies within a factor of $\sim$2. This gives the maximal difference of UV intensity between clear and cloudy sky conditions during flight. Ray trace simulations were then performed using the OffLine code to translate $\hat{N}_0$ in I_0 values. In the area with no artificial light sources, based on the airglow and starlight models, the measured count rate from the diffuse light under clear atmosphere conditions corrsponds to I_0 = 300 - 320 photons m^{-2} sr^{-1} ns^{-1} in the 300 – 500 nm band. This value is consistent with previous measurements and confirms a good understanding of the detector performance also in this respect, which is very important in view of JEM-EUSO.

The helicopter events revealed to be extremely useful to understand the system's performance and to test the capability of EUSO-Balloon to detect and reconstruct EAS-like events. Laser tracks are used to test the reconstruction algorithms [14]. The analysis is based on the geometry of the triggered pixels. The typical time fit of a laser event and the direction reconstruction are shown in Fig. 3 after requiring basic quality cuts: a) a set of events of same energy (15 mJ, equivalent to $\sim 10^{20}$ eV EAS); b) a track lasting at least 4 GTUs. It is important to remember that the read-out period of 2.5 μ is optimised for JEM-EUSO, which is expected to detect EAS at $\sim$400 km distance, instead of $\sim$35 km as in case of EUSO-Balloon. The fact that EAS-like tracks can be reconstructed in EUSO-Balloon is quite promising in view of JEM-EUSO.

EUSO-SPB was launched on April 25th, 2017, from Wanaka New Zealand as a mission of opportunity on a NASA Super Pressure Balloon (SPB) test flight planned to circle the southern hemisphere [5]. The primary scientific goal was to make the first observation of UHECR-EASs by looking down on the atmosphere with an optical fluorescence detector from the near-space altitude of 33 km. After 12 days 4 hours aloft, the flight was terminated prematurely in the Pacific Ocean about 300 km SE of Easter Island due to a leak in the balloon. The telescope was similar to EUSO-Balloon. An autonomous internal trigger was implemented according to [15] to detect UHECRs. About 30 hours of useful data were collected during night time.

In October 2016, the fully assembled EUSO-SPB detector was tested for a week at the EUSO-

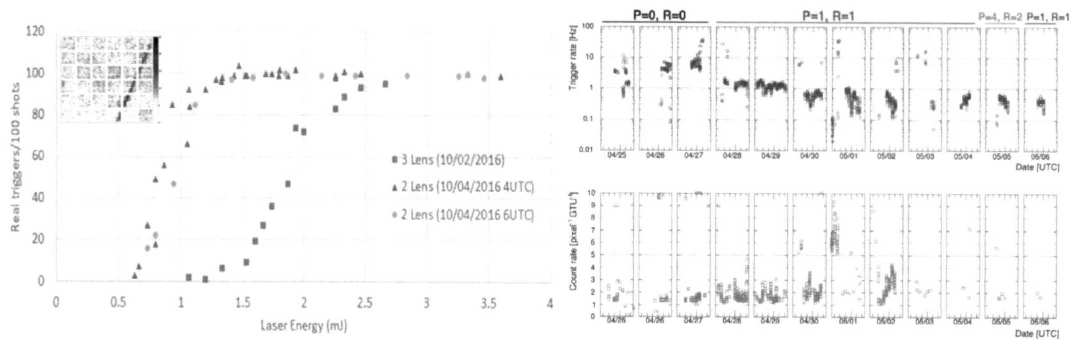

Figure 4. Left: trigger efficiency as a function of GLS energy with 2 and 3 lens systems for vertical shots. An example of GLS track (2 mJ energy) is displayed on the top-left corner. Right: Trigger rate (top) and average count rate measured at pixel level (bottom) in flight. P and R define variations of the trigger logic that were tested on flight.

TA site to measure its response and to calibrate it by means of a portable Ground Laser System (GLS). Observations of CLF, stars, meteors were performed. The left side of Fig. 4 shows an example of a GLS track detected by EUSO-SPB as well as the trigger efficiency curve of the 2 and 3 lens system as a function of GLS energy. With the 2 lens system, which is the one that flew from Wanaka, the $\sim$50% trigger efficiency is reached at laser energies whose luminosity is equivalent to $\sim$45° inclined EAS of E $\sim 3 \times 10^{18}$ eV seen from above by a balloon flying at 33 km altitude. This condition is represented by the blue point displayed in Fig. 2.

Right side of Fig. 4 shows the average pixel count and trigger rates during flight. The pixel count rate will be used to determine the exposure. The different intensity levels are due to changing conditions, such as clear atmosphere, clouds and presence of moon light. The trigger rate was on average compliant with JEM-EUSO requirements. The analysis of the collected data is ongoing. Tracks of cosmic rays directly crossing the detector have been recognized. However, no EAS track has been clearly identified yet. Simulations post-flight indicate that the number of expected events is $\sim$1 in the available data sample, confirming pre-flight assumptions for such a flight duration. Analyses are on still on going.

A more ambitious mission is currently under development, EUSO-SPB2 [6]. It will be equipped with 3 telescopes to observe UHECRs and try to catch ν_τ induced EAS, employing a Schmidt camera optics in view of POEMMA. The flight is expected by 2022.

4. TUS & Mini-EUSO

The Track Ultraviolet Setup (TUS) detector was launched on April 28th, 2016 as a part of the scientific payload of the Lomonosov satellite. TUS [7] is the world's first orbital detector aiming at detecting EECRs. The satellite has a sun-synchronous orbit with an inclination of 97.3°, a period of $\sim$94 min, and at altitude of 470 - 500 km. The telescope consists of two main parts: a modular Fresnel mirror-concentrator with an area of $\sim 2\times2$ m^2 and 256 PMTs arranged in a 16×16 photo-receiver matrix located in the focal plane of the mirror. The pixel's FoV is 10 mrad, which corresponds to a spatial spot of $\sim$5 km $\times$ 5 km at sea level from a 500 km orbit altitude. Thus, the full area observed by TUS at any moment is $\sim$80 km $\times$ 80 km. TUS is sensitive to the near UV band with a time resolution of 0.8 μs in a full temporal interval of 256 time steps. During its operation TUS has detected about 8×10^4 events that have been subject to an offline analysis to select among them those satisfying basic temporal and spatial criteria of EECRs. A few events passed this first screening. In order to perform a deeper analysis of such candidates, a dedicated version of ESAF as well as a detailed modeling of TUS optics and detector are being

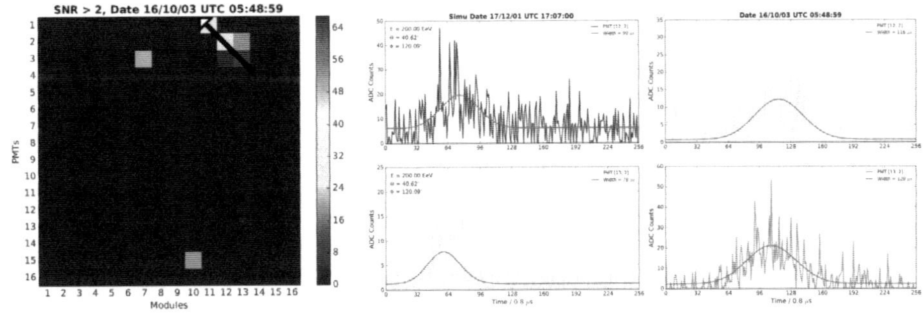

Figure 5. Left: Event 161003 occurring over Minnesota on October 3rd 2016. Right: Comparison between signals from ESAF simulation and those from the 161003 event for corresponding PMTs. Left panels: Signals in PMTs from ESAF simulation (EAS angle from the zenith $\theta = 41°$). Arbitrary units are used as the calibration of the detector response in simulation is still in progress. Right panels: Signals in the same PMTs taken from the candidate event.

developed. Fig. 5 shows an example of event which has passed first EECR selection criteria. A deeper analysis and comparison with simulations is still on going. A detailed study of this event is presented elsewhere [16] which includes also the study of weather conditions and presence of artificial lights. This example shows the importance of TUS measurements to determine analysis strategies in view of K-EUSO.

Mini-EUSO [8] is a UV telescope to be placed in 2019 inside the ISS, looking down on the Earth from a nadir-facing window in the Russian Zvezda module. Mini-EUSO will map the earth in the UV range (290 - 430 nm) with a spatial and temporal resolutions of ~5 km (like TUS) and 2.5 μs, respectively. Mini-EUSO has a FS similar to EUSO-TA. The optical system consists of 2 Fresnel lenses of 25 cm diameter with a large FoV of ~19°. A multi-level trigger [17] will allow the measurement of UV transients at different time scales, complementing TUS observations. Laboratory experiments with Mini-EUSO engineering model and simulations confirm the sensitivity of Mini-EUSO to EECR-like transients around 10^{21} eV.

4.1. Acknowledgments
This work is supported by the Italian Ministry of Foreign Affairs and International Cooperation.

References
[1] Dawson B.R., Fukushima M. and Sokolski P. 2017 *Prog. Theor. Exp. Phys.* **12** 12A107.
[2] Adams J.H. et al. (JEM-EUSO Coll.) 2015 *Experimental Astronomy* **40** 3.
[3] Abdellaoui G. et al. (JEM-EUSO Coll.) 2018 *Astroparticle Physics* **102** 98.
[4] Adams J.H. et al. (JEM-EUSO Coll.) 2015 *Experimental Astronomy* **40** 281.
[5] Wiencke L. and Olinto A. for the JEM-EUSO Coll. 2017 *PoS(ICRC2017)* 1097.
[6] Adams J.H et al. 2017 *ArXiv e-prints* [[arXiv]1703.04513.
[7] Klimov P. et al. (TUS Coll.) 2017 *Space Science Reviews* **8** 1.
[8] Capel F. et al. 2017 *Advances in Space Research* 10.1016/j.asr.2017.08.030.
[9] Casolino M. et al. (JEM-EUSO Coll.) 2017 *PoS(ICRC2017)* 368.
[10] Olinto A. et al. (POEMMA Coll.) 2017 *PoS(ICRC2017)* 542.
[11] Bisconti F. et al. (JEM-EUSO Coll.) 2017 *PoS(ICRC2017)* 463.
[12] Adams J.H. et al. (JEM-EUSO Coll.), 2013 *Astroparticle Physics* **44** 76.
[13] Bertaina M. et al. (JEM-EUSO Coll.) 2017 *PoS(ICRC2017)* 445.
[14] Abdellaoui G. et al. (JEM-EUSO Coll.), 2018 *J. of Instrumentation* **13** 05023.
[15] Abdellaoui G. et al. (JEM-EUSO Coll.) 2017 *Nucl. Instr. & Meth. A* **866** 150.
[16] Klimov P. et al. 2018 *This Conference Proceedings.*
[17] Belov A. et al. 2017 *Advances in Space Research* 10.1016/j.asr.2017.10.044.

IOP Conf. Series: Journal of Physics: Conf. Series **1181** (2019) 012075 doi:10.1088/1742-6596/1181/1/012075

Detector developments for a hybrid particle and radio array for cosmic-ray air-shower detection

Max Renschler[1], Aswathi Balagopal[1], Andreas Haungs[1], Thomas Huber[1,2], Tim Huege[1], Timo Karg[2], Marko Kossatz[2], Samridha Kunwar[2], Agnieszka Leszczynska[1], Marie Oehler [1], Harald Schieler[1], Frank G. Schröder[1,3], Karl-Heinz Sulanke[2], Andreas Weindl[1], Mark Weyrauch[1]

[1] Karlsruhe Institute of Technology, Kaiserstraße 12, 76131 Karlsruhe, Germany

[2] DESY, Platanenallee 6, 15738 Zeuthen, Germany

[3] Department of Physics and Astronomy, University of Delaware, Newark, DE, USA

E-mail: Max.Renschler@kit.edu

Abstract. Large-scale air-shower arrays could profit from a radio sub-detector as the radio emission in air showers is sensitive to the electromagnetic component. In particular, with an array of radio antennas in combination with particle detectors highly inclined EAS can be detected. This gives rise to new science cases, e.g. the search for PeV gamma rays coming from the Galactic Center at the South Pole, from where it is visible all over the year at an inclination of 61° degree. We report on progress and plans in the development of an optimized hybrid detector.

1. Introduction

The planned upgrade of the IceCube experiment to IceCube-Gen2 foresees a considerable extension of the in-ice detector volume [1]. With this, also the surface-detector array IceTop has to be extended to fulfill its primary goal as a veto detector regarding the search for ultra-high energy neutrinos. With an extended surface array also the detection and precise measurement of cosmic-ray air showers can be lifted up to a new level. As a first step towards a IceCube-Gen2 surface array, new detector designs and detection techniques are tested within the IceTop Enhancement. This is an upgrade of the already existing IceTop detector array with new detector prototypes of different kind [2]. Aim of the IceTop Enhancement is not only a study for the future Gen-2 surface array but also to establish IceTop as a world-leading detector for measuring cosmic rays.

A new hybrid detector consisting of a scintillation particle detector and radio detection is at the moment under development. When compared to the already existing ice-Cherenkov detectors of IceTop, scintillation detectors have several advantages like an easier and cheaper deployment. This enables a denser spacing of detectors with the effect of a lower energy threshold to detect and reconstruct cosmic-ray air shower. Exploiting also the radio signal of cosmic rays with a hybrid detector improves reconstruction accuracy of cosmic-ray events and opens whole new physics studies and science cases which will be discussed in the next section.

Regarding the scintillation detectors, a first version prototype detector has been developed and a first station consisting of seven detectors and the associated DAQ has successfully been deployed at the South Pole in the summer season of 18/19 [2]. The radio extension is at the moment under development and the deployment of two antennas at the South Pole has been proposed for the season 18/19. The future plan for the IceTop Enhancement is a deployment of 37 detector stations consisting of 259 scintillation detectors potentionaly enhanced by 74 radio antennas inside the footprint of the existing IceTop detector array [3].

2. Benefits of a radio surface array at the South Pole

Radio measurements with first prototype antennas a decade ago confirmed excellent background conditions at the South Pole [4]. Meanwhile the radio technique for cosmic-ray air showers has matured, and its potential for precise measurements has been experimentally demonstrated [5, 6]. A deployment of radio antennas as part of a hybrid detector has practical advantages, such that both detectors can share a common DAQ and that the deployment infrastructure can be used for the simultaneous deployment of both detectors. Beside this practical reasons there are multiple advantages to the measurement of cosmic-ray air shower as well as to the search for neutrinos and even new science cases which can only be reached with a combination of radio and particle detectors.

Figure 1. SKALA-2 antenna in the South-Pole design with a height of 0.3 m from ground to the lowest lobe.

The main purpose of the IceTop array is vetoing atmospheric neutrinos to support the search for galactic and extra-galactic neutrinos with the IceCube detector. Due to the sensitivity of radio to highly inclined events up to an inclination of $\Theta = 70°$, the veto capabilities of the IceTop veto detector will increase with the use of radio antennas. Due to the increase of the radio footprint with inclination of the shower, inclined showers can be reconstructed successfully even if the shower core is hundreds of meters outside of the area of IceTop. Of course, this high sensitivity for inclined events is also a benefit for the search for cosmic-ray events in general and turns the IceTop detector into a high-potential cosmic-ray observatory.

With the high sensitivity for inclined shower, radio enables the search for PeV gamma rays from the galactic center which is not possible with particle detectors alone. The IceCube site has the very unique feature that the center of our Galaxy is visible all over the year in an inclination of 61°. Measurements performed with the High Energy Stereoscopic System (H.E.S.S.) telescope suggest that our Galactic Center houses a Pevatron [7]. Due to the high inclination in which the Galactic Center is seen from the IceCube site, the detection of these PeV gamma rays is only possible with a combination of radio antennas and particle detectors. An energy threshold of below 1 PeV with radio detection can be achieved by measuring in a higher frequency range of 100 - 190 MHz [8].

In addition to this, radio measurements together with particle-detector measurement and in-ice measurements of atmospheric muons will probe and improve present hadron interaction models used in air-shower simulations [3].

3. The scintillation detector and the data acquisition system

The IceCube-Gen2 Collaboration has designed, tested, and deployed a scintillation detector at the South Pole [2].

Each scintillation detector consists of 16 scintillator bars manufactured by Fermi Lab. Two

IOP Conf. Series: Journal of Physics: Conf. Series **1181** (2019) 012075 doi:10.1088/1742-6596/1181/1/012075

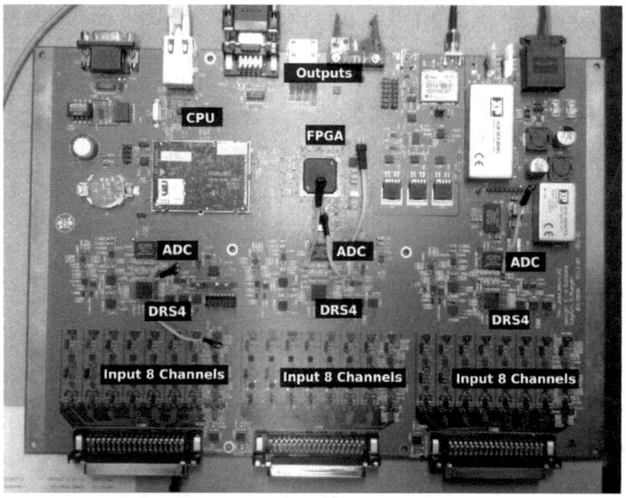

Figure 2. The IceTAXI DAQ equipped with three DRS4 sampling chips.

Wavelength shifting fibers go through each bar and return though the next but one bar back to the optical readout. All fiber ends are bundled and optically coupled to a 6 × 6 mm^2 Silicon Photomultiplier *S13360-6025* from Hamamatsu [9]. The electronics inside of the detector houses the Hamamatsu SiPM power supply *C11204-02* which includes a temperature control loop to stabilize the over-voltage of the SiPM at variable ambient temperatures. The SiPM signal is amplified in two amplification stages in a high-gain channel (10x) and a low-gain channel (1x) and transformed to a differential signal. The differential analog signal is send via 70 m of cable to the DAQ (IceTAXI). Only the high-gain or the low-gain signal can be transferred at one time. The purpose of two different amplifications of the SiPM signal is doing the in-field characterisation of the SiPM with the high-gain channel and use the low-gain channel for cosmic-ray measurements.

The DAQ of the scintillation detectors samples the incoming analog signals with DRS4 (Domino Ring Sampler) sampling chips [10]. Figure 2 shows a picture of the IceTAXI DAQ. The IceTAXI DAQ can be equipped with three DRS4 chips. Each DRS4 has 8 input channels and can sample the incoming signal with a variable sampling rate of 0.7 - 5 GHz. For sampling the seven scintillation detector signals, one DRS4 chip is used. A trigger signal is generated with a discriminator of variable threshold. When a trigger occurs, the samples stored with capacities inside of the DRS4 are read out with a 33 MHz ADC and send to an FPGA which processes the data and sends it to a UNIX based operation system running on an ARM chip on the TAXI board. Data transfer and time synchronization are realized with a White Rabbit Network [2].

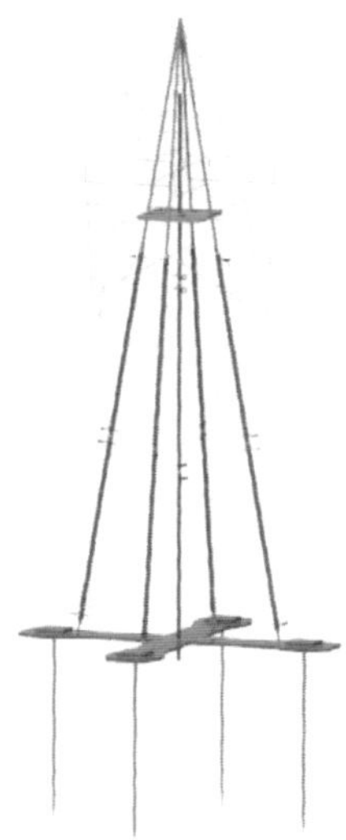

Figure 3. CAD drawing of the highest South-Pole antenna design. The distance of the lowest antenna lobe to the ground level is 1.8 m.

4. The radio antenna candidate SKALA

The SKALA antenna developed for the SKA (Square Kilometer Array) obervatory is considered as antenna [11]. The SKALA antenna is at the moment under development for the low-frequency part of the SKA experiment [12]. First measurements have been made with the first prototype version of the SKALA antenna at KIT. With four

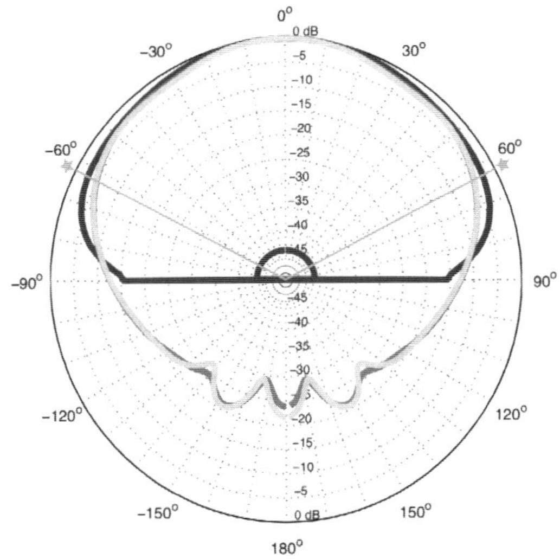

Figure 4. Simulated direction pattern of the SKALA antenna. The blue line shows the response for infinite ground, the red line for soil and the green line for a mesh over soil ground. The important angle of 61° is emphasized in the plot for better orientation. (Adopted from [14])

antennas of the second version prototype kindly provided to KIT, a small test array has been build up together with scintillation detectors and particle detectors of the former KASCADE experiment [13]. Two of these version two prototype antennas are considered to be deployed at the South Pole. Figure 1 shows a picture of the SKALA antenna.

The SKALA antenna meets the requirements for the deployment and use at the South Pole. The frequency bandwidth of the antenna is 50 - 350 MHz which contains the target frequency band of 100 - 190 MHz. Especially regarding the aim of measuring PeV gamma rays from the Galactic Center a good sensitivity for inclined signals is of great importance. Figure 4 shows the simulated directive pattern of the SKALA antenna for a signal frequency of 150 MHz and a 45° cut [14]. For all simulated grounds the attenuation for an inclination of $\Theta = 61°$, which would be the inclination at which the Galactic Center is visible, is only about $5\,dB$ compared to the zenith, which is an acceptable value. The simulated LNA noise with 30 - 40 K is sufficiently low to detect cosmic-ray signals down to the PeV regime. The LNA noise becomes even more important since it is the limiting factor for measurements in higher frequency bands like it is planned for the measurements at the South Pole.

5. Preparations of the antenna for a deployment at the South Pole

It is foreseen to deploy in a first step two version-two prototype SKALA antennas at South Pole. The antennas will have different heights to study the effect of snow accumulation around the antenna. To prepare the antennas for a deployment at the South Pole, special care has to be taken regarding the stability of the used material and the antenna electronics at temperatures of lower than -70°C and a high radiation of UV light.

Figure 3 shows a CAD drawing of the antenna design. The antenna legs as well as the middle pole and the snow spikes are made out of glass-fiber reinforced plastic material. For the antennas which are to be deployed at the South Pole, three mountings have been assembled with antenna legs of a length of 0.5 m, 1.0 m and 2.0 m. This resembles a distance of the lowest antenna lobe to the snow level of around 0.3 m, 0.8 m and 1.8 m. The middle pole stabilizes the antenna structure but serves primarily as a cable guide. The antenna sits on a base made out of multiplex plywood. The base is attached with snow spikes made out of glass-fiber reinforced plastic poles to the snow and ice surface.

All original plastic parts of the antenna like the LNA housing or the stabilizing plate in the middle

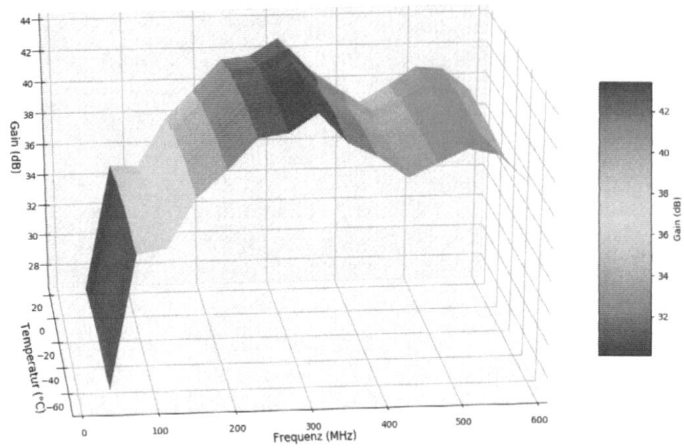

Figure 5. Measurement of the SKALA LNA gain for a temperature range from 20°C to -72°C.

of the metal antenna legs have been stress-tested down to a temperature of -70°C. During these tests no cracks or deformations have been observed.

Special care was taken about the low-temperature test and characterization of the antenna LNA. The LNA has been characterized down to a temperature of -72°C within a wide frequency range of up to 1 GHz. For this, a frequency generator was attached to the inputs of the LNA. During temperature cycles in the temperature chamber, the LNA response to signal pulses from the frequency generator have been recorded. The gain of the LNA depending on the frequency and the temperature has been received by comparing the Fourier spectra of the pulse from the frequency generator and the corresponding LNA response. Figure 5 shows the result of the LNA gain measurement depending on the temperature and the frequency. The gain measurements of the LNA show no significant dependence of the gain regarding the temperature.

6. Modifications to the IceTAXI DAQ for radio

To include the sampling and digitization of the radio signals into the current DAQ of the scintillation detectors, IceTAXI, some modifications have to be applied to the IceTAXI DAQ. Figure 6 shows a sketch of the planned layout of the DAQ. The IceTAXI DAQ is equipped with three DRS4 sampling ships of which one is sampling the scintillation detector signals. It is foreseen to operate two antennas with two polarizations each with one IceTAXI DAQ. For this, each of the remaining two DRS4 will sample one antenna with two polarizations. For this, the DRS4 chips are operated in channel cascading mode, which means that the incoming signal is given to multiple input channels of the DRS4 chips. The DRS4 chip cascades the sampling of the incoming signal which results in longer trace length. With this, a trace length of 4.1 µs at a sampling frequency of 1 GHz can be achieved.

Additional electrical parts to process the radio signal are added to the IceTAXI by including additional PCBs. These PCBs house the bias-tees for powering LNAs in the antennas, filters and amplifiers. Since the DRS4 is operated in channel cascading mode, the signal line of each antenna polarization has to be fanned out into 4 DRS4 inputs. Additionally the incoming 50 Ω single-ended signal from the antenna LNA has to be transformed to a 100 Ω differential signal to be processed with the DRS4. For this, the antenna signal is fanned out into four identical amplifiers which amplify the antenna signal by a factor of 10dB to compensate attenuation by the bias-tee, filters and the fan-out of the signal. At the same time, the amplifiers transform the incoming single-ended signal into the needed differential form.

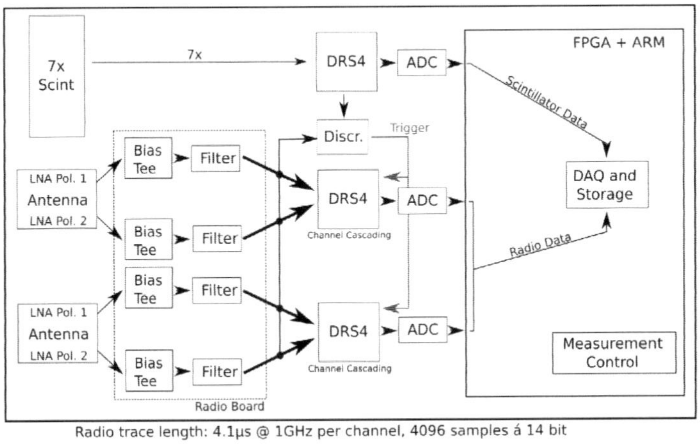

Figure 6. Block diagram of the hybrid IceTAXI DAQ digitizing seven scintillation detectors and two radio antennas with two polarizations each.

7. Conclusion

A scintillator-based particle detector with silicon photomultiplier (SiPM) has been developed and deployed by the IceCube-Gen2 Collaboration. We have designed a radio extension and test measurements with prototype radio antennas are ongoing. The SKALA antenna developed by the SKA observatory is considered as candidate. For a deployment at the South Pole a low-temperature and UV-light resistant antenna structure based on glass-fiber reinforces plastic material has been designed and successfully tested. Processing of the radio signal will be integrated into the existing scintillator DAQ. The DAQ will sample the signals of two antennas with two polarization channels each. Sampling is realized with DRS4 sampling chips triggered by the particle detectors. This enables the measurement of radio signals in the target frequency range of 100-190 MHz with a sampling rate of 1 GHz and a sampling depth of 14 bit. This will be used to test the predicted low PeV threshold of the radio measurement in radio-quiet environments.

References

[1] IceCube-Gen2 Collaboration 2014 *arXiv* **1412.5106**
[2] IceCube-Gen2 Collaboration 2017 *PoS (ICRC2017)* **301** The IceTop Scintillator Upgrade
[3] Schröder F G for the IceCube-Gen2 Coll. *EPJ WoC (Proc. of ARENA 2018)* submitted
[4] Böser S 2012 *NIM A* **662** 66
[5] Schröder F G, 2017, *Progress in Particle and Nuclear Physics* **93** 1
[6] Huege T 2016 *Physics Reports* **620** 1
[7] HESS Colaboration 2016 *Nature* **531** 476
[8] Balagopal V A, Haungs A, Huege T, Schröder F G 2018 *The European Physical Journal C* **78** 111
[9] Renschler M et al. A 2018 *NIM A* **888** 257
[10] Karg T et al. 2014 *ARENA 2014 [Book of Abstracts]* **arXiv 1410.4685**
[11] de Lera Acedo E, Drought N, Wakley B, Faulkner A 2015 *ICEAA* Evolution of SKALA (SKALA-2) 839
[12] Huege T et al. 2017 *EPJ WoC* **135** 02003
[13] KASCADE Collaboration 2003 *NIM A* **513** 490
[14] de Lera Acedo E, Razavi-Ghods N, Troop N, Drought N, Faulkner A J 2015 *Experimental Astronomy* **39** 3

Simulation study for the IceCube IceTop enhancement with a scintillator array

Agnieszka Leszczyńska[1] **for the IceCube-Gen2 Collaboration**
[1]Institut of Nuclear Physics, KIT - Karlsruhe Institute of Technology, Germany
E-mail: agnieszka.leszczynska@kit.edu

Abstract. The IceTop upgrade foresees the deployment of a scintillation-detector array placed within the IceTop footprint. This will improve the current measurements of cosmic rays by providing a complementary signal to the IceTop Cherenkov tanks. A detailed simulation study of the proposed array was performed to show its capabilities of air-shower detection. An optimisation of the reconstruction was investigated using different lateral particle distribution functions and adjusting the reconstruction parameters. The reconstruction efficiency of the new array as well as its accuracy were evaluated for a chosen configuration.

1. Introduction

The IceTop surface array of the IceCube Observatory measures cosmic rays in the transition region from galactic to extra-galactic sources [1, 2]. The high altitude of the array places the observation level close to the shower maximum enabling more accurate determination of the properties of the primary. However, snow accumulation on the Cherenkov tanks leads to signal attenuation [3, 4] influencing the electron to muon number ratio and thus the cosmic-ray composition analysis. Installing an almost twice as dense array of cost-effective scintillation modules (for a similar total instrumented area) will mitigate this effect. The proposed enhancement foresees the deployment of 37 stations within the IceTop instrumented area, where one station comprises of 7 detectors [5] as illustrated in Fig. 1. This configuration will also improve the veto capability for in-ice neutrino measurements by lowering the detection threshold for cosmic rays.

The difference in response to shower particles between scintillation detectors and Cherenkov tanks requires accurate studies of air-shower characteristics reconstructed by the new detection device. A complete simulation chain combining air-shower cascades and interactions within the detectors has been established to estimate the capabilities of the enhancement and create a preliminary tool for future data analyses.

2. Simulation structure

The final scintillation detector design is under investigation. Two different prototype stations were deployed at the South Pole in the austral summer of 2017/18 for on-site evaluation and testing. The following simulation study is based on a simplified model of the prototype detectors, combining an adequate level of specificity with the possibility to investigate the response of the full planned array to extensive air-showers.

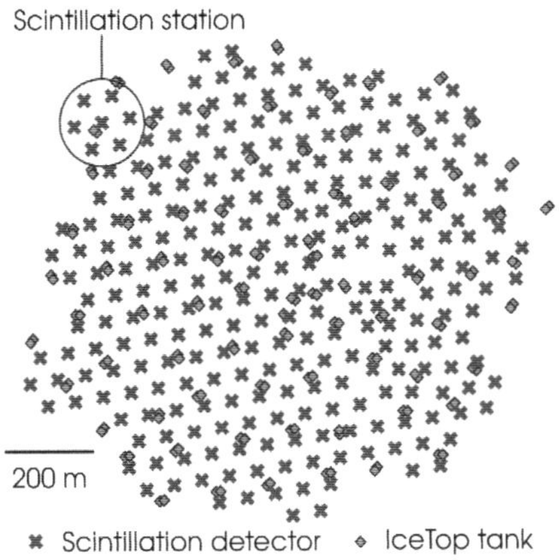

Figure 1. The scheme of the scintillation upgrade. Red crosses show planned positions of scintillation modules and blue squares show positions of IceTop tanks.

Figure 2. The scintillator bars with out-coming wavelength-shifting fibres and readout system used in one of the prototype stations. The scintillator bars are covered with styrofoam supported by plywood.

2.1. Scintillation detector geometry and definition of signal

The prototype design was inherited from AugerPrime [6]. The scintillation detector consists of 16 polystyrene bars coated with a reflecting TiO_2 layer of 250 μm thickness creating 1.5 m^2 of sensitive area (0.8 m $\times$ 1.875 m). Scintillation photons are guided through the wavelength-shifting fibres to a SiPM-based readout system. To provide light-tightness, the scintillator is wrapped in a polyethylene foil with high static dissipative properties. One of the prototypes is shown in Fig. 2.

The core of the simulations lies in the incorporation of the Geant4 toolkit [7] within the framework of the IceCube standard software. The implemented geometry treats the scintillator bars as the sensitive boxes coated with a reflective layer and placed in an 1 mm aluminium casing, padded with 20 mm plywood support structure and styrofoam. The enclosing details are descriptive of one of the two sets of panels but the differences of the other set of panels are not expected to change the simulations results.

The simulations do not include the fibres and the fibre holes. The output of the detector simulation is the energy deposited within the sensitive material as well as a number of produced scintillation photons calculated by means of Birks' law [8] (assuming a scintillation yield of 11200 photons/MeV and Birks' constant of 0.111 mm/MeV). The simulated signal is defined as the sum of photons generated by all particles crossing the detector and by convention is normalised to the average number of photons generated by vertically passing muons of 3 GeV kinetic energy (VEM = vertical equivalent muon) randomly distributed across the detector. In this case, the reference value used for such normalisation is the peak of the Landau distribution parameterisation [9] fitted to the number of photons (Fig. 3).

2.2. Air-shower simulations

The extensive air-showers are simulated with CORSIKA v7.5600 [10] using Sybill 2.3 [11] (high energy hadronic interaction model) and FLUKA [12] (low energy interaction model). The particles are read out at the observation level of 2838 m (Atmosphere model 13, October 1997 [13]) and further propagated to the scintillation array placed 30 cm lower. The threshold on the secondary particle energy in the air-shower is set to 0.1 GeV for hadrons, 0.1 GeV for muons, 0.3 MeV for electrons and 0.3 MeV for photons. In order to investigate the capabilities

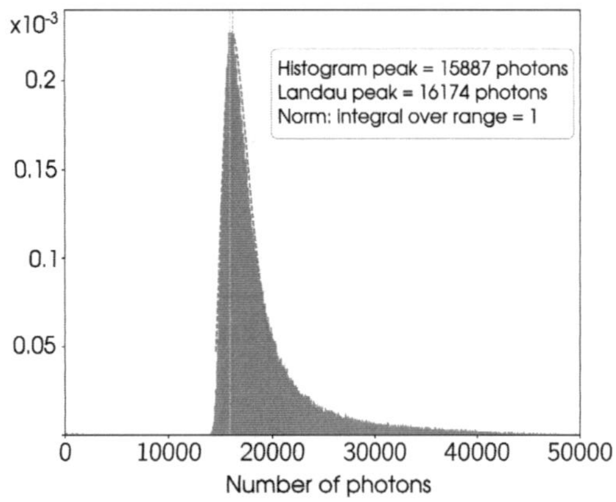

Figure 3. Number of generated photons produced in the scintillation module by 3 GeV vertically muons randomly distributed across the detector. The parametrisation of Landau distribution was taken to estimate the peak value.

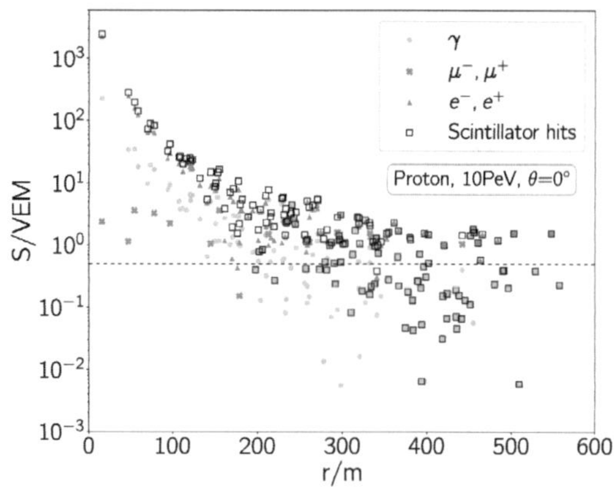

Figure 4. Lateral distribution of air-shower signals. The signals as well as partial contributions from different particle species are shown. The horizontal line indicates the cut (> 0.5 VEM) to be considered in the reconstruction.

of detection in the "knee" region of the cosmic-ray energy spectrum, as well as of primary mass identification, the simulations were conducted for proton and iron primaries of energies 100 TeV–100 PeV with incoming zenith angles from 0° to 40° and azimuth -180° to 180°, and further resampled within the array.

3. Array response

Energy and direction of the primary cosmic ray can be estimated from spatial and temporal distributions of the secondary particle signals which depend on the detector geometry and spacing, as well as on the atmospheric conditions. Therefore the proper description for a given experimental site is essential. Fig. 4 shows an example simulated response of the scintillation array to an air-shower induced by a high-energy proton. The shape of the lateral distribution indicates the size of the shower which is directly connected to the primary energy. The contributions from different particle species indicate the leading signals at different distances to the shower axis. The ratio between the electromagnetic and muonic component reveals the primary particle mass, since heavier primaries generate a higher number of muons.

3.1. Arrival time distributions

The arrival times of the particles hitting each detector allow tracing back the direction of an incoming cosmic ray. This information is included in the curvature function, which represents the signal arrival delay with respect to the plane front assumption. The accuracy of the reconstructed direction is shown in Fig. 6 using two different lateral distribution functions.

3.2. Lateral distributions

The importance of a proper definition of the lateral distribution function and at the same time its challenge lies in the need of simultaneous estimation of core, age and size of the air-shower. Originating from the theoretical Nishimura-Kamata-Greisen description [14, 15], its different modifications (like in [16, 17]) are most frequently used in various experiments like for example Eq. 1 [18]. In addition some groups reported a good agreement of scintillation array results with a function like Eq. 2 accounting for higher scaling with radial distance [19, 20, 21].

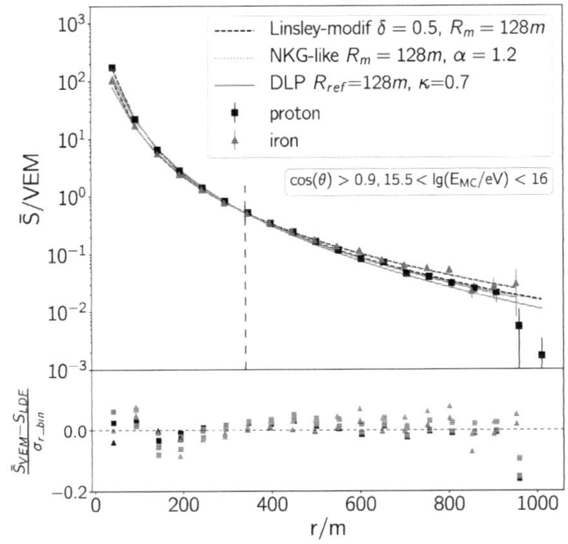

Figure 5. Average lateral distribution of signals for proton (black squares) and iron (blue triangles) primaries with energies $10^{15.5}$–10^{16} eV and arrival directions smaller than 25°. The crossing region for different primaries is marked with a vertical line. Lines and residual markers represent respectively: black — modified Linsley, blue — NKG-like, red — DLP functions. The fit was obtained with MCMC sampler [23]. The shown parameters are exemplary and R_m refers to the theoretical Molière radius for IceTop atmosphere. Even though average distributions show small deviations from different functions, the stability of single fits vary significantly.

The function used in the standard reconstruction with IceTop tanks is represented by Eq. 3 (in the logarithmic form this equation is called a double logarithmic paraboloid — DLP), which takes into account the flatter shape of the tank response distribution [22].

$$S_{\text{NKG-like}}(r) = \frac{N}{2\pi R_m^2} \frac{\Gamma(\beta - s)}{\Gamma(s - \alpha + 2)\Gamma(\alpha + \beta - 2s - 2)} \left(\frac{r}{R_m}\right)^{s-\alpha} \left(1 + \frac{r}{R_m}\right)^{s-\beta} \quad (1)$$

$$S_{\text{Linsley}_{\text{modif}}}(r) = \frac{N}{R_m^2} \left(\frac{r}{R_m}\right)^{-\alpha} \left(1 + \frac{r}{R_m}\right)^{\alpha-\beta} \left[1 + \left(\frac{r}{10R_m}\right)^2\right]^{-\delta} \quad (2)$$

$$S_{\text{DLP}}(r) = S_{ref} \left(\frac{r}{R_{ref}}\right)^{-\beta - \kappa \log_{10}\left(\frac{r}{R_{ref}}\right)} \quad (3)$$

In the above functions: the parameters α, β, δ, κ and s influence the slope of the function and thus the age of shower development, where R_m and R_{ref} indicate scaling and reference distances. The signal at R_{ref} distance and size of the air-shower, N, correspond to the primary energy.

Fig. 5 shows the average lateral distribution of proton- and iron-induced showers, where the detectors without signals are accounted for. The distance at which two distributions cross is marked. It is important to notice that this point strongly depends on the primary energy and the zenith angle. This distance is of particular interest as it could provide a mass-independent estimate of the primary energy. The choice of proper parameters for a particular function is a challenging task and was evaluated for single air-showers by minimising the negative logarithm of the Poisson likelihood. Based on this the best function was chosen for further reconstruction of individual events.

3.3. Reconstruction performance

The reconstruction performed within the IceTop framework [22, 2] handles the parametrisation of the lateral distribution function and shower-front function in three steps. The distribution of arrival delays with respect to the plane-front is described by the sum of a parabolic and a

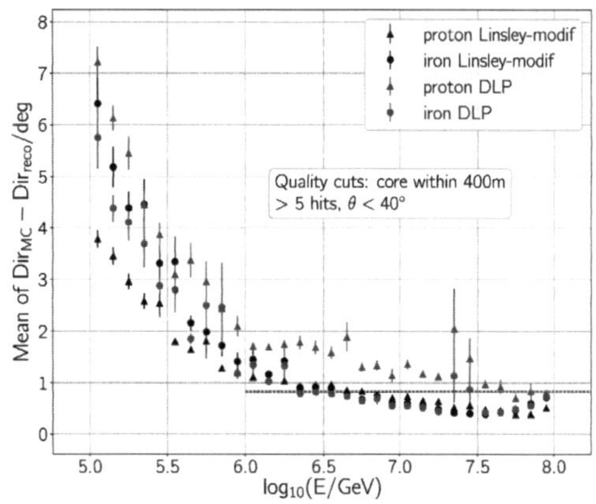

Figure 6. Difference between simulated and reconstructed directions. The blue points correspond to reconstruction with the DLP function and the black ones to the modified Linsley function. The average difference above 1 PeV is smaller than 1° and it is marked in the picture as the horizontal dashed lines for both cases.

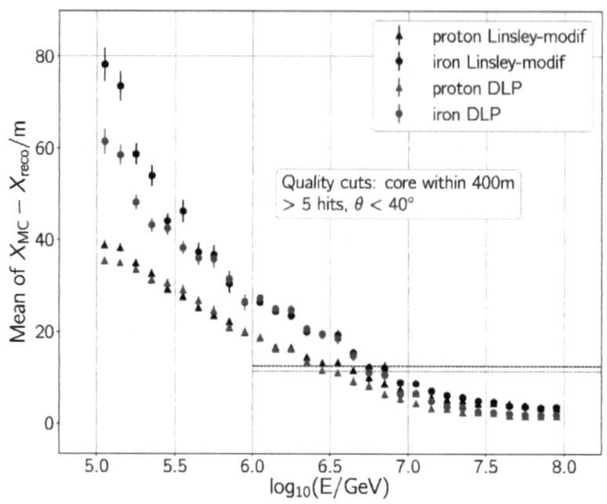

Figure 7. Difference between simulated and reconstructed core positions. The blue points correspond to reconstruction with the DLP function and the black ones to the modified Linsley function. The average difference above 1 PeV is around 12 m and it is marked in the picture as the horizontal dashed lines for both cases.

Gaussian function for a single event. Within the reconstruction the negative log-likelihood is minimised, taking the Poisson distribution for non-zero signals, the Poisson cumulative distribution for silent stations, and a Gaussian distribution for time delays.

The overall reconstruction shown in Fig. 8 was performed using the modified Linsley function with parametrisation on β parameter (the form of parameterisation was adopted from [24, 25]). It shows full efficiency for air-showers even below 1 PeV, which can significantly improve detailed studies of the transition region. The average accuracy of reconstruction above 1 PeV is smaller than 1° in direction (Fig. 6) and around 12 m in core position (Fig. 7). These results are however obtained without refined quality cuts which will improve the accuracy. It is important to note that despite our choice of the function the overall behaviour of reconstruction does not change significantly, but does change across energies.

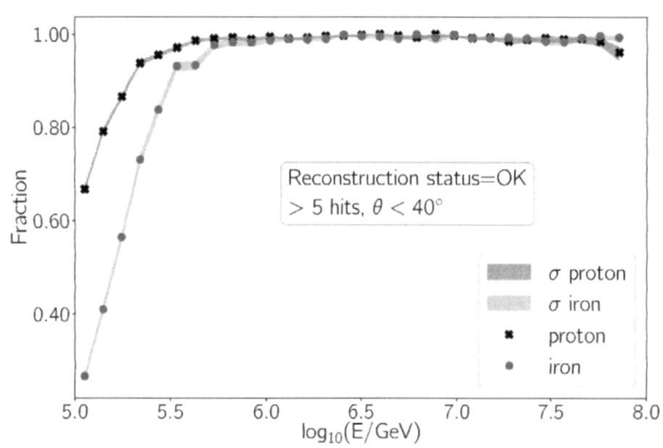

Figure 8. The efficiency of air-shower reconstruction for proton and iron primaries using the modified Linsley function. The proposed array can reconstruct air showers with full efficiency even below 1 PeV. At lower energies the reconstruction will improve the veto efficiency for in-ice measurements. Sigma is defined as the Wald interval [26].

4. Concluding remarks

The scintillation enhancement of the IceTop array was investigated using a full simulation framework. The fundamental distributions and array response to the air-showers were obtained together with studies of lateral distribution functions. After adjusting the existing reconstruction routine, the efficiency of air-shower reconstruction has shown a possibility to lower the measured energy threshold, going below the "knee" region. The results qualitatively indicate the upgrade's capabilities and reveal its challenges for future combined analyses of scintillation and the Cherenkov signals.

Specifics of the new array require further detailed studies, especially in order to account for different behaviours of fluctuations and therefore the proper choice of a likelihood function for the minimisation procedure. Further studies and improvements are ongoing, including an evaluation of the validity of the functions employed in order to obtain a more reliable estimate of the primary energy.

References

[1] Abbasi R *et al* (IceCube Collaboration) 2013 *Nucl. Instrum. Meth. A* **700** 188-220 [arXiv:1207.6326]
[2] Abbasi R *et al* (IceCube Collaboration) 2013 *Astropart. Phys.* **44** 40-58 [arXiv:1202.3039]
[3] Rawlins K *et al* (IceCube Collaboration) 2013 *PoS(ICRC2013)* 1106 37-40 [arXiv:1309.7006]
[4] Rawlins K *et al* (IceCube Collaboration) 2015 *PoS(ICRC2015)* 0628 68-75 [arXiv:1510.05225]
[5] Huber T *et al* (IceCube-Gen2 Collaboration) *PoS(ICRC2017)* 401 62-68 [arXiv:1710.01207]
[6] Aab A *et al* (Pierre Auger Collaboration) 2016 The Pierre Auger Observatory Upgrade - Preliminary Design Report [arXiv:1604.03637]
[7] Agostinelli S *et al* (GEANT4 Collaboration) 2003 *Nucl. Instrum. Meth. A* **506** 250-303
[8] Birks J B 1951 *Proc. Phys. Soc. A* **64** 874-877
[9] Moyal J E 1955 *Phil. Mag.* **46** 263-280
[10] Heck D *et al* 1998 Report FZKA 6019 [https://publikationen.bibliothek.kit.edu/270043064]
[11] Riehn F *et al* 2015 A new version of the event generator Sibyll *PoS(ICRC2015)* 558 [inspirehep.net/record/1395980/files/ICRC2015_558.pdf]
[12] Ferrari A *et al* 2005 *FLUKA: A Multi-Particle Transport Code* CERN-2005-10, INFN/TC_05/11, SLAC-R-773
[13] https://ccmc.gsfc.nasa.gov/modelweb/models/msis_vitmo.php
[14] Greisen K 1956 *Progress in Cosmic Ray Physics* vol 3 ed. by Wilson J G and Robinson G (Amsterdam: North Holland Publ.Co.) chapter 1 pp 19-37
[15] Kamata K and Nishimura J 1958 *Prog. Theor. Phys. Suppl.* **6** 93-155
[16] Linsley J 1962 *J. Phys. Jpn. Suppl. A-III* **17** 148
[17] Capdeviellev J N and Cohen F 2005 *J. Phys. G: Nucl. Part. Phys.* **31** 507-524
[18] Apel W D *et al* (KASCADE Collaboration) 2006 *Astropart. Phys.* **24** 467-483 [arXiv:astro-ph/0510810]
[19] Lagutin A A and Raikin R I 2001 *Nucl. Phys. B (Proc. Suppl.)* **97** 13
[20] Lagutin A A, Raikin R I, Inoue N and Misaki A 2002 *J. Phys. G: Nucl. Part. Phys.* **28** 1259-1274
[21] Yoshida S 1995 *J. Phys. G: Nucl. Part. Phys.* **20** 651-665
[22] Klepser S 2008 Reconstruction of Extensive Air Showers and Measurement of the Cosmic Ray Energy Spectrum in the Range of 1-80PeV at the South Pole. PhD thesis, Humboldt-Universitt zu Berlin
[23] Foreman-Mackey D *et al* 2013 *Publ. Astron. Soc. Pac.* **125** 306-312 [arXiv:1202.3665]
[24] Grieder P K F 2010 *Extensive Air Showers: High Energy Phenomena and Astrophysical Aspects — A Tutorial, Reference Manual and Data Book* (Springer-Verlag Berlin Heidelberg) chapter 8 pp 390-391
[25] Glushkov A V, Makarov A K and Makarov I T 2011 *Proceedings 32nd International Cosmic Ray Conference (ICRC 2011)* **6** 153-155
[26] Brown L D *et al* 2001 *Statist. Sci.* **16** (2) 101-133

IOP Conf. Series: Journal of Physics: Conf. Series **1181** (2019) 012077 doi:10.1088/1742-6596/1181/1/012077

Status of the Lunar Detection Mode for Cosmic Particles of LOFAR

T. Winchen[1] A. Bonardi[2], S. Buitink[1], A. Corstanje[2], H. Falcke[2,3,4], B. M. Hare[5], J. R. Hörandel[2,3], P. Mitra[1], K. Mulrey[1], A. Nelles[6], J. P. Rachen[2], L. Rossetto[2], P. Schellart[2,7], O. Scholten[5,8], S. ter Veen[2,4], S. Thoudam[2,9], T. N. G. Trinh[5]

[1] Astrophysical Institute, Vrije Universiteit Brussel, Pleinlaan 2, 1050 Brussels, Belgium,
[2] Department of Astrophysics/IMAPP, Radboud University, P.O. Box 9010, 6500 GL Nijmegen, The Netherlands,
[3] NIKHEF, Science Park Amsterdam, 1098 XG Amsterdam, The Netherlands,
[4] Netherlands Institute of Radio Astronomy (ASTRON), Postbus 2, 7990 AA Dwingeloo, The Netherlands,
[5] KVI-CART, University Groningen, P.O. Box 72, 9700 AB Groningen,
[6] Institut für Physik, Humboldt-Universität zu Berlin, Unter den Linden 6, 10099 Berlin, Germany,
[7] Department of Astrophysical Sciences, Princeton University, Princeton, NJ 08544, USA,
[8] Interuniversity Institute for High-Energy, Vrije Universiteit Brussel, Pleinlaan 2, 1050 Brussels, Belgium,
[9] Department of Physics and Electrical Engineering, Linnéuniversitetet, 35195 Växjö, Sweden

E-mail: `tobias.winchen@rwth-aachen.de`

Abstract. Cosmic particles hitting Earth's moon produce radio emission via the Askaryan effect. If the resulting radio ns-pulse can be detected by radio telescopes, this technique potentially increases the available collective area for ZeV scale particles by several orders of magnitude compared to current experiments. The LOw Frequency ARray (LOFAR) is the largest radio telescope operating in the optimum frequency regime for this technique. In this contribution, we report on the status of the implementation of the lunar detection mode at LOFAR.

1. Introduction

The low fluxes of the particles of interest, which consequently require huge detectors at the highest energies, is one of the major challenges in astroparticle physics. As a possible solution it has already been proposed in the 1960s [1] to use the Moon as a particle detector by searching for the ns radio pulses generated by a charge excess in the shower developing in the lunar rock. Several searches for these signals have been conducted with earth-bound radio telescopes (see reference [2] for a review), but so far none of these searches has been sensitive enough to expect to detect the observed flux of charged cosmic rays. The main reason here is that most of these previous searches used telescopes operating at GHz frequencies. While the amplitude of the emitted pulse reaches a maximum in the GHz range, the effective area increases towards lower frequencies as the Cherenkov cone becomes broader and thus also inclined events become detectable. The resulting optimum frequency band for corresponding observations is found to be above approximately 100 MHz [3].

The currently largest telescope operating in this optimal frequency range is the LOw Frequency ARray (LOFAR) [4]. It is composed of more than 50 stations located throughout Europe with 24 stations placed in a dense core in the Netherlands. Each of these core-stations contains 768 high band antennas (HBA) operating in the frequency range 110–190 MHz organized in two fields of 24 tiles of 16 antennas each. The signal of the individual stations are filtered into sub-bands by a polyphase filter (PPF) and combined into a beam of approximately 5° width at 120 MHz before it can be accessed by users of the telescope with dedicated observation pipelines.

2. The NuMoon Pipeline

A pipeline to detect ns pulses with LOFAR has to perform three different tasks to reconstruct the pulses before triggering.

1. Form tied array beams. To maximize the collective area, the signals of as many stations as possible have to be combined coherently. As this reduces the beam-width, several beams have to be formed to cover the surface of the Moon.
2. Invert the PPF. The PPF on station level reduces the time resolution by a factor of 1024 from approximately ns to approx. μs. To recover ns time resolution, the PPF has to be inverted.
3. Ionospheric dedispersion. By travelling through Earth's ionosphere, the signal is dispersed proportional to the electron content of the ionosphere. The amplitude of the pulse must consequently be reconstructed to detect the pulses.

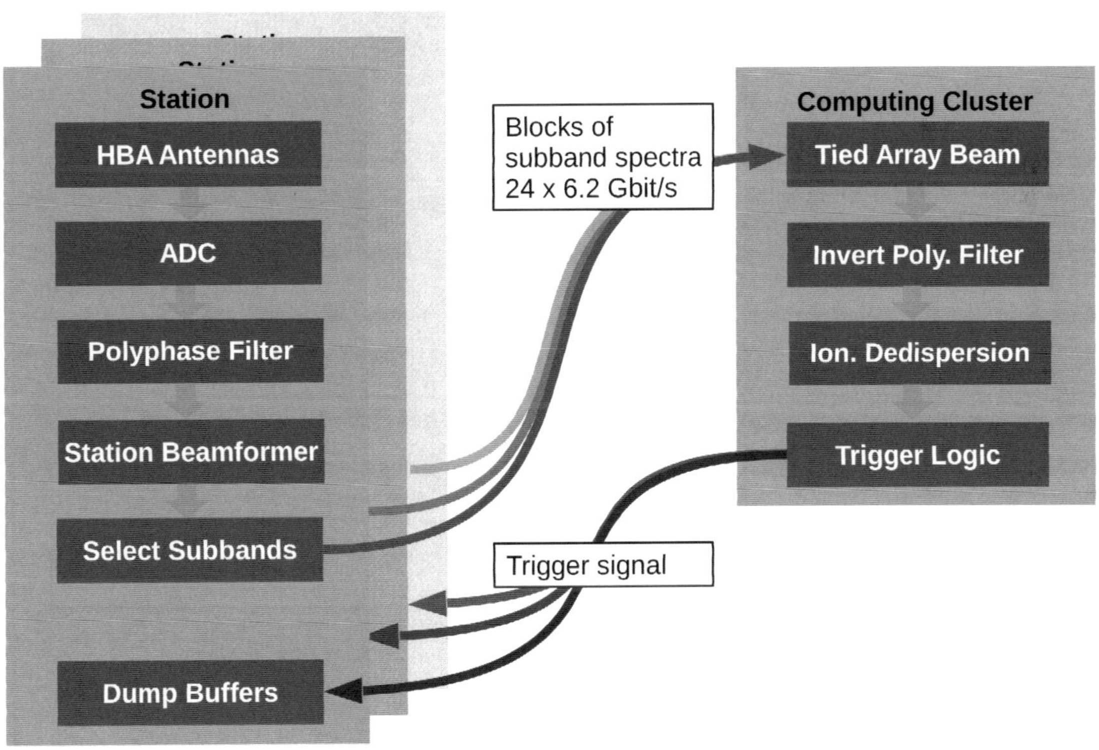

Figure 1. Data processing steps in the individual stations and central computing system for the LOFAR pipeline to observe pulses on ns timescales.

The corresponding data processing procedure for LOFAR is sketched in fig. 1. In particular the inversion of the PPF is compute-intensive. A direct inversion of the PPF produces echos of short pulses as e.g. shown in [5] that limits the usability here, as we thus expect a large number of false triggers generated from RFI by this method. Instead, we use an iterative method that requires solving a linear system [6] which requires $\mathcal{O}(1000)$ GFLOPs^{-1} of computing power [7] per beam.

The required computing resources are, unfortunately, not available on the default LOFAR computing system. Instead, we will use the DRAGNET cluster [8], a computing cluster build for realtime searches of pulsars located in the same computing center next to the default LOFAR facilities. DRAGNET is connected to the LOFAR network with a 56 Gbit/s Infiniband network only, that limits realtime processing of the data of about five stations. The selection of five out of the 24 stations is non-trivial as the geometry of the stations results in very different side lobe patterns of the final beams [9].

With five stations and the available processing power we expect to be able to cover the full Moon with a grid of up to 49 analysis beams. The data-stream in these will be constantly monitored for ns pulses during an observation and eventually trigger a read-out of the transient buffers of all core stations to obtain a large unfiltered data set for offline analysis.

3. Trigger Concept

We expect the data to contain many RFI pulses originating from the horizon. A background pulse originating from the horizon will produce a signal in every analysis beam, while a signal pulse from the target direction on the sky will be, ideally, only visible in the beam pointing to the direction of its origin. RFI suppression can thus be, in principle, easily be achieved by requiring an anti-coincidence between the individual analysis beams. However, due to the circular geometry of the beams, and also due to side-lobes the beams overlap, which prevents a simple anti-coincidence and requires a more sophisticated trigger. An example for the reconstructed pulse amplitude for simulated signals originating from the Moon and the horizon are shown in

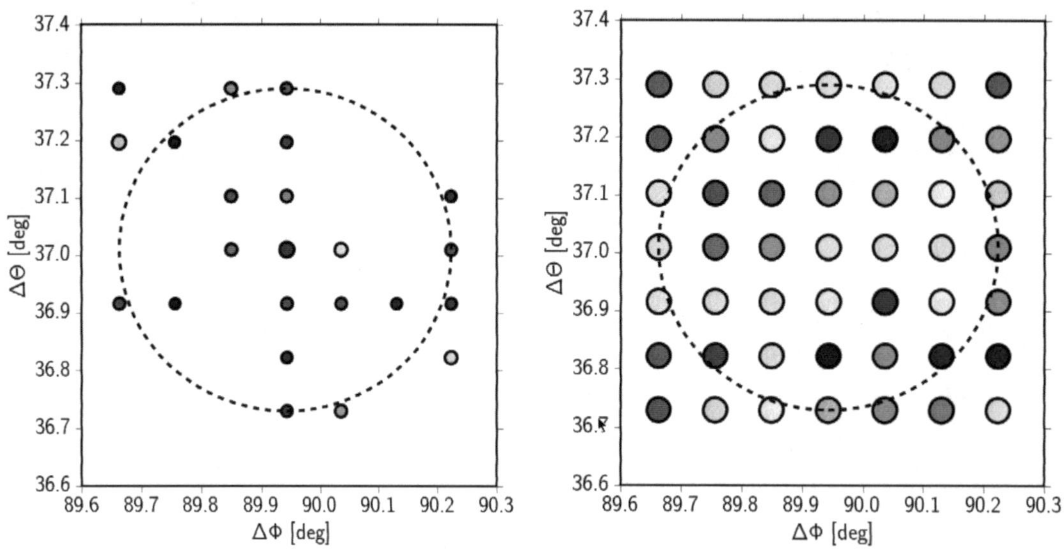

Figure 2. Amplitude above threshold of reconstructed pulses in a grid of 7×7 beams covering the area around the Moon. The dashed circle indicates the position and extent of the Moon. **(left)** Pulse originating from the center of the Moon. **(right)** Pulse originating from the horizon.

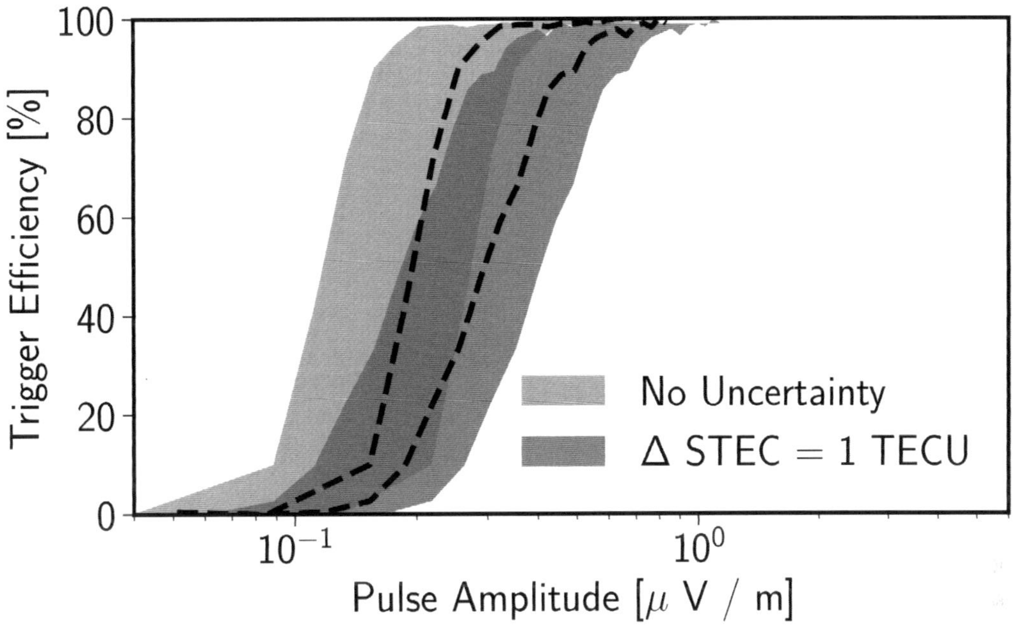

Figure 3. Trigger efficiency as function of pulse amplitude for simulated pulses added to measured background. Pulses have been dedispersed assuming an ionospheric electron content of 20 TECU and dispersed by the same amount (orange) respectively a random value of 20 ± 1 TECU (blue). The bands denote the uncertainty on the pulse amplitude given by the uncertainty of the antenna calibration.

figure 2.

We consequently implement the following prototype trigger strategy. First, the signal in individual analysis beams is monitored and a first level trigger is emitted when the signal passes a threshold corresponding to an acceptable trigger rate from thermal noise of one event per minute. The number of beams above threshold in a given time window, and the RMS of the signal amplitudes of all beams above threshold, is then used as second criterion for signal / background separation.

We evaluated this procedure using simulated events of varying strength originating from the horizon and from a random position on the Moon. For weak pulses that are visible in less than 10 beams we achieve a 90% efficiency and 90% purity using a simple cut on the normalized RMS of the amplitudes. For strong pulses that are visible in every analysis beam we achieve 100% efficiency at 95% purity. The anti-coincidence criterion thus does not limit the method to low energy events as has been previously remarked [2].

4. Expected Sensitivity

The expected sensitivity of the pipeline is estimated using simulated signal pulses of varying strength that are added to observed noise traces. The simulated pulses are dispersed assuming an exact slant total electron content (STEC) of 20 TECU, respectively a normal distributed STEC value centered around 20 TECU with width of 1 TECU to account for imperfect reconstruction. All traces were then processed by the observation pipeline with an dedispersion corresponding

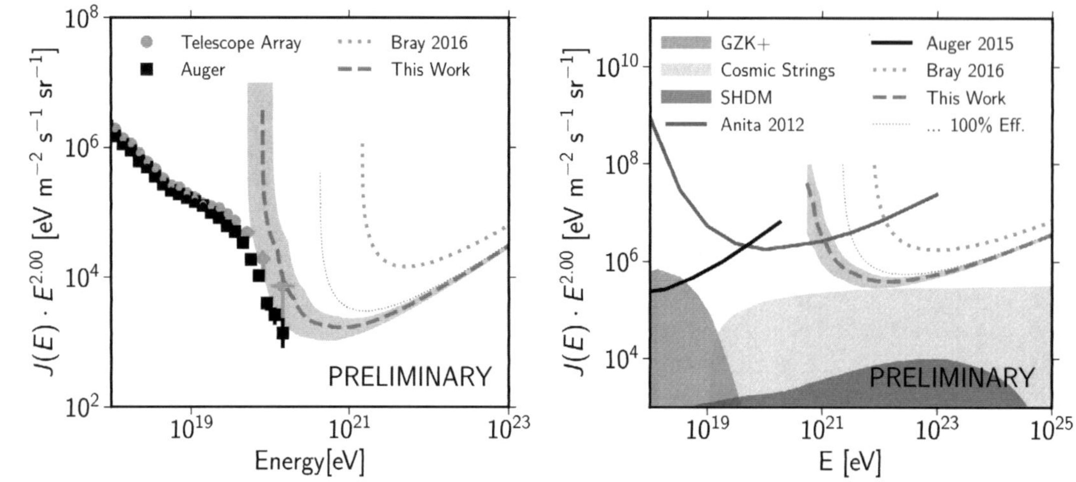

Figure 4. Expected limits for UHECR (left) and neutrinos (right) with available data [10, 11] for the UHECR flux respectively model predictions for the flux of neutrinos [12, 13, 14] for 200 hours of observation. The dashed line labeled 100% Eff. corresponds to a minimum detectable field strength of $\mathrm{E} = 0.8\,\mu\mathrm{V\,m^{-1}}$.

to 20 TECU. We require one sample above a threshold value in at least one out of 49 traces for analysis beams forming a grid covering the Moon as trigger condition. The threshold value is chosen to yield a trigger rate of 1 / min from thermal noise only. The resulting trigger efficiency is shown in figure 3 as a function of the pulse amplitude. Full efficiency is reached for pulses with amplitude $E \approx 0.8\,\mu\mathrm{V\,m^{-1}}$ and approximately 20% efficiency is reached for pulses with $E \approx 0.2\,\mu\mathrm{V\,m^{-1}}$ if the STEC is known to 1 TECU.

Following the procedure described in references [15, 2] we calculate the sensitivity to charged cosmic rays and neutrinos as shown in figure 4. For this calculation we assume that any reduction in observation time due to the RFI trigger rate is negligible. With the pipeline described here we expect an improved sensitivity compared to previous results [2], because we can use a lower trigger threshold and an increased band-width. With the full simulation we were also, for the first time, able to include the effect of the smooth cut-off of sensitivity to low amplitude pulses. This softens the low-energy cut off of the sensitivity compared to previous estimates.

5. Conclusion

We developed a complete prototype for the online and offline analysis pipeline for the Lunar detection of cosmic particles with the LOFAR radio telescope. For the first time the sensitivities of corresponding observations have been estimated based on full simulations of the radio telescope, data processing, and trigger. These simulations enabled design and evaluation of new trigger concepts and consequently lead to an improved estimated sensitivity to cosmic particles. Compared to previous semi-analytical estimates the full detector simulations reveal an increased sensitivity to low energetic particles of the prototype pipeline. However, with the available resources a detection of cosmic rays with LOFAR remains unlikely, but it will become possible by implementing the here described prototype at the upcoming SKA telescope. Preparations for first observations to demonstrate the capabilities of the prototype pipeline are currently ongoing.

Acknowledgements
The LOFAR cosmic ray key science project acknowledges funding from an Advanced Grant of the European Research Council (FP/2007-2013) / ERC Grant Agreement n. 227610. The project has also received funding from the European Research Council (ERC) under the European Union's Horizon 2020 research and innovation programme (grant agreement No 640130). We furthermore acknowledge financial support from FOM, (FOM-project 12PR3041-3) and NWO (Top Grant 614- 001-454, and Spinoza Prize SPI 78-409). TW is supported by DFG grant WI 4946/1-1. LOFAR, the Low Frequency Array designed and constructed by ASTRON, has facilities in several countries, that are owned by various parties (each with their own funding sources), and that are collectively operated by the International LOFAR Telescope foundation under a joint scientific policy.

References

[1] Askaryan G 1962 *Sovjet Physics J.E.T.P* **14** 441
[2] Bray J 2016 *Astroparticle Physics* **77** 1–20 (*Preprint* 1601.02980)
[3] Scholten O *et al.* 2006 *Astroparticle Physics* **26** 219 – 229
[4] van Haarlem M P *et al.* 2013 *Astronomy and Astrophysics* **556** A2 (*Preprint* 1305.3550)
[5] Singh K *et al.* 2012 *Nucl. Instrum. Meth.* **A664** 171–185 (*Preprint* 1108.5745)
[6] Winchen T *et al.* (LOFAR Cosmic Ray KSP) 2017 *EPJ Web Conf.* **135** 04003 (*Preprint* 1609.06590)
[7] Winchen T *et al.* 2016 *22nd International Conference on Computing in High Energy and Nuclear Physics (CHEP 2016)* (*Preprint* 1612.06592)
[8] Hessels J 2015 DRAGNET: A high-speed, wide-angle camera for catching extreme astrophysical events URL `http://www.astron.nl/dragnet`
[9] Winchen T *et al.* 2017 *Proceedings of the 35th International Cosmic Ray Conference (ICRC)* (*Preprint* 1711.07749)
[10] Fenu F and for The Pierre Auger Collaboration 2017 *Proceedings of the 35th International Cosmic Ray Conference (ICRC)*
[11] Ivanov D 2012 *Energy Spectrum Measured by the Telescope Array Surface Detector* Ph.D. thesis Rutgers, The State University of New Jersey
[12] Roulet E, Sigl G, van Vliet A and Mollerach S 2013 *Journal of Cosmology and Astroparticle Physics* **1** 028 (*Preprint* 1209.4033)
[13] Berezinsky V, Sabancilar E and Vilenkin A 2011 *Phys. Rev.* **D84** 085006 (*Preprint* 1108.2509)
[14] Aloisio R, Matarrese S and Olinto A V 2015 *JCAP* **1508** 024 (*Preprint* 1504.01319)
[15] Gayley K G, Mutel R L and Jaeger T R 2009 *Astrophys. J.* **706** 1556–1570 (*Preprint* 0904.3389)

IOP Conf. Series: Journal of Physics: Conf. Series **1181** (2019) 012078 doi:10.1088/1742-6596/1181/1/012078

Filters for RFI suppression in AERA radio detection of cosmic rays

Zbigniew Szadkowski

Faculty of Physics and Applied Informatics, Department of High-Energy Astrophysics, 90-236 Łódź, Poland, tel: +48 42 635 56 59

E-mail: zbigniew.szadkowski@fis.uni.lodz.pl

for the Pierre Auger Collaboration

Observatorio Pierre Auger Av. San Martn Norte 304, 5613 Malarge, Argentina,
Full author list: http:\\www.auger.org\archive\authors_2018_07.html

E-mail: auger_spokespersons@fnal.gov

Abstract.

Radio stations can observe radio signals caused by coherent emissions due to geomagnetic radiation and charge excess processes mainly in the frequency band from 30 to 80 MHz. This range is often highly contaminated by human-made RFI. In order to improve the signal to noise ratio, RFI filters have to suppress this contamination.

The paper compares several (currently in use and proposed) RFI filters used in the Auger Engineering Radio Array (AERA), being a part of the Pierre Auger Observatory. We analyzed both: the non-adaptive filter (IIR notch) as well as adaptive filters: IIR notch supported by the virtual NIOS® processor, FIR based on the linear prediction supported by the virtual NIOS® processor, the FIR fully implemented into FPGA fabric used the Levinson algorithm without any processor (all coefficients fully calculated in the FPGA logic elements) and finally the Delayed Least Mean Square filter also with a modification proposed by Harris, variable step size and Instantaneous LMS filter.

We investigated: the efficiency of the several mono-carriers suppression, the distortion factors of cosmic rays pulses after filtering, the speed of the filter response as well as the power consumption crucial for systems supplied by solar panels.

The 200 MW solar farm with 250 kV power line is anticipated to be built in the Southern part of the Auger array. High-power DC/AC converter for sure introduce a huge noise in the electromagnetic spectrum being a topic of interest of AERA. Very efficient, adaptive filter becomes a crucial issue. Simulations and laboratory measurement suggest that the most promising will be the Instantaneous Least Mean Square filter with dynamically modified input range.

1. Introduction

Particles coming from the Universe with ultrahigh energy, known as cosmic rays (UHECRs), penetrate the Earth's atmosphere and generate avalanches of charged and neutral particles known as Extensive Air Showers (EAS) discovered in the late 1930s [1]-[5]. The origin of UHECRs still remains a mystery. Several experiments (e.g. the Pierre Auger Observatory

[6]) investigate EAS to estimate the energy, arrival direction of each cosmic ray as well as to analyze a distribution of primary masses (cosmic ray composition).

Radio detection of EAS started in the 1960s [7]. In the last decade, the LOPES [8] and CODALEMA [9] Collaborations continued radio detection together with other radio array, such as LOFAR [10] and Tunka-Rex [11]. The Auger Engineering Radio Array [12]-[13] is the radio detector in the Pierre Auger Observatory (together with the fluorescence and surface ones) to measure EAS with an energy larger than 10^{17} eV.

The geomagnetic radiation [14] and the Askaryan charge-excess [15] are the main responsible for the radio emission. The geomagnetic effect is a consequence of the induction of a transverse electric current in the shower front generated by the Earth's magnetic field. The charge excess in the shower front is to a large extent due to the knock-out of fast electrons from the ambient air molecules by high-energy photons in the shower. The refractive index of the air is slightly bigger from unity. This causes a Cherenkov-like time compression [16], which affects both the radiation mentioned above.

The crucial condition for a radio detection is the radio signals coherence, which is investigated in a frequency range of 30 - 80 MHz (analog electronics uses a band-pass filter). The coherence appears also above 80 MHz, but the upper coherence frequency depends strongly on the distance to the shower axis reaching even GHz frequencies. Additionally, 80 MHz cut-off removes also the FM radio band. The lower frequencies usually reflect frequently in the atmosphere, so detectors can register emission from very distant storms. The lower limit is optimized for the signal-to-noise ratio of EAS pulses vs. Galactic noise. The selected frequency band contains mostly Galactic noise and human-made RFI. In order to assure a high quality self-trigger, radio signals should be cleaned from RFI.

2. Filters currently in use and proposed

The environment is mostly contaminated by narrow-band mono-carriers of short-wave transmitters. The narrow peaks in the frequency domain have to be strongly suppressed before the trigger is generated. The improvement of the signal to noise ratio is crucial to avoid spurious triggers and a registration of empty events. The AERA used two types of the RFI filters:

- **1**) an adaptive one based on the FFT conversion from time to frequency domains filtered afterwards by the median filter and the second conversion from the frequency to the time domain by the Hilbert transform [18].
- **2**) a non-adaptive IIR notch filter with fixed coefficients [17], which suppress four narrow bands. Coefficients calculated in an external C program have been implemented into the FPGA code as constants [17]. If the frequencies of the contaminations are stable (e.g., correspond to short-wave transmitter carriers) a non-adaptive filter is very efficient. However, if the frequencies of contamination do not match the IIR coefficients, the IIR filter is practically useless: the output signals are approximately the same as the input ones.

The adaptive FFT-Median-iFFT filter (**1**) was installed in the first generation of radio stations with Cyclone®III FPGA and 180 MHz sampling frequency. The relatively low frequency of the ADC clock (providing a relatively narrow safety margin to the Nyquist condition) was a consequence of used over-clocking necessary to avoid aliasing. Even the fastest Cyclone®III FPGA did not allow on higher clock frequency. Two polarization channels would require straightforward to use two FFT engines for calculating the frequency domain signal, while setting their imaginary input to zero. A trick was used, two channels were connected to a single FFT engine (real and imaginary inputs), outputs were next disentangled to the Fourier components. Although the FFT approach was working correctly, it has not been implemented in the second generation design with Cyclone®IV due to high power consumption and very complicated FPGA code.

The non-adaptive IIR notch filter (**2**) were implemented in the next generation of electronics - in radio stations with Cyclone®IV and 200 MHz sampling frequency. Adaptive filters allowing an RFI suppression independently of the frequencies of the contamination, are more flexible than non-adaptive filters with fixed parameters usually tuned for the local transmitter frequencies. Such a solution is efficient: however, it cannot react to any additional RFI sources. Nevertheless, for well defined short-wave carriers in AERA region the IIR filter has been working extremely efficient and with very low power consumption. If the risk of contamination by the solar farm is low, the IIR filter would be still the best solution.

A proposed filter was already tested in Argentinean pampas:

- **3a**) The adaptive 32-stage FIR filter based on the linear prediction (LP) of the signal supported by the NIOS® processor. The NIOS® processor had to solve 32 linear equations in the FPGA. Calculations in the NIOS® provided the same precision as in the Microsoft Visual C++. This filter was tested in real Argentinean pampas conditions and proved its efficiency and stability [19][20]. Coefficients calculation in the NIOS® processor take ~600 ms. This refresh time is considered too long. The filter cannot react on short-time spikes.

The filter **3a**) fulfilled our expectations, however further work was done and new filters with interesting and promising features are proposed:

- **3b**) The variant with the Hardcore Processing System, with embedded micro-controller in the FPGA. The design was developed on Cyclone®V SoC Development Kit and SoC Embedded Design Suite with 5CSXFC6D6F31C6N (SoC) FPGA. FIR coefficients are calculated in the embedded micro-controller with 925 MHz clock (instead of 100 MHz in NIOS® allowed decreasing the refresh time from ~600 ms to ~20 ms only. This design as it required FPGA with SoC could not be tested in the field. Its feature was the same as **3a**) with 30 times shorter calculation of FIR coefficients.
- **3c**) The linear prediction method generates the auto-covariance Toeplitz matrix calculated from ADC samples. Linear equations are next solved by much faster Levinson procedure instead of relatively slow Gaussian elimination. The algorithm of solving 32 linear equations based on the Levinson procedure has been implemented directly in the FPGA fabric [22]. The refresh time was only ~600 μs. Implementation of parallel processes directly in the FPGA fabric reduces the refresh time. However, it increases the power consumption, a crucial factor for applications supplied from solar panels. This variant will be tested in the field, especially in seriously contaminated regions.
- **4a**) The adaptive Normalized Least Mean Square (NLMS) filter with a canonical FIR section [23]. This is the first design based on the LMS approach. It is relatively slow as using a division operation for a normalization. Unfortunately, the Altera Mega-function implementing a division operation does not use the fast multipliers from DSP blocks. In order to assure a sufficient speed of 200 MHz (required for ADCs) the length of the FIR filter should not be longer than 16. It seems to be too short.
- **4b**) The adaptive Delayed Least Mean Square (DLMS) filter with non-canonical FIR section [24]. It is significant modification of the NLMS algorithm without any division operation. It provides much faster registered performance and allows an implementation of 32- or 64-stage FIR.
- **4c**) The DLMS modification with variable step size [26] proposed by Harris [25]. Laboratory measurements did not confirm a superiority of variable step approach in comparison to the fixed learning factor design.
- **4d**) The Instantaneous LMS modification (ILMS) without a delay line as in the DLMS design. An analysis showed that the DLMS algorithm distorted cosmic rays signal introducing echoed artifacts. Removing a delay line significantly improved the shape of

IOP Conf. Series: Journal of Physics: Conf. Series **1181** (2019) 012078 doi:10.1088/1742-6596/1181/1/012078

signal after filtering. The most efficient filtering is for very large contaminated signal. The investigated cosmic ray signals deeply hidden in the background are very efficiently extracted from the noise.

- **4e**) The ILMS algorithm can perfectly extract signals from a large noise but it does not show so high efficiency for smaller contamination. In order to improve the RFI suppression also for relatively low RFI levels, the dynamical shift of the ADC input has been proposed (an artificial amplification and reduction after filter). This LMS variant seems to be most optimal and will be tested in the field.

3. Suppression efficiency

The suppression efficiency has been tested with four mono-carriers, a condition the IIR notch filter was optimized for. Selected frequencies were as follows: 29.17, 41.11, 59.37 and 72.43 MHz, respectively. Amplitudes were tuned to be at a similar level.

3.1. IIR notch filter

The IIR notch filter was tested for 4 frequencies of mono-carriers (Figure 1A) exactly matching the assumed condition of perfect suppressions (Figure 1b) and with a single frequency evidently out of the suppression window (Figure 1c). If at least a single mono-carrier does not match the optimal IIR notch suppression window, a full filter is practically useless. For this mono-carrier the IIR notch filter is practically transparent and a suppression of the rest of mono-carriers does not matter. However, if all frequencies of mono-carriers match the assumed coefficients the RFI suppression is extremely efficient (spectral amplitudes are suppressed by a factors of ~30 - 100).

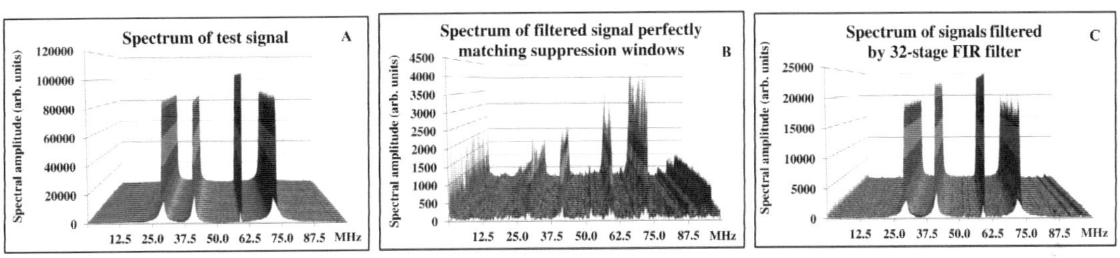

Figure 1. Spectra of signals: original from 4 external generators (test signal with 4 carriers with 29.17, 41.11, 59.37 and 72.43 MHz, respectively - left plot: A), filtered by the IIR notch filter with perfectly matching of all frequencies for a maximal suppression (middle plot: B) and filtering by the LP FIR filter (right plot: C).

3.2. FIR filter based on the linear prediction

The 32-stage FIR filter based on the linear prediction [19] has been successfully tested in Argentinean pampas [20]. In comparison to IIR notch filter its suppression efficiency is much smaller (Figure 1C) - the suppression factors are at the level of ~5 while for IIR notch filter the suppression factors are above 20 and sometimes reach even ~100. The main advantage of the FIR filter is its adaptivity, which allows a suppression of either new sources or the same source with moved frequency. Adaptive filters are especially useful in environment with non-stationary RFI sources.

The filter based on the hardware implementation of the Levinson procedure directly into the FPGA fabric will not be analyzed as it is based on the same algorithm which does not provide a sufficient RFI suppression. The filter with Levinson algorithm implemented directly in the FPGA fabric is much faster than FIR filter supported by the NIOS® processor, but the power consumption is much higher due to parallel processing and much bigger resources occupation.

3.3. LMS32 filters

The NLMS filter solves the problem of normalizing by the learning factor updated in every iteration. The division operation requires a 24-stage pipeline LPM_DIVIDE Mega-function, which is rather slow since it is not supported by DSP blocks as multipliers. The filter uses the canonical FIR section requiring additionally 5 pipeline stages for a 32-stage length.

Much faster than using the non-canonical FIR algorithm is the DLMS32 filter showing a very good suppression efficiency, practically for all allowed learning factors. Optimization of the DLMS32 for elimination echoed artifacts removed the delay line. The ILMS32 provides the same suppression features as the DLMS32 (Figure 2). Variable step size LMS [25][26] as nor indicating superiority than the fixed learning factor approach are not developed anymore. The last design and most promising, the ILMS32 with dynamical input range (VRILMS) seems to be the the optimal for very wide range of potential contaminations [27].

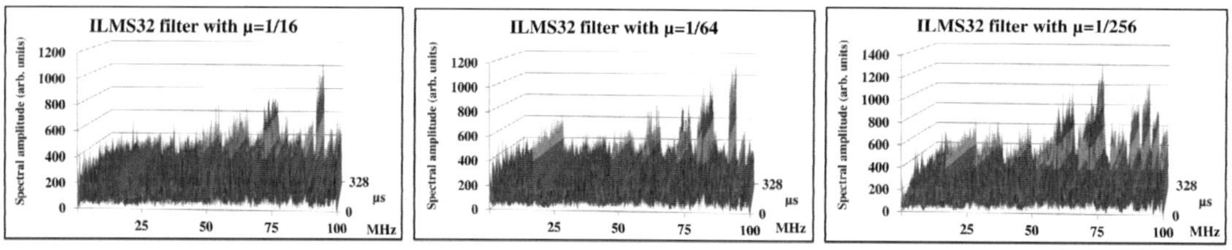

Figure 2. Spectra of signals filtered by the 32-stage DLMS filter with various learning factors.

4. Distortion factors

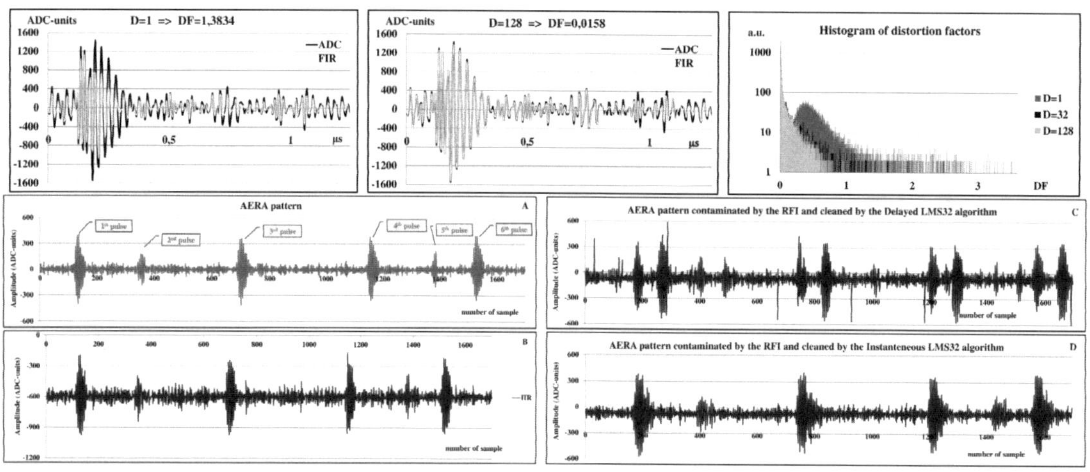

Figure 3. Example of signal more or less distorted depending on the length of the delay line for the LP FIR filter (the first row) and six original AERA pulses (A) contaminated by four sine RFI and next cleaned the IIR filter (B), the DLMS (C) and the ILMS algorithms (D).

Distortion factors in the LP FIR filter were reduced by an adjustment of the delay line length to D=128. However, this approach for the LMS filters shows echoes signals. The minimum of distortion factors was obtained in the case of absence of any delays. Figure 3 shows an example of real AERA events contaminated artificially by four mono-carriers and next cleaned by IIR, DLMS and ILMS filters. The IIR provides the minimal distortion factors, however, the ILMS algorithm assures a sufficiently good signal reconstruction for the self trigger circuit.

5. Power consumption

As expected, the IIR filter shows the lowest power consumption for both FPGAs. However, Cyclone® V indicates rather a large static power dissipation. For potentially future applications Cyclone®10 is definitely the favorite. Unexpectedly, the power consumption for VRILMS32 is twice bigger than for the FIR32 (for Cyclone®10). Let us notice that the power consumption for Cyclone®10 and Cyclone®IV E are comparable. This simplifies tests in real pampas conditions where EP4CE75F29C6 FPGA is now used (Figure 4).

Power in mW	EP4CE75F29C6				5CEFA9F31I7				10CL120YF780I7G			
	IIR	FIR32	Levinson	VRILMS	IIR	FIR32	Levinson	VRILMS	IIR	FIR32	Levinson	VRILMS
Total Thermal Power Dissipation	152.8	189.6	297.8	361.3	543.8	772.3	1134.9	739.9	140.1	170.7	280.8	362.9
Core Dynamic Thermal Power Dissipation	19.2	39.2	146.1	215.0	15.2	200.4	562.4	172.9	24.2	39.1	148.1	228.3
Core Static Thermal Power Dissipation	90.9	95.2	97.7	91.7	518.7	527.9	530.3	521.1	74.4	78.1	80.0	75.2
I/O Thermal Power Dissipation	42.7	55.2	54.1	54.7	10.0	44.0	42.3	45.9	41.5	53.5	52.7	59.4

Figure 4. Power consumption for various filters and three FPGAs from Cyclone®IV E, Cyclone®V and Cyclone®10 families estimated by the Quartus® Power Analyzer Tool.

6. Conclusions

The analysis of the RFI filters shows that the most promising for the future working in an very contaminated environment is the VRILMS32 based on the Cyclone®10 FPGA.

References

[1] Rossi B 1934 *Supplemento a la Ricerca Scientifica* **1** 579.
[2] Schmeiser K and Bothe W 1938 *Annals of Physics* **424** 161
[3] Kolhörster W, Matthes I and Weber E 1938 *Naturwissenschaften* **26** 576
[4] Auger P, Maze R and Robley A F 1939 *Comptes Rendus* **208** 1641
[5] Auger P, Ehrenfest P, Maze R, Daudin J and Robley A F 1939 *Rev. of Modern Physics* **11** 288
[6] Aab A et al., (Pierre Auger Collaboration) 2015 *Nucl. Instrum. Meth.* A798 (2015) 172.
[7] Allan H R 1971 *Progress in Part. and Nucl. Phys: Cosmic Ray Physics (North-Holland Publishing Company (Amsterdam)* **10**, 169
[8] Falcke H et al. 2005 *Nature (London)* **435** 313
[9] Ardouin D et al. 2006 *Astropart. Phys.* **26** 341
[10] Schellart P et al. (LOFAR Collaboration) 2013 *Astron. Astrophys.* A **98** 560
[11] Bezyazeekov P A et al., (Tunka-Rex Collaboration) 2015 *Nucl. Instrum. and Meth.* A **802** 89.
[12] Huege T (Pierre Auger Collaboration) 2010 *Nucl. Instrum. Meth.* A **617** 484
[13] Aab A et al. (Pierre Auger Collaboration) 2012 *JINST* **7** P10011, 2016 *JINST* **11** P01018
[14] Kahn F D and Lerche I 1966 *Proc. of the Royal Society of London Series A-Math. and Phys, Sci.* **289** 206
[15] Askaryan G A 1962 *Sov. Phys. JETP* **14** 441
[16] James C W, Falcke H, Huege T and Ludwig M 2011 *Phys. Rev.* E **84** 056602
[17] Kelley J L for the Pierre Auger Collaboration 2013 *Nucl. Instrum. Meth.* A **725** 133
[18] Schmidt A, Gemmeke H, Haungs A, Kampert K-H, Rühle C and Szadkowski Z 2011 *IEEE Trans. on Nucl. Science* **58** 4 1621
[19] Szadkowski Z, Fraenkel E D and van den Berg A M 2013 *IEEE Trans. Nucl. Science* **60** 5 3483
[20] Szadkowski Z, Głas D, Timmermans C and Wijnen T for the Pierre Auger Collaboration 2015 *IEEE Trans. Nucl. Science* **62** 3 977
[21] Szadkowski Z and Głas D 2016 *IEEE Trans. Nucl. Science* **63** 3 1455
[22] Szadkowski Z 2017 *IEEE Trans. Nucl. Science* **64** 11 1455
[23] Szadkowski Z and Głas D 2017 *IEEE Trans. Nucl. Science* **64** 6 1304
[24] Szadkowski Z and Szadkowska A 2018 *Real-Time and Embedded Systems and Technologies (RTEST) (Tehran)* DOI: 10.1109/RTEST.2018.8397081
[25] Harris R W et. al, 1986 *IEEE Trans. on Acoustics, Speech and Signal Processing* **ASSP-34(2)** 309
[26] Szadkowski Z 2018 *International Conference on Advances in Big Data, Computing and Data Communication Systems (icABCD 2018) (Durban)*
[27] Szadkowski Z 2018 submitted to *IEEE Trans. Nucl. Science*

IOP Conf. Series: Journal of Physics: Conf. Series **1181** (2019) 012079 doi:10.1088/1742-6596/1181/1/012079

Multiple Linsley method for EAS energy determination in LAAS

Hiroki Matsumoto[1], Atsushi Iyono[2], Saya Yamamoto[2], Kazuhide Okei[1], Shuhei Tsuji[1], Takao Nakatsuka[3] and Nobusuke Takahashi[4]

[1] Kawasaki Medical School, Kurashiki, Japan
[2] Okayama University of Science, Okayama, Japan
[3] Okayama Shoka University, Okayama, Japan
[4] Hirosaki University, Hirosaki, Japan

E-mail: h.matsumoto@med.kawasaki-m.ac.jp

Abstract. The observation of the primary energy spectrum in its energy region of 10^{16} eV to 10^{19} eV by using compact Extensive Air Shower (EAS) arrays and the time structure of EAS particles which was studied by Linsley (Linsley method) have been carried out by the Large Area Air Shower group. We have estimated the EAS core distance by using only an EAS array and the Linsley method. To improve the determination accuracy of the primary energy, we consider the estimation method of the core distance by using two EAS arrays (multiple Linsley method). We report on the idea of the multiple Linsley method and the comparison of simulation results in various observation modes.

1. Introduction

Cosmic rays have been studied in order to understand the origin of the primary cosmic rays, its propagation mechanism and many other phenomena, and various experiments have been carried out. The Large Area Air Shower (LAAS) experiment is also one of them and has been carried out for the observation of Extensive Air showers (EASs) since 1996 [1, 2]. This experiment has researched EAS events observed simultaneously by EAS arrays located at distance of several 100 km to about 900 km. Each EAS array which have eight scintillation counters typically in the LAAS experiment, and they are installed on several observation sites in Okayama University of Science (OUS), Nara Sangyo University and Hirosaki University in 2018. The experimental results in large scale observations are reported on GZ-effect events [3, 4] by the photon disintegration in Ultra High Energy cosmic rays, and anisotropy of cosmic rays [5]. Additionally, Okayama group in LAAS has studied of the primary energy spectrum, a correlation between solar modulation and cosmic rays, and a comparison of hadronic interaction models.

The LAAS EAS array is compact one which have eight scintillation counters (up to eight scintillation counters) and covers an area of several 100 m^2, because the array is designed for the observation of simultaneous and large scale EAS events. Thus, the observation of the lateral distribution of electrons and muons is difficult, and the primary energy of EASs is not able to be obtained. However, ideas and experimental results obtained by Linsley [6] allowed the estimation of the primary energy of EASs in the primary energy region of 10^{16} eV to 10^{19} eV by using even a compact EAS array. The estimation of the EAS core distance has been carried out by using only an EAS array. In addition, we plan to estimate the core distance by using more than an EAS array.

We report on the simulation results of the acceptance and the accuracy of the primary energy determination in the estimation method in the EAS core distance by using two arrays.

2. EAS arrays

There are five EAS arrays which are described as OUS1, OUS2, OUS3, OUS4 and OUS5, respectively, in Okayama University of Science in Okayama, Japan. Figure 1 shows the arrangement of OUS arrays in the campus of Okayama University of Science. The OUS1~OUS3

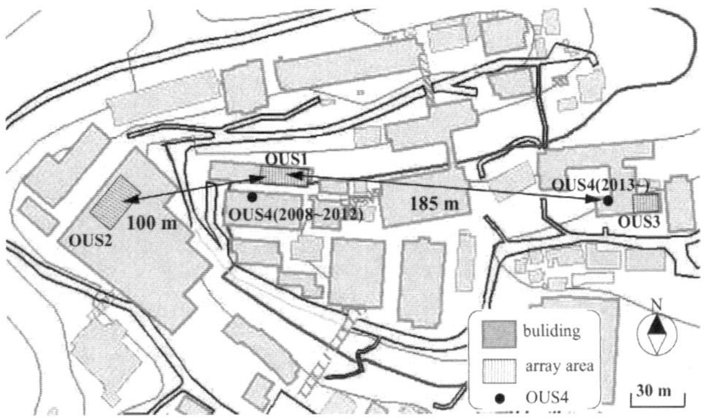

Figure 1. The arrangement of OUS arrays. This campus map is made by using a data in the Geospatial Information Authority of Japan [7].

are installed on the rooftop of buildings in OUS, and the OUS5 is installed on the ground in OUS's botanical garden located at ~ 2km distance from OUS. The OUS4 array is installed in the same building set on the OUS3. The distance between the OUS1 and the OUS2 is about 100 m, and the distance between the OUS1 and the OUS4 is about 185 m.

OUS arrays equip eight plastic scintillation counters typically, of which size is 50 cm × 50 cm × 5 cm and a PMT. Each counter is covered a stainless-steel case of 0.5 cm thickness. Data acquisition system consists of a TDC, a ADC and a shift register module which is only equipped with the OUS1. Timing information of the event trigger of EAS is obtained by using a GPS module. Data of the TDC, the ADC and the GPS, and the shift register are stored in a Linux OS PC and a Windows OS PC, respectively. The EAS event trigger signal is generated by hitting more than three scintillation counters within 2.5 μs.

Event signals generated with counters are digitalized by a discriminator and are temporarily stored in the shift register. When the shift register receives the event trigger signal, timing information of event signals within ± 2.5 μs time window from the trigger timing is stored to the Windows PC. The time accuracy of the shift register is 5 ns.

A GPS module is used for obtaining timing information of the event trigger and for synchronizing observation time among EAS arrays. The time accuracy of this module is maintained at 1 μs by receiving 1 pps GPS signal.

The OUS4 is used for obtaining the zenith angle of EAS. The OUS4 equips scintillation counters for the event trigger at the top layer and the bottom layer. The event trigger signal is generated by the coincidence of both counters. There are four scintillation counters as anticounters on each side of the OUS4. The trigger counters and the anticounters allow the restriction for the EAS zenith angle. When only the counters for the event trigger hit, a EAS zenith angle is obtained as within 30 degree. The OUS4 had been installed in the building located at distance of $\sim$ 10 m from the OUS1, and was relocated to the near site of the OUS3 in 2013. The update of the OUS4 was reported on reference [8].

3. Estimation of EAS core distance

3.1. Linsley method

The relation between the EAS thickness and its core distance r had been studied by Linsley. His results allow the estimation of the EAS core distance with a compact EAS array. We described the estimation method of the EAS core distance by using EAS thickness as the Linsley method.

We defined a dispersion of the arrival time distribution of particles in each EAS as σ_t. The average of σ_t in observation data is described as $\langle\sigma_t\rangle$. In the Linsley method, $\langle\sigma_t\rangle$ is described as

$$\langle\sigma_t\rangle = \sigma_{t0}\left(1+\frac{r}{r_t}\right)^b, \qquad (1)$$

where $\sigma_{t0} = 1.6$ ns, $r_t = 30$ m, $b = (2.08 \pm 0.08) - (0.4 \pm 0.06)\sec\theta + (0 \pm 0.06)\log(E/10^{17}\ \mathrm{eV})$, θ and E are the zenith angle and the primary cosmic ray energy [6], respectively. The parameter b depends on θ and E, however, is used by the averaged value. The averaged b is 1.65 and is nearly equal to when $\theta \sim 30$ degree. We rewrite Eq. 1 as

$$r = r_t\left(\frac{\sqrt[b]{\langle\sigma_t\rangle}}{\sigma_{t0}} - 1\right), \qquad (2)$$

and plot $\langle\sigma_t\rangle$ against r in figure 2.

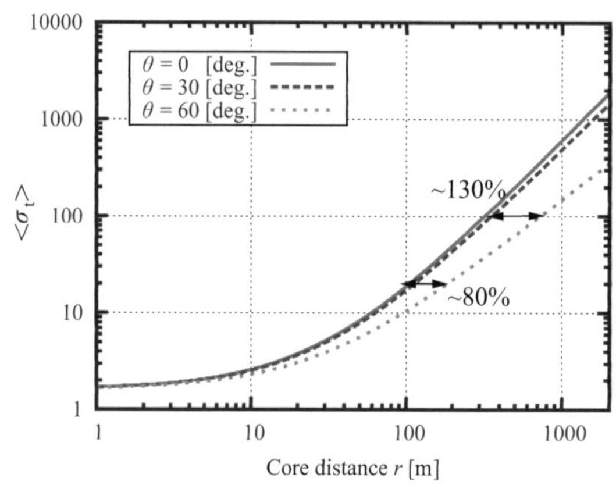

Figure 2. The relation of r and $\langle\sigma_t\rangle$, when $\theta = 0,\ 30,\ 60$ degree.

The systematic error of estimated r increase in increasing a difference between the true zenith angle and the averaged one. When $\theta = 60$ degree, the systematic error are 80 % ($\sigma_t = 20$) and 130 % ($\sigma_t = 100$), respectively.

The arrival time distribution of EAS is assumed to be a gamma distribution by Linsley. An estimator of the dispersion of the distribution is

$$\sigma_t = \frac{\langle t\rangle^2}{2}, \qquad (3)$$

where t is the arrival time of EAS particles. We have adapted the median to the estimator, because of suppressing noise events. The discussion for estimators is reported on Okita [9]. Thus, σ_t is described as

$$\sigma_t = \frac{\sqrt{2}}{1.67} t_{\mathrm{m}}, \tag{4}$$

where t_{m} is the median of the arrival time distribution of EAS particles. Using Eq.2 and Eq.4, the core distance r is described as

$$r = 30\left((1.35 t_{\mathrm{m}})^{\frac{1}{1.65}}\right). \tag{5}$$

3.2. Multi Linsley method

We have estimated the EAS core distance by using only an array. Additionally, we approach the estimation method by using two arrays and the Linsley method. This method is described as the Multiple Linsley method in this paper.

The core distance of EAS estimated independently by using the OUS1 and the OUS2 are described as r_1 and r_2, respectively. An EAS core position is estimated by the position with an overlap between r_1 and r_2. When there is not the overlap, we estimate the core position from the relation between r_1 and $r_2 + \Delta r_2$ ($r_1 > r_2$), or $r_1 + \Delta r_1$ and r_2 ($r_1 < r_2$), where Δr_1 and Δr_2 are systematic errors of r_1 and r_2, respectively. The selection criterion for the systematic error is within 100 % in consideration of the distance of two arrays.

4. Simulation result

The database of the lateral distribution and the shower size of electrons and muons required for carrying out the detector simulation is made by using Air shower simulation program AIRES [10]. In order to making the database, we assumed primary nuclei to be protons and irons, and hadronic interaction models to be QGSJET II-3 [11] /Hillas Splitting Algorithm [12]. The sampling of the primary energy E_0 is every $10^{0.1}$ eV in the primary energy region of 10^{15} eV to 10^{20} eV. The sampling of the EAS zenith angle θ is every 10 degree from 0 degree to 60 degree. The detector simulations of the OUS1, OUS2 and OUS4 are carried out by using this database. The procedures of the database simulation and the detector simulation are described in the reference [13].

4.1. Acceptance

Figure 3 shows the acceptance of each observation mode in case of proton primaries assumed. The OUS1 and the OUS1+2 represent the single observation of the OUS1 array only and the synchronized observation mode between the OUS1 array and the OUS2 array, respectively. And for the synchronized observation among above arrays and the OUS4, the OUS1+4 and the OUS1+2+4 are described, respectively. In a comparison of the OUS1 and the OUS1+2, the ratio of the acceptance of OUS1+2/OUS1 is about $1/2 \sim 1/3$ above $E_0 \geq 10^{17}$ eV, is $\sim 1/50$ in the low energy region. Similarly, in a comparison of the OUS1+4 and the OUS1+2+4, a ratio of the acceptance of OUS1+2+4/OUS1+4 is $\sim 1/5$ above $E_0 \geq 10^{17}$ eV, is $\sim 1/50$ in the low energy region. Event rates are expected to $\sim$ 120 /day and $\sim$ 1 /day in the OUS1+2 and the OUS1+2+4, respectively, because the average event rate of the OUS1 is $\sim$ 6000 /day.

Figure 4 shows the acceptance of each observation mode in case of iron primaries assumed. The acceptance of each observation mode in iron primaries shows similar trend in case of proton primaries.

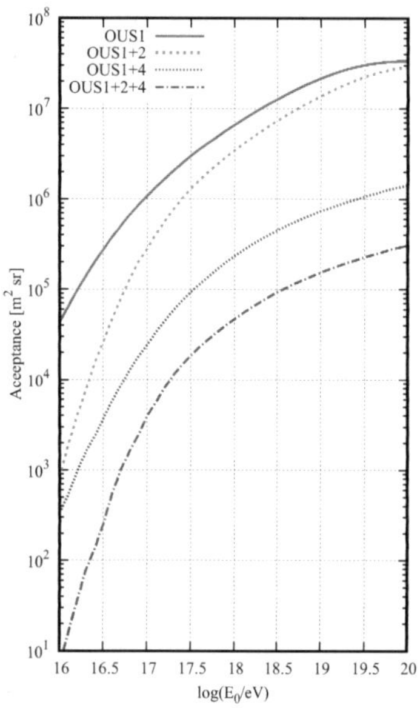

Figure 3. The comparison of four observations in the acceptance in the primary nucleus assumed to be proton.

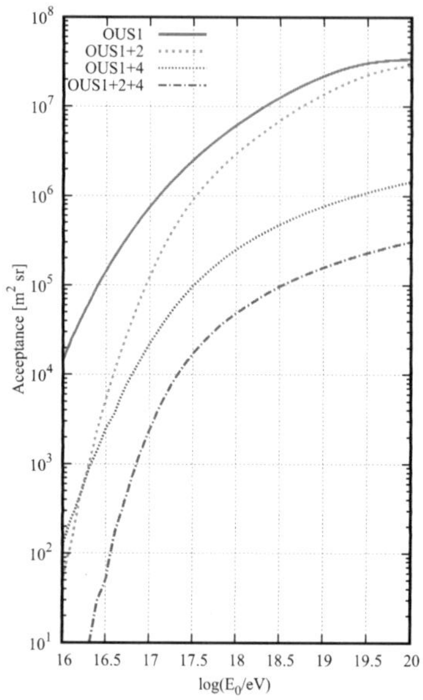

Figure 4. The comparison of four observations in the acceptance in the primary nucleus assumed to be iron.

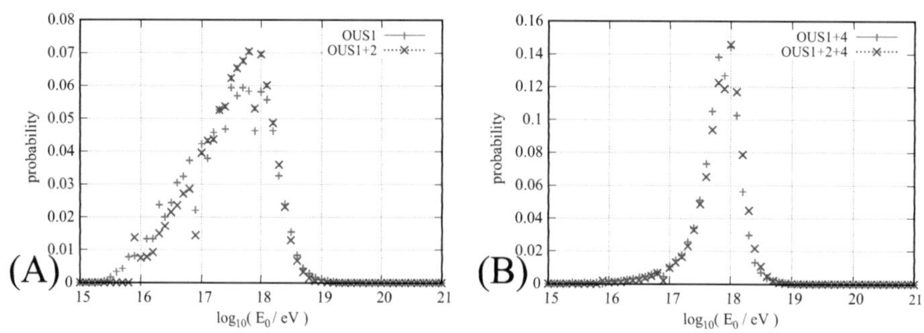

Figure 5. The primary energy energy distribution obtained by detector simulations assumed to be proton primaries at $E_0 = 10^{18}$ eV. (A) The comparison of the OUS1 and OUS1+2. (B) The comparison of the OUS1+4 and OUS1+2+4.

4.2. Accuracy of the primary energy determination

In the primary energy distribution obtained by detector simulations, figure 5 shows comparisons of distributions assumed to be proton primaries at $E_0 = 10^{18}$ eV. In the distribution of the OUS1, the peak energy of the distribution is $10^{17.7}$ eV and the FWHM is from $10^{16.5}$ eV to $10^{18.4}$ eV. The peak energy and the FWHM of the OUS1+2 are $10^{17.8}$ eV and from $10^{16.9}$ eV to $10^{18.4}$ eV, respectively. The primary energy distribution of the OUS1+2 is sharper than that of the OUS1 in the lower energy region than the peak energy in figure 5-(A). On the other hand, the primary energy distributions of the OUS1+2+4 is similar to that of the OUS1+4 in figure 5-(B), because the effect of the restriction of the EAS zenith angle is dominant.

In comparisons of each observation mode in iron primaries, the primary energy distributions assumed to be iron primaries show similar trends in case of proton primaries in figure 6.

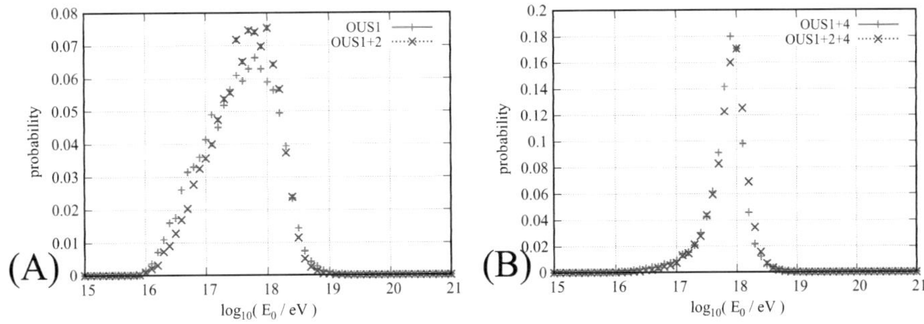

Figure 6. The primary energy energy distribution obtained by the detector simulation assumed to be iron primaries at $E_0 = 10^{18}$ eV. (A) The comparison of the OUS1 and OUS1+2. (B) The comparison of the OUS1+4 and OUS1+2+4.

5. Conclusion

Linsley method allows the estimation of the EAS core distance and the primary energy by using a compact EAS array in the LAAS experiment. As the results, we have been observed the primary energy spectrum in its energy region of 10^{16} eV to 10^{17} eV since 2008. In order to improve the accuracy of the primary energy estimation, we carried out the simulation for the Multi Linsley method. In addition we will develop an observation apparatus for the Multiple Linsley method.

References

[1] Wada T, Ochi N, Kitamura T, Unno W, Chikawa M, Kato Y, Konishi T, Tsuji K, Ohara S, Takahashi T, Takahashi N, Ohmori N, Sasaki H, Hasebe N, Yamamoto I, Nakatsuka T and Large Area Air Shower group (LAAS group) 1999 *Nucl. Phys.* B *(Proc. Suppl.)* **75** (issue 1-2) 330-2

[2] Ochi N, Iyono A, Kimura H, Konishi T, Nakamura T, Nakatsuka T, Ohara S, Ohmori N, Okei K, Saitoh K, Takahashi N, Tsuji S, Wada T, Yamamoto I, Yamashita Y, Yanagimoto Y and the Large Area Air Shower (LAAS) group 2003 *J. Phys.* G: *Nucl. Part. Phys.* **29** 1169-80

[3] Gerasimova N M and Zatsepin G T 1960 *JETF* **11** 899

[4] Iyono A, Matsumoto H, Okei K, Tsuji S, Ohara S, Ochi N, Konishi T, Takahashi N, Yamamoto I, Nakatsuka N, Nakamura T, Ohmori N and Saitoh K 2011 *Astrophysics and Space Sciences Transactions* **7** 327-33

[5] Noda C, Iyono A, Matsumoto H, Masuda M, Okita M, Okei K, Morita T, Takahashi N, Ochi N, Konishi T, Nakatsuka T, Ohara S, Takahashi N, Tsuji S, Wada T, Yamamoto I, Yamashita Y, Nakamura T and Saitoh K 2008 *Nucl. Phys.* B *(Proc. Suppl.)* **175-176** 459-62

[6] Linsley J 1986 *J. Phys.* G: *Nucl. Phys.* **12** 51-7

[7] Geospatial Information Authority of Japan *www.gsi.go.jp*

[8] Matsumoto H, Iyono A, Okei K, Tsuji S, Ohara S, Ochi N, Takahashi N, Nakatsuka T and Yamamoto S 2017 *Proc. of 35th International Cosmic Ray Conference* (Busan: PROCEEDINGS OF SCIENCE)

[9] Okita M, Wada T, Yamashita Y, Okei K, Morita T, Liang S, Takahashi N, Iyono A, Matsumoto H, Noda C, Masuda M, Yamamoto I, Kohata, Ochi N, Nakatsuka T and Tsuji S 2008 *Nucl. Phys.* B *(Proc. Suppl.)* **175-176** 322-25

[10] S. J. Sciutto 1999 AIRES *Preprint* astro-ph/9911331

[11] S. Ostapchenko 2006 *Phys. Rev.* D **74** 014026

[12] Hillas A M 1997 *Nucl. Phys.* B *(Proc. Suppl.)* **52** (issue 3) 29-42

[13] Matsumoto H, Iyono A, Yamamoto I, Kohata M, Okei K, Tsuji S, Nakatsuka T and Ochi N 2010 *Nucl. Instr. & Meth.* A **614** 475-82

Search for simultaneous and parallel cosmic gamma rays in the balloon-borne emulsion telescope experiments (GRAINE2015)

Saya Yamamoto[1], Atsushi Iyono[1], Shigeki Aoki[2], Satoru Takahashi[2], Hiroki Rokujo[3], Fukashi Mizutani[2], Kaname Hamada[4], Toshio Hara[2], Tatsuki Inoue[2], Katsumi Ishiguro[3], Hiroaki Kawahara[3], Koichi Kodama[5], Ryosuke Komatani[3], Masahiro Komatsu[3], Tetsuya Kosaka[2], Motoaki Miyanishi[3], Kunihiro Morishima[3], Misaki Morishita[3], Mitsuhiro Nakamura[3], Toshiyuki Nakano[3], Akira Nishio[3], Kimio Niwa[3], Naoto Otsuka[3], Keita Ozaki[2], Osamu Sato[3], Emi Shibayama[2], Atsumu Suzuki[2], Ryo Tanaka[2], Yurie Tateishi[2], Shuichi Tawa[2], Misato Yabu[2], Kyohei Yamada[2], Masahiro Yoshimoto[3]

[1] *Okayama University of Science*
[2] *Kobe University*
[3] *Nagoya University*
[4] *Institute of Space and Astronautical Science / Japan Aerospace Exploration Agency*
[5] *Aichi University of Education*

E-mail: iyono@das.ous.ac.jp, s_yamamoto@das.ous.ac.jp

Abstract. The emulsion chamber telescopes for observing 10 MeV - 100 GeV gamma rays have been developed by GRAINE (Gamma Ray Astro-Imager with Nuclear Emulsion) projects. We have performed the balloon filight in May, 2015 at Alice Springs, Australia, with 14.4 hours duration (including 11.5 hours of level flight) and 0.378 square meter aperture (GRAINE2015). GRAINE experiments utilized the multiple stage systems of emulsion films for timestamper of registered tracks.

In this paper, we report the potential of GRAINE2015 experiments to detect the parallel and simultaneous cosmic gamma rays such as charged particles due to hadronic interactions, EAS (Extensive Air Shower) and GRBs (Gamma Ray Bursts).

1. GRAINE project in 2015 (GRAINE2015)

1.1. GRAINE: Gamma Ray Astro-Imager with Nuclear Emulsion

GRAINE project is joint experiment by using the balloon-borne gamma ray telescope with high spatial resolution and large area emulsion films. The purpose of GRAINE project is to carry out the precise gamma ray imaging of high energy (energy range of 10 MeV - 100 GeV) astrophysical objects by utilizing milliradian angular resolution and to explore the origins and the emission mechanisms of gamma rays.

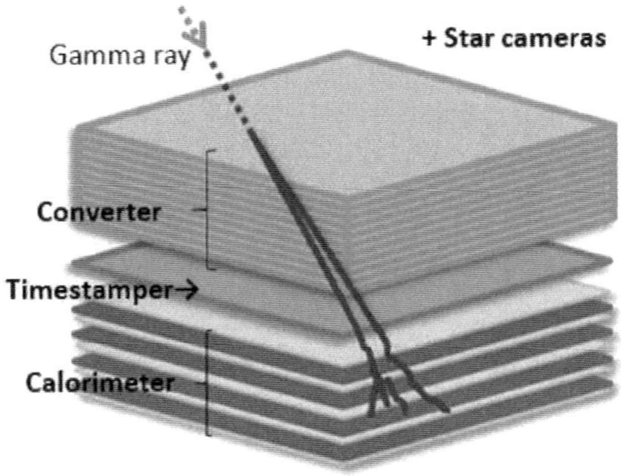

Figure 1. Constitution of main detector in gondola of GRAINE2015. This detector consists of converter, timestamper, calorimeter and three star cameras. Three star cameras are out of gondola.

1.2. Flight of GRAINE2015

GRAINE2015 flight in Australia, from Alice Springs to Longreach (about distance of 1000 km). Flight time is 14.4 hours in 37.2 km (including 11.5 hours of level flight). Expore area is 0.38 m^2. Purposes of GRAINE2015 flight were to confirm the overall system functions and to observe the morphology of the well-known and brightest gamma ray object, vela pulsar.

The main detector of GRAINE2015 consists (Fig. 1) of converter, timestamper[3], calorimeter and three star cameras. The converter's consist is stacking of one hundred emulsion films. Gamma rays were converted into electron-positron pairs in converter. Electron momenta were measuring with multiple scattering deflection angles. The timestamper's consist is three stages (top stage, middle stage, bottom stage) configuration using emulsion films. By move differently each stages, it can provide the track arrival time information (see 2.2). The calorimeter's consist is stacking of seventeen emulsion films and sixteen stainless steel plates. It measures the total energy of particles. And the three star cameras registeres the stellar objects. It evaluates the attitude of gondola in the flight.

2. Track data for analyze

2.1. Scan of emulsion filmes

After flight completed, developing emulsion films in Sydney University, scanning with HTS (Hyper Track Selector) were performed in Nagoya University. Scanning with HTS, the tracks in emulsion film are convert to data. Scanning and analyzing were done by dividing the emulsion film into nine areas (Fig. 2). In this paper, we used one scan area (about 9 cm $\times$ 13 cm including overlap) for this analysis.

Track data contain as coodinate (x, y, z) in each area, projected zenith angles ($\tan\theta_x$, $\tan\theta_y$), darkness of track called PH (Pulse Height) and PHV (Pulse Height Volume) about one emulsion film. HTS scan sixteen layer in emulsion, and PH is layer count of HTS can scanning. PHV is total density in sixteen layer of PH seven or more.

2.2. Time data

Arrival time data are obtained from timestamper. The timestamper utilized three motorized stages with different speeds (Fig. 3). The top stage moves 18.8 min/step, the middle stage moves 9 sec/step, and the bottom stage moves 500 um/s in vela mode. And vela mode is observable

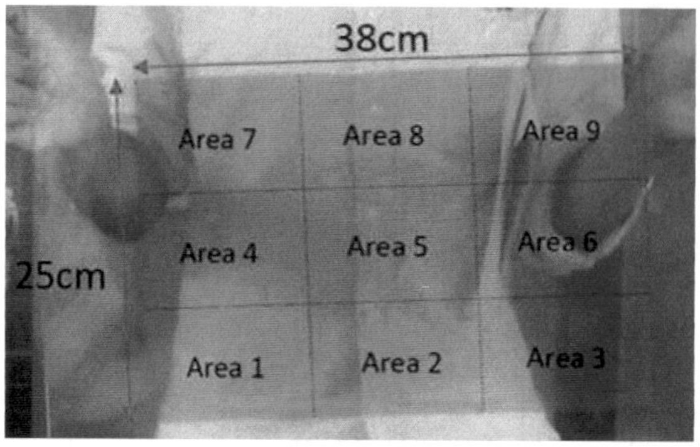

Figure 2. One of the emulsion film of GRAINE2015's converter and dividing into nine areas. In this paper, we used one scan area (about 9 cm × 13 cm including overlap) for this analysis.

time of vela pulsar. Determining the arrival time of tracks displacements by connecting them at all stages (Fig. 4). For example, in Fig. 4, the track of orange line passed in this time, and other tracks (green and blue lines) passed in other time.

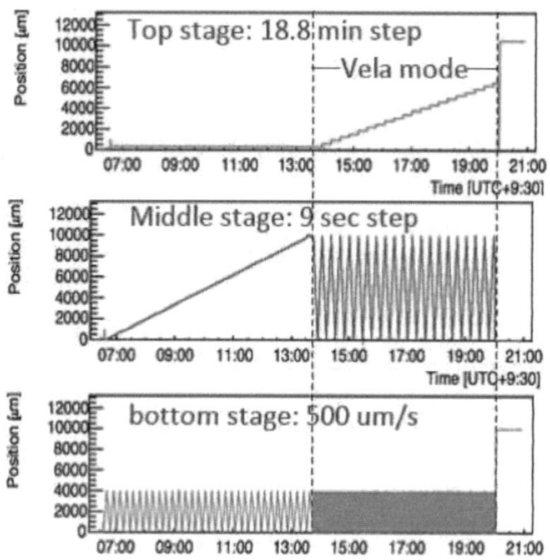

Figure 3. Movement of timestamper. Vertical axis is the shift position, and horizontal axis is the time in ACST (UTC+9:30). Vela mode is observable time of vela pulsar.

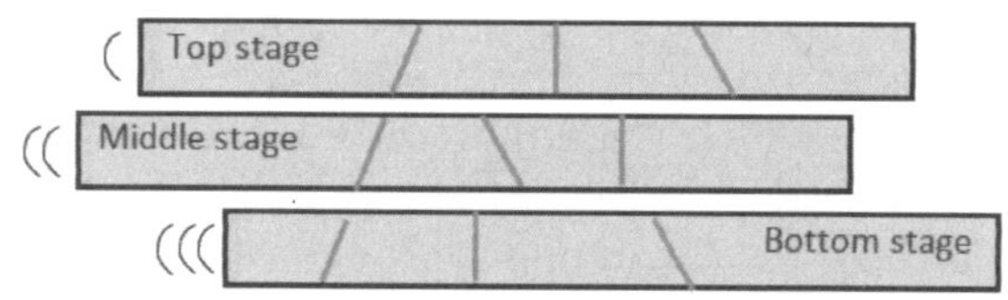

Figure 4. The track of orange line passed in this time. And other tracks (green and blue lines) passed in other time.

3. Selection of electron pair and gamma ray from data

3.1. Selection of electron pair

The method of electron pair selection is to eliminate tracks unrelated to pair-production. Tracks unrelated to pair-production are defined as penetrated track in converter, not found midway track due to energy threshold, and track of scattered in large angle. The track penetrating converter is not to make pair-production in converter. And the track scattered in large angle is too small energy to tracking. Finally, the track of non pair were eliminated.

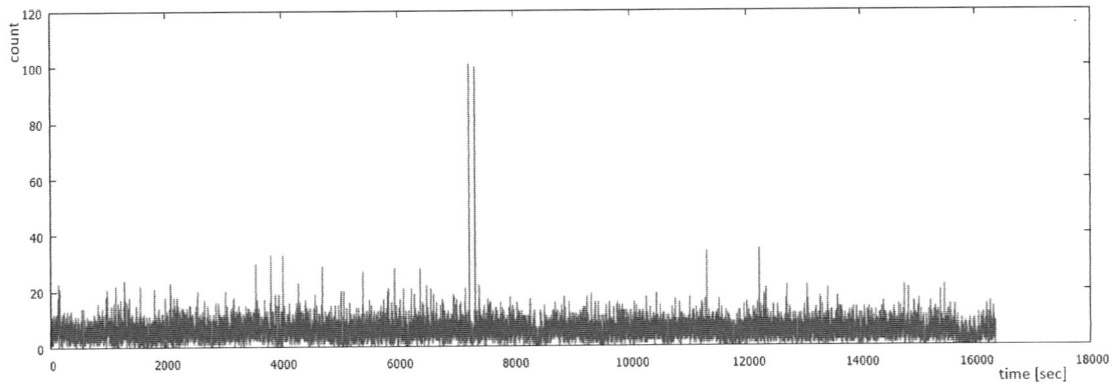

Figure 5. Timeline of count rate for timestamper data that could be decide a track time and that was able to see vela pulsar within zenith angle of 40 degrees. The horizontal axis is elapsed time from could see vela pulsar within zenith angle of 40 degrees to observated end time. The bin is one second.

3.2. Time stamp of electron data

Timestamper data at three stages data (top stage, middle stage, bottom stage) and the bottom layer in converter. Converter data and timestamper data can be combined by using the bottom data in converter. Fig. 5 is histogram of timestamper data that could be decide a track time and that was able to see vela pulsar within zenith angle of 40 degrees. And the bin is one second. The horizontal axis is elapsed time from could see vela pulsar within zenith angle of 40 degrees to observated end time. The start time of Fig. 5 is about 15:30 in ACST. Highest could comes from hadron interaction.

3.3. Selection of electron pair with timestamper data

The method of electron pair selection with timestamper data is as follows. At first, combine electron pair data and timestamper data. Next, selected the electrons that could determine the time. And third, selected the electrons that could determine the time two or more. This order select both electrons at the same time. And finaly, selected the electrons that have PHV<150. This order select the electrons that low charge. The electron pairs selected would come from in this method as gamma rays conversion.

Gamma ray counts in electron pair time is shown in Fig. 6. Fig. 6 left graph shows gamma ray data which can timestamp in bottom stage and which can connect to converter, and right graph shows the last two bins in top stage. Also, both graph in Fig. 6, the upper graph is histogram of top stage, the middle graph is histogram of middle stage, and the lower graph is histogram of bottom stage. In Fig. 6 right graph, we can see bottom stage can distribute among more detailed time than Middle stage. And Fig. 6, we could not show the simultaneous gamma rays in this analyze area. Also GRB also did not occur in the catalog of website of [5] observing from 10 MeV to 100 GeV gamma rays.

Summary

GRAINE carried out the emulsion gamma ray telescope balloon flight in Australia in 12th May 2015 (GRAINE2015). Converter and timestamper data were combined and track's arrival time were able to be determined. We could not find the simultaneous and parallel gamma rays in this analysis so far. Also GRB also did not occur in the catalog of website of [5] observing from 10 MeV to 100 GeV gamma rays.

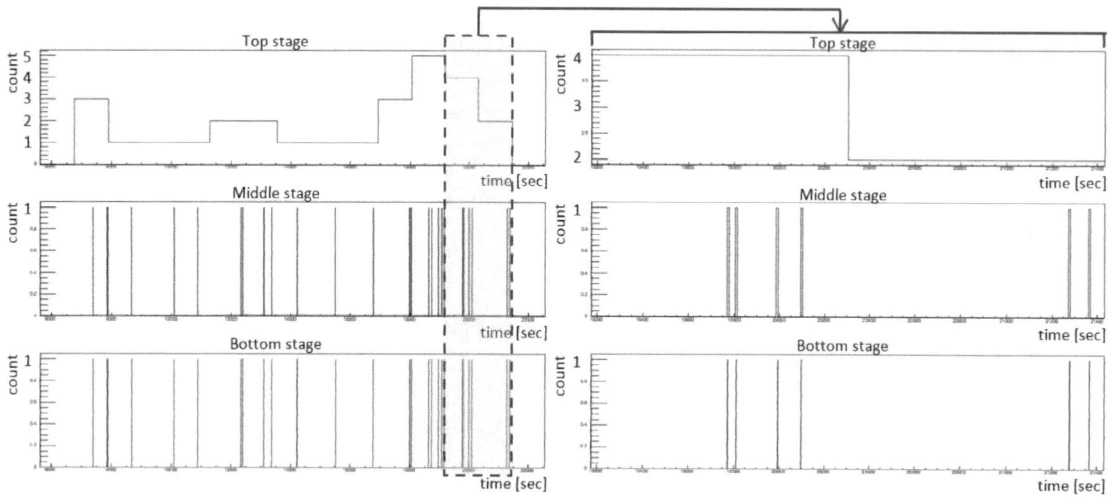

Figure 6. Left graph is Histogram of detected gamma ray times using all gamma ray data. And right graph is Histogram of detected gamma ray times using last two bins in top stage.

References

[1] Hiroki Rokujo, Shigeki Aoki, et al., "First demonstration of gamma-ray imaging using balloon-borne emulsion telescope", PTEP, 2018
[2] Satoru Takahashi, Shigeki Aoki, et al., "GRAINE2015, aballoon-borne emulsion γ-ray telescope experimet in Australia", Prog. Theor. Exp. Phys. 2016, 073F01, 2016
[3] F. Mizutani, et al., NIM A (submitted)
[4] Masahiro Yoshimoto, et al.,"Hyper-track selector nuclear emulsion readout system aimed at scanning an area of one thousand square meters",PTEP, 2017
[5] GRBs localized with BSAX or BATSE/RXTE or ASM/RXTE or IPN or HETE or INTEGRAL or Swift or AGILE or Fermi or MAXI, http://www.mpe.mpg.de/ jcg/grbgen.html

Status of the URAN array for detection of EAS neutron component

I I Yashin[1], F A Bogdanov[1], D M Gromushkin[1], R P Kokoulin[1], A A Petrukhin[1], P A Semov[1], I A Shulzhenko[1], Yu V Stenkin[1,2], K O Yurin[1]

[1] National Research Nuclear University MEPhI (Moscow Engineering Physics Institute), Moscow 115409, Russia
[2] Institute for Nuclear Research of the Russian Academy of Sciences, Moscow 117312, Russia

E-mail: IIYashin@mephi.ru

Abstract. First prototypes of PRISMA-32 neutron detectors based on the ZnS (Ag) scintillator demonstrated the perspectives of their application for detection of neutrons in extensive air showers. Detected thermal neutrons are generated due to interactions of the hadronic component with the nuclei of the atmosphere or the matter surrounding the detector and carry important information about the energy of primary particles. A mixture of zinc sulfide with natural boron (ZnS + B_2O_3) is used in the URAN setup. Detectors of the URAN array are located on the roofs of NEVOD and neighboring laboratory buildings; the area of each detector is 0.36 m^2. The array consists of independent clusters of 12 detectors in each and is united by a central station of collection and processing of information. Two new clusters of the array were put into operation in 2017, synchronization with other NEVOD setups (NEVOD-EAS array and NEVOD-DECOR-SCT) was provided, and collection of experimental data was started with the area of about 10^3 m^2. The first events of the EAS with neutrons have been registered and data on the temporal distribution of EAS neutrons have been obtained.

1. Introduction

In recent years, a new method for the study of EAS, based on the detection of thermal neutrons, which are generated due to interactions of the hadron component of EAS with the nuclei of the atmosphere and the matter near the facility, has been widely used [1]. This method allows detect the hadronic component throughout the facility area. The first results on EAS neutrons obtained at the PRISMA-32 facility [2] demonstrated the feasibility of this research method [3-5]. The next step in this direction was the creation of the URAN installation (an installation for registration of atmospheric neutrons), whose detectors are located on the roofs of the buildings on the territory of the MEPhI campus [6]. The location outside the building allows to minimize the influence of various structural elements on the registration of neutrons in the EAS, and to reach a much larger area in comparison with the PRISMA-32 installation. For the URAN setup, new en-detector with new scintillator and cluster type electronics that allows digitizing the signals with different frequencies was created.

2. Design of en-detector and registration system of the URAN installation

Design of the novel type en-detector is based on that developed for the PRISMA-32 array [4]. The construction of the en-detector is presented in figure 1. We use a black cylindrical (Ø740 x 570) standard commercial plastic water tank of 200 l volume as a light protecting housing. To improve light collection, a reflecting cone is used. On the top of the cone, the 6" photomultiplier tube (FEU200) looking to the scintillator at the cone bottom is located. The scintillator area is equal to 0.36 m^2. Zinc coated 0.9 mm iron cylinder (Ø880 x 1100) with a conic roof is used for environmental protection.

A special inorganic compound scintillator ($ZnS(Ag)+B_2O_3$) with natural boron is used for the URAN array construction. It is made of grains of the compound alloy LRB-1 (Russian production) in a transparent silicon disk of 70 cm diameter and 5 mm thickness. Mass content of boron in the compound is equal to ~ 13%, while content of ^{10}B is 2.6%. Total compound thickness is 50 mg/cm^2.

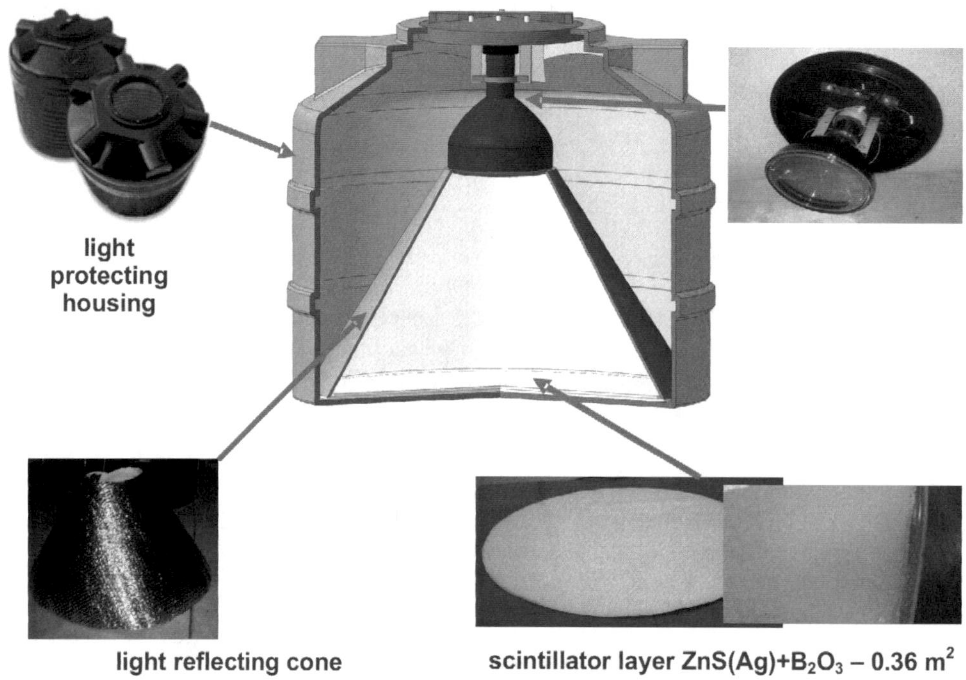

Figure 1. The en-detector design.

The signal is produced by heavy charged particles (^4He and ^7Li) which lose their energy inside one grain. Scintillator ZnS is effective enough for heavy particles and has a record α/e separation ratio.

The recording system of the URAN array clusters consists of twelve boards of amplitude analysis with two integrated controllers BAAC12. Structural scheme of the URAN array electronics is shown in figure 2. The use of two boards per cluster is necessary to increase the dynamic range of the registration of the electromagnetic component.

BAAC12 board has six dual-channel 12-bit ADC with 200 MHz sampling, it receives signals from the detectors (D#1-12) and implements the processing at a given logic. Boards are installed in the Local Post (LP) of the primary data processing, which, like the detectors, is located on a building roof. A controllable unit of the high-voltage power supply and a data transmission system are also located in the LP. Each local post is equipped with a thermostabilization system consisting of a temperature sensor, a fan and a heater to ensure stable operation temperature of the electronics. Transferring of analog signals from the detectors to the amplitude analysis boards is carried out by means of coaxial cables of the same length (25 m).

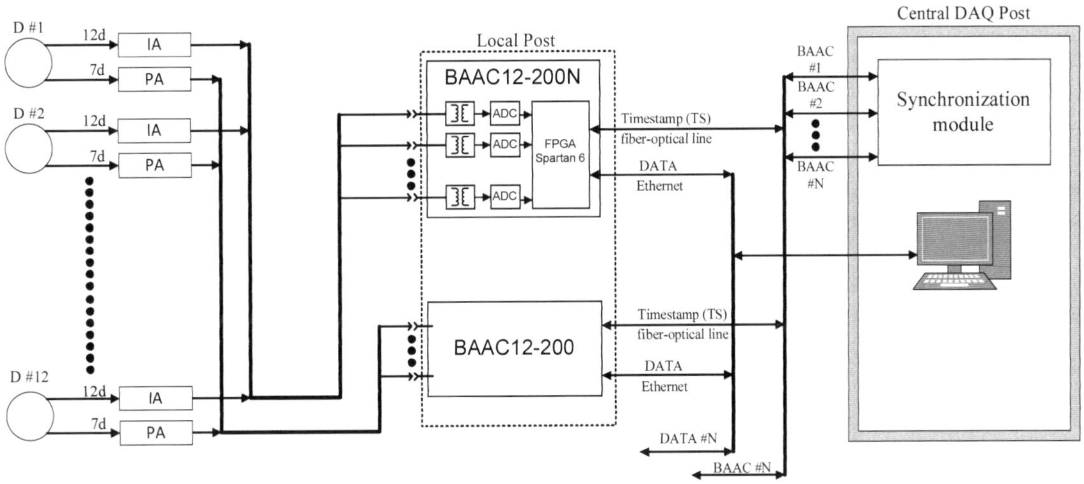

Figure 2. Structural scheme of the URAN array registration system.

BAAC12 boards are combined in the local post. Data transmission from the LP to the central post of data acquisition and processing (CDP) is carried out by means of fiber optic lines. CDP consists of the PC on which a client program, allowing to carry out a set-up and start-up of the boards is installed. The synchronization module and data storage and data transfer systems are also located in the CDP. In addition, the CDP is receiving and storing data coming from the boards.

All 12 cards work synchronously of the same frequency, due to the synchronization module. The input signal from the PMT of the detector passes through the integrator-amplifier (IA) for the 12th dynode and through the integrating pre-amplifier (PA) for the 7th dynode, in both cases the time-integration constant is 1 μs, comes to BAAC12-200. It passes through a matching transformer used for galvanic isolation, and then is fed to the ADC. Then the signal is digitized with a step of 5 ns and enters the digital comparator, wherein the data are also written to the buffer. If incoming signal exceeds the threshold of the digital comparator, the "REQUEST" signal is generated, and the value of the internal timer is stored. If the number of "REQUEST" signals for a certain period is greater than or equal to the number specified by trigger conditions, than the "TRIGGER" signal is generated. The "TRIGGER" signal stops recording of the data in the buffer and forms a data packet. In the data packet recorded digitized signals from 12 channels and time of the internal timer are stored: for BAAC12-200N card 1024 samples with steps of 5 ns and 20,000 samples with steps of 1 ms, and from BAAC12-200 card 1024 samples in steps of 5 ns are recorded. The boards implement the internal buffer of 50 packets for security of information in the case of heavy traffic when transmitting to the CDP. To send data, the TCP / IP protocol is used.

Time synchronization of clusters with an accuracy of 10 ns is performed using global positioning systems (GPS/GLONASS). Each BAAC12 board has an internal timer whose value is saved and passed along with the event. Also synchronization module ensures simultaneous starting of this timer in order to set the initial value. The synchronization module performs periodic polling of internal time of boards (once per second), and in case of difference aligns their values.

3. Registration of EAS

Registration of the EAS takes place on the leading edge of the shower and then the oscilloscope is open for writing in the time window of 20,000 microseconds, where the neutrons accompanying the shower are detected. Clusters work independently and have the same criteria for triggering and recording of events. During the series of measurements, it was required that the signals from two or more detectors have amplitudes at least 16 mV. Figure 3 shows the oscillogram obtained by recording the electromagnetic component of the EAS in four detectors of the cluster, the step of digitization is 5 ns, and the oscillogram of the detected delayed neutron with a sampling step of 1 μs. The signals from neutrons of the EAS whose amplitude is ≥ 10 mV are registered after the electromagnetic component. Also, the signal duration from neutron should be at least 4 μs. Figure 3 shows pulses of electron component and five neutrons detected during 20 ms time window (right panel).

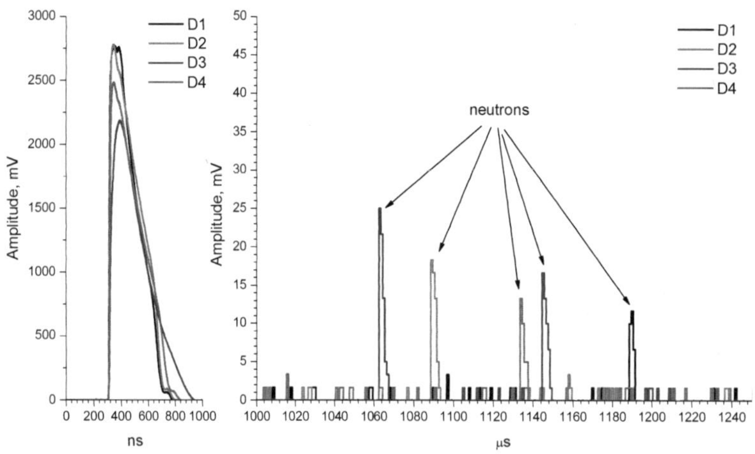

Figure 3. Waveforms from the charged component (left) of the EAS and neutrons accompanying the shower (right).

4. Results

The current configuration of the URAN setup consists of 5 clusters. The temporal distribution obtained with the 5 clusters (Cl1-Cl5) of the URAN array and their total amount (AllCl) of the EAS detection is shown in the figure 4. The time step is 100 μs. The distribution obtained for events with a trigger condition ≥ 6 detectors was fitted by an exponential function:

$$F(t) = A_1 \times \exp(-t/t_1) + y_0 \quad (1)$$

with the parameter of the function $t_1 = 646.8 \pm 19.8$ μs.

The obtained value can be compared with the data on the temporal distribution of neutrons in the EAS measured with the PRISMA-32 array. This distribution can be described with a double exponential function:

$$F(t) = A_1 \times \exp(-t/t_1) + A_1 \times \exp(-t/t_2) + y_0 \quad (2)$$

with the parameters $t_1 = 0.49 \pm 0.01$ ms. and $t_2 = 3.44 \pm 0.2$ ms.

As it is seen, the function parameter in the URAN array is consistent with the parameter of the first exponent in the PRISMA-32 array. The obtained result confirms the previously made assumption that in the PRISMA-32 array the second exponent ($t_2 = 3.44 \pm 0.2$ ms) has a connection with the neutrons generated due to interactions of the hadronic component with the roof or walls of the building (the

PRISMA-32 detectors are located inside the building under concrete overlap). And the first exponent is related to the average lifetime of locally born neutrons in the matter under the detector.

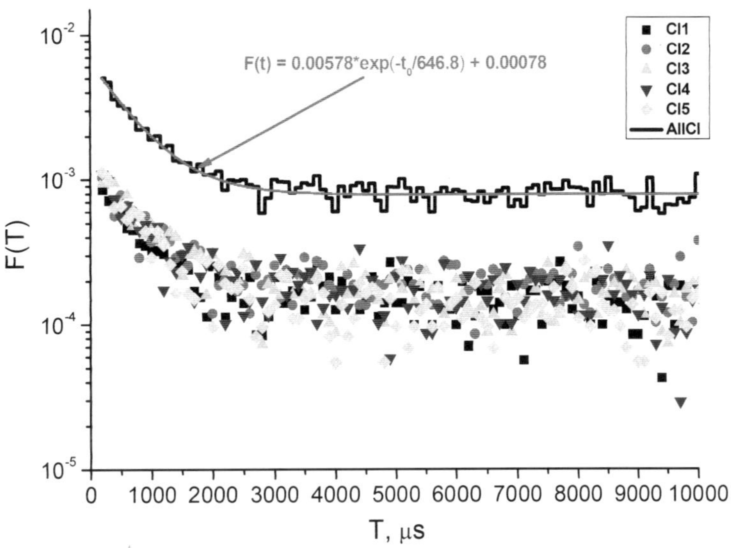

Figure 4. Temporal distribution of neutrons measured with the URAN array.

5. Conclusion

The first stage of the URAN installation for registration of the neutron component of the EAS was created. The installation consists of five clusters of 12 en-detectors capable to detect electrons and thermal neutrons of extensive air showers. The first series with five clusters of the URAN installation were carried out. The first results of data processing showed the possibility of the URAN setup for reliable registration of the neutron component of the EAS. The temporal distribution of the EAS neutrons in the URAN facility is described by a one exponential function whose behavior is comparable to that of one of the exponents obtained by measuring the time distribution of neutrons in the PRISMA-32 facility. The presence of the first exponent in PRISMA-32 data indicates the influence of the roof over the detector. This indicates that when measuring the neutron component of the EAS, it is necessary to take into account the location of the detectors and the distribution of matter around them, in which the neutrons are thermalized.

Acknowledgments

The work was performed at the Unique Scientific Facility "Experimental complex NEVOD" with the support of the Ministry of Education and Science of the Russian Federation (project RFMEFI59114X0002, government task, MEPhI Academic Excellence Project 02.a03.21.0005) and the Russian Foundation for Basic Research (grant 16-29-13028-ofi-m).

References

[1] Gromushkin D M et al. 2015 *Phys. Atom. Nucl.* **78** 379
[2] Gromushkin D M et al. 2014 *Journal of Instrumentation* **9** C08028
[3] Gromushkin D M et al. 2013 *J. Phys.: Conf. Ser.* **409** 012044
[4] Gromushkin D M et al. 2011 *Bull. RAS. Physics* **75** 374
[5] Stenkin Yu V 2009 *Nucl. Phys.* **109** 293
[6] Gromushkin D M et al. 2016 *J. Phys.: Conf. Ser.* **675** 032043

IOP Conf. Series: Journal of Physics: Conf. Series **1181** (2019) 012082 doi:10.1088/1742-6596/1181/1/012082

SENSE - Ultimate Low Light-Level Sensor Development

G. Bonanna[1], A. Haungs[2], K. Henjes-Kunst[3], T. Huber[2,3], K. Link[2], A. Nagai[4], R. Mirzoyan[5], T. Montaruli[4], G. Romeo[1], D. Strom[5] and H. Tajima[6]

[1] INAF- Catania Astrophysical Observatory, Italy
[2] Institut für Kernphysik, Karlsruhe Institute of Technology, Germany
[3] DESYDeutsches Elektronen-Synchrotron, Hamburg, Germany
[4] Départment de physique nucléaire et corpusculaire, Université de Genève, Switzerland
[5] Max-Planck-Institute for Physics, Munich, Germany
[6] Nagoya University, Japan

E-mail: `sensepro@desy.de`

Abstract. SENSE, a roadmap for the ideal low light-level sensor development is a project funded by the European Commission under Future and Emerging Technologies (FET) Open Coordination and Support Action (CSA). It aims at coordinating, monitoring, and evaluating the R&D efforts of research groups and industry in advancing low light-level (LLL) sensors and liaise with strategically important European initiatives and research groups and companies worldwide.

The projects objectives are: (1) to conduct the development of a European R&D roadmap towards the ultimate LLL sensors, and to monitor and evaluate the progress of the development with respect to the roadmap, (2) to coordinate the R&D efforts of research groups and industry in advancing LLL sensors and liaise with strategically important European initiatives and research groups and companies worldwide, (3) to transfer knowledge by initiating information and training events and material, (4) to disseminate information by suitable outreach activities.

The consortium has four partners: the Deutsches Elektronen Synchrotron (Coordinator), Germany; the Université de Genève, Switzerland; the Max-Planck Institute for Physics, Germany and Karlsruhe Institute of Technology, Germany. Several international experts for all areas of LLL development are involved in SENSE, some advise the project n the Experts Group and the others in the working group of the project which is regulated by a Cooperation Agreement.

1. Introduction

The SENSE project started its activities in September 2016 and received funding for three years. The goal of the project is the coordination and evaluation of the R&D developments in the field of low light-level sensoring.

To achieve these objectives, the SENSE project working plan is structured in five work packages (WPs), which are lead by different partners of the consortium: **WP1** (lead by Max-Planck Institute for Physics) deals with the creation of the roadmap for LLL sensor R&D and the monitoring and evaluation of the subsequent R&D work.

WP2 (lead by Université de Genève) focuses on working on an agreement on R&D cooperation between Research Institutes in Europe, linking to other European initiatives as well as fostering the exchange between academia and industry. A laboratory set up has been created and validated at Ideasquare at CERN for measuring properties of silicon photosensors.

WP3 (Karlsruhe Institute of Technology) takes care of outreach of the project and developing the Technology Exchange Platform (TEP/ SENSE Forum), as major part to exchange between developers.

Three tasks are planned for **WP 4** (Deutsches Elektronen Synchrotron) all around training and learning: the organisation of technology training sessions during existing summer schools, establishing a training section on the TEP and to organise LLL technology training events. All groups participated to outreach through public events.

WP5 (Deutsches Elektronen Synchrotron) describes the management duties of the overall project.
All work is carried out in close cooperation and with mutual assistance.

2. Roadmap

For preparation of the first draft of the roadmap the different developments concerning LLL were discussed and evaluated within the consortium and the experts. A first document was drafted based on this. It is focused on developments that are crucial for two photo-sensing technologies: silicon photomultipliers (SiPMs) and photomultipliers (PMTs).

Three major sectors of development were identified for each technology:

- the performance of the sensors (which usually depends on the application);
- the readout and control electronics;
- the integration of such electronics into the sensor.

For each sector, the specifications required to individual fields of application and which challenges must be overcome are pointed out. In addition, the results of ongoing specific R&D activities taking place in line with the roadmap idea within SENSE are shortly presented. Until the end of the project a final document with clear recommendations and also some concrete examples will be published.

3. Cooperation Agreement

A cooperation agreement for providing a framework for the testing and performance studies of photosensors was developed. Various institutions and industrial partners, have expressed their interest in carrying out together a long-term detector R&D for future ultimate LLL photosensor, which also includes an educational and technology training component. This agreement should lead to open access publications which will be approved by the SENSE consortium.

At the moment the document is signed by the following institutions:
University of Geneva (UNIGE), Max-Planck-Institute for Physics (MPI), Karlsruhe Institute of Technology (KIT), Deutsches Elektron Synchrotron (DESY), Nagoya University (Nagoya), INAF-Osservatorio Astrofisico di Catania (INAF-CT), University of Heidelberg, Institute for High Energy Physics of Barcelona (IFAE), University of Barcelona ICCUB.

An example of measurements that were produced between cooperating institutes is shown in figure 1, which shows the optical cross talk as a function of the over-voltage for some silicon photomultipliers which were compared. These measurements were carried out under the direction of the University of Geneva and are published in [1].

4. SiPM Characterisation

Initiated by SENSE a collaboration between several labs, experienced in measuring photosensors, was developed. The aim is to characterize LLL sensors and standardize measurements and analysis procedures. Each lab has slightly different test facilities for the measurement of

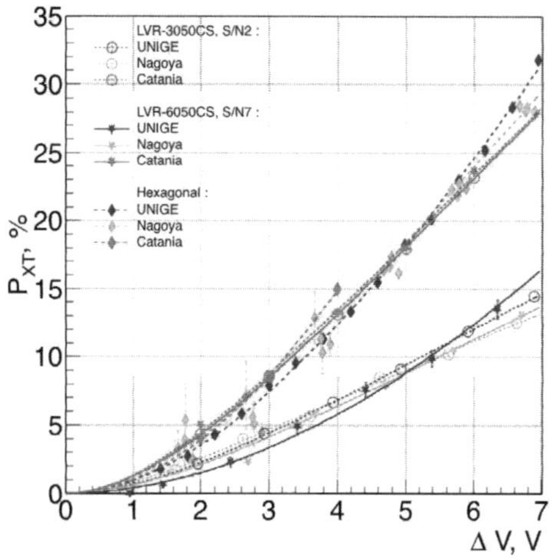

Figure 1. Cross talk probability of three compared silicon photomultipliers by three institutions in the cooperation agreement as a function of the overvoltage, from [1].

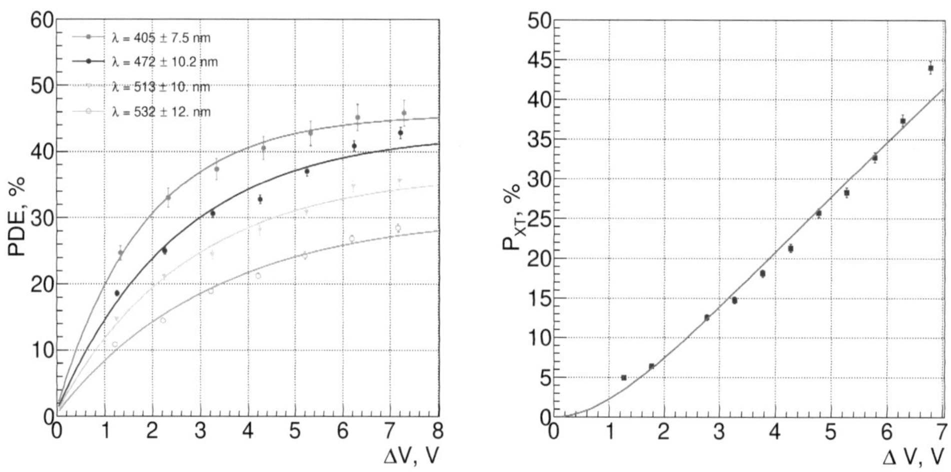

Figure 2. Photon detection efficiency and cross talk probability of SensL MicroFJ-SMTPA-30035 as a function of the overvoltage.

photosensors, so in a first step each platform is characterized and systematic errors relevant for various measurements are compared. This minimizes the double effort for the characterization of sensors and common precision on measured quantities are established. First results are published jointly under open access in [1]. An example of the current measurement is shown in Fig. 2.

5. Public Relations

In terms of outreach activities the project aims to inform, to disseminate results and to allow for open communication between all involved partners, for the people concerned and also for the general public. Therefore it is important to offer information material at different levels and to

Figure 3. SENSE booth at the spring meeting of the German Physical Society

use various channels.

A broad audience with different interest and knowledge can be reached online, e.g. via a website or using social media, while the participation at conferences is dedicated to a specific audience. In order to address as many people as possible we at SENSE therefore use both online and offline outreach opportunities.

5.1. Website, Newsletter & TEP

The SENSE website `www.sense-pro.org` provides information about the project itself and the people behind as well as general informations on different photosensors and their usage. Under the menu item *Portraits*, test stands and experiments working with photosensors are presented. This category is regularly extended by further articles, which are also sent out in a newsletter.

Reports about current actions of SENSE, references to conferences and published articles as well as other events of interest to the community are also sent in the newsletter and can be found on the website as *News* and *Events*. Information on conferences are also available in the calendar.

For the future a sophisticated section for training and learning shall be developed and made available on the website.

Furthermore, the results obtained in the activities on SiPM characterisation will be collected in a database and published on the website.

5.2. Presentation at Conferences

For the presentation of SENSE at conferences various information materials are produced: A general brochure on SENSE containing information about the project and general on low light-level sensors, a flyer dedicated to the roadmap and posters or talks, which are adapted to the respective needs and the respective audience. As an example, figure 3 shows the SENSE booth at the spring meeting of the German Physical Society.

In addition to making SENSE generally known, the scientific results from lab measurements are also presented at conferences. However, personal contact and exchange and the opportunity to establish new contacts with various community working on photosensors are one of the main reasons for attending conferences.

6. Training Activities

In order to continue advancing the development of photosensors in the future, interested young scientists are needed. That is why we want to introduce young people to the topic and inspire them with enthusiasm.

A series of lectures on different topics like detector principles, light interactions with matter or signal processing, are developed and will be made available on the website. These lectures should be held additionally at a summer school dedicated to photosensor development.

We already held lectures and practical sessions at the MAGIC hardware school and the TESHEP summer school, see Fig. 4- on the left. But also already high school students got in touch with SENSE during the TecDay in Geneva, see Fig. 4- on the right, or the IceCube Masterclass.

Figure 4. Lecture during TESHEP summer school (left) and TecDay (right)

In this context, two show case experiments were developed, one by UNIGE and the other one still in the design phase at KIT.

6.1. Show Case Experiments

Both experiments aim at the measurement of cosmic muons once with a thermos flusk used as water Cherenkov detector and once using a scintillator bar as detector. Both setups use a SiPM as light detector, operated without high voltage, which is of special importance for a student hands-on experiment.

Figure 5 shows the experimental setup of the experiment currently developed at KIT. The idea of using thermos flusk is already realized in many student experiments but so far the light pulses are read out using a PMT. The biggest advantage of this improved setup is the elimination of high voltage due to the use of SiPMs for light detection. For the analysis of the measured signal the free and open source software picoCosmo [2], also developed at KIT, is used. This software also includes the possibility to define different trigger conditions and to search for double pulses which allows the measurement of the lifetime of muons.

The setup developed at UNIGE is shown in figure 6. Muons are measured using a 1m plastic scintillator bar that is coupled on both ends to SiPM detectors. The measured signal is visualized using an oscilloscope. To separate the signal of cosmic ray muons from detector noise signals the detectors are read out in coincidence mode. Then the rate of cosmic ray muons can be measured, and also the muon rate as a function of the inclination can be studied.

7. Summary

SENSE was proposed to European Commission by experts who spent a long time understanding photosensors - SENSE should create a common voice towards industry. Therefor, many institutes working in the LLL development area have to be convinced to work together with SENSE. A

IOP Conf. Series: Journal of Physics: Conf. Series **1181** (2019) 012082 doi:10.1088/1742-6596/1181/1/012082

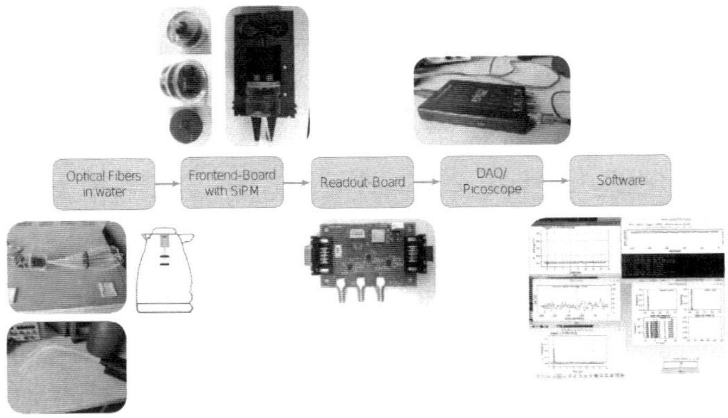

Figure 5. Setup of the show case experiment developed at KIT

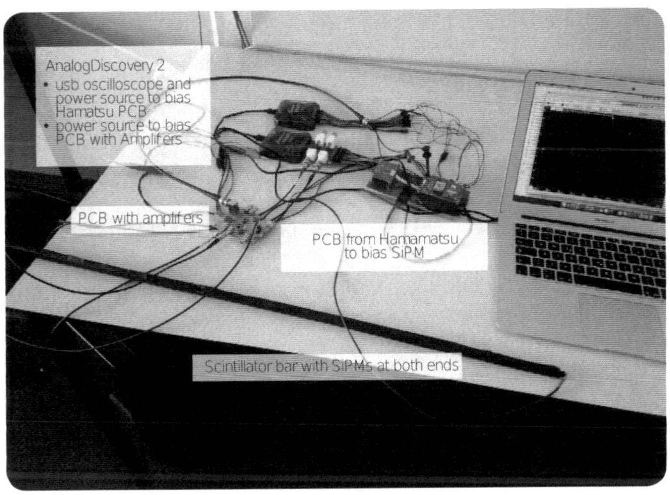

Figure 6. Setup of the show case experiment developed at UNIGE

first step in this direction was done with the agreement of common testing in the cooperation agreement. During many conferences and trainings SENSE was made public. The publicity which is reached now in different application disciplines of LLL sensors should be now be used for the last year of SENSE to develop a strong final roadmap introducing directions in which development of LLL sensors should go in future.

Acknowledgments
This project received funding from the European Unions Horizon 2020 research and innovation programme under grant agreement no. 713171.

References
[1] Nagai A et al. 2017 SENSE: A comparison of photon detection efficiency and optical crosstalk of various SiPM devices *NIM* A
[2] Software picoCosmo, available at *https://github.com/GuenterQuast/picoCosmo*

Additional aperture detectors of gamma-telescope GAMMA-400 calibrations on synchrotron "PAKHRA": possibility of temporal profiles fractal analysis

I V Arkhangelskaja[1], A I Arkhangelskiy[1, 2], E N Chasovikov[1], S I Suchkov[2], M F Runtso[1], Y T Yurkin[1], A M Galper[1, 2], N P Topchiev[2], A A Leonov[1, 2], A E Egorov[2]

[1]National Research Nuclear University MEPhI (Moscow Engineering Physics Institute), Kashirskoe highway 31, Moscow, 115409, Russia
[2]Lebedev Physical Institute, Russian Academy of Sciences, Leninskiy prospekt 53, Moscow, 119991, Russia

E-mail: irene.belousova@usa.net

Abstract. GAMMA-400 (Gamma Astronomical Multifunctional Modular Apparatus) will be the new generation satellite gamma-observatory. The gamma-ray telescope GAMMA-400 consists of anticoincidence system (top and lateral sections - ACtop and AClat), the converter-tracker (C), time-of-flight system (two sections S1 and S2), position-sensitive calorimeter CC1, electromagnetic calorimeter CC2, scintillation detectors of the calorimeter (S3 and S4) and lateral detectors of the calorimeter LD.Three apertures provide events registration both from upper and lateral directions. The main aperture provides the best angular (all double (X, Y) tracking coordinate detectors layers information analysis) and energy resolution (energy deposition in the all detectors studying). The main aperture created firstly due to converter-tracker (C): gammas converted in tungsten conversion foils are registered. Triggers in the main aperture will be formed using information about particle direction provided by time of flight system and presence of charged particles or backsplash signal formed according to analysis of energy deposition in combination of both layers anticoincidence systems ACtop and AClat individual detectors. Other two apertures used for observation of transient events do not require best angular resolution as gamma-ray bursts and solar flares both from upper and lateral directions. Additional aperture allows particles registering from upper direction, which don't interact with converter-tracker and don't formed TOF signal. Particles detection in additional aperture starts with signal of CC1 fast discriminators in anticoincidence with TOF. Energy band for gammas registration in this aperture is similar to the main one. In the lateral aperture low energy (0.2-100 MeV) photons classified by using simple anticoincidence signals from the individual detectors of LD and CC2. Higher energies γ-quanta (E>100 MeV) recognized using energy deposition analysis in the individual detectors of S3, S4, LD and CC2. Prototype of additional aperture functioning of GAMMA-400 contains two detectors. One of them AC/LD prototype based on BC-408 scintillator with dimensions of 128x10x1 cm^3. Other is CC1 prototype composed of CsI(Tl) crystal with dimensions of 33x5x2 cm^3. The positron beam with energies 100-300 MeV was used for calibration of prototypes of GAMMA-400 detectors on synchrotron "PAKHRA". We calculate fractal dimension of temporal profiles measured during calibrations of AC/LD and CC1 prototypes. Preliminary results are 1.50±0.05 and 1.48±0.08 correspondingly. This is similar to Poisson statistics or Erlang one with coefficient up to 10.

1. GAMMA-400 apertures short description.

The gamma-ray telescope GAMMA-400 (Gamma Astronomical Multifunctional Modular Apparatus) consists of three types of detectors: double (X, Y) tracking coordinate detectors, used in the converter-tracker (C) and position-sensitive calorimeter CC1, plastic and non-organic scintillators [1, 2]. Following detectors based on BC-408 plastics: time-of-flight system TOF (two sections S1 and S2), top (ACtop) and lateral (AClat) sections of anticoincidence system, scintillation detectors of the calorimeter (S3 and S4) and lateral detectors of the calorimeter (LD) (its installation required for particles registration from lateral directions). All detector systems ACtop, AClat, S1-S4 and LD consist of two sensitive layers of 1 cm thickness each [3]. Two calorimeters made of CsI(Tl): position-sensitive (CC1) and electromagnetic (CC2) ones. CC1 contain of 2 strips layers and 2 scintillation layers. The thickness of CC1 and CC2 is ~2 X_0 (~0.1 λ_0) and ~20 X_0 (~0.9 λ_0) respectively (where λ_0 is nuclear interaction length). The total calorimeter thickness is ~22 X_0 or ~1.0 λ_0 for vertical incident particles registration and ~54 X_0 or ~2.5 λ_0 for laterally incident ones [3]. The silicon photomultipliers (SiPM) are used in all scintillation detectors instead of vacuum PMT for minimization of power consumption. The physical scheme of the under consideration variant of gamma-ray telescope GAMMA-400 construction and its three apertures are shown at figure 1.

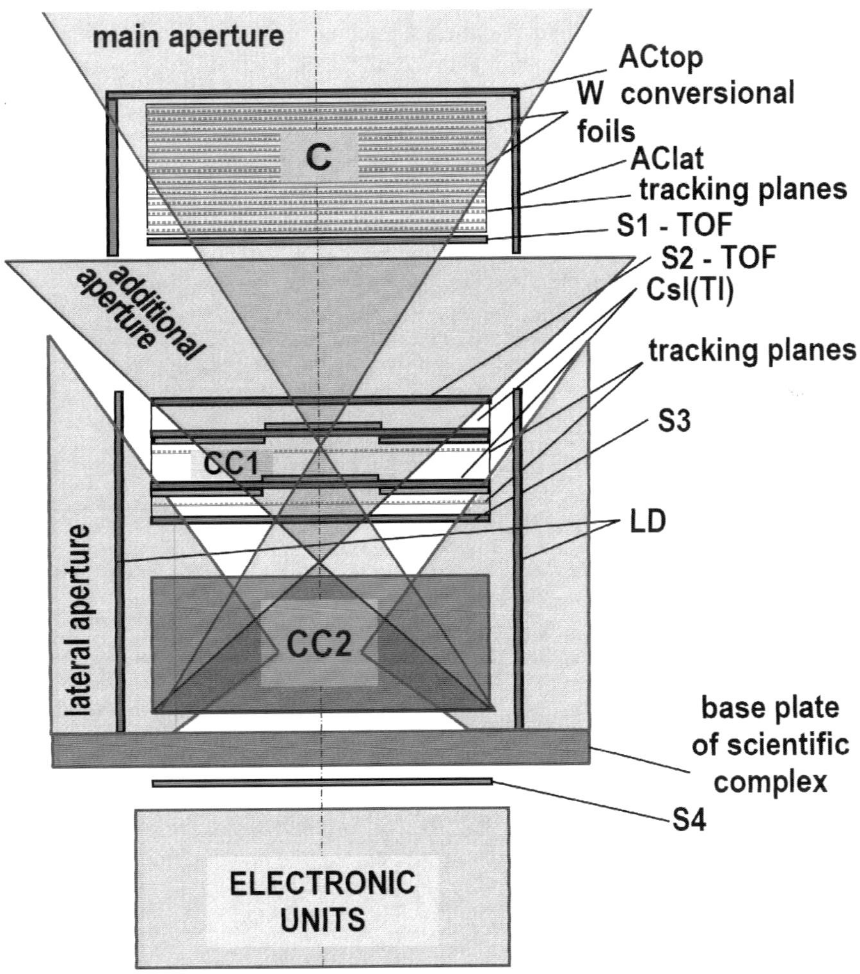

Figure 1. The physical scheme of the under consideration variant of GAMMA-400 gamma-ray telescope construction and its three apertures.

The gamma-ray telescope is optimized for registration of γ-quanta and charged particles with energy above 100 GeV with the best parameters in the main aperture from upper direction. The main aperture is created firstly due to converter-tracker (C): gammas converted in tungsten conversion foils are registered. Triggers in the main aperture will be formed using information about particle direction provided by TOF system and about presence of charged particles or backsplash obtained from ACtop and AClat anticoincidence detectors in energy band of 20 MeV-1.0 TeV for gammas and E>100 MeV for electrons [4]. Gamma-telescope operated in event by event registration mode in this aperture.
Other two apertures are used for observation of transient events as gamma-ray bursts and solar flares both from upper and lateral directions do not require best angular resolution. Additional aperture allows to registered particles from upper directions which don't interact with converter-tracker and don't formed TOF signal. Particles detection in additional aperture starts with signal of CC1 fast discriminators in anticoincidence with TOF [4]. Energy band for gammas registration in this aperture is similar to the main one. In the lateral aperture low energy (0.2-100 MeV) photons classified by using simple anticoincidence signals from the individual detectors of LD and CC2. Higher energies γ-quanta (E>100 MeV) are recognized using energy deposition analysis in the individual detectors of S3, S4, LD and CC2. The angular resolution is provided by double (X, Y) tracking coordinate detectors layers in the CC1. Electromagnetic shower starting point position id defined due to methods analogues to using in accelerator technique (so-called «center-of-gravity technique») allow accuracy ~1 mm for electrons (positrons) with E~8 GeV – for example, in experiments BTeV [5] and PANDA [6].

2. Prototype of additional aperture of GAMMA-400 calibration on synchrotron "PAKHRA"

High-energy γ-quanta are registered in GAMMA-400 mostly by formation of electron-positron pairs in converter-tracker C. The positron beam with energies 100-300 MeV [7] was used for calibrations of prototypes of GAMMA-400 detectors on synchrotron "PAKHRA". Scheme of beam forming and apparatus installation on synchrotron C-25P "PAKHRA" is presented at figure 2.

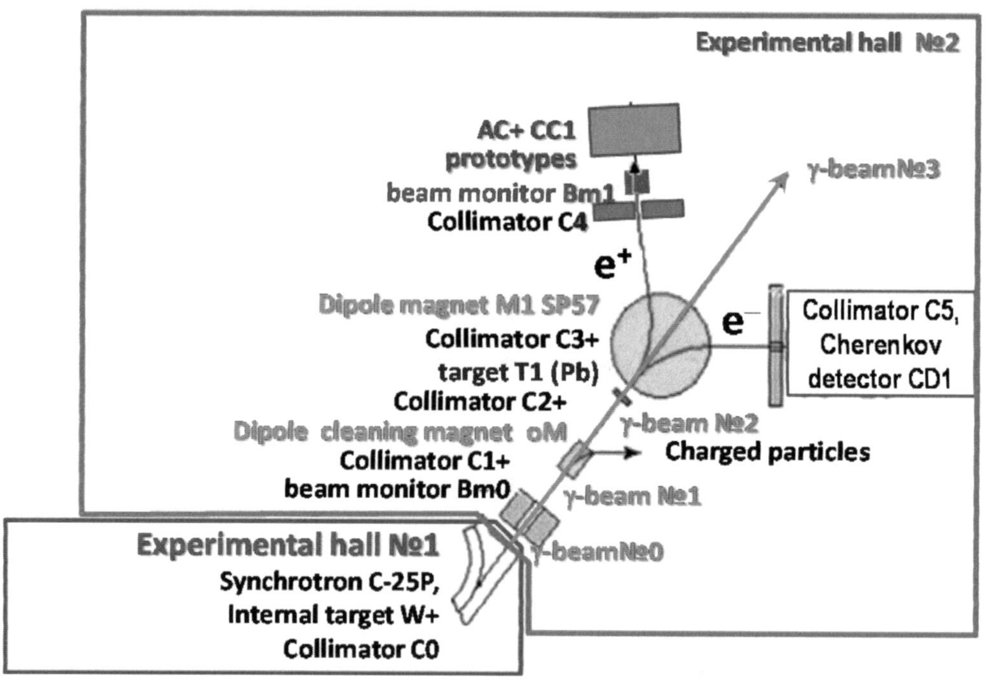

Figure 2. The scheme of beam forming and apparatus installation on synchrotron C-25P "PAKHRA".

GAMMA-400 prototype of additional aperture functioning consist of two detectors. One of them is BC-408 based with dimensions of 128×10×1 cm^3 (one detecting strip from AC prototype) and other composed of CsI(Tl) crystal with dimensions of 33×5×2 cm^3 (one block of CC1 prototype).

3. Fractal dimension of temporal profiles measured during calibrations of AC and CC1 prototypes

For investigation of time series corresponding to solar flares, gamma-ray bursts and other transient events fractal analysis is often applied. It has some features that allow it to be used to study sets with characteristics varying over a wide range: scaling (two events with similar temporal profiles but with different durations have a fractal dimensions) and the possibility to process simultaneously the fractal dimension distributions obtained by using data from different detectors if the background fractal dimensions for these detectors are the same. Moreover, the fractal dimensions must be different for the temporal profiles of events caused by different physical processes [8, 9]. Thus background fractal dimension is useful characteristic of detector.

Fluctuations of count rate registered in scintillation detector during satellite experiments caused due to three reasons. First is fluctuations background caused by gammas and charged particles cosmic and magnetosphere origin. Statistical fluctuations of such background are described by Poisson or Gauss statistics outside the radiation belts and other disturbed regions of magnetosphere. Second reason is fluctuations of produced scintillation photons and photoelectron number in photomultiplier. Corresponding statistical fluctuations are poissonian or gaussian in the first approximation in the linear region of SiPM. Other reason is transient processes in electronic system.

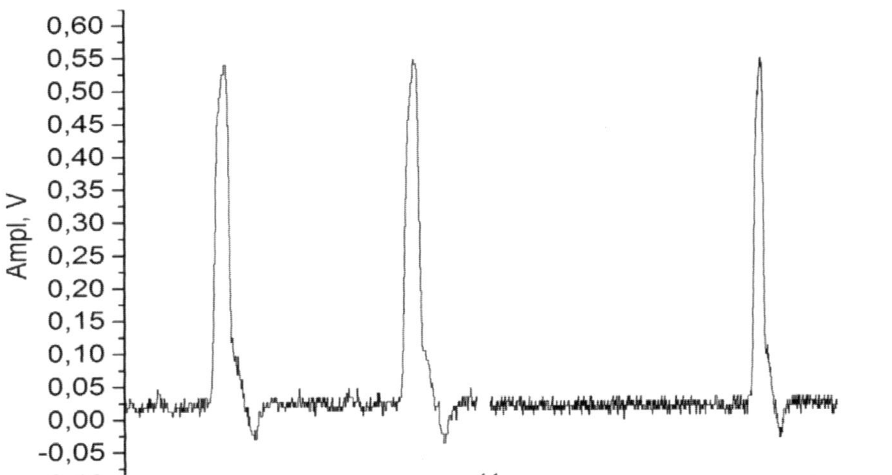

Figure 3. The example of trend of CC1 prototype, $E_{positrons}$=300 MeV, run#5, December 6, 2017.

There are some methods for time profile fractal dimension definition. In the main, these methods based on dissection of a time profile on bins and analysis of count rate statistical fluctuations in each bin [8, 9]. If amount of experimental points in bin k is not enough for statistical analysis (usually if k≤20) then cell algorithm of fractal dimension definition is used [8, 9]: the part of plane in which analyzable curve is locate covers by cells with side δ. Let N(δ)-amount of cells, which has one generic point with this curve even if. Then we define certain gauge for this curve:

$$L = N(\delta) \times \delta^{D} \quad (1)$$

For usual (non fractal) curve L=0 for $\delta \to 0$ but for fractal curve gauge (1) is nonzero for some $D \neq 1$. For practical application it is more suitable to plot dependence of $N(\delta)$ for set of different δ. If it looks

$$N(\delta) = a \times \delta^D \tag{2}$$

for a>0, then fractal dimension is equal *D*.

We have measured continuous temporal profiles of signals from CC1 and AC prototypes and then simulate time target events operation modes of apparatus using time intervals between signals as marker of event registration (single count). Than we modeled background temporal profiles using summation of "registered counts" in time intervals defined as integer numbers of signal duration $\Delta tsig_{AC}$ ~20 ns and $\Delta tsig_{CC1}$ ~5 μs and use cell algorithm for simulated temporal profile fractal dimension calculation - see, for example, [8]. In our work minimal sides of the cell are 20 ms and 5 μs for CC1 and AC prototypes correspondingly. Than we used values of 30, 40, 50 and 60 ms for CC1 prototype and 10, 20, 30, 40 and 50 μs for AC one.

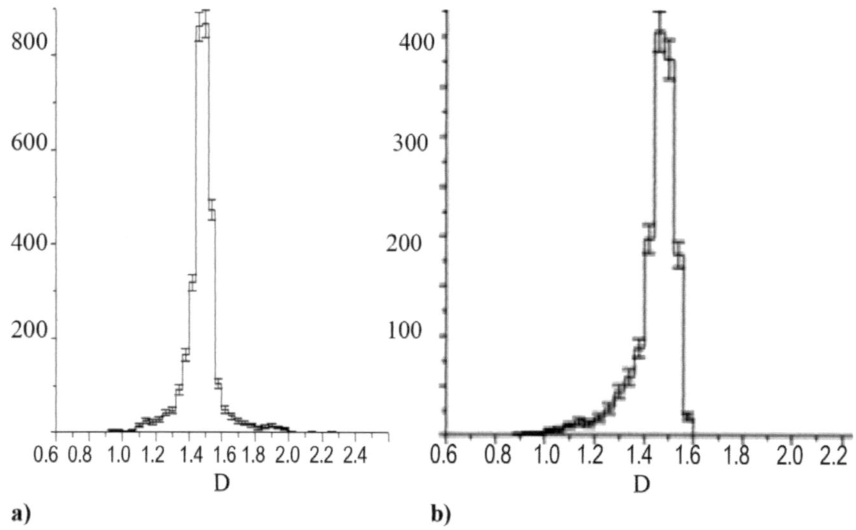

Figure 4. Fractal dimensions distributions of temporal profiles: a) obtained from trends of CC1 prototype, in particular, one shown at figure 3; b) LAD detectors of BATSE experiment.

The whole curve of time profile was divided to *k* strips. In every strip we have calculated maximum and minimum values of the count rate σ: σ_{max} and σ_{min}. Sum of the cells with the width δ_j in strip *k* that have at least one common point with our curve were calculated by:

$$N_k = \frac{\sigma_{max} - \sigma_{min}}{\delta_j} + 2; \qquad N_j = \sum_k N_k \tag{3}$$

If we build a line through all points of the graphic of $\lg N_j(\lg \delta_j)$ and its angle α don't equal to 135° ($D \neq 1$) than we determine the fractal dimension of our time profile: D = - tg α. In common, the received value D is not a real fractal dimension and our time profile is just – prefractal [9] - because really fractal is a limit of some endless process but we have a minimal value – time resolution of the detector. The example of trend of CC1 prototype is presented at figure 3, its fractal dimensions distribution – at figure 4a. Analysis of this distribution gives fractal dimension D=1.48±0.08.

For comparison let's consider fractal dimension of real temporal profiles obtained in orbital experiment, for example BATSE [11, 12] (Burst and Transient Source Experiment) onboard the Compton Gamma Ray Observatory (CGRO). CGRO [13] was launched on April 5, 1991 and finished its functioning on June 4, 2000. BATSE registered γ-emission temporal profiles in four energy bands: 25-50 keV, 50-100 keV, 100-300 keV and > 300 keV. For obtaining the event temporal profiles with high resolution Large Area Detectors (LAD - NaI(Tl) scintillation detector 50.8 cm in diameter and 1.27 cm thick) were used [11, 12]. No any onboard and ground data filtering were used in this experiment. Fractal dimensions of BATSE background and events temporal profiles were investigated in [14]. Following this work, in our examination we used data with the time resolution of 64 ms, sides of the cell were 200 ms, 300, 400, 500 and 600 ms. Its fractal dimensions distribution is presented at figure 4b. Analysis of this distribution gives fractal dimension D=1.47±0.06. Both results are similar to Poisson statistics or Erlang one with coefficient up to 10.

4. Conclusion

The additional aperture of GAMMA-400 gamma-ray telescope allows registering particles from upper directions which don't interact with converter-tracker and don't formed TOF signal. Prototype of additional aperture functioning of GAMMA-400 contains two detectors. One of them is single strip of AC prototype based on BC-408 scintillator with dimensions of $128\times10\times1$ cm^3. Other is CC1 prototype composed of CsI(Tl) crystal with dimensions of $33\times5\times2$ cm^3. We calculate fractal dimension of AC/LD and CC1 prototypes temporal profiles measured during calibrations and obtain preliminary results of 1.50±0.05 and 1.48±0.08 correspondingly. Analysis of LAD detectors of BATSE experiment fractal distribution gives fractal dimension D=1.47±0.06. Both results are similar to Poisson statistics or Erlang one with coefficient up to 10.

Acknowledgments

Authors thank for the support from National Research Nuclear University MEPhI in the framework of the Russian Academic Excellence Project (contract No. 02.a03.21.0005, 27.08.2013)

References

[1] Galper A M *et al* 2013 *Advances in Space Research* **51** 297
[2] Galper A M *et al* 2015 *Physics Procedia* **74** 177
[3] Topchiev N P *et al* 2016 *J. of Phys.: Conf. Ser.* **675(3)** 032009
[4] Arkhangelskaja I V *et al* 2015 *Physics Procedia* **74** 212
[5] Sokol G A et al 2000 *Letters to Exp. Particles At. Nucl.* **102(5)** 71
[6] Batarin V A et al 2006 *Nucl. Phys.B (Proc)* **150** 262
[7] Semenov P A et al 2009 *NIM A* **598** 224
[8] Crownover R M 1995 *Introduction to Fractals and Chaos*. (Boston: Jones and Bartlett)
[9] Mandelbrot B B 1977 *Fractals, Form, Chance, and Dimension* (San Francisco: Freeman)
[10] Feder E *Fractals* 1991 (Moscow: Mir)
[11] Fishman G J et al 1994 *Astrophys J. Suppl. Series* **92** 229
[12] Briggs M S 1999 et al *Astrophys. J. SuppL.* **122** 468
[13] Varendorff M 1995 *SIF Conf. Proc.* **47** 183
[14] Arkhangelskaja I V 2004 *Advances in Space Research* **34(12)** 2723

IOP Conf. Series: Journal of Physics: Conf. Series **1181** (2019) 012084 doi:10.1088/1742-6596/1181/1/012084

A Fundamental of the LPM showers in water up to 10^{21}eV

K Kato[1], T Tanemori[2], N Takahashi[3] and A Misaki[4]

[1] Kyowa Interface Science Co.,Ltd, Niiza, Saitama, 352-0011 Japan
[2] Department of Physics, Saitama University, 255 Shimo-Okubo, Sakura-ku, Saitama, 338-8570 Japan
[3] Faculty of Science and Technology, Hirosaki University,Hirosaki,036-8561 Japan
[4] Graduate School of Science and Engineering, Saitama University, 255 Shimo-Okubo, Sakura-ku, Saitama, 338-8570 Japan

amisaki@sand.ocn.ne.jp

Abstract. In order to pursue the highest energy of neutrino events in lepton physics, say, ~10^{21} eV, as attained in extensive air showers, in hadron physics, the fundamental understanding towards electron-initiated cascade showers, namely, the LPM showers at extremely high energies, is strongly requested, because the LPM showers at these energies behave in completely different manner against those of the conventional BH showers. For the purpose, we need, at first of all, start from the morphological study of the LPM showers from 10^{16} eV to 10^{21} eV in water and 10^{23} eV in the atmosphere. The different characteristics of the LPM shower are clarified in different energy region.

1. Introduction

The final goal of the present paper, not at present, but in future, is to offer the fundamental materials for the design study for the electron neutrino astrophysics above 10^{21} eV, which is attained already by extensive air showers. For the purpose, we should carefully investigate big diversity of the LPM showers at extremely high energies, which should be regarded the strongest characteristics of the LPM showers [1].

In figure 1, the overall average pictures for the LPM showers over wider primary energies are shown. To make it possible to understand characteristics of the LPM showers over wider energy ranges, we show the cascade curves for shower particles with different primary energies, keep $E_{prim}/E_{min}=10^5$, for the LPM showers, where E_{prim} and E_{min} denote the primary energy and the minimum energy for the LPM showers, respectively. Primary energies are covered with 10^{15} eV to 10^{21} eV and the primary particles are electrons. If we assume the absence of the LPM effect, we expect all the cascade showers show the same structures as the BH shower shown in the figure, irrespective of primary energies as the results of under Approximation A [2].

2. Morphology of the LPM showers from mono- peak to multi-peak structure

As emphasized in Introduction, the characteristics of the LPM showers at extremely high energy lies in the existence of the extraordinary diversities among the LPM showers. Our final aim is the establishment of the reliable energy determination of LPM showers at extremely high energies which behave in very complicated manner and nobody could understand their characteristics in detail. In such the situations, the first of all, we must get accurate information around the LPM showers as much as possible. This is the reason why we need morphological studies around the LPM showers.

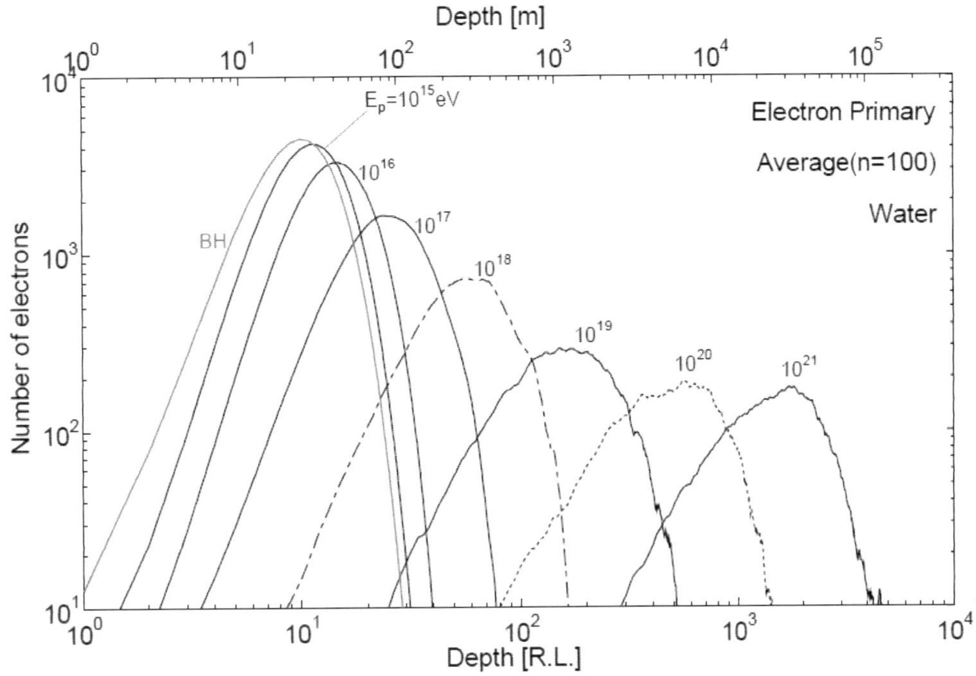

Figure 1. Cascade curves for electrons by the averaged LPM showers with different primary energies from 10^{15} eV to 10^{23} eV for keeping of $E_{prim}/E_{min}=10^5$. The average BH shower with 10^{15} eV is attached for readers' reference.

2.1. *The LPM effect begin to be effective ~10^{15} eV in water and become clear ~10^{16} eV*

The LPM effect begins to influence over the development of cascade showers in water ~10^{15} eV, as shown in figure 1. However, the LPM effect becomes effective essentially at ~10^{16} eV. In figure 2, we show clear indication of the LPM effect in both aspects, namely, one is the prolongation of the cascade shower compared with the BH shower and the other is the clear fluctuations around their average values. Here, [r.l] (radiation length) is the same as [c.u.] (cascade unit).1 r.l=36cm in water.

2.2. *A LPM shower with primary energies above 10^{17} eV. The appearance of multi-peak structure*

The most distinguished character of the LPM showers which we never expect in the usual cascade showers (BH showers) lies in the existence of the multi-peak structure [1]. It is the direct reflection of the strong deviations among shower particles produced in both bremsstrahlung and pair production with LPM effect. This characteristic is firstly found by Konishi et al in the LPM shower with 10^{17} eV in lead [1]. Compared the LPM shower in water shown in figure 3 with the corresponding one in lead with the same primary energy [1], "the degree" of multi-peak structure in figure 3 is rather weaker than that in lead, because the LPM effect is a kind of density effect which is stronger in higher density.

2.3. *The appearance of multi-peak structure (> 4) ~10^{18} eV*

The characteristic of the LPM showers ~10^{18} eV is the appearance of clear multi-peak (>4) in addition to large prolongation of the cascade curve as shown in figure 4.

2.4. *Appearance of the sub-showers with multi-peak structure ~10^{19} eV*

In these energy regions, as the energy distribution among shower particles may lean on one particle in both bremsstrahlung and pair production, some of shower particles penetrate into deeper depth without energy loss. In figure 5, it is understood that the first sub-shower is initiated at 20 r.l. (10 meter) by the primary electron penetrate there. The second sub-shower is generated at the point ~70 r.l. (25 meter) after the first sub-shower terminates. Such phenomena are never imagined in the BH showers.

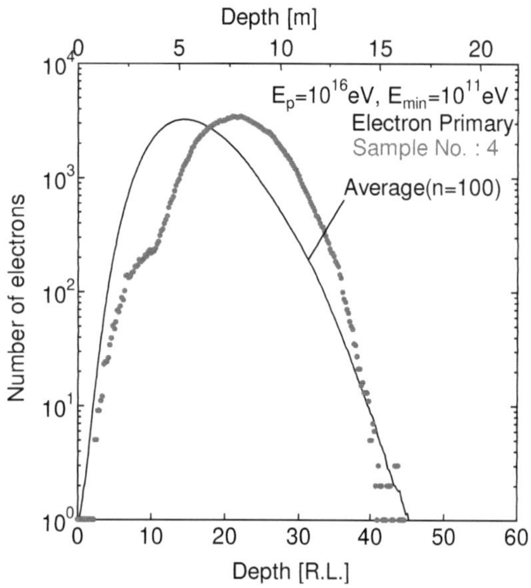

Figure 2. In two reasons, LPM effect appear clear, (1) average picture is clearly different from that of the BH shower(see, Figure 1,also), (2) the real one shows clear fluctuation compared with the average picture.

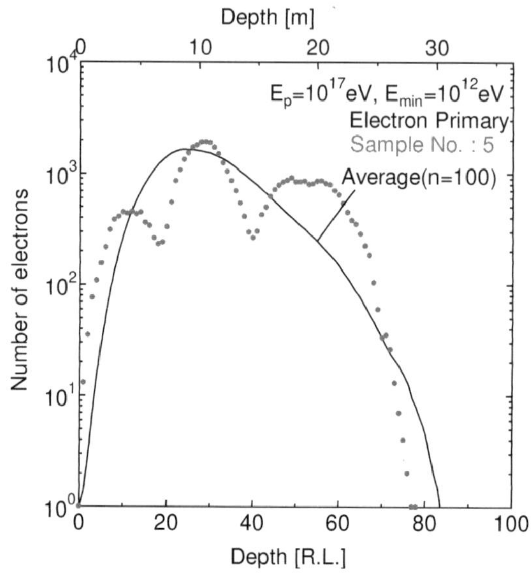

Figure 3. The multiple peak structure of the LPM shower in water clearly appears ~10^{17} eV.

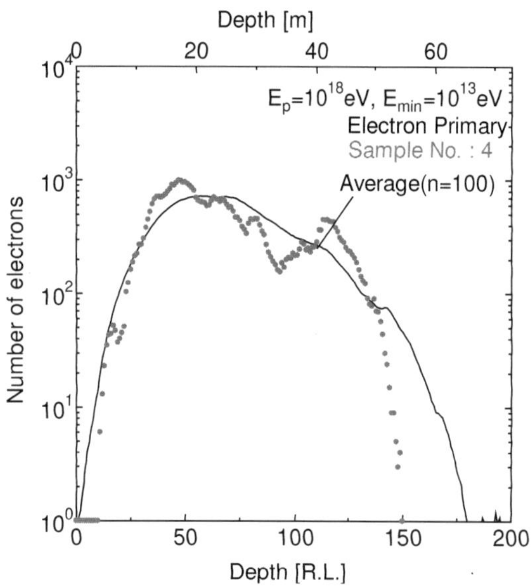

Figure 4. The LPM shower with 10^{18} eV. The typical structure of the multi-peak structure.

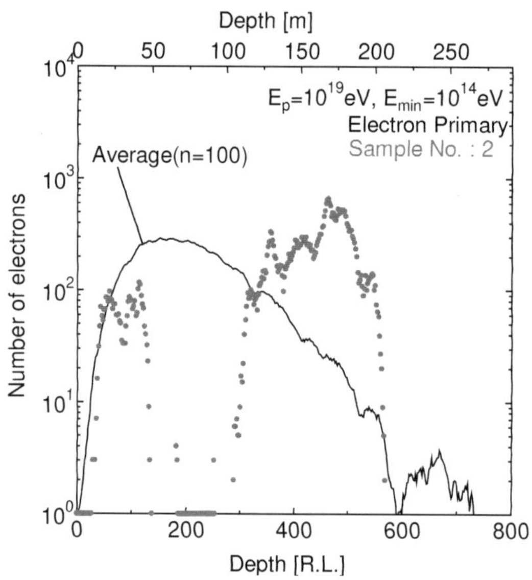

Figure 5. The separated sub-showers appear in ~10^{19}eV. A deformed kind of multi-peak structure.

2.5. Appearance of the sub-showers with longer traversed length ~10^{20} eV.
It should be noticed that the LPM shower traverses too long distance. In figure 6, we show one example with 10^{20} eV which shows the total length ~1790 r.l. (640 meter). The extension of the first sub-shower is 390 r.l. (140 meter) and that of second is 1250 r.l. (450 meter), which of the separation between two is 150 r.l. (60 meter). The second sub-shower is composed of many peaks.

2.6. Appearance of single cascade shower with 10^{21} eV over 6000 r.l.(2100 meter)
We show a single LPM shower with many peaks but continuously developed over 6000 r.l. (2100 meter) in figure 7. Namely, this LPM shower could not be stored within the interior of the KM3 detectors. Nobody has imagined such a characteristics of cascade showers.

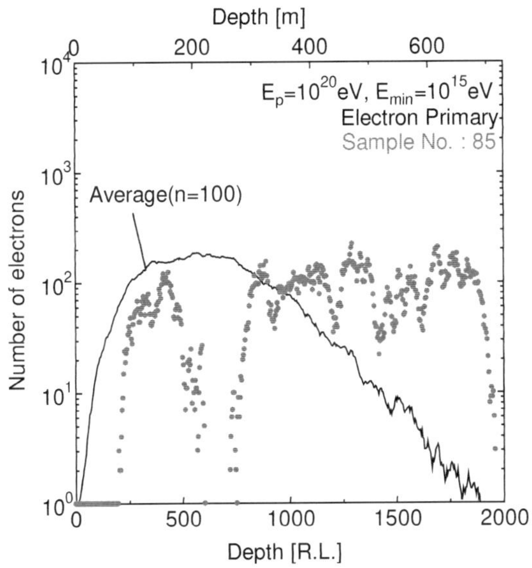

Figure 6. An example of two separated sub-showers with 10^{20} eV.

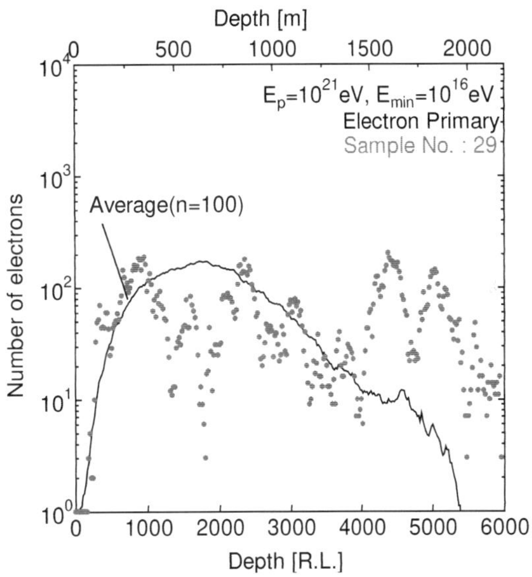

Figure 7. The LPM shower with 10^{21} eV traverses with many multi-peak without producing sub-shower. Notice that shower traverse 6000 c.u. (~ 2100 meter!!).

3. Track length distributions from the view point of energy estimations of the LPM showers

All reliable energy estimation of cascade showers is finally reduced to the exact estimation of track lengths of electrons in the corresponding cascade showers, irrespective of the different means for the measurements, namely by optical measurements for Cherenkov lights or Radio pulse measurements from cascade showers.

The fluctuation in the track lengths for shower particles as the function of the depth traversed by cascade particles is rather small compared with the corresponding ones due to cascade curves which are shown in figures 2 to 7. Therefore, the track length might be a measure for more reliable estimation.

In figure 8, one example of the cascade curves for track length with 10^{20} eV is shown. It is seen from the figure that the track lengths saturate roughly at ~1000 r.l. From this fact, one may conclude that the length for reliable energy estimation is roughly, 1000 r.l (360 meters). In similar manner, one may conclude from figure 9 that we need ~3000 r.l. (~1 km) for 10^{21} eV. Of course, one may say that the present calculation is limited to $E_{prim}/E_{min}=10^5$, however, in the real experiment, $E_{prim}/E_{min}=10^{14}$ for 10^{20} eV, where one adopts 1 MeV as the minimum energy. In spite of big difference between 10^5 and 10^{14}, the depth where the saturation is attained is independent on the value of E_{prim}/E_{min}. (See, the next section)

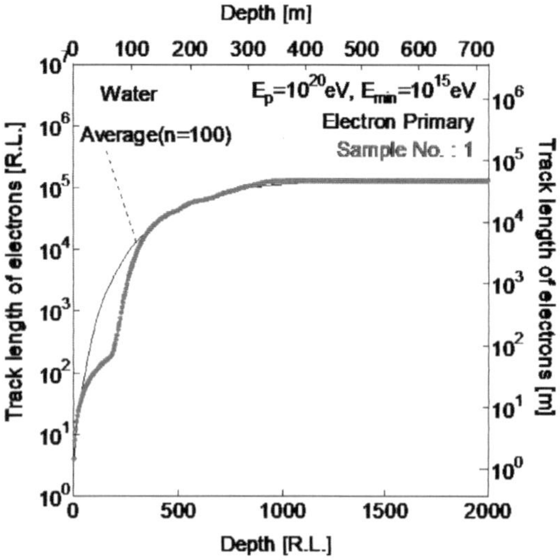

Figure 8. Track length of a LPM shower with primary energy $E_p=10^{20}$ eV.

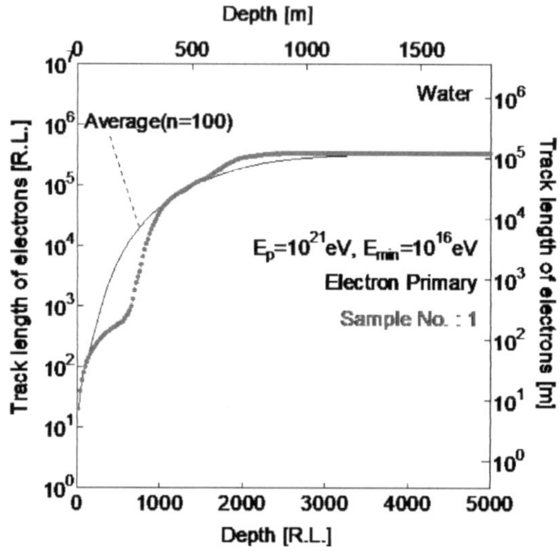

Figure 9. Track length of a LPM shower with primary energy $E_p=10^{21}$ eV.

4. Could the LPM showers at extremely high energy keep their morphological features down to lower energy, say, 1 MeV?

The minimum energies of the LPM showers under consideration are rather high because of $E_{prim}/E_{min}=10^5$. On the other hand, really observed LPM showers include many shower particles whose energies are so low (~1 MeV) that the LPM effect can be completely neglected there. Consequently, it is natural for someone to have the same opinion as the conjecture by Kamata and Nishimura [3]. However, this conjecture is finally excluded [3]. Namely, the LPM showers with extremely higher primary energies keep their essential features even if the energy of each shower particle become so low that the LPM effect does not influence over them.[3] .

In figures 10 and 11, we give two examples among the LPM showers in the atmosphere whose primary energies are 10^{23} eV. The showers whose zenith angles are 90 degrees traverse through the deepest atmosphere (on the ground) whose total lenght is ~7000g/cm^2. Here, we take into account of the change of the density in the atmosphere which influence over the LPM effect.

In these calculations, we utilize the hybrid method for calculation of the total number of electrons. Namely, we calculate cascade shower down to 10^{16} eV where the LPM effect is still effective in the atmosphere by the exact Mote Carlo method and we utilize the NKG function for the connection with results by the Monte Carlo results. In the energy region below 10^{16} eV in the atmosphere, the cascade showers concerned are regarded as BH showers in which fluctuation effect is really can be neglected.

In the NKG function, we assume no fluctuation. Therefore, in figures 10 and 11, the fluctuation effect appears only in the LPM shower, namely in the exact Monte Carlo part. The NKG function is an approximation to the BH showers. In real BH shower, of course, the fluctuation effect exists but its effect is very small compared with the case in the LPM shower. Therefore, for the rough estimation of the reliable energy of the LPM shower, it may be allowed to connect NKG function with the exact Monte Carlo method for the LPM shower. It is shown in figure 10 and 11 that green lines represent cascade curves by the exact Monte Carlo method and red lines represent cascade curves by hybrid method. From both figures, we conclude that the features in the green curve keep even red curves.

This can be applied to the discussion around figure 8 and 9 directly. It is clear from those figures that track lengths of shower particles whose minimum energy ~1 MeV saturate at nearly the same points shown as those of figures 8 to 9 in spite of big numerical difference in track lengths themselves.

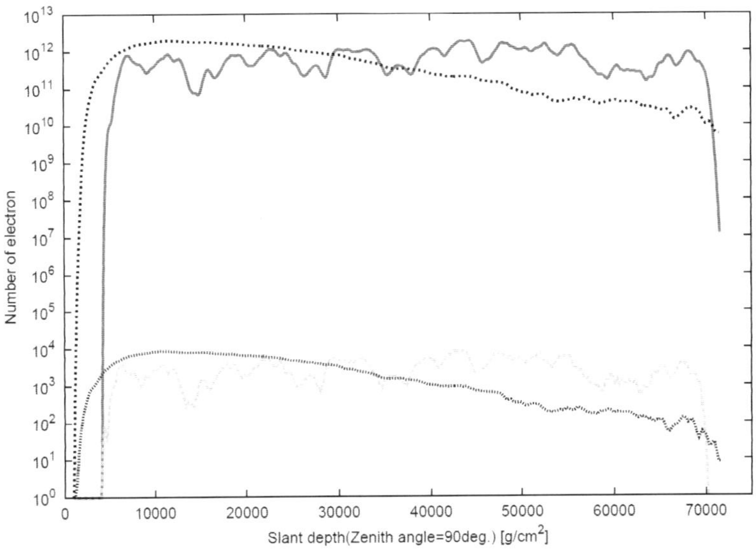

Figure 10. Cascade curve of LPM shower (No.1) with 10^{23} eV in the atmosphere

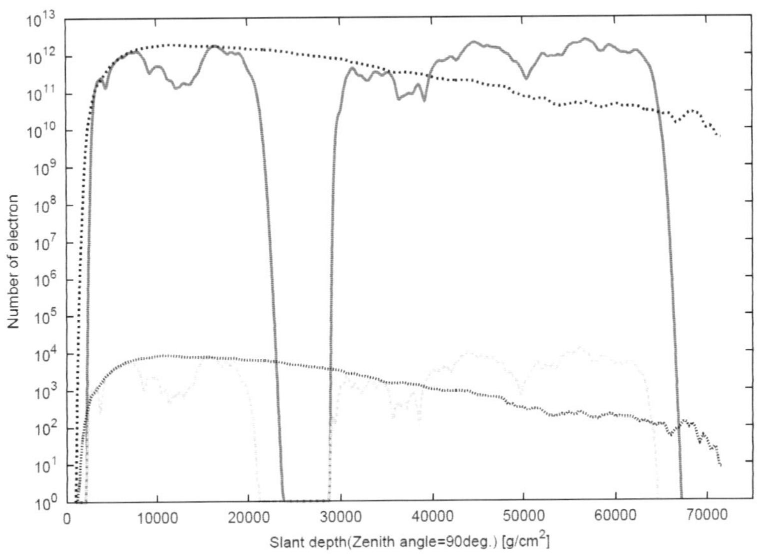

Figure 11. Cascade curve of LPM shower (No.2) with 10^{23} eV in the atmosphere

5. Conclusion
For the moment, the most high energy cosmic ray events have been found in extensive air shower. We should try to find higher events in electron neutrino astrophysics.

Acknowledgement
One of the authors (A.M) should like to appreciate LOC of (26thE+CRS 35 th RCRC) for him to be the chairman of the workshop1: [the LPM effect and its applications for extremely high energy cosmic ray research]. The present paper is a part of his key note report to the workshop together with [3].

References
[1] Konishi E, Adachi A, Takahashi N and Misaki A 1991 *J.Phys.G:Nucl.Part.Phys* **17** 719-732
[2] Rossi B and Gresisen K 1941 *Rev.Mod.Phys.* **13** 249-309
[3] Misaki A, in this workshop

IOP Conf. Series: Journal of Physics: Conf. Series **1181** (2019) 012085 doi:10.1088/1742-6596/1181/1/012085

A historical Introduction to the LPM shower

A Misaki

Graduate School of Sciece and Engineering, Saitama University, 255 Shimo-Okubo, Sakura-ku, Saitama, 338-8570 Japan

akeobarn@gmail.com

Abstract. The matter before and after the birth of the LPM showers is outlined. The most important tasks for the LPM shower studies are thoroughgoing examinations on the validity of the methods around calculations of the cascade showers.

1. Introduction

The LPM effect is a kind of density effect, which brings about the decrease of the bremsstrahlung due to multiple scattering at higher energies. Landau-Pomeranchuk treat this problem in semi-classical manner [1] and afterwards, Migdal formulates it in quantum electrodynamical manner [2].

Now, it is named as the LPM effect [3, 4]. Nowadays, it is well known that the LPM effect makes influence on the cosmic ray phenomena greatly at extremely high energies and the cascade showers characterized by the LPM effect are called as the LPM showers.

2. The history of the LPM showers

2.1. The pioneering work in the LPM showers by A.A.Pomansky

In 1968, Pomansky represented a paper entitled as Consideration of the cascade curve at super high energies [5]. This is the first application of the LPM effect to the electromagnetic cascade showers which is shown in figure 1. In his paper, he did not use the physical term of the LPM effect, only cited Migdal's name in his paper, because at that time the physical term of the LPM effect did not exist.

Unfortunately, his results were not correct [6]. However, his work should be highly appreciated due to its pioneering nature in the history of the LPM showers. We quote a part of the first page of his paper in figure 1.

2.2. The first result of the correct LPM showers and their subsequent developments

In 1974, Misaki et al. calculated cascade curves where the cross sections obtained by Migdal were utilized by an exact Monte Carlo method. Their cascade showers were clearly different from the usual cascade showers under Bethe-Heitler's cross section [7]. Then, they named this effect as L-P-M effect, because it made possible to calculate cascade showers in quantitative manner only by the advent of Migdal's cross section. The figure 2 is reproduced from original one.

When their paper was presented at an international conference, where Khristiansen casted strong doubt around the reliability of their results. His strong criticism toward their results seemed to be quite reasonable at that time, because Kamata and Nishimura made the following comment in their famous paper on three-dimensional cascade theory [8].

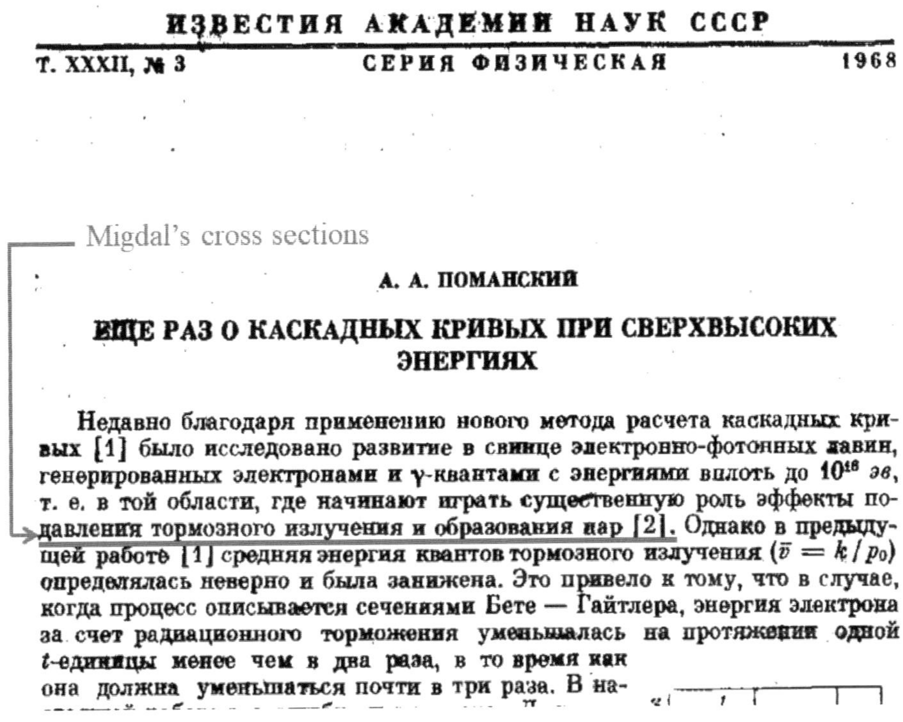

ИЗВЕСТИЯ АКАДЕМИИ НАУК СССР

Т. XXXII, № 3 СЕРИЯ ФИЗИЧЕСКАЯ 1968

А. А. ПОМАНСКИЙ

ЕЩЕ РАЗ О КАСКАДНЫХ КРИВЫХ ПРИ СВЕРХВЫСОКИХ ЭНЕРГИЯХ

Недавно благодаря применению нового метода расчета каскадных кривых [1] было исследовано развитие в свинце электронно-фотонных лавин, генерированных электронами и γ-квантами с энергиями вплоть до 10^{16} *эв*, т. е. в той области, где начинают играть существенную роль эффекты подавления тормозного излучения и образования пар [2]. Однако в предыдущей работе [1] средняя энергия квантов тормозного излучения ($\bar{v} = k/p_0$) определялась неверно и была занижена. Это привело к тому, что в случае, когда процесс описывается сечениями Бете — Гайтлера, энергия электрона за счет радиационного торможения уменьшалась на протяжении одной t-единицы менее чем в два раза, в то время как она должна уменьшаться почти в три раза. В на-

Figure 1. The pioneering work by A.A.Pomansky. From Izv.Akad.Nauk SSSR,Ser.Fiz.,32,496(1967)

"Recently, the limit of application of the complete screening cross sections at very high energies has been critically discussed by Landau-Pomeranchuk. They pointed out that, at very high energies, the path length pf an electron which is the effective in the collision processes become so long that the interference effects of the adjacent atoms should be taken into account and that the cross section should decreases in the high energy regions. Even if his cross section should fall off at very high energies, the average energy of shower particles would soon decrease by order of magnitude by the rapid multiplication of particles. Thus these effects, if exist any, should concern only the beginning stage of shower development and the overall behavior should suffer little change." [8]

In order to overcome Khristiansen's bitter remark, and, furthermore, to exclude the conjecture by Kamata-Nishimura, they had to verify the validity of their calculation as rigorously as possible. The validity of the Monte Carlo method utilized was verified in the following ways.

i. First of all, they calculated cascade showers within the framework of the one-dimensional analytical theory under Approximation A [7]. The physical quantities, such as, total track lengths of shower particles and cascade curves for shower particles were obtained by the analytical theory, the accuracies of which are clarified [7].
ii. The total track lengths for shower particles and the cascade curves by the Monte Carlo method are compared with the corresponding ones by the analytical theory [9]. Both quantities obtained by mathematically different methods showed an excellent agreement. This fact confirms that the validity of their Monte Carlo method in one-dimensional was established.
iii. They extended their method to three-dimensional case. For the purpose, they built the effect of the multiple scattering due to shower electrons into their Monte Carlo procedure in one dimensional case.

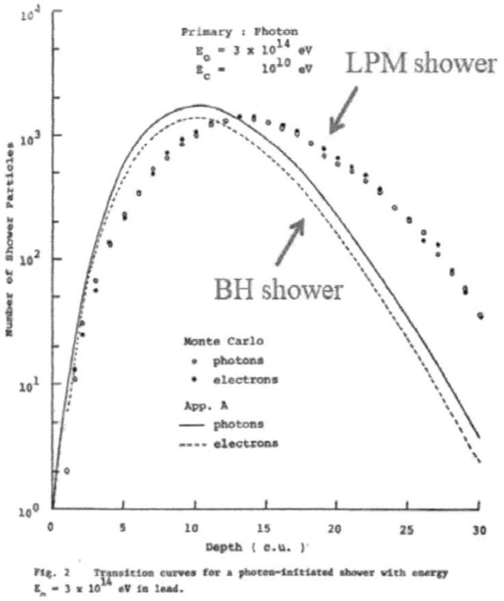

Figure 2. Transition curves for a photon-initiated shower with energy $E_0 = 3 \times 10^{14}$ eV in lead. Reproduced from Proc.Inter. Cosmic Ray Symp. on Hugh.Ener.Phys. 148(1974),Tokyo

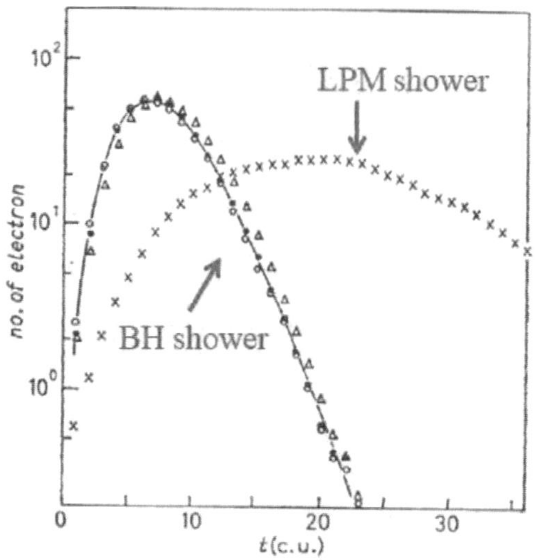

Figure 3. Transition curves of mean number of electrons in lead. Open circles show the results with Bethe-Heitler cross-sections ($E_0/E_{min}=10^3$) and closed circles, triangles and crosses show the results including the LPM effect initiated by photons with energies $E_0=10^{11}$, 10^{13}, 10^{15} eV and $E_m=10^8$, 10^{10}, 10^{12} eV, respectively.

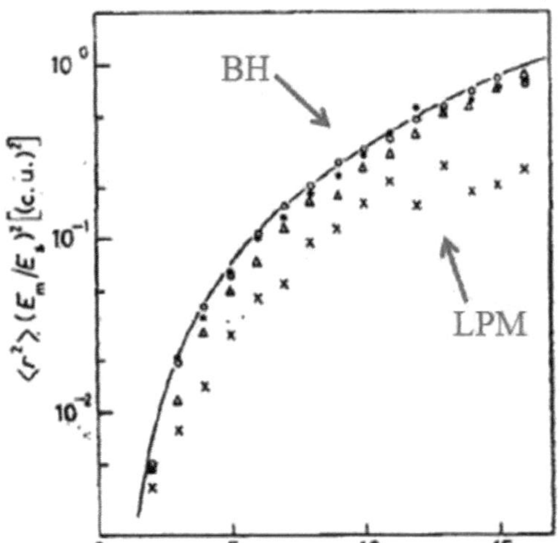

Figure 4. Mean square lateral spread of electrons in lead. Marks are the same as figure 3. $< r^2 >$ is normalized by multiplying by the factor $(E_m/E_s)^2$, where E_m is the minimum energy of observation and E_s is the scattering energy E_s=21.2 MeV. The solid curve shows analytical results under approximation A.

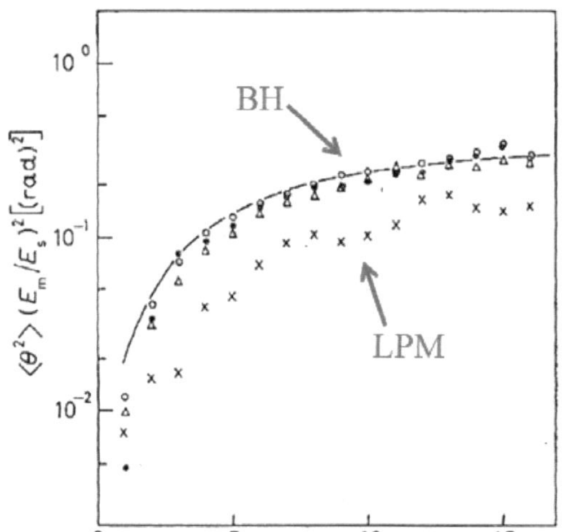

Figure 5. Mean square angular spread of electrons in lead. Marks are the same as figure 3. $< \theta^2 >$ is normalized by multiplying by the factor $(E_m/E_s)^2$, where E_m is the minimum energy of observation and E_s is the scattering energy E_s=21.2 MeV. The solid curve shows analytical results under approximation A.

The validity of their three-dimensional procedure in the Monte Carlo method was finally checked by the comparison of the mean square lateral and angular spreads of shower particles by the Monte Carlo method with corresponding ones by the analytical procedure [9]. The corresponding quantities by the analytical theory had been proved to be right mathematically. They confirmed that the agreement between them were excellent (See, figures 3 to 5). As the results of it, they confirmed the validity of their Monte Carlo procedure in three dimensional cases. Figure 3 to 5 are reproduced from E.Konishi et al., Il Nuovo Cimento, 44A, (1978)509-503. The agreements between the Monte Carlo results and analytical ones in these figures under Approximation A guaranteed finally the validity of their three-dimensional Monte Carlo procedure.

iv. As the validity of the general structures of the three-dimensional Monte Carlo procedure in BH showers had been verified, the applications of this procedure to the LPM showers were automatically guaranteed, because the difference between two procedures was merely the difference in the numerical values between respective cross sections.

2.3. The final denial toward Khristiansen's comment and the partially -denial towards Kamata-Nishimura 's conjecture

The title of the paper in which figures 3 to 5 are included is named as, "Three-Dimensional Cascade Showers in Lead taking account of the Landau-Pomeranchuk-Migdal effect". Namely, the physical term of the LPM effect appeared firstly in a peer-reviewed journal [4].

It is clear from these figures that LPM effect influence really on the development of the cascade showers. Consequently, the criticism by Khristiansen should be finally excluded. However, it should be noticed that the conjecture by Kamata-Nishimura cited above are not completely excluded, because the minimum energies of shower particles in figures 3 to 5 are rather high where the LPM effect is still effective. In order to exclude their conjecture finally, they need to develop a new technique by which it makes possible to pursue shower particles down to lower energies where the LPM effect is completely neglected. Because, it was impossible to utilize the exact Monte Carlo method for the purpose due to huge time consuming at that time and even now. Thus, a new technique was forced to be developed for the purpose [10] and its details was given in [11].

In figure 6, the validity of the matrix method was verified through the agreements among cascade curves obtained by the methodologically independent means. This figure is reproduced from A.Misaki, Physical Rev.D49 (1989)3086—3096 by the matrix method [12]. The validity of this method was verified by other two independent methods, namely, the analytical and Monte Carlo methods on which are methodologically independent, as shown in figure 6. Based on the confirmation of the validity of the matrix method, the transition curves for shower particles both the cases without LPM effect (figure 7) and with LPM effect (figure8) are calculated.

In these numerical calculations, the effect of the ionization loss is included (in the sense of "under Approximation B"). It is clear from the comparison between figure 7 and figure 8 that the developments of the LPM showers are completely different from corresponding ones of the BH showers. Here, shower particle are pursued down to almost zero energies where LPM effect is completely neglected. Thus, it is clear from the figures that the conjecture by Kamata-Nishimura should be finally excluded.

In table 1, the physical quantities which characterize the behaviors of cascade showers are compared between the LPM showers and BH showers. Namely, *Nmax*, the maximum number for cascade showers, *Tmax*, the depth where the cascade showers attain at the maximum, and their *FWHM* , are compared between LPM showers and BH showers. It is clear from the table that the LPM showers are quite different from the BH showers.

The comparison of cascade curves (average pictures) without the LPM effect with the corresponding ones with LPM effect is made in figure 7 and 8. It is clear that the LPM showers are quite different from the BH showers as shown in figures 7 and 8, and table1, whichi are reproduced from A. Misaki, Fortschritte der Physik, **38** (1990)413-446.

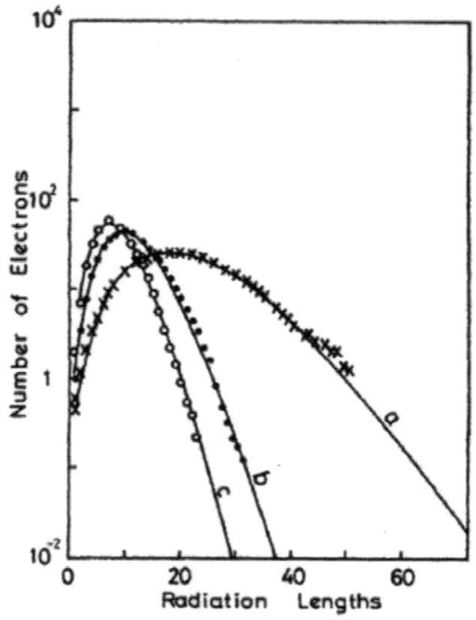

Figure 6. Comparison between the present results and the results obtained by Konishi et al. [4]. The comparison is made for LPM showers without ionization loss in lead, keeping E_0/E_{th} fixed and values of $E_0=10^{11}$, 10^{13}, 10^{15} eV. The crosses, closed circles and open circles denote $E_0=10^{15}$, 10^{13}, 10^{11} eV respectively. From A.Misaki, Phys.Rev,D,40, (1989)3086-3096

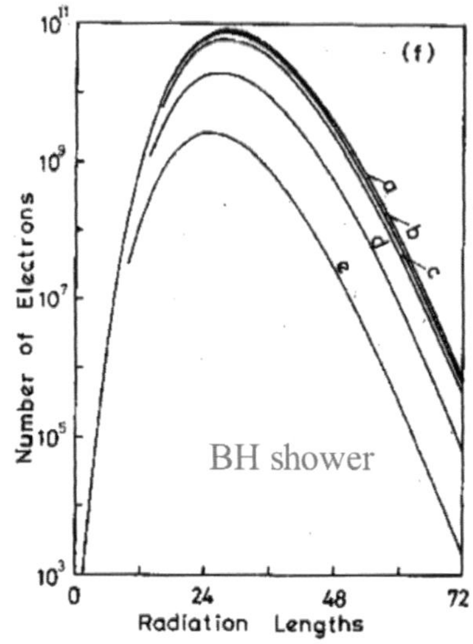

Figure 7. Transition curves of electron numbers in water under Approximation B initiated by a photon of 10^{20} eV for various threshold energies. The letter attached to each curve denotes the threshold energies: a=10^3 eV, b=10^6 eV, c=10^7 eV, d=10^8 eV, e=10^9 eV. From: A.Misaki, Fortschr. Phys. 38 (1990) 413-446

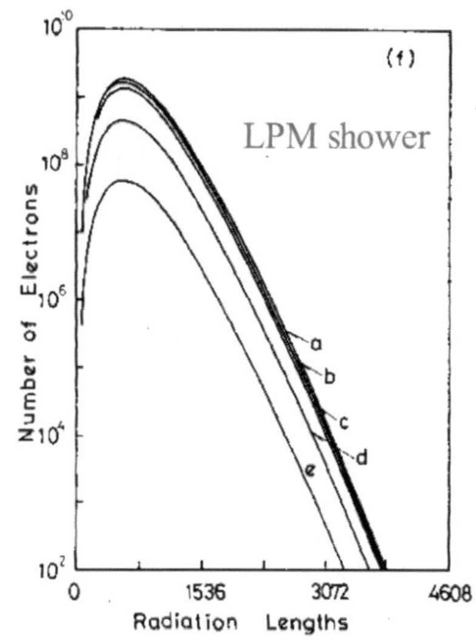

Figure 8. Transition curves of electron numbers in water in the presence of the LPM effect including ionization loss. The primary energy is 10^{20} eV and the threshold energies, *a* to *e* are the same as in figure 7. From: A.Misaki, Fortschr. Phys. **88** (1990) 413-446

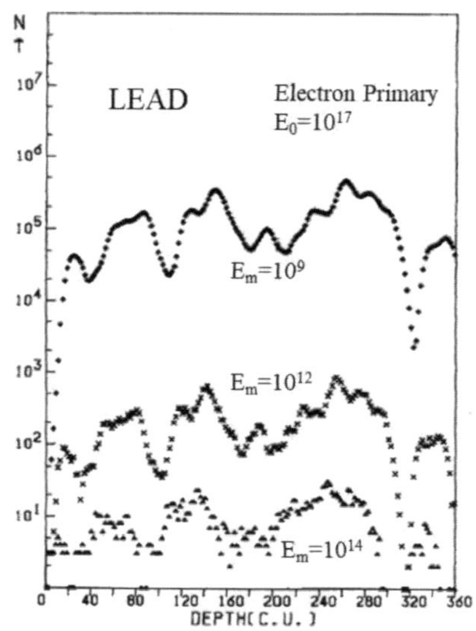

Figure 9. Number N of electrons in an electron-initiated shower of primary energy $E_0=10^{17}$ eV in lead as a function of depth in the absorber. The three curves refer to three different values of cut off energy $E_m=10^9$, 10^{12}, 10^{14} eV. From: Konishi et al, Phys.G: Nucl.Part.Phys. **17**(1991)719-732

Table 1. Numerical values for electron number at shower maximum, N_{max}, depth of shower maximum, T_{max}, and full width half maximum, FMHM, in both BH and LPM showers including ionization loss in water. In each row, numerical values on upper row are due to the LPM shower, while the corresponding ones on lower row are due to the BH shower. Reproduced from: A.Misaki, Fortschr. Phys. 38 (1990) 413-446.

E_0/eV	10^{15}	10^{16}	10^{17}	10^{18}	10^{19}	10^{20}	10^{21}
N_{max}	1.04×10^6	8.31×10^6	4.58×10^7	1.73×10^8	5.82×10^8	1.91×10^9	6.16×10^9
	1.07×10^6	1.00×10^7	9.48×10^7	9.04×10^8	8.63×10^9	8.26×10^{10}	7.97×10^{11}
T_{max}	18 (r.l.)	24 (r.l.)	38 (r.l.)	78 (r.l.)	200 (r.l.)	569 (r.l.)	1696 (r.l.)
	17 (r.l.)	19 (r.l.)	22 (r.l.)	24 (r.l.)	26 (r.l.)	29 (r.l.)	31 (r.l.)
FWHM	13 (r.l.)	16 (r.l.)	28 (r.l.)	75 (r.l.)	223 (r.l.)	676 (r.l.)	2049 (r.l.)
	12 (r.l.)	13 (r.l.)	14 (r.l.)	15 (r.l.)	15 (r.l.)	16 (r.l.)	17 (r.l.)

2.4. New aspect in the structure of the cascade showers

In 1991, E.Konishi et al. [13] found the quite new phenomena in cascade showers which are called as multi-peak structure now. In figure 9, one example of an individual shower with multiple peaks is shown. In the figure, there are three different cascade curves with different cut off energies for the same showers with the primary energy of 10^{17} eV in lead.

It should be noticed that smoothed cascade curves for different cut off energies in figure 9 are the resultants of the average of the individual cascade shower with multi-peak structures for different cut off energies., The existence of similarities in the cascade curves for different cut off energies in figure 9 suggests us that the essential energy configuration among shower particles in the LPM shower at extremely high energies are accidentally determined in higher energy particles. This is the origin of variety of the diversities among the LPM showers. Figure 9 is reproduced from E.Konishi et al, J.Phys.G:Nucl.Part.17(1991) 719-932.

3. Conclusion

The diversity of the LPM showers at extremely high energies is a quite new phenomenon which never appears in the usual BH cascade showers. The characteristic of the multi-peak structure at extremely high energies among the LPM showers will be clarified in another paper in this workshop, which play decisively important role in the study for extremely high energy electron neutrino astrophysics.

References

[1] Landau L D and Pomeranchuk I J 1953 *Dokl. Akad. Nauk. SSSR* **92** 586-588, 589-593
[2] Migdal A 1956,*Phys.Rev.***103** ,1811-1820
[3] Misaki A, Shibata M, Yuda T, Shirai T and Mori K 1974 *Proc.Inter.Cosmic Ray Sympo on High Ener.Phys.* **142 ,**Institute for Cosmic Ray Reseach,University of Tokyo
[4] Konishi E, Misaki A and Fujimaki N 1978 *Nuov Cimento* **44** 509-530
[5] Pomansky A A 1947 *Uzb.Akad.Nauk SSSR,Ser.Fiz.* **32** 497
[6] Misaki A 1988 *ICR-Report* 169-88-15, July, (Tanashi: Institute for Cosmic Ray Research, University of Tokyo)
[7] Rossi B High Energy Particles 1952 Prentice Hall Incoporated New York
[8] Kamata K and Nishimura J 1958 *Suppl.Prog.Theor.Phys.***6** 93-155
[9] Misaki A,1964 Suppl.**32** *Prog.Theor.Phys*.82-103
[10] Fujimaki N and Misaki A 1988 *ICR-Report* 165-88-14 (Tanashi: Institute for Cosmic Ray Research, University of Tokyo)
[11] Misaki A 1990 *Fortschr.Phys.* **88** 413-446
[12] Misaki A 1989 *Phys.Rev.* D **40** 3086-3096
[13] Konishi E, Adachi A, Takahashi N and Misaki A 1991 *J.Phys.Nucl.Part.Phys.* **17** 719-732

IOP Conf. Series: Journal of Physics: Conf. Series **1181** (2019) 012086 doi:10.1088/1742-6596/1181/1/012086

LPM effect in cosmic-ray electron observations with emulsion chambers

Kenji Yoshida[1], **Tadashi Kobayashi**[2], **and Jun Nishimura**[3]

[1] Shibaura Institute of Technology, 307 Fukasaku, Minuma-ku, Saitama 337-8570, Japan
[2] Aoyama Gakuin University, 5-10-1 Fuchinobe, Chuo, Sagamihara 252-5258, Japan
[3] ISAS, JAXA, 3-1-1 Yoshinodai, Chuo, Sagamihara 252-5210, Japan

E-mail: yoshida@shibaura-it.ac.jp

Abstract. We have performed a series of cosmic-ray electron + positron observations using balloon-borne emulsion chambers since 1968, together with electron beam tests at CERN-SPS. In emulsion chamber experiments, we can measure the location of shower tracks in each emulsion plate with a precision of 1 μm. Because of the high position resolution, we can identify the first electron-positron pairs of electron-induced showers. The LPM effect predicts a reduction of amplitude for bremsstrahlung photon emission by a high-energy electron. It causes the optical depth of the first electron-positron pair to increase. From the measurements of the first pair depths with 200 GeV and 250 GeV beam electrons at CERN, and cosmic ray electrons + positrons from 400 GeV to 3 TeV (on average at 900 GeV), we found direct evidence of suppression of the bremsstrahlung cross section due to the LPM effect. In this work, we present the experimental results compared to analytic approximations and Monte-Carlo simulations.

1. Introduction

Landau, Pomeranchuk, and Migdal (LPM) predicted that the interaction cross sections of bremsstrahlung and pair creation by the Bethe-Heitler formulas [1] are suppressed at high energy and in dense media [2, 3, e.g.]. The LPM effect for bremsstrahlung is the suppression of photon production due to multiple scattering of the electron. In multiple scattering of an electron while traversing the formation zone, the probability of bremsstrahlung photon emission is reduced due to the interference of the bremsstrahlung amplitudes from before and after the scattering [3, e.g.]. A similar suppression of the cross section occurs for pair creation.

The existence of the LPM effects affects high energy phenomena involving electro-magnetic showers in thick targets with bremsstrahlung radiation and pair creation when the energy of the incident electron is in high enough on transversing material. If the LPM effect does exist, the phenomena involving such processes would look very different from those without such an effect. Therefore, it is important to verify the LPM effect in high energy region. Since the LPM effect will cause the elongation of high-energy electromagnetic showers with large fluctuations of the number of shower particles. In fact, Konishi *et al.* (1991) demonstrated that such behaviors are observed in their Monte-Carlo simulations [4].

Below a few TeV region, although the reduction of the pair creation cross section due to the LPM effect is expected to be negligible small, the LPM effect of the bremsstrahlung radiation is expected to be significantly appeared. So far, the experimental verification of the LPM effect have been achieved only by few accelerator experiments, which measure bremsstrahlung gamma rays from electron beams of 8 GeV – 290 GeV traversing thin targets of several % r.l. [5, 6]. In this paper, we present the results of the verification of the LPM effect with emulsion chambers by using electron beams of 50 GeV, 200 GeV and 250 GeV, and cosmic ray electrons of 400 GeV to 3 TeV that is 900 GeV on average. We confirmed that the experimental results were consistent with the expected results from the LPM effect by means of analytical calculations and Monte-Carlo simulations.

2. Cosmic ray electron observations

We have carried out fourteen balloon flights from 1968 to 2001 for cosmic ray all-electron (electrons + positrons) observations with emulsion chambers. The geometrical factor is $S\Omega_e T = 8.19$ m^2 sr day and the total exposure is 11.8 day. Figure 1 shows the cosmic ray all-electron spectrum from 30 GeV to 3 TeV observed with emulsion chambers [7], compared to our SNR model [8]. In Fig. 1, the recent electron + positron spectra observed with H.E.S.S. on the ground [9, 10] and AMS-02, Fermi-LAT, DAMPE, and CALET in space are also plotted [11, 12, 13, 14]. Our all-electron spectrum is in good agreement with the recent spectra from the experiments in space.

In addition to the balloon observations, we have performed accelerator beam tests at CERN-SPS with electron beams of 50 GeV, 200 GeV and 250 GeV.

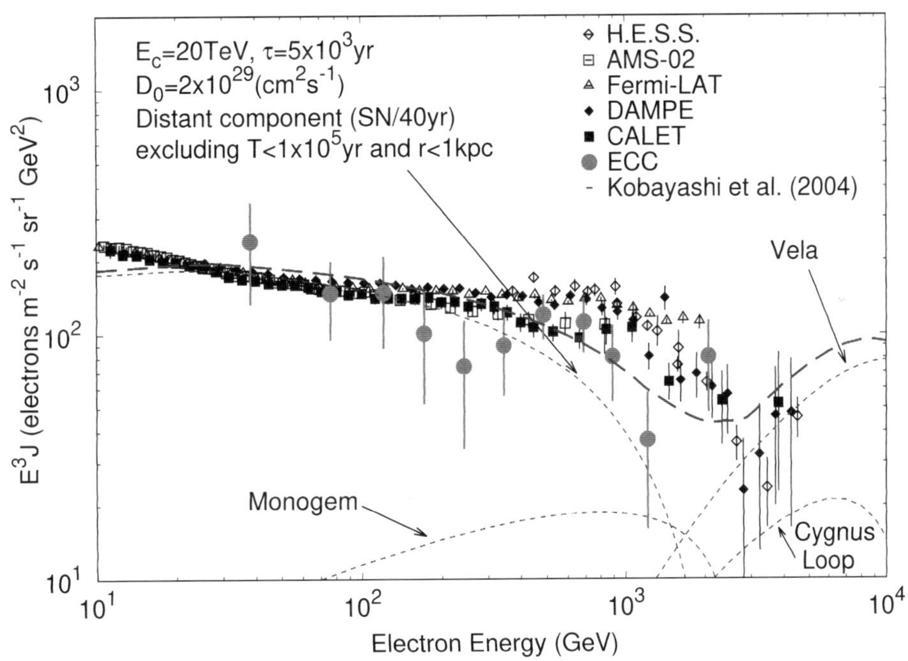

Figure 1. Cosmic ray electron + positron spectrum from 30GeV to 3TeV observed with balloon-borne emulsion chambers [7], compared to the recent observations on the ground and in space [9, 10, 11, 12, 13, 14], and our SNR model spectrum [8].

3. Verification of the LPM effect

3.1. Emulsion chambers

Emulsion chambers consist of nuclear emulsion plates, X-ray films, and lead plates, which are stacked alternately. The typical size and thickness are 40 cm × 50 cm and about 9 cm (∼9 r.l.),

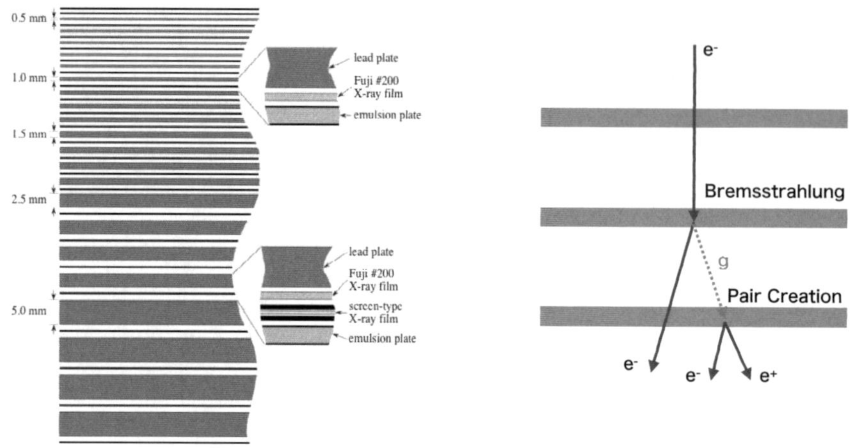

Figure 2. Schematic side view of emulsion chambers (left) and conceptual diagram of the first electron-positron pair creation in emulsion chambers (right).

respectively. A left panel of Fig. 2 shows a schematic side view of emulsion chambers. In emulsion chambers, it is possible to measure the location of shower tracks in each emulsion plate with a precision of 1 μm. The high position resolution enables us to inspect the shower starting points in detail, which are used for identification of electrons, gamma rays, and other hadronic interaction events.

Electron energies were determined by comparing the number of shower tracks within a circle of 100 μm radius from shower axis at various depths with the calculated transition curves, and fitting to the integrated track length, used to estimate total ionization in the cascade [15].

3.2. Measurements of the first electron-positron pairs

An incident electron generates gamma-ray photons through bremsstrahlung radiation, and these gamma rays generate electron-positron pairs in emulsion chambers. Because of the high position resolution, we can identify the location of the first electron-positron pair that is produced by a gamma ray from bremsstrahlung of an incident electron. A right panel of Fig. 2 shows the conceptual diagram of the first pair creation in emulsion chambers. The first pair depths directly depend on the cross sections of bremsstrahlung and pair creation. Since the LPM effect predicts the suppression of the Bethe-Heitler cross section, the first pair depths are expected to become deeper than that of the Bethe-Heitler cross section. We derived the distributions of the first electron-positron pair depths from the electron beam tests at CERN-SPS and the electron observations with balloon-borne emulsion chambers.

3.3. Calculations of the first electron-positron pairs

Monte-Carlo simulations can represent the distributions of the first electron-positron depths in emulsion chambers, which have the complicated structure of nuclear emulsion plates, X-ray films, and lead plates. We derived the distributions of the first electron-positron pair depths by using two completely different Monte-Carlo simulation codes of Epics [16] and Geant4 [17, 18]. All essential cross sections of the electromagnetic processes are implemented in Epics and Geant4, respectively.

In addition to the Monte-Carlo simulations, we derived analytic approximations of the distributions of the first pair depths. Not only do the analytic approximations allow a good qualitative understanding of the the features of the first pair depth distributions, but they are valuable for the direct comparison to the observed distributions of the first pair depths. Letting g be bremsstrahlung gamma-ray emission rate per r.l. and σ be pair creation rate per r.l., in the case of σ and g are constant, the probability density of the first electron-positron pair is derived to be

$$P(t) = ge^{-gt}(1 - e^{-\sigma t})e^{\frac{g}{\sigma}(1-e^{-\sigma t})}, \tag{1}$$

where t is the depth of the first pair in radiation length (r.l.). In this calculation of the first pair depths, the energy dependence of g and σ is taken in, including the effect of complete screening and electron straggling.

4. Results

We compared the experimental distributions of the first pair depths with the calculations. Figure 3, 4, 5 and 6 present the first pair depth distributions with electrons of 50 GeV, 200 GeV, and 250 GeV at CERN-SPS, and cosmic-ray electrons + positrons from 400 GeV to 3 TeV, which is 900 GeV on average, compared to the calculations with the analytic approximations, Epics, and Geant4 Monte-Carlo simulations. In Table 1 the comparisons among the experimental and calculated results are summarized with reduced-χ^2 values. As for 50 GeV electrons, since there are almost no differences between the Bethe-Heitler predictions and the LPM predictions, the first pair depth distribution of 50 GeV electrons is well represented by both the cross sections. Above 200 GeV electrons, the differences between the Bethe-Heitler predictions and the LPM predictions become more pronounced. As shown in Fig. 4, 5, 6, and Table 1, the first pair depth distributions of 200 GeV electrons, 250 GeV electrons, and electrons from 400 GeV to 3 TeV reject the Bethe-Heitler cross sections for the all calculations. On the contrary, they are well represented by the LPM predictions for the all calculations.

5. Conclusions

For the verification of the LPM effect, we applied the different approach from the previous accelerator experiments [5, 6]. By using the first pair measurements with emulsion chambers, we found direct evidence of the LPM effect in 200 GeV, 250 GeV, and 400 GeV–3 TeV (on average at 900 GeV) electrons. This verification of the LPM effect up to 3 TeV is the highest energies so far.

Table 1. Summary of comparisons between the experimental and calculated results

	χ^2_ν (Prob.)		
	Analytic	Epics	Geant4
50 GeV (225 events, 9 d.o.f.)			
BH	0.746 (66.7%)	0.694 (71.5%)	0.551 (83.8%)
LPM	0.513 (86.7%)	0.319 (96.9%)	0.371 (94.9%)
200 GeV (289 events, 8 d.o.f.)			
BH	3.015 (2.2%)	2.051 (3.7%)	1.814 (6.9%)
LPM	0.467 (88.0%)	0.801 (60.1%)	1.422 (18.1%)
250 GeV (363 events, 9 d.o.f.)			
BH	5.46 (2×10^{-5}%)	4.56 (5×10^{-4}%)	4.46 (7×10^{-4}%)
LPM	0.791 (62.5%)	0.655 (75.0%)	0.932 (49.5%)
400GeV–3TeV (113 events, 5 d.o.f.)			
BH	3.550 (0.33%)	2.591 (2.4%)	2.325 (4.0%)
LPM	0.870 (48.9%)	1.676 (13.6%)	2.103 (6.2%)

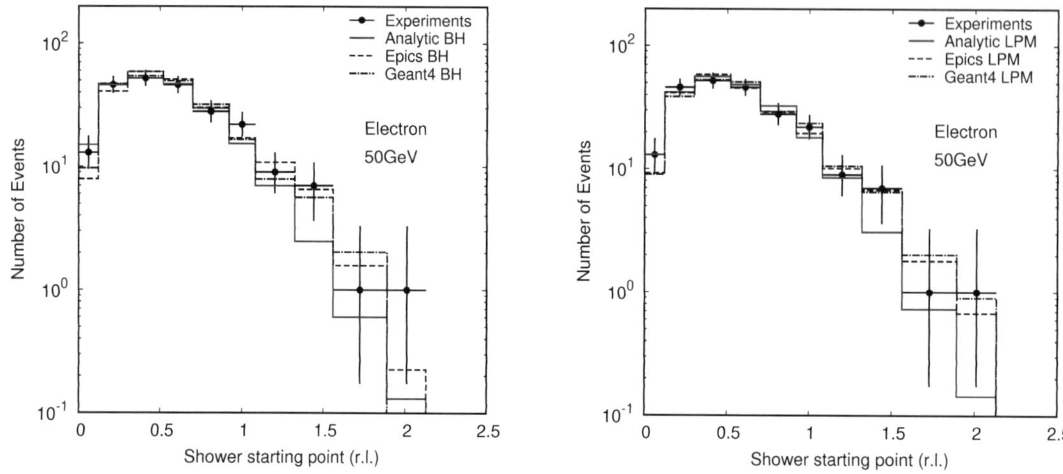

Figure 3. First pair depth distributions of 50 GeV electrons at CERN-SPS, compared to calculations of the Bethe-Heitler cross section (left) and the LPM effect (right). The calculations are carried out with the analytic approximations, Epics and Geant4.

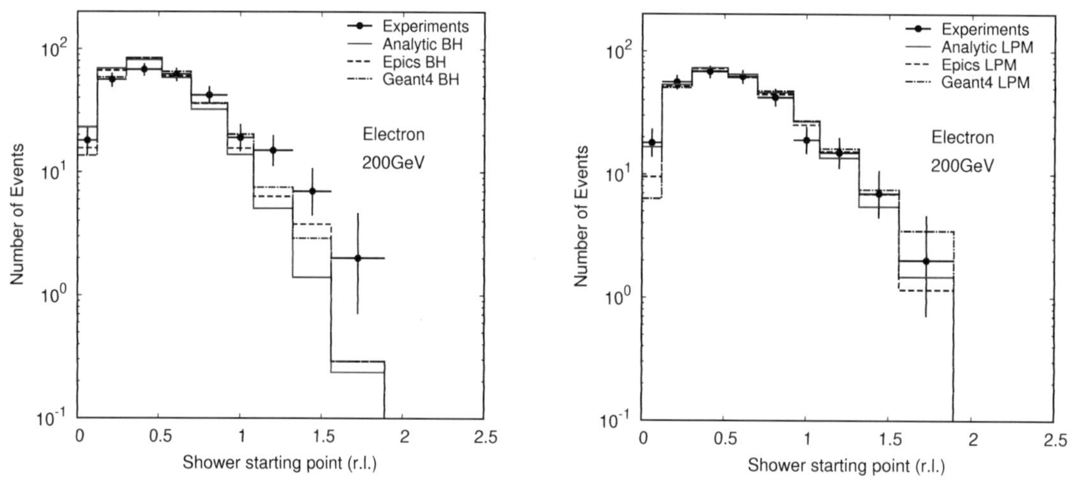

Figure 4. First pair depth distributions of 200 GeV electrons at CERN-SPS, compared to calculations of the Bethe-Heitler cross section (left) and the LPM effect (right).

Acknowledgments

We are grateful to Prof.Misaki for his inviting us to the workshop 1 of the 26th Extended European Cosmic Ray Symposium (E+CRS 2018).

References

[1] Tsai Y S 1974 *Rev. Mod. Phys.* **46** 815
[2] Klein S 1999 *Rev. Mod. Phys.* **71** 1501
[3] Baier V N & Katkov V M 2005 *Phys. Rep.* **409** 261
[4] Konishi E, Adachi A, Takahashi N, and Misaki A 1991 *J. Phys. G: Nucl. Part. Phys.* **17** 719
[5] Anthony P *et al.* 1997 *Phys. Rev. D* **56** 1373
[6] Hansen H *et al.* 2004 *Phys. Rev. D* **60** 032001

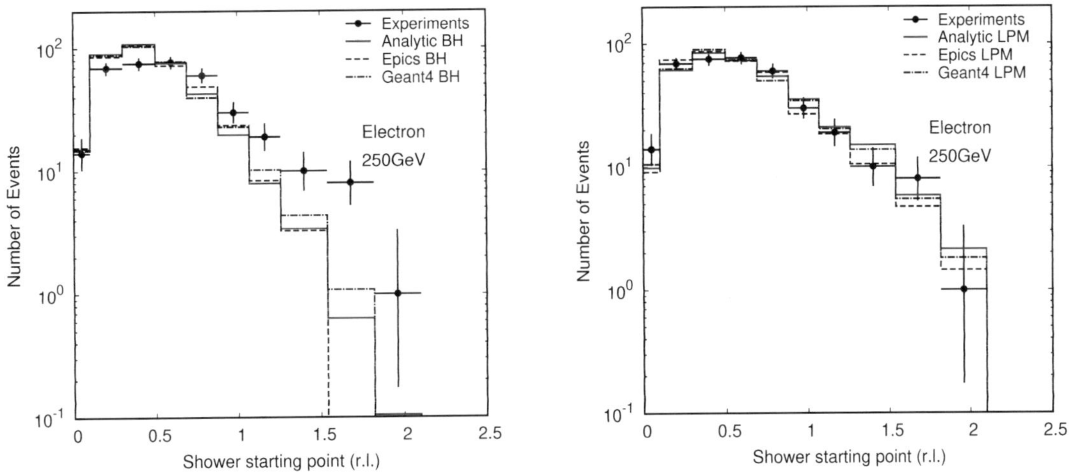

Figure 5. First pair depth distributions of 250 GeV electrons at CERN-SPS, compared to calculations of the Bethe-Heitler cross section (left) and the LPM effect (right).

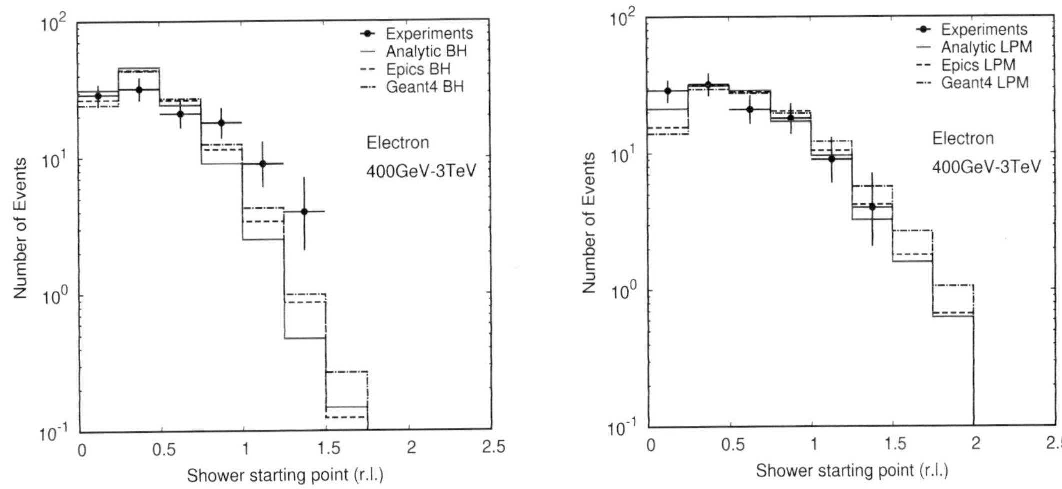

Figure 6. First pair depth distributions of all cosmic ray electrons from 400 GeV to 3 TeV (on average at 900 GeV), compared to calculations of the Bethe-Heitler cross section (left) and the LPM effect (right).

[7] Kobayashi T *et al.* 2012 *Astrophys. J.* **760** 146
[8] Kobayashi T *et al.* 2004 *Astrophys. J.* **601** 340
[9] Aharonian F *et al.* 2008 *Phys. Rev. Lett.* **101** 261104
[10] Aharonian F *et al.* 2009 *Astron. Astrophys.* **508** 561
[11] Aguilar S *et al.* 2014 *Phys. Rev. Lett.* **113** 221102
[12] Abdollahi S *et al.* 2017 *Phys. Rev. D* **95** 082007
[13] Ambrosi G *et al.* 2017 *Nature (London)* **552** 63
[14] Adriani O *et al.* 2018 *Phys. Rev. Lett.* **120** 261102
[15] Nishimura J *et al.* 1980 *Astrophys. J.* **238** 394
[16] Kasahara K 2018 http://cosmos.n.kanagawa-u.ac.jp/EPICSHome/index.html
[17] Agostinelli S *et al.* 2003 *Nucl. Instrum. Methods Phys. Res. A* **506** 250
[18] Allison J *et al. IEEE Trans. Nucl. Sci.* **53** 270

Application of an Improved Molière Theory to the Description of the Landau–Pomeranchuk–Migdal Effect

O O Voskresenskaya and H T Torosyan

Joint Institute for Nuclear Research, 141980, Dubna, Moscow region, Russia

E-mail: voskr@jinr.ru

Abstract. We present exact analytical and numerical results for high-energy Coulomb corrections (CCs) to the parameters of an improved Molière multiple scattering theory over a wide range of the nuclear charge number Z of the target atom and show that the values of these corrections increase up to $40\% - 45\%$ for $Z = 92$. We also report our results of applying this improved Molière multiple scattering theory for calculating CCs to the quantities of the classical Migdal theory of the Landau–Pomeranchuk–Migdal (LPM) effect, and we demonstrate that the improved Migdal LPM effect theory allow one to completely eliminate the discrepancy between the predictions of the LPM effect theory and their measurement at least for high-Z targets. We also obtained analytical and numerical results concerning high-energy CCs to the quantities of the quantum Migdal theory of the LPM effect in the ranges of Z from 6 to 92 and the Molière expansion parameter B from 4.5 to 8.5, and we managed to show that the magnitude of relative CC to the spectral radiation rate reaches -19% in these ranges considered and must be borne in mind, e.g., in the Monte-Carlo analysis of electromagnetic cascade LPM showers in extremely high-energy region.

1. Introduction

The theory of the multiple scattering of charged particles has been treated by several authors. However, the most widely used at present is the multiple scattering theory of Molière [1]. The multiple scattering theory is of interest for numerous applications related to particle transport in matter [2] and is widely used in the cascade shower theory [3]. It can also be useful in applications motivated by exploring the extremely high-energy Landau–Pomeranchuk–Migdal (LPM) showers [4] and IceCubes neutrino-induced showers with energies above 1 PeV [5, 6]. The neutrino-oscillation experiments [7], the astronomical cosmic-ray observations [8], as well as the cosmic-ray experiments in ultrahigh-energy region [9], etc., face a problem of taking into account the multiple scattering effects.

As the Molière theory can be currently used in ultrahigh-energy region, the role of the high-energy Coulomb corrections to the parameters of this theory becomes significant. Of especial importance is the Coulomb correction to the screening angular parameter, as this parameter enters into other important quantities of the Molière theory and the Migdal theory of the LPM effect [10]. It is shown that the Coulomb corrections to the spectral bremsstrahlung rate of the LPM effect theory for the regime of small LPM suppression allow completely to eliminate the discrepancy between the predictions of the LPM effect theory and its measurement at least for

IOP Conf. Series: Journal of Physics: Conf. Series **1181** (2019) 012087 doi:10.1088/1742-6596/1181/1/012087

high-Z targets and also to further improve the agreement between the predictions of the LPM effect theory analogue for a thin layer of matter and experimental data [10]. The aim of the present work is to estimate analytically and numerically the CCs to some important quantities in the quantum LPM effect theory, especially to the Migdal function $G(s)$ and the spectral bremsstrahlung rate which are of special interest in describing the shower production over the energy range $10^{16} \leq E \leq 10^{20}$ eV [5, 6].

The outline of this papers is as follows. In Section 2 we present briefly our results for the high-energy Coulomb corrections[1] to the parameters of an improved Molière multiple scattering theory [11]. Then, in Section 3 we report our results of applying this improved Molière multiple scattering theory for calculating CCs to the quantities of the classical Migdal theory of the LPM effect in the regime of small LPM suppression. In Section 4 we obtain analytical and numerical results for high-energy CCs to some quantities of the quantum Migdal theory of the LPM effect in the regime of strong LPM suppression. Finally, in Sec. 4 we state our conclusions.

2. Coulomb corrections to some parameters of the Molière theory

In this section we present an exact analytical result for the screening angular parameter θ_a ($\theta_a' = \sqrt{1.167}\theta_a$) of the Molière theory valid to all orders in the Born parameter $\xi = Z\alpha/\beta$,

$$\Delta_{CC}[\ln(\theta_a')] \equiv \ln(\theta_a') - \ln(\theta_a')^B = f(\xi), \tag{1}$$

with the Bethe–Maximon function $f(\xi) \equiv \xi^2 \sum_{n=1}^{\infty}[n(n^2+\xi^2)]^{-1}$, instead of an approximate one for the exact (θ_a^M) and first-order (θ_a^B) values of the screening angle (θ_a) valid to second order in the parameter ξ,

$$\theta_a^M = \theta_a^B\sqrt{1+3.34\cdot\xi^2}, \quad \theta_a^B = 1.20\cdot\alpha\cdot Z^{1/3}, \tag{2}$$

that was obtained in the original paper of Molière [1]. Here, α is the fine structure constant and $\beta = v/c$ is the velocity of a projectile in units of the velocity of light.

We also present the analytical and numerical results for the Coulomb corrections to the parameters b, B, and $\overline{\vartheta^2}$ of the Molière expansion method for the angular distribution function

$$w_M(\vartheta, L) = \sum_{n=0}^{\infty}\frac{1}{n!}\frac{1}{B^n}w_n(\vartheta, L), \quad w_n(\vartheta, L) = \frac{1}{\overline{\vartheta^2}}\int_0^{\infty} y dy J_0\left(\frac{\vartheta}{\bar{\vartheta}}y\right)e^{-y^2/4}\left[\frac{y^2}{4}\ln\left(\frac{y^2}{4}\right)\right]^n, \tag{3}$$

where $\vec{\vartheta}$ is a two-dimensional particle scattering angle in the plane orthogonal to the incident particle direction, $\overline{\vartheta^2} = \theta_c^2 B$, θ_c is the characteristic angle, L denotes the thickness of an absorber, $J_0(\vartheta\eta)$ presents the Bessel function, $y = \theta_c\eta$, and the expansion parameter B is defined by the transcendental equation $B - \ln B = b$ with $b = \ln(\theta_c/\theta_a')^2$.

For the Coulomb corrections to the Born parameters b^B, B^B, and $\overline{\vartheta^2}^B$, we get

$$\Delta_{CC}[b] = -f(\xi), \ \Delta_{CC}[B] = \frac{f(\xi)}{1/B^B - 1}, \ \Delta_{CC}\left[\overline{\vartheta^2}\right] = \theta_c^2\cdot\Delta_{CC}[B]. \tag{4}$$

The relative Coulomb corrections become

$$\delta_{CC}\left[\overline{\vartheta^2}\right] = \delta_{CC}[B] = f(\xi)/(1-B^B), \ \delta_{CC}[\theta_a] = \exp[f(\xi)] - 1\,. \tag{5}$$

The relative difference between the approximate θ_a^{M} and exact θ_a results reads

$$\delta_{\mathrm{CCM}}[\theta_a] \equiv \frac{\theta_a - \theta_a^{\mathrm{M}}}{\theta_a^{\mathrm{M}}} = \frac{\theta_a}{\theta_a^{\mathrm{M}}} - 1 = R_{\mathrm{CCM}}[\theta_a] - 1\,. \tag{6}$$

[1] The Coulomb correction (Δ_{CC}) is the difference between the exact Born parameter ξ result and the Born approximation.

Table 1. The Z dependence of the Coulomb corrections and differences defined by Eqs. (1), (4), (5), and (6) for $\beta = 1$ and $B^B = 8.46$.

Target	Z	$f(\xi)$	$\delta_{CC}[\theta_a]$	$\delta_{CCM}[\theta_a]$	$\delta_{CC}\left[\overline{\vartheta^2}\right]$	$\Delta_{CC}[b]$	$\Delta_{CC}[B]$
W	74	0.2813	0.3249	−0.0565	−0.0377	−0.2813	−0.3190
Pt	78	0.3067	0.3590	−0.0582	−0.0411	−0.3067	−0.3478
Au	79	0.3125	0.3670	−0.0585	−0.0419	−0.3125	−0.3545
Pb	82	0.3316	0.3930	−0.0600	−0.0445	−0.3316	−0.3760
U	92	0.3951	0.4845	−0.0622	−0.0530	−0.3951	−0.4481

The Z dependence of these corrections and differences for some high-Z targets presents Table 1. It shows that while the modules of the quantities $\delta_{CC}\left[\overline{\vartheta^2}\right]$ and $\Delta_{\rm CCM}[\theta_a]$ reach about 6% over the range $74 \le Z \le 92$, the maximal value of the relative correction $\delta_{CC}[\theta_a]$ amounts approximately to 50% for $Z = 92$. The modules of the Coulomb corrections to the parameters b and B are: $\left|\Delta_{CC}[B]\right| \sim 0.45$ and $\left|\Delta_{CC}[b]\right| \sim 0.40$, such as $\Delta_{CC}\ln(\theta_a')] \sim 0.40$ for $Z = 92$. These corrections will be used in the next section to obtain Coulomb corrections to the quantities of the Landau–Pomeranchuk–Migdal effect theory.

3. Applying the improved Molière theory to the description of the LPM effect

Based on the Coulomb corrections found in the previous section, we first get the analytical and numerical results for the Coulomb corrections to the quantities of the classical Migdal theory of the LPM effect for sufficiently thick targets, the basic formulae of which in the units $\hbar = c = 1$, $e^2 = 1/137$ read [12, 13]

$$\left\langle \frac{dI}{d\omega}\right\rangle = \Phi(s)\left(\frac{dI}{d\omega}\right)_0, \quad \left(\frac{dI}{d\omega}\right)_0 = \frac{2e^2}{3\pi}\gamma^2 q\, L, \quad q = \overline{\vartheta^2}/L, \tag{7}$$

$$\Phi(s) = 24 s^2\left[\int_0^\infty dx\, e^{-2sx}\mathrm{cth}(x)\sin(2sx) - \frac{\pi}{4}\right], \quad s^2 = \frac{\lambda^2}{\overline{\vartheta^2}}, \quad \lambda^2 = \gamma^{-2}\,. \tag{8}$$

Here, $\langle dI/d\omega\rangle$ is the electron spectral bremsstrahlung intensity averaged over various trajectories of electron motion, $(dI/d\omega)_0$ presents the spectral bremsstrahlung rate without accounting the multiple scattering effects in the radiation, and γ denotes the Lorentz factor of the scattered particle.

The analytical result for the Coulomb correction Δ_{CC} to the Born spectral bremsstrahlung rate $(dI/d\omega)_0$ is as follows:

$$\Delta_{CC}\left[\left(\frac{dI}{d\omega}\right)_0\right] \equiv \left(\frac{dI}{d\omega}\right)_0 - \left(\frac{dI}{d\omega}\right)_0^B = \frac{2e^2}{3\pi}\gamma^2 L\cdot\Delta_{CC}[q], \quad \Delta_{CC}[q] = \frac{1}{L}\cdot\Delta_{CC}\left[\overline{\vartheta^2}\right]$$

with

$$\Delta_{CC}\left[\overline{\vartheta^2}\right] = \theta_c^2\cdot\Delta_{CC}[B], \quad \Delta_{CC}[B] = f(\xi)/(1/B^B - 1)\,, \tag{9}$$

according to (4) and (5). In doing so, $\Delta_{CC}\left[(dI/d\omega)_0\right]$ and $\delta_{CC}\left[(dI/d\omega)_0\right]$ become

$$\Delta_{CC}\left[\left(\frac{dI}{d\omega}\right)_0\right] = \frac{2(e\gamma\theta_c)^2}{3\pi\,(1/B^B - 1)}\cdot f(\xi), \quad \delta_{CC}\left[\left(\frac{dI}{d\omega}\right)_0\right] = \frac{f(\xi)}{1 - B^B}\,. \tag{10}$$

IOP Conf. Series: Journal of Physics: Conf. Series **1181** (2019) 012087 doi:10.1088/1742-6596/1181/1/012087

Table 2. The dependence of the relative Coulomb correction $-\delta_{CC}\left[\langle dI/d\omega\rangle\right]$ value (%) on the Migdal parameter s in the regime of small LPM suppression for high-Z targets.

Target	Z	s = 1.0	s = 1.1	s = 1.2	s = 1.3	s = 1.5	s = 2.0	s = ∞
Au	79	4.32	4.28	4.26	4.24	4.22	4.21	4.19
Pb	82	4.58	4.54	4.51	4.49	4.47	4.46	4.45
U	92	5.45	5.41	5.36	5.34	5.33	5.31	5.30

$$\delta_{CC}\left[\langle dI/d\omega\rangle\right] = -4.50 \pm 0.05\%\ (Z=82)\ ,\ \delta_{CC}\left[\langle dI/d\omega\rangle\right] = -5.35 \pm 0.06\%\ (Z=92)\ ,\ \text{and}$$
$$\bar{\delta}_{CC}\left[\langle dI/d\omega\rangle\right] = -4.70 \pm 0.49\%\ .$$

For the regime of small LPM suppression, $\Phi(s) \approx 1 - 0.012/s^4$, the analytical result for the relative Coulomb correction to the Migdal function $\Phi(s)$ in the entire range $1 \leq s \leq \infty$ reads

$$\delta_{\rm CC}\left[\Phi(s)\right] = \frac{0.012}{s^4}\,\delta_{\rm CC}\left[s^4\right] \cdot \frac{\left(s^4\right)^{\rm B}}{\left(s^4\right)^{\rm B} - 0.012} = 0.012\,\frac{\delta_{\rm CC}\left[s^4\right]}{\delta_{\rm CC}\left[s^4\right]+1} \cdot \frac{1}{\left(s^4\right)^{\rm B} - 0.012}\ , \tag{11}$$

where

$$\delta_{\rm CC}\left[s^4\right] = 1 - \left(\frac{\left(\overline{\vartheta^2}\right)^{\rm B}}{\overline{\vartheta^2}}\right)^2 = 1 - \frac{1}{\left(\delta_{\rm CC}[\overline{\vartheta^2}]+1\right)^2}\ . \tag{12}$$

Table 2 presents the average values of the corrections $-\delta_{CC}\left[\langle dI/d\omega\rangle\right]$ (%) for some high-Z targets and also the sampling mean $-\bar{\delta}_{CC}\left[\langle dI/d\omega\rangle\right]$ (%) over the entire range $1.0 \leq s \leq \infty$ of the parameter s. It shows that $\delta_{CC}\left[\langle dI/d\omega\rangle\right] = (-4.50 \pm 0.05)\%$ $(Z = 82)$ and $\delta_{CC}\left[\langle dI/d\omega\rangle\right] = (-5.35 \pm 0.06)\%$ $(Z = 92)$ coincide within the experimental error with the sizes of the normalization corrections $(-4.5 \pm 0.2)\%$ for $2\%L_R$ lead target and $(-5.6 \pm 0.3)\%$ for $3\%L_R$ uranium target, respectively. The average value $\bar{\delta}_{CC}\left[\langle dI/d\omega\rangle\right] = (-4.70 \pm 0.49)\%$ excellently agrees with the weighted average value $(-4.7 \pm 2)\%$ of the normalization correction obtained for 25 GeV data of the SLAC E-146 experiment [14].

This means that applying the improved multiple scattering theory by Molière allows one to avoid multiplying the results of predictions of the conventional Migdal LPM effect theory by the above normalization factor and leads to agreement between the improved Migdal theory of the LPM effect and experimental data. We believe that this allows one to understand the origin of the normalization problem for high-Z targets discussed in [14].

4. Coulomb corrections to some quantities of the quantum LPM effect theory

We have also obtained analytical and numerical results for the Coulomb corrections to the quantities of the quantum Migdal LPM theory

$$\left\langle\frac{dI}{d\omega}\right\rangle = \frac{1}{4}\left(\frac{dI}{d\omega}\right)_0 \left\{\varepsilon^2 G(s) + 2[1+(1-\varepsilon)^2]\Phi(s)\right\}\ , \tag{13}$$

with the energy of radiation $\varepsilon = \omega/E$ in units of the incident particle energy E, Migdal's parameter $s = \sqrt{(\omega/q)}/8\gamma^2$, and the Migdal functions G(s) and $\Phi(s)$

$$G(s) = 12\pi s^2 - 48s^2\sum_{k=0}^{\infty}\frac{1}{(k+s+1/2)^2+s^2},\quad \Phi(s) = 6s - 6\pi s^2 + 24s^2\sum_{k=1}^{\infty}\frac{2}{(k+s)^2+s^2} \tag{14}$$

for the regime of strong LPM suppression, $G(s)_{s\to 0} \to 12\pi s^2$ and $\Phi(s)_{s\to 0} \to 6s$ [13, 15].

IOP Conf. Series: Journal of Physics: Conf. Series **1181** (2019) 012087 doi:10.1088/1742-6596/1181/1/012087

Table 3. Coulomb corrections to the quantities of the quantum Migdal LPM theory, $\delta_{CC}[G(s)]$, $\delta_{CC}[\Phi(s)]$, $\delta_{CC}[(dI/d\omega)_0]$, and $\delta_{CC}[\langle dI/d\omega\rangle]$ in the regime of strong LPM suppression

1. $B^B = 8.46$, $s = 0.05$, and $\beta = 1$

Target	Z	$f(\xi)$	$\delta_{CC}[G(s)]$	$\delta_{CC}[\Phi(s)]$	$\delta_{CC}\left[\left(\frac{dI}{d\omega}\right)_0\right]$	$\delta_{CC}\left[\left\langle\frac{dI}{d\omega}\right\rangle\right]$
C	6	0.0041	−0.0005	−0.0003	−0.0005	−0.0008
Al	13	0.0107	−0.0014	−0.0007	−0.0014	−0.0023
Fe	26	0.0430	−0.0058	−0.0029	−0.0058	−0.0094
W	74	0.2810	−0.0392	−0.0194	−0.0377	−0.0621
Au	79	0.3130	−0.0438	−0.0216	−0.0419	−0.0698
Pb	82	0.3320	−0.0466	−0.0230	−0.0445	−0.0738
U	92	0.3950	−0.0560	−0.0276	−0.0530	−0.0887

2. $B^B = 4.50$, $s = 0.05$, and $\beta = 1$

Target	Z	$f(\xi)$	$\delta_{CC}[G(s)]$	$\delta_{CC}[\Phi(s)]$	$\delta_{CC}\left[\left(\frac{dI}{d\omega}\right)_0\right]$	$\delta_{CC}\left[\left\langle\frac{dI}{d\omega}\right\rangle\right]$
C	6	0.0041	−0.0012	−0.0006	−0.0012	−0.0019
Al	13	0.0107	−0.0031	−0.0015	−0.0031	−0.0050
Fe	26	0.0430	−0.0125	−0.0062	−0.0123	−0.0201
W	74	0.2810	−0.0873	−0.0427	−0.0803	−0.1349
Au	79	0.3130	−0.0982	−0.0479	−0.0894	−0.1507
Pb	82	0.3320	−0.1049	−0.0511	−0.0949	−0.1603
U	92	0.3950	−0.1273	−0.0617	−0.1129	−0.1921

For the relative CC to the Born spectral bremsstrahlung rate (13), one can get

$$\delta_{CC}\left[\left\langle\frac{dI}{d\omega}\right\rangle\right] = \delta_{CC}\left[\left(\frac{dI}{d\omega}\right)_0\right] + \frac{\Delta_{CC}[G(s)] + \Delta_{CC}[\Phi(s)]}{|G^B(s) + \Phi^B(s)|}$$

$$= \delta_{CC}\left[\left(\frac{dI}{d\omega}\right)_0\right] + \frac{G^B}{|G^B + \Phi^B|}\delta_{CC}[G] + \frac{\Phi^B}{|G^B + \Phi^B|}\delta_{CC}[\Phi]\,, \tag{15}$$

$$\delta_{CC}\left[\left(\frac{dI}{d\omega}\right)_0\right] = \frac{\Delta_{CC}[\ln(\theta_a')]}{1 - B^B}\,. \tag{16}$$

For the regime of strong LPM suppression, we have obtained

$$\delta_{CC}\,[G(s)] = 1 - \frac{1}{\delta_{CC}[\overline{\vartheta^2}] + 1}\,, \quad \delta_{CC}\,[\Phi(s)] = 1 - \frac{1}{\sqrt{\delta_{CC}[\overline{\vartheta^2}] + 1}}\,. \tag{17}$$

Table 3 and Fig. 1 demonstrate the values of $\delta_{CC}\,[G(s)]$, $\delta_{CC}\,[\Phi(s)]$ (17), $\delta_{CC}\,[(dI/d\omega)_0]$ (16), and $\delta_{CC}\,[\langle dI/d\omega\rangle]$ (15) in the regime of strong LPM suppression over the ranges $0 \le Z \le 100$ and $4.5 \le B^B \le 8.5$ at $s = 0.05$. It can be seen from Table 3 that $\delta_{CC}\,[G(s)]$ is comparable with $\delta_{CC}\,[(dI/d\omega)_0]$ in the case of strong suppression. The value of $|\delta_{CC}\,[G(s)]\,|$ is twice as large than $|\delta_{CC}\,[\Phi(s)]\,|$ and reaches about 4.7% at $B^B = 8.46$ and approximately 10.5% at $B^B = 4.50$ for $Z = 82$ (Pb). The modulus of $\delta_{CC}\,[\langle dI/d\omega\rangle]$ acquires a value of 7.4% at $B^B = 8.46$ and 16.0% at $B^B = 8.46$ for $Z = 82$ ($s = 0.05$). Its upper limit is 19% over the ranges considered [16].

Thus, we can conclude that such corrections as $\delta_{CC}\,[\langle dI/d\omega\rangle]$, $\delta_{CC}\,[G(s)]$, and $\delta_{CC}\,[(dI/d\omega)_0]$ become significant in the regime of strong LPM suppression and must be borne in mind, e.g., in the investigations of the LPM showers in extremely high-energy region.

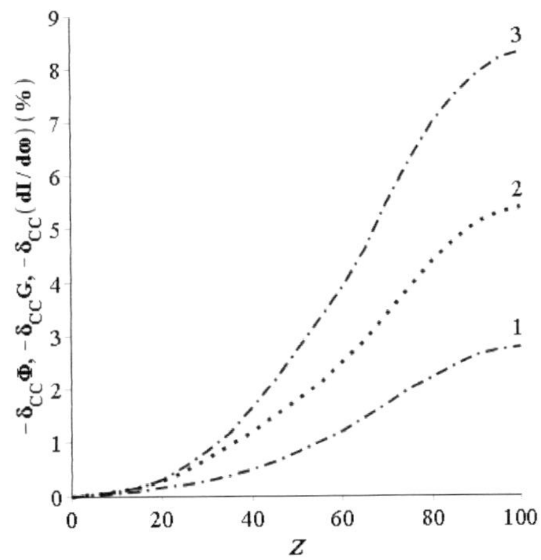

Figure 1. The Z dependence of relative CCs to the quantities of the quantum LPM effect theory for the regime of large LPM supression: 1 – $\delta_{CC}[\Phi(s)]$, 2 – $\delta_{CC}[G(s)]$, and 3 – $\delta_{CC}[\langle dI/d\omega\rangle]$ at $s = 0.05$ and $B^B = 8.46$.

5. Conclusion

The developed approach can be useful for the analysis of cosmic-ray experiments in ultrahigh-energy region, where the LPM effect becomes significant (for instance, in the applications related to investigations of extremely high-energy LPM showers, in exploring their structure and behavior, computing their characteristics, etc. [17, 18]).

Acknowledgments

The authors express their gratitude to Prof. Akeo Misaki (Saitama University, Japan) for his interest to the work and stimulating discussions.

References

[1] Molière G 1947 *Z. Naturforsch.* **2 a** 133
[2] Nakatsuka T, Okei K, Iyono A and Bielajew A F 2015 *Eur. Phys. J.* A **51** 161
[3] Kamata K and Nishimura J 1958 *Suppl. Prog. Theor. Phys.* **6** 53
[4] Misaki A 1989 *Phys. Rev.* D **40** 3086
[5] Klein S 2013 private communication
[6] Gerhardt L and Klein S R 2010 *Phys. Rev.* D **82** 074017
[7] Ambrosio M et al 2002 *Nucl. Instrum. Methods Phys. Res.* A **492** 376
[8] Okei K, Takahashi N and Nakatsuka T 2007 *Proc. 30th Int. Cosmic Ray Conf.* vol 4 (Merida, Mexico: Yucatan Autonoma Univ.) p 157
[9] Aartsen M G et al (IceCube Collaboration) 2018 *Phys. Rev* D **98** 062003
[10] Voskresenskaya O, Kuraev E and Torosyan H 2014 *Phys. Part. Nucl. Lett.* **11** 366
[11] Kuraev E, Voskresenskaya O and Tarasov A 2014 *Phys. Rev.* D **89** 116016
[12] Migdal A B *Dokl. Akad. of Science of SSSR* 1954 **96** 49
[13] Akhiezer A I and Shul'ga N F 1996 *High-Energy Electrodynamics in Matter* (Amsterdam: Gordon and Breach Science Publishers SA)
[14] Anthony P L et al 1997 *Phys. Rev.* D **56** 1373
[15] Migdal A B *Phys. Rev.* 1956 **103** 1811
[16] Voskresenskaya O 2018 Coulomb corrections to the spectral bremsstrahlung rate in the quantum Migdal theory of the Landau–Pomeranchuk effect *Preprint* arXiv:submit/2413019 [hep-ph]
[17] Konishi E, Adachi A, Takahashi N and Misaki A 1991 *J. Phys.* G **17** 719
[18] Misaki A 2018 *26th Extended European Cosmic Ray Symposium and 35th Russian Cosmic Ray Conference. Book of Abstracts* ed A A Lagutin, R I Raikin et al (Barnaul: Altay State University Press) p 207

Lateral distributions of electrons in air showers initiated by ultra-high energy gamma quanta taking into account LPM and geomagnetic field effects

Tatyana Serebryakova[1], Alexander Goncharov[1], Anatoly Lagutin[1], Roman Raikin[1] and Akeo Misaki[2]

[1] Altai State University, 61 Lenin Ave., Barnaul 656049, Russia
[2] Saitama University, 255 Shimookubo, Saitama-city 338-8570, Japan

E-mail: serebryakova@theory.asu.ru

Abstract. Lateral distributions of electrons in air showers initiated by photons of ultra high energies ($10^{18} - 10^{22}$ eV) obtained on the basis of numerical solution of adjoint cascade equations are presented. An extended analysis has been made considering separately the Landau-Pomeranchuk-Migdal (LPM) effect and the interaction of photons and electrons with the geomagnetic field (GMF) with respect to the scaling formalism for lateral distributions. It is shown that one-parametric scaling description of the lateral distribution of electrons remains valid up to the highest energies considering the LPM and GMF effects, that allows effective primary particle type discrimination using the surface detectors data of largest ground-based air shower arrays.

1. Introduction

The determination of the ultra-high energy (UHE) cosmic ray composition, including identification of primary photons and setting up high-precision limits on their fluxes, is a vital effort for understanding the origin of cosmic rays [1–5]. UHE photons could originate either in the interactions of primary cosmic-ray nuclei with the cosmic microwave background radiation or in the exotic phenomena, e.g. resulting from decay of hypothetical super-massive particles. Though the latter scenarios are highly disfavoured by the results of recent studies, the reliable identification of primary photons from the experimental data of giant air shower arrays on both the average and the event-by-event basis remains an actual problem.

The limits on UHE photons fluxes established so far are very stringent, well below the predictions of the most of the exotic models. Thus, extracting the information about the primary particle type from the data of surface detectors, which provides an order of magnitude larger statistics in comparison with atmospheric telescopes, plays an important role.

While simulations of nuclei-induced air showers at ultra-high energies rely on hadronic interaction models fundamentally uncertain in the relevant energy range, in case of primary photons the following two effects should be properly taken into account: the Landau-Pomeranchuk-Migdal (LPM) effect [6–8] and the interaction of EHE photons and electrons with the geomagnetic field (GMF) [9]. In this paper we present the results of the extended analysis of lateral distributions of electrons in photon-generated air showers of $10^{18} - 10^{22}$ eV

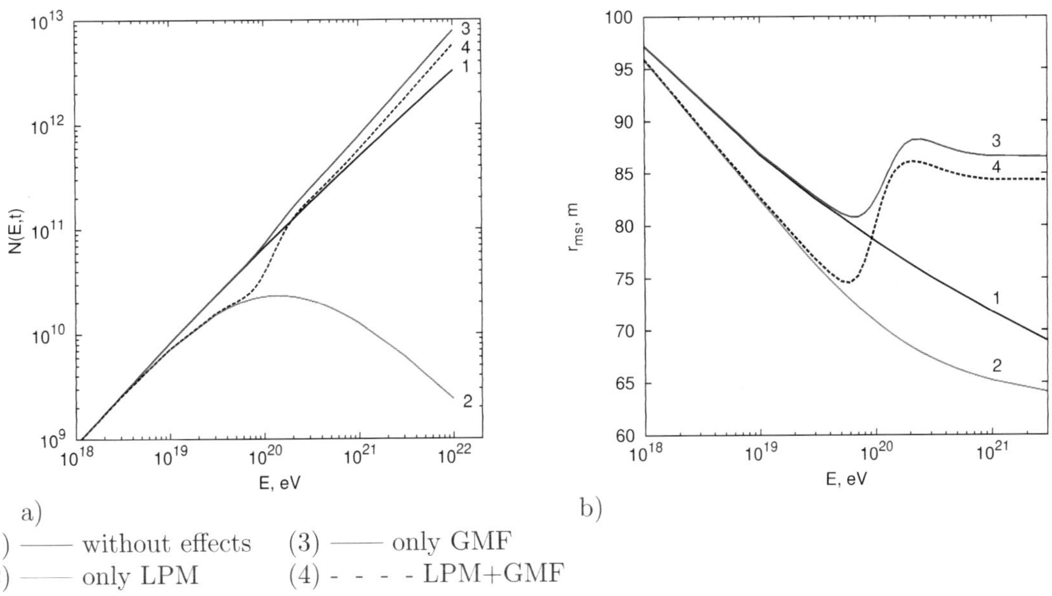

Figure 1. Energy dependences of the total number of electrons (a) and root mean square radius of electron component (b) in vertical electromagnetic air showers produced by photons with $E_\gamma = 10^{18} - 10^{22}$ eV at the observational level of 890 g/cm^2 with and without taking into account the UHE-effects (LPM and GMF).

considering separately the LPM and GMF effects with respect to the scaling formalism for lateral distributions. We show that one-parametric scaling description of the lateral distribution of electrons remains valid up to the highest energies considering the LPM and GMF effects. Effectiveness of primary particle type discrimination using the scale factors of lateral distributions measured by surface detectors of largest ground-based air shower arrays is discussed.

2. Calculation methods and results of analysis

The numerical solution of adjoint cascade equations was used to calculate characteristics of electromagnetic cascades in the atmosphere and magnetosphere of the Earth [10–13].

We consider all the essential processes of cascade particle interaction with matter at energy region $E \geq 10^4$ eV. Calculations were performed for the US standard atmosphere. The GMF intensity profile corresponds to the southern location of the Auger Observatory (Mendoza, Argentina). The primary energies range is $E_\gamma = 10^{18} - 10^{22}$ eV.

In Figure 1 energy dependences of the total number of electrons $N(E,t)$ (shower size) and root mean square radius of electron component in vertical electromagnetic air showers produced by photons with $E = 10^{18} - 10^{22}$ eV at the observational level of $t = 890$ g/cm^2 are shown with and without taking into account the UHE-effects (LPM and GMF).

It can be seen from Figure 1 that the interaction with GMF counteracts the LPM effect by converting the primary photon into a bunch of synchrotron photons with energies effectively below the threshold of the LPM effect (see [11] for details).

The root mean square radius of electron component $r_{\rm ms}$ is defined in a standard way as a second moment of electron lateral distribution function (LDF):

$$r_{\rm ms}(E,t) = \left(\frac{2\pi}{N(E,t)} \int_0^\infty r^2 \rho(r; E,t) r dr \right)^{1/2}, \tag{1}$$

where $\rho(r; E, t)$ is the local particle density at radial distance r from the core position in shower with primary energy E observed at the depth t in the atmosphere.

According to the scaling approach [14], when radial distance is rescaled with respect to the variable $x = r/r_{\rm ms}$, the shape of electron LDF doesn't depend on primary energy and shower age for $x = 0.05 - 20$:

$$\rho(r; E, t) = \frac{N(E,t)}{r_{\rm ms}^2} F\left(\frac{r}{r_{\rm ms}}\right). \tag{2}$$

Here $F(x)$ is the invariant function. Thus, $r_{\rm ms}$ could be considered as a single parameter describing variations of the shape of electron LDF with energy and observation depth. It is important that, according to our calculations, the scaling property (2) remains valid up to the highest energies regardless of what UHE-effects were taken into account. Numerical values of the ratios of $r^*_{\rm ms}$ with taken into account UHE-effects to $r_{\rm ms}$ are presented in Table 1.

Later it was shown [15–17] that the scaling property of the same type extends to electron LDF in extensive air showers generated by primary nuclei.

In order to examine the possibility for effective primary particle discrimination with identifications of photons on average and event-by-event basis we have simulated vertical air showers generated by photons, protons and iron nuclei of 10^{18} eV and evaluated root mean square radii from lateral distributions of electrons and also depths of maximum $t_{\rm max}$. Simulations were performed using CORSIKA v.7.4100 [18] with EPOS LHC v.3400 (FLUKA 2011.2c.2) hadronic interaction models. In order to get reliable data on local particle densities at very large distances from the shower core the thinning level and particle weight limit were set as $\varepsilon_{\rm th} = 10^{-8}$ and $\omega = 10^2$ respectively.

The distributions of $r_{\rm ms}$ and $t_{\rm max}$ for different primaries are displayed in Figure 2. 200 showers were included in each data set. One can see that the root mean square radius of electron component is generally more sensitive to the primary particle type in comparison with the depth of shower maximum.

Table 1. The behavior of the ratio of $r^*_{\rm ms}$ with taken into account effects to $r_{\rm ms}$ without effects for vertical cascade on the observation level 890 g/cm^2.

E_γ, eV	$\frac{r_{\rm ms}^{LPM}}{r_{\rm ms}}$	$\frac{r_{\rm ms}^{GMF}}{r_{\rm ms}}$	$\frac{r_{\rm ms}^{LPM+GMF}}{r_{\rm ms}}$
10^{18}	0.987	1.000	0.987
10^{19}	0.950	1.002	0.953
$3 \cdot 10^{19}$	0.924	1.003	0.931
10^{20}	0.902	1.053	1.023
$3 \cdot 10^{20}$	0.897	1.172	1.142
10^{21}	0.909	1.207	1.176
$3 \cdot 10^{21}$	0.930	1.255	1.223
10^{22}	0.958	1.309	1.274

3. Conclusion

Using the numerical solution of adjoint cascade equations we analysed characteristics of air showers produced by ultra-high energy photons taking into account the Landau-Pomeranchuk-Migdal effect and the interaction of UHE photons and electrons with the geomagnetic field. An extended analysis has been made considering separately the LPM and GMF effects with respect to the scaling formalism for lateral distributions of electrons. It was obtained that

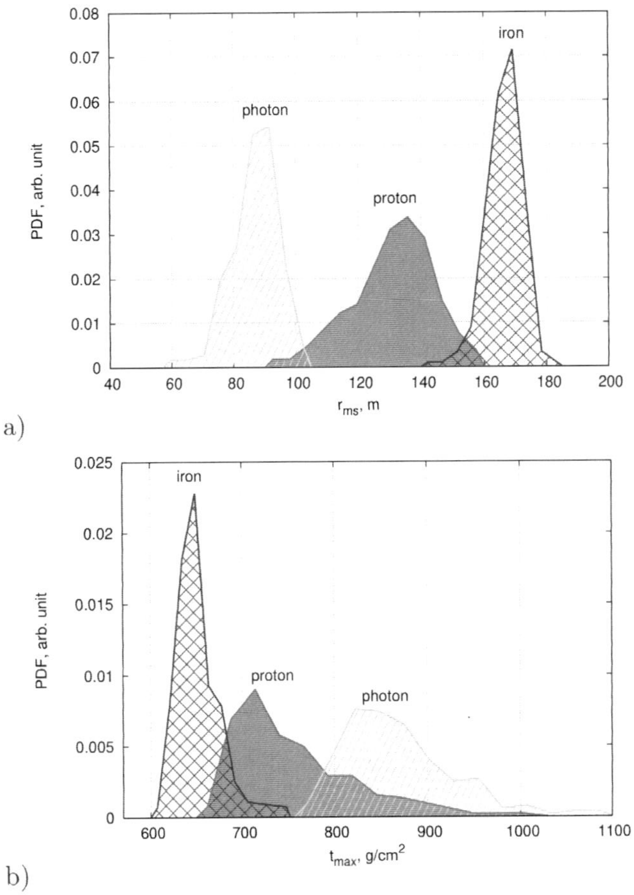

Figure 2. Distributions of primary particle discriminating observables: root mean square radius of electron component r_{ms} at sea level (a) and depth of shower maximum $t_{\max}$ (b) for vertical air showers initiated by photons (green), protons (red) and iron nuclei (blue) with primary energy 10^{18} eV.

one-parametric scaling description of the lateral distribution of electrons remains valid up to the highest energies. CORSIKA Monte-Carlo simulations of air showers initiated by primary photons and nuclei showed that using root mean square radius of electron component under the framework of scaling approach allows effective primary particle type discrimination from surface detectors data of largest ground-based air shower arrays.

Acknowledgments

This work was supported by Russian Foundation for Basic Research (grant 16-02-01103 a).

References

[1] Aloisio R, Berezinsky V and Blasi P 2014 *Journal of Cosmology and Astroparticle Physics* **2014** 020
[2] Aloisio R 2017 *Progress of Theoretical and Experimental Physics* **2017**
[3] Risse M, Homola P 2007 *Modern Physics Letter A* **22** No11 749
[4] Homola P 2018 *CERN Proceedings, [S.l.]* **1** 283
[5] Karpikov I, Rubtsov G, Zhezher Y 2018 *Phys. Rev. D* **98** 103002
[6] Landau L D and Pomeranchuk I J 1953 *Dokl. Akad. Nauk SSSR* **92** 535
[7] Landau L D and Pomeranchuk I J 1953 *Dokl. Akad. Nauk SSSR* **92** 735
[8] Migdal A B 1956 *Phys. Rev.* **103** 1811
[9] McBreen B and Lambert C J 1981 *Phys. Rev.* D **24** 2536
[10] Goncharov A I 1991 *Spatial characteristics of electron-photon showers in the Earths atmosphere* (PhD Thesis, Tomsk) 146 p. (in Russian)
[11] Plyasheshnikov A V and Aharonian F A 2002 *J. Phys. G: Nucl. Part. Phys.* **28** 267

[12] Goncharov A et al 2003 Proc. 28 ICRC (Tsukuba) **1** 575
[13] Goncharov A I, Lagutin A A, Plyasheshnikov A V and Serebryakova T L 2005 *Izv. RAN. Ser. Fiz.* **69** 369
[14] Lagutina A, Plyasheshnikov A, Goncharov A 1997 *Nucl. Phys. Proc. Suppl.* 60 Issue 3 161
[15] Lagutin A, Misaki A, Raikin R 1997 *Proc. 25 ICRC (Durban)* **6** 285
[16] Lagutin A A and Raikin R I 2001 *Nucl. Phys. Proc. Suppl.* **97** 274
[17] Lagutin A A et al. 2002 J. Phys. G **28** 1259
[18] Heck D et al. (Report FZKA 6019. Forschungszentrum Karlsruhe, 1998)

IOP Conf. Series: Journal of Physics: Conf. Series **1181** (2019) 012089 doi:10.1088/1742-6596/1181/1/012089

Do LHC data contradict superhigh-energy cosmic-ray coplanarity of most energetic particles?

R Mukhamedshin[1] and T Sadykov[2]

[1] Institute for Nuclear Research of RAS, Moscow, Russia

[2] Institute for Physics and Technology, Almaty, Kazakhstan

E-mail: rauf_m@mail.ru

Abstract. To research for a coplanarity of most energetic subcores of γ-ray–hadron families observed in cosmic-ray experiments at $E_0 \gtrsim 10^{16}$ eV ($\sqrt{s} \gtrsim 5$ TeV), a new phenomenological FANSY 2.0 model is designed. The model makes it possible to simulate interactions for both traditional and coplanar hadron generation options, reproduces LHC data, including high-η and x_F data on γ-rays and neutrons. An experiment at the LHC is proposed to test the model.

1. Introduction

The paper considers one of phenomena observed in mountain-based and stratospheric X-ray–emulsion chamber (XREC) experiments, namely, a strong azimuthal effect that manifests itself as a tendency to a coplanarity of most energetic subcores of γ-ray–hadron ($\gamma - h$) families, i.e., groups of high-energy ($\gtrsim 1$ TeV) particles in EAS cores initiated mainly by protons and light nuclei of the primary cosmic radiation (PCR) with energies $E_0 \gtrsim 10^{16}$ eV.

The effect has been found by the *Pamir* Collaboration [1, 2, 3, 4, 5] and observed in other experiments [6, 7, 8, 9] with different XREC types, i.e., so-called "carbon", "lead", "iron" (C, Pb, Fe) XRECs. Five sets of experimental data were accumulated. This is, first of all, data of mountain-based *Pamir* Collaboration's C- and Pb-XRECs on $\gamma - h$ families with observed energies of γ-rays and $e^{\pm}$, $\sum E_\gamma \geq 700$ TeV. Close results at $\sum E_\gamma \geq 500$ TeV were obtained by Mt.Canbala Collaboration using Fe-XRECs [6]. Besides, only two stratospheric events with $\sum E_\gamma > 1$ PeV (the *Strana* [7, 8] and *JF2af2* [9]) have been observed at very high altitudes, and both these events demonstrate a very high coplanarity of most energetic particles (MEP).

A new parameter λ_N was proposed [1] to calculate the degree of a so-called *alignment* of N point-like tracks on a target plane, namely, $\lambda_N = 1$ for N points aligned along a straight line and $\lambda_N \approx -1/(N-1)$ in the isotropic cases.

Table 1 presents experimental data sets and types of used XRECs; criteria used to select experimental coplanar events (λ_N and $\sum E_\gamma$); number of coplanar families, $N_{\rm exp}^{\rm copl}$, and total family number $N_{\rm tot}$; fraction of coplanar families, $F_{\rm exp}^{\rm copl}$; references (Columns 1–6, respectively). Simulated background fractions of coplanar γ-ray families, $F_{\rm sim}^{\rm copl}$ obtained with the use of the QGSM-based FANSY 1.0 model [12] are given in Column 7. Although the statistical security of experimental data is not too high, each of these data sets shows an phenomenon excess over the simulated value. Both stratospheric events are unique in energy and coplanarity.

This phenomenon cannot be studied in EAS experiments because of their poor coordinate and energy resolution. Fortunately, it could be investigated in LHC experiments (Sect. 4).

The probability for the total set of these experimental results to be produced by cascade fluctuations is much lower than $\sim 10^{-10}$ [10, 11, 12].

The phenomenon was initially interpreted as a manifestation of coplanar generation of MEPs ($x_{Lab} = E/E_0 \gtrsim 0.01 - 0.05$) characterized by relatively large transverse momenta, $p_t^{copl} \gtrsim 1$ GeV/c. However, as regards γ-ray–hadron families, an accurate measurement of p_t^{copl} values was not made. The average ratio of p_t components lying in the coplanarity plane, p_t^{copl}, to p_t components directed normally to the coplanarity plane, p_t^{norm}, was only estimated, namely, $\langle p_t^{copl}/p_t^{norm}\rangle \sim 10$ [3]. It is also very important that the coplanar particle generation (CPG) is characterized by a relatively large cross section (comparable with σ_{prod}^{p-air}) [10].

A few theoretical ideas were proposed to explain this phenomenon [14, 15, 16], which include large p_t as an almost mandatory element. Besides, the most extraordinary and speculative idea was proposed which assumes a connection between the coplanarity phenomenon and the hypothesis of so-called "crystal world" [17, 18] which implies a gradual reduction of space dimensionality from three to two dimensions at sufficiently high energies and entails p_t localization in some plane and a reduction of transverse momenta directed perpendicularly to this plane. Unfortunately, all of these ideas are not quantitatively developed. In any case, this experimental result shows that strong "forward-physics" interactions at superhigh energies are not well-described with the concept used by QGS models.

The coplanar generation of MEPs can be phenomenologically realized in the framework of two modes, i.e., initial generation of (a) quark-gluon strings with the following hadron production; (b) an energetic leading system (diffraction cluster, e.g.) with a lower multiplicity and higher energies of hadrons. The first approach is considered in this work.

The phenomenological model FANSY 1.0 [11, 12] was developed a dozen of years ago. It described main characteristics of γ-ray families and helped to understand general features of the experimental coplanar events. However, FANSY 1.0 cannot properly describe some features of γ-ray families [19] and LHC data. Therefore, a new phenomenological model FANSY 2.0 [21] is designed. The model makes it possible to simulate both QGSM-based traditional interactions and coplanar particle generation (QGSJ and QGSCPG versions, respectively). All interaction characteristics (excluding azimuthal ones) simulated with QGSJ and QGSCPG versions, are similar. Both QGSJ and QGSCPG versions become identical at $\sqrt{s} \lesssim 2$ TeV. They reproduce high-$x_{\rm F}$ and high-η data on γ-rays and neutron-like hadrons accumulated by the LHCf experiment (Sect. 3) as well as low-rapidity data. FANSY 2.0 reproduces a lot of low-energy data as well.

A detailed comparison of experimental data with results of FANSY 2.0 QGSJ simulation at energies $\sqrt{s} = 17$ GeV – 13 TeV are given in Ref. [21].

Sect. 2 presents a short discussion on simulation problems related to the initial concept of coplanarity features and describes shortly the FANSY 2.0 QGSCPG's algorithm of hadron coplanarization. Sect. 3 compares high-$x_{\rm F}$ and high-η experimental and simulated results. Sect. 4 considers proposed experiment at the LHC.

2. Coplanarity simulation

2.1. Coplanarity concepts

Simulations using tentative FANSY 2.0 QGSCPG versions show the following problem. The initial concept exploited earlier [11, 12] qualitatively explains the observed coplanarity with assuming *high* p_t^{copl} of MEPs in a *coplanarity* plane. However, a significant p_t^{copl} growth suppresses $d\sigma/dy$ and $d\sigma/d\eta$ cross sections of hadron generation at highest $|y|$ and $|\eta|$ values and creates robust peaks at $2 \lesssim |y,\eta| \lesssim 4$ which are contrary to LHC data. This is an unsoluble problem for all considered high-p_t^{copl} QGSCPG versions.

Table 1. Summary of experimental data. Columns: 1. Data set and type of used XRECs; 2–3. Criteria for selection of coplanar events; 4. Number of coplanar families, $N_{\rm exp}^{\rm copl}$, and total family number $N_{\rm tot}$; 5. Fraction of coplanar families, $F_{\rm exp}^{\rm copl}$. 6. References. 7. Simulated background fraction of coplanar γ-ray families, $F_{\rm sim}^{\rm copl}$ [12].

Experimental data set	Selection criteria λ_N	$\sum E_\gamma$,TeV	$N_{\rm exp}^{\rm copl}$ / $N_{\rm tot}$	$F_{\rm exp}^{\rm copl}$	Ref.	$F_{\rm sim}^{\rm copl}$
Pamir (Pb)	$\lambda_4 \geq 0.8$	≥ 700	6 / 14	0.43 ± 0.13	[3]	0.041 ± 0.004
Pamir (C)	$\lambda_4 \geq 0.8$	≥ 700	5 / 35	0.15 ± 0.05	[5]	0.049 ± 0.004
Mt.*Kanbala* (Fe)	$\lambda_3 \geq 0.8$	≥ 500	6 / 12	0.50 ± 0.13	[6]	0.180 ± 0.010
the *Strana*	$\lambda_4 = 0.99$	$\simeq 1500$	1	1	[13]	$(26 \pm 3) \cdot 10^{-4}$
the *JF2af2*	$\lambda_4 = 0.998$	$\simeq 1500$	1	1	[9]	$(9 \pm 3) \cdot 10^{-4}$

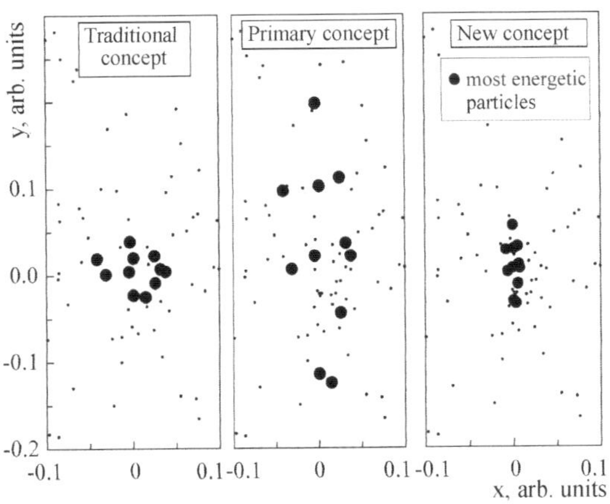

Figure 1. Particle tracks on a target plane in cases of traditional particle generation (left), initial large-p_t CPG model (mid) and new traditional-p_t CPG concept (right).

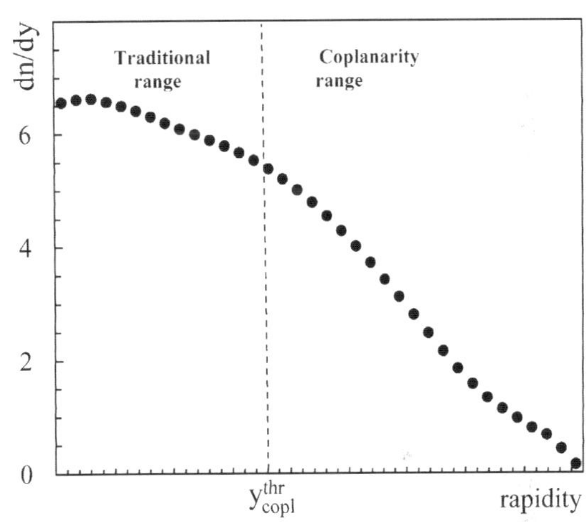

Figure 2. Qualitative scheme of traditional and coplanarity ranges of dn/dy distribution.

However, the coplanarity is actually observed in cosmic-ray experiments. Is it possible to reconcile this result with the LHC data?

Simulation shows that an agreement of LHC data and idea of coplanar generation becomes real only using a new concept assuming some *decrease* of p_t components directed *normally* to the coplanarity plane, while p_t becomes (almost) invariant simultaneously with the appearance of azimuth effects. It could be provided with some turning of MEPs' $\vec{p}_t$ vectors with no change of its absolute value.

Figure 1 shows three versions of observed tracks of particles, generated in the same imaginary interaction and arriving on a target plane placed at some distance from the interaction point: (1) traditional QGSM-like interaction (left); (2) initial-concept coplanar interaction with high-p_t^{copl} (mid); (3) new-concept coplanar interaction with traditional p_t^{copl} values (right). MEPs are shown with large black symbols. The geometric scale is given in arbitrary units.

2.2. Coplanarity simulation

All characteristics of the main bulk of secondary particles are primarily simulated with the traditional QGSJ option through the QGS fragmentation. After this, c.m.s. longitudinal momenta of hadrons are listed in decreasing order of magnitude, i.e., $p_{max} > p_1 > p_2 > p_3... > -p_{max}$, where $p_{max} = \sqrt{s/4 - m_p^2}$. Rapidities of these hadrons are denoted as y_1, y_2, y_3....

Figure 2 shows a schematic drawing of traditional and coplanarity ranges of dn/dy distributions. A "coplanarization" algorithm rotates the $\overrightarrow{p}_t$ vector of each of MEPs with $|y| > y_{thr}^{copl}$ towards the coplanarity plane along the shortest path, if the total energy of secondary particles in their c.m.s., $\sqrt{s^{eff}}$, is higher than a fixed value, $\sqrt{s_{thr}^{eff}}$. Here $y_{thr}^{copl} = y_2 - \Delta y^{copl}$, where a free parameter $\Delta y^{copl} \approx 3 - 5$, y_{thr}^{copl} varies from $3 - 5$ for hadrons with $p_z > 0$ due to fluctuations of hadron rapidity values. Similar procedure is carried out for hadrons with $p_z < 0$. Fluctuations are mainly determined by variations of $\sqrt{s^{eff}}$ and y_2 values. The coplanarity-plane orientation is determined by the total $\overrightarrow{p}_t$ of two leading hadrons. The azimuthal-angle distribution of $\overrightarrow{p}_t^{copl}$ of MEPs with respect to the coplanarity plane is described by a Gaussian distribution with a standard deviation σ^{copl}. Results are obtained at $\sqrt{s_{thr}^{eff}} = 250$ GeV/c, $\Delta y^{copl} = 4$, $\sigma^{copl} = 0.05$ rad, $\langle y_2 \rangle = 7.15$, $\langle y_{thr}^{copl} \rangle = 3.15$.

The described coplanarization algorithm is highly arbitrary and can be used as a zero-order approximation tool only to estimate the influence of CPG processes on interaction characteristics.

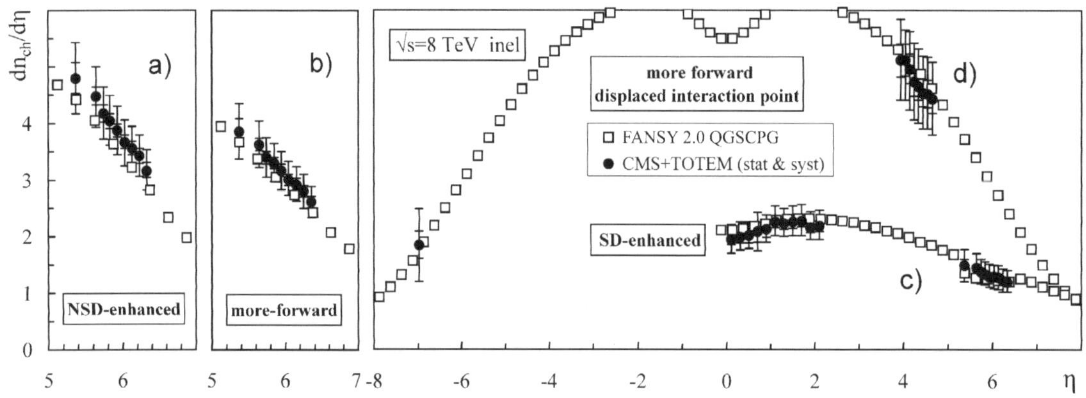

Figure 3. CMS+TOTEM and FANSY 2.0 QGSCPG $dn_{ch}/d\eta$ data at $\sqrt{s} = 8$ TeV. (a) "NSD-enhanced" events: $n_{ch} \geq 1$ at $-6.5 < \eta < -5.3$ **and** $5.3 < \eta < 6.5$; (b) "more-forward" events: $n_{ch} \geq 1$ at $-6.5 < \eta < -5.3$ **or** $5.3 < \eta < 6.5$; (c); "SD-enhanced" events (lower symbols): $n_{ch} \geq 1$ **only** in the ranges $-6.5 < \eta < -5.3$ **or only** $5.3 < \eta < 6.5$ (c); "more-forward" data (upper symbols): $n_{ch} \geq 1$ at $-7.0 < \eta < -6.0$ **or** $3.7 < \eta < 4.8$.

3. LHC data and FANSY 2.0 QGSCPG results

As the coplanarity is associated with MEPs, only high η and x_F data of the CMS+TOTEM [22] on charged particles at $\sqrt{s} = 8$ TeV and LHCf experiments [20] at $\sqrt{s} = 7$ TeV are considered below.

Figure 3 shows CMS+TOTEM [22] and FANSY 2.0 QGSCPG $dn_{ch}/d\eta$ data at $\sqrt{s} = 8$ TeV.

Figure 4 shows FANSY 2.0 QGSCPG and LHCf "neutron" ($n, \bar{n}$, Λ^0, $\bar{\Lambda}^0$, K_L^0) $d\sigma/dE$ cross sections.

Figure 5 shows FANSY 2.0 and LHCf γ-ray n_γ/N_{inel}/GeV energy spectra [20] at $\sqrt{s} = 7$ TeV. The upper spectrum is multiplied by 10.

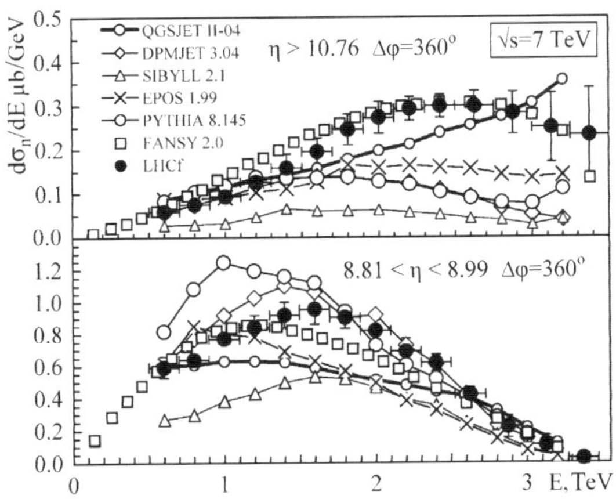

Figure 4. LHCf and FANSY 2.0 "neutron" $d\sigma/dE$ cross sections at $\eta > 10.76$ (top) and $8.81 < \eta < 8.99$ (bottom), $\Delta\phi = 360°$.

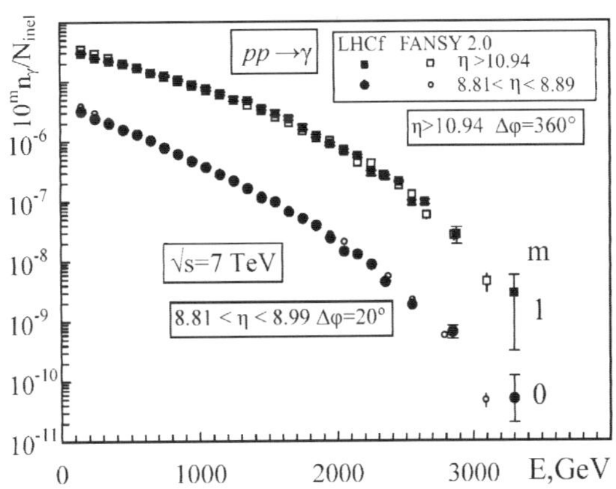

Figure 5. LHCf and FANSY 2.0 γ-ray n_γ/N_{inel}/GeV energy spectra at $\eta > 10.94$, $\Delta\phi = 360°$ (upper symbols) and $8.81 < \eta < 8.99$, $\Delta\phi = 20°$ (lower symbols).

4. Search for coplanarity at the LHC

The LHCf detector aimed at the study of high x_F and η values is capable of detecting only one neutral particle for several dozens of interactions. This makes it impossible to study azimuthal correlations of particles. Let us estimate potentialities of the CASTOR detector, which seems to be more promising, taking into account that much more detailed simulation is undoubtedly required. CPG effects could manifest themselves in its operation pseudorapidity range, i.e. $5.25 < \eta < 6.5$. The detector consists of 16 segments. Particles are assumed to be detected if their η values are in the range $5.3 < \eta < 6.5$ and they do not fall into the vertical gap.

Figure 6 shows a simplified cross section scheme of CASTOR and an example of detection of one coplanar interaction. Black circles show tracks of particles. The larger the circle size, the higher the energy of the particle. High-energy particles show a tendency to some coplanarity. Low-energy particles form a more or less azimuthally isotropic background.

To analyze observed interactions, energy values "measured" in each of the segments, E_i, are used. Events with total energy $\sum E_i \geq 1$ TeV are only analyzed, in which the number of segments with measured energy $E_i > 100$ GeV is $N_s \geqslant 2$. The first number is assigned to a segment with the maximum release of energy, E_{max}. Other segments are numbered clockwise. Here and below, $E_1 = E_{max}$, $E_{\rm copl} = E_1 + E_9$; $E_{tr} = E_5 + E_{13}$, i.e., E_{tr} is the energy measured in 9th and 13th segments, perpendicular to the first segment. A simple parameter is applied, namely, $\varepsilon_{\rm copl} = E_{\rm copl}/(E_{\rm copl} + E_{tr})$. The degree of event coplanarity is maximum at $\varepsilon_{\rm copl} = 1$. Figure 7 shows $d\omega/d\varepsilon_{\rm copl}$ probability distributions (normalized to unity) reproduced with FANSY 2.0 QGSJ and QGSCPG versions, which obviously, predict different probability magnitudes to observe coplanar interactions at $\varepsilon_{\rm copl} \to 1$.

5. Conclusion

The FANSY 2.0 model is developed to study superhigh-energy cosmic-ray "forward physics" interactions in general and the coplanar particle generation (CPG) phenomenon in particular.

FANSY 2.0 includes the traditional QGSM-based QGSJ version and unconventional QGSCPG one with a CPG mode. Results of the QGSCPG version do not contradict to LHC data if to move from the primary concept of *growth* of p_t of MEPs in the coplanarity plane to a new concept of *reduction* of transverse momentum component directed normally to the coplanarity plane so that the average p_t values remain traditional.

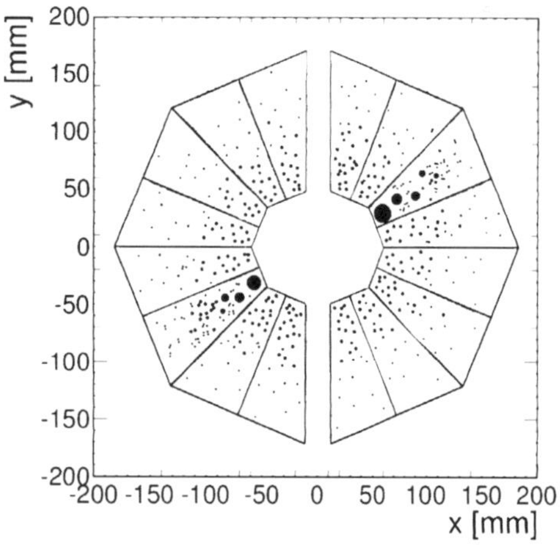

Figure 6. CASTOR's simplified cross section with an example of a coplanar interaction.

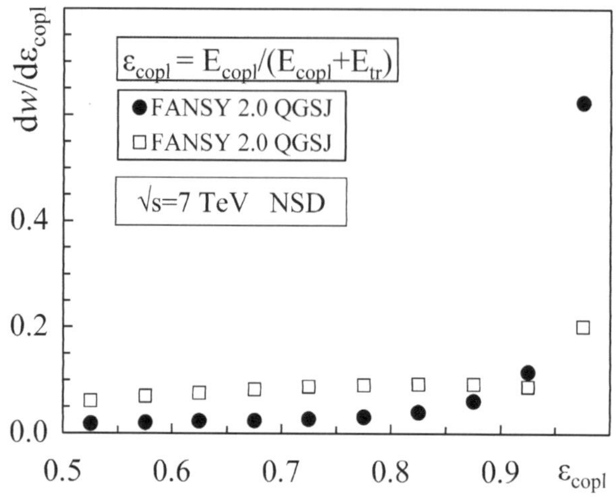

Figure 7. Probability $d\omega/d\varepsilon_{\rm copl}$ distributions predicted by FANSY 2.0 QGSJ and QGSCPG versions.

It is proposed to test the FANSY 2.0 QGSCPG model with using the CASTOR detector.

References

[1] Borisov A *et al* (Pamir Collaboration) 1986 in *Proc. 4th Int. Symp. on Very High Energy Cosmic Ray Interactions (Beijing)* p 4-29
[2] Ivanenko I P *et al* 1992 *JETP Lett.* **50** 2125
[3] Kopenkin V V *et al* 1995 *Phys. Rev.* D **52** 2766
[4] Pamir Collaboration 1989 *Preprint INP MSU 89-67/144*
[5] Borisov A S *et al* 2001 *Nucl. Phys.* B (Proc. Suppl.) **97** 118
[6] Xue L *et al* 1999 *Proc. 26th Int. Cosmic Ray Conf. (Salt Lake City)* vol 1 p 127.
[7] Apanasenko A V *et al* 1977 *Proc. 15th Int. Cosmic Ray Conf. (Plovdiv)* vol 7 p 220
[8] Managadze A K *et al* 2001 *Proc. 27th Int. Cosmic Ray Conf. (Hamburg)* vol 1 p 1426
[9] Capdevielle J N 1988 *J.Phys.* **G 14** 503
[10] Mukhamedshin R A 2005 *J. High Energy Phys.* JHEP05(2005)049
[11] Mukhamedshin R A 2009 *Nucl. Phys.* B (Proc. Suppl.) **196C** 98
[12] Mukhamedshin R A 2009 *Eur.Phys.J.* C **60** 345
[13] Managadze A K *et al* 2005 *Proc. 29th Int. Cosmic Ray Conf. (Pune)* vol 9 p 81
[14] Wibig T 2000 Alignment in hadronic interactions *preprint arXiv:hep-ph/0003230*
[15] Royzen I I 1994 *Mod. Phys. Lett.* **A9** no.38 3517
[16] Capdevielle J N *et al* 2003 *Proc. 28th Int. Cosmic Ray Conf. (Tsukuba)* vol 1 p 1599
[17] Anchordoqui L A *et al* 2010 Vanishing Dimensions and Planar Events at the LHC *preprint arXiv:1003.5914 [hep-ph]*
[18] Anchordoqui L A *et al* 2011 Searching for the Layered Structure of Space at the LHC *preprint arXiv:1012.1870v2 [hep-ph]*
[19] Kempa J, Mukhamedshin R and Tamada M 2013 *J. Phys.: Conf. Series* **409** 012109
[20] Adriani O *et al* 2011 *Phys. Lett.* B **703** 128
[21] Mukhamedshin R A FANSY 2.0: a Monte Carlo tool for study of superhigh-energy cosmic-ray interactions. Proton-proton interactions (sent to *Eur.Phys.J.* C)
[22] The CMS collaboration, The TOTEM Collaboration 2014 *Eur.Phys.J.* C **74** 2053

IOP Conf. Series: Journal of Physics: Conf. Series **1181** (2019) 012090 doi:10.1088/1742-6596/1181/1/012090

Possible approach to the analysis of nucleus-nucleus interactions at very high energies

A A Petrukhin, E S Sozinov and V V Shutenko

National Research Nuclear University MEPhI
(Moscow Engineering Physics Institute), Moscow, 115409, Russia

E-mail: AAPetrukhin@mephi.ru

Abstract. Interactions of various nuclei in the frame of a simple geometrical approach are considered with the purpose to determine the number of nucleons participating in these interactions. It is shown that on average a part of interacting nucleons does not exceed one quarter of the total number of nucleons of both nuclei.

1. Introduction

Nucleus-nucleus interactions at very high energies (> 10^{15} eV in cosmic rays, > 1 TeV in colliders) are interesting and important from at least two points of view. Firstly, in these interactions the transition from nucleon-nucleon or/and quark-quark to interactions of many nucleons or/and quarks appears. Secondly, nucleus-nucleus interactions play an important role in cosmic ray investigations since cosmic rays consist mainly of nuclei (~ 60% at energies < 10^{15} eV), the fraction of which is increased at higher energies.

But namely at energies about 10^{16} eV, which correspond to several TeV in the center-of-mass system, various unusual events as in cosmic rays so in LHC experiments are observed [1, 2]. Numerous attempts of explanation of these results in the frame of existing models did not lead to success. And only a model of production of quark-gluon blobs with a large orbital momentum allows to explain in principle all observed events and phenomena from a single point of view [3, 4]. But their quantitative explanation requires more detailed information about nucleus-nucleus interactions.

In this paper a simple approach to nucleus-nucleus collisions is considered. In this approach, it is assumed that the nucleus is a hard sphere of radius R with a constant density. Since in the considered model the interactions between nucleons are not taken into account and their concentration in nucleus volume is constant, collisions between nuclei can be considered in a pure geometrical approach. For transition to the number of nucleons in interacting parts of nuclei it is sufficient to multiply the volume by the density of nucleons.

The basic parameter of collisions of two nuclei is the impact parameter b: the distance between nuclei centers. Its value determines the volume of interacting parts of two nuclei. It can change between 0 and (R_1+R_2), for equal nuclei up to $2R_0$. Correspondingly, the total number of interacting nucleons is equal to $\Delta A=\Delta A_1+\Delta A_2$. Below, two cases are discussed: collisions of two equal nuclei (A_0) and collisions of two different nuclei (A_1 and A_2).

2. Collisions of equal nuclei

The general scheme of interacting nuclei with the number of nucleons A_0 and radius R_0 moving along z-axis is shown in figure 1(a). Though colliding nuclei are spheres, interacting volume is described by two cylindrical surfaces of radius R. In figure 1(b) a nucleus without interacting part is shown. Therefore for the number of interacting nucleons we can obtain:

$$\Delta A = 2A_0 \frac{\Delta V_1}{V_1}, \text{ where } \Delta V(b) = 2\int_{-x_0}^{x_0} dx \int_{b-\sqrt{R_0^2-x^2}}^{\sqrt{R_0^2-x^2}} \sqrt{R_0^2 - x^2 - y^2}\, dy, \quad (1)$$

and $x_0 = \frac{\sqrt{4R_0^2-b^2}}{2}$ is the point of intersection of the circles (see figure 1).

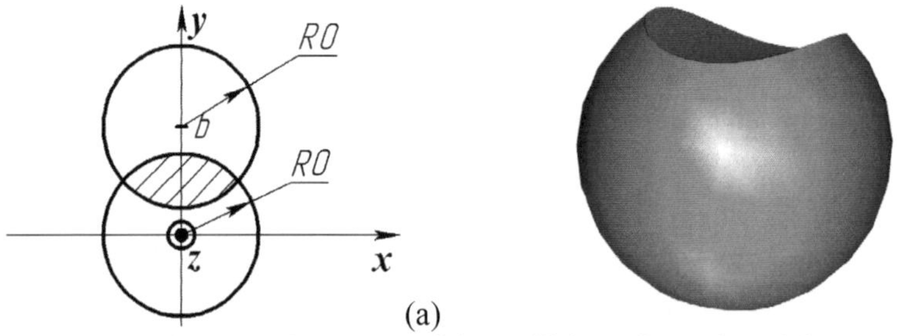

(a) (b)

Figure 1. Geometry of the collision of equal nuclei.

The dependence $\Delta A(b)$ calculated by the formula (1) for $A_0 = 1$ *(1 is a normalization value, but not one nucleon!)*, $R_0 = 1$ is presented in figure 2(a). It is interesting, that one half of colliding nucleons $\Delta A = A_0$ will interact when $b = 0.86R_0$, and $\Delta A = 0.8A_0$ when $b = R_0$.

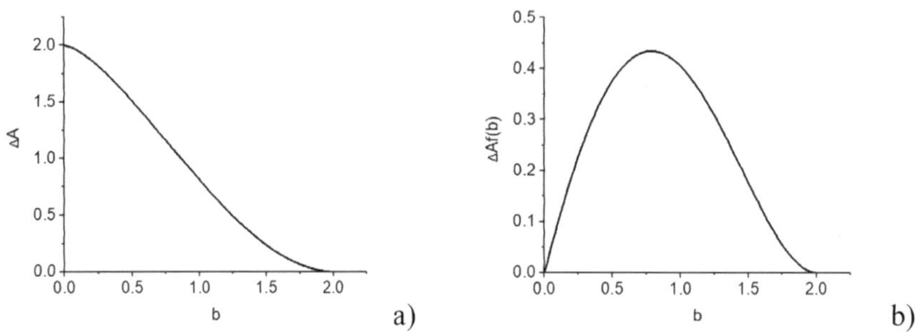

Figure 2. Curves corresponding to collision of equal nuclei for $A_0 = 1$, $R_0 = 1$.

In order to calculate the average quantity of interacting nucleons we assume that the probability distribution law depends on the parameter b only. In this case the effective cross-section is

$$d\sigma = 2\pi \cdot b \cdot db \quad (2)$$

and the total cross-section corresponds to the integral from 0 to $2R_0$; that is

$$\sigma_{tot} = \pi \cdot (2R_0)^2 = 4\pi \cdot R_0^2. \quad (3)$$

Then we can determine:

the probability $dP = \frac{d\sigma}{\sigma_{tot}} = \frac{b}{2R_0^2} db$, (4)

the probability density $f(b) = \frac{b}{2R_0^2}$ (5)

as a function of the impact parameter b.

The dependence of the number of interacting nucleons $\Delta Af(b)$ on the impact parameter is shown in figure 2(b). According to figure 2(b), the most probable value of the impact parameter is

$b = 0.78R_0$. The value ΔA corresponding to it is $\Delta A(0.78R_0) = 1.12\,A_0$. Using the obtained dependence, one can calculate the average value $< \Delta A >$:

$$< \Delta A > = \int \Delta A dP = \int_0^{2R_0} \Delta A f(b) db. \tag{6}$$

After calculations, we will get that

$$< \Delta A > = \frac{A_0}{2}.$$

So, on average only one quarter of nucleons of two colliding nuclei participate in interactions.

3. Collisions of different nuclei

Now we consider collisions of nuclei with different radii R_1 and R_2, ($R_1 > R_2$ for definiteness) depending on the parameter b. There are three areas:

$$0 < b < R_1 - R_2 < b < \sqrt{R_1^2 - R_2^2} < b < R_1 + R_2 \tag{7}$$

3.1. *The first area of interaction*: $0 < b < R_1 - R_2$ (figure 3).

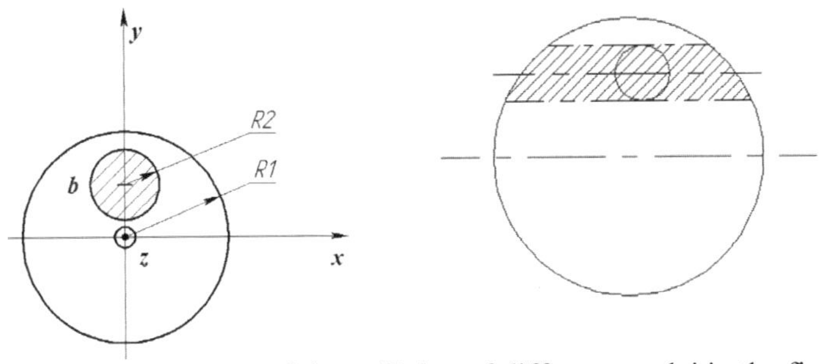

Figure 3. Geometry of the collision of different nuclei in the first area $(0 < b < R_1 - R_2)$.

We define two numbers of particles A_1 and A_2, which correspond respectively to nuclei of radius R_1 and R_2. It can be seen from figure 3 that the number of colliding particles is equal to:

$$\Delta A = A_2 + A_1 \frac{\Delta V_1}{V_1}, \tag{8}$$

where ΔV_1 is the volume of intersection of a cylinder of radius R_2 with a sphere of radius R_1, and

$$\Delta V_1 = 2 \int_{-R_2}^{R_2} dx \int_{b-\sqrt{R_2^2 - x^2}}^{b+\sqrt{R_2^2 - x^2}} \sqrt{R_1^2 - x^2 - y^2}\, dy\,. \tag{9}$$

3.2. *The second area of interaction*: $R_1 - R_2 < b < \sqrt{R_1^2 - R_2^2}$ (figure 4).
In this case, the number of particles:

$$\Delta A = \frac{A_2}{2} + A_2 \frac{\Delta V_2}{V_2} + A_1 \frac{\Delta V_1}{V_1}, \tag{10}$$

where

$$\Delta V_1 = 2 \int_{b-R_2}^{y_1} dy \int_{-x_1}^{x_1} \sqrt{R_1^2 - x^2 - y^2} dx + 2 \int_{y_1}^{R_1} dy \int_{-x_2}^{x_2} \sqrt{R_1^2 - x^2 - y^2} dx,$$

$$\Delta V_2 = 2 \int_b^{y_1} dy \int_{-x_1}^{x_1} \sqrt{R_2^2 - x^2 - (y-b)^2} dx + 2 \int_{y_1}^{R_1} dy \int_{-x_2}^{x_2} \sqrt{R_2^2 - x^2 - (y-b)^2} dx, \tag{11}$$

$$y_1 = \sqrt{R_1^2 - R_2^2 + \frac{(R_1^2 - R_2^2 - b^2)^2}{4b^2}},\ x_1 = \sqrt{R_2^2 - (y-b)^2},\ x_2 = \sqrt{R_1^2 - y^2}\,. \tag{12}$$

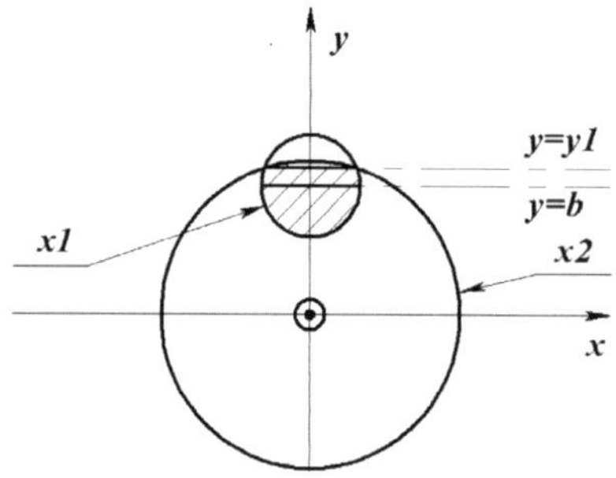

Figure 4. Geometry of the collision of different nuclei in the second area ($R_1 - R_2 < b < \sqrt{R_1^2 - R_2^2}$).

3.3 *The third area of interaction:* $\sqrt{R_1^2 - R_2^2} < b < R_1 + R_2$ (figure 5).

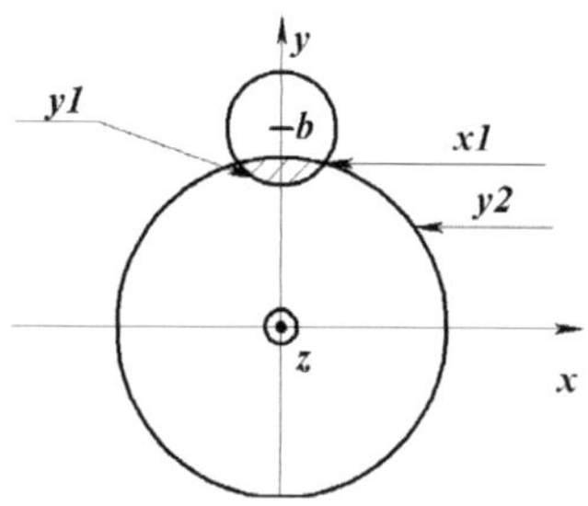

Figure 5. Geometry of the collision of different nuclei in the third area ($\sqrt{R_1^2 - R_2^2} < b < R_1 + R_2$).

Here the number of particles

$$\Delta A = A_1 \frac{\Delta V_1}{V_1} + A_2 \frac{\Delta V_2}{V_2} \tag{13}$$

where ΔV_1 and ΔV_2 are the volumes, corresponding to intersections of spheres of radii R_1 and R_2 with cylinders of radii R_2 and R_1, respectively:

$$\Delta V_1 = 2\int_{-x_1}^{x_1} dx \int_{b-\sqrt{R_2^2-x^2}}^{\sqrt{R_1^2-x^2}} \sqrt{R_1^2 - x^2 - y^2}\, dy\,, \tag{14}$$

$$\Delta V_2 = 2\int_{-x_1}^{x_1} dx \int_{b-\sqrt{R_2^2-x^2}}^{\sqrt{R_1^2-x^2}} \sqrt{R_2^2 - x^2 - (y-b)^2}\, dy\,, \tag{15}$$

where x_1 is the point of intersection of the circles on the plane:

$$\mathrm{x}_1 = \sqrt{R_2^2 - \frac{\left(R_1^2 - R_2^2 - b^2\right)^2}{4b^2}}. \tag{16}$$

The curve of $\Delta A(b)$ for $R_1 = 1,\ R_2 = 0.5,\ A_1 = 1,\ A_2 = 0.5^3$ is shown in figure 6(a).

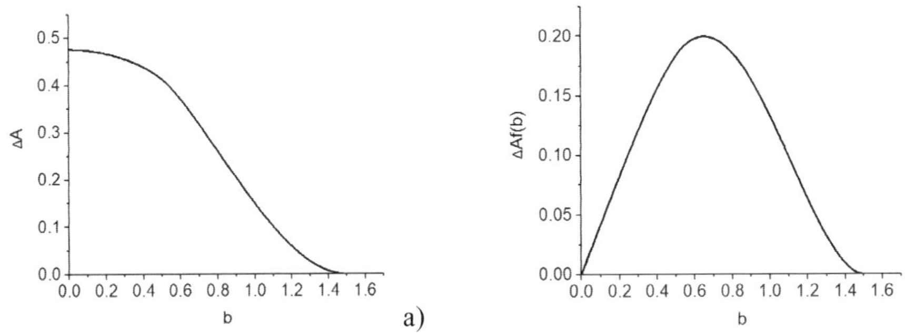

Figure 6. Curves corresponding to collision of nuclei with $R_1 = 1$, $R_2 = 0.5$, $A_1 = 1$, $A_2 = 0.5^3$.

Here we define $f(b)$ as:

$$f(b) = \frac{2b}{(R_1 + R_2)^2}.$$

From the curve of $\Delta A(b)f(b)$ (figure 6(b)) for such radii and numbers of particles we get the value: $< \Delta A > = 0.17$. Using the curve we can get that the most likely value of the impact parameter is $b = 0.65$ and a value of ΔA, corresponding to $\Delta A(0.65) = 0.345$ (see figure 6(b)).

4. Conclusion

The average part of the number of interacting nucleons at collisions of nuclei depends on their sizes. The maximal average value corresponds to one quarter at collisions of two equal nuclei. At interactions of cosmic ray nuclei with atomic nuclei of the atmosphere ($A_2 = 14.5$) this value varies from 0.2 (helium and iron) to 0.25 (nitrogen and oxygen).

Acknowledgements

Authors thank A. G. Bogdanov, R. P. Kokoulin, V. A. Okorokov for fruitful discussions. The work is supported by Ministry of Science and Higher Education of the Russian Federation (State task and MEPhI Academic Excellence Project).

References

1. Slavatinski S A 2003 *Nucl. Phys. B (Proc. Suppl.)* **122** 3
2. Aamodt K *et al* 2010 *Phys. Rev. Lett.* **105** 252301
3. Petrukhin A A 2008 *Nucl. Phys. B (Proc. Suppl.)* **175-176** 125
4. Petrukhin A A and Bogdanov A G 2016 *Frascati Physics Series* **64** 291

AUTHOR INDEX